Lecture Notes in Computer Science 7529

Lecture Notes in Computer Science 7529

Commenced Publication in 1973
Founding and Former Series Editors:
Gerhard Goos, Juris Hartmanis, and Jan van Leeuwen

Editorial Board

Fu Lee Wang Jingsheng Lei Zhiguo Gong
Xiangfeng Luo (Eds.)

Web Information Systems and Mining

International Conference, WISM 2012
Chengdu, China, October 26-28, 2012
Proceedings

Volume Editors

Fu Lee Wang
Caritas Institute of Higher Education, Department of Business Administration
18 Chui Ling Road, Tseung Kwan O, Hong Kong, China
E-mail: pwang@cihe.edu.hk

Jingsheng Lei
Shanghai University of Electric Power
School of Computer and Information Engineering
Shanghai 200090, China
E-mail: jshlei@shiep.edu.cn

Zhiguo Gong
University of Macau, Department of Computer and Information Science
Av. Padre Tomás Pereira, Taipa, Macau, China
E-mail: fstzgg@umac.mo

Xiangfeng Luo
Shanghai University, School of Computer
Shanghai 200444, China
E-mail: luoxf@shu.edu.cn

ISSN 0302-9743 e-ISSN 1611-3349
ISBN 978-3-642-33468-9 e-ISBN 978-3-642-33469-6
DOI 10.1007/978-3-642-33469-6
Springer Heidelberg Dordrecht London New York

Library of Congress Control Number: 2012946752

CR Subject Classification (1998): H.4.1-3, H.3.4-5, H.5.3, H.2.7-8, K.4.4, K.6.4, E.3, I.2.6, C.2.0, D.4.6, C.2.4

LNCS Sublibrary: SL 3 – Information Systems and Application, incl. Internet/Web and HCI

Typesetting: Camera-ready by author, data conversion by Scientific Publishing Services, Chennai, India

Printed on acid-free paper

Springer is part of Springer Science+Business Media (www.springer.com)

Preface

The 2012 International Conference on Web Information Systems and Mining (WISM 2012) was held during October 26–28, 2012 in Chengdu, China. WISM 2012 received 418 submissions from 10 countries and regions. After rigorous reviews, 87 high-quality papers were selected for publication in the WISM 2012 proceedings. The acceptance rate was 21%.

The aim of WISM 2012 was to bring together researchers working in many different areas of web information systems and web mining to exchange new ideas and promote international collaboration. In addition to the large number of submitted papers and invited sessions, there were several internationally well-known keynote speakers.

On behalf of the Organizing Committee, we thank Xihua University and Leshan Normal University for its sponsorship and logistics support. We also thank the members of the Organizing Committee and the Program Committee for their hard work. We are very grateful to the keynote speakers, session chairs, reviewers, and student helpers. Last but not least, we thank all the authors and participants for their great contributions that made this conference possible.

October 2012

Fu Lee Wang
Jingsheng Lei
Zhiguo Gong
Xiangfeng Luo

Preface

October 2012

Organization

Organizing Committee

General Co-chairs

WeiGuo Sun	Xihua University, China
Qing Li	City University of Hong Kong, Hong Kong, China

Program Committee Co-chairs

Zhiguo Gong	University of Macau, Macau, China
Xiangfeng Luo	Shanghai University, China

Steering Committee Chair

Jingsheng Lei	Shanghai University of Electric Power, China

Local Arrangements Co-chairs

Yajun Du	Xihua University, China
MingXing He	Xihua University, China
Jin Pei	LeShan Normal University, China

Proceedings Co-chairs

Fu Lee Wang	Caritas Institute of Higher Education, Hong Kong, China
Ting Jin	Fudan University, China

Sponsorship Chair

Zhiyu Zhou	Zhejiang Sci-Tech University, China

Program Committee

Ladjel Bellatreche	ENSMA - Poitiers University, France
Sourav Bhowmick	Nanyang Technological University, Singapore
Stephane Bressan	National University of Singapore, Singapore
Erik Buchmann	University of Karlsruhe, Germany
Jinli Cao	La Trobe University, Australia
Jian Cao	Shanghai Jiao Tong University, China

Badrish Chandramouli	Microsoft Research, USA
Akmal Chaudhri	City University of London, UK
Qiming Chen	Hewlett-Packard Laboratories, USA
Lei Chen	Hong Kong University of Science and Technology, China
Jinjun Chen	Swinburne University of Technology, Australia
Hong Cheng	The Chinese University of Hong Kong, China
Reynold Cheng	Hong Kong Polytechnic University, China
Bin Cui	Peking University, China
Alfredo Cuzzocrea	University of Calabria, Italy
Wanchun Dou	Nanjing University, China
Xiaoyong Du	Renmin University of China, China
Ling Feng	Tsinghua University, China
Cheng Fu	Nanyang Technological University, Singapore
Gabriel Fung	The University of Queensland, Australia
Byron Gao	University of Wisconsin, USA
Yunjun Gao	Zhejiang University, China
Bin Gao	Microsoft Research, China
Anandha Gopalan	Imperial College, UK
Stephane Grumbach	INRIA, France
Ming Hua	Simon Fraser University, Canada
Ela Hunt	University of Strathclyde, Glasgow
Renato Iannella	National ICT, Australia
Yan Jia	National University of Defence Technology, China
Yu-Kwong Ricky	Colorado State University, USA
Yoon Joon Lee	KAIST, Korea
Carson Leung	The University of Manitoba, Canada
Lily Li	CSIRO, Australia
Tao Li	Florida International University, USA
Wenxin Liang	Dalian University of Technology, China
Chao Liu	Microsoft, USA
Qing Liu	CSIRO, Australia
Jie Liu	Chinese Academy of Sciences, China
JianXun Liu	Hunan University of Science and Technology, China
Peng Liu	PLA University of Science and Technology, China
Jiaheng Lu	University of California, Irvine, USA
Weiyi Meng	Binghamton University, USA
Miyuki Nakano	University of Tokyo, Japan
Wilfred Ng	Hong Kong University of Science and Technology, China
Junfeng Pan	Google, USA

Zhiyong Peng	Wuhan University, China
Xuan-Hieu Phan	University of New South Wales (UNSW), Australia
Tieyun Qian	Wuhan University, China
Kaijun Ren	National University of Defense Technology, China
Dou Shen	Microsoft, USA
Peter Stanchev	Kettering University, USA
Xiaoping Su	Chinese Academy of Sciences, China
Jie Tang	Tsinghua University, China
Zhaohui Tang	Microsoft, USA
Yicheng Tu	University of South Florida, USA
Junhu Wang	Griffith University, Australia
Hua Wang	University of Southern Queensland, Australia
Guoren Wang	Northeastern University, USA
Lizhe Wang	Research Center Karlsruhe, Germany
Jianshu Weng	Singapore Management University, Singapore
Raymond Wong	Hong Kong University of Science and Technology, China
Jemma Wu	CSIRO, Australia
Jitian Xiao	Edith Cowan University, Australia
Junyi Xie	Oracle Corp., USA
Wei Xiong	National University of Defence Technology, China
Hui Xiong	Rutgers University, USA
Jun Yan	University of Wollongon, Australia
Xiaochun Yang	Northeastern University, China
Jian Yang	Macquarie University, Australia
Jian Yin	Sun Yat-Sen University, China
Qing Zhang	CSIRO, Australia
Shichao Zhang	University of Technology, Australia
Yanchang Zhao	University of Technology, Australia
Sheng Zhong	State University of New York at Buffalo, USA
Aoying Zhou	East China Normal University, China
Xingquan Zhu	Florida Atlantic University, USA

Table of Contents

Applications of Web Information Systems

Applications of Web Mining

E-government and E-commerce

Information Security

Intelligent Networked Systems

Management Information Systems

Mobile Computing

Semantic Web and Ontologies

Web Information Extraction

Web Intelligence

Web Interfaces and Applications

XML and Semi-structured Data

Study on the Design of Automatic Cotton Bale Inspecting Management System

Jie Zhao[1], Fang Yu[1], Minfeng Tang[1], Wenliao Du[2], and Jin Yuan[2]

[1] Shanghai Entry-Exit Inspection and Quarantine Bureau, Shanghai, China
[2] School of Mechanical Engineering, Shanghai Jiao Tong University, Shanghai, China
{zhaojie2012126,fangyu126,yumintang126}@126.com,
{wenliaodu,j.yuan72}@gmail.com

Abstract. The study object in this paper is the cotton bale's high-speed weighing and sampling system. Firstly it introduces the requirements of the cotton bale weighing and sampling system. And then the control system is discussed including simplified schematic, system composition motion control. Motion control is the focal point of this paper. One of the emphases is the choice and installation of sensors, limit switches and execution units in this system, the other is the principle, procedure and flow chart of the automatic deviation correction. Finally, it discusses the real-time solution for multi production lines convey by using the VxWorks real-time kernel and NTP to do time synchronization to ensure the control units meet the time requirements of high-precision.

Keywords: cotton bale, Cotton fiber sampling, VxWorks.

1 Introduction

In the inbound and outbound of the imports and expert cotton and domestic cotton trade, the warehouse management process includes cotton bale weighing, sampling and marking work, due to the low degree of automation of existing technology, to take a more discrete and manual way. Customs cotton bale warehouse are usually more dispersed, especially cotton commodities, large volume, variety and large quantities of cotton bale, its regulatory and inspection work need to spend a lot of manpower, working hours, which not only give a great deal of pressure to inspection and quarantine departments for storage and transportation, but also increased trade delivery. Cotton bale test and management system is developed for continuous, automated which requires high security, reliability, ease maintenance[1-2], while the distributed automated system control and man-machine interface separation can be achieved using modern network communication technology remote real-time data acquisition, monitoring in remote. A fully functional distributed systems and highly specialized, high reliability and enables distributed human-machine dialogue system with flexible expansion, its management features simple and convenient. Distributed systems through the communication network to a host of large complete sets of equipment and a number of auxiliary interconnected, sharing resources to work together to achieve distributed monitoring and centralized operation and management diagnosis.

F.L. Wang et al. (Eds.): WISM 2012, LNCS 7529, pp. 1–8, 2012.

The use of high reliability PLC and the SCADA monitoring software make it easy to build a control system with a data acquisition, control circuit, automatic sequence, calculation and other major operations of equipment. It can realize the automatic control of packaging process and transportation process with a real-time monitoring, fault diagnostic and quality evaluation, which meets the reliability, stability and real-time requirements of industrial system [3].

The cotton bale's high-speed weighing and sampling has been chosen as a study object in this paper. Firstly it introduces the requirements of the cotton bale belt conveyor and sampling system. And then the control system is discussed including simplified schematic, system composition motion control. Motion control is analyzed in details. One of the emphases is the choice and installation of sensors, limit switches and execution units in this system, the other is the principle, procedure and flow chart of the automatic deviation correction. In the final, it also discusses the real-time solution for multi production lines convey by using the VxWorks real-time kernel and NTP to do time synchronization to ensure the control units can meet the time requirements of high-precision.

2 Requirement Analysis

As combined transportation and packaging technology has been applied much widely in materials packaging conveyor system, the daily output from conveyor has also increased greatly and hence belt conveyor with middle-long distance and huge traffic becomes one of the main delivery equipment in cotton bale sampling and qualify-test production. Now, the safety, reliability and efficiency of such conveyor are very important in cotton packaging production [4].

However, due to the complexity of production conditions, centralized belt conveyor in many packaging production line has to suffer such problems as impact of big discharge and uneven loads on the belt during the process of package feeding, thus probably leads to decrease the strength of belt, in worse to destroy the driving motor or break the belt, which means great loss in economy and safety. Toward such problems, this paper, on the background of the production conditions of the field point of our investigation, designs an auto control system based on rotating arm, belt feeder and programmable logic controller. And also the belt-conveyor can be strengthened to use the chain-conveyor to enhance the carrying power and stability.

The cotton bale roller conveyor in the field point is multi-segment transport equipment, with four conveyors (roller conveyors) and one motor-less conveyor distributed regularly to form the whole roller conveyor of production line. The conveyor 1 is the entry conveyor which gets the feed of cotton packet. The method of feeding cotton bale on the belt in the entry port is that workers open the port manually to let cotton bale on the belt freely without accurate interval, and there is no coordination between the two discharge workers, which may cause overload on the belt. The conveyor 2 is to regulate the array of the cotton bales of transferred by conveyor 1, which is prepared to be weighed in the dynamic scale pan. If there have a cotton bale on the scale pan, the counting roller conveyer (conveyer 2) will stop to ensure there can only be one cotton bale on the scale pan. After being weighed in the scale pan, there is a branch conveyor (conveyor 4) which does the functions of sampling and qualify test. When the counter reaches a predefined value, the rotating arm 1 put down. The chain-conveyor pulls up, and the cotton bale runs into roller conveyor 4. After sampling, if the cotton bale passes the test, it will be returned to the main line (conveyor 3). The cotton bale passing the test will go to the exit port, which is made up of motor-less conveyor in the right outlet. The main part of cotton bale production line is shown in Fig. 1.

3 System Design

According to the above requirements, we can simplify the mechanical parts to be a transportation system with five conveyors (four motor conveyors and one motor-less conveyor).

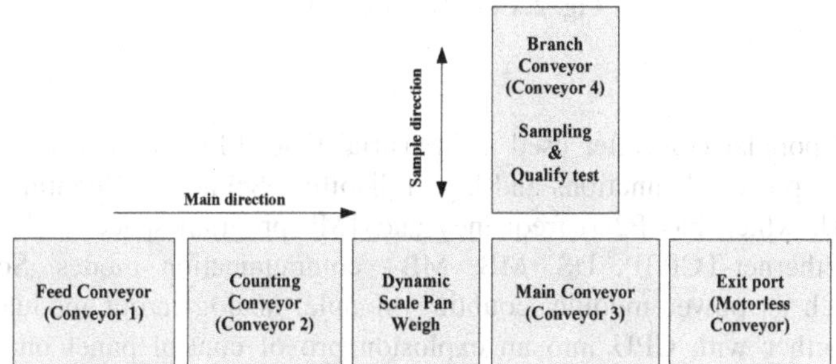

Fig. 1. The main part of cotton bale production line schematic

3.1 Simplified Schematic

The design in this next is to use controllable devices to regulate and to control the feeders automatically according to the real-time load on the weigh belt conveyor, and then finish the work of sampling and qualify test. All the transportation system can be simplified as four conveyors and one motor-less conveyor.

3.2 System Composition

Cotton Bale Feeder

As in above, manually feed is shown in the figure 1 to load cotton bale on the belt. In fact, the conveyor 1 is the preparation feed to the production line. The conveyor 2 can be looked as the actual feeder to the main line (conveyor 3). Both conveyors are the microscale belt conveyor with counter in them. It can be installed below the discharge port closely and to carry cotton bale out of from the port to the main belt conveyor (conveyor 3). The feeder is driven by motor with current frequency altered by corresponding transducer. The quantity of cotton bale loaded by the feeder, proportional to run time and the feeder's velocity determined by the current frequency, thus can be calculated according to the relationship among the above factors after testing.

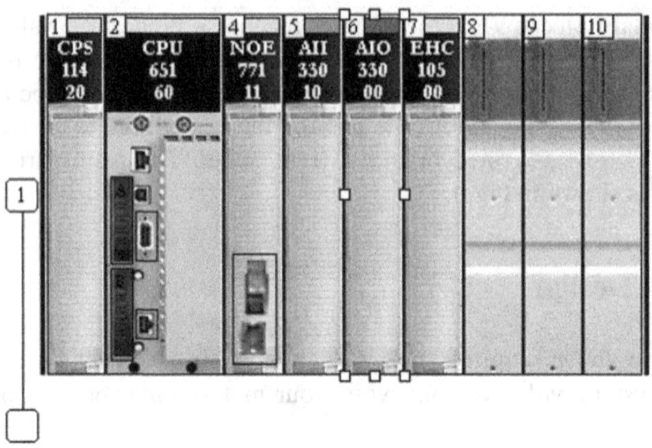

Fig. 2. Configuration of PLC

Controller

As the most popular controller used in industrial field, PLC is chosen in this paper because of its powerful functions and high reliability. Schneider Quantum 651 60 is used as CPU, which has P266 frequency and 1Mb program space, and it supports PCMCIA, Ethernet-TCP/IP, US, MB, MB+ communication modes. Some other modules, such as power module, counting module, analog in/out module, etc. are installed together with CPU into an explosion pro of control panel on which are corresponding functional key and screen in order to control or adjust the system manually if need. The configuration of PLC is shown in Fig. 2.

Other Modules

140-CPS-114 power supplies are designed to not require external EMI filter, ferrite bead and Ol flex cable. Quantum power supplies include on-board early power detection logic which is used to signal other modules on the rack that input power is lost.

The Ethernet communication module NOE 77111complies with the following standards: UL 508, CSA 22.2-142, CE, EMI, EN55011, EN61131-2, FM 3611 Class 1 Div2 Groups ABCD, IEC61131-2, IEEE 802.3 2002, ODVA. It supports both the I/O communication and the explicit messaging capacities.

The AII 330 10 is the analog input module to get the input of the photo electric device, sensor or any other input elements. It is configured by the function block AII33010_Instance, which is used to edit the configuration data of an AII 330 10 Quantum module for subsequent use by the scaling EFBs. The module has 8 unipolar intrinsically safe channels. The following ranges can be selected: 0-20 mA, 0-25 mA and 4-20 mA. For the configuration of an AII 330 10 the function block in the configuration section is connected to the corresponding SLOT output of the QUANTUM or DROP function block. The %IW references specified in the I/O map are automatically assigned internally to individual channels. The channels can only be occupied by unallocated variables. The analog values can be further processed in

Scaling Sections using the I_NORM, I_PHYS, I_RAW and I_SCALE functions. EN and ENO can be configured as additional parameters [5].

The AIO 33000 is for analog output. This module has 8 intrinsically safe symmetrical output channels for controlling and supervising the currents in the ranges 0 to 20 mA, 0 to 25 mA and 4 to 20 mA. For the configuration of an AIO 330 00 the function block in the configuration section is connected to the corresponding SLOT output of the QUANTUM or DROP function block. The MW references specified in the I/O map are automatically assigned internally to individual channels. The channels can only be occupied by unallocated variables.

The EHC-105-00 module is a high-speed counter module for the Modicon Quantum controller. The module includes 5 independent counters. Each one can be operated with either 5 or 24 VDC pulse input signals. Counting frequencies of up to 100 kHz can be monitored, depending upon cable length, transmitter type and voltage. These I/O points can be assigned to the cotton bales counters.

The rack has used a 10-rack extensible back plane. There still have 3 free slots to be install extent module.

PLC is connected with belt feeders by optical fiber. Each transducer of the feeder is equipped with a photoelectric converter and Modbus is chosen as the communication protocol.

3.3 Motion Control

The motion control includes control actions of the motors, limit switches and some other motion elements or devices in the process of conveying, sampling and quality-test. Such as rotating arm, air cylinder, sampling cutter. The automatic degree of controlling these things will greatly influence the efficiency of the whole production line.

According to the desired function, the system is designed as a closed loop control system in which current value of belt feeder motor and of belt conveyor driving motor is taken as feedback parameters to judge whether the load on belt conveyor is in the set range or not.

Requirements of Motion Control

The six main limit position switched are triggered under the following conditions.

- Limit switch 1 is in the front of cutter, it indicates the limit position of cutting feed.
- Limit switch 2 is in the end of cutter, it indicates the limit position of reverse cutting feed.
- Limit switch 3 in on the roller path 4, it indicates the transportation limit of the sampling cotton bale.
- Limit switch 4 is on the roller path 3, it indicates the space limit after sampling.
- Limit switch 5 is on the rotating arm 1, it indicates sampling with a predefined proportion.
- Limit switch 6 is on the rotating arm 2, it is for remove the wire-steel strapping band.

Control Logic

The motion logic mainly includes two actions: weigh and sample. It is shown in Fig. 3. The part of Ladder Diagram of PLC programming is shown in Fig.4.

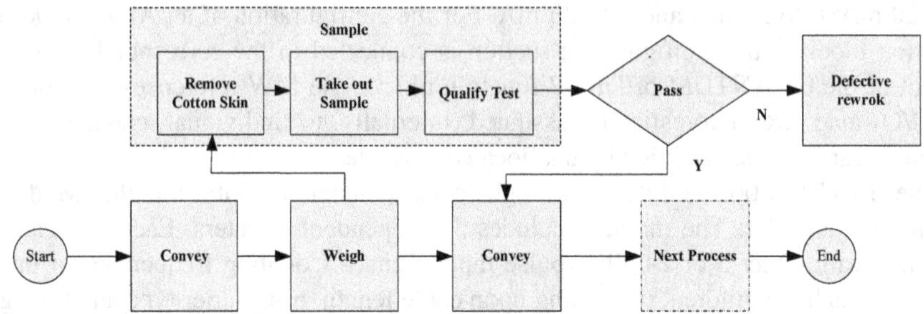

Fig. 3. The control logic process

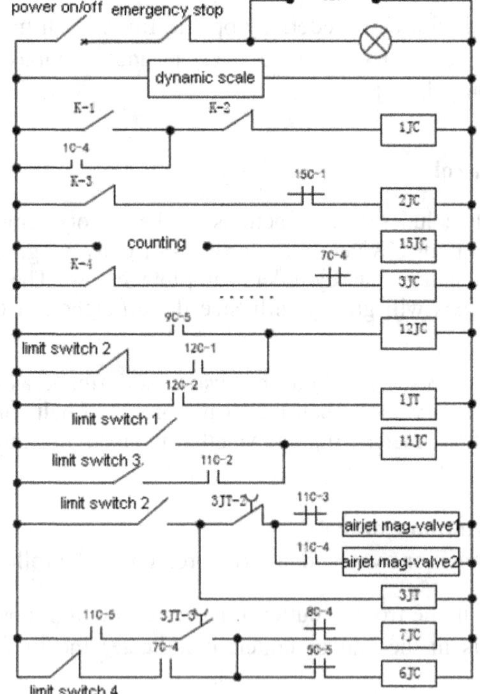

Fig. 4. Part of PLC program

1. Weigh: Weigh cotton bale one by one. If there has a cotton bale on the scale pan, the counting control roller conveyer (conveyor 2) will stop to ensure there is only one cotton bale on the scale pan.
2. Sample: Sample with the predefined proportion.

- When the counter reach a predefined value, the rotating arm 1 is put down. The chain-conveyor pulls up, the cotton bale runs into roller path 4 (conveyor 4).
- Put down the rotating arm 2, the cotton bale is waiting for removing the strapping wire-steel.
- Remove cotton bale cover. The sampling cutter forward rotates, and cuts into cotton bale with predefined feeding time to restrict cutting depth. Grab the cover cotton, the gliding block air cylinder move to exit. Then the sampling cutter passively rotates, runs until reach the limit position 2 and stop. The magnetic valve 1 opens to blow off the cover cotton. The gliding block air cylinder move forward.
- Take out the sample. The sampling cutter positively rotates, and cuts into cotton bale (the limit switch 1 restricts cutting space). Grab the cotton skin, the gliding block air cylinder move off and exit. Then the sampling cutter back rotates, runs until reach the limit position 2 and stop. The magnetic valve 2 will open and blast air to blow off the sample. The gliding block air cylinder move forward.
- The cotton bale is send back to the main line (conveyor 3).

4 Real-Time Solution for High-Speed Multi-line Convey

If there have a group of cotton bale transportation and sampling production lines which runs at the mean time. It requires high-precision time synchronization of each device. And every device must have the good real-time capability and performance to react to the request from the main control center. So the VxWorks is adopted as the kernel of the intelligent terminal. And NTP (Network Time Protocol) is used to synchronization for each communication module of each production line.

VxWorks is a real-time operating system with high performance that has a leading position in industry field developed by WindRiver company, which is telescopic, scalable and having high reliability, at the same time, it is applicable to all main stream CPU target platform. VxWorks has historically supported a kernel execution mode only. Both user applications and central support functionality, such as that provided by the network stack, all ran in the same memory space. With Base 6, VxWorks now supports a Real Time Process (RTP) mode in addition to kernel mode. This RTP mode is based on RTP Executable and Linking Format (ELF) object files. These object files are fully linked reloadable executables with full name space isolation [6].

The VxWorks kernel has included the Simple Network Time Protocol (SNTP) client service. And NOE 77111 has the NTP, which can realize the time synchronization from a NTP sever for devices of multi production line. SNTP client uses INCLUDE_SNTPC to include an SNTP client implementation in a VxWorks image. If a VxWorks image includes an SNTP client, sntpcTimeGet( ) can be use to retrieve the current time from a remote SNTP/NTP server. The sntpcTimeGet( ) routine reports the retrieved value to a format suitable for POSIX-compliant clocks. To get time information, sntpcTimeGet( ) either sends a request and extracts the time from the reply, or it waits passively until a message is received from an SNTP/NTP server executing in broadcast mode [7].

The NTP client service is to obtain the time from the provider and to pass it to the repository of the local time, the PLC. The repository of the local time provide the time to the end user, the application, that can then use it for its own purpose, provided that the time is maintained by the repository with enough accuracy to meet its needs. In our project, the time quality (the difference between the Time in the PLC and the NTP Server) can reach 1 ms typically, and the worst case is not greater than 4 ms.

5 Conclusions and Perspective

The cotton bale's high-speed multi-line belt conveyor and sampling has been chosen as the study object in this paper, the configuration and process of real-time control have been analyzed also. It analyzed the requirements of the cotton bale belt conveyor and sampling system. Then the design control system is discussed including simplified schematic, system composition and motion control. Motion control is analyzed in details. During the controlling process, one of the emphases is the choice and installation of sensors, limit switches and execution units in this system, the other is the principle, procedure and flow chart of the automatic deviation correction. Finally, it also discussed the real-time solution for multi production lines convey by using the VxWorks real-time kernel and NTP to do time synchronization to ensure the control units can meet the time requirements of high-precision.

Using the automatic materials packaging conveyor system can solve the problem of many packaging conveyor is set up in bad environment device, such as a lot of dust, relative humid air, decentralized operating position. The safety, reliability, and maintainability are enhanced greatly with the automatic transportation packaging control system. And roller conveyor is one the most important equipment for bulk material transportation. Traditional control methods cannot meet the design requirements of the heavy-duty belt conveyor. The method and controlling process can be used in other normal big workload production. It can remove the bottleneck which constrains the production in material feeding control system.

References

1. Abrahamsen, F., Blaabjberg, Pedersen, J.K., Grabowski, P.Z., Hogersen, P.: On the energy optimized control of standard and high-efficiency induction motors in CT and HVAC applications. IEEE Transaction on Industry Applications 34(4), 822–831 (1998)
2. Abdel-Halim, M.A., Badr, M.A., Alolah, A.I.: Smooth starting of slip ring induction motors. IEEE Transaction on Industry Applications 12(4), 317–322 (1997)
3. Rodriguez, J., Pontt, J., Becker, N., Weinstein, A.: Regenerative rives in the megawatt range for high-performance downhill belt onveyors. IEEE Transaction on Industry Applications 38(1), 203–210 (2002)
4. Hu, K.: The research on virtual prototype of belt conveyor with twin driving respective drums. Master degree dissertation. Anhui University of Science &Technology (2008)
5. Guo, Y.C., Hu, K., Xu, Z.Y.: Research and design on virtual prototype of belt conveyor. Hoisting and Conveying Machinery (8), 64–66 (2009)
6. Hu, W.: Research of numerical simulation technique of belt conveyor dynamic character. Xi'an University of Science and Technology (2005)
7. Zhang, X., Yang, W.: Application of fuzzy control technology in coal feeder control. Coal 14(1), 35–36 (2005)

The Smallest Randić Index for Trees*

Bingjun Li and Weijun Liu

College of Mathematics, Central South University , Changsha, 410000, China
Department of Mathematics, Hunan Institute of Humanities
Science and Technology, Loudi city, Hunan 417000, China
ydlbj@yahoo.com.cn

Abstract. The general Randić index $R_\alpha(G)$ is the sum of the weight $d(u)d(v)^\alpha$ over all edges uv of a graph G, where α is a real number and $d(u)$ is the degree of the vertex u of G. In this paper, for any real number $\alpha > 0$, the first three minimum general Randić indices among trees are determined, and the corresponding extremal trees are characterized.

Keywords: extremal graph, tree, the general Randić index. 2010 Mathematics Subject Classification: 05C70.

1 Introduction

In 1975, Randić [8] introduced the branching index as the sum of $d(u)d(v)^{-1/2}$ over all edges uv of a (molecular) graph $G = (V, E), i.e.$,

$$R(G) = \sum_{uv \in E(G)} (d(u)d(v))^{-1/2},$$

where $d(u)$ denotes the degree of $u \in V(G)$.

Randić noticed that there is a good correlation between the Randić index R and several physic-chemical properties of alkanes: boiling points, chromatographic retention times, enthalpies of formation, parameters in the Antoine equation for vapor pressure, surface areas, etc[8]. So finding the graphs having maximum and minimum Randić indices attracted the attention of many researchers. The Randić index has been extensively studied by both mathematicians and theoretical chemists in the past 30 years, see [1, 9].

Later, in 1998 Bollobás and Erdös [2] generalized this index by replacing $-\frac{1}{2}$ with any real number α, which is called the general Randić index, *i.e.*,

$$R_\alpha(G) == \sum_{uv \in E(G)} (d(u)d(v))^\alpha.$$

For a survey of results, we refer to the reader to a book, which is written by X. Li and I. Gutman [7].

* A Project Supported by Scientific Research Fund of Hunan Provincial Education Department and Central South University Postdoctoral Science Foundation Research.

F.L. Wang et al. (Eds.): WISM 2012, LNCS 7529, pp. 9–14, 2012.

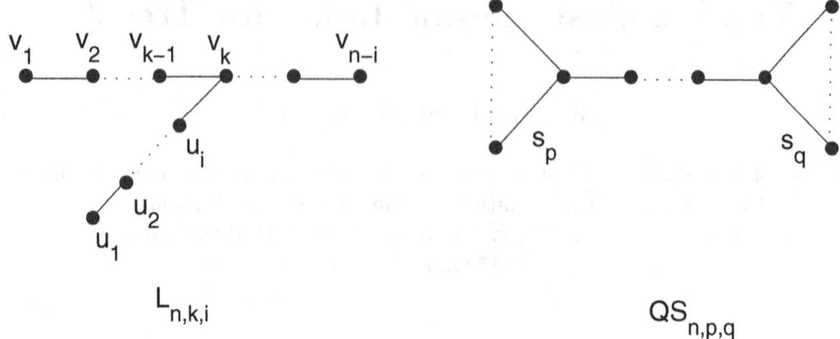

Fig. 1. $L_{n,k,i}$ and $QS_{n,p,q}$

Let $d(u)$ and $N(u)$ denote the degree and neighborhood of vertex u, respectively. A vertex of degree i is also addressed as an $i-$ *degree vertex*. A vertex of degree 1 is called a pendent vertex or a leaf. A connected graph without any cycle is a tree. The path P_n is a tree of order n with exactly two pendent vertices. The star of order n, denoted by S_n, is the tree with $n-1$ pendent vertices. The double star $S_{p,q}$ is the tree with one p-degree vertex, one q-degree vertex and $p+q-2$ pendent vertices. If $|p-q| \leq 1$, then the double star is said to be balanced. A comet is a tree composed of a star and a pendent path. For any number n and $2 \leq n_1 \leq n-1$, we denote by P_{n,n_1} the comet of order n with n_1 pendant vertices, *i.e.*, a tree formed by a path P_{n-1} of which one end vertex coincides with a pendant vertex of a star S_{n_1+1} of order n_1+1. Let $L_{n,k,i}$ $(1 \leq k, i \leq n-i)$ denote the tree obtained from a path $v_1v_2\cdots v_{n-i}$ of length $n-i$ by attaching a suppended path of length i rooted at v_k. A quasi double star graph $QS_{n,p,q}$ is a tree obtained from connecting the centers of S_p and S_q by a path of length $n-p-q$ as shown in Figure 1.

The question of finding the minimum general Randić index and the corresponding extremal graphs also attracted much attention of many researchers. Hu, Li and Yuan [6] showed that among trees with $n \geq 5$ vertices, the path P_n for $\alpha > 0$ and the star S_n for $\alpha < 0$, respectively, has the minimum general Randić index. And in the same paper the trees of order $n \geq 6$ with second minimum and third minimum general Randić index for $0 < \alpha \leq -1$ are characterized. Chang and Liu [5] showed that among trees with $n \geq 7$ vertices, the comet $P_{n,3}$ has second minimum general Randić index for $\alpha > 0$.

In this paper, we discuss the first three minimum general Randić index for trees. We use a relatively simple way to prove results mentioned above, and determine trees with the third minimum general Randić index for $\alpha > 0$ (answering a question posed by Chang and Liu [5]).

2 The Case for $\alpha > 0$

Lemma 2.1 Let T_1 be a tree of order $n \geq 7$ which is not a star, $\alpha > 0$. Assume that $P = v_1v_2\cdots v_t$ is a longest path in T_1 with $d = d(v_2) > 2$, and

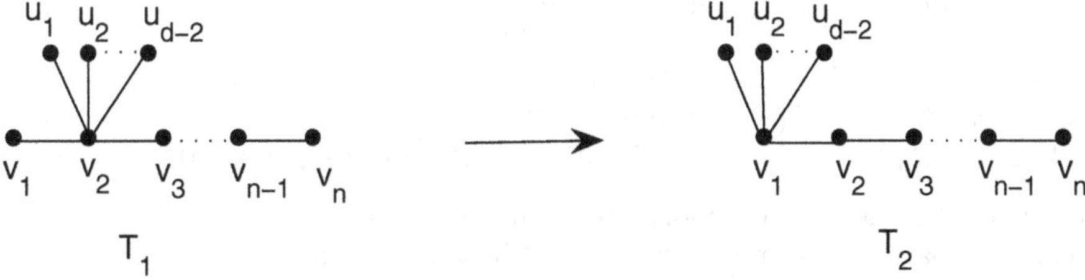

Fig. 2. Transform T_1 to T_2

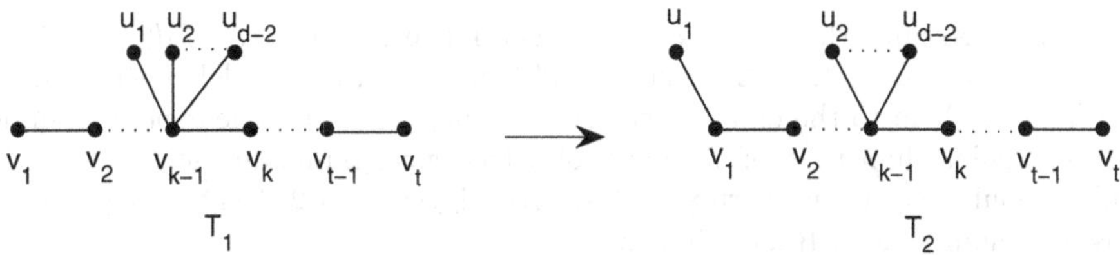

Fig. 3. Transform T_1 to T_2

$u_1, u_2, \cdots, u_{d-2}$ are neighbors of v_2 other than v_1, v_3. T_2 is formed by deleting the edges $v_2u_1, v_2u_2, \cdots, v_2u_{d-2}$ and adding the edges $v_1u_1, v_1u_2, \cdots, v_1u_{d-2}$

Then $R_\alpha(T_1) > R_\alpha(T_2)$.

Proof. Since T_1 is not a star and P is a longest path, then $d(u_1) = d(u_2) = \cdots = d(u_{d-2}) = 1$ and $d(v_3) > 1$.

$R_\alpha(T_1) - R_\alpha(T_2)$
$= d(v_3)^\alpha(d^\alpha - 2^\alpha) - 2^\alpha(d-1)^\alpha + (d-1)d^\alpha - (d-2)(d-1)^\alpha$
$\geq 2^\alpha(d^\alpha - 2^\alpha) + (d-1)d^\alpha - 2^\alpha(d-1)^\alpha - (d-2)d^\alpha$
$= 2^\alpha(d^\alpha - (d-1)^\alpha) + d^\alpha - 4^\alpha$.

If $d \geq 4$, $R_\alpha(T_1) - R_\alpha(T_2) > 0$; if $d = 3$,
$R_\alpha(T_1) - R_\alpha(T_2) > 2^\alpha(3^\alpha - (d-1)^\alpha) + 3^\alpha - 4^\alpha$.
$= 6^\alpha + 3^\alpha - 2 * 4^\alpha > 2 * \sqrt{18^\alpha} - 2 * \sqrt{16^\alpha} > 0$. □

Corollary 2.1 The general Randić index of a comet P_{n,n_1} is monotonously decreasing in n_1 for $\alpha > 0$.

Lemma 2.2 Let T_1 be a tree of order $n \geq 7$, $\alpha > 0$. Assume that $P = v_1v_2 \cdots v_t$ is a longest path in T_1 with $d(v_2) = d(v_{t-1}) = 2$, and $d = d(v_k) > 2$ for some $3 \leq k \leq t-2$. $u_1, u_2, \cdots, u_{d-2}$ are neighbors of v_k other than v_{k-1}, v_{k+1}. T_2 is formed by deleting the edge v_ku_1 and adding a new edge v_1u_1, as shown in Figure 3.

Then $R_\alpha(T_1) > R_\alpha(T_2)$.

Proof. $R_\alpha(T_1) - R_\alpha(T_2)$
$\geq (d * d(u_1))^\alpha + (d * d(v_{k-1}))^\alpha + 2^\alpha - (2 * d(u_1)^\alpha + (d(v_{k-1})(d-1))^\alpha - 4^\alpha)$

$= d(u_1)^\alpha(d^\alpha - 2^\alpha) + d(v_{k-1})^\alpha(d^\alpha - (d-1)^\alpha) + 2^\alpha - 4^\alpha$
$\geq d^\alpha - 2^\alpha + 2^\alpha(d^\alpha - (d-1)^\alpha) + 2^\alpha - 4^\alpha$
$= d^\alpha - 4^\alpha + 2^\alpha(d^\alpha - (d-1)^\alpha)$.
If $d \geq 4$, $R_\alpha(T_1) - R_\alpha(T_2) > 0$; if $d = 3$,
$R_\alpha(T_1) - R_\alpha(T_2) = 6^\alpha + 3^\alpha - 2 * 4^\alpha > 2 * \sqrt{18^\alpha} - 2 * \sqrt{16^\alpha} > 0$.

Theorem 2.1 Let $\alpha > 0$, among trees with $n \geq 7$ vertices,

(1) The path P_n has minimum general Randić index;

(2) The comet $P_{n,3}$ has second minimumm general Randić index;

(3) Let μ proximately equal to 3.6475, if $\alpha > \mu$, $L_{n,k,1}(k = 3, 4, \cdots, n-3)$ has third general Randić index. If $\alpha < \mu$, the quasi double star $QS_{3,3}$ has third general Randić index.

Proof. (1) Assume that $P = v_1 v_2 \cdots v_t$ is a longest path in T. If $d(v_i) > 2$ for some $i = 2, 3, \cdots, n-1$, by Lemma 2.1 and Lemma 2.2, we can delete some edges and adding them to the end vertices to get a new tree with the general Randić index smaller than that of T. Therefore, tree with minimum general Randić index should contain no vertex with degree bigger than 2. Hence the path P_n has minimum general Randić index.

(2) Let T' be a tree with second minimum general Randić index. We assert that T_1 contains no vertex with degree bigger than 3 or two 3-degree vertices. Assume the contrary, if $d(v_i) \geq 4$ or $d(v_j) = d(v_k) = 3$ for some $1 \leq i, j, k \leq n$. then we can delete one edge adjacent to v_i or v_j and add it to a pendent vertex to get a new tree T'. By Lemma 2.1 and Lemma 2.2, $R_\alpha(T'') > R_\alpha(T')$. Obviously, T'' is not a path, a contradiction. Hence, T' contains exactly one 3-degree vertex. Note that $R_\alpha(L_{n,3,1}) = R_\alpha(L_{n,4,1}) = \cdots = R_\alpha(L_{n,n-2,1})$ and $R_\alpha(L_{n,k,l})$ are the same for $3 \leq k \leq n-2$ and $2 \leq l \leq k$. Thus it suffice to compare the general Randić index of the following trees.

$R_\alpha(H_2) - R_\alpha(H_1)$
$= 2 * 6^\alpha + 3^\alpha + 2 * 2^\alpha - 2 * 3^\alpha - 2^\alpha - 6^\alpha - 4^\alpha$
$= 6^\alpha + 2^\alpha - 3^\alpha - 4^\alpha$
$= 3^\alpha * 2^\alpha + 2^\alpha - 3^\alpha - 2^\alpha * 2^\alpha$
$= (3^\alpha - 2^\alpha)(2^\alpha - 1) > 0$
$R_\alpha(H_3) - R_\alpha(H_2)$
$= 3 * 6^\alpha + 3 * 2^\alpha - 2 * 6^\alpha - 3^\alpha - 4^\alpha - 2 * 2^\alpha$
$= 6^\alpha + 2^\alpha - 3^\alpha - 4^\alpha$
$= 3^\alpha * 2^\alpha + 2^\alpha - 3^\alpha - 2^\alpha * 2^\alpha$
$= (3^\alpha - 2^\alpha)(2^\alpha - 1) > 0$.

Hence H_1(the comet $P_{n,3}$) has the second general Randić index.

(3) Let $\overline{T}$ be a tree with third minimum general Randić index. By Lemma 2.1 and Lemma 2.2, we assert that $\overline{T}$ contains no vertex with degree bigger than 4, or two 4-degree vertices, or one 4-degree vertex and one 3-degree vertex. Hence $\overline{T}$ must be one of the following:

If $d(u_1) > 1$ or $d(u_2) > 1$, by Lemma 2.2, $R_\alpha(T_2) > R_\alpha(T_5)$. If $d(u_1) = d(u_2) = 1$, by Lemma 2.2, $R_\alpha(T_2) > R_\alpha(T_3)$. By the proof of (2), $R_\alpha(T_5) > R_\alpha(T_3)$. For T_6, if $d(v_1) > 1$ or $d(v_2) > 1$, by Lemma 2.2, $R_\alpha(T_6) > R_\alpha(T_5)$. So If

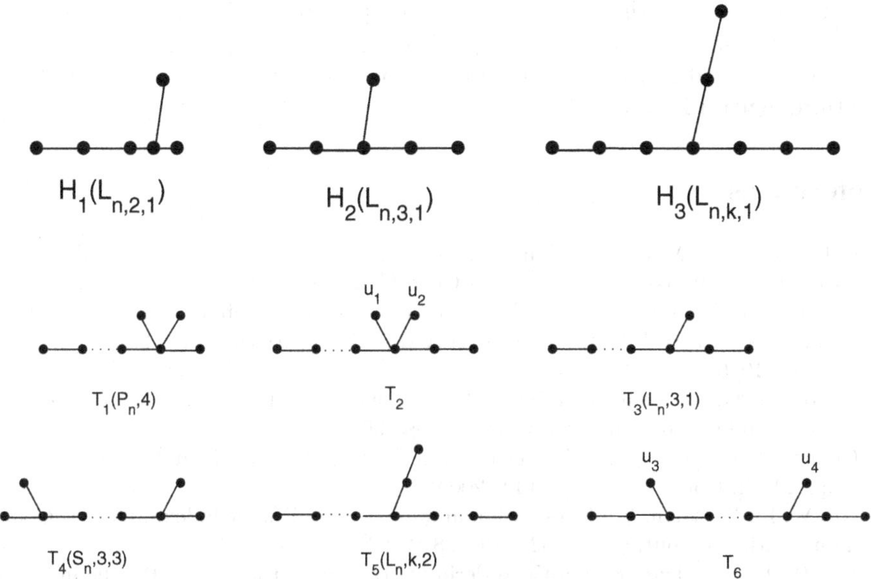

Fig. 5. Trees possible with third minimum general Randić index

$d(v_1) = d(v_2) = 1$, by Lemma 2.2, $R_\alpha(T_6) > R_\alpha(T_3)$. Therefore, $\overline{T}$ must be one of T_1, T_3, T_4.

$R_\alpha(T_1) - R_\alpha(T_3) = 3 * 4^\alpha + 8^\alpha + 2^\alpha - 2 * 6^\alpha - 3^\alpha - 2 * 2^\alpha$
$= 8^\alpha + 3 * 4^\alpha - 2 * 6^\alpha - 3^\alpha - 2^\alpha$
Let $f(x) = 8^x + 3 * 4^x - 2 * 6^x - 3^x - 2^x, \quad x \geq 0$
$f(0) = 0$
$f'(x) = 3 * 8^x \ln 2 + 6 * 4^x \ln 2 - 2 * 6^x \ln 6 - 3^x \ln 3 - 2^x \ln 2$
$> 3 * 8^x \ln 2 + 6 * 4^x \ln 2 - 2 * 6^x \ln 6 - 3^x \ln 6 = g(x)$.
$g(0) = 3 * \ln 2 + 6 * \ln 2 - 2 * \ln 6 - \ln 6 = \ln(2^9) - \ln(6^3) > 0$.
Since $(\ln 6)^2 < 7(\ln 2)^2$ and $\ln 6 * \ln 3 < 6 * (\ln 2)^2$, then
$g'(x) = 9 * 8^x(\ln 2)^2 + 12 * 4^x(\ln 2)^2 - \ln 6(2 * 6^x \ln 6 - 3^x \ln 3)$
$> (9 * 8^x + 12 * 4^x - 14 * 6^x - 6 * 3^x \ln 3)(\ln 2)^2$
$> (9 * 8^x + 6 * 4^x - 14 * 6^x)(\ln 2)^2$
Let $h(x) = 9 * 8^x + 6 * 4^x - 14 * 6^x$, $h(0) > 0$.
$h'(x) = 27 * 8^x * \ln 2 + 12 * 4^x \ln 2 - 14 * 6^x * \ln 6$
$h''(x) > 81 * 8^x * (\ln 2)^2 + 24 * 4^x * (\ln 2)^2 - 14 * 6^x * (\ln 6)^2$.
$h''(0) = 81 * (\ln 2)^2 + 24 * (\ln 2)^2 - 14 * (\ln 6)^2$
$> 105 * (\ln 2)^2 - 98(\ln 2)^2 > 0$.
$h'''(x) > 243 * 8^x * (\ln 2)^3 - 14 * 6^x * (\ln 6)^3$.
$> (243 * (\ln 2)^3 - 14 * (\ln 6)^3) * 8^x > 0$.

Therefore, $h(x) > 0$ for $x > 0$, thus $f(x) > 0$ for $x > 0$. Hence $R_\alpha(T_1) > R_\alpha(T_3)$.
$R_\alpha(T_4) - R_\alpha(T_3)$
$= 4 * 3^\alpha + 2 * 6^\alpha - (2 * 6^\alpha + 3^\alpha + 2 * 2^\alpha + 4^\alpha)$
$= 3 * 3^\alpha - 4^\alpha - 2 * 2^\alpha$

It is easy to calculate that for μ proximately equal to 3.6457, if $\alpha > \mu$, $R_\alpha(T_4) > R_\alpha(T_3)$; otherwise, $R_\alpha(T_3) > R_\alpha(T_4)$. Therefore, if $\alpha > \mu$, $L_{n,k,1}(k = 3, 4, \cdots, n-3)$ has third general Randić index. If $\alpha < \mu$, the quasi double star $QS_{3,3}$ has third general Randić index.

References

1. Balaban, A.T., Motoc, I., Bonchev, D., Mekenyan, O.: Topological indices for structure-activity correlations. Topics Curr. Chem. 114, 21–55 (1983)
2. Bollobás, B., Erdös, P.: Graphs of extremal weights. Ars Combin. 50, 225–233 (1998)
3. Bondy, J.A., Murty, U.S.R.: Graph Theory with Applications. Macmillan Press Ltd., London (1976)
4. Caporossi, G., Gutman, I., Hansen, P., Pavlović, L.: Graphs with maximum connectivity index. Comput. Biol. Chem. 27, 85–90 (2003)
5. Chang, R., Liu, G.: Trees with second minimum general Randić index for $\alpha > 0$. J. Appl. Math. Comput. 30, 143–149 (2009)
6. Hu, Y., Li, X., Yuan, Y.: Trees with minimum general Randić index. MATCH Commun. Math. Comput. Chem. 52, 119–128 (2004)
7. Liu, H., Lu, M., Tian, F.: On the ordering of trees with the general Randić index of the Nordhaus-Gaddum type. MATCH Commun. Math. Comput. Chem. 55, 419–426 (2006)
8. Li, X., Gutman, I.: Mathematical Aspects of Randić-Type Molecular Structure Descriptors. Mathematical Chemistry Monographs No.1, Kragujevac (2006)
9. Randić, M.: On characterization of molecular branching. J. Amer. Chem. Soc. 97, 6609–6615 (1975)

Design of Meridian and Acupoints Compatibility Analysis System

Jingchang Pan and Guangwen Wang

School of Mechanical, Electrical & Information Engineering
Shandong University at Weihai
Weihai, China, 264209
pjc@sdu.edu.cn, guangwenweihai@163.com

Abstract. Based on traditional Chinese meridian theory, this paper discussed the design and establishment of a complete meridian database system, including various information about meridian, acupoints, health knowledge, and acupoints compatibility, by means of computer database, image processing and data mining technologies. We provided a tool which can be used to analyze the relations between symptoms and the acupoints compatibilities through graphical and interactive queries. The system is very useful for auxiliary treatment, and popularizing the Chinese traditional meridian knowledge.

Keywords: meridian, acupoint, acupoint compatibility, diagnosis.

1 Introduction

Meridian theory is one of the core contents of Chinese traditional theoretical system [1]. It is an important adjustment and control system for a human body[2-3].

In current clinical studies of meridian and acupuncture, the main approaches remain the way of looking up the wall charts or referencing meridian text materials manually, which has its limitations. Literature [4] proposed a good idea to establish graphic information system about human's meridian system. However, there are many improvements to be made, such as lacking interactive queries between meridian and acupoints. Also, it only made prescription for a certain symptom, without the statistical and intuitional analysis about the meridians and related acupoints information.

This paper designed and established a complete meridian database system, by means of computer database and image processing technologies. The system is based on the traditional Chinese meridian theory, providing various information about meridian, acupoints, health knowledge, and acupoints compatibility. It also provided a tool to analyze the relations between symptoms and the acupoints compatibilities using data mining technology and graphical interactive queries, which is meaningful for medical research. The system is very useful for auxiliary treatment, and popularizing the Chinese traditional meridian knowledge.

The paper is structured in the following manner. A simple review on the meridian and acupoint theory is given in Section 2. In Section 3, we showed how to design the meridian information system, based on which, the analysis of acupoints compatibility issues are discussed in Section 4. The summary and future work are discussed in Section 5.

F.L. Wang et al. (Eds.): WISM 2012, LNCS 7529, pp. 15–21, 2012.

2 Introduction to Meridian and Acupoint

2.1 Meridian System

Meridians are distributed throughout the human's body. They are the main channels for the flow of Qi and blood in the human body. They are also the basic routes connecting every part of human's body and play an important roll in the transform and delivery of substance, energy and information in human's body. Because of these connections, the interrelations, the intercoordinations, the mutual promotions and restrictions are made possible between parts of the living organism [5].

Meridian system is a complex network in human's body, which is composed of the meridians, collaterals, muscles along the regular meridians, dermal parts. Meridian system includes twelve meridians, eight extra channels and twelve divergent channels. Collateral system includes fifteen collaterals channels and countless minute collaterals, float windings. Muscles along the regular meridians include twelve muscles meridian channels. Dermal parts include twelve parts [6]. Figure 1 shows an example of some meridians in human's body.

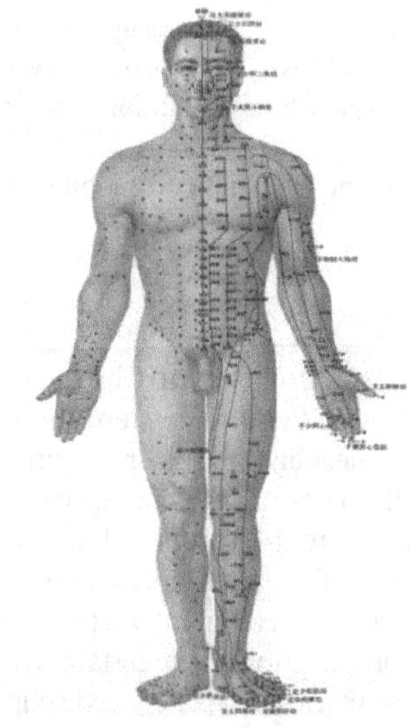

Fig. 1. Meridians in human body

2.2 Acupoint

Acupoints are the concentration parts for essence of QI and blood fluid. They are special zones for connecting body surface, Zang-fu organs and related parts, and mainly distributed among the meridians. They are attached to meridians, and internally connected to Zang-fu organs.

Acupoints have the function to fight disease, reflect sickness and feel stimulation. When inflicting acupuncture and massaging on the acupoint, the stimulations can be transferred into the internal and the power to cure disease can be inspired. There are many acupoints in the body, 361 of which are commonly used. The two red points in Figure 2 show the examples of acupoints.

2.3 Acupoints Compatibility (AC)

In Chinese acupuncture and massage therapy, it is often the case that several acupoints on different meridians are touched together. Stimulating these acupoints simultaneously can cure some diseases. The combination of these interrelated acupoints is called acupoints compatibility (AC). Through practice of thousand years of Chinese medical treatment, many ACs are found and proved efficient, hundreds of ACs being very common. For example, one AC treatment for headache is to massage the St36 acupuncture point in liver meridian and the Close valley acupuncture point in Yangming Large Intestine Meridian of Hand, as shown in Figure 2.

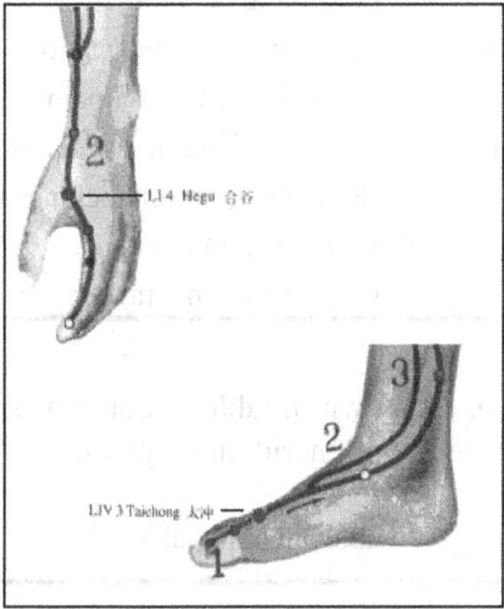

Fig. 2. Example of AC

3 Design of Meridian Information System

3.1 The Design of Meridian Database

The meridian database is composed of the following tables. These tables lay a foundation for our further queries and diagnosis analysis.

(1) Meridian table, as shown in table 1, contains the detailed information about the commonly used meridians. 19 meridians are recorded in this table.

Table 1. Meridian table

Field	Meaning
ID*	Unique ID, keyword
TitleC	Title in Chinese
TitleE	Title in English
Number	Acupoints in this meridian
Scope	Distribution in body
Profile	Profile of this meridian

(2) Acupoint table, as shown in table 2, contains the detailed information about commonly used acupoints. 407 acupoints are stored in this table.

Table 2. Acupoint table

Field	Meaning
ID*	Unique ID, keyword
LabelC	Title in Chinese
LabelE	Title in English
Meridian	Meridian this acupoint belongs to
Place	Position in the body
Anatomy	Local anatomy information
Effect	The effect of this acupoint
Function	Related disease and symptom
GraphID	Graph this acupoint belongs to
X	x coordinate in Graph
Y	y coordinate in Graph

(3) Meridian graph table, as shown in table 3, contains specific graphs relating to each meridian in table 1. There are 20 meridian graphs in this table.

Table 3. Graph table

Field	Meaning
ID*	Unique ID, keyword
Graph	Meridian Graph in OLE format

(4) AC table, as shown in table 5, contains the detailed information of ACs available. Note that AID and SN combine together to form the keyword. We have stored 973 ACs in this table.

Table 4. AC table

Field	Meaning
AID*	Acupoint ID
SN*	Serial number of AC
Effect	Disease this AC can cure

3.2 Function Modules of the System

The system function chart is shown in Figure.3. It's developed using C# and Microsoft Access 2007.

(1) Meridian module provides the information about the meridian's function, text and graph information and interactive query facilities. In the graph interface, the user can inquiry any acupoint information interactively by clicking the acupoint. The interface also provides the zooming and shifting function to let the user browse the whole graph to explore the meridians and acupoints.

(2) Acupoint module provides the queries to acupoints by selecting meridians in the pull list menu or by inputting the acupoint name. Detailed in formation, such as the meridian it belongs to, the position where it resides, its treatment effects, and so on can be obtained easily.

(3) Aided Medical module provides the treatment plan for most of the symptoms specified by the user. It supplies a set of candidate ACs for the user graphically.

(4) Analysis module provides the statistical information about the symptom specified by user. This tool is especially useful for medical research. It supplies the information of how many meridians and acupoints are involved in the ACs for that symptom. Information is displayed in a graphical interface and further inspection can be carried out in this interface.

(5) Medical knowledge provides abundant know ledges about meridians and health care.

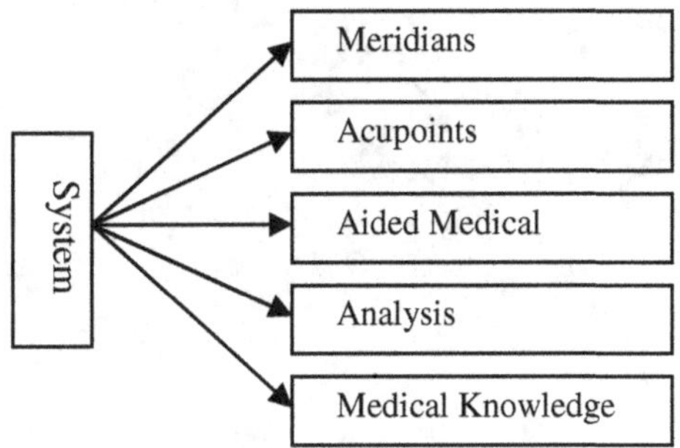

Fig. 3. System structure

4 Acupoints Compatibility Analysis

4.1 AC Statistics

This module is to analyze dynamically the ACs generated for the specific symptom specified. Based on the database we built, we can analyze the frequency of the meridians and acupoints that have effects on that symptom. In this way, the most important meridians and acupoints can be obtained, which are very useful for medical research purposes. The pseudo code for analysis algorithm is as follows.

```
Algorithm Acu-Comp-Analysis
Input: Symptom // input by user
       Data   // information stored in the above defined tables
Output: Analysis results
Method:
```

(1) Find all the records containing the symptom in AC table.
(2) Record the accounts of related acupoints in the AC
(3) Record the accounts of meridians containing the acupoints in (2)
(4) Show the results as network in figure 5.
(5) Show statistic information in the bottom of figure 5.
(6) React to user click-query and show related acupoint and meridian information in new windows.

4.2 Explanations of the Operations

The analysis interface is given in Figure 4. The meridians are listed in the left column of Figure 4. The acupoints that belong to each meridian are listed right after the meridian on the right, represented by a solid circle. For the limitation of space, the name of each acupoint is not displayed directly. However, when you move the mouse to a certain circle, the name of the acupoint displays instantly. Clicking each acupoint, all its information can be displayed.

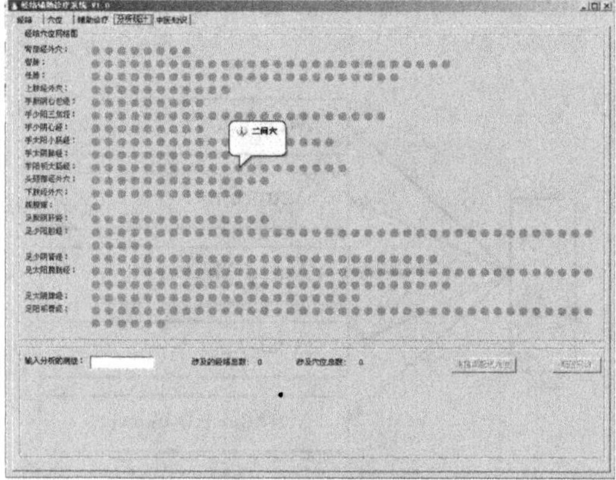

Fig. 4. Analysis interface

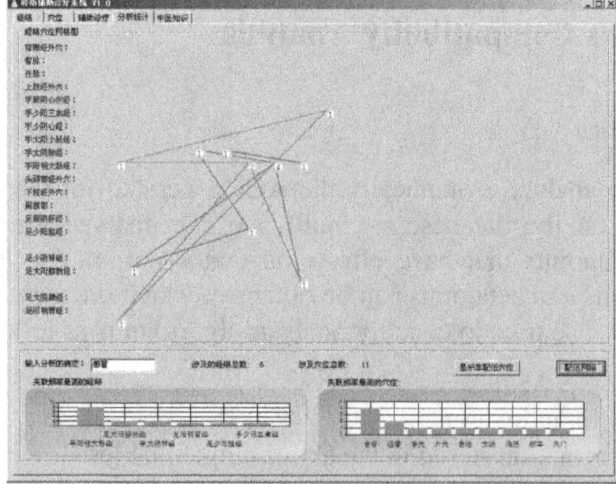

Fig. 5. Statistics interface

When user inputs a certain symptom in the textbox and presses enter, all the ACs information and be summed up and shown in Figure 5. Each AC is represented by a closed polygon with acupoints involved being its vertexes, in a unique color. At the bottom of interface shown in Figure 5, two bar graphs are displayed, showing the most frequently used meridians and acupoints respectively (9 at most if any). When clicking the ACs network, the related AC graphical information will be displayed, as shown in Figure 6.

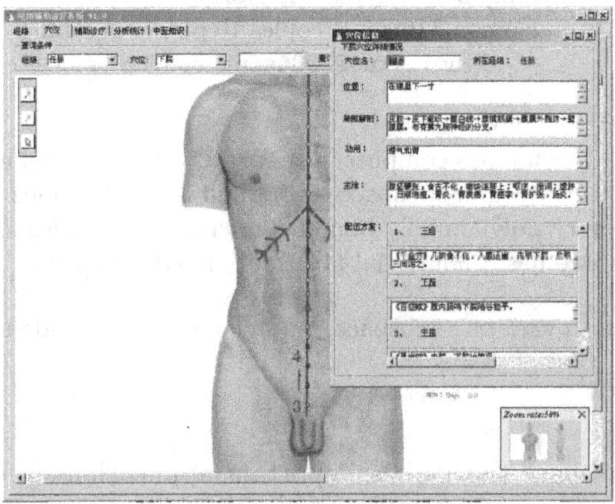

Fig. 6. Acupoints interactive query

5 Conclusion and Future Work

In this paper, the meridian database is designed and established and an information and analysis system is developed based on the Chinese medical meridian theory. It provides a useful tool to access meridian information about medical treatment plans and to analyze the meridians and acupoints for medical research. However, all the work is done only based on 2-D body graphs, which is not too intuitive. To make the system more visualized, we will improve the system on 3-D image platform in the future.

References

1. Hu, X.: The Main Achievements and Basic Experiences in Meridian Research in Recent 15 Years. Acupuncture Research 3, 165–172 (1987)
2. Niu, H., Jiang, S.: Exploration of The Nature of Meridian. Journal of Xi'an Medical University 18(3), 409–419 (1997)
3. Ni, F.: TCM Meridian Theory and Modern Meridian Research. Journal of Fujian College of Traditional Chinese Medicine 9(3), 31–41 (1999)
4. Cai, G., Yu, L.: The Human Body Meridian Graphic Information Query System Development. Journal of Fujian College of Traditional Chinese Medicine 3, 32–34 (1995)
5. The Human Body Meridian System Knowledge [EB/OL], `http://www.360doc.com/content/10/0210/21/76929_15632212.shtml`
6. Huang, J., Li, R., Zhang, J.: Meridian and Acupoint Theory. Beijing University of Chinese Medicine (2007)

Invariant Subspaces for Operators with Thick Spectra*

Mingxue Liu

School of Computer Science and Technology
Guangdong Polytechnic Normal University
Guangzhou 510665, Peopleś Republic of China
liumingxue9698@sina.com.cn

Abstract. The famous computer scientist J. von Neumann initiated the research of the invariant subspace theory and its applications. This paper show that every polynomially bounded operator with thick spectrum on a Banach space has a nontrivial invariant closed subspace.

Keywords: Invariant subspace, polynomially bounded operator, spectrum.

The invariant subspace theory has important applications in computer science and information science (see [1-16] and so on).

As stated in [3], it was the famous computer scientist J. von Neumann who initiated the research of the invariant subspace theory for compact operators and its applications. To be more specific, J. von Neumann showed that every compact operator on a Hilbert space H has a nontrivial invariant closed subspace.

The famous mathematician P. R. Halmos said ([9] p.100), "one of the most important, most difficult, and most exasperating unsolved problems of operator theory is the problem of invariant subspace".

S.Brown,B.Chevreau and C.pearcy [7] showed that every contraction operator with thick spectrum on a Hilbert space has a nontrivial invariant closed subspace.

In this paper, we show that every polynomially bounded operator with thick spectrum on a Banach space has a nontrivial invariant closed subspace.

It is well known that by the von Neumann inequality every contraction operator is polynomially bounded, but the converse is not true. For example, G. Pisier [12] given an example of a polynomially bounded operator that is not similar to a contraction operator (and so that is not a contraction operator).

It is also well known that every Hilbert space is a Banach space, but the converse is not true (for example, $l_p(p \neq 2)$-space, $L_p(p \neq 2)$-space, $C[a, b]$and so on).

Let G is a nonempty open set in the complex plane C. Let us denote by $H^\infty(G)$ the Banach algebra of all bounded analytic functions on G equipped with the norm $\|f\| =$ sup $\{|f(\lambda)|; \lambda \in G\}$.

* The research was supported by the Natural Science Foundation of P. R. China (No.10771039).

F.L. Wang et al. (Eds.): WISM 2012, LNCS 7529, pp. 22–25, 2012.

A subset σ of the complex plane C will be called dominating in G if $\|f\| = \sup\{|f(\lambda)|; \lambda \in \sigma \cap G\}$ holds for all $f \in H^\infty(G)$,

A compact subset σ of the complex plane C is called thick if there is a bounded open set G of the complex plane C such that σ is dominating in G.

An operator T on a Banach space X is said to be polynomially bounded if there is a constant k such that the inequality

$$\|p(T)\| \leq k\|p\|$$

holds for every polynomial.

For the notation and terminology not explained in the text we refer to [1], [4] and [16].

Lemma 1. Let T be a polynomially bounded operator on a Banach space X. If the set

$$\Gamma = (\sigma(T) \cap G) \cup \{\lambda \in G \backslash \sigma(T); \|(\lambda - T)^{-1}\| \geq \frac{3}{1-|\lambda|}\}$$

is dominating in G, then the functional calculus of T is isometric. That is, $\|f(T)\| = \|f\|_\infty$ for every $f \in H^\infty(G)$. Where $\sigma(T)$ denotes the spectrum of T.

Proof. The proof of Lemma 1 is similar to that of Theorem 3.3 in [6] and is therefore omitted.

Lemma 2. Let T be a polynomially bounded operator on a Banach space X. If the functional calculus of T is isometric, then T has a nontrivial invariant closed subspace.

Proof. the proof of Lemma 2 is identical with that of contraction operators (cf. [5], [6]) and is therefore omitted.

Now we are in a position to give the main result.

Theorem 1. Let T be a polynomially bounded operator on a Banach space X such that the set

$$S = \{\lambda \in G;\ \text{there is a vector}\ x \in X,\ \text{such that}\ \|x\| = 1,\ \text{and}\ \|(\lambda - T)x\| < \frac{1}{3}(1-|\lambda|)\}$$

is dominating in G, then T has a nontrivial invariant closed subspace.

Proof. Set

$$\Gamma = (\sigma(T) \cap G) \cup \{\lambda \in D \backslash \sigma(T); \|(\lambda - T)^{-1}\| \geq \frac{3}{1-|\lambda|}\}.$$

We now show $S \subset \Gamma$. In fact, for every $\lambda \in S$, if $\lambda \in \sigma(T)$, then $\lambda \in \Gamma$; if $\lambda \in \rho(T)$, then by the definition of S, there is a vector $x \in X$, such that $\|x\| = 1$, and

$$\|(\lambda - T)x\| < \frac{1}{3}(1 - |\lambda|).$$

Therefore we have

$$\begin{aligned}1 = \|x\| &= \|(\lambda - T)^{-1}(\lambda - T)x\| \\ &\leq \|(\lambda - T)^{-1}\|\|(\lambda - T)x\| \\ &\leq \frac{1}{3}\|(\lambda - T)^{-1}\|(1 - |\lambda|).\end{aligned}$$

This shows that $\|(\lambda - T)^{-1}\| \geq \frac{3}{1-|\lambda|}$, and so $\lambda \in \Gamma$. From the above we have $S \subset \Gamma$. Since the set S is dominating in G, the set Γ is dominating in G. It follows from Lemma 1 that the functional calculus of T is isometric. Thus by Lemma 2 T has a nontrivial invariant closed subspace. This completes the proof of Theorem 1.

Corollary 1. Let T be a polynomially bounded operator on a Banach space X. If the spectrum $\sigma(T)$ of T is dominating in G, then T has a nontrivial invariant closed subspace.

Proof. If $\sigma_r(T) \neq \{0\}$, then $\overline{\text{range }(\lambda - T)}$ is a nontrivial invariant closed subspace for T.

We now can assume with loss of generality that $\sigma(T) = \sigma_a(T)$. Thus for every $\lambda \in \sigma(T)$, there is a sequence $\{x_n\}$ in X such that $\|x_n\| = 1$ and $\|(\lambda - T)x_n\| \to 0 (n \to \infty)$. This shows that there is an n such that

$$\|(\lambda - T)x_n\| < \frac{1}{3}(1 - |\lambda|).$$

That is, $\lambda \in S$, where

$$S = \{\lambda \in G;\ \text{there is a vector } x \in X,\ \text{such that } \|x\| = 1,\ \text{and } \|(\lambda - T)x\| < \frac{1}{3}(1 - |\lambda|)\}.$$

Therefore we have $\sigma(T) \subset S$. Since the set $\sigma(T)$ is dominating in G, the set S is dominating in G. Thus by Theorem 1 T has a nontrivial invariant closed subspace. This completes the proof of Corollary 1.

Corollary 2. Let T be a contraction operator on a Hilbert space H. If the spectrum $\sigma(T)$ of T is dominating in G, then T has a nontrivial invariant closed subspace. That is, every contraction operator with thick spectrum has a nontrivial invariant closed subspace.

Remark 1. Corollary 2 is the main result of [7].

References

1. Abramovich, Y.A., Aliprantis, C.D.: An Invitation to Operator Theory. Amer. Math. Soc., Providence (2002)
2. Apostol, C.: Ultraweakly closed operator algebras. J. Operator Theory 2, 49–61 (1979)

3. Aronszajn, N., Smith, K.T.: Invariant subspaces of completely continuous operators. Ann. of Math. 60, 345–350 (1954)
4. Beauzamy, B.: Introduction to Operator Theory and Invariant Subspacs. North-Holland (1988)
5. Bercovici, H.: Factorization theorems and the structure of operators on Hilbert space. Ann. of Math. 128(2), 399–413 (1988)
6. Bercovici, H.: Notes on invariant subspaces. Bull. Amer. Math. Soc. 23, 1–36 (1990)
7. Brown, S., Chevreau, B., Pearcy, C.: Contractions with rich spectrum have invariant subspaces. J. Operator Theory 1, 123–136 (1979)
8. Foias, C., Jung, I.B., Ko, E., Pearcy, C.: Hyperinvariant subspaces for some subnormal operators. Tran. Amer. Math. Soc. 359, 2899–2913 (2007)
9. Halmos, P.R.: A Hilbert Space Problem Book, 2nd edn. Springer, Heidelberg (1982)
10. Kim, H.J.: Hyperinvariant subspaces for operators having a normal part. Oper. Matrices 5, 487–494 (2011)
11. Kim, H.J.: Hyperinvariant subspaces for operators having a compact part. J. Math. Anal. Appl. 386, 110–114 (2012)
12. Pisier, G.: A polynomially bounded operator on Hilbert space which is not similar to a contraction. J. Amer. Math. Soc. 10, 351–369 (1997)
13. Liu, M., Lin, C.: Richness of invariant subspace lattices for a class of operators. Illinois J. Math. 47, 581–591 (2003)
14. Liu, M.: Invariant subspaces for sequentially subdecomposable operators. Science in China, Series A 46, 433–439 (2003)
15. Liu, M.: Common invariant subspaces for collections of quasinilpotent positive operators on a Banach space with a Schauder basis. Rocky Mountain J. Math. 37, 1187–1193 (2007)
16. Radjavi, P., Rosenthal, P.: Invariant subspaces. Springer, New York (1973)

Voronoi Feature Selection Model Considering Variable-Scale Map's Balance and Legibility

Hua Wang*, Jiatian Li, Haixia Pu, Rui Li, and Yufeng He

Faculty of Land Resource Engineering,
Kunming University of Science and Technology, Kunming, China
{yiniKunming2011,ljtwcx}@163.com,
{pugongying928,heyufeng8805}@126.com,
zisefeiyang2010@hotmail.com

Abstract. Variable-scale map because of its variability, destroys the constant of original scale, causing the map enlarge regional information easy to read, other regions are severely compressed and difficult to identify, reducing the map legibility. In this paper, we proposed a new pattern called Voronoi Feature Selection to solve the problem of information compression, considering map's legibility and equilibrium. The main idea is that we use voronoi adjacency relationship model to select features, instead of traditional euclidean distance model, use voronoi influence ratio to determine the feature whether or not to remain, and remove the small influence features to reduce the loading of extrusion area, as well as improve the legibility of map. The comparative experiment results show that our methods make the transformed map readability and clearness to express, and it has a good feasibility.

Keywords: variable-scale map, information balance, legibility, voronoi.

1 Introduction

Magnifier map is a typical delegate to variable-scale map, whose important objects on the map are represented by the larger-scale, while the rest are displayed on a small-scale, in this "distortion", readers can focus on the key objects. In form, scale regulation is close to a continuous change, has no big jump, so it is better to meet the demands of continuity of map reading. In operation, it can get more information but avoid the frequent conversion of map zoom in or zoom out. Thus, the magnifier map is considered to be a very practical mapping form for years.

In recent years, the research on variable-scale map is mainly concentrated in following three aspects: (a) Projection method: it starts from the perspective of mathematical, to transform the flat map into the variable-scale map. Wang et al. [1, 2] scientifically summarized this projection method, and proposed an adjustable

* Corresponding author.

F.L. Wang et al. (Eds.): WISM 2012, LNCS 7529, pp. 26–35, 2012.

magnifier projection method aiming at the problem of lack of effective control of scale change. This map projection deformation can draw up various map of differences scale, Yang et al. [3] designed combination projection method, made local-scale everywhere on the map to differ a fixed multiple. (b) Visualization method: Harrie [4] demanded for mobile mapping, researched the deformation of variable-scale map based on the thought of coordinate transformation. AI et al. [5] applied the variable scale method in navigation electronic map, realized the effect of "near larger, far smaller". (c) Self-adaption method: Fairbairn et al. [6, 8] presented a linearly reduced-scale method from city center to edge area, and it solved the problem of crowded of space target preferable. According to the target number equilibrium, Chen et al. [7] devised mobile mapping clip model under voronoi adjacency relationship.

2 Question Describe

In physics, magnifier has three elements: object, image, focal length. The distance from objects to lens is called object distance(u), as well, the distance from image to lens is called image distance(v), they have a relationship with focal length: $u^{-1} + v^{-1} = f^{1}$. If put the object in focus, at this time, through the lens, we can see a image which the object is magnified, this is the basic theory of magnifier imagery. Because it is not a actual convergence point of refraction lights, but their back lines intersected, so it can't received by the physical screen, it's a virtual image. Thus, projection method is used to simulate the process of magnifier imaging. Fig. 1(a) shows the grid lines corresponding relations of pojection transform before and after. Regard the x, y axis increasing at same time as the enlarged area, and the rest for extrusion area, the transformed region can be divided into enlarged and extrusion area, shown as in Fig.1(b). Firstly, the projection transformtation process does not take the non-uniform distribution of spatial objects into account, and can't guarantee the integrity and outstanding expression of adjacent information, namely the problem of information balance. Secondly, the legibility of map has been proposed qualitatively by scholars, but still lack effective calculation model to support it. These two are the key and core problem of adaptive method [6-8]. In essence, magnifier map is the results for select partial targets of flat map and transform. This paper proposed a variable-scale map feature select medol based on Voronoi Adjacency relationship. Voronoi diagram is a geometry structure of spatial partition, it assumes that a group of growth points expand around at same time, until meeting, to form the spatial coverage of every growth point [9, 10]. On the basic of Voronoi adjacent relation of the key target, we can determine the set of targer selection, and establish a legibility evaluation which measured by voronoi influence scope of each target, then we can remove poor legibility targets from their selection.

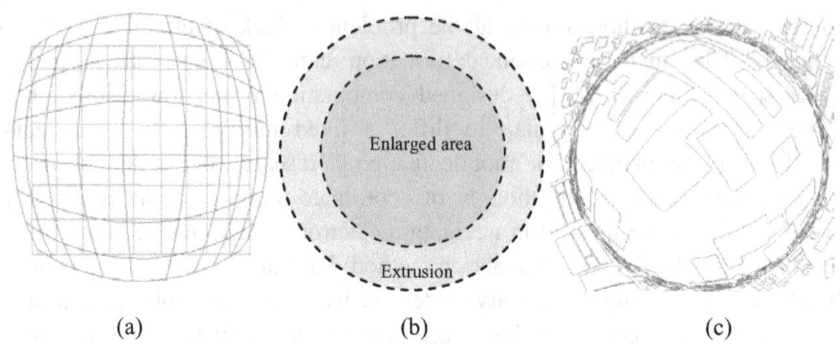

Fig. 1. The Projection Transformation for Magnifier Map

3 Feature Select Model

3.1 Modeling Process Describe

The general process of variable-scale map projection transformation can be describe as: $f(\mathbf{A}) \rightarrow \mathbf{A}'$, that is, to vary the scale under projection rule f, the target dataset **A** in the planimetric map project turn into **A′** in variable-scale plane, f is a mapping function among targets in mathematics, show as formula (1):

$$\begin{cases} x' = x_0 + f(r)(x - x_0) \\ y' = y_0 + f(r)(y - y_0) \\ f(r) = \alpha(1+r)^{\beta} \\ r = \sqrt{(x-x_0)^2 + (y-y_0)^2} \end{cases} \quad (1)$$

Where, (x_0, y_0) is focus point of magnifier map, (x, y) is any point of original map; (x', y') is the point (x, y) after transformed. α reflects the level of scale change and control the size of scale. β reflects the form of scale change: center focus can be zoomed in if β takes a positive value, while, center focus can be zoomed out if takes a negative value. The physical meaning of formula is that using the distance r (from point to focus) to adjust the focal length $f(r)$, the larger the distance is, the smaller the focal length is, and the smaller the imaging is, on the contrary, imaging is greater. Because of the impact of f, $a(a\in\mathbf{A})$ converts to $a'(a'\in\mathbf{A}')$, lead to a' product some changes in size or shape, thus, we can get a prominent effect on the expression of key target. From the result of variable-scale map we can see, the final effect is focus on the target-centric information balanced and legibility, hence, our study views the maximal change rate of projection area as the key target, for relative to the general objects, the key object is closer to the projection center. If the key target is represented by a_{max}, it can be calculated as:

$$a_{max} = \boldsymbol{max}\{a_i \mid \boldsymbol{area}(a_i')/\boldsymbol{area}(a_i),\ a_i\in\mathbf{A},\ a_i'\in\mathbf{A}',\ f(\mathbf{A})\rightarrow\mathbf{A}'\} \quad (2)$$

all above is ecumenical process how constant-scale turns into the variable-scale. We select part of elements $\mathbf{A}(\mathbf{A} \subseteq \mathbf{M})$ from the whole dataset $\mathbf{M}$, and transform them to $\mathbf{A}'$, after, find the key element. Figure 1 shows the process of modeling, see as in Fig. 2.

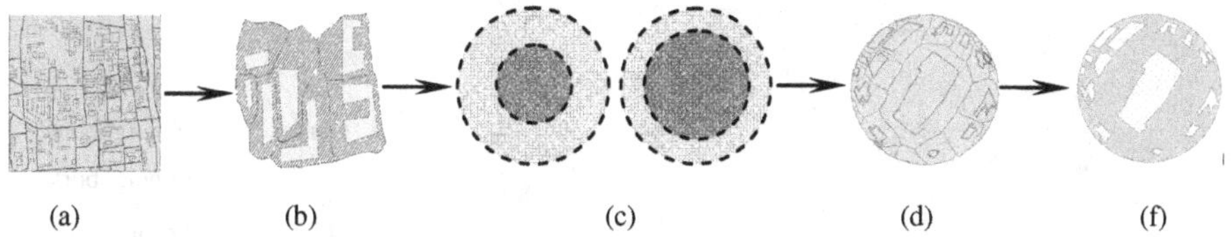

Fig. 2. This figure shows the voronoi select process of Magnifier Map. There, (a) is the original map, (b) is target object and its voronoi neighbors, (c) adjust focal length, (d) remove non-legibility objects, (f) is the result objects.

3.2 Balance Modeling

Legibility refers that the number of features after the projection ransformation should not be too much or too little, and the features are distributed around the key target more evenly. Voronoi nerghbors which is defined based on Voroni adjacency has been shown to have a uniform distribution characteristic[7, 9, 10]. Legibility modeling is that focusing on the target-centric then selecting features among its Voronoi neighbours, and making these selection transform to meet the magnifier effect.

Based on Voronoi adjacency relationship, we look for the voronoi neighbors collection $\mathbf{A}_{vor}$ of target-centric $a_{\max}$. Now, the establishment of $\mathbf{A}_{vor}$ is based on the full Voronoi adjacency, so the number of features within the expression scope is not sure, if use this scope to inverse compute projection scale, the expression of key target will be reduced. In order to highlight the amplification of key target, we select 1-order neighbours MBR (Minimum Enclosing Rectangle) as transformation experimental objects, sign as MBR_v. The feature contained by Voronoi polygon which has same edge with the $a_{\max}$'s Voronoi polygon, has one Voronoi distance from $a_{\max}$, called 1-order nature neighbour of $a_{\max}$. All these which have same edge with 1-order neighbours are called 2-order nature neighbor of $a_{\max}$, and they have two voronoi distance. Thus, the features who have k voronoi distance are called k-order nature neighbor of target-centric. Fig. 3 shows the key target and its voronoi k-order neighbors.

Magnifier planes are generally roundness or square, the random distribution of map features makes MBR_v not necessarily square, when MBR_v project to magnifier, it can not overspread all magnifier plane and leave some space. We utilize affine transformation to slove this problem. Affine transformation can maintain the basic shape of feature, and does not disorganize the topological relation between features. Suppose the size of MBR_v is $w \times h (w > h)$, after the affine transformation is $w \times w$, the point original coordinate is(X,Y,1), after transformed is (X′,Y′,1), if the side of MBR_v on x axle is longer than y axle, so the tranformation formula is:

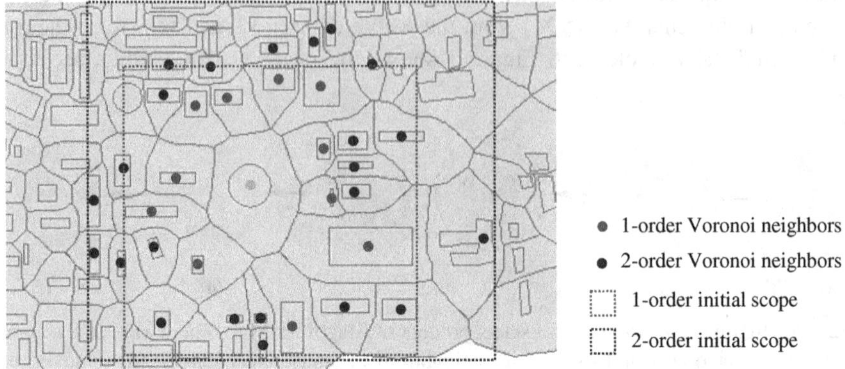

Fig. 3. Key target, Voronoi k-Order Neighbors and Its Initial Expression Scope

$$\begin{bmatrix} X' \\ Y' \\ 1 \end{bmatrix} = \begin{bmatrix} 1 & 0 & 0 \\ 0 & b/a & 0 \\ 0 & 0 & 1 \end{bmatrix} \bullet \begin{bmatrix} X \\ Y \\ 1 \end{bmatrix} \tag{3}$$

In contrast, the side on *y* axle is longer than *x* axle,its transition matrix is:

$$\begin{bmatrix} b/a & 0 & 0 \\ 0 & 1 & 0 \\ 0 & 0 & 1 \end{bmatrix} \tag{4}$$

Then, project the features to magnifier plane, as shown in figur 4:

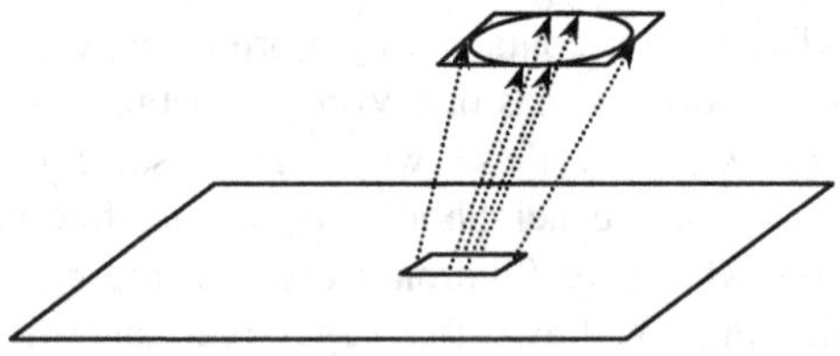

Fig. 4 Affine Transformation

To reduce the amout of compulation in the middle, above process can be simplified one step, transform a map coordinate system to the device coordinate system directly. We can presume the map affine coordinate system is I:$[O;e_1,e_2,e_3]$, and the device affine coordinate system is I':$[O;e_1',e_2',e_3']$, in I and I', their respective coordinate is (X,Y,1)、$(x',y',1)$, if the transition matrix is C, and the transformation formula is:

$$\begin{bmatrix} X \\ Y \\ 1 \end{bmatrix} = \begin{bmatrix} c_{11} & c_{12} & c_{13} \\ c_{21} & c_{22} & c_{23} \\ 0 & 0 & 1 \end{bmatrix} \bullet \begin{bmatrix} x' \\ y' \\ 1 \end{bmatrix} \tag{5}$$

Easily, we can accquire the coordinate of four vertexs and geometrical center of the MBR of magnifier to calculate **C**. Then,we get the definite formula, project all the features in MBR_*v* to magnifier plane.

3.3 Legibility Modeling

After transformation, the magnifier can be divided into enlarged area and extrusion area. Expected result is the key target appears in the zoom area and express an enlarged effect, and its Voronoi neighbors distribute in the extrusion or a little appears zoom area. 1-order Voronoi select model can make the feature extract balance, however, the map information after transformation is too poor to disturb the spatial analysis for key target. Using k-order Voronoi feaure set for the selection can add more informaiton for analysis, nonetheless, features in the extrusion area suddenly increase,affect the readability of map. In essence, extrusion area is the change from large-scale to small-scale for the map. In the map generalization process, it needs to romove or fuse some minor information, to meet small-scale display dimension. Therefore, the features in extrusion area should be reclassify, filter some feature which has small influence, to improve the legibility of map.

The map features are mostly discrete distributed, and each feaure has a certain influence scope. This influence scope is not only related to the size of feature on map, but also related to the distribution of other features around. The Voronoi diagram of feature can effectively measure its spatial neighbour relation with other's, as well the influence scope.

Voronoi diagram has such properties: a) In the context of same density, the greater the growth factor is, the bigger the Voronoi region will be. b)In the same capacity space, the smaller the density of growth factor is, the bigger the Voronoi region will be, conversely, the Voronoi region will decresent when density goes to largen. From the properties of Voronoi,we can see, the Voronoi diagram of some feature group (such as city business center) is very small, but the spatial information they contained have a materialeffect for human, if only consider featuer's Voronoi region as the legibility extraction factor, these features will be removed, so that some effect infomation of map will be lost. Therefore, we choose the ratio of the feature area into its Voronoi region as the measurement factor of feature's selection. The formula is as follows:

$$p_i = \frac{S_{oi}}{S_{vi}} \times I_i \tag{6}$$

There, P_i is the selection possibility of feature, S_{oi} is the feature's area and S_{vi} is the Voronoi area, I_i stands for the importance degree (the Voronoi distance from this feature to the key target). Note that, for a wide line (such as roads, rivers), let its width instead of S_{oi}, while a thin line, let S_{oi} be 1, namely, only make the Voronoi area as the judgement factor, the amended formula is:

$$p_i = \frac{1}{S_{vi}} \times I_i \tag{7}$$

The modeling process describes simply as follows:

a) Create Voronoi diagram on the magnifier map after balance modeling.
b) Find the k-order feature of $a_{\max}$ and label it, that is, I_i=k.
c) Layering these features according to their different type (point, line, polygon), calculate their possibility value with (6)(7), and calculate the average value upon the different levels.(note: $a_{\max}$ does not participate in the average.)
d) Traveral every feature on its level, the featuer will be removed to meet both of the following conditions: ① P_i<$P^{\wedge}$; ②Its 1-order feature has not been removed. Repeat the operation, until there are no features can be removed. Here, P_i is the possbility value of each feature, $P^{\wedge}$ is the average.
e) Determine the final features remained on the map, complete the legibility adujstment of extrusion area.

4 Experiment and Conclusions

4.1 Experimental Result

we utilized a city map (residential building and street) as the experimental data, in the environment of Visual Studio 2008 and C#+ArcEngine 9.3, we accomplished the proposed method and related algorithm. Illustrations of the partial experimental result are shown in Fig. 5.

Fig. 5(b) shows the initial variable-scale map transformed through formula, for failing to consider the legibility and balance of information display, the distortional map is hard to read than the original map, and lose the significance of variable-scale. Fig. 5(d) shows the final result for adjusting the legibility using Voronoi feature area ratio, comparing with Fig. 5(b), it is obviously to improve the readability. Just because the feature's influence weight was considered with Voronoi adjacency, features that have important influence to the key target are remained.

4.2 Evaluation of Legibility

The amendment of extrusion area on variable-scale map, in essence, is the cartographic generalization from a large-scale to a small-scale. In the process of rarefying features, the more effective information transformed, the less information loss, so the better the result of map generalization is.

Entropy is one of the main theories for quantitative measurement of vector maps. The maximum entropy theory: when the first n messages occur with equal probability (i.e. the Voronoi area of each feature is equal), the information entropy has the maximum value. Therefore, for a map with the same number of features, if these

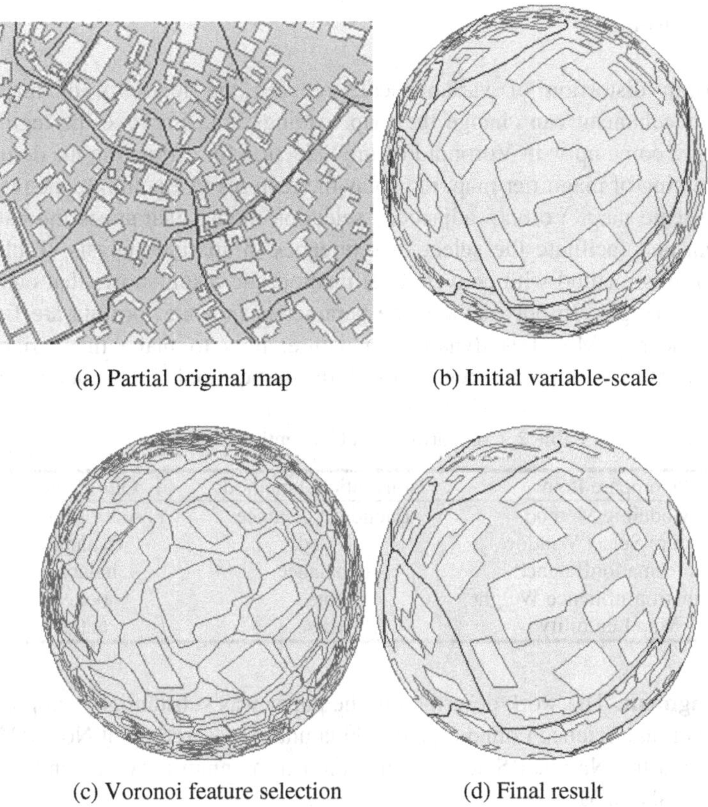

(a) Partial original map (b) Initial variable-scale

(c) Voronoi feature selection (d) Final result

Fig. 4. Experimental Result

features distribute more uniform, the geometric information entropy they contain will be greater. We use Voronoi feature area ratio to dilute the map, considering each feature's influence weight to the key target in magnification center, and the terrain distribution around, to make the results more uniform. Even though the decreasing number of map features reduces the maximum entropy, however, the more uniform distribution make the geometric entropy increase, the relative entropy of map is also ascending, see as Table 1:

Table 1. Entropy Calculate[1]

	Maximum entropy	Geometry entropy	Relative entropy
Fig.5(b)	6.233	4.984	72%
Fig.5(d)	5.977	5.012	83%

[1] Maximum entropy= $\log_2(n)$ Relative entropy= Geometry entropy/Maximum entropy

Geometry entropy $= -\sum_{i=1}^{n}(s_{vi}/s)(\log_2 s_{vi} - \log_2 s)$

5 Conclusions

The extrusion distortion of variable-scale reduces the legibility of map; using reasonable mechanism can change the map loadings, and improve the legibility. In this paper, we come up with Voronoi Feature Selection for estimating the deformation and compressing of magnifier map, redisplay the features to ameliorate the readability of variable-scale map. Voronoi adjacency selection model guarantees the balance of information, and facilitate the select of references in spatial analysis. Table 2 is a compare for this method with conventional method, the result shows that our method are more advantages. In our method, the parameters in scale formula are fixed, but Vonoroi adjacency Model is dynamic in nature, how to make the variable-scale parameter adapt to Vonoroi adjacency model, this issue should be further researched.

Table 2. Comparing with Conventional Method

Compare Item	Conventional method	Our method
Modeling Method	Euclidean distance	Voronoi adjacency
Feature Select Window	Fixed	dynamic
InformationBalance	unbalance	balance
InformationInfluence Weight	worse	better
Map Legibility	worse	better

Acknowledgment. The work described in the paper was substantially supported by a grant the National Science Foundation of China under research grant No. 40901197 & 41161061, and the National Science Foundation in Yunnan province under research grant No. 2008D032M.

References

1. Wang, Q., Hu, Y.: A CAC-based General Multi-focal Pojection. Joural of Wuhan Technical University of Surveying and Mapping 17, 18–25 (1992) (in Chinese)
2. Wang, Q., Hu, Y.: A Kind of Adjustable Map Projection With "Magnifying Class" Effect. Agta Geodaetioa et Cartographica Sinica 22, 270–278 (1993) (in Chinese)
3. Yang, X., Yang, Q., Zhao, Q.: A New Kind of Variable-scale Map Projection Research on Composite Projection. Journal of Wuhan Technical University of Surveying and Mapping 24, 162–165 (1999) (in Chinese)
4. Harrie, L., Tiina Sarjakoski, L., Lahto, L.: A Variable-Scale Map for Small-Display Cartography. In: Proceeding of the Joint International Symposium on "Geospatial Theory, Processing and Applications," (ISPRS/Commission IV, SDH 2002), Ottawa, Canada (2002)
5. Ai, T., Liang, R.: Variable-Scale Visualization in Navigation Electronic Map. Geomatics and Infromation Science of Wuhan University 32, 127–130 (2007) (in Chinese)
6. Hampe, M., Sester, M., Harrie, L.: Multiple Representation Databases to Support Visualization on Mobile Devices. In: Proceeding of the XXth ISPRS Congress, Istanbul, Turkey (2004)

7. Chen, J., Zhao, R.: Voronoi-based k-order neighbour relations for spatial analysis. ISPRS Journal of Photogrammetery & Remote Sensing 59, 60–72 (2004)
8. Fairbairn, D., Taylor, G.: Developing a Variable-Scale Map Projection for Urban Areas. Computers & Geosciences 21, 1053–1064 (1995)
9. Gold, C.M.: Review.: Spatial Tessellations-Concepts and Application of Voronoi Diagrams. International Journal of Geographical Information Science 8(2), 237–238 (1994)
10. Okabe, A., Boots, B., Sugihara, K.: Nearest neighborhood operations with generalized Voronoi diagram: a review. International Journal of Geographical Information System 8(1), 43–71 (1994)
11. Agrawala, M., Stolte, C.: Rendering Effective Route Maps: Improving Usability through Generalization. In: Proceeding of SIGGRAPH 2001, pp. 241–249 (2002)
12. Guerra, F., Boutoura, C.: An Electronic Lens on Digital Tourist City-Maps. In: Proceedings of the 20th International Cartographic Conference, Beijing, China, p. 1151 (2001)
13. Snyder, J.P.: Magnifying-Class Azimuthal Map Projection. The American Cartographer 14, 61–68 (1987)
14. Sukhov, V.I.: Information capacity of a map entropy. Geodesy and Aerophotography X, 212–215 (1967)
15. Sukhov, V.I.: Application of information theory in generalization of map contents. In: International Yearbook of Cartography, X, pp. 41–47 (1970)

A Code Dissemination Protocol of Low Energy Consumption

Haiyong Wang[1,2], Geng Yang[1], Jian Xu[1], Zhengyu Chen[1], and Zhen Yang[1]

[1] Key Lab of Broadband Wireless Communication & Sensor Network Technology of Ministry of Education,Nanjing University of Post &Telecommunications, Nanjing, 210003, China
[2] College of Internet of Things, Nanjing University of Posts and Telecommunications, Nanjing 210003,China
{why,yangg,xuj,chenzy,yangz}@njupt.edu.cn

Abstract. Dissemination and update of the code image plays an important role in application of the Internet of Things (IoT). As a part of the IoT, Wireless sensor networks should provide the code dissemination. The commonly used method for periodical broadcasting metadata packages to determine whether code dissemination is needed consumes excessive energy. So a low energy consumption code dissemination protocol is proposed for this issue. That is whether code dissemination is necessary can be determined when the sensor node is under normal communication. The simulation result shows that the proposal in this article has reduced the energy consumption when the network needs code dissemination.

Keywords: Code dissemination, Low energy consumption, Code image, Wireless sensor networks, Metadata.

1 Introduction

Wireless sensor networks(WSNs) is comprised of a large number of sensing nodes with limited energy, communicating capability and hardware resources. The sensing nodes are deployed at target areas to monitor、sense and collect information of the object. The collected information from sensing node will be transmitted to sink node via organization network for user to analyze, process and make a decision.

In practical application, when versions of sensing nodes within a network are different, sensing data format acquired with sensing node of older version will result in an error during the data aggregation, and lead the sink node receives inaccurate data. This result will affect the user's analyzing process, sometimes will even result in the network be separated into several sub-networks.

On one hand, with the change of network topology and demands and the development of technique, the existing program is found to need modifications, this will inevitably modify the applied program carried by the deployed sensor node or to add new application program. Being deployed in severe environment with rare signs of human habitation or even in hostility controlled area, it is impractical to reclaim sensor nodes to load the updated program and re-deploy them. This requires the new

F.L. Wang et al. (Eds.): WISM 2012, LNCS 7529, pp. 36–43, 2012.

code image to be disseminated to carry out unified network online code update. For example, sensor nodes can be deployed in nuclear leakage area to carry out the costly task to evaluate radiation situation after the nuclear leakage in 2011 Japanese earthquake through monitor, but such task is unlikely to be wholly predicted in advance, it is impractical to preload necessary software completely before the nodes deployed, but a code update of the sensor node can be realized through code dissemination.

On the other hand, product development of WSNs is a progressive process, which needs continual code debugging and modifying operation. By using technique of code dissemination and code update will be able to reduce the repetition work of the researchers and improve the work efficiency.

2 The Proposed Encryption Algorithm

The latest version of TinyOS2.1.1[1] published in April, 2010 has self-contained Trickle, Deluge code dissemination protocol. In which the Trickle[2] is a kind of standard one to multiple code dissemination algorithm in operation status of WSNs. In the Trickle algorithm, the sensor node sets a threshold value k (to suppress the broadcast redundancy of code metadata), and an update cycle τ (for the time interval of code dissemination). Once a sensor node receives a metadata the same as itself, a variable c with initial value is zero will start a plus 1 operation. Within an update cycle τ, when c>k, the sensor node suppress itself make no operation, otherwise the metadata will be broadcasted once by random selection ($\tau/2, \tau$) and set variable c with the initial value zero within the consequent time interval τ. It is found that the smaller τ is, the higher energy consumption is needed for code dissemination, while the bigger τ is, the longer time delay is needed. Under normal conditions, a relatively bigger τ is set for network operation status to reduce the network energy consumption. But when the metadata discrepancy is found and an update is needed, τ should be reset immediately to a very small value to accelerate the network code dissemination speed.

Deluge[3] extends Trickle and supports more effective code dissemination of big data quantity, and divides the updated codes into N pages with pagination technique, each page will be divided into a group of data packages with a fixed size. Each page will be determined whether an update is necessary according to Age Vector encapsulated in metadata to improve parallel transmission capability through pipeline technique, data transmitting is realized through 3-handshake (ADV-REQ- DATA) mechanism and the whole process is divided into 3 statuses: maintenance status, requesting status and transmitting status. Where the maintenance status uses a maintenance mechanism similar to Trickle protocol, the requesting status adds a pause process when requesting failed, the transmitting status follows page bottom priority principle when the data is transmitted.

IDEP[4] adopts metadata negotiation and message suppress mechanism, packet loss detection mechanism, spatial multiplexing mechanism, distance and energy based node choosing algorism to reduce redundant message transmission and realizes rapid and reliable mirror distribution.

The literature[5] proposed a kind of code dissemination model suitable for delay tolerant mobile sensor network. During the general running time of network, sink node maintains the statistic of contact rates between nodes. By taking advantage of the information, it can predict a node sequence of code acquisition before the dissemination stage, then the distribution time can be calculated out to ensure that all the nodes boot the code at same time. The literature[6] improves the code dissemination performance via random linear code and pagination method.

Within the lifetime of WSNs, the sensor node consumes most energy when transmitting and receiving messages. During the network operation phase, a periodic broadcast is used in algorithm[2-5], energy consumption will be increased with the operation time. As a matter of fact, the energy consumed in metadata broadcast dissemination is much more than code dissemination, this results energy consumption of metadata broadcast dissemination during operation is as much as tens to hundreds times of that of the actual code dissemination. For example, Deluge is a default code dissemination protocol of TinyOS2.1.1 system, the advertisement data package is broadcasted periodically every 2 minutes during network operation phase, a broadcast of 24 hours corresponds with the energy consumption of disseminating 2.5 KB codes. During the network operation phase, energy consumption of code dissemination can be reduced by increasing broadcast metadata time interval. Yet because of the checking delay, this time interval cannot be increased obviously.

3 Numerical Simulation

Communication between nodes is always accompanied with plenty of energy consumption in the wireless sensor network. To reduce the energy consumption caused by determining if a code distribution is needed, the broadcast metadata data package items sent for determining a necessity for code distribution should be reduced as much as possible. The author proposes a Code Dissemination of Low Energy Consumption (LECCD) protocol in view of the network node periodic broadcast data package to determine code dissemination issue in the algorithm[2-5]. Here is the detailed description of LECCD protocol.

3.1 Construction of Node Neighbor List

After deployment of wireless sensor network node, the location information of the nodes and the connectivity between nodes are unknown, but can be achieved with equipped GPS for each node or orientation algorithm. Because of the price and the accuracy of orientation, methods from SEEM algorithm are sometimes applied in practice. That is to transmit a “Neighbors Discovery (ND)” broadcast package through sink node to carry out flooding broadcast and in this way to obtain the neighboring node location information, the obtained ID of the neighbor node is then saved in the neighbor list.

Here is the description in detail:

1) Sink node sends ND broadcast data package first by broadcast. The data package introduces random diffuse mechanism to select the next hop node for every ND. The source node ID is included in ND information.

2) When the middle node S_k receives the ND_j data package transmitted from S_j, the suffix j denotes the ND data package transmitted from the node j.

Firstly check if S_j is enlisted in its own neighbor list. If not, then S_j is added into the neighbor list and broadcast the ND data package. Otherwise no operation will be carried out.

3) When there is no ND data package to be transmitted in the network or the value exceeds the setting time value, the neighbor node ID information is stored in each node.

In the above neighbor list construction, each network node has been included its own ID designation into the broadcast data package and included the source node ID of the received ND broadcast data package into its own neighbor list. So when the network completes the initialization, every node has stored the neighbor node ID information in the list. Meanwhile, during the network operation, the neighbor list will be cleared when a node reset or update the code.

3.2 Determine the Code Dissemination Process

In order to reduce the energy consumption caused by determine code dissemination, the status of network node are characterized as operation state, ready state and dissemination state, Figure 1 illustrates the node state conversion. Where the operation state means action status of node acquisition, transmission, receiving and convergence; ready state means status of the node is likely need to perform code dissemination, it's an intermittent status between operation and dissemination states; dissemination state means status of the node to run the dissemination protocol to carry out code dissemination. The following passages describe the conversion process under relevant status.

When the node is under operation state:

(1) When node n1 receives data package from node n2, whether n2 is included in neighbor list is checked firstly.

A. If n2 is in the list, n1 receives the data package

B. If n2 is not in the list then n1 will enter the ready state. A broadcast data package with objective node ID of n2 will be transmitted at the same time. Among those, the broadcast data package includes source node metadata, source node ID and objective node ID.

(2) When node n2 receives broadcast data package with an objective node ID of n2, metadata version will be compared. If the version of itself is lower, then a broadcast data package will be transmitted to a destination address of NULL.

(3) When node n3 receives a broadcast data package with an objective address of NULL or none n3, it will compare with the metadata in the received data package.

A. When the two have the same version, the received broadcast data package will be discarded;

B. When the metadata version of its own is lower, a requesting code dissemination broadcast data package will be transmitted and enters a dissemination state;

C. When the metadata version of its own is found higher than the one received, it means the source node needs an update, so a broadcast data package will be transmitted continuously with a destination address as null till the suppress condition is satisfied, which means k number of the same version metadata broadcast package have been received. Thus the metadata versions of n1 and n2 are assured to be the same and with a lower version, both nodes need update can also be detected.

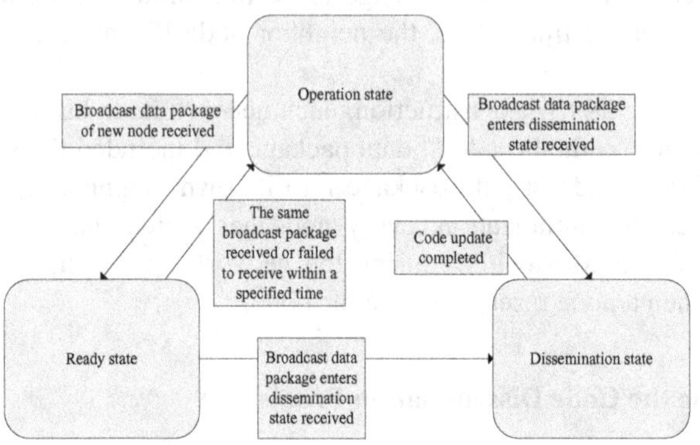

Fig. 1. State conversion of network nodes

(4) A request code dissemination broadcast data package is received at any node and enters dissemination state.

When a node is in a ready state:

Suppose node n1 receives the data package transmitted from n2, while n2 is not included in the neighbor list of n1 and enters ready state.

(1) Node n1 transmits broadcast data package at a frequency of stipulated time T with a destination node ID of n2. When the suppress condition is satisfied or there is no broadcast data package transmitted from node n2 is received, it will return to operation state and discard the data package received.

(2) Node n1 receives the broadcast data package transmitted from node n2, if the two have the same metadata, n2 will be added into the neighbor list and return to operation state, accepts the data package from n2. Otherwise a request of code dissemination broadcast data package will be transmitted and enters dissemination state.

(3) Any code receives a request of code dissemination broadcast data package will enter dissemination state.

When a node is in the state of dissemination:

Any existing dissemination protocol can be used in dissemination state. This proposed code dissemination protocol design philosophy is orthogonal with the existing dissemination protocols at present and any kind of them can be used to perform code dissemination.

LECCD protocol no longer needs a node broadcast metadata periodically but change the node status to realize reducing the quantity of broadcast metadata data package transmitted to determine the need of code dissemination, the communication quantity between nodes are reduced and thus a low energy consumption code dissemination target is achieved.

4 Experiment Result and Analysis

We performed simulation experiment with TinyOS2.1.1 with a self-contained simulating platform TOSSIM. 100 nodes were taken during the experiment and GRID structure was adopted in topological structure, the nodes were distributed in a square of 10×10, gaps between the nodes were the same. 5 feet and 10 feet gap between nodes were taken with a transmitting radius of 50 feet each.

4.1 Energy Consumption Compare

The broadcast data package quantity transmitted with Deluge protocol and LECCD protocol within a network operation cycle. In Fig. 2, when the network is in operation, the quantity of broadcast data package transmitted with Deluge protocol from nodes in the network shows a linear increase with time; when LECCD protocol is used, the quantity of broadcast data package transmitted from network nodes remained unchanged when reaching a certain value. This is because the broadcast data package will be transmitted at update cycle for metadata comparing in Deluge protocol, to determine if a code dissemination is needed till the suppress condition is satisfied. LECCD protocol uses a neighbor list of every node. When the network initialization completed and starts a normal operation state, only new nodes are added into the network or the network nodes are restarted will cause relevant nodes in the network to transmit broadcast data package for a metadata check, in order to determine if a code dissemination is needed. In other cases, a neighbor list stores the neighbor node IDs whose data packages have been received, the periodic transmission of broadcast data package is no longer needed, the energy consumption caused by determine the necessity of code dissemination is reduced.

When the node gap is 10 feet, both Deluge protocol and LECCD protocol have more quantity of broadcast data package than the node gap of 5 feet. This is because the smaller the node gap is, the easier the suppress conditions can be satisfied.

4.2 Code Dissemination Time Compare

Suppose the gap between nodes is 10 feet, load new code to the network topological structure at the top-left node and measure the time for the whole network completing the code dissemination,10 measurements have been taken for compare, Fig. 3 shows the result.

In Fig. 3, the time difference for completing the whole network node code dissemination between using LECCD protocol and Deluge protocol is very small, this difference is almost negligible. So when a network needs to proceed code dissemination, LECCD protocol can none the less ensure its real-time property with little energy consuming.

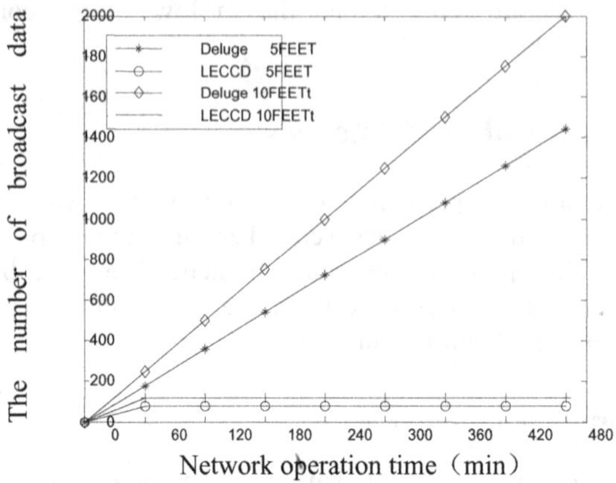

Fig. 2. The number of broadcast data package within different time

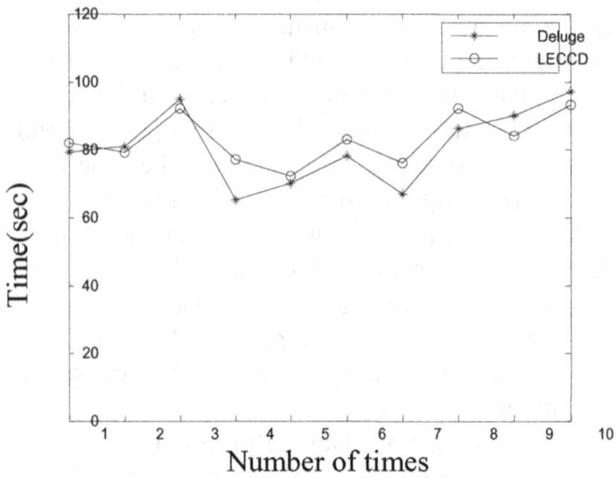

Fig. 3. Code dissemination time

5 Result Analysis Summary

Code dissemination is an important issue of wireless transducer network. The LECCD protocol proposed in this article take full advantage of network node neighbor list in node communication, which will not create excessive affects to the network performance, realizes node version difference check during normal communication. Compare with present dissemination protocol such as Trickle, Deluge, where the node version difference are detected via transmitting and receiving broadcast metadata package periodically, LECCD protocol reduces the periodic broadcast data package and thus reduces the energy consuming greatly. Compare with the existing relevant work, the complexity of the proposed LECCD protocol is lower, the design philosophy is orthogonal with the existing dissemination protocol, which can be used in any of them to carry out code dissemination, and can also be applied to the limited character of the wireless sensor network resource. According to the simulating result, the proposed method has superior performance to other methods in energy consumption.

The future study will be majored on the following two aspects: (1) take into account of the new nodes included in the network and nodes exit situation, (2) to improve the model proposed under more complicated network environment.

Acknowledgments. This work was supported by the National Basic Research Program of China (973 Program) under Grant No.2011CB302903, the National Science Foundation of China under Grant No.60873231, the Natural Science Foundation of Jiangsu Province under Grant No.BK2009426, the Natural Science Research Project of Jiangsu Education Department under Grant No.09KJD510008, the Natural Science Foundation of the Jiangsu Higher Education Institutions of China under Grant No.11KJA520002, Research Fund for the Doctoral Program of Higher Education of China No.20113223110003. the Innovation Project for postgraduate cultivation of Jiangsu Province under Grant No.CXLX11_0415.

References

1. http://docs.tinyos.net/tinywiki/index.php/Getting_started
2. Levis, P., Patel, N., Shenker, S., et al.: Trickle: a self-regulating algorithm for code propagation and maintenance in wireless sensor networks. Technical report, University of California at Berkeley, pp. 15–28 (2004)
3. Hui, J.W., Culler, D.: The dynamic behavior of a data dissemination protocol for network programming at scale. In: Proc. ACM SenSys, pp. 81–94 (2004)
4. Huang, R.H., Zhang, X.-Y., Wang, X.-M., et al.: Efficient and Energy-Saving Network Reprogramming Dissemination Protocol: IDEP. Journal of Information Engineering University 11(4), 504–507 (2010)
5. Huang, Y.F., Lin, F., Zheng, L., et al.: Code Distribution Model in DT-MSN with Variable Contact Rate. Computer Engineering 37(19), 101–103 (2011)
6. Dong, W., Chen, C., Liu, X., et al.: A lightweight and density-aware reprogramming protocol for wireless sensor networks. IEEE Transactions on Mobile Computing 10(10), 1403–1415 (2011)

Dynamic Spectrum Analysis of High-Speed Train Passenger Compartment Luggage Rack Noise*

Chuanhui Wu, Xiangling Gao, and Pinxian Gao

School of Mechanical Engineering
Southwest Jiaotong University
Chengdu 610031, China

Abstract. In order to understand the dynamic changes of vibration radiation noise of high-speed train's passenger compartment luggage rack, the dynamic spectrum is used for analysis and research. Dynamic spectrum is a three-dimensional spectrum which is build on the time - frequency analysis, the research results show that the dynamic spectrum can fully reflect the time varying characteristic, spectral structure and dynamic range of luggage rack vibration radiation noise, and it is sensitive to the changes of train operation status, operation speed and track condition, thus this method can be extended to noise measurement and analysis of the passenger compartment, at the same time, it has important application value on the monitoring of train operation status and comfort level evaluation of passenger compartment noise.

Keywords: luggage rack vibration radiation noise, dynamic spectrum, condition monitoring, noise comfort level.

1 Introduction

Luggage rack vibration radiation noise is structure-borne noise, which belongs to radiation noise generated by luggage rack resonance when the train running at high speed and it has significant impact on the overall noise of passenger compartment.

Noise analysis methods are of two main kinds: time domain analysis method and frequency domain analysis method, time domain analysis method is used for assess noise level, which reflects the intensity of noise; frequency domain analysis method is used to reflect the spectrum structure of noise, namely frequency components. For stationary random noise, it is able to evaluate noise characteristics through the two methods above mentioned. When the train speed is low, the vibration radiation noise is generally weakly stationary, conventional noise analysis method is feasible, but when the train running in high speed or disturb by external environment, the vibration radiation noise is typical non-stationary random noise, especially in the tunnel. At present, the commonly used method for non-stationary random noise is time - frequency analysis, such as short time Fourier transform and wavelet analysis, etc [1]. In this paper, dynamic spectrum method which is based on short time Fourier transform is put forward

* The work is supported by Natural Science Supported Planning of China (2009BAG12A01).

F.L. Wang et al. (Eds.): WISM 2012, LNCS 7529, pp. 44–50, 2012.

to analysis luggage rack vibration radiation noise, it is simple, has clear physical meaning and strong intuitive sense, and application convenience, which was used to analyze friction noise effectively[2], this paper employs it to vibration radiation noise of high-speed train's passenger compartment luggage rack and receives good effect.

2 Dynamic Spectrum

At present, dynamic spectrum is mainly used in speech signal processing, known as the spectrogram [3], namely the language spectrum analysis. Actually dynamic spectrum is a three-dimensional frequency spectrum, the vertical axis is time and the abscissa is the frequency. Spectrum amplitude is expressed by chromaticity level of the image, that is to say, the size of spectrum diagram amplitude presents though color depth, this paper uses white to indicate the maximum noise decibel value and minimum noise value is black, other noise values change from black to white (chroma changes). In the language spectrum analysis, various features of the spectrogram have clear physical meaning. In fact, the dynamic spectrum has long been applied in the random signal analysis, such as: the time spectrum, vehicle speed, traffic spectrum, waterfalls maps, etc [4]. Only in these dynamic spectrums, the spectrum amplitude is described by curve, for a certain time-varying narrowband random signal appear more regularity "mountains" like peak in the time spectrum, its magnitude and direction have a clear physical meaning. The passenger compartment luggage rack vibration radiation noise is a special non-stationary random noise, the particular order statistical parameters is basically a periodic function of time at resonance, in modern signal processing, such signals is called as cyclostationary signals [5]. Trains running at high speed or in tunnel, the luggage rack vibration radiation noise generally exhibit non-stationary characteristics, in the normal road condition exhibit weakly stationary features, for the feature this paper considers that use chroma described dynamic spectrum to analyze have significant superiority.

3 Luggage Rack Vibration Radiation Noise Dynamic Spectrum Acquirement

Dynamic spectrum is a digital image, in the passenger compartment noise measurement, the microphone output signal is an analog signal, after amplification, conditioning and A / D convert to digital signal, sampling frequency fs of A / D convert and data length N is selected in accordance with the sampling theory and the spectral resolution, that is, the sampling theory:

$$f_s \geq 2f_m \tag{1}$$

Spectral resolution:

$$\Delta f = \frac{1}{T} = \frac{f_s}{N}, \quad N = \frac{f_s}{\Delta f} \tag{2}$$

fm is the highest frequency component of the signal, Δf is spectral resolution T is the total sampling time. Train passenger compartment noise frequency analysis range is between 31.5Hz and 8000Hz [6][7], in order to get a better analysis results, according to theory (1), fs = 30kHz, continuous sampling and storage, refer to the ISO3381 standard, short time Fourier transform sequence width of the window is one second, namely 30000 data, assuming that one second time noise is stable or weakly stable, discrete short time Fourier transform is

$$X_i(k) = \sum_{i=-\infty}^{\infty} x(m)w(i-m)e^{-j\frac{2\pi km}{N}} \quad 0 \le k \le N-1 \tag{3}$$

In which, Xi(k) is the noise digital spectrum, x(m) is noise sequence, w(m) is the window sequence, in actual application, i value is of finite length. Rectangular window function has great impact on spectral leakage, so we use Hanning window

$$w(m) = \begin{cases} \frac{1}{2}\left(1-\cos\frac{2\pi m}{N-1}\right) & 0 \le m \le N-1 \\ 0 & n < 0, \ m \ge N \end{cases} \tag{4}$$

The short time Fourier transform Equation (3) describes is easy to obtain though FFT fast algorithm according to window length.

After the noise digital spectrum obtained, chromaticity of the spectrum according to image quantization theory, commonly adopt uniform quantization [8], chroma level selected between 0 and 255, then get chrominance images from the chroma level using the RGB function. Analysis time is 10 seconds (evaluation time), the noise spectrum chromaticity diagram matrix as follows:

$$F_{i,j} = \begin{bmatrix} X_{11} & X_{12} & \cdots & X_{1N} \\ X_{21} & X_{22} & \cdots & X_{2N} \\ \cdots & \cdots & \cdots & \cdots \\ X_{M1} & X_{M2} & \cdots & X_{MN} \end{bmatrix} \tag{5}$$

$$i = 0,1,\cdots,9, \qquad j = 0,1,\cdots,29999$$

In virtual instrument programming language, such as LabWindow/CVI, has the function of composed of two-dimensional matrix and displayed as two-dimensional chromaticity diagram from a set of one-dimensional array [9], and the plot is very convenient. The graphics in this article are all produced by CVI controls. Figure 1 shows the noise dynamic spectrum near passenger compartment luggage rack when the train running at the speed of 300km/h in the normal state, the noise contains the compartment average noise and luggage rack vibration radiated noise. We can clearly see the time variability of the noise, from the statistical parameters analysis point of view, this is a weakly stationary random noise and it is the typical noise dynamic spectrum the luggage rack around.

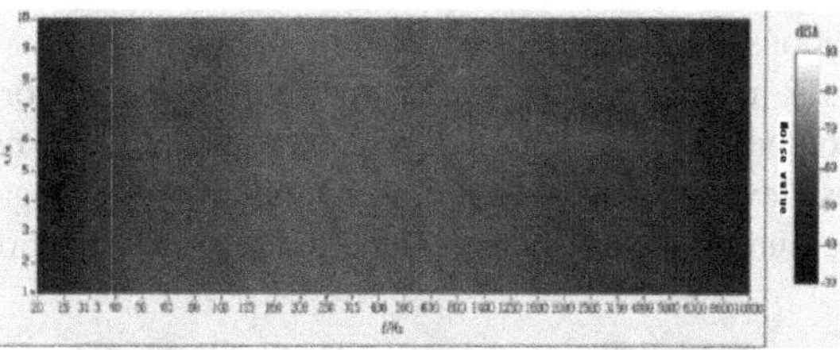

Fig. 1. Noise dynamic spectrum near luggage rack when running at 300km/h

4 Luggage Rack Vibration Radiation Noise Dynamic Spectrum Analyze

Because of the dynamic spectrum combined time domain analysis and frequency domain analysis organically, so dynamic spectrum owns rich noise time - frequency characteristics, figure 1 is the dynamic spectrum, although containing the passenger compartment average noise, this paper uses sound intensity measurement technique [10], the measuring point is close to the luggage rack, the main impact factor is the luggage rack vibration radiation noise which has the following characteristics:

4.1 Luggage Rack Noise Is Non-stationary Random Noise

Along timeline of figure 1, the composition and magnitude of the spectrum is changing which shows this is a non-stationary random noise, there are several characteristics of this time-varying:

In the frequency range (160-800Hz) is basically stationary, spectrum amplitude is almost constant within 10 seconds;

Low frequency (below 160Hz) and high frequency (above 800Hz) is time-varying, but the amplitude is small;

The most sensitive band is 630-5000 Hz for vibration changes.

4.2 Energy Distribution of the Luggage Rack Noise

Along frequency axis of figure 1, we can see that the spectrum mainly in the 150Hz-400Hz which constitute the main body of the noise energy; the amplitude is small below 100Hz and above 500Hz, which have little effect on the whole condition, therefore, luggage rack noise reduction measures should focus on the main energy band.

4.3 Luggage Rack Resonance Noise

The noise near luggage rack is around 70dBA (including the passenger compartment average noise) in normal condition, but when the resonance of luggage rack, the luggage rack noise could reach 80dBA or higher. Figure 2 is a typical radiated noise dynamic spectrum of luggage rack resonance, the main part especially the 600Hz-200Hz, the chroma level significantly increases, reflecting the noise level magnitude and the middle to high frequency components increase, and appear periodic fluctuations, which resulting in the passenger compartment average noise level increase and the comfort level of the passenger compartment noise decrease.

Periodic fluctuations frequency is low, from the relationship of sound and vibration point of view, this is luggage rack vibration frequency range, in normal condition, the modes are not excited, but once excited, which would produce resonance radiation noise, and the noise level increases rapidly, so inhibition of the luggage rack resonance is an important measure of the passenger compartment noise reduction.

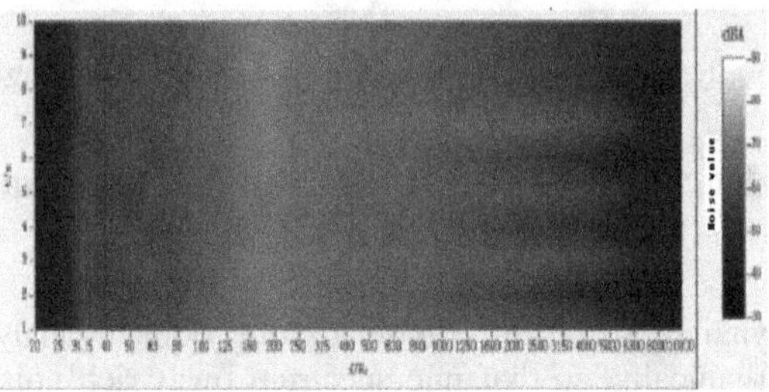

Fig. 2. Noise dynamic spectrum near luggage rack when luggage rack resonance running at 300km/h

5 Luggage Rack Dynamic Spectrum Sensibility for Speed and Railway Condition

Luggage rack dynamic spectrum is sensitive to train speed and changes of line condition (mainly refers to the tunnel), chroma value (median or mean value) and deviation or standard deviation can be used as monitoring parameters and noise comfort evaluation impact indicators, in which the standard deviation is preference, this method can be extended to the passenger compartment noise measurement and analysis.

5.1 Noise Dynamic Spectrum at Various Speed

Figure 3 is noise dynamic spectrum that the measuring point is near luggage rack at the speed of 350km/h, the volatility of spectrum structure is similar to 350km/h, but the spectrum amplitude (chromaticity level) is significantly increase, especially in the 160Hz-200Hz band, A sound level is generally above 70dBA, indicating that with the increase of the speed, luggage rack vibration intensify.

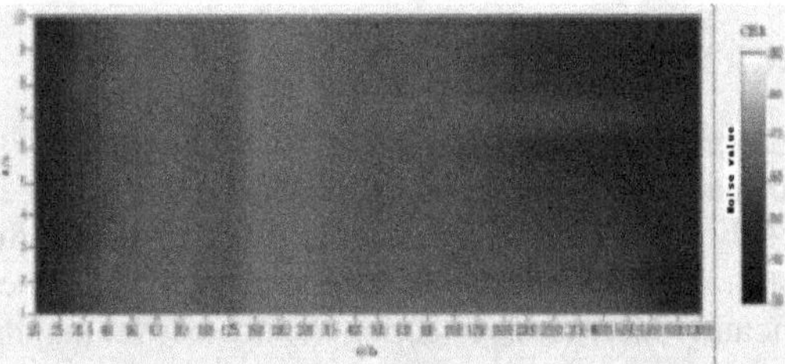

Fig. 3. Noise dynamic spectrum near luggage rack when running at 350km/h

5.2 Noise Dynamic Spectrum When Running in the Tunnel

Figure 4 is noise dynamic spectrum of the measurement point near luggage rack at 330km/h from the tunnel to the open line, the first 6 seconds running in the tunnel, the follow 4 seconds running in the open line, the boundaries are very clear, so it is very favorable for the operating state monitoring. It can be seen that the noise dynamic spectrum has the following characteristics when the train running in the tunnel:

The amplitude was significantly increased up to more than 80dBA;

The main energy band extended to 160Hz-1000Hz;

The amplitude significantly increased compared with open line below 100Hz and above 500Hz;

Cyclical fluctuations basically eliminated.

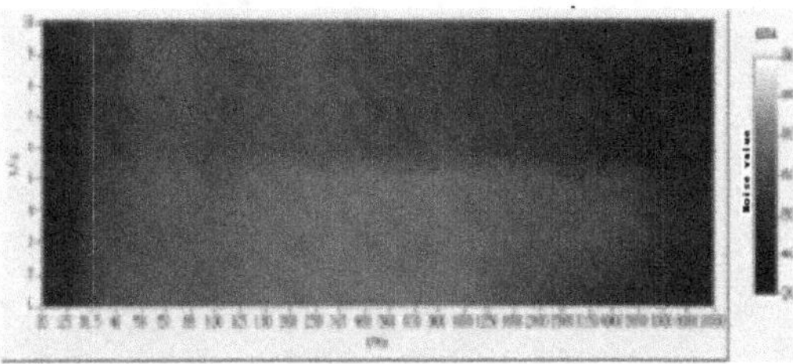

Fig. 4. Noise dynamic spectrum near luggage rack when running in the tunnel at 300km/h

6 Conclusions

Through the analysis of dynamic spectrum of luggage rack vibration radiation noise, we get the following conclusions:

Dynamic spectrum can clearly reflect the time-varying characteristics of the passenger compartment noise, both understand the time domain changes and reflect changes in the frequency domain, has the advantages of time-frequency analysis;

The method is simple which do not need to make complex calculations. Chroma level is used to describe the noise changes in different frequency, intuitive sense strong and convenient.

Chroma level (amplitude) in the dynamic spectrum is very sensitive to changes of the luggage rack vibration state, therefore use some of the parameters of the dynamic spectrum, such as standard deviation is feasible to the luggage rack vibration state monitoring.

The method can be generalized to the entire passenger compartment noise and exterior noise measurement and analysis, and used to train operation state monitoring and comfortable evaluation of passenger compartment noise.

References

1. Gao, P.X.: Test and Measurement Technology of vibration, shock and noise, pp. 11–12. Southwest Jiaotong University Press, China (2010)
2. Wang, H.: The non-stationary random signal analysis and processing, pp. 1–29. National Defence Industry Press (1999)

3. Hu, H.: Speech Signal Processing, pp. 46–47. Harbin Institute of Technology Press (2000)
4. Si, Z.: Vibration test and analysis technology, pp. 392–394. Tsinghua University Press (1992)
5. Huang, Z.: Smooth signal processing and application cycle, pp. 1–15. Science Technology Press (2005)
6. ISO 3381.2005(E), Railway applications-Acoustics- Measurement of noise inside railbound vehicles, 4 (2005)
7. PRC "Industrial noise testing norms": Fifth
8. Gonzalez: Digital Image Processing, pp. 224–259. Electronic Industry Press (2004)
9. Liu, J.H.: Virtual instrument programming language LabWindows/CVI course, pp. 87–91. Electronic Industry Press (2001)
10. Zhou, G.L.: Scanning sound intensity technology, pp. 28–32. Harbin Engineering University Press (2007)

Port-Based Composable Modeling and Simulation for Safety Critical System Testbed

Yujun Zhu, Zhongwei Xu, and Meng Mei

School of Electronics & Information Engineering, Tongji University, Cao'an highway 4800, 201804 Shanghai, China
{dyzhuyujun,xuzhongweish,mei_meng}@163.com

Abstract. While there has been much attention paid to the applications of Modeling and Simulation (M&S) for safety critical system testbed lately, little has been done to address related technology areas that enable M&S to be more easily constructed.Model Composabilty is the ability to compose component across a variety of application domains. This contribution discusses basic researches for a port-based object (PBO) approach to integrated M&S of testbed. We give the formal description of PBO with the safety property and describe the method of composable design based on PBO, and then illustrate this approach by a simple example .

Keywords: Modeling and Simulation, Safety critical system, Testbed, Port.

1 Introduction

Safety critical systems are complex systems, that if major failures can render adverse consequences for health, safety, property, and the environment[1]. Tests for safety critical systems require considerable effort and skill and consume large of cost. Due to the growing complexity of such systems it has to be expected that their trustworthy test will become unmanageable in the future if only conventional techniques. For these reasons a simulation testbed is increasingly recognized as an effective approach to testing such system. Tested enables a process to be operated with simulation environment.

Development of testbed for safety critical system is an important and emerging area of system or software testing research. The complexity of these systems leads to requirements for new techniques. Simulation prototypes need to model the behavior of the equivalent physical prototype adequately accurately, otherwise, the predicted behavior does not match the actual behavior resulting in poor design decisions. However, not always are the most detailed and accurate simulation models also the most appropriate; sometimes it is more important to evaluate many different alternatives quickly with only coarse, high-level models. At this stage, the accuracy of the simulation result depends more on the accuracy of the parameter values than on the model equations; simple equations that describe the high-level behavior of the system are then most appropriate. Equally important to accuracy is the requirement that simulation models be easy to create. Creating high-fidelity simulation models is a

F.L. Wang et al. (Eds.): WISM 2012, LNCS 7529, pp. 51–58, 2012.

complex activity that can be quite time-consuming[2].This article introduces a multi-level composable modeling and simulation (M&S) method based on port object. This method has the ability to generate system-level simulations automatically by simply organizing the system components. Although [3] and [4] have discussed the port-based approach to integrated M&S of physical systems and their controllers, they consider the CAD system more without the property of safety, real-time and was not suitable to build the safety critical systems testbed. We further the evolution towards a seamless integration of simulation for safety critical system testbed with the idea of a port object.

2 Safety Critical System Testbed

Many safety critical systems are tested by putting them into their designated working environment and seeing whether they perform according to expectation. Given the importance and complexity of modern digital control systems, including a lot of code per project and distributed over several controller boards with multiple processors each and real-time communication between them, the traditional approach is no longer adequate. One of the most powerful, but also most demanding, tests for safety critical systems is to connect the inputs and outputs of the control system under test to a real-time simulation of the target process[8]. this test method is often called simulation testing and this environment where the real system tested working called testbed. A testbed is generally composed of four parts: simulation system, interface, controller and support service system including display, record, plotting and analysis. The safety critical system testbed diagram is depicted in Figure 1. The testbed has the potential to provide significant reductions in the cost and time than a physical mock-up. However, creating high fidelity simulations for complex safety critical system can be quite a challenging because of the fowling reasons:

(1) Cost. Although the use of physical prototypes for testing or verification is a very clstly, creating simulations including seamless connection with the real interface also costs a pretty penny.
(2) Real-Time. One of the attribute of safety critical system is real-time constraints. Tasks in hard real-time systems must be scheduled to ensure that these timing constraints are met. The testbed in correspondence with the such system uder test must ensure to be scheduled ,that is difficult.
(3) Development Time. Increasingly complex applications of testbed gives the significant pressure to reduce development time.
(4) Safety & Reliability. Safety and reliability in safety critical systems are very important, the failure of such fuctions can cause damage or injury.The testbed must satisfy test condition for safety attribute.

Therefore,we propose a composable M&S framework based on the concepts of port-based object.

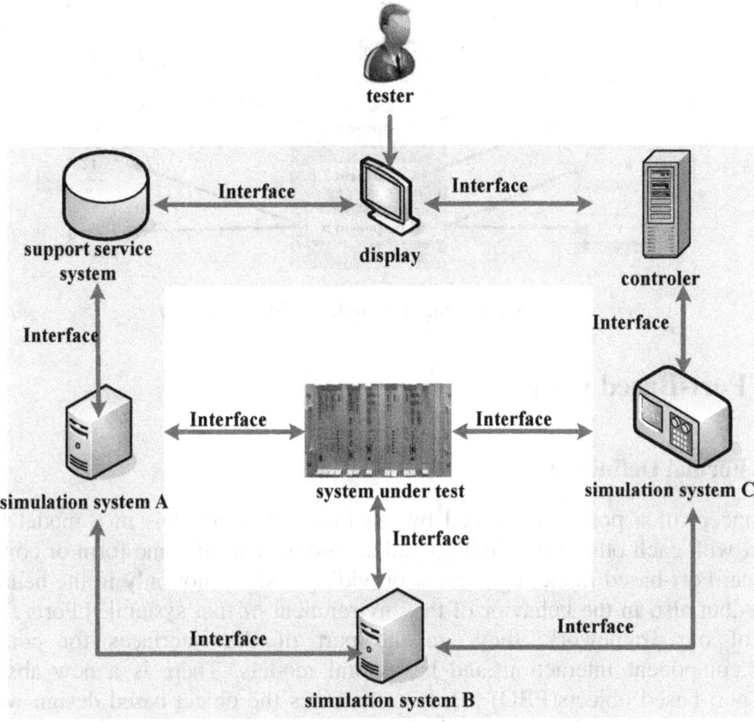

Fig. 1. Testbed Diagram for Safety Critical System

3 Composable Modeling and Simulation

Composability is the capability to select and assemble components of a model or simulation in various combinations to satisfy specific user requirements meaningfully. It has sometimes been seen as the elusive holy grail of M&S[5]. The science of composable M&S is substantial and growing. Some relevant theory and technology are increasingly proposed or applied. For example, understanding languages and notations—e.g., unified modeling language(UML) and discrete-event system specification(DEVS), expressing models—e.g.,agent-based and object-oriented methods. There are a few of advantages of composability that the safety critical system testbed need[6,7]:

(1) Reduce costs and allow us to do things once , don't need all the many models that now exist.
(2) Allow us to combine models from different disciplines into integrated system-level models.
(3) Allow models of sub-systems to evolve throughout the design process.

We can creat the testbed like the diagram depicted in Figure 2.

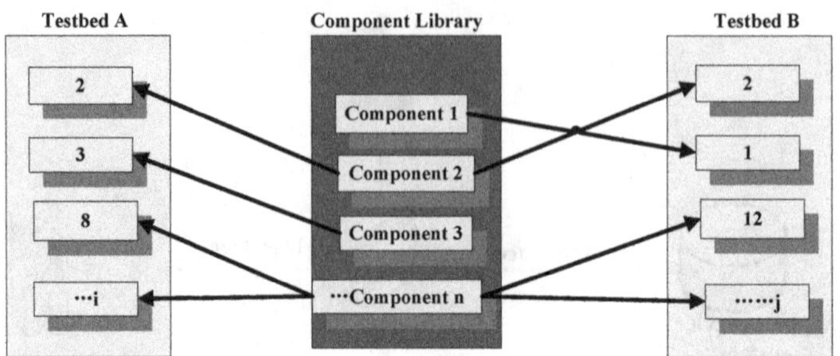

Fig. 2. Conceptual Example of Composability

4 Port-Based Object

4.1 Formal Definitions

The concept of a port is generated by the fact that submodels in a model have to interact with each other by definition and accordingly need some form of conceptual interface. Port-based modeling aims at providing insight, not only in the behavior of systems,but also in the behavior of the environment of that system[3].Ports form the basis of our framework, they are the part of the interfaces for component objects,component interactions,and behavioral models. There is a new abstraction called port-based objects(PBO) [9], that combines the object-based design with port automaton design. A PBO is not only defined as an object, but also has various ports for real-time communication. Each PBO is a real-time task that communicates only through its ports. A PBO gets data through its input ports, shares its results with other PBOs with its output ports, and the constraints constants are used to keep specific hardware or applications within bounds. This interface is depicted in Figure 3.

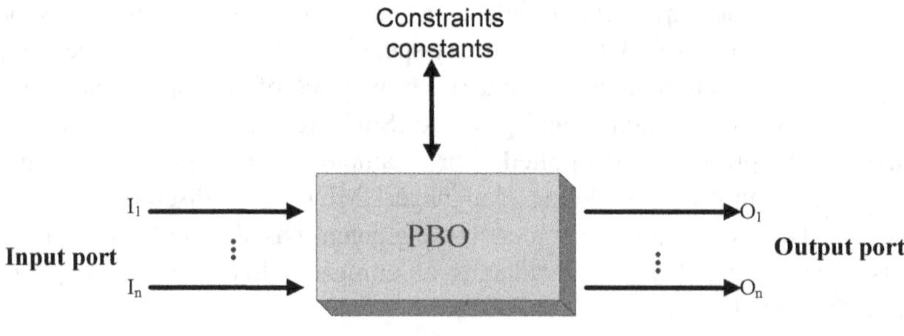

Fig. 3. A PBO Module

Here are the formal definition of the PBO. A $PBO = (S, I, O, C, \rightarrow, S_0)$ consists of a set of states S, a set of input ports I, a set of output ports O, a set of constraints constants C, a transition relation $\rightarrow \subseteq \mathrm{S} \times 2^N \times \mathrm{S}$ and a set of initial states $S_0 \subseteq S$ .We write transitions as $q \xrightarrow{n} p$ with $q, p \in S$ and

$n \subseteq N$ .The interpretation is that there is data flow at the ports N and no data flow at the rest of the ports $N \setminus n$.

4.2 PBO Design

Data transmitted over the ports can be any type in a PBO, it can be raw data such as input from an A/D or D/A converter, user operated data or processed data. Constraints constants are important for the safety critital system testbed, they are the key to connect seamlessly with the real system. We could come up with to illustrate the detail design.Trains need to run in the tracks, the track needs communication with the train control system. We creat a track PBO instance shown in table 1.

Table 1. A Track PBO Instance

PBO name	Description	Input	Output	Constraints	State
$Track_i$	Track circuit code sender	$Iport_1$: coding data(m_1) $Iport_1$: Synchronous data(m_2) $Iport_2$: the state of PBO $Track_{i-1}$(m_4)	$Oport_1$: state data(m_3) $Oport_2$: current state(m_5)	25ms	HU,U, UU,U US,U2 ,U2S, LU,L, L2,L3, L4,L5

The upgrade relations of track circuit code is the following order HU→UU→UUS→U→U2→U2S→LU→L→L2→L3→L4→L5.The state L5 is the highest, that means the train in this track can run to the highest spped.If the tracki get the state LU from the tracki-1, it will update the own state to L higher a level than LU,then send the current state to tracki+1.The communication sequences of PBO tracki shown in Figure 4.

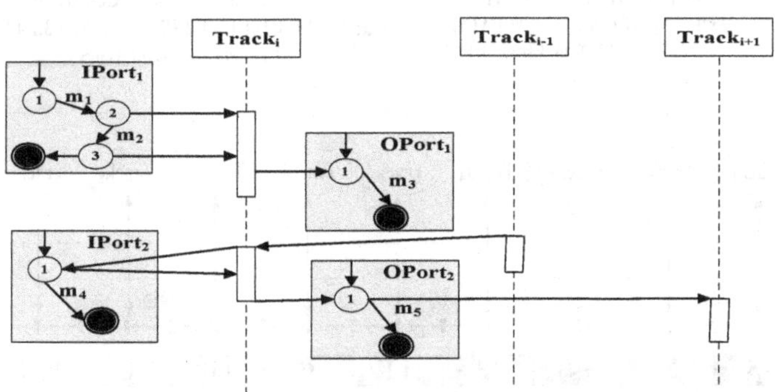

Fig. 4. Communication Sequences of PBO $Track_i$

5 Port-Based Composable Modeling and Simulation

5.1 Component Represent

The basis of composition is the components which is the key for composable modeling and simulation. We use the PBO to represent the components and give the definition as two-tuples:

$$CM = (PBO, C)$$

with a set of PBO, and the conditions for the composition C. we now use conditions to define given context of a component. The composition conditions can be Orderly sequence of a PBO, they define the environment for PBO that reside in the environment. Create a context in the activation process of the PBO, the PBO is configured to require certain services such synchronization, transactions, real-time activation, safety and so on. Multiple PBOs can be retained within a context.

5.2 Composable Design

To provide better support for composable design of testbed, we build blocks, PBO componet, within our composable design and simulation environment. We divided the design process into three layers shown in in Figure 5. First layer is presentation layer,we represent our componet grammars using XML, an XML DTD is a representation of attribute grammar.We can replace the componet in this layer.The Second is PBO modeling Layer, PBO can set up the dynamic modle of componet. The last one is simulation layer , PBOs are combined in the context and designed.

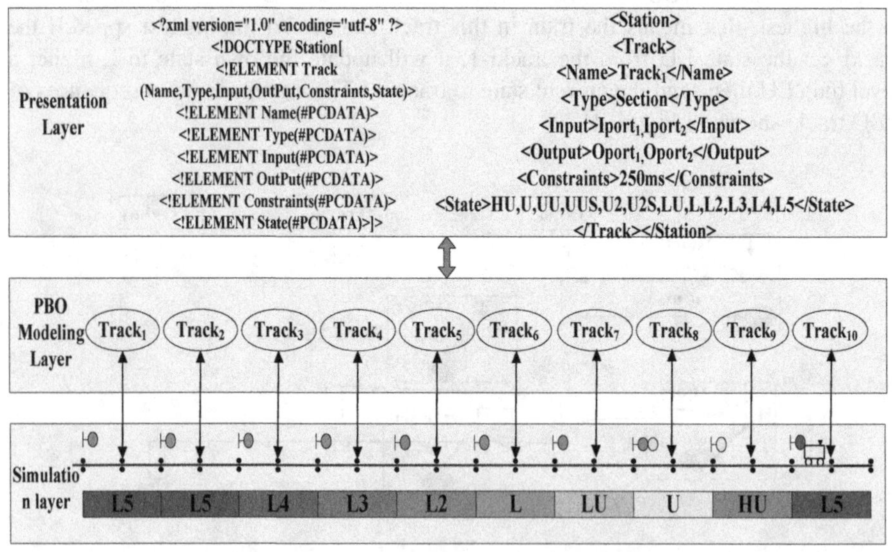

Fig. 5. A Three-layer Composable Model and Simualtion

5.3 Application

In order to facilitate fast and efficient train traffic across borders in our country, a unified standard of Chinese Train Control System(CTCS) is proposed. Some of the safety critical system of CTCS are under development in several companies. In order to test those system, it is necessary to build the laboratory simulation environment. We take the testbed for Train Control Center(TCC) which is one sub-system of CTCS for example to illustrate the port-based composable modeling and simulation.

TCC is a classic safety critical system, it should control of code of the track circuit based on the information of the train route and the section of track state.TCC needs obtain states information of the track circuit every 250ms. We should build the simulation of track circuit as a part of the testbed. For the dynamic nature of track circuit,we use the PBO composable method. We creat different track circuit test scenarios through the different PBO configuration. The detail design has been described in section 3.2 and 4.2. The general process is depicted in Figure 6.

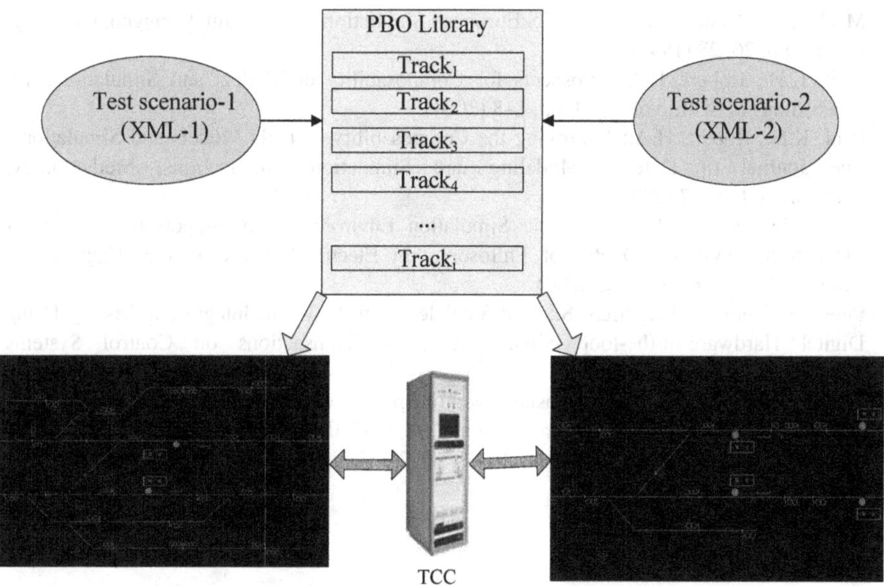

Fig. 6. PBO testbed for TCC

6 Conclusion

The presented PBO composable modeling and simulation for safety critical system testbed offers an integrated system approach to many of the requirements of simulations. The scope of this paper did not allow for the discussion of the current state of technology, but that technology is sufficiently advancedto allow for pursuit of some of the issues as recommended.This paper has shown how modern PBO composable method can help to make proper design of testbed and to create more insight in the simulation. An example demonstrated that one of the major achievements is that this approach enables to easily change PBO components acording to the changed requirements, then creats some different integrated test systems. The discussion of how best to pursue this research should begin now.

References

1. Bin, L., Xin, W., Hernan, F., Antonello, M.: A Low-Cost Real-Time Hardware-in-the-Loop Testing Approach of Power Electronics Controls. IEEE Transactions on Industrial Electronics 57, 919–931 (2007)
2. Paredis, J.J., Diaz-Calderon, A., Sinha, R., Khosla, P.K.: Composable Models for Simulation-Based Design. Engineering with Computers 17, 112–128 (2001)
3. Peter, C.B.: Port-based Modeling of Mechatronic Systems. Mathematics and Computers in Simulation 66, 99–127 (2004)
4. Diaz-Calderon, A., Paredis, J.J., Khosla, P.K.: A Composable Simulation Environment for Mechatronic Systems. In: 1999 SCS European Simulation Symposium, Erlangen, Germany, vol. 10, pp. 26–28 (1999)
5. Paul, K.D., Robert, B.A.: Prospects for Composability of Models and Simulations. In: Proceedings of SPIE, vol. 5423, pp. 1–8 (2004)
6. Paul, K.D., Robert, H.A.: Improving the Composability of DoD Models and Simulations. The Journal of Defense Modeling and Simulation: Applications, Methodology, Technology 1, 5–17 (2004)
7. Diza-Calderon, A.: A Composable Simulation Environment to Support the Design of Mechatronic Systems. Doctor of Philosophy in Electrical and Computer Engineering, Carnegie Mellon University (2000)
8. Peter, T., Thomas, K., Erich, S.: Rail Vehicle Control System Integration Testing Using Digital Hardware-in-the-loop Simulation. IEEE Transactions on Control Systems Technology 7, 352–363 (1999)
9. Stewart, D.B., Volpe, R.A., Khosla, P.K.: Integration of real-time software modules for reconfigurable sensor-based control systems. In: 1992 IEEE/RSJ International Conference on Intelligent Robots and Systems (IROS 1992), Raleigh, North Carolina, pp. 325–333 (1992)

Risk Assessment Method of Radio Block Center in Fuzzy Uncertain Environment

Qiuxiang Tao[1], Wei Nai[2], Haiming Gao[3], Jiliang Tu[2,1]

[1] School of Information Engineering, Nanchang Hangkong University, Nanchang Jiangxi 330063, China
[2] School of Transportation Engineering, Tongji University, Shanghai 201804, tujiliang@yahoo.com.cn
[3] Institute of Physical Education, Nanchang Hangkong university, Nanchang, Jiangxi 330063, China

Abstract. In this paper, a risk assessment method of radio block center (RBC) for train control based on fuzzy multi-criteria decision-making theory is proposed. By a thorough study of the system risk assessment problem of RBC including Fault Mode and Effects Analysis (FMEA) in fuzzy uncertain environment, in the proposed method, 4 indices including occurrence possibility, severity, detectability and maintainability are chosen for risk assessment and then fuzzificated by triangular fuzzy number, their weights are calibrated respectively to show their relative importance by information entropy weight method; and the final risk assessment result is acquired by sorting the fault mode employing the fuzzy multi-criteria decision-making theory. An application case is studied in this paper to verify the effectiveness and feasibility of the proposed method, and the assessment process and result indicates its convenience for application and its suitableness for popularization.

Keywords: radio block center, fuzzy multi-criteria decision-making, uncertain environment, risk assessment, entropy weight.

1 Introduction

Radio Block Center (RBC) is the key equipment of Chinese Train Control System level 3 (CTCS-3) for train operation and has significant influence on train safety. The central processing system of RBC connects numerous external equipments (including OBU, computer based interlocking (CBI) equipment, temporary speed restriction server (TSRS), centralized traffic control (CTC) and centralized signal measuring (CSM) system, etc.) and makes them work coordinately [1-3], so its safety quality can be affected by various factors. The system failures of RBC could not be specified or identified between each other, because in some cases the same reason may cause different kind of failures and it is also possible that different reason can cause the same kind of failure to happen. Therefore, there are many uncertainties and fuzziness in the risk assessment process. At present, some related research has been done on risk assessment and its countermeasure for safety[4-7]. Aims at dealing with the flaws that discussed in the literature review above, in this paper, a novel risk assessment method is proposed based on fuzzy multi-criteria decision-making theory.

F.L. Wang et al. (Eds.): WISM 2012, LNCS 7529, pp. 59–66, 2012.

2 Fault Mode and Effects Analysis (FMEA) of RBC

Table 1. FMEA of RBC system (MA: memory administration; CEM: common equipment module; SMA: Shared memory architecture; UEM: universal energy management; O: occurrence possibility; S: severity; D: detectability; M: maintainability)

No.	*Fault modes*	*Fault phenomenon*	*Fault reason*	*Fault effects*	*Solving Methods*	*O*	*S*	*D*	*M*
1	Hardware fault and system resources error	MA or CEM generating error	CPU or memory error	Train overspeed or overrun	Introducing 2 out of 3 or 2 out of 2×2 safety platform in	2	5	4	2
2	Basic data error	Railroad line data error	Railroad line data source error or data generating error	Train overspeed or overrun	Testing data strictly according to the safety process	1	4	4	2
3	Intra-system communication error	SMA, CEM and UEM sending delay	Intra-system communication link or equipment failure	Train overspeed or overrun	Introducing universal computing security or fail-safe security platform in	2	4	3	3
4	Inter-system communication error	Receiving information of external equipment error	Communication Channel interference or external attack	Train overspeed or overrun	Strictly according to the security communication protocol	2	3	3	4
5	Inter-system communication failure	The status of TSR and railway track approaching status changed	External communication channel failure or equipment connection error	Train overspeed or overrun	Adding redundant communication channel or set operating rules	3	3	2	4

Fault Mode and Effects Analysis (FMEA) is a reliability analysis technique which comes from practice [8]. Based on the research result of paper [4], combined with a large amount of data including operating history, maintenance history, equipment fault history, and accident analysis history of urban metro system, intercity rail transit, maglev lines which serves as research basis, according to the expert experience and research summary, FMEA analysis has been done in this paper only considering the structure and function of RBS system, and the results is shown in Table 1. It should be pointed out that the risks caused by equipment failure like OBU or CBI are not been considered in this paper.

In order to enhance the credibility and reliability of the assessment result, the value of indices for FMEA is divided into 5 levels and is shown in number 1, 2, 3, 4 and 5, the value for each index in Table 1 is set according to the expert experience.

3 Fuzzy Multi-criteria Based Risk Assessment

3.1 Fuzzification of the Assessment Indices

In order to reduce the effects from the uncertain factors in indices choosing, triangular fuzzy number is used to fuzzificate the 4 indices according to their characteristics. The triangular fuzzy number can be defined as [9]: if in real number field, a fuzzy set $\tilde{A}$ is a triangular fuzzy number, then its subordinate function can be written as:

$$\mu_A(x) = \begin{cases} 0, & x < a \\ \dfrac{x-a}{m-a}, & a \le x < m \\ \dfrac{b-x}{b-m}, & m \le x < b \\ 0, & x \ge b \end{cases} \tag{1}$$

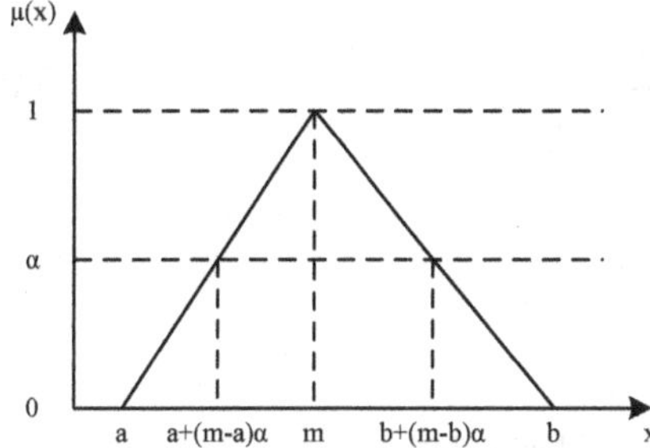

Fig. 1. Triangular fuzzy number

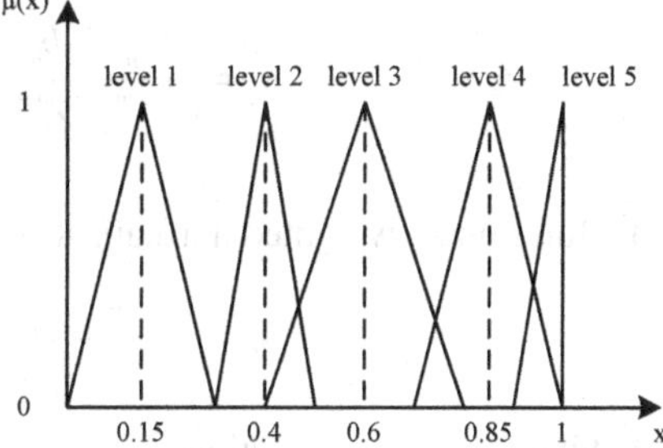

Fig. 2. The subordinate function for each level

As shown in Fig. 1, the subordinate function is determined by a, b and m, which can be denoted by (a, b, m). According to the expression of subordinate function and the characteristics of 4 indices chosen for risk assessment, the subordinate function of the triangular fuzzy number which is used to fuzzificate each level for risk assessment of the 4 indices is shown in Fig. 2. It can be seen from the figure that the subordinate function for each level are as follows: level 1 - (0, 0.15, 0.3) level 2 - (0.3, 0.4, 0.5); level 3 - (0.4, 0.6, 0.8); level 4 - (0.7, 0.85, 1.0); level 5 - (0.9, 1.0, 1.0).

3.2 Process of Risk Assessment

Fuzzy multi-criteria decision-making theory [10-11] is formed by setting fuzzy ideal solution and fuzzy ill-ideal solution as reference datum. Provided that there are m fault modes, n assessment indices, and there are p experts in relevant fields when making the assessment, the assessment value can be expressed as matrix X, where Xij represents the value of jth assessment index of the ith fault mode. The assessment steps can be deduced as follows:

1) Calibrate the weights of each assessment indices

In information theory, the concept of information entropy depicts the disorder of a system, it reflects the uncertainty of a certain system, the formula for calculating information entropy [12] is shown as follows:

$$H(p_j) = -k\sum_{i=1}^{m} p_{ij} \ln p_{ij}, \text{ where } 0\ln 0=0,\ p_{ij} = x_{ij} / \sum_{i=1}^{m} x_{ij},\ k=1/\ln m$$

The weights of each assessment indices wj can be calculated by formula:

$$w_j = (1-H(p_j)) / \sum_{i=1}^{n}(1-H(p_j)),\ j=1, 2, \ldots, n \tag{2}$$

2) Normalize fuzzy indices value matrix

The xij discussed here is triangular fuzzy number,

$$\widetilde{x_{ij}} = \left(\frac{a_{ij}}{a_{ij}^{\max}}, \frac{b_{ij}}{b_{ij}^{\max}}, \frac{c_{ij}}{c_{ij}^{\max}} \wedge 1 \right) \tag{3}$$

3) Build fuzzy weighted normalization matrix

$$\widetilde{r_{ij}} = w_j \widetilde{x_{ij}} \tag{4}$$

4) Find fuzzy ideal solution $\widetilde{V}^+$ and fuzzy ill-ideal solution $\widetilde{V}^-$

$$\widetilde{V}^+ = (\widetilde{V_1}, \widetilde{V_2}, \ldots, \widetilde{V_n}), \text{ where } \widetilde{V_j} = \left\{ (\max_i \widetilde{r_{ij}} \mid j \in J_1), (\min_i \widetilde{r_{ij}} \mid j \in J_2) \mid i = 1, \ldots, m \right\} \tag{5.1}$$

$$\widetilde{V}^- = (\widetilde{v_1}, \widetilde{v_2}, \ldots, \widetilde{v_n}), \text{ where } \widetilde{V_j} = \left\{ (\min_i \widetilde{r_{ij}} \mid j \in J_1), (\max_i \widetilde{r_{ij}} \mid j \in J_2) \mid i = 1, \ldots, m \right\} \tag{5.2}$$

In equation (5.1) and (5.2), J1 is indicator for benefits, J2 is indicator for costs.

5) Calculate the Hamming Distance between r_{ijL} and V_{jL}, r_{ijR} and V_{jR}, $\widetilde{r_{ijL}}$ and v_{jL}, r_{ijR} and v_{jR}

$$d(\widetilde{r_{ij}}, \widetilde{V_j}) = \int_{S(\widetilde{r_{ij}} \cup \widetilde{V_j})} \left| \mu_{\widetilde{r_{ij}}}(x) - \mu_{\widetilde{V_j}}(x) \right| dx \tag{6.1}$$

$$d(\widetilde{r_{ij}}, \widetilde{v_j}) = \int_{S(\widetilde{r_{ij}} \cup \widetilde{v_j})} \left| \mu_{\widetilde{r_{ij}}}(x) - \mu_{\widetilde{v_j}}(x) \right| dx \tag{6.2}$$

6) Calculate the Hamming Distance between the assessment object and the fuzzy ideal solution or fuzzy ill-ideal solution

$$d_i^+ = (\sum_{j=1}^{n}[d(\widetilde{r_{ijL}},\widetilde{V_{jL}})+d(\widetilde{r_{ijR}},\widetilde{V_{jR}})]^2)^{\frac{1}{2}} \tag{7.1}$$

$$d_i^- = (\sum_{j=1}^{n}[d(\widetilde{r_{ijL}},\widetilde{v_{jL}})+d(\widetilde{r_{ijR}},\widetilde{v_{jR}})]^2)^{\frac{1}{2}} \tag{7.2}$$

7) Specify the relative proximity

$$d_i = d_i^- /(d_i^+ + d_i^-)\ ,\ \ i = 1,2,...,n \tag{8}$$

According to the value of relative proximity, the risk priority of fault modes can be acquired.

4 Case Study

In the assessment system discussed in this paper, the assessment result can be strongly affected by the weight of each assessment index, different weights of the indices would cause different assessment results. In this part, using the FMEA results of RBC system discussed above as data reference, a risk assessment of the fault modes in a certain example has been done by employing the assessment method in this paper based on fuzzy multi-criteria decision-making theory which calibrate the weights of indices by information entropy weight method.

1) From the FMEA result shown in Table 1, an assessment matrix X can be formed by 4 risk assessment indices of 5 fault modes, which can be normalized as matrix D.

$$X = \begin{bmatrix} 2 & 5 & 4 & 2 \\ 1 & 4 & 4 & 2 \\ 2 & 4 & 4 & 3 \\ 2 & 3 & 3 & 4 \\ 3 & 3 & 2 & 4 \end{bmatrix},\ D = \begin{bmatrix} 0.2 & 0.263 & 0.250 & 0.133 \\ 0.1 & 0.211 & 0.250 & 0.133 \\ 0.2 & 0.211 & 0.188 & 0.200 \\ 0.2 & 0.158 & 0.188 & 0.267 \\ 0.3 & 0.158 & 0.125 & 0.267 \end{bmatrix}$$

2) The weight of each assessment index can be calculated by formula (2), the results are shown in Table 2.

Table 2. The information entropies and weights of assessment indices

	Occurrence Possibility	*Severity*	*Detectabi lity*	*Maintainabili ty*
H_j	0.967	0.807	0.982	0.752
w_j	0.033	0.392	0.037	0.504

3) According to formula (3), fuzzificate all the elements in assessment matrix X based on the subordinate function of all the assessment level shown in Fig. 2, and normalize the acquired fuzzy indices matrix.

$$\widetilde{X} = \begin{bmatrix} (0.38,0.67,1) & (0.9,1,1) & (0.7,1,1) & (0.3,0.47,0.71) \\ (0,0.25,0.75) & (0.7,0.85,1) & (0.7,1,1) & (0.3,0.47,0.71) \\ (0.38,0.67,1) & (0.7,0.85,1) & (0.4,0.71,1) & (0.4,0.71,1) \\ (0.38,0.67,1) & (0.4,0.6,0.89) & (0.4,0.71,1) & (0.7,0.85,1) \\ (0.5,1,1) & (0.4,0.6,0.89) & (0.4,0.47,0.71) & (0.7,0.85,1) \end{bmatrix}$$

4) According to formula (4), weight all the fuzzy indices in matrix $\widetilde{X}$, thus a fuzzy weighted assessment matrix $\widetilde{R}$ can be acquired.

$$\widetilde{R} = \begin{bmatrix} (0.013,0.022,0.033) & (0.353,0.392,0.392) & (0.026,0.037,0.037) & (0.151,0.237,0.358) \\ (0.000,0.008,0.025) & (0.274,0.333,0.392) & (0.026,0.037,0.037) & (0.151,0.237,0.358) \\ (0.013,0.022,0.033) & (0.274,0.333,0.392) & (0.015,0.026,0.037) & (0.202,0.358,0.504) \\ (0.013,0.022,0.033) & (0.157,0.235,0.349) & (0.015,0.026,0.037) & (0.353,0.429,0.504) \\ (0.017,0.033,0.033) & (0.157,0.235,0.349) & (0.015,0.017,0.026) & (0.353,0.429,0.504) \end{bmatrix}$$

5) According to formula (5.1) and (5.2), the fuzzy ideal solution and the fuzzy ill-ideal solution can be calculated.

$$\widetilde{V}^{+} = [(0.017,0.033,0.033) \quad (0.353,0.392,0.392) \quad (0.026,0.037,0.037) \quad (0.353,0.429,0.504)]$$
$$\widetilde{V}^{-} = [(0.000,0.008,0.025) \quad (0.157,0.235,0.349) \quad (0.015,0.017,0.026) \quad (0.151,0.237,0.358)]$$

6) According to formula (6.1) and (6.2), the Hamming Distance between $\widetilde{r_{ijL}}$ and $\widetilde{V_{jL}}$, $\widetilde{r_{ijR}}$ and $\widetilde{V_{jR}}$, $\widetilde{r_{ijL}}$ and $\widetilde{v_{jL}}$, $\widetilde{r_{ijR}}$ and $\widetilde{v_{jR}}$ can be calculated.

7) Based on step 6), the Hamming Distance between the assessment object and the fuzzy ideal solution, the Hamming Distance between the assessment object and the fuzzy ill-ideal solution can be calculated according to formula (7.1) and (7.2) respectively:

$d_1^+ =0.366,\ d_2^+ =0.38,\ d_3^+ =0.209,\ d_4^+ =0.277,\ d_5^+ =0.279,$

$d_1^- =0.28,\ d_2^- =0.367,\ d_3^- =0.285,\ d_4^- =0.367,\ d_5^- =0.367.$

8) the relative proximity can be calculated according to formula (8):
d_1=0.433, d_2=0.491, d_3=0.577, d_4=0.567, d_5=0.568.

9) It can be seen from the relative proximity results that the risk priority of the 5 fault modes is: d3 > d5 > d4 > d2 > d1.

5 Conclusion

1) By a thorough study of the system risk assessment problem of RBC including Fault Mode and Effects Analysis (FMEA) in fuzzy uncertain environment, 4 indices

including occurrence possibility, severity, detectability and maintainability are chosen in this paper to depict the risk assessment. These 4 indices are research objects for risk assessment of RBC. The fuzzification of the 4 indices by triangular fuzzy number has eliminated the effects of the uncertain factors to certain extent; while the weight calibration for the 4 indices by information entropy weight method has effectively solved the low accuracy problem caused by treating each index in the same way.

2) RBC has longer authorized distances for wireless communication and more complex methods for information exchanging, moreover, the value determination of assessment indices has strong fuzziness, so fuzzy mathematic analysis is an effective way to do the risk analysis of RBC, and worth to be popularized.

3) Further study for the risk assessment of RBC system, such as changing fuzzy uncertainty into fuzzy randomness or fuzzy roughness, or considering the necessary condition of equipment maintenance, etc., these are the future work which require great effort to solve.

Acknowledgement. This paper is partly supported by National Basic Research Program of China under Grant No. 2007AA11Z247, Science Foundation for the Youth (Department of Education, Jiangxi Province) under Grant No. GJJ11175.

References

[1] Ji, X., Tang, T.: Study on the Integrated Testing Platform for CTCS Level 3 Train Operation Control System. Railway Signaling & Communication 43(7), 1–3 (2007) (in Chinese)
[2] Lu, J., Tang, T., Jia, H.: Modeling and Verification of Radio Block Center of CTCS-3 Train Control System for Dedicated Passengers Lines. Journal of the China Railway Society 32(6), 35–40 (2010) (in Chinese)
[3] Liang, N., Wang, H.: Real-time Performance Analysis of RBC System for CTCS Level 3 Using Stochastic Petri Networks. Journal of the China Railway Society 33(2), 67–68 (2011) (in Chinese)
[4] Zhang, Y., Liu, C., Li, Q., et al.: Analysis of the Safety Related Risks of Radio Block Center and Countermeasures. China Railway Science 31(4), 112–117 (2010) (in Chinese)
[5] Armin, Z., Gunter, H.: A Train Control System Case Study in Model-Based Real Time System Design. In: IEEE International Parallel and Distributed Processing Symposium, pp. 118–126 (2003)
[6] Xu, T., Li, S., Tang, T.: Dependability Analysis of Data Communication Subsystem in Train Control System. Journal of Beijing Jiaotong University (Natural Science Edition) 31(5), 24–27 (2007) (in Chinese)
[7] Tommaso, P.D., Flammini, F., Lazzaro, A., et al.: The simulation of an alien in the functional testing of the ERTMS/ETCS trackside system. In: The 9th IEEE International Symposium on High Assurance Systems Engineering, pp. 131–139 (2005)
[8] Guo, Y.: Principles of Reliability Engineering, pp. 56–89. Tsinghua University Press, Beijing (2002) (in Chinese)
[9] Wang, Y., Zhou, J., Chen, W., et al.: Risk Assessment Method of Power Transformer Based on Fuzzy Decision-making. Chinese Journal of Scientific Instrument 30(8), 1662–1666 (2009) (in Chinese)

[10] Xie, X., Zhou, Y., Zhang, C., et al.: Reinforcing effect assessment of soft foundation based on fuzzy multi-attribute decision method. Journal of Natural Disasters 19(2), 62–64 (2010) (in Chinese)
[11] Chen, T.-Y., Tsao, C.-Y.: The interval-valued fuzzy TOPSIS method and experimental analysis. Fuzzy Sets and Systems 159(11), 1410–1428 (2008)
[12] Li, R.: Fuzzy Multiple Criteria Decision Making Theory and Its Application. Science Press, Beijing (2002)

Learning Research in Knowledge Transfer

Bing Wu and Pingping Chen

School of Economics and Management, Tongji University
200092 Shanghai, China

Abstract. This study is a productivity review on the literature gleaned from science citation index expanded (SCI-EXPANED) database on web of science, concerning learning research in knowledge transfer. The result indicates that the number of literature productions on this topic mainly distributes in recent years, reaching climax of 5 in 2010 and then followed by 2008. The main research development country is USA accounting for 27%. And from the analysis of institutions, HONG KONG UNIV SCI TECHNOL and UNIV OTTAWA rank parallel top one. As for source title, management science is in the first place. The related research can be classified into three branches, including effects of learning, approaches to knowledge transfer and modeling of knowledge transfer.

Keywords: knowledge transfer, network, computer science, management, expert system.

1 Introduction

Learning and the acquisition of new knowledge are preconditions for learning, continuous improvement and when new tasks have to be carried out. High learning rates of employees can translate into lower costs and quality improvements, both highly relevant in competitive manufacturing environments. Although learning in an organization is stimulated by a number of factors, task understanding and skill development are the main components for learning a new task and improving task performance. While both are regarded as being relevant for learning a new task, for an in-depth analysis the components have to be disaggregated into their underlying mechanisms.

The main objective of this paper is to analyze learning research in knowledge transfer on Science Citation Index Expanded (SCI-E) database from web of science. As a result the related work in this area can be thoroughly explored. The rest of this article is organized as follows: Section 2 surveys the related research in this topic. Section 3 briefly introduces research focuses of them. Last, we conclude our article in Section 4.

2 Analyze Result of Learning Research in Knowledge Transfer

Knowledge transfer has long occupied a prominent, if not always explicit, place in research on strategic management and corporate expansion. Deploying and extending

F.L. Wang et al. (Eds.): WISM 2012, LNCS 7529, pp. 67–72, 2012.

productive knowledge to new facilities is inherent in corporate growth. The speed and effectiveness of that process can determine ability of one firm to penetrate new markets, preempt and respond to rivals, and adapt to market changes. In sum, existing research from various perspectives underscores the importance of knowledge implementation and transfer to firm growth and performance.

According to Science Citation Index Expanded and Social Sciences Citation Index Database in web of science, only 22 records were found in related discipline when "knowledge transfer" and "learning" as a combined search title.

2.1 Country/Territory Analyze

According to country/territory, analyze results can be shown as table 1. USA accounts for 27%, ranking in the top one, following that is GERMANY accounting for 22%.

Table 1. Analyze Results of Country/Territory

Country/Territory	Record Count	% of 22
USA	6	27.273%
GERMANY	5	22.727%
CANADA	3	13.636%
FRANCE	3	13.636%
PEOPLES R.CHINA	3	13.636%
ENGLAND	2	9.091%

2.2 Publication Year

Almost all the papers were published mainly in last ten years. As shown in table 2, there are 5 papers in publication year 2010, reaching climax. And accordingly citations in year 2010 reach more than 50, ranking top one as well.

Table 2. Analyze Results of Publication Year

Publication Year	Record Count	% of 22
2010	5	22.727%
2008	3	13.636%
2000	2	9.091%
2003	2	9.091%
2009	2	9.091%
2011	2	9.091%
2012	2	9.091%

2.3 Institutions

As far as institutions are concerned, as shown in table 3, HONG KONG UNIV SCI TECHNOL and U UNIV OTTAWA rank parallel top one, accounting for 9% respectively.

Table 3. Analyze Results of Institutions

Institutions	Record Count	% of 22
HONG KONG UNIV SCI TECHNOL	2	9.091%
UNIV OTTAWA	2	9.091%

2.4 Source Title

There are three main sources, one is management science, and the other two is lecture notes in artificial intelligence and implementation science respectively. They are sources for one third of all papers, as shown in Table 4.

Table 4. Analyze Results of Source title

Source Titles	Record Count	% of 22
MANAGEMENT SCIENCE	3	13.636%
LECTURE NOTES IN ARTIFICAL INTELLIGENCE	2	9.091%
IMPLEMENTATION SCIENCE	2	9.091%

3 Focuses of Learning Research in Knowledge Transfer

3.1 Effects of Learning

Knowledge Stickiness. Increased outsourcing yields less vertically-integrated firms, suppliers have to rely on different buyers and interdisciplinary teams for acquired and utilized knowledge to improve performance. However, knowledge transfer from buyers to suppliers is not always successful.

Accordingly, Li Chia-Ying (2012) assesses perception of suppliers regarding how specific knowledge transfer stickiness influences their manufacturing capabilities based on the perspectives of the contextual aspects of learning capability and social embeddedness [1]. The results showed the influence of knowledge stickiness on manufacturing capability would be enhanced by the moderating variables of social embeddedness and learning capability. This study is expected to provide valuable information to both academics and practitioners in terms of designing appropriate integrative mechanisms for supplier performance improvement.

Ranking Problem. Learning to rank employs machine learning techniques to automatically obtain the ranking model using a labeled training dataset. Much contribution has been made in developing advanced rank learning approaches. Motivated by the labeled data shortage in real world learning to rank applications, a challenging problem is addressed Chen Depin et al (2010) [7]. Starting from a heuristic method introduced in previous work, the knowledge transfer for learning to rank at feature level and instance level was theoretically studied with two promising methods proposed, respectively. In future work, the cross domain learning to rank problem was planned to further study under the scenario where the source and target domain data originally have different feature sets.

Implementation Performance. Although scholars from various disciplines recognize the potential impact of knowledge transfer, deployment, and implementation on organizational success and viability little attention has been devoted to how such effects manifest in the time it takes to make new facilities viable. This gap is filled by analyzing how competitive, firm, and technology characteristics combine to affect the time it takes firms to get their facilities fully operational [8].

Prior Activities. Daghfous A (2004) provides more insight into the value and workings of firm knowledge and organizational learning by producing empirical evidence on the relationships between these activities and their benefits [10]. This study developed reliable and valid measures related to knowledge, learning activities, and especially learning benefits. Although the nature of this study is exploratory, the results obtained provide valuable prescriptions for more successful technology transfer projects.

3.2 Approaches to Knowledge Transfer

Reusing Structured Knowledge. The outcomes of Letmathe Peter et al (2012) have shown that the transfer of explicit knowledge in combination with autonomous learning and self-observation or with autonomous learning, self observation and additional outcome feedback is most beneficial with respect to manufacturing performance in terms of quality and assembly time when a new task is introduced [2]. This analysis has focused on a well-structured and semi-complex manufacturing task and revealed the benefits of self-observation in combination with explicit knowledge transfer. Research could prove whether these results also hold in a more complex setting.

Transfer Learning. Transfer learning aims at solving exactly these types of problems, by learning in one task domain and applying the knowledge to another. An overview by Yang Qiang et al (2011) is given on three research works that transferred progressively more sophisticated structured knowledge from a source domain to a target domain to facilitate learning [3]. The conclusion is that structured knowledge transfer can be useful in a wide range of tasks ranging from data and model level transfer to procedural knowledge transfer, and that optimization method can be applied to extract and transfer deep structural knowledge between a variety of source and target problems. How to identify good source domains and to come up with a good quantitative measure of similarity between different domains at declarative and procedural levels of learning is the future research direction.

Advice Using. Reinforcement learning is a continual learning process in which an agent navigates through an environment trying to earn rewards [5]. A novel technique is presented of extracting knowledge gained on one task and automatically transferring it to a related task to improve learning [9]. Key idea is that the models learned in an old task as a source of advice for a new task can be viewed. In future work, the sensitivity of algorithm to errors and omissions in mapping advice should be evaluated.

3.3 Models of Knowledge Transfer

Structural Knowledge Transfer. An essential quality of a cognitive being is its ability to learn, that is, to gain new knowledge or skills as well as to improve existing knowledge or skills based on experience. The role of abstraction principles for knowledge transfer in agent control learning tasks is investigated by Bajoria Rekha et al (2011) [4]. The use of so-called structure space aspectualizable knowledge representations that explicate structural properties of the state space is proposed and a posteriori structure space aspectualization (APSST) as a method to extract generally sensible behavior from a learned policy was presented. This new policy can be used for knowledge transfer to support learning new tasks in different environments.

Bridging Domains. Traditional supervised learning approaches for text classification require sufficient labeled instances in a problem domain in order to train a high quality model. However, it is not always easy or feasible to obtain new labeled data in a domain of interest. The lack of labeled data problem can seriously hurt classification performance in many real world applications. To solve this problem, transfer learning techniques, in particular domain adaptation techniques in transfer learning are introduced by capturing the shared knowledge from some related domains where labeled data are available, and use the knowledge to improve the performance of data mining tasks in a target domain. A novel transfer learning approach, called BIG (Bridging Information Gap) is designed by Xiang Evan Wei et al (2010) to effectively extract useful knowledge in a worldwide knowledge base [6]. Research work is planned to continue in the future by pursuing several avenues.

Learning by Hiring. To investigate the conditions under which learning-by-hiring is more likely, Song J et al (2003) study the patenting activities of engineers who moved from United States (U.S.) firms to non-U.S. firms [11]. The results support the idea that domestic mobility and international mobility are similarly conducive to learning-by-hiring.

Learning Negotiation Skills. Review of the learning and training literature revealed four common methods for training people to be more effective negotiators: didactic learning, learning via information revelation, analogical learning, and observational learning. Each of these methods is tested experimentally in an experiential context and found that observational learning and analogical learning led to negotiated outcomes that were more favorable for both parties, compared to a baseline condition of learning through experience alone [12]. Interestingly, negotiators in the observation group showed the largest increase in performance, but the least ability to articulate the learning principles that helped them improve, suggesting that they had acquired tacit knowledge that they were unable to articulate.

4 Conclusions

Knowledge-based competition has magnified the importance of learning alliances as a fast and effective mechanism of capability development. The early literature on strategic alliances focused primarily on alliance structuring and outcomes. Starting in the late 1980s, more research attention has been directed to the processes of learning and knowledge transfer, especially in the context of technology transfer, R&D collaboration, and strategic alliances. This growing attention, fueled by the global race for a sustainable competitive advantage through superior dynamic capabilities, has produced significant conceptual work and anecdotal evidence, but relatively few empirical studies.

Acknowledgement. This work is supported by the National Nature Science Foundation of China (No. 71071117).

References

1. Li, C.-Y.: Knowledge stickiness in the buyer-supplier knowledge transfer process: The moderating effects of learning capability and social embeddedness. Expert Systems with Applications 39(5), 5396–5408 (2012)
2. Letmathe, P., Schweitzer, M., Zielinski, M.: How to learn new tasks: Shop floor performance effects of knowledge transfer and performance feedback. Journal of Operations Management 30(3), 221–236 (2012)
3. Yang, Q., Zheng, V.W., Li, B., et al.: Transfer Learning by Reusing Structured Knowledge. AI Magazine 32(2), 95–106 (2011)
4. Bajoria, R., Shah, F., Rodeck, C.H., et al.: Virtual Learning Using Interactive Multimedia-Based Tools for Knowledge Transfer and Development of Global Patient Care Pathway in Hemoglobinopathies. Hemoglobin 35(5-6), 643–652 (2011)
5. Frommberger, L., Wolter, D.: Structural knowledge transfer by spatial abstraction for reinforcement learning agents. Adaptive Behavior 18(6), 507–525 (2010)
6. Wei, X.E., Bin, C., Hao, H.D., et al.: Bridging Domains Using World Wide Knowledge for Transfer Learning. IEEE Transactions on Knowledge and Data Engineering 22(6), 770–783 (2010)
7. Depin, C., Yan, X., Jun, Y., et al.: Knowledge transfer for cross domain learning to rank. Information Retrieval 13(3), 236–253 (2010)
8. Salomon, R., Martin, X.: Learning, knowledge transfer, and technology implementation performance: A study of time-to-build in the global semiconductor industry. Management Science 54(7), 1266–1280 (2008)
9. Torrey, L., Walker, T., Shavlik, J., Maclin, R.: Using Advice to Transfer Knowledge Acquired in One Reinforcement Learning Task to Another. In: Gama, J., Camacho, R., Brazdil, P.B., Jorge, A.M., Torgo, L. (eds.) ECML 2005. LNCS (LNAI), vol. 3720, pp. 412–424. Springer, Heidelberg (2005)
10. Daghfous, A.: An empirical investigation of the roles of prior knowledge and learning activities in technology transfer. Technovation 24(12), 939–953 (2004)
11. Song, J., Almeida, P., Wu, G.: Learning-by-hiring: When is mobility more likely to facilitate interfirm knowledge transfer? Management Science 49(4), 351–365 (2003)
12. Nadler, J., Thompson, L., Van Boven, L.: Learning negotiation skills: Four models of knowledge creation and transfer. Management Science 49(4), 529–540 (2003)

A Post-filtering Technique for Enhancing Acoustic Echo Cancelation System

Yaxun Wang

School of Computer, Dongguan University of Technology, Dongguan, 523808, China
yaxunwang@yahoo.com.cn

Abstract. A post-filtering technique for enhancing acoustic echo cancelation (AEC) system is proposed in this paper, the proposed filter takes in the output of the conventional AEC system and estimates the percentage of the residual error, with respect to the reference signal. This percentage of residual error is terms as "leakage" and is used in adjusting adaptation step size and calculating the post-filter gain. The proposed technique improves the convergence speed and performance of the conventional AEC systems and is simple to implementation.

Keywords: Post-filtering, AEC, DTD, FFT.

1 Introduction

Over the past years, hands-free communication has found increasing applications in areas such as videoconferencing systems and mobile phones. The acoustic coupling between the loudspeaker and the microphone at the near end will cause the far end speech to be transmitted back to the far end and form the acoustic echo. This will severely deteriorate the quality of communication. The cornerstone of a normal acoustic echo cancellation (AEC) system is the adaptive filter. Approaches combining echo cancellation and noise reduction have been proposed to achieve sufficient quality of the transmitted speech [1]. The realization of such a combined system is, however, difficult. The difficulties in cancelling acoustic echo are caused by high computational complexity and some influences such as background noise, near-end speech, and variations of the acoustic environment which disturb the adaptation of the echo canceller. In practice, especially in mobile hands-free applications where all these factors play a significant role, the residual echo remains at the output of an adaptive filter because of the constraint of finite filter length and background noise [2].

A perfect post-filtering technique should not alter the formant information and should attenuate null information in the speech spectrum in order to achieve noise reduction and hence produce better speech quality. The use of post-filters can significantly improve the performance of the echo canceller by suppressing the residual echo. Park et al. proposed a residual-echo cancellation scheme based on the autoregressive (AR) analysis [3] and [4]. Myllyla proposed a method to suppress

F.L. Wang et al. (Eds.): WISM 2012, LNCS 7529, pp. 73–80, 2012.

the residual echo by estimating the power spectral density of the echo [5]. However, these post-filters require much computational complexity for the step-by-step process on fullband signals. There are also several post-filters proposed for the frequency-domain AEC (FAEC) algorithms [6] and [7]. In order to suppress the ambient noise as well, some post-filters combined the residual-echo suppression with the noise reduction were discussed [8] and [9]. Furthermore, psychoacoustic schemes [10] and comfort noise fill [11] were also adopted in FAEC's post-filters to improve the comfortability of the results.

It commonly happens in practice that as long as an adaptive filter is not perfectly adapted, the error signal always contains some information about the echo that is supposed to be canceled. The amount of echo is often referred to as "residual". Moreover, the residual decreases when the filter weights misadjustment decreases (or as the filter converges). Therefore, a time-varying step size and gain control mechanism should be developed as supplement to further attenuate the residual. The main function of this proposed mechanism lies in the two aspects.

1) Attenuate the residual caused by the misadjustment of the adaptive filter.

2) Control the step size of the commonly used normalized least mean squares (NLMS) algorithm.

The paper is organized as follows: In Section 1, the conventional post-filter structure is addressed briefly. In Section 2, an improved post-filter combined with background noise suppression is proposed for AEC system. In Section 4, some simulation results and discussions are given. The conclusion of this paper is drawn in Section 5.

2 An Acoustic Echo Canceller with Post Filter

The system performs acoustic residual echo cancellation with post filter. Fig.1 shows the scheme including the conventional double talk detector (DTD) based on normalized cross-correlation between far-end talker signal and microphone input signal, adaptive echo cancellation filter, and NR post filter.

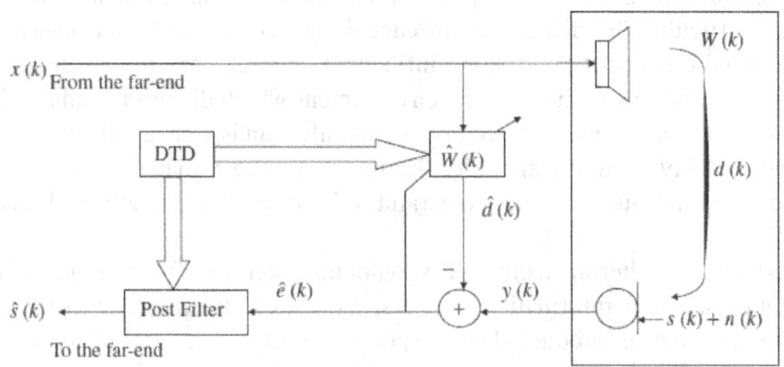

Fig. 1. The structure of an acoustic echo canceller with post filter

Referring to Fig.1, the adaptive echo cancellation filter creates a replica $\hat{d}(k)$ of echo signal $d(k)$. If this replica is identical with the original echo, the echo can be removed by simple subtraction from the near-end signal $y(k)$:

$$y(k) = s(k) + n(k) + d(k) \tag{1}$$

Here, $s(k)$ is the near-end talker signal and $n(k)$ is ambient noise. The output of acoustic echo canceller or the error signal $e(k)$ is used to adjust the coefficients $\hat{W}(k)$ of the adaptive filter so that the coefficients converge to a close representation of the echo path $W(k)$:

$$e(k) = s(k) + n(k) + d(k) - \hat{d}(k) = s(k) + n(k) + r(k) \tag{2}$$

where $r(k) = d(k) - \hat{d}(k)$ indicates residual echo.

In the experiments, we set the order of an adaptive FIR filter 256 and adopt normalized least mean square (NLMS) algorithm. The coefficients of the NLMS adaptive filter $\hat{W}(k)$ are updated according to

$$\hat{W}(k+1) = \hat{W}(k) + \frac{\mu}{\hat{\sigma}_X^2 + \gamma} X(k)e(k) \tag{3}$$

where $\hat{W}(k) = [\hat{w}_0(k), \hat{w}_1(k), \ldots, \hat{w}_{N-1}(k)]^T$ is a $N\times 1$ coefficient vector, $X(k) = [x(k), x(k-1), \ldots, x(k-N+1)]^T$ is a $N\times 1$ excitation vector, $\hat{\sigma}_X^2$ is an estimated input power, μ is a constant controlling the convergence and γ is a stabilization factor.

In practice, residual echo remains at the output of an echo canceller due to an error of the adaptive filter. Therefore noise reduction technique has been used for reducing the residual echo as well as ambient noise [12] and [13]. However, the post filters only based on noise reduction algorithms rarely reduce the residual echo, since the dominant characteristic of residual echo is also speech. Although the power of the residual echo is small, a voice activity detector (VAD) used in NR may fall into an error in determining residual echo as transmitting speech, which results in transmission of the residual echo to a far-end talker.

3 Enhancing Acoustic Echo Cancelation System with Post Filter

The system performs acoustic residual echo cancellation with post filter. Fig.2 shows the diagram of a conventional AEC system with the proposed post-filter block included. All the signals are processed within the frequency domain so the blocks of Fast Fourier Transform (FFT) and IFFT are omitted. $X(k, l)$, $Y(k, l)$, $V(k, l)$, $D(k, l)$ are the reference signal, output of the acoustic system (local room), noise signal and the microphone signal, respectively. $H(k,l)$ and $\hat{H}(k,l)$ are the actual impulse response of the local room and the approximated one (weights of the adaptive filter). In all the above variables, k is the frequency index and l is the frame index.

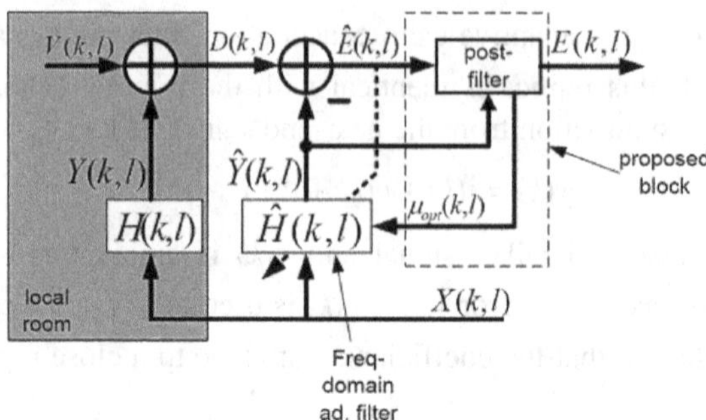

Fig. 2. Block diagram of a frequency domain AEC system with proposed post-filter. Frame index l, frequency index k

The first variable that the proposed post-filter calculates is the optimum step size that is used in LMS adaptive algorithm, given by

$$\mu_{out} = \frac{\sigma_r^2(k,l)}{\sigma_e^2(k,l)} \tag{4}$$

where $\sigma_r^2(k,l)$ and $\sigma_e^2(k,l)$ are the energy of residual and filter output, respectively. The values of $\sigma_r^2(k,l)$ and $\sigma_e^2(k,l)$ are estimated by using the introduction of the "leakage" $\eta(k,l)$ factor which is defined as the ratio of the residual energy over the output of the adaptive filter. The estimated values of $\sigma_r^2(k,l)$ and $\sigma_e^2(k,l)$ are thus given by

$$\hat{\sigma}_r^2(k,l) = \eta(k,l)\sigma_{\hat{Y}}^2(k,l) = \eta(k,l)\,\|\,\hat{Y}(k,l)\,\|^2 \tag{5}$$

$$\hat{\sigma}_e^2(k,l) = \|\,\hat{E}(k,l)\,\|^2 \tag{6}$$

Combining the above two equations, the optimum step size is given by

$$\hat{\eta}_{out}(k,l) = \hat{\eta}(k,l)\frac{\|= \hat{Y}(k,l)\,\|^2}{\|= \hat{E}(k,l)\,\|^2} \tag{7}$$

Moreover, the estimated leakage factor $\hat{\eta}(k,l)$ can be obtained by

$$\hat{\eta} = (k,l) = \frac{R_{\hat{E}\hat{Y}}(k,l)}{R_{\hat{Y}\hat{Y}}(k,l)} \tag{8}$$

where $R_{\hat{E}\hat{Y}}(k,l)$ and $R_{\hat{Y}\hat{Y}}(k,l)$ are given by

$$R_{\hat{E}\hat{Y}}(k,l) = (1-\beta(l))R_{\hat{E}\hat{Y}}(k,l-1) + \beta(l)P_{\hat{Y}}(k,l)P_{\hat{E}}(k,l) \quad (9)$$

$$R_{\hat{Y}\hat{Y}}(k,l) = (1-\beta(l))R_{\hat{Y}\hat{Y}}(k,l-1) + \beta(l)P_{\hat{Y}}(k,l)^2 \quad (10)$$

where $\beta(l)$ is a smoothing factor and is given by

$$\beta(l) = \beta_0 \min(\frac{\sigma_{\hat{Y}}^2(k,l)}{\sigma_{\hat{E}}^2(k,l)},1) \quad (11)$$

The quantities $P_{\hat{Y}}(k,l)$ and $P_{\hat{E}}(k,l)$ are also estimated by

$$R_{\hat{Y}}(k,l) = (1-\alpha)P_{\hat{Y}}(k,l-1) + \alpha \| \hat{Y}(k,l) \| \quad (12)$$

$$R_{\hat{E}}(k,l) = (1-\alpha)P_{\hat{E}}(k,l-1) + \alpha \| \hat{E}(k,l) \| \quad (13)$$

where α is also a constant smoothing factor. The second variable that the proposed post-filter calculates is the gain $G(k,l)$ for further attenuating the residual, given by

$$G(k,l) = e^{-r\hat{\eta}}(k,l) \quad (14)$$

where γ is a predefined constant for controlling the attenuation level of the post-filter. The final output signal $E(k,l)$ is thus given by

$$E(k,l) = G(k,l)\hat{E}(k,l) \quad (15)$$

Therefore, the operations that are carried out within the proposed post-filter block are summarized as follows.

1) Get input signals $\hat{Y}(k,l)$ and $\hat{E}(k,l)$
2) Calculate $P_{\hat{Y}}(k,l)$ and $P_{\hat{E}}(k,l)$ using (12) and (13).
3) Calculate $R_{\hat{E}\hat{Y}}(k,l)$ and $R_{\hat{Y}\hat{Y}}(k,l)$ using (10) and (9).
4) Calculate leakage factor $\hat{\eta}(k,l)$ using (8).
5) Calculate optimum step size $\hat{\mu}_{out}(k,l)$ using (7).
6) Calculate filter gain $G(k,l)$ using (14).
7) Calculate fitter output $E(k,l)$ using (15).

The flowchart of the operations for the post-filter is depicted in Fig.4. It should be noted that this only shows the processing of frequency bin k. All the other frequency bins should be processed in the same way as shown in Fig.2 and Fig.4. In practice, the microphone signal *d*(*n*), reference signal *x*(*n*) should be first converted into frequency domain by using FFT and the error signal E(*k*, *l*) is then converted back into time domain by using IFFT. This is shown in Fig.3.

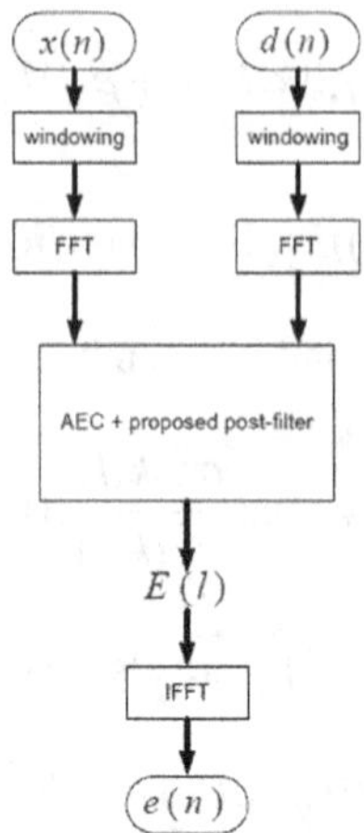

Fig. 3. Block diagram of converting time domain signals into frequency domin

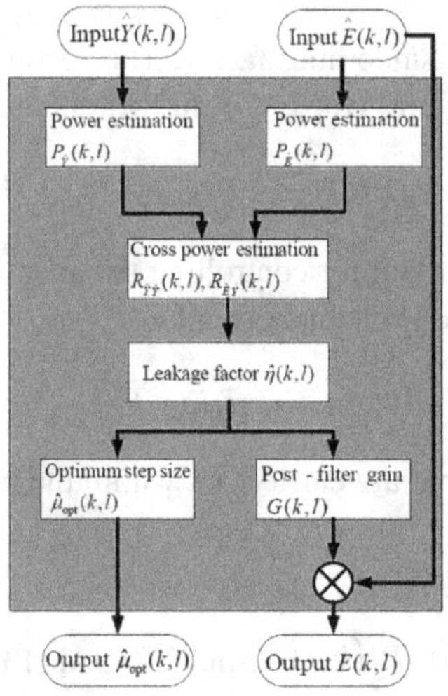

Fig. 4. Block diagram of the proposed post-filter

4 Simulation Experimental Results

The AEC system in the following simulations uses the structure based on blocks of Fast Fourier Transform. The length of the prototype low-pass filter is 160. The sampling rate of the signal is set to be 8KHz. Since it is the post-filter we are investigating, we assume the double-talk detector to be ideal. This is easily achieved in simulations because the near-end speech can be added to the input separately. The normalized least mean square (NLMS) algorithm is implemented in the given structure, with the stepsize 0.3. To generate strong residual echo, the filter length in the band is limited at 20 taps, which can only provide a rough echo path model. We compared the performance of the proposed system with that of the conventional AEC system, these results are represented in Fig.5 and Fig.6 show the waveforms of the related signals.

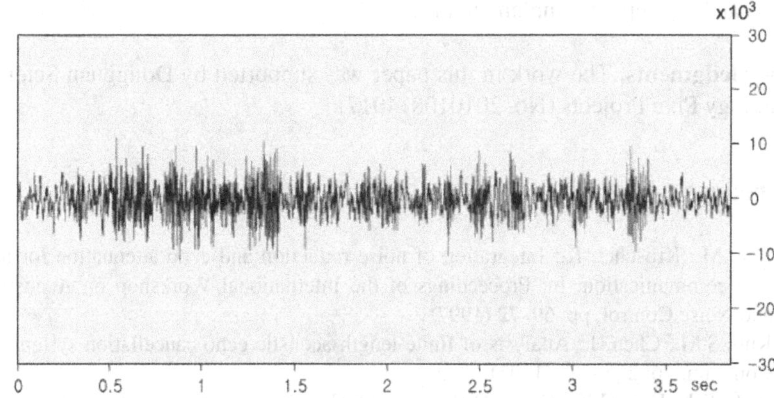

Fig. 5. Waveforms of the related signals in the conventional AEC system

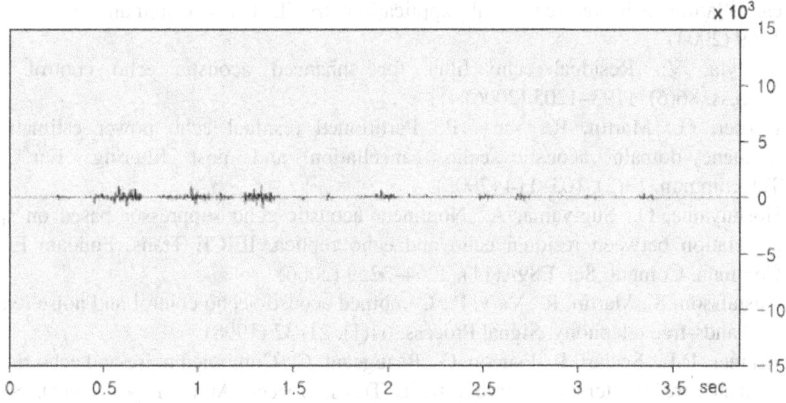

Fig. 6. Waveforms of the related signals in the proposed AEC system with post-filtering

Based on the results above, we can see that the proposed AEC system is the proper choice for reducing background noise and residual echo.

5 Conclusion

In this paper, we proposed a novel post-filtering for AEC systems. The post-filtering takes in the output of the conventional AEC system and estimates the percentage of the residual error, with respect to the reference signal. Compared with the conventional AEC system, the proposed post-filtering can effectively attenuate the residual echo and the background noise, while keeping the speech distortion and musical noise at a low level. The simulation results show that the post-filtering achieve superior performance both in the speech quality and the robustness and are particularly effective when noise reduction is imperative. And the proposed post filters improve the convergence speed and performance of the conventional AEC systems and is simple to implementation.

Acknowledgments. The work in this paper was supported by Dongguan Science and Technology Plan Projects (No. 201010814017).

References

[1] Ihle, M., Kroschel, K.: Integration of noise reduction and echo attenuation for handset-free communication. In: Proceedings of the International Workshop on Acoustic Echo and Noise Control, pp. 69–72 (1997)
[2] Kuo, S.M., Chen, J.: Analysis of finite length acoustic echo cancellation system. Speech Commun. 16, 255–260 (1995)
[3] Park, S.J., Lee, C.Y., Youn, D.H.: A residual echo cancellation scheme for hands-free telephony. IEEE Signal Process Lett. 9(12), 397–399 (2002)
[4] Kang, S., Baek, S.J.: A new post-filtering algorithm for residual acoustic echo cancellation in hands-free mobile application. IEICE Trans. Commun. E87B(5), 1266–1269 (2004)
[5] Myllyla, V.: Residual echo filter for enhanced acoustic echo control. Signal Process. 86(6), 1193–1205 (2006)
[6] Enzner, G., Martin, R., Vary, P.: Partitioned residual echo power estimation for frequency-domain acoustic echo cancellation and post filtering. Eur. Trans. Telecommun. 13(2), 103–114 (2002)
[7] Hoshuyama, O., Sugiyama, A.: Nonlinear acoustic echo suppressor based on spectral correlation between residual echo and echo replica. IEICE Trans. Fundam Electron. Commun. Comput. Sci. E89A(11), 3254–3259 (2006)
[8] Gustafsson, S., Martin, R., Vary, P.: Combined acoustic echo control and noise reduction for hands-free telephony. Signal Process. 64(1), 21–32 (1998)
[9] Jeannes, R.L., Scalart, P., Faucon, G., Beaugeant, C.: Combined noise and echo reduction in hands-free systems: a survey. IEEE Trans. Speech Audio Process. 9(8), 808–820 (2001)
[10] Gustafsson, S., Martin, R., Jax, P., Vary, P.: A psychoacoustic approach to combined acoustic echo cancellation and noise reduction. IEEE Trans. Speech Audio Process. 10(5), 245–256 (2002)
[11] Benesty, J., Morgan, D.R., Gay, S.L.: Advances in network and acoustic echo cancellation. Springer, New York (2001)
[12] Basbug, F., Swaminathan, K., Nandkumar, S.: Integrated noise reduction and echo cancellation for IS-136 systems. In: Proceedings of the IEEE International Conference Acoustic, Speech Signal Processing, pp. 1863–1866 (2000)
[13] Beaugeant, C., Turvin, V., Scalart, P., Gilloire, A.: New optimal filtering approaches for hands-free telecommunication terminals. Signal Processing 64, 33–47 (1998)

Packet Dropping Schemes and Quality Evaluation for H.264 Videos at High Packet Loss Rates

Yuxia Wang, Yi Qu, and Rui Lv

School of Information Engineering
Communication University of China
10024, Beijing, China
yuxiaw@cuc.edu.cn

Abstract. Packet loss may lead to serious degradation of video quality in network communication. Sometimes a router must drop quite a lot of packets to alleviate the congestion. This paper designs multiple dropping schemes for H.264 videos, considering varying packet loss rates. We perform packet dropping experiments for three video sequences. Experiment results show that both the packet type and the packet distribution have an impact on the video quality. Given higher packet loss rates, dropping a few B frames in a GOP results in less quality degradation than dropping packets randomly from all the B frames.

Keywords: Objective Evaluation, Packet Dropping, Packet Loss Rate, PSNR, VQM.

1 Introduction

The video transmission is the main traffic sources in current networks. When sending compressed video across communication networks, packet losses may occur due to the congestion. Thus the decoder could not receive all the encoded video data because of the losses, video quality will degrade more or less. Actually many factors can affect the decoded video quality at the receiver, including channel loss characteristics, video encoder configurations, and error concealment methods of video decoders, etc.

Considerable research has been done to explore how packet losses impact video quality. Traditional approaches of queue management assume that all packet losses affect quality equally and the traffic is not differentiated, so each packet is treated identically. But as for video resources, the impact on perceptual quality of each packet or frame may be quite different. S. Kanumuri and T.-L. Lin modeled packet loss visibility for MEPG-2 and H.264 videos based on subjective experiments [1, 2, 3, 4]. Using these kinds of models, the router can drop the least visible packets or frames to achieve the required bit reduction rates. The relation between PSNR and perceptual quality scores is considered in [5]. This work found that packet losses are visible when the PSNR drop is greater than a threshold, and the distance between dropped packets is crucial to perceptual quality. The above work considered the perceptual quality of a decoded video subjected to a single transmission loss (which can cause the loss of multiple consecutive frames).

F.L. Wang et al. (Eds.): WISM 2012, LNCS 7529, pp. 81–89, 2012.

In [6], the issue of temporal degradations in transmission of digital video is addressed. A psychophysical experiment is conducted to study the impact of jerkiness and jitter on perceived video quality. A methodology of subjective evaluation was proposed in [7], which assess the perceptual influence of frame freezes and blockiness in full length movies. Through the subjective experiments they found that usually frame freezes are less noticeable than blockiness. Our prior work [8, 9] focused on visual quality of video degraded by both packet losses and compression artifacts. We developed a network-based model for predicting the objective VQM quality scores so that video packet importance can be obtained. We validated this model by performing packet dropping experiments for multiple combinations of video streams at different bit rates.

Now we are interested in another question: if a router must drop multiple packets and when packet loss rate (PLR) is high (for example: 10% to 30%), which packets or frames should be dropped to obtain better quality of video at the decoder side? In other words, with a packet loss visibility model mentioned above, we can probably conclude that a certain packet is less important than another in a GOP due to varying contents, and a packet from I or P frame is more important than that from B frame. But when a hundred B packets need to be dropped, which packets should we select? We randomly drop the packets from all B frames in the GOP or just drop one or two whole B frames in that GOP? Also when dropping whole frames, should we select the continuous ones or just randomly pick some of them? This paper gives a detailed experiment implementation for varying packet-dropping methods. By objective evaluation algorithms PSNR and VQM [10, 11], we examine the different influences on video quality of each dropping method for H.264 videos.

This paper is organized as follows. In Section 2, objective quality evaluation methods PSNR and VQM are introduced. Section 3 describes the video sequences for the experiment and codec configurations, followed by the design of the packet dropping experiments. Section 4 presents simulation results at various target packet dropping rates. Section 5 concludes the paper.

2 Methods of Objective Quality Evaluation

As we all know, subjective assessment methods attempt to evaluate the perceived quality by asking viewers to score the quality of a series of test scenes, which have been a fundamental and reliable method to characterize video quality. However, the implementation of subjective evaluation is expensive and time-consuming, which brings a lot of attention to the research on objective evaluation metrics. Peak Signal-to-Noise Ratio (PSNR) is widely used as a video quality metric or performance indicator. Typical values for the PSNR of lossy videos are between 20 and 50dB, where higher is better. For video transmission in a network, video quality at the receiver can be highly affected by packet losses. VQM is a standardized full-reference (FR) method of objectively measuring video quality. It has been adopted by the ANSI as a U.S. national standard and as international ITU Recommendations for closely predicting the subjective quality ratings. The General Model has objective parameters for measuring the perceptual effects of a wide range of impairments such as blurring, block distortion, jerky/unnatural motion, noise (in both the luminance and chrominance channels), and error blocks. The values of VQM are usually in the range [0, 1], where 0 corresponds to best quality, and 1 is the worst quality. VQM may occasionally exceed one for video scenes that are extremely distorted [10].

3 Sequences Coding and Packet Dropping Schemes

3.1 Sequences Selection and Codec Configurations

We used three original video sequences: *news*, *foreman* and *coastguard* which present fixed shot and camera motion. Each sequence is of 300 frames and encoded at 600 kbps, which we assume there is no compression artifacts for the videos. As shown in Table1, the videos are compressed by the H.264 JM9.3 encoder in CIF resolution (352 by 288). The GOP structure is (IDR)BBPBBPBB… with 15 frames per GOP. The first frame of each GOP is an IDR frame which prevents a packet loss in the previous GOP from propagating errors into the current GOP. Rather than using a fixed quantization parameter, we use the default rate control of the JM9.3 encoder. The values of the quantization parameter can vary from frame to frame. One packet consists of one horizontal row of macroblocks (MBs), so there are 18 packets in one frame (288/16).

Table 1. Video coding configurations

	Parameters
Spatial resolution	352*288
Bit rate	600kbps
Duration (frames)	300
Compression	H.264 JM
GOP length	IDR BBP/15
Frame rate	30
Rate control	on

The lossy videos after packet dropping are decoded by FFMPEG [12], which uses the default error concealment algorithm. There are two kinds of algorithms corresponding to two different packet loss patterns: packet losses and frame losses. The error concealment for packet loss is as follows: for the MBs which are estimated to be intra coded, FFMPEG takes a weighted average of the uncorrupted neighboring blocks for error concealment, and for inter coded MBs, it performs bi-directional motion estimation to conceal the MBs. On the other hand, FFMPEG conceals whole frame losses using temporal frame interpolation. A lost B frame is concealed by temporal interpolation between the pixels of the previous and the future frames [11]. Once the packet dropping is performed for each GOP, the FFMPEG decoding and error concealment are run.

3.2 Packet Dropping Schemes

We would like to see the influences on video quality of different frame types and different packet dropping methods. We design three high PLRs: 10%, 20%, and 30%. For each sequence, we drop packets within one GOP to obtain the three given PLRs. We do this for each GOP so that the impact on quality degradation of video content can be explored.

3.2.1 The Influence of Frame Types

Usually, the packet losses from I and P frames will lead to more subjective quality degradation than those from B frames because of the error propagation. We'd like to validate this phenomenon through objective evaluation. Given a lower PLR (5%), dropping packets from I, P and B frames separately within a GOP. We use a lower PLR because there is only one I frame in a GOP, the rate of packets from I frame is no more than 1/15. For each case, we randomly select the packets from that type of frame. For example, there is only one I frame in a GOP, so we randomly drop the packets from this frame until the PLR is obtained. But for B frame, there are 10 B frames in that GOP, thus we can select the packets from all the 10 B frames randomly.

3.2.2 The Influence of Dropping Methods for B Frames

Given the three PLRs (10%, 20% and 30%) we drop packets only from B frames, there are three dropping methods:

a) Scatter

We average the total number of packets need to be dropped by the number of B frames in a GOP, then for each B frame, the packets number is randomly generated. We denote this method as 'scatter'.

b) Focus-random

We calculate the total number of packets for each PLR, then drop all the packets within one B frame. If the packets in the frame are not enough for the PLR, packets from a second B frame are selected. Here the frame number of each B frame is randomly generated in a GOP. We denote this method as 'focus-random'

c) Focus-tail

As the second method above, we drop whole B frames. Differently, the last B frame of the GOP is firstly selected, and then the second B from the last, the rest can be done similarly until the PLR is met. We denote this method as 'focus-tail'.

4 Experiment Results and Discussion

There are 20 GOPs for each sequence, and packet dropping experiment is performed for GOP2 to GOP20. Based on the decoded videos and original videos, we can get a PSNR value and a VQM score for each GOP. Also we can average the quality scores over all GOPs for each sequence.

4.1 The Influences of Frame Types

When PLR is 5%, the quality scores of three sequence are obtained. Figure 1 gives the three PSNR values versus GOP number for video *news*. Figure 2 gives the three VQM scores versus GOP number. Both present the quality scores for packet dropping from I, P and B frames. We can see the PSNR values for B are much larger than those for P and I frames, while the VQM scores for B are clearly smaller than those for P and I frames. That means given a certain PLR, packet dropping from I frame results in much more seriously quality degradation than from P frames and B frames, as one would expect. Similar results can be obtained from the videos *foreman* and *coastguard.* Table 2 gives the average quality scores over all GOPs for all sequences.

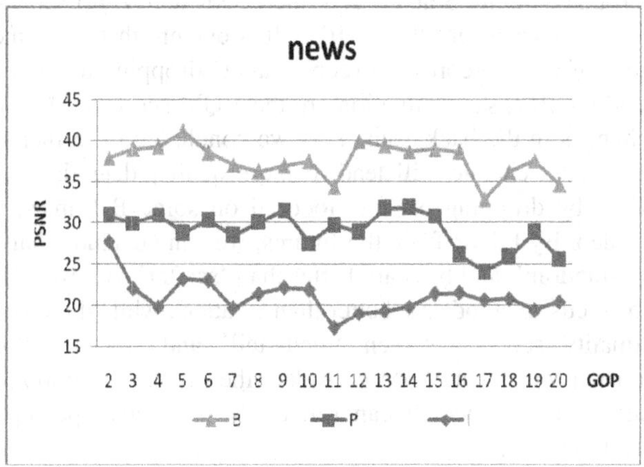

Fig. 1. PSNR values versus GOP number for video *news*

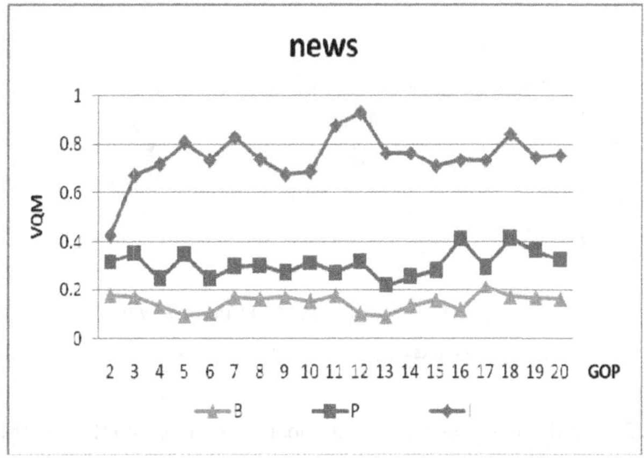

Fig. 2. VQM scores versus GOP number for video *news*

Table 2. Average quality scores over all GOPs

	PSNR			VQM		
Frame type	I	P	B	I	P	B
news	21.12	29.07	37.58	0.7439	0.3077	0.1498
foreman	21.29	26.03	33.19	0.8016	0.4187	0.2207
coastguard	19.33	25.20	30.45	0.8900	0.4627	0.3050

4.2 The Influences of Dropping Methods for B Frames

As discussed in section 3, for each of the three high PLRs, we adopt three methods to drop packets or frames. Figure 3 and 4 give the PSNR and VQM scores versus GOP number for the video *news* when PLR is 10%. It is evident that for some GOPs there are big differences of PSNR scores between 'scatter' dropping method and the other two for GOP7 and GOP17, same situation are the VQM scores in. It is because there are two scene changes in the background. As we concluded in former studies, larger motion or more content change will lead to more quality degradation when packet loss happens. Here by dropping packets focused on some B frames, we get better quality for that video. By PSNR from the figures, we can conclude that 'focus-tail' is better than 'focus-random', and both are better than 'scatter'. But by VQM scores, we can see that both 'focus' methods are better than 'scatter', while there is no significant difference of quality scores between 'focus-tail' and 'focus-random' dropping methods. A main reason is that VQM considers the effects of temporal information besides the spatial artifacts, which can reflect the subjective perception of video quality more effectively.

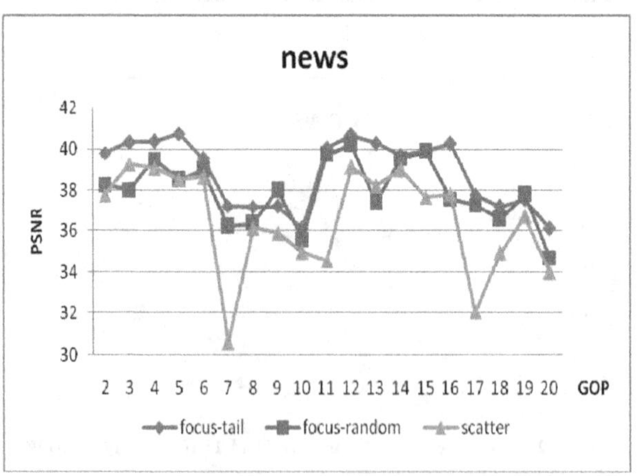

Fig. 3. PSNR values versus GOP number for *news* video (PLR=10%)

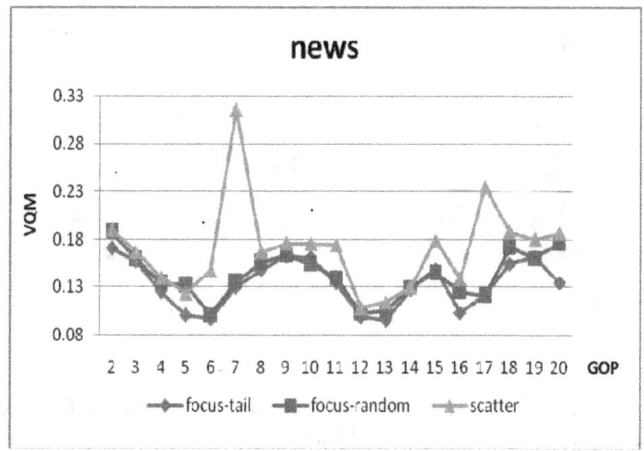

Fig. 4. VQM scores versus GOP number for *news* video (PLR=10%)

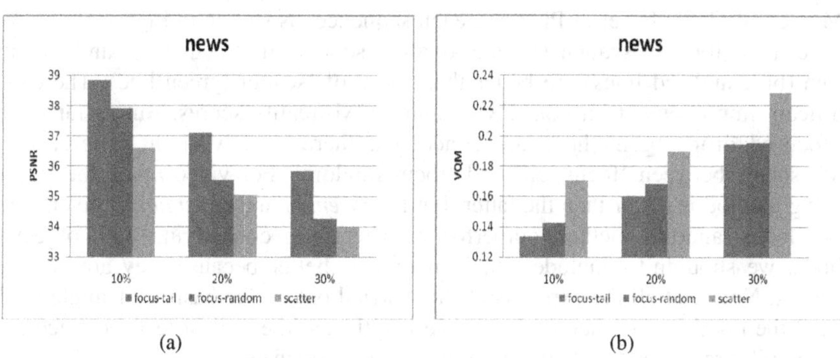

(a) (b)

Fig. 5. Video *news* (a) average PSNR values over all GOPs for three PLRs. (b) average VQM scores over all GOPs for three PLRs.

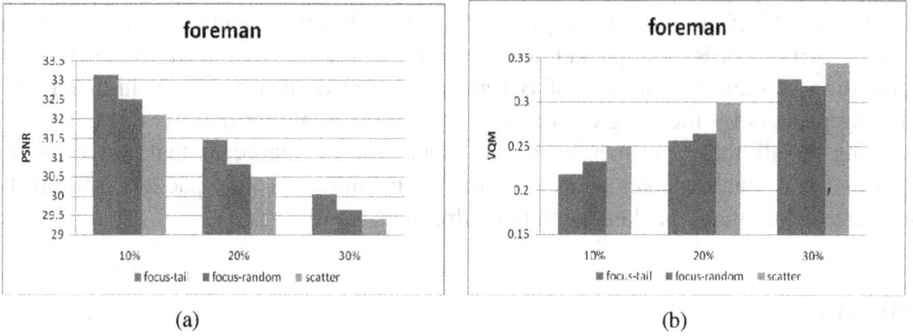

(a) (b)

Fig. 6. Video *foreman* (a) average PSNR values over all GOPs for three PLRs. (b) average VQM scores over all GOPs for three PLRs.

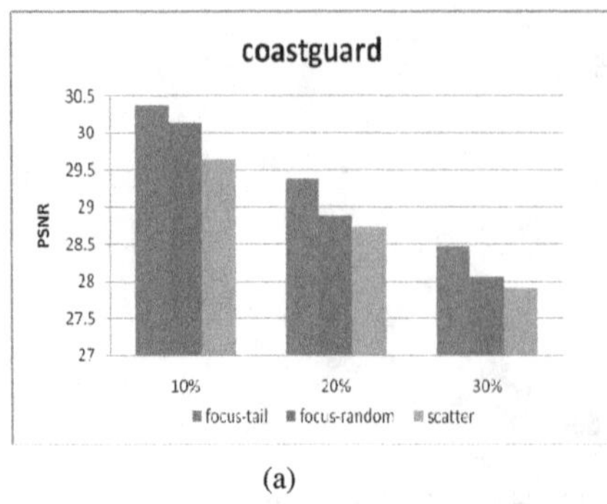

(a)

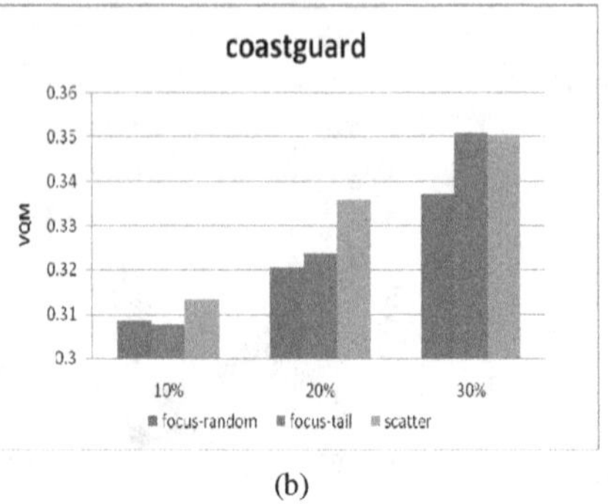

(b)

Fig. 7. Video *coastguard* (a) average PSNR values over all GOPs for three PLRs. (b) average VQM scores over all GOPs for three PLRs.

To present the results more directly, we calculate the average scores of PSNR and VQM over all GOPs for each PLR and each sequence, as shown in Figure 5 to 7. We can see that for all sequences, the quality scores of 'focus-tail' and 'focus-random'(blue and red lines) are better than those of 'scatter'(green line). There is a siginificant improvement on both PSNR and VQM quality scores. But as far as the two focused dropping methods are concerned, there is only slight difference on quality socres between 'focus-tail' and 'focus-random'. For video *news*, the former dropping method is better than the latter, but for *foreman* and *coastguard*, given some PLRs, 'focus-random' method outperforms. Therefore, considering both objective methods, we shouldn't conclude which method is better because they are content-dependent. Note that all the experiments are carried out on the basis of a single GOP, to avoid the impact of sequence length, we lengthened the unit of test sequence as 4 GOPs. Similar results were obtained for all the test sequences.

5 Conclusion

By objective metrics PSNR and VQM, we studied the influences on video quality degradation of different frame types and packet dropping methods in this paper. Given a certain PLR, the packet loss of I or P frames will lead to much more degradation of quality than that of B frames. It is also shown that, at higher PLRs, dropping packets by focusing on a few B frames gives higher quality scores than by scattering in all the B frames randomly. Our findings indicate that not only the importance of each packet, but also the distribution of the packets should be considered when many packets need to be dropped.

References

1. Kanumuri, S., Cosman, P.C., Reibman, A.R., Vaishampayan, V.: Modeling Packet-Loss Visibility in MPEG-2 Video. IEEE Trans. Multimedia 8, 341–355 (2006)
2. Kanumuri, S., Subramanian, S.G., Cosman, P.C., Reibman, A.R.: Packet-loss Visibility in H.264 Videos Using a Reduced Reference Method. In: IEEE ICIP (2006)

3. Lin, T.-L., Zhi, Y., Kanumuri, S., Cosman, P., Reibman, A.: Perceptual Quality Based Packet Dropping for Generalized Video GOP Structures. In: ICASSP (2009)
4. Lin, T.-L., Cosman, P.: Network-based packet loss visibility model for SDTV and HDTV for H.264 videos. In: International Conference on Acoustics, Speech, and Signal Processing, ICASSP (2010)
5. Liu, T., Wang, Y., Boyce, J., Yang, H., Wu, Z.: A Novel Video Quality Metric for Low Bitrate Video Considering both Coding and Packet Loss Artifacts. IEEE J. of Selected Topics in Signal Processing 3(2), 280–293 (2009)
6. Huynh-Thu, Q., Ghanbari, M.: Impact of jitter and jerkiness on perceived video quality. In: VPQM 2006 Proceedings, Scottsdale, AZ (2006)
7. Staelens, N., Vermeulen, B., Moens, S., Macq, J.-F., Lambert, P., Van de Walle, R., Demeester, P.: Assessing the influence of packet loss and frame freezes on the perceptual quality of full length movies. In: Fourth International Workshop on Video Processing and Quality Metrics for Consumer Electronics, VPQM 2009 (2009)
8. Wang, Y., Lin, T.-L., Cosman, P.: Network-based model for video packet importance considering both compression artifacts and packet losses. In: IEEE Globecom (2010)
9. Wang, Y., Lin, T.-L., Cosman, P.: Packet dropping for H.264 videos considering both coding and packet-loss artifacts. In: IEEE Packet Video Workshop (2010)
10. Pinson, M., Wolf, S.: A New Standardized Method for Objectively Measuring Video Quality. IEEE Transactions on Broadcasting 50(3), 312–322 (2004)
11. `http://www.its.bldrdoc.gov/n3/video/index.php`
12. The Official website of FFMPEG, `http://ffmpeg.org`

A Blocked Statistics Method Based on Directional Derivative

Junli Li[1], Chengxi Chu[2], Gang Li[2], and Yang Lou[2]

[1] Visual Computing and Virtual Reality Key Laboratory of Sichuan Province, Sichuan Normal University, Chengdu 610066, China
[2] Information Science and Engineering College, Ningbo University, Ningbo 315211, China

Abstract. The basic idea of identifying the motion blurred direction using the directional derivative is that the original image be an isotropic first-order Markov random process. However, the real effect of this method is not always good. There are many reasons, of which the main is that a lot of pictures do not meet the physical premises. The shapes of objects and texture of pictures would be vulnerably influenced for identifying. In this paper, according to the image characteristics of the local variance, we extract multiple blocks and identify the motion directions of the blocks to identify the motion blurred direction. Experimental results show that our method not only improve the identification accuracy, but also reduce the amount of computation.

Keywords: Motion Blur, Markov Process, Directional Derivative, Weighted Average Method, Blocked Statistics Method.

1 Introduction

The uniform linear motion blur motion blur and defocus blur have good "prior knowledge". So if we can identify the blurred parameters and obtain the point spread function (PSF), it can make the image restoration of this NP problem into P. The scope of this article is to identify the problem of linear motion blur direction.

Previous studies have been proposed including a variety of methods to identify blur direction. These methods can be generally divided into two types, i.e., the transform domain algorithms [1-4] and the airspace algorithms [5-10]. Ref. [1] proposed the periodic zero values stripes in the Fourier spectrum to identify the blur direction, but this method is only applicable to uniform linear motion blurring, and also very sensitive to noise. Ref. [2] has proposed that the above method is sensitively affected by noise, and practical difference. Ref. [3] uses the amplitude spectrum of the gradient image to improve the accuracy of identification. Ref. [4] compares several different advantages and disadvantages of blur parametric methods, including the cepstrum method and the Radon transform method. Overall, the most significant drawback of the transform domain methods is that it is more sensitive to noise, while

F.L. Wang et al. (Eds.): WISM 2012, LNCS 7529, pp. 90–97, 2012.

the airspace algorithm has better resistance to noise. The "error-parameter analysis" given by Ref. [5] can identify the horizontal direction of the uniform linear blur and defocus blur point spread function. This method can improve resistance to noise, but the process is of computational complexity, and human judgment may cause some errors as well. Y. Yitzhaky et al. [6-8] use a 2×2 differential operator to identify the blur direction, but the scope is limited to -45 degree to 0 degree, and the accuracy is not high. Chen Qianrong et al. [9] use 3×3 differential operator to identify, which the angle range extends to -90 degrees to 90 degrees, and also the accuracy is improved. It should be a good physical idea, the original image power spectrum as for an isotropic first-order Markov random process, but does not achieve good results in practical application. Because a lot of don't meet the physical premises. Ref. [10] has given some improvement, but only the operator is changed. And the practical application is not good.

In our view, to select some characteristics blocks can improve the accuracy and reduce computation. This paper will give why and how we select them. In addition, in order to further reduce the amount of calculation, we give two different steps to search the data.

2 Directional Derivative Method

2.1 Point Spread Function

It can cause motion blurring in the relative movement between the object and the imaging system. As the exposure time is very short, the motion can be regarded as a uniform motion. So we have the blur length L=vt, where t is the exposure time and v is the constant speed. θ is the angle along the horizontal direction.
We have the point spread function (PSF)

$$h(x,y)=\begin{cases}1/L, 0\le |x| \le L\cos\theta, y=L\sin\theta \\ 0, otherwise\end{cases} \tag{1}$$

From the above, we can get the point spread function h(x, y), if we know the blur direction θ and the blur length L. Under normal circumstances, seek first the blur direction and then the blur length. Once the blur direction is known, this two-dimensional problem becomes one-dimensional problem. So the accuracy of identification of the blur direction is very important.

2.2 3×3 Differential Multiplied Operator

Usually the original image is a first order Markov isotropy in our method, that is the original image autocorrelation and power spectrum is isotropic. As motion blur reduces the high frequency components in the image, the greater the departure of the direction is, the smaller it impacts. That denotes the loss of the high frequency components on the direction of motion is the most, and on this direction the image gray value is the smallest. Y Yitzhaky et al. [6-8] and Chen Qianrong et al. [9] use the same physical idea. But the former uses a 2×2 differential multiplier, while the latter uses a 3×3 differential multiplier, which can improves the accuracy of identification of the motion direction.

Figure 1 shows the direction differential schematic, where we set a point g(i, j), the size of the blur image is M×N. g(i′, j′)is the point on the arc, where the radius is Δr, and the center point is g(i′, j′). Δr is the micro-element. The value of Δr can be flexibly selected, such as to take 1, 1.5 or 2 pixels, according to the movement type (uniform linear motion, acceleration, vibration and etc). ∂ is the direction of the differential, define the range [−π/2, π/2), ∂ is 0 on the horizontal direction and negative under the horizontal direction, non-negative above . We use bilinear interpolation to get the value of g(i′, j′), where i′, j′ satisfy the condition

$$\begin{cases} i' = i + \Delta r * \sin\partial \\ j' = j + \Delta r * \cos\partial \end{cases} \tag{2}$$

So we get the image gray value on the direction ∂

$$\Delta g(i,j)_\partial = g(i',j') - g(i,j) = g(i,j) \otimes D_\partial \tag{3}$$

Where D_∂ is the differential multiplier on the direction ∂.

$$D_\partial = \begin{bmatrix} -1 & 2-4\sin\partial-2\cos\partial+4\sin\partial\cos\partial & -1+2\sin\partial+2\cos\partial-4\sin\partial\cos\partial \\ 0 & 4\sin\partial-4\sin\partial\cos\partial & -2\sin\partial+4\sin\partial\cos\partial \\ 0 & 0 & 0 \end{bmatrix} \tag{4}$$

Then get the gray sum of absolute

$$I(\Delta g)_\partial = \sum_{i=0}^{N-1}\sum_{j=0}^{M-1} |\Delta g(i,j)_\partial| \tag{5}$$

In the interval [−π/2, π/2), we can get the value of ∂ according to a certain step. Then find the value ∂ which makes the value of $I(\Delta g)_\partial$ minimum, and this ∂ is the motion blur direction. Use the similar methods, the other five regional operator derived results can also be calculated.

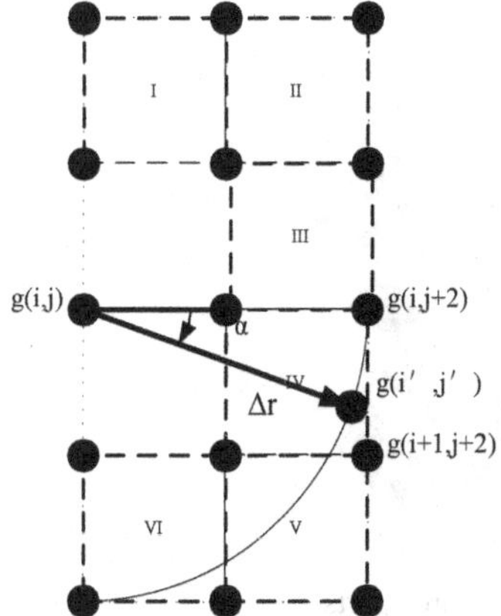

Fig. 1. Direction differential schematic

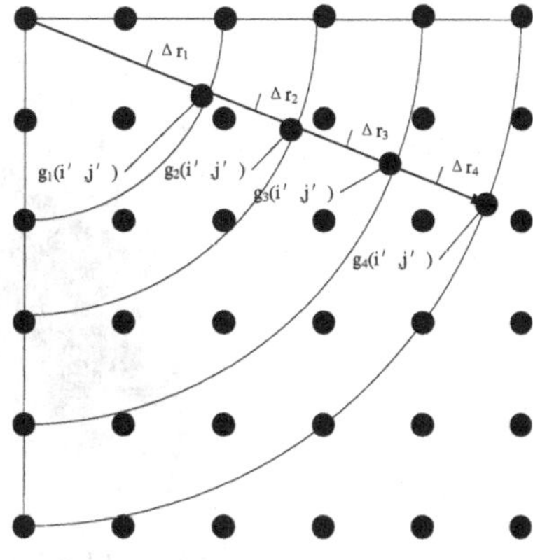

Fig. 2. Weighted average method

2.3 The Weighted Average Method Based on Directional Derivative

There will be some local deviations during the identification. To average the random factors and to improve the accuracy, the weighted average method is proposed on the basis of the directional derivative.

Shown in Figure 2, first to obtain the absolute value of the differential image gray $I(\Delta g)_{\Delta r_1,\partial}, I(\Delta g)_{\Delta r_2,\partial}, I(\Delta g)_{\Delta r_3,\partial}, I(\Delta g)_{\Delta r_4,\partial}$ using several micro-element (Δr_1, Δr_2, Δr_3, Δr_4), then get the weighted summation

$$I(\Delta g)_{\partial} = \sum_{i=1}^{4} w_i * I(\Delta g)_{\Delta r_i,\partial} \tag{6}$$

where $w=[w_1,w_2,w_3,w_4]=[1,1,1,1]$. And at last we get the differential curve $I(\Delta g)_{\partial}$ - ∂,

$$\hat{\partial} = \underset{\partial\in[-\pi/2,\pi/2)}{\arg\min} \left(\sum_{i=1}^{4} w_i * I(\Delta g)_{\Delta r_i,\partial}\right) \tag{7}$$

3 The Blocked Statistical Method Based on the Directional Derivative

Usually the whole blur image should have the same PSF, but the identifications of the various parts are different. Figure 5 shows the identification of three parts of Camera. The identification of the whole image is 48 degrees, 46 degrees in Zone 1, Zone 2 and 3 deviate large. The edge of the image contains important information in a blur image. We can get a good result if we select the strong edge region.

Fig. 3. Identification of three parts

In most cases, due to the intricacies of the texture, the strong edge region is difficult to extract. To simplify it, we select some regions of large local variance. First calculate the local variance of the blur image, and then select some large variance points to determine the blocks. Specific process is as below:

Step 1: De-noising pretreatment, if there is some noise, this step is necessary;
Step 2: Calculate the local variance of the whole image, and sort the variances;
Step 3: Randomly select several big ones, and then get the blocks of which the center points are the selected variance points;
Step 4: Identify the blurred direction for every block using the weighted average method based on directional derivative;
Step 5: Get the results and average them.

4 Experimental Results and Analysis

4.1 The Limitation of the Identification for the Whole Image Using Directional Derivative

Motion blur can be seen as the superposition of the pixel gray, that is the process of pixel linear superposition on the blur direction. In visual performance, it is blur ghosting in some direction [11]. That the original image is regarded as "isotropic first-order Markov process" has its limitations, because it is susceptible to the influence of the factors susceptible to the influence of the factors such as the shape and texture of the objects. We take the image Baboo, Camera and Lena as examples. The experimental results are shown in Table 1.

It can be seen from the results, the weighted average method based on directional derivative can basically identify the blur direction of the image Baboo. The error of the identification of Lena is very large, followed by the image Camera. Analysis of the three images, the intensity change of Baboo is uniform, isotropic performance.

Table 1. Blur direction identification results of the images

Blur length (/pixels)	Blur direction (/degrees)	Direction identification (/degrees)							
		(a)	(b)	(c)	(d)	(e)	(f)	(g)	(h)
10	-82	-81	-89	-89	-89	-85	-89	-84	-89
-20	-65	-66	-68	-84	-68	-72	-86	-69	-84
25	-36	-36	-37	-74	-38	-45	-70	-38	-40
30	0	0	0	72	0	0	45	0	35
32	45	45	48	54	45	48	50	45	52
24	75	76	76	81	76	82	80	76	82

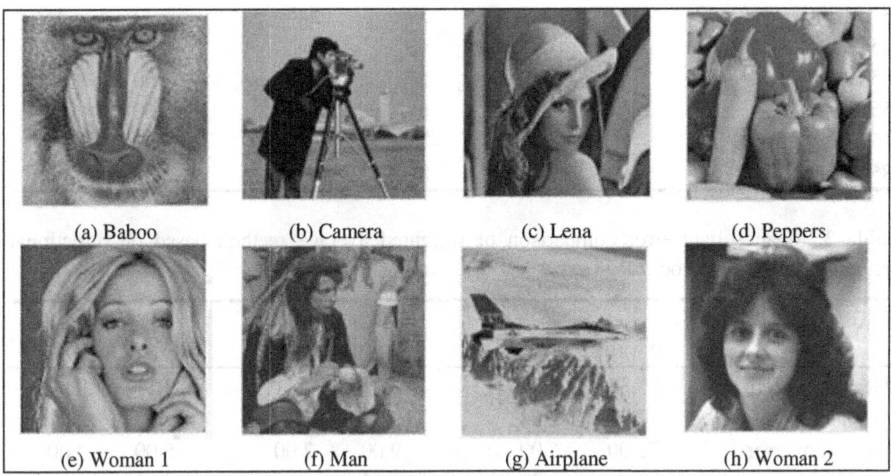
(a) Baboo (b) Camera (c) Lena (d) Peppers
(e) Woman 1 (f) Man (g) Airplane (h) Woman 2

Fig. 4. Some experimental images

Texture of Lena is complex, and a large range of gray change. Even in non-noise case, it cannot get accurate results. In Camera, there is a large black region, and the identification error of certain angles. That denotes object shapes can have some impacts on the identification. In Figure 4, (d), (e), (f), (g) and (h) are images of which the identification errors are large using weighted average method based on directional derivative for the whole image. Analysis of the experimental results can be drawn, and the shapes of the objects and the texture can affect the identification results.

4.2 Comparison of Identification Accuracy

Table 2 shows identification results of Lena and Camera and Baboo. Table 3 is the comparison of the errors. It can be seen from the two tables, our method can indeed improve the accuracy of the identification of blur direction.

Table 2. Results comparison of weighted average method based on directional derivative and our method (/degrees)

Original	Lena		Camera		Baboo	
orientation	Ref.[9]'s	Ours	Ref.[9]'s	Ours	Ref.[9]'s	Ours
-90	-89	-89	-89	-89	-89	-89
-80	-89	-83	-89	-82	-80	-80
-60	-85	-65	-64	-62	-60	-60
-50	-64	-55	-54	-52	-50	-50
-30	-70	-35	-36	-32	-30	-30
-10	20	-4	-15	-8	-15	-10
0	72	0	0	0	0	0
20	36	22	23	21	22	20
35	42	37	37	35	32	35
45	54	46	48	46	45	45
55	59	56	57	55	50	52
65	81	68	68	66	60	62
75	81	76	75	76	80	75
85	89	87	89	86	89	85

Table 3. Mean square error comparison of weighted average method based on directional derivative and our method (/degrees)

Errors	Lena		Camera		Baboo	
	Ref.[9]'s	Ours	Ref.[9]'s	Ours	Ref.[9]'s	Ours
Mean square error	25.35	3.32	4.13	1.41	3.11	0.94
Average error	17.36	2.64	2.94	1.14	2.14	0.43
Maximum error	72.00	6.00	9.00	2.00	5.00	3.00
Minimum error	1.00	0.00	0.00	0.00	0.00	0.00

4.3 Blur Direction Search Optimization

In order to speed up the search time, step-by-step search is proposed. The first step is set 10 degrees, and the second step is set 1 degree. After the first step search, we choose the angle whose gray absolute value is the minimum. If the angle is -90 degrees, the search interval is [-90, 90), otherwise the search interval is the angle around 10 degrees extension. So to maintain the necessary accuracy, we can reduce the time of the blind search greatly.

5 Conclusion

Experimental results show that texture and shapes of the objects will affect the results of the identifications. The identification using directional derivative for the whole image sometimes makes large errors. Blocked statistics method is proposed, and usually our method can give a good result. In addition, step-by-step search is given to speed up the search time. In experiments, we find the results are not good in case of strong noise. We will focus on how to use directional derivative to accurately identify the blur direction in strong noise.

References

[1] Cannon, M.: Blind deconvolution of spatially invariant image blurs with phase. IEEE Transactions on Acoustics, Speech and Signal Processing 24(1), 58–63 (1976)

[2] Fabian, R., Malah, D.: Robust identification of motion and out-of-focus blur parameters form blurred and noisy image. CVGIP: Graphical, Models and Image Processing (53), 403–412 (1991)

[3] Ji, H., Liu, C.: Motion blur identification from image gradients. In: IEEE Conference on CVPR, pp. 1–8 (2008)

[4] Krahmer, F., Lin, Y., McAdoo, B., Ott, K., Wang, J., Widemannk, D., Wohlberg, B.: Blind image deconvolution: motion blur estimation. Tech. Rep., University of Minnesota (2006)

[5] Zou, M.-Y.: Deconvolution and Signal Recovery. National Defence Industry Press, Beijing (2001)

[6] Yitzhaky, Y., Mor, I., Lantzman, A., et al.: Direct method for restoration of motion-blurred images. Journal of Optical Society of American 15(6), 1512–1519 (1998)

[7] Yitzhaky, Y., Kopeika, N.S.: Vibrated image restoration from a single frame. In: The International Society for Optical Engineering, SPIE, vol. 3808, pp. 603–613 (1999)

[8] Yitzhaky, Y., Milberg, R., Yohaev, S., et al.: Comparison of direct blind deconvolution methods for motion-blurred images. Applied Optics 38(20), 4325–4332 (1999)

[9] Chen, Q., Lu, Q., Chen, L.: Identification of the motion blurred direction based on the direction of differential. Chinese Journal of Graphics 10(5), 590–595 (2005)

[10] Yan, X., Liu, H., Liu, B.: Motion blur airspace estimation method based on the rotation difference operator. In: 15th National Image and Graphic Conference, pp. 18–23 (2011)

[11] Lokhande, R., Arya, K.V., Gupta, P.: Identification of parameters and restoration of motion blurred images. Chinese Journal of Computers 30(4), 686–692 (2007)

[12] Yitzhaky, Y., Kopeika, N.S.: Identification of blur parameters from motion blurred images. Graphical Models and Image Processing 59(5), 310–320 (1997)

[13] Chen, Y.S., Choa, I.S.: An approach to eastimating the motion parameters for a linear motion blurred image. IEICE Transactions on Information System E83-D(7), 1601–1603 (2000)

Study on Cooperative Game Model of Talent Training through School-Enterprise Coalition

Jun Li, Yan Niu, and Jingliang Chen

Computer School, Hubei University of Technology, Wuhan, China

lj_py@126.com, {ny,cjl}@mail.hbut.edu.cn

Abstract. Joint talent training is an important part of school-enterprise cooperation. The success of school-enterprise cooperation and cooperative strategy is closely related to the training cost of the school or the enterprise. This paper proposes a student-, school- and enterprise-based three-party cooperative game model to rationally quantify relevant parameters such as the training cost of the school and enterprise. Simulation of the three-party cooperative game model is also studied in the paper. Results of simulation will help the school and enterprise develop corresponding strategies. The model can also give a reasonable explanation on realities of school-enterprise coalition.

Keywords: school-enterprise coalition, talent training, game, Shapley value, cooperation.

1 Introduction

School-enterprise cooperation is a new cooperation model created for the school and enterprise to realize resource sharing and mutual benefit. The school has advantageous talents to tackle key technological problems in scientific research and industrial upgrading. In addition, the school is also a cradle for talents with broad knowledge background, strong knowledge-update ability and innovative ideas. The enterprise is the innovation subject, i.e. the input subject, research subject, benefit distribution subject and risk and responsibility undertaking subject. Enterprises and universities have heterogeneous resource and capability as well as inter-organizational complementarity, allowing the creation of linkage between universities and enterprises and synergistic effect as well as effective innovation and integration. The combination of school and enterprise not only can bring their respective advantages into full play, but also can jointly train talents meeting social and market demand, making the school-enterprise model a win-win model for both the school and enterprise (society) [1]. Globally, “dual education system”, “sandwich placement”, “TAFE”, “industry-academic cooperation” and other models are common forms of school-enterprise cooperation. Meanwhile, with the development of demand characteristics and increased innovation risk, school-enterprise cooperation tends to be further strengthened and intensified. Talent training is an important joint program of school-enterprise cooperation. As the smallest unit of market economy, the enterprise regards profit and value as its ultimate goals and talents as the core competitiveness. Therefore, the enterprise must pay high attention to talent training.

F.L. Wang et al. (Eds.): WISM 2012, LNCS 7529, pp. 98–104, 2012.

On the other hand, the goal of a school is to train and provide talents. In particular, training of technical talents and skilled talents with innovation ability should be the prominent feature of a school. For both the school and student, firstly, school-enterprise cooperation allows the school and student to make full use of education resources and environment inside and outside the school. Secondly, the student can combine knowledge learned in class with specific operation in actual working environment. Through front-line internship or placement, students can broaden their knowledge scope and have a good understanding of their future job as well as build comprehensive ability [3, 4]. Industry-academic cooperation and combination of working and learning bring obvious benefits to the school and student. Schools and enterprises have to pay costs for cooperation. However, on the whole, through coalition, schools and enterprises, and even the policy provider, the government can obtain high payoff, realizing win-win for the school, enterprise and society. Therefore, school-enterprise cooperation represents the maximization of cooperation and collective payoff. No matter what kind of cooperation model is adopted, school-enterprise cooperation is an organization built for the purpose of mutual benefit.

2 Cost Sharing-Based Cooperative Game Model

Through cooperation to realize mutual complementarity, synergistic effect produced by the school and enterprise through cooperation is higher than the sum of values the two created for the society. Therefore, the cooperative game is a non-zero sum cooperative game. As for results of school-enterprise cooperation, cooperation should be regarded as the precondition or premise for distribution of cooperation payoff and each player should make joint efforts with other players or even competitors to strive for more payoff. Cooperative game generally can be divided into transferable utility game (utilities of players can be superposed) and non-transferable utility game (utilities do not have marginal payment nature). Cooperative game highlights collective rationality, efficiency, fairness and equality. Win-win is possible only under the framework of cooperative game which generally gains high efficiency and payoff. Considering the positive linear relation between cost and payoff under non-cooperative game, the model is applicable to transferable utility. As cooperation lowers cost and increases total profit, a mathematical model is established to find a scheme for rational distribution of the profit increment within the coalition.

2.1 Characteristic Function

Player set mainly includes the school, enterprise and student. N= {1,2,……n}. We define v(S) as player's coalition and $S \subseteq N$ can generate the maximum worth. Call $v(\bullet)$ as a characteristic function and assume it has the following properties [6]:

1) $v(\phi) = 0, \phi$ indicates an empty set.

2) If R and S are two non-intersecting player subset, then $v(R \cup S) \geq v(R) + v(S)$

From the assumption we can see that school-enterprise cooperative game is changed to characteristic function game with $<N,v>$ as the model. In terms of value quantification, student's biggest goal is to obtain corresponding innovation ability and skills. In terms of measurement, economic indicators such as credit hour and tuition can be quantified. Payoff obtained by the school can also be measured by the input cost of laboratory, teacher's expense, credit hour and tuition of the school. Payoff obtained by the enterprise can be measured by work load and staff training cost. Different from the three-party game, the school should, under the premise that the budget for running a school is not exceeded, choose the maximum value for student as the optimal solution. Formula (2) can be changed to a cost sharing problem, i.e. the total cost of running two combined coalitions is lower than that of running two independent coalitions. Value will not be produced by a coalition which has no player. Formula 2 shows superadditivity. Assume x_i can represent payoff the player i can get, then an n-dimension real vector consisting of payoffs of n players is formed as $x=(x_1,x_2,\dots,x_n)$, which is called a characteristic vector. The following assumptions are satisfied:

$$v(\{i\}) \le x_i, i \in N, S \subseteq N \tag{1}$$

$$\sum_{i \in N} x_i = v(N) \tag{2}$$

$$v(S) \le \sum_{i \in N} x_i, S \subseteq N \tag{3}$$

The above assumptions show the conditions of personal rationality, Pareto efficiency and collective rationality of game players [6]. It is easy to see that enterprise and school can be regarded as rational individual and can form a rational group.

2.2 Parameter Quantification

The skill a talent possesses is proportional to the input of credit hour. If a talent wants to acquire the same skill, a certain number of credit hour should be spent on learning. Assume the completed skill as A_{skill}. Talent training cost paid by a school is reflected in credit hour, lecture fee and construction of a laboratory. Construction cost of the laboratory is one-time investment and the laboratory has a long service life, thus, construction cost of the laboratory can be regarded as a constant μ. Only course offering, credit hour, and lecture fee should be taken into consideration. To simplify the talent training cost, the cost of credit hour * lecture fee/credit hour can be assumed as $x * pay$. Therefore, talent training cost of a school is $v\{school\} = x * pay$. The main cost of the enterprise is the cost of skill training for new recruits such as skill training and internship. However, as the enterprise is not a professional educational organization, the cost of training is higher than that of the school. We can use m to represent the cost increase coefficient of the school and enterprise, and the m is called multiplication factor. $0 \le m \le 1$, without loss of generality, staff training cost of an enterprise should not be higher than twice of school training cost. $v\{\text{company}\} = x * pay * (1+\text{m})$. Students can only require a few skills if they do not participate in these courses and choose self-learning which is assumed as $v\{student\} = \beta$. Assume that corresponding skills can be acquired under the premise that three parties exist. We can make the following assumptions that the payoff of talent standard jointly reached by three parties is 100 and the payoff (cost deducted) of school to reach such talent standard independently is $v(stu, school) = 100 - x * pay$. While the payoff (cost deducted) of school to reach such talent standard independently is $v(stu, company) = 100 - x * pay * (1+m)$. Independent enterprise, school and student produce no payoff. No payoff is produced by school-enterprise coalition if students do not participate in the coalition. Therefore, characteristic function of the three parties can be defined in this way. However, there are constraint conditions for such function, i.e. total credit hours should not be exceeded and the training cost of an enterprise should not be too high, which is $0 \le m \le 1$.

2.3 Calculation of Shapley Value

According to Shapley theorem, there is an only value function

$$\varphi_i(v) = \sum_{S \subseteq N\{i\} i \in S} \gamma_i(S)(v(S) - v(S - \{i\})) \tag{4}$$

$i \in N$, where, $\gamma_i(S) = \dfrac{(s-1)!(n-s)!}{n!}$

And corresponding characteristic function is

$x = [1,1,1]$ $v = 100$;

x=[1,0,1] x=100-x*pay*(1+m);

x=[0,1,1] x=0;

x=[1,1,0] x=100-x*pay;

x=[1,0,0] x=0;

x=[0,1,0] x=0;

x=[0,0,1] x=0;

3 Simulation and Analysis of Cooperative Game Model

We can see from Formula 1 and assumptions that the solution is in line with the Pareto optimal solution. Core configuration only exists when there is a big coalition formed by all players [7]. Payoff of the coalition can be understood as the weighted evaluation of marginal contribution made by players to the coalition.

Situation one: the training cost of an enterprise is $x^{*}pay^{*}(1+\mathrm{m})=80$, in which m changes within 0.1~0.9, and the following Shapley can be obtained.

Table 1. m value and Shapley value

m	stu	college	company
0.1	38.16667	33.16667	28.66667
0.2	39.66667	34.66667	25.66667
0.3	41.16667	36.16667	22.66667
0.4	42.66667	37.66667	19.66667
0.5	44.16667	39.16667	16.66667
0.6	45.66667	40.66667	13.66667
0.7	47.16667	42.16667	10.66667
0.8	48.66667	43.66667	7.666667
0.9	50.16667	45.16667	4.666667
1	51.66667	46.66667	1.666667

It can be seen that if the training cost of an enterprise is fixed, for the training cost of a school, which is $\frac{60}{1+m}$, the payoff of a school increases with the increase of the multiplication factor. Payoff obtained by the student also increases accordingly, while the same obtained by the enterprise will decrease.

From Fig. 1, we can see that if the enterprise trains its own staff, the payoff can be 20 units. When the multiplication factor excesses 0.4, i.e. the training cost of the enterprise is more than 1.4 times of the teaching cost of the school, the payoff of the enterprise in school-enterprise coalition will be lower than 20 units, thus, the enterprise will not be enthusiastic in school-enterprise coalition. The optimal strategy of the enterprise is to reject the coalition or to keep a negative attitude and wait to be given a free ride by the school. Or the enterprise will recruit talents directly from a job fair and the cost will be lower than that of establishing cooperation with the school and training talents jointly. It is understandable that the enterprise makes such choice. The success of joint talent training through school-enterprise coalition depends on the condition that full attention is given to the interests of each party, in particular, the enterprise has discovered the value of such coalition so as to ensure the student obtains the optimal local payoff. Of course, it takes a long time and long-term investment to train skilled talents and not all enterprise can afford the cost. Although the school has to pay certain costs, it should actively promote school-enterprise cooperation to avoid negative attitude from the enterprise.

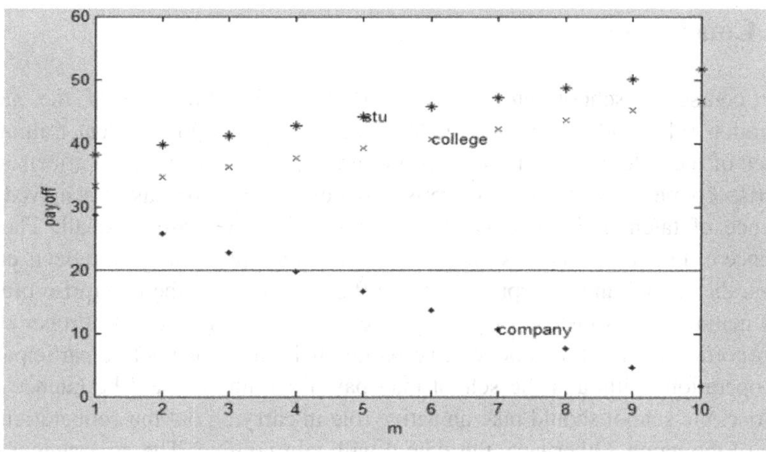

Fig. 1. Payoff trend of each party with different m values. Payoff obtained by the student and college will increase with the increase of the multiplication factor. However, the enterprise's payoff will decrease.

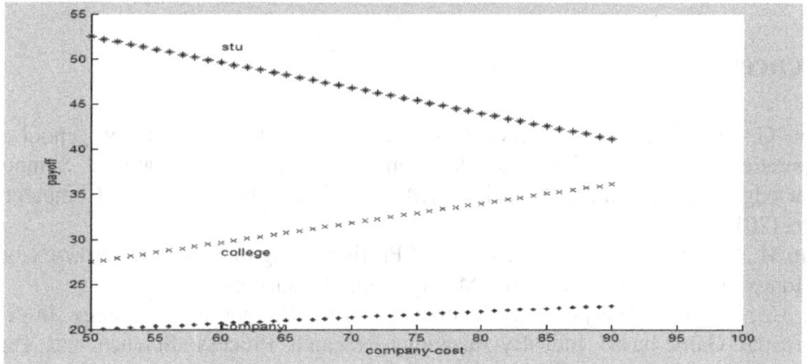

Fig. 2. Payoff trend of each party with the change of training cost and fixed m value

On the other hand, if the multiplication factor is fixed, for example, when $m = 0.2$, as training costs of the enterprise and school increases, the payoff obtained by school in the coalition will decrease, which is in line with the actual situation. If the enterprise increases the cost of staff training, the payoff shows no obvious increase (see the slope of company in the Fig. 2). The school will save more cost to accomplish the same goal. Under such circumstance, the enterprise will not actively take part in the coalition. The payoff obtained by the school is higher than that of the enterprise. Therefore, when the multiplication factor is fixed, the most reasonable cost paid by the enterprise in the coalition should not be higher than the product of training cost of the school multiplying multiplication factor, i.e. $x * pay * (1+m)$.

4 Conclusion

In the course of school-enterprise cooperative game, the core of the game is cooperation rules and profit-sharing. Multiplication factor m of talent training cost variance of the enterprise and school is an important parameter which affects school-enterprise cooperation. School-enterprise cooperation can be easily achieved if the difference of talent training cost of the school and enterprise is small. The small difference of talent training cost of the school and enterprise can also reflects equality between the school and enterprise. If the difference is large, the enterprise prefers to recruit mature talents instead of putting investment in talent training. Proper strategy and proportion of the three cooperative parties will encourage active participation in the cooperation. Although the school may pay a certain cost and be hitched by the enterprise, the school should take an active role in carrying out the cooperation in the course of the game. Otherwise, the school will gain nothing. The government should take the role of cooperative game designer and issue incentive policies so as to grant preferential treatment to enterprise which is willing to and actively takes part in school-enterprise cooperation when pursuing payoff maximization. The government should also improve public opinion and promote healthy development of social assessment.

References

1. Chen, G.: Innovation and practice of training high-skilled workers by school-enterprise cooperation in higher vocational education. In: 2011 4th International Symposium on Knowledge Acquisition and Modeling, KAM 2011, October 8-9. IEEE Computer Society, Sanya (2011)
2. Zhan, M., Pan, J.: Analysis on the Game of Profit Sharing of School-Enterprise Cooperation and Innovation. Scientific Research Management (January 2008)
3. Liu, C., Ye, C.: Analysis on School-Enterprise Cooperation Strategy in Two-stage Sequential Game-based Industry-Academia-Research Process. Science and Technology Management Research (June 2011)
4. Ding, X.: Training high vocational talents with school-enterprise cooperation: A case study. In: 2nd International Conference on Information Science and Engineering, ICISE 2010, December 4-6. IEEE Computer Society, Hangzhou (2010)
5. Gu, Y., Tang, G.: Sequential Game: New Development of Cooperative Game. Journal of Chongqing Normal University (Natural Science Edition) (February 2012)
6. Zhao, Y.: Game Theory and Economic Model, Pu, Y. (ed.). China Renmin University Press, Beijing (2010)
7. Yu, X., Zhang, Q.: Analysis on Interval Shapley Value-based Profit-sharing of Production and Cooperation. Journal of Beijing Institute of Technology (July 2008)

Research on Internet Public Opinion Detection System Based on Domain Ontology

Chunming Yang

School of Computer Science and Technology, Southwest University of Science and Technology, Mianyang 621010, China
yangchunming@swust.edu.cn

Abstract. How to get the Internet public opinion timely and comprehensively is a key problem for local governments. The method of keyword matching lack of analysis and statistics on semantic, reduced the precision and recall. In this paper, the detection algorithm of semantic expansion is proposed by utilizing *Integrated E-Government Thesaurus (Category Table)* to construct domain ontology and through expanding the area knowledge. The experimental results show that the precision rate and recall rate of public opinion detection have all been improved obviously.

Keywords: Internet public opinion, domain ontology, Thesaurus.

1 Introduction

With the sharp increase of the popularization and application of mobile terminals, such network media as blog, forum, micro-blog and instant messaging become the major media for netizens to express their own opinions and to communicate information. Internet public opinion consists of the opinions generated from the stimulation of all sorts of events, the integration of those knowledge, attitude, emotion and behavior disposition on the event of people who communicate it through internet, and the common opinion with certain influences and disposition[1].

Through internet public opinion detection, governmental departments can promptly learn public feelings, integrate civil wisdoms, and timely discover the information that could lead to emergencies so as to make decision quickly and adopt corresponding measures to control and guide the development of the event and thus to reduce the social influences of this event.

2 Related Work

The target of Internet Public Opinion Detection (IPOD) is to discover the sensitive information, hotspot, new topics in designated public opinion sources, while follow up the topics and analyze the emotional disposition so as to convenient the government to master key information at first time to implement specialized work.

F.L. Wang et al. (Eds.): WISM 2012, LNCS 7529, pp. 105–110, 2012.

Currently, some scientific research institutes, colleges and universities and enterprises have developed some internet public opinion analysis methods and implemented systematic research and design work.

The major systems include the analysis on the occurrence and change rules of internet public opinion based on automatically collected public opinion information with court system as the application background and the classified standard catalog of news information as the default public opinion theme[2].

TRS[3] public opinion detection system adopts the analysis on similarity of content to discover new hot news, and then filters the hot information which is not concerned by users based on automatic classification; to the already known hot information, follow up it according to the thematic character and make analysis on communication paths; timely detect the sensitive information concerned by users and provide warnings for the negative information.

In terms of analysis methods, [4]utilizing Chinese word segmentation technology to statistic the term frequency in internet information so as to discover hot information, [5]utilizing fuzzy reasoning to provide internet public opinion warning, [6]utilizing the theory and method of signal analysis to provide the internet public opinion warning model for emergency and [7]utilizing related technology for opinions mining to construct the original model of urban image internet detection system so as to identify the emotional disposition and evolution rule in comments. And then make primary research on the internet detection of city impression.

The government usually hopes to get the related sensitive information released on those news websites, blogs and forums in their administrative regions. Although above mentioned systems and methods might be used for the IPOD of local governments, and some of them have been put into trial run, the precision and completeness of public opinion detection are to be improved.

It mainly utilizes keyword matching method to retrieve whether the collected information contains designated sensitive word information in current systems. As netizens are free on expression, they might usually use different words to express the same concept (topic). For instance, the evaluation on Secretary of Municipal Party Committee might use such terms as "市委书记", "×书记" or a name rather than adding regional restriction like "××市" to it. This kind of phenomenon of "default region" and "multi-word for one meaning" based on utilizing keyword matching method lack of analysis and statistics on semantic, so it reduced the precision and completeness of public opinion detection.

Based on above problems, this paper constructed the Internet Public Opinion Detection System (IPODS), based on domain ontology according to governments' realistic demands on IPOD by utilizing *Integrated E-Government Thesaurus (Category Table)* to construct domain ontology of public opinion detection and through expanding the area knowledge.

3 Discovery of Public Opinion Based on Domain Ontology

3.1 Construction of Domain Ontology Based on Thesaurus

Domain Ontology refers to the conceptual model used for describing domain knowledge. It defines the basic terms and the relations that consist of vocabulary of some certain domain, and then truly reflect the substantive characteristics of the object by combining these terms and relation definition[8].

There are mainly three types of semantic relations between terms in domain ontology: synonyms, hypernym and hyponym. These semantic relations can solve the phenomenon of "one meaning multi words" and "default administrative region" in public opinion detection. The thesaurus is the Chinese phrase library with semantic relation. It shows the theme of some certain subjects and fields through integrating normalized words. *Integrated E-Government Thesaurus (Category Table)* [9]collected the words of all subjects in E-government field, and standardized them. And the effective category and non-official theme words (synonym) are confirmed. Therefore, the thesaurus can be utilized to construct the public opinion detection domain ontology and regular words and administrative region knowledge can be added according to realistic demands, such as local administrative region distribution, leaders name and institutions etc.

When implementing specific construction, find the grade one and grade two categories from the *Integrated E-Government Thesaurus (Category Table)*, and then extract the core hierarchy of conception type and analysis type. And then take subject word as the sample of category, and then confirm the synonymy among samples and the hierarchy relation between categories. And finally add in the regional information, like the terms including administrative region portion, administrative institute and leaders' names as well as the corresponding hierarchy and semantic relation like synonymy, thus the needed domain ontology can be formed. The domain ontology will change with the change of demands. Therefore, the public opinion detection domain ontology shall be maintained and expanded, mainly including adding terms and confirming the relationship between terms.

The system adopts protégé 4.02 to construct ontology, and then lead in the constructed ontology into database for storage after checking the consistency, so as to be used for discovering public opinion. A total of 17 class one categories are created with a total of more than 3800 subject words (exclude thesaurus). And the domain ontology maintenance tool is developed, and it can be added according to demands.

3.2 Public Opinion Detection Based on Semantic Expansion

The detection of internet public opinion refers to that with users setting up corresponding sensitive words, the system utilizes domain ontology to make semantic expansion on the words, and then return the information that satisfies searching conditions from the collected information to users. Here's semantic expansion is to find the synonyms, hypernymy and hyponym of the words at its concept node from the domain ontology, and then respectively transfers it into *OR* and *AND* logical search. The definition of semantic expansion is as follows:

> *W* refers to word, *O* represents domain ontology, *N* means concept node. So to word *W*, if *W* matches the *N* in *O*, or *W* and *N* has child of relation, or *W* and *N* has equivalent relation, so *W* totally matches with the *N* in *O*; if there are many *N*, then matching conceptual set is formed. The formula is showed as follows:

$$C(W,O)=\{N_1,N_2,\ldots,N_n\},\ (N_i\square O)$$

According to above definitions, after finding the conceptual set of sensitive words *W* in domain ontology, turn *hasChildOf* relation to and relation, and turn equivalent relation to or relation, then the logical expression formula after semantic expansion can be acquired. Utilize this formula to make logic search in the collected information and then send the information that satisfies logic condition to users. The algorithm as follows:

> Input: sensitive word, domain ontology, and document set *D*
> Output: the document set T that satisfies conditions
> Step 1: Find *W* in *O*, if found, then record as *N*, or record as *Q*, and then turns to Step5;
> Step 2: Find words $N_1,N_2,\ldots,N_k$ that *hasChildOf* relation with *N* in *O*, the formula is $Q_1= (N_1 \text{ or } N_2 \text{ or } \ldots N_k)$;
> Step 3: Find the words $M_1, M_2,\ldots, M_j$ that has equivalent relation with *N* in O, express it as $Q_2= (M_1 \text{ or } M_2 \text{ or } \ldots M_j)$;
> Step 4: Assemble *W* with Q_1 and Q_2 to form the inquiry formula $Q=W$ and $(N_1 \text{ or } N_2 \text{ or } \ldots N_k) \text{ or } (M_1 \text{ or } M_2 \text{ or } \ldots M_j)$;
> Step 5: Utilize *Q* to make logic inquiry in *D*, and then return the documents that satisfy conditions to users according to time descending order.

In order to improve the precision and timeliness of public opinion detection, the system first makes subject indexing and extraction of named entity on collected information, and then mainly matches from the indexed subject and named entity when making logical inquiry. Subject indexing adopts TF-IDF method [10], namely add domain ontology and network buzzword at Chinese participle stage. Meanwhile corresponding weighing is made to the positions of part of speech, term length and word. 5~20 subject words are confirmed for each information according to the length of the context, namely, at least 5 subject words for the information less than 100 words, and then increase one subject word for the increase of every 100 words, until the subject words achieves 20.

The named entity only extracts the regional name information by utilizing domain ontology according to the results after Chinese participle so as to mark the region described in the information.

4 Analysis on Experiments and Results

The system is realized by adopting J2EE. It can effectively detect those influential news websites, forums and blogs that are related to the administrative regions and concerned by local governments, quickly collect the information of these websites and analyze whether they contain any sensitive information concerned by government, and then form corresponding public opinion report. The system provides such functions as discovery of sensitive information, full text search, meta-search, information collection, public opinion journal and hot term statistics.

In order to verify the precision and completeness of public opinion detection, the precision rate and recall rate are adopted for measurement. Precision rate refers to the percentage of correct information in returned information while the recall rate means the percentage of correct information in all of the information related to this term. Meanwhile, F-measures are adopted to make evaluation on general performance. Over 2000 pages are collected in the experiment. To the 8 categories (many concerning terms are set in each category) concerned by users are compared by using the results of traditional keyword matching and semantic expansion respectively. The results are shown in the Table 1:

Table 1. Statistics of Public Opinion Detection Performance

Concerning	Keywords matching			Semantic Expansion		
	Precision	Recall rate	F-measures	Precision	Recall rate	F-measures
就业	81.82%	72.58%	76.92%	89.09%	79.03%	83.76%
***[1]	92.86%	76.47%	83.87%	93.33%	82.35%	87.50%
***	100.00%	60.00%	75.00%	100.00%	80.00%	88.89%
***	90.91%	90.91%	90.91%	100.00%	100.00%	100.00%
长虹	85.29%	76.32%	80.56%	85.29%	76.32%	80.56%
九州	66.67%	66.67%	66.67%	83.33%	83.33%	83.33%
三台县	83.33%	61.40%	70.71%	93.33%	73.68%	82.35%
游仙区	83.33%	55.56%	66.67%	92.00%	85.19%	88.46%
Average value	85.53%	69.99%	76.41%	92.05%	82.49%	86.86%

The statistical result shows that the precision rate and recall rate after adopting semantic expansion have been greatly improved, and the rates can achieve 100% to the names of people or institutes. This might because of that the name of people or institution formed logical search with the hypernym and hyponym of administrative region in domain ontology, so the precision rate is improved. The systematical average precision rate achieved 92.05%, average recall rate 82.49% and F-measures 86.86% which could basically satisfy users' demands. It can be shown that after utilizing domain ontology to make semantic expansion on concerning terms, the public opinion information concerned by users can be discovered more comprehensively and precisely.

5 Conclusion

Detection on internet public opinion has become the hot issue concerned by all levels of governmental departments. Fast information communication and the freedom of

[1] The *** in the table refers to the names of people or institutes.

language expression by netizens have brought challenges to the timeliness and correctness of public opinion detection. Therefore, this paper effectively improved the precision rate of public opinion detection and basically satisfied users' demands by utilizing the domain ontology of public opinion detection according to the demands of local governments on public opinion detection. At present, the system timely discovers related information only according to the terms set by users, but there are more tasks on the detection and analysis on internet public opinion, such as discovery of hotspot information, theme detection and follow-up and analysis on public opinion disposition etc.

The next step of work shall be to further complete the public opinion detection ontology to improve the precision rate of public opinion detection, while implement in-depth public opinion analysis so as to provide more valuable information for local governments.

References

1. Zeng, R.: Research on Information Resource Sharing of Network Opinion. Journal of Intelligence 28, 187–191 (2009)
2. Qian, A.: A Model for Analyzing Public Opinion under the Web and Its Implementation. New Technology of Library and Information Service 24, 49–55 (2008)
3. Du, Y., Wang, H., Wang, H.: The Solutions of Network Public Opinion Monitoring. Netinfo Security 6, 69–70 (2008)
4. Zheng, K., Shu, X., Yuan, H.: Hot Spot Information Auto-detection Method of Network Public Opinion. Computer Engineering 36, 4–6 (2010)
5. Li, B., Wang, J., Lin, C.: Method of Online Public Opinions Pre-warning based on Intuitionistic Fuzzy Reasoning. Application Research of Computers 27, 3312–3315, 3325 (2010)
6. Xu, X., Zhang, L.: Research on Early Warning of Network Opinion on Emergencies Based on Signal Analysis. Information Studies: Theory & Application 33, 97–100 (2010)
7. Li, G., Chen, J., Cheng, M., Kou, G.: Study on the City Image Network Monitoring System Based on Opinion-mining. New Technology of Library and Information Service 2, 56–62 (2010)
8. Missikoff, M., Navigli, R., Velardi, P.: The Usable Ontology: An Environment for Building and Assessing a Domain Ontology. In: Horrocks, I., Hendler, J. (eds.) ISWC 2002. LNCS, vol. 2342, pp. 39–53. Springer, Heidelberg (2002)
9. Integrated E-Government Thesaurus (Category Table). Science and Technology Literature Press, Beijing (2005)
10. Yang, C., Han, Y.: Fast Algorithm of Keywords Automatic Extraction in Field. Computer Engineering and Design 32, 2142–2145 (2011)

A Security Analysis Model Based on Artificial Neural Network

Rongqi Liu and Shifen Zhong

School of Mathematics and Computer Engineering, XiHua University
610039 Chengdu, Sichuan Province, China
394431956@qq.com

Abstract. Current problems in investment market are discussed and the classic artificial neural network is introduced. Then, it mainly introduces a security analysis model based on the artificial neural network, including how to choose neurons, calculate weights and make decision etc. Finally, a decision-making system based on the model is introduced. Some possible problems about the system and strategies on how to invest are also discussed.

Keywords: Security Analysis, Artificial Neural Network.

1 Introduction

Investment is an issue that every one must face. Since everyone is unwilling to devalue your assets, you must invest and finance. According the rule of the currency issue, the total amount of money will grow larger steadily. Then it decides that inflation must be tendency in the long run. The monetary assets will be devalued ceaselessly. This point can be viewed by the tendency of gold price.

The current investment markets include stock, future, interest, art and foreign exchange. People are more interested in stock and future markets. To grasp the rule of the market is to understand what the participators are. For example, there are four types of investor in the stock market: value investors, tendency investors, technique investors and ordinary investors.

Value investors are that once they think the specific share is valuable to invest, they will hold it until they think it worthless or not worth to hold. These investors often have some vision better than anyone else and always earn the largest sum of money.

Tendency investors are that once they think the share is in a good tendency, they will not sell the share that they hold until they think the tendency of share bad. When common market is not well, they may rest and do not participate the market. These investors always can earn money from the market.

Technique investors often sell or buy according to the technique about specific share. Some of them maybe earn money, but most of them will lose money.

The ordinary investors do not know anything about market well, so they lose money in the market frequently.

F.L. Wang et al. (Eds.): WISM 2012, LNCS 7529, pp. 111–116, 2012.

According to the statistic data, there is an proportion relation: 70 : 15 : 15. That is to say among 100 persons in stock market, about 70 persons will lose money, 15 persons will not earn or lose money, only 15 persons can earn money[1]. Meanwhile, in future market, this relation should be 90 : 5 : 5. Since there is a leverage effect in future market, the risk is larger than that in stock market. How to success in the market? The value and tendency investment may be the best choices. If activities are allowed to a bull or bear market with the advantage of value and tendency investment, they will have favorable odds to success in the market.

How to invest with the advantage of value and tendency ? There are so many factors to affect market. How to evaluate these factors is the key to success. The paper introduces a tendency analysis model of investment market based on the artificial neural network. It can be used to find out the tendency of market and the valuable share, to support decision-making and supervise the investment behavior.

Security analysis and speculation is not same. Do not put your whole money on investment. Investment is not gambling and its success depends on security analysis[2].

2 The Artificial Neural Network

The classic artificial neural network is developing on the basis of modern neuroscience. It is the abstract mathematical model that reflects the structure and function of the human brain. Since 1943, American psychologist W. McCulloch and mathematician W. Pitts had given the abstract mathematical model of the formal neuron; the theory of the artificial neural network has been developed for 50 years. Especially in 1980s, it has made a big progress and become an interdiscipline among physics, mathematics, computer science and neuron-biology. It is applied in many fields, such as: pattern recognition, image processing, intelligence control, combinatorial optimization, financial forecast and application, communication, robot and expert system etc. There are more than 40 types of models of neural network. The basic model and ways of the artificial neural network will be discussed.

Fig 1 shows the basic neural model of the artificial neural network. It has three factors.

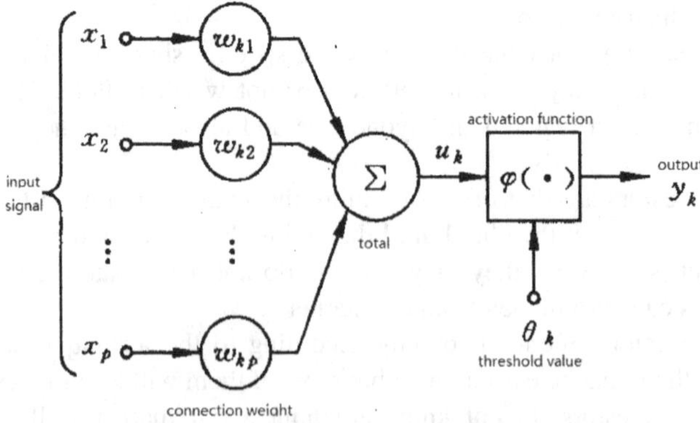

Fig. 1. Three factors of a basic neural model

(1) A group of connection weight: intensity of connection can be expressed by weight. The positive value means active and negative value means constrain.
(2) A cell to total can be used to calculate the sum about all input signals' weights (linear combination).
(3) A nonlinear active function can be used to confine the output value in (0,1)[3].

3 An Investment Analysis Model

An investment analysis model based on the artificial neural network can be defined:

$$f(x_1,x_2,x_3,\ldots,x_n) = \sum_{i=1}^{n} w_i x_i \tag{1}$$

Above, w_i is the weight of number i neuron and xi is the observed value for number i neuron. In principle, this analysis model has regulations as following:

(1) $\sum_{i=1}^{n} w_i = 1$， 且$0 \le w_i \le 1$。 (2)

(2) Number n+1 neuron can be allowed to add and assign weight: w_{n+1}, In order to meet (**2**), all previous $w_i(1 \le i \le n)$ must be replaced with $(w_i)'$ as (**3**).

$$(w_i)' = (1-w_{n+1})*w_i \tag{3}$$

(3) Number j ($1 \le i \le n$) neuron can be allowed to delete, In order to meet (**2**), all other $w_i(1 \le i \le n$ and $i <> j)$ must be replaced with $(w_i)'$ as (**4**).

$$(w_i)' = (1+w_i)*w_j \tag{4}$$

(4) Number j ($1 \le i \le n$) neuron can be allowed to adjust. The new weight for Number j neuron is w_j'.In order to meet (**2**), all other $w_i(1 \le i \le n$ and $i <> j)$ must be replaced with $(w_i)'$. If $w_i' > w_i$, then $(w_j)' = (1-(w_i'-w_i)) * w_j$; if $w_i' < w_i$, then $(w_j)' = (1+(w_i-w_i')) * w_j$.
(5) x_i is observed value, and confined in $-100 \le x_i \le 100$.
(6) $f(x_1,x_2,x_3,\ldots,x_n)$ is the value used to make decision and defined [-100,100]. It meets the demand (**5**)

$$f(x_1,x_2,x_3,\ldots,x_n) = \begin{cases} [-100,-5) & \textit{bear} \\ [-5,5) & \textit{balance} \\ [5,100] & \textit{bull} \end{cases} \tag{5}$$

When market is bull, investors can make money by buying shares and hold the shares until the market is bear.

In this model, x_i is observed value which is usually defined for the specific share by the specific user. w_i is weight of number i neuron, which is defined by experts. Though the values of w_i is same to everyone, the values of x_i observed by each people may be difference. Then values of $f(x_1,x_2,x_3,\ldots,x_n)$ may be difference.

4 How to Define Neurons

In author's opinion, there are three tendencies in investment market: long, medium, short. Investors can be aggressive when long tendency of the market is good. Short tendency can be used to predict the specified share whose price will make a u-turn. Medium tendency can be used to predict the index of stock market which will change its direction. There is a famous saying in the market: If mansion will collapse, no eggs in it can be well preserved. If stock market index can be predicted, especially when stock market makes u-turn, investors can earn a large sum of money.

By using investment analysis model based on the artificial neural network to make decision on investment, investors can avoid emotional behaviors, analysis the current market rationally, and then investment behaviors can be made regularly and scientifically. By using this model, tendency change about the investment market can be predicted. To some degree, this can make person successful.

Taking stock market as example, the author supposes, all neurons in market can be divided two types: one is for common market, the other is for the specified share. Each type can be divided into three subtypes: long, medium and short tendencies. Among them, value investors can choose long tendency neurons, tendency investors can choose medium neurons, and technique investors can choose short tendency neurons.

Long tendency neurons of share include: property features, whether the s price of product is controlled by state, year line, half-year line of Moving averages system etc. Among these, the meaning of observed value about the 'property features' are following: whether property can be duplicated or not (e.g.: famous scenery site, celebrated brand), whose observed value is 80 means that the company has a good earning power and never be challenged in the future; 30 means that the future may be good but may be challenged to some extent; 0 or below means confined or restricted in the future), etc.

Medium tendency neurons for share include: book value per share, earning power per share, comparability etc. Among these, the meaning and observed values about the comparability are following: One share has higher comparability means that comparing with other shares in the same fields, this share has lower price, higher earning power or other material can be publicized. Thus, this share may be valuable to invest. The highest observed value for this is 80, medium is 20, common is 0, below common is -20, the lowest is -50, etc.

Short tendency neurons for share include: active degree, hot degree, K-line indicator. Among these, the meaning and observed values about the active degree are that whether the share has experienced of huge fluctuation in price, reaches its' limit-up, has a large trading volume recently. The observed value can be given more than 50 if the share reaches the limit-up or has huge price amplitude recently. In general, the shares in highway and steel sections are the most inactive; on the other hand, the shares in nonferrous metals or security company are the most active. The observed value for this neuron can be defined: active, 60; common, 0; inactive, -60.

There are about 55 neurons being used in the model. More neurons and their meaning will be discussed in author other paper.

5 Construction and Evaluation

Using the analysis model, the investment decision-making system is constructed. In order to use w_i simply, w_i' is used to replace with w_i. The meaning of w_i' is the importance degree of the neuron and its value is between 0 and 1, then $w_i' = \frac{w_i}{\sum_{j=1}^{n} w_j}$.

The system gave the evaluation of the market share price of China Shenhua Energy Company Limited whose stock number is 601088 from May to June in 2011. The long, medium, short tendency lines are shown in the system as following:

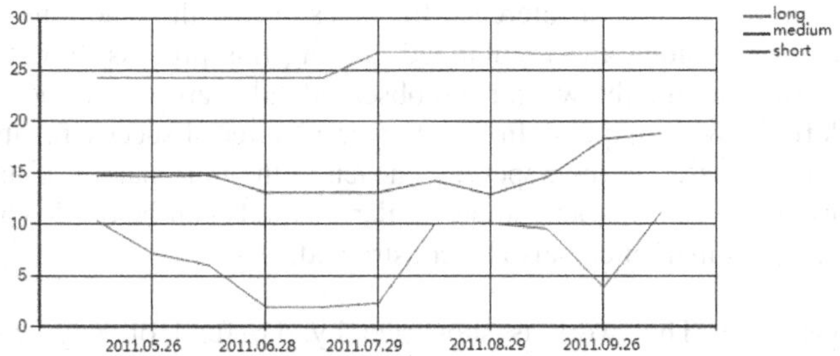

Fig. 2. Evaluation of 601088

Relatively the long tendency line changes a little, meanwhile, short tendency line changes huge. During this period, the market price of 601088 steadily went up from the lowest 26 Yuan to the highest 32.5Yuan. Due to the time, the author can not trace the price tendency of 601088 in half or year until the paper was published.

6 Conclusion

Using the analysis model, the investment decision-making system is constructed. In Just as the beginning of the paper, investment and financing will be the issues that everyone must face. However, Investment is also taken a big risk. So, investor must be prudent. They prefer to miss the chance rather than do it wrongly. How to invest reasonable and properly? The paper gave some strategies which can be used as references.

When market is in bull market, the 30 percent of personal assets can be used in value investment. In principle, the shares can be chosen if the related companies have predominant influence in its industry, they can control their products' price, their developing field is supported by the state policy, they have something for speculation, their future earning powers are high, and the current market prices are far less from the top prices. In short their intrinsic values are good. When investors choose shares, they must pay more attention where the long tendency is upward.

When market is in bull market, the 40 percent of personal assets can be used in tendency investment, in which, 60 percent can be for fixed investment after tendency for common market is worthy for investment. That is to say, there must some features show that the common market will be upward in the future. The other assets can be used for earning short-term interest. When tendency for the common market changes, state policy must be paid more attention, shares that have good earning powers and future prospects must be hold steadily. When ratio of earning money is more than 20 percent, the 30 percent shares for tendency investment can be sold for cash.

The 30 percent of personal assets can be saved at bank or for other purpose.

If market is bear, investors can take rest.

By using the methods of the neural artificial network, factors which can affect the market can be used as every neuron, and then weight is assigned. This weight is the impact degrees for the market. Then, the total result can reflect the tendency of the market. That can be used as a reference for investor and the emotional behavior for investment can be avoided and regularized. During this process, how to choose the neurons and how to assign the weight and observed value are very important. The first two can be defined by the experts. In fact, they are matter of successful application.

The neurons discussed in the paper are chosen in the bull market, if investors want to earn money in the bear market, some similar methods can be used. The models for the future or interest market can also be constructed.

Acknowledgements. This work is supported by the fund of Key Disciplinary of Computer Software and Theory, Sichuan (SZD0802-09-1).

References

1. http://finance.ifeng.com/people/detail/comchief/linyuan.shtml
2. Graham, B., Dodd, D.L.: Security Analysis. McGraw-Hill Companies
3. http://wenku.baidu.com/view/6b53ba0d6c85ec3a87c2c567.html

Social Network Analyses on Knowledge Diffusion of China's Management Science

Hongjiang Yue[1,2]

[1] School of Management,Nanjing University,
Nanjing,China, 210093
[2] Institute of Public Administration and Performance Evaluation,
Nanjing Audit University, Nanjing, China, 210029
yuehj@nau.edu.cn

Abstract. There are two prospective goals for this research, firstly, bring in society network analytic technique to research the communication mode of the management science's researcher community. Secondly, describe the citation network's variation of management science journal. Using social network analyses on the citation network—measures of centrality, multivariate measures, position in the network—an examination of a management science community was discussed by promulgating the two-period citation modes' variation to describe the management science's evolution and maturity. Results show that the exchange of knowledge of management science journals in the network density increases and community was more direct and open. The subject tends to cross in the field of fusion. The network has maintained relative stability, but it also has some obvious changes. These journals can be considered to be the "invisible college" that has core members.

Keywords: journal citation, social network analysis, centrality, visualization.

1 Introduction

Management science is an interdisciplinary science which aims at revealing and applying the basic laws of various management activities, which combines natural science, engineering science, technology science and social science, is an integrative decussate science.

To see from world wide, the management science has undergone hundred years of developmental history. In China, as an academic subject, management science got development after the Reform and Opening-up. Especially in 1997, The Academic Degrees Committee of the State Council and the Council Education Committee published Academic Subjects and Disciplines Catalogue for Doctor & Master Degree Grant and Master Education, which independently takes the management science as an academic subject, by utilizing mathematical tool and combining basic theories such as economic science and behavior science, etc..

F.L. Wang et al. (Eds.): WISM 2012, LNCS 7529, pp. 117–124, 2012.

The destination of this paper's research is to track and describe China management science knowledge's evolution. In order to get a better realization to the evolution features of management science's research fields, it is selected to research the management science journals' citation network. The journals' citation networks can reflect the management science knowledge's resource, fluxion and extension [1].

There are two prospective goals for this research, firstly, bring in society network analytic technique to research the communication mode of the management science's researcher community. Secondly, describe the citation network's variation of management science journal. By promulgating the two-period citation modes' variation to describe the management science's evolution and maturity.

2 Methods and Data

2.1 Methods

The main method for this research is the society network analytic technique with journal cited material as basis.

The citation analysis is a science measurement method [2] for researching science documents' citation relation. In academic subjects, the science documents' mutual connections express in the documents' mutual citation, we usually call the mutual citation relation as citation network, the science documents' mutual citation is a presentation of science development rule, and it reflects science knowledge's accumulation, continuity and succession.

Society network analysis is a demonstration research method with relation as basic unit [3], mainly analysis the relation data. "Society network" means society activists and the aggregation of their relations. That is to say, a society network is an aggregation which is composed with many points and lines between the points. Points can be individuals, units, companies, cities and countries, etc. The relation types are multiple, which can be the relations of friends, foreign trades, science cooperation or journals' citation and cited-by, etc. the society network analysis method has got extensive application in many science fields such as society science, psychology and management science, etc. This paper mainly analyzes the information communication networks in and out of the management science journals, to promulgate knowledge's transmission feature among members, is also beneficial for getting a better realization to the science communication mode. The society network analysis is also beneficial for getting a better realization to the academic subjects' relations. This analysis method also gets application in science research management field, such as analyzing the science research's cooperative relationship network [4], [5].

Citation analysis only provides basic data for analyzing academic subject's development, while society network analysis can utilize citation data to promulgate citation network's structure feature, and the combination of both can be a powerful analysis tool.

2.2 Data Source

The management science journals' sample selection is the precondition for analysis. In China, management science has developed for more than twenty years, has shown the emerging academic subject's feature of rapid increased document quantity. However, comparing with the other academic subjects, management science is still a developing subject, has not formed any critical journal groups [6] which get common view in the this academic science for now, even the relatively important journals also don't have unified standard and realization, expresses in the management departments/faculties list different critical journals for postgraduate education. Several important internal document data bases transplant management science into their academic subject ranges, respectively select out critical journals for management science. After eliminating the repeated journals, we select 18 kinds of management science journals with high impact factor (IF), these journal cover each field of Chinese management science, and own extensive representativeness in respective management science field.

Utilizing the document data base to search and establish a citation data matrix for the management science journals. For now, the domestic document data bases which possess citation search function respectively are: China National Knowledge Infrastructure (CNKI), Chinese Sciences Citation Database(CSCD), Chinese Social Science Citation Information (CSSCI) and Chinese Science and Technology Paper and Citation Database (CSTPCD), etc. Through comparison, this paper selects CNKI as statistics resource, and utilizes the journal citation search function of this data base's systematic reference document, to search out the citation data for these 18 kinds of journals.

In order to research the management science's evolution, we select a long enough periods to research, i.e. from 1996 to 2006; the long period can guarantee to promulgate the variation features of academic communication mode for management science. For eliminating the other factors' influence, the data are separated into two periods, respectively are 1996-2000 and 2002-2006. Five-year period like this can much clearly show the journals' citation relation.

Search out the cite and cited-by data for the 18 kinds of journals through on-line handling, form two 18×18 citation matrixes which are non-symmetrical and with directed relation, in order to conveniently utilize the society network analysis tool, at the same time, to eliminate the unbalancedness of the journals' citation data, according to different network features, we can transform the multi-value matrix into adjacent, normalized or symmetrical matrix. The normalized matrix taken use by this paper is with row normalization, the symmetrical normalization utilizes the cosine function of included angle to take similarity measure and give normalization to the matrix. X and Y are respectively two vectors for matrix, the calculating formula is:

$$Cosine = \frac{\sum_{i=1}^{n} x_i y_i}{\sqrt{\sum_{i=1}^{n} x_i^2}\sqrt{\sum_{i=1}^{n} y_i^2}} = \frac{\sum_{i=1}^{n} x_i y_i}{\sqrt{(\sum_{i=1}^{n} x_i^2)(\sum_{i=1}^{n} y_i^2)}} \quad (1)$$

The value is between 0 and 1. The bigger the cosine values become, means the citation modes of these journals are much closer, it is opposite in the contrary situation.

The journal citation matrix shows the citation and cited-by data among different journals, and presents the citation relation network among different journals, forms a culminating point and brim network in the citation relation network, the culminating point is those journals, brims present the journals' citation relations, in this way, we can form a journal citation relation network with directed multi-value relation, in this network, we can take individual or integral network analysis, through the integral network analysis we can promulgate this network's structure feature, such as the network's centrality, cohesive subgroups, block-models structure, etc, a series of quantized features.

By throwing the original and the normalized matrixes into the society network analysis software UCINET [7], we get the following result analysis with the software's calculation result as basis.

3 Result Analysis

In the data matrixes of the two periods, diagonal data presents journal self-citing frequency, because each journal publishes different quantity of papers every year, it is unfair to compare the journals' self-citing data. We take the ratio of self-citing frequency and the quantity of papers published by each journal every 5 years as journal's self-citing index, by comparison, we can see the journal's self-citing ratio level. High ratio means this journal's paper adoption has continuity, the front and the back publications connect well, gradually forms and maintains its own academic style and feature, also explains the journal's team in this field is in stable development stage, but also means the specialized fields of the papers published by this journal are narrow, the relative independence (encapsulation) is high, the capability (openness) to absorb the achievements of the other fields is low, in the meanwhile, means its influence to the other fields is small, without high position in science field.

From the two periods' comparison to see, the self-citation ratio of all the journals in the second period is higher than that in the first period, explains all journals are strengthening to keep their academic style and feature. From the comparison of the journals' sequencing in the two periods to see, the journals' self-citation frequency occurs huge variation. In the first period, the self-citation frequency of FEM, NBR and CIE is low, while that of SETP, SSTI and CSS is high; in the second period, the self-citation frequency of FEM, CSS and IEM is low, while that of MSC, CJMS and AR is high.

In the society network analysis, the relation totality which actually exists in the network divided by the theoretical maximum possible relation totality [8] is equal to the integral density of the two-value relation network. Give two-value treatment to the journals' citation relations in the two periods, in the network, for any two journals that exist citation relation, the code is 1. For the journals which don't exist citation relation, the code is 0; we can get a contiguous matrix in this way. The density of journals' citation relation network is 0.5686 in 1996-2000, while that is 0.8627 in 2002-2006. The integral density of the second period is bigger than that of the first period, this explains that, the mutual citation relations of the 18 kinds of management science journals in the second period is bigger than that in the first period, the extension and fluxion of the journal's knowledge is more extensive, the academic subjects' fields are tending to mergence.

Utilizing Quadratic Assignment Procedure (QAP) theory, which is a method used to compare the similarity of each lattice value in the two matrixes, we can calculate the relevance modulus in the two matrixes. The calculation result of the original data matrixes in the two periods shows the relevance modulus is 0.755. The journals' citation networks of the two periods keep relative stability, in the mean while, also occur some obvious variations.

3.1 Centrality

To the relation network, estimate its centrality is an effective method to depict society activist's "power". In the relation network, if an activist directly connects with the other activists in its local environment, the measured centrality is called as local centrality. Measure the centrality of the directed relation network, find each point has two local centrality, one is corresponding to the Indegree, and the other one is corresponding to Outdegree. In the journal citation network, indegree is the local network centrality which is formed by a journal cites the other journals, while outdegree is the local network centrality which is formed by a journal cited by the other journals. The results are listed in the table 2.

From the table 2 we can see that, both the standard indegree and centrality of the journals' networks in the two periods are bigger than the standard outdegree and centrality, this shows that, the journals' cite and cited-by relations have huge asymmetry. The closer the centrality approaches to 100%, explains the network owns more central tendency, the result shows the central tendency of the citation network in the second period is bigger than that of the first period. From the integral comparison to see, the integral centrality of the citation relation network is relative big.

Table 1. The indegree & outdegree of journals' citation networks

	1996-2000		2002-2006	
Journal	Indegree	Outdegree	Indegree	Outdegree
CSS	0.447	1.476	0.688	1.099
CJMS	0.439	0.615	0.408	0.823
CIE	0.830	0.121	0.557	0.616
RDM	0.349	0.532	0.500	0.725
JSE	0.586	0.328	0.661	0.373
SETP	0.326	1.199	0.588	0.825
FEM	0.696	0.289	0.838	0.033
SR	0.360	0.186	0.492	0.159
SSTI	0.045	0.090	0.175	0.144
NBR	0.250	0.384	0.663	0.408
CD	0.328	0.120	0.285	0.196
SRM	0.553	0.805	0.689	1.047
SSS	0.517	0.382	0.499	1.014
AR	0.197	0.065	0.269	0.232
MW	0.587	0.370	0.694	0.560
MSC	0.674	0.649	0.619	0.536
IEEM	0.699	0.569	0.799	0.589
IEM	0.421	0.123	0.423	0.468
Centrality	18.2%	6.6%	15.2%	8.0%

To observe journals' local centrality, the diversity between indegree and outdegree is bigger than zero means the journal's cited-by centrality is bigger than its cite centrality, contrast if the diversity value is smaller than zero. In the first period, CSS and SETP have comparatively stronger centrality, in the second period, SSS and CSS have comparatively stronger centrality. The integral diversity comparison of the two periods shows, the network centrality of FEM and MSC becomes much smaller (respectively from -0.439 to -0.805, from 0.264 to -0.210), while the centrality of CJMS and SSS become much bigger (respectively from 0.070 to 0.415, from -0.098 to 0.515), the influence of SETP is declining (respectively from 1.029 to 0.411, from 0.873 to 0.237). From the journals' indegree diversity comparison of two periods to see, the diversity of NBR and SETP is becoming bigger, while that of CIE and MSC is becoming smaller. From the journals' outdegree diversity comparison of two periods to see, the diversity of SSS and CIE is becoming bigger, while that of CSS and SETP is becoming smaller.

3.2 Network Role Position

According to the cosine value of journal's citation vector, draw a journal citation network chart. Thereby realize the similarity of journals' knowledge communication modes, the line's size corresponds with journals' similarity, the rougher is the line, the more similar is the connected journals' citation mode, and the stronger is the similarity of journals' fields or specialized subjects. Visualization can much directly realize the similarities and dissimilarities of the knowledge communication of management science journals. Fig. 3 and 4 give visualization results.

The lines in the citation network chart are only visualization expression of the journal citation mode's similarity, when different journal groups which cluster according to research fields appear in the network, it is difficult to identify the journal groups only from the lines' size, at this time, we need to take "simplify" to the complex network.

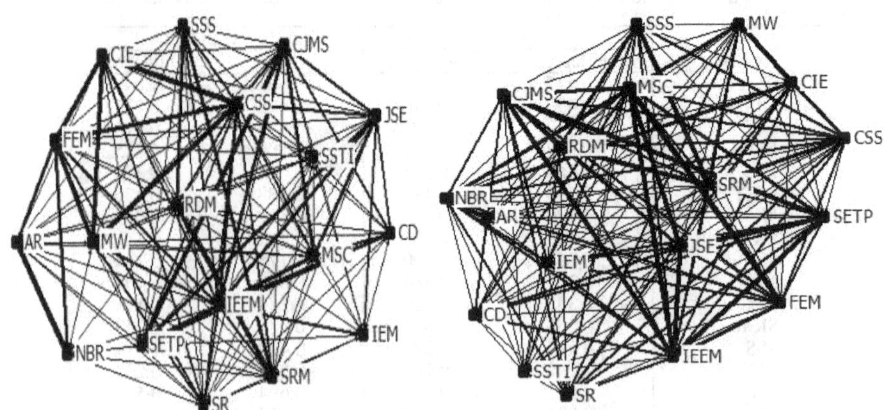

Fig. 1. Journal's citation network in 1996 - 2000

Fig. 2. Journal's citation network in 2002-2006

The block-models in the society network analysis can sort the network activists according to Structural Equivalence, which is a method to research the network position mode and a descriptive quantitative analysis to society roles [8]. In the journal citation network, the structural equivalence can be used to measure to what an extent one or more kinds of journals play the same role or perform the same function. Structural Equivalence journals take mutual citations with the other journals of the same role. Although sometimes two journals don't mutually cite another journal, they still play the same role in the network, but the roles played by the journal members maybe multiple and in variation. The Concor program in the Ucinet software is according to the structural equivalence method to identify all points. Through transplanting journal citation network data, we can identify which journals are playing the same roles. According to the strict equivalence standard, we can divide the points into four blocks, and the results of the two periods are listed in table 2.

Table 2. The papartion of citation networks

	1996-2000	2002-2006
Zone 1	AR, CIE, MW, NBR, CSS, SR	AR, CIE, MW, NBR, FEM, CSS, SR
Zone 2	SSTI, CJMS, SSS, RDM	SSTI, CJMS, SSS, RDM
Zone 3	MSC, FEM, IEM	MSC, SRM, IEEM
Zone 4	SRM, JSE, SETP, CD, IEEM	JSE, SETP, CD

To see from the overall, although the journals of the two periods are both divided into 4 zones, except the journals in the second area keep constant, the journal formation of the rest areas occur some variation. There are 6 kinds of basic journals in the first area, respectively are AR, CIE, MW, NBR, CSS and SR. In the second period, the FEM joins into the first area, the journals in this area mainly research macroscopic management and policy; There are 4 kinds of basic journals in the second area, respectively are SSTI, SRM, SSS and RDM, mainly take technology management research; the basic journal in the third area is MSC, in the second period, CJMS and IEEM joined in, this group mainly research management science in a narrow sense; the basic journals in the forth area are JSE, SETP, CD, mainly research system engineering, the CJMS and IEEM of the first period also belong to this area, which are divided into the third area in the second period.

Through visual direct analysis and quantitative role partition, we can see the citation role and function of the management science journals keep stability on the whole, but occurred some variation.

4 Conclusion and Discussion

By transplant society network analysis methods – centrality, subgroup structure and block-models – to research the management science's citation network, we successfully accomplish the analysis goal of this paper. Firstly, various society network analysis methods respectively promulgate and describe the knowledge citation network features of the management science community; secondly, the promulgated features of the network structure is beneficial to build a management science communication system.

As a whole, along with time changes, the knowledge communication density of management science journals is increasing, the knowledge communication among journals become more and more direct and open, the academic branch fields are tending to mergence. The journals' whole citation and cited-by scale of the two periods keep relative stability and with some variation. Many researches show that, the network's centric position is very important for accepting and diffusing information. China Soft Science Magazine and MSC lie in the network's center, play important roles in the knowledge communication network. Journal's cluster and partition also show that, the journal's citation exists "invisible college", i.e. exists journal community. Their critical members exist in all communities.

Acknowledgments. This work was supported by supported by the Humanities and Social Sciences Planning Fund of Chinese Ministry of Education under grant No. 11YJC630272, supported by China Postdoctoral Science Foundation, under grant No. 20110491375, supported by the Social Science Foundation of Jiangsu Province, China, under grant No. 10ZZC005. supported by A Project Funded by the Priority Academic Program Development of Jiangsu Higher Education (Auditing Science and Technology) , supported by Jiangsu Province College of Humanities and Social Science Fund, under grant No. 2011SJB630037

References

1. Borgman, C.L.: Scholarly communication and bibliometrics. Sage, Newbury Park (1990)
2. Price, D.J.: Networks of scientific papers. Science 149, 510–514 (1965)
3. Scott, J.: Social network analysis: a handbook. Sage, London (1991)
4. Wagner, C.S., Leydesdorff, L.: Mapping the network of global science: comparing international co-authorships from 1990 to 2000. International Journal of Technology and Globalisation 1, 185–208 (2005)
5. Newman, M.E.J.: The structure of scientific collaboration networks. Proceedings of the National Academy of Sciences of the United States of America 98, 404–409 (2001)
6. Chen, X.T., Jin, B.H., Yang, L.X., et al.: Select important journals of management sciences in China. Journal of Manegement Sciences in China 2, 70–76 (1999)
7. Borgatti, S.P., Everett, M.G., Freeman, L.C.: Ucinet for windows: software for social network analysis. Analytic Technologies, Harvard (2002)
8. Hanneman, R.A., Riddle, M.: Introduction to social network methods. University of California, Riverside (2005)

A Parallel Association-Rule Mining Algorithm

Zhi-gang Wang[1,2] and Chi-she Wang[1,2]

[1] School of IT, Jinling Institute of Technology, Nanjing 211169, China
[2] Information Analysis Engineering Laboratory of Jiangsu Province, Nanjing 211169, China

Abstract. Although the FP-Growth association-rule mining algorithm is more efficient than the Apriori algorithm, it has two disadvantages. The first is that the FP-tree can become too large to be created in memory; the second is the serial processing approach used. In this paper, a kind of parallel association-rule mining algorithm has been proposed. It does not need to create an overall FP-tree, and it can distribute data mining tasks over several computing nodes to achieve parallel processing. This approach will greatly improve efficiency and processing ability when used for mining association rules and is suitable for association-rule mining on massive data sets.

Keywords: Data mining, Association rules, Frequent pattern, FP-tree, Parallel algorithm.

1 Introduction

Association rules were first proposed by Agrawal in 1993 [1] and are an important data mining research topic. The Apriori algorithm was proposed in 1994 by Agrawal and Srikant as an original algorithm for mining Boolean-type association rules [2]. Many improved algorithms have been derived from the Apriori algorithm. However, all Apriori-like algorithms have the following deficiencies [3]: (1) they must spend a large amount of time processing a large number of candidate item sets, and (2) they must repeatedly scan the database to match candidate item sets to find frequent patterns. To avoid the deficiencies of the Apriori-like algorithms, the FP-growth algorithm was proposed by Han et al. [3] for mining association rules. Research has shown that the FP-growth algorithm is approximately one order of magnitude faster than the Apriori algorithm.

In recent years, along with the digitalization of society, massive data applications have become more and more prevalent. Although the FP-growth algorithm is more effective than Apriori, it is still unsuitable for performing association-rule mining tasks on massive data sets. The reasons for this are that the FP-growth algorithm performs serial processing and that the FP-tree may become too large to be created in memory.

Association-rule mining problems based on applications of distributed data structures have made apparent the practical need for a distributed data mining algorithm to avoid the transmission of large amounts of data in networks.

F.L. Wang et al. (Eds.): WISM 2012, LNCS 7529, pp. 125–129, 2012.

To achieve parallel association-rule mining or association-rule mining on distributed data structures, approaches reported in the literature include data division [4], multithreaded memory-sharing parallel algorithms [5-6], and division of the overall FP-tree into small FP-trees for parallel processing once the overall FP-tree has been obtained [7]. Reference [8] suggested that in the FP-growth algorithm, the processes of mining every conditional-pattern base are independent, and therefore frequent pattern mining of every conditional-pattern base can be regarded as an independent subtask, and the whole task of frequent pattern mining can be divided into a set of subtasks which can be assigned to different nodes of a computer cluster.

This paper proposes an improved algorithm based on the FP-growth algorithm. The algorithm is designed to operate in a distributed application data framework, does not need to create an overall FP-tree, and uses parallel processing in all its principal steps.

2 Description of the Problem

Let I={i1,i2,…,in} be the set of all items, and let transaction T be a set composed of several items from I, T$\subseteq$I. Each transaction has a unique Tid identifier.

Let D be a global database composed of multiple local databases, D=$\sum_{j=1}^{m} d_j$, where d_j ($j=1,2,...m$) are distributed in different storage nodes M_j.

Let C_j (j=1,2…p) be a group of computing nodes with powerful calculation capabilities. In a physical sense, a storage node and a computing node can completely coincide, partially overlap, or be completely distinct.

The algorithm will attempt to make full use of the computing nodes C_j (j = 1, 2... p) to discover the frequent patterns as quickly as possible from the overall transaction database D to obtain association rules.

3 Algorithm Description

(1) For each storage node M_j: obtain the count of all items in d_j.

(2) In the central computing node: collect and sum the counts of all the items in each d_j to obtain the count of all items in the overall database D. Order the items in descending order. Delete items for which the count is less than the specified minimum count supported. The result is the overall 1-item frequent set L.

(3) In the central computing node: to distribute the frequent pattern mining task over many computing nodes by items to realize parallel processing in the following step, the central computing node needs to assign a computing node to every item in L and to add its assigned results to L. Finally, L is obtained as shown in Table 1.

Table 1. Overall 1-item frequency set and computing node assignment results

Item	Count	Computing node
...	...	...
i2	3510	C3
i5	2580	C4
i9	2500	C5
...	...	...

The central computing node distributes L over all storage nodes.

Many strategies are available to assign a computing node to each item. The simplest way is to assign computing nodes to items by their count order. For example:

Items are designated as 1,...,n according to their count in descending order. The number of computing nodes is j, and the nodes in turn are called C1,...,Cj. Assign computing node Ci+1 for item n when n mod j = i.

(4) For each storage node M_j: process every transaction in 6 according to the overall 1-item frequent set L. The main steps of the procedure are:

① Filter out the items that are not in L.

② Order the rest of the items by descending count order in L.

③ Add the transactions to the local FP-tree.

After this step, there can be several local FP-trees and several local header tables that are bound by the overall 1-item frequency set L.

(5) For each storage node M_j: according to the additional computing-node assign information in L, send the items in the local header tables with their conditional-pattern base in the local FP-trees to their corresponding computing nodes.

(6) For each computing node C_j: aggregate the conditional-pattern base group by item and perform frequent pattern mining. If the conditional-pattern base of an item is very large, it is possible to invoke this algorithm recursively to disaggregate the frequent-pattern mining task for this item to make it suitable for parallel processing.

(7) For each computing node C_j: send the frequent patterns to the central computing node, which summarizes them and generates the overall association rule set.

4 Discussion

In step (3), more than one strategy is available to assign a computing node to every item in L, for example, static distribution which reduces communication overhead and is suitable for remote distributed computing. If communications are readily available, a dynamic distribution strategy can be used. One item is initially assigned to every computing node, beginning from the end of the overall 1-frequency set L. After a computing node has finished its task, it is assigned another, as yet unassigned item from the end of L.

5 Test Results

The computer configuration used for the experiment was the following: CPU: Intel Celeron D Processor 3.456 GHz, RAM: 1GB, OS: Microsoft Windows XP Professional.

The data used for the experiment are found in the T10I4D100K data set (http://fimi.ua.ac.be/data/T10I4D100K.dat). The data file is 3.93 MB in size and contains 100,000 transactions. The minimum serial number of the items is 0, and the maximum serial number is 999. The number of items which actually appear in all transactions is 871. This means that there are 29 items which do not appear in all transactions. The total number of times that an item appears in a transactions is 1,010,221, and the average length of a transaction is 10.1.

To simulate parallel processing, at each parallel step of the algorithm, the task was disaggregated into several subtasks. Then all the subtasks were executed on a single computer, the maximum processing time of all subtasks was taken to be the execution time of this step, and the execution times for all steps were added together to obtain the total execution time of the algorithm.

The minimum support value was set to 1%, which means that the minimum support count was 1000. Test results from the algorithm are shown in Fig. 1.

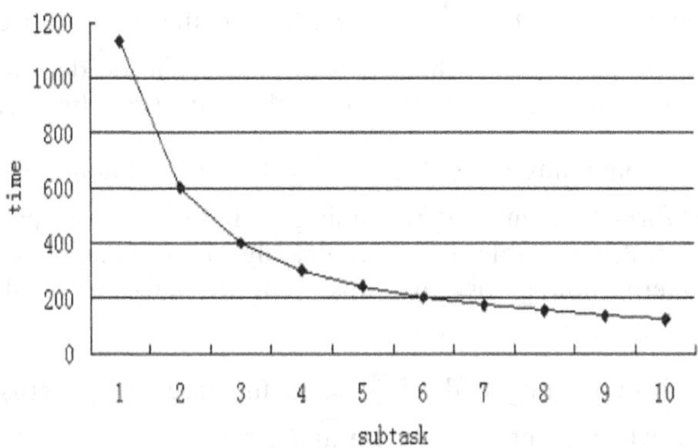

Fig. 1. Test results

6 Conclusions

The proposed algorithm is based on a distributed application data framework and does not need to create an overall FP-tree. This can avoid the problem that the overall

FP-tree may become too large to be created in RAM. The algorithm uses parallel processing in all its principal steps. It can greatly improve the efficiency and processing ability of the association-rule mining algorithm. It is suitable for association-rule mining on massive data sets which the traditional FP-growth algorithm cannot handle.

Experiments have shown that this algorithm is faster than the FP-growth algorithm for association-rule mining on problems at the same data scale. Because it does not need an overall FP-tree, this algorithm can deal with larger data sets than the FP-growth algorithm.

References

1. Agrawal, R., Imielinski, T., Swami, A.: Mining association rules between sets of items in large databases. In: Proceedings, ACM SIGMOD International Conference on Management of Data, Washington, DC, pp. 207–216 (1993)
2. Agrawal, R., Srikant, R.: Fast algorithms for mining association rules. In: Proceedings, 1994 International Conference on Very Large Data Bases (VLDB 1994), Santiago, Chile, pp. 487–499 (1994)
3. Han, J., Pei, J., Yin, Y.: Mining frequent patterns without candidate generation. In: Proceedings, ACM-SIGMOD International Conference on Management of Data, pp. 1–12. ACM Press, Dallas (2000)
4. Pramudiono, I., Kitsuregawa, M.: Parallel FP-Growth on PC cluster. In: Proceedings, International Conference on Internet Computing, Seoul Korea, pp. 467–473 (2003)
5. Zaïane, O.R., Mohammad, E.H., Lu, P.: Fast parallel association rule mining without candidacy generation. In: Proceedings, 1st IEEE International Conference on Data Mining, pp. 665–668. IEEE Computer Society Press, San Jose (2001)
6. Liu, L., Li, E., Zhang, Y., et al.: Optimization of frequent item-set mining on multiple-core processors. In: Proceedings, 33rd International Conference on Very Large Data Bases, pp. 1275–1285. VLDB Endowment, Vienna (2007)
7. Tan, K.-L., Sun, Z.-H.: An algorithm for mining FP-trees in parallel. Computer Engineering and Applications (13), 155–157 (2006) (in Chinese)
8. Chen, M., Li, H.-F.: FP-Growth parallel algorithm in cluster system. Computer Engineering (20), 71–75 (2009) (in Chinese)

An Algorithm of Parallel Programming Design Based on Problem Domain Model

Xin Zhang, Guangnan Guo, Lu Bai, Yu Zhang, and Hao Wang

Department of Fundamental Courses, Air Force Logistics College, Xuzhou, China
zhangxin_0609@163.com

Abstract. Parallel algorithm in parallel program design has close relationship with problem type to be solved. Correct utilization of problem type can effectively improve parallelism of program. The paper proposed parallel programming selection algorithm based on problem domain. It models and matches program structure to find appropriate parallel program design method. The practical application example of proposed algorithm was provided to illustrate it can effectively improve working efficiency of parallel program. The performance analysis result also shows that the method can adequately give full play of parallel nature of program.

Keywords: Problem Domain, Classification, Parallel Decision-making, Speedup.

1 Introduction

With the era of multi-core CPU, many emerging computer application fields have brought out higher requirements on computing performance of computers [1, 2]. It is difficult for traditional serial programs to meet needs of high performance computation. In addition, the potential execution of multi-core CPU has also some inadequacies. Therefore, it has become real needs to transform and design new parallel programs.

The popular parallel program design schema includes parallelization method based on task decomposition, parallelization method based on data decomposition as well as that based on data flow decomposition and etc. Each parallel algorithm has its own appropriate problem domain [3, 4]. In case of parallel code generation, the programmer mainly complete it depend on personnel experience or system with parallel compilation functions. The process is often certain blindness, namely if programmer or complier system has not select appropriate parallel strategies, it will lead to parallel program cannot achieve higher improvement in efficiency, even occur error. Therefore, it is the premise of play parallel program efficiency by classify problem domain, analyze features of different domain and design appropriate parallel programming strategy of different domain. The paper proposed parallel programming selection algorithm based on problem domain. It models and matches program structure to find appropriate parallel program design method. The paper is organized

F.L. Wang et al. (Eds.): WISM 2012, LNCS 7529, pp. 130–135, 2012.

as follows: section 2 introduces several common classification methods of problem domain; section 3 gives parallel program design algorithm based on problem domain; section 4 performs performance analysis on parallel program design algorithm; section 5 concludes our work.

2 Common Problem Domain Classifications and Related Definition

2.1 Related Definitions

Definition 1: Define program structure as *P*(*C*, *D*), where *C* is control module set of program *P* and *D* is data set of *P*. The *p* is used to represent sub-task of *P*.

Definition 2: Set program control module as *C*(*I*, *O*, *L*). Where, *I* is input of module *C*; *L* is sentence set of *C* and *O* is output of *C*. The *c*(*i*, *o*, *l*) is used to represent control module of sub-task *p*.

Definition 3: $p_1 \cup p_2 = (c_1 \cup c_2, d_1 \cup d_2)$, $p_1 \cap p_2 = (c_1 \cap c_2, d_1 \cap d_2)$.

In order to simply representation, the paper uses $c_1 \cdot o_1$ to show output o_1 of control structure c_1. In this way, $p_1 \cdot c_1 \cdot l_1$ is the sentence set l_1 in the control unit c_1 of sub-task p_1.

2.2 Common Problem Classification

(1) Transaction independent problem domain

Set *n* is an arbitrary natural number. There are $p_1, p_2, p_3, \cdots, p_n$ belong to *P*, when $p_1 \cup p_2 \cup p_3 \cup \cdots \cup p_n = P$ and $p_1 \cdot c_1 \cdot l_1 \cap p_2 \cdot c_2 \cdot l_2 \cap \cdots \cap p_n \cdot c_n \cdot l_n = \varphi$, the program *P* is transaction independent problem domain.

The task to be completed of transaction independent problem domain can be divided into independent sub-tasks. The data to be processed by each sub-task is same, but operations on these data are varying from each other.

(2) Data independent problem domain

Set *n* is an arbitrary natural number. There are $p_1, p_2, p_3, \cdots, p_n$ belong to *P*, when $p_1 \cdot d_1 \cup p_2 \cdot d_2 \cup \cdots p_n \cdot d_n = P \cdot D$, $p_1 \cdot d_1 \cap p_2 \cdot d_2 \cap \cdots p_n \cdot d_n = \varphi$ and $p_1 \cdot c_1 \cdot l_1 = p_2 \cdot c_2 \cdot l_2 = \cdots = p_n \cdot c_n \cdot l_n$, the program *P* is data independent problem domain.

The task to be completed of data independent problem domain can be divided into several different sub-tasks. The data to be processed by each sub-task is quite different, but operations of which are identical.

(3) Transaction associated or data associated problem domain

Assume *m*, *n*, *o* are arbitrary natural numbers and $m < n$, $o < n$. There is $p_1, p_2, p_3, \cdots, p_n$ belongs to *P*, it meet $\forall p_m \in P$, $\exists p_o \in P$ and

$p_m \cdot c_m \cdot i_m = p_o \cdot c_o \cdot i_o$, the program P is transaction associated or data associated problem domain. Where, p_m and p_o are component of a data flow.

The task to be completed of transaction associated or data associated problem domain can be divided into associated bus-tasks. The input and output among sub-tasks are depending on each other. The associated transactions constitute an integral data flow.

2.3 Conclusion of Common Problem Domain

(1) The transaction independent problem domain matches parallel algorithm based on task decomposition. Each independent sub-task cannot be executed in parallel. It needs to carry out synchronization operation on common data.

(2) Data independent problem domain matches parallel algorithm based on data decomposition. Each independent sub-task can execute in parallel.

(3) Transaction associated or data associated problem domain matches parallel algorithm based on data flow decomposition. Each independent sub- task cannot be executed in parallel.

(4) Problem domain of other types uses non-parallel algorithms.

3 Parallel Programming Design Algorithm Based on Problem Domain

3.1 Main Idea

As different problem domain has different parallel strategy, the algorithm firstly models program structure according to representation method in the definition using set data structure. Utilizing above four conclusions, the sentence-level parallel solution can be arrived by sub-task division, problem matching and algorithm recursive.

3.2 Algorithm Description

Step 1: Establish structural mode *P*(*C*, *D*) of program *p*. Number each sentence in the program. The sentence set constitutes set *L* of control module *C* and all data constitutes set *D*.

Step 2: Divide program into several sub-tasks $p_1, p_2, p_3, \cdots, p_n$ based on logical function.

Step 3: Construct program structure model *P*(*c*, *d*) of each sub-task.

Step 4: Detect control module and data module of each sub-task, and then determine which problem domain of the relationship between it and other modules.

Step 5: Utilizing former mentioned four conclusion, we can obtain parallel algorithms matching to each sub-task and sub-task level parallel solution.

Step 6: Repeat Step 1 to Step 6 on each sub-task till sentence level parallel solution was determined.

3.3 Algorithm Application Example

Parallel program design algorithm based on problem domain can be used to transform existing serial program, or to design new program solution. Matrix operations are problem objects to be solved in the parallel program. The iterative sentence is also usually used as optimization processing object [3]. In the next, the program example is used to illustrate how to perform assignment and operation with the algorithm to conduct parallel operation. The code segmentation is as follows:

/*Save value of matrix *A* with three-dimensional array and assign value to array *a*[3][3]*/

$$p_1 \begin{cases} for(i=0;i<3;i++) & (1) \\ for(j=0;j<3;j++) & (2) \\ A[i][j]=inputA[i][j]; & \end{cases}$$

/*Save value of matrix *B* with three-dimensional array and assign value to array *b*[3][3]*/

$$p_2 \begin{cases} for(m=0;m<3;m++) & (3) \\ for(n=0;n<3;n++) & (4) \\ B[m][n]=inputB[m][n]; & \end{cases}$$

/* Multiplication operation between matrix *A* and *B**/

$$p_3 \begin{cases} for(p=0;p<3;p++) & (5) \\ for(q=0;q<3;q++) & (6) \\ output[p][q]=A[p][0]\bullet B[0][q]+ & \\ A[p][1]\bullet B[1][q]+A[p][2]\bullet B[2][q]; & \end{cases}$$

Construct mathematical model for the whole above code segment and mark it with *P(C, D)*, Where *C* is {{input*A* input*B*}, {output}, {(1) (2) (3) (4) (5) (6)}}, *D* is {input*A*, input*B*, output, *i*, *j*, *m*, *n*, *p*, *q*}.

Based on logical functions of program, the code segment can be divided into three sub-tasks, namely assign value to matrix *A* and matrix *B* as well as conduct multiplication operation on *A* and *B*, which is respectively marked by p_1, p_2 and p_3.

Construct structural model of each sub-task:

$$p_1(\{\{inputA\},\{A[3][3]\},\{(1),(2)\}\},\{i,j,inputA,A[3][3]\})$$

$$p_1(\{\{inputB\},\{B[3][3],\{(3),(4)\}\},\{m,n,inputB,B[3][3]\})$$

$$p_1(\{\{A[3][3],B[3][3]\},\{output\},\{(5),(6)\}\},\{p,q,A[3][3],B[3][3],output\})$$

Based on problem domain concept and four conclusions, detect the problem domain and corresponding parallel program of p_1, p_2 and p_3 as well as other two sub-tasks.

The p_1 and p_2 are data independent problems, so p_1 and p_2 cannot compute in parallel. The p_1 and p_3 as well as p_2 and p_3 are data or transaction associated problem, so p_1 and p_2 cannot compute with p_3 in parallel. We can only firstly compute p_1, p_2, and then compute p_3.

For the sub-task p_1, p_2 and p_3, build model with same method and examine problem type among sentences, so as to obtain execution program in parallel. The sentence (1) and (2) in p_1 belong to data independent problem domain, so (1) and (2) can execute in parallel. Similarly, (3) (4) in p_2 and (5) (6) in p_3 can be executed in parallel. In summary, the parallel solution firstly perform (1)-(4) in parallel. After completion, the (5) and (6) will be executed.

4 Performance Analysis

After parallel transform on the program with proposed algorithm, the program performance can be improved. A complete program in the algorithm is divided into several different sub-tasks. The parallel execution of these sub-tasks improves efficiency. As long as the processor number meets requirements of parallel execution, the efficiency will improve significantly. Here, the concept of speedup is used to quantitatively analyze on performance improvement in the detection.

Based on modified Amdahl, the speedup= $\dfrac{1}{s+(1-s)/n+H(n)}$.

Where, s is the proportion of serial program in the total sentences; n is the number of processors; $H(n)$ is system overhead in case of parallel execution of program [5-7]. When $n \to \infty$, the speedup can reach maximum $1/s$. Therefore, the key to improve performance is to decrease proportion of serial part in program. In the above example, as (1)-(4) can be executed in parallel and (5) (6) can also be executed in parallel. The proportion of serial program occupies 2/27. As the parallel code itself has certain system overhead, its speedup may be less than 27/2. It is visible that majority sentences of program has been parallelized, which play fully parallelism of program.

5 Conclusion

The parallel program design algorithm based on problem domain models program structure on problem to be solved and determines problem type that function modules belongs to, so as to complete decision-making on parallel algorithm. In order to implement algorithm, the set data structure is used to establish program structure model. It can clearly reflect relationship in program structure and help for determining problem domain type. The example application shows that the algorithm can design parallel program with high parallelism and improves algorithm performance. It taps the potential of multi-core processors in high performance computing. The downside is that the algorithm can only make a sentence-level parallel program decision-making, but not perform detail operation into depth. Therefore, the parallelization inside sentence can be further improved to improve program design efficiency, which is our focus in the future.

References

1. Liu, Q.-S., Yang, H.-B., Wu, Y.: Simultaneous multithreading technology. Computer Engineering and Design 29(4), 963–967 (2008)
2. Le, X.-B., Huang, M.: Apply Petri net to analyze the parallelism of serial program. Mini-Micro System 22(11), 1391–1395 (2001)
3. Nasser, G., Oliver, S.: Parallel iterator for parallelizing object oriented applications. In: Proceedings of 14th IEEE International Conference on Parallel and Distributed Systems, pp. 113–130. IEEE Computer Society Press (2008)
4. Bischof, H., Gorlatch, S., Leshchinskiy, R.: Generic Parallel Programming Using C++ Templates and Skeletons. In: Lengauer, C., Batory, D., Blum, A., Vetta, A. (eds.) Domain-Specific Program Generation. LNCS, vol. 3016, pp. 107–126. Springer, Heidelberg (2004)
5. Giacaman, N., Sinnen, O.: Task Parallelism for Object Oriented Programs. In: Proceeding of the International Symposium on Parallel Architectures, Algorithms and Networks, pp. 13–18. ACM Press, USA (2008)
6. Wang, L., Franz, M.: Automatic Partitioning of Object-Oriented Programs for Resource-Constrained Mobile Devices with Multiple Distribution Objectives. In: Proceedings of 14th IEEE International Conference on Parallel and Distributed Systems, pp. 369–376. IEEE Computer Society Press (2008)
7. Liu, X.-L.: Exploiting Object-Based Parallelism on Multi-Core Multi-Processor Clusters. In: Proceedings of Eighth International Conference on Parallel and Distributed Computing, Applications and Technologies, pp. 78–89. ACM Press, USA (2007)

Metadata-Aware Small Files Storage Architecture on Hadoop

Xiaoyong Zhao*, Yang Yang, Li-li Sun, and Han Huang

School of Computer and Communication Engineering, University of Science and Technology Beijing, Beijing, China
yongxiaozhao@163.com

Abstract. The ZB (trillion GB) scales of data produced globally each year, making the distributed data storage become a trend. Research and application on Hadoop which is the most representative open source distributed file system is increasing. However, Hadoop is not suitable for handling massive small files, this paper presents a metadata-aware storage architecture for massive small files, taking full advantage of the metadata of file, merging the small files into Sequence File by the classification algorithm of merge module, and the efficient indexing mechanism be introduced, make a good solution to the problem about the bottleneck of NameNode memory. Taking MP3 files as an example, the experiments show that the architecture can obtain good results.

Keywords. Hadoop, metadata aware, small files, storage architecture, MP3.

1 Introduction

According to the report from IDC, the amount of data produced from all over the word in 2010 is 1.2ZB（1ZB=1 trillion GB）,and in 2011there will be 1.8ZB[1]. Jim Gray, the winner of Turing Award, put forward an experiential law: In networks environment, every 18 months, the number of data produced equals to all the data counted together before. Until now the growth in data rates is true about this law. Traditional methods for data storage and handling are increasingly hard to handle the massive data, so the distributed storage becomes inevitable. The two most representative of them is the distributed data storage (GFS/BigTable) [2] and management architecture (Map-Reduce) from Google, and the open source Hadoop architecture.

Hadoop is a software architecture that can manage the massive data distributive, by a reliable, high-efficient, scalable way. Hadoop contains multiple components, and Hadoop Distributed File System (HDFS) is the most basic one. HDFS is high fault-tolerance, designed to deploy on low-cost hardware, providing extremely high data throughput and suitable for those with large data set of applications [3].

Essentially, HDFS is a kind of streaming block the file system. It is First designed to handle big files (the file size is usually above hundreds of MB). When tackle with small ones, it will bring some problems. Small file means those whose file size smaller than HDFS block (64MB default). This kind of files will bring serious problems to Hadoop's expansibility and performance. First of all, too many small files will impact

F.L. Wang et al. (Eds.): WISM 2012, LNCS 7529, pp. 136–143, 2012.

on the use of NameNode RAM and DataNode RAM [4]. In HDFS, every block, file or catalog are store as objects, every object occupies 150 byte.

If there are 10,000,000 small files, every file occupy one block, it will take almost 2G. That means we store ten million small files, the NameNode needs 20GB RAM. The RAM of NameNode will seriously restrict the cluster's extensibility. Secondly, the speed of visiting massive small files is far slower than some of big files. Visiting massive small files need jump from one DataNode to another. It also affects the system's performance. Finally, the speed of dealing with massive small files is slower than dealing with the same size of big files. Every small file takes place a slot, and the start of task will cost a lot of time, so far as to spend most of time on start and release the task. If the small file is too many, it also need many times of interacting with NameNode, which will enhance the load of NameNode and add many times of seeking and jumping between DataNode.

But in daily-use files, small files take a great percentage, such as Word docs, PowerPoint, MP3 and so on. These kinds of files are smaller than 64M (HDFS block default). At present, there are seldom researches aiming at the management of small files. [5] use HAR to realize the combination of small files to improve the metadata storage ability. [6] and [7] propose a storage mechanism adapt to WebGIS and PPT files, improving the storage and access efficiency of small files. Above all, they all focus on special field of application, usually aiming at specific files and could not be used universally.

This text brings up a kind of metadata-aware small files storage architecture on Hadoop, fully using the information included in the file itself. Through the preprocessing module, using classify-algorithm, the strong relative files are merged to the Sequence File[8]. This method reduces the number of files in the HDFS; Meanwhile, it brings into a high efficient extended first level index mechanism, which can locate the small files in the Sequence File and its offset; Moreover, Using the metadata management module, the meta-information can be indexed and manage centrally and could support all kinds of application scenarios; At last, take MP3 file for example, confirm this architecture is effective.

2 The Overview of Storage Architecture

We can see from figure 1, this architecture expands the traditional HDFS. Besides the traditional HFDS NameNode and DataNode, it brings in the extended first class index and metadata node. It could store richer meta-data and provide stronger support on applications.

2.1 File Write

After the preprocessing of small files, except writing the merged SequenceFile into HDFS file system, It also writes the relation between small files and serial files into the extended first class index. Meanwhile, it will write richer metadata (Like the ID3V1 of MP3 File) into metadata node.

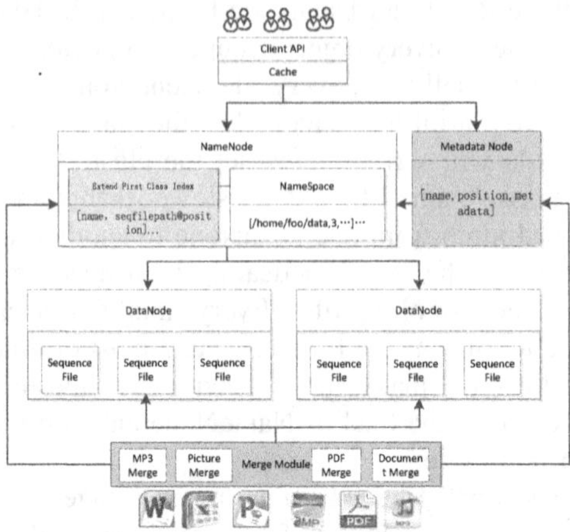

Fig. 1. The storage architecture

2.2 File Access

When visiting a small file, firstly it will find the name of SequenceFile in extended first class index and the offset position of this small file. And then, it uses HDFS metadata to search the information of the SequenceFile. Client read the corresponding small file from the offset position. Meanwhile it puts the extended first class index and metadata into Cache for future using.

3 Module Design

3.1 Merge Module

The file accesses of users usually have local characteristics, namely they are considerable relevance. According to these characteristics, we can use the classify algorithm to deal with all kinds of files. The common algorithm is show as follow:

Step 1. Read the meta-information and put it into metadata node;

Step 2. According to the file type, use the relevant module to conduct and merge to a SequenceFile;

Step 3. Record the classify results in metadata node;

Step 4. Store the relevance between small files and SequenceFile into the extended first class index;

3.2 Index Module

The module saves the information as [the small file name, the sequence file path@ offset position] in NameNode and uses the same mechanism as metadata, initially loads into NameNode RAM. Meanwhile, the client use the cache mechanism to store the files index that read before in order to search quickly and reduce the communicating times with the NameNode. There into, the sequence file path starts directly from the root directory of HDFS, using 6 characters as the auto incremented generated file names(every character can use a-z0-9, 2,100,000,000 files all together). The average length of small file name is 10 characters, the initial position use 36 notations (the max number system by Java default supported) , occupying 6 byte totally. So every small file needs 23(10+6+1+6) byte to store this index and 2G RAM can store one hundred million files.

3.3 Metadata Manage Module

Due to the metadata in traditional way of Hadoop NameNode is mainly for file management, cannot deal with richer metadata. That is why the metadata manage module is worth bringing into the system to index and manage all metadata intensively. The metadata manage module extends the NameNode, using Key-Value document database CouchDB[9]. And considering the performance, Memcache[10] is used as cache. Based on these ample metadata, it can provide strong support to upper level applications. The structure of metadata is[File name, Position, meta-information]; As to MP3 file, the structure is [File name, Position, Song name, Artist, Album, Year, Genre].

4 Simulation

4.1 Experimental Environment

The experimental platform is Hadoop cluster including five IBM X3200M3 PC-server , every node have quad-core intel Xeon X3430 CPU ,2.4GHz,4G RAM, 2T SATA hard disk. The OS is 64 bit CentOS 5.5 (kernel version:2.6.18) , Hadoop version is 0.20.2, JDK version is 1.6.0。

In the five server, one as NameNode. three as DataNode. The last one is metadata Node.

As MP3 is the de facto standard in the digital music era, there are amount of MP3 files and the need of access by users is more and more rapidly (top100.cn includes 4000,000 MP3 files[11]). This paper chooses MP3 files to test. During testing, there are 100,000 MP3 files and we simulate the condition of 100 concurrent users access.

4.2 MP3 Metadata

This paper uses the ID3V1 information as the metadata. The ID3V1 has a simple structure, storing at the end of MP3 files and containing the information of title, artist, album etc.[12] The structure is showing as follow:

```
typedef struct tagID3V1{
char Header[3];
char Title[30];
char Artist[30];
```

```
char Album[30];
char Year[4];
char Comment[28];
char reserve;
char track;;
char Genre;
} ID3V1;
```

The album is the major way of release to modern music, It usually contains 3-15 songs together and about 12-74 minute of time. The size of MP3 files contained in one album are commonly smaller than 64M(HDFS default).This paper use the classify algorithm as follow:

Step 1. Put all the metadata of MP3 (IDV3)into metadata Node;

Step 2. Read the files belonged to the same album and merge it as a Sequence File. Record the classify information in storage and store its correlations as the the extended first class index;

Step 3. Classify all files which could not be classified at Step 2 by artist and put the same class MP3 files together as a Sequence File. Record the classify information in storage and store its correlations as the extended first class index;

Step 4. Classify all files which could not be classified at Step 3 by year and put the same class MP3 files together as a Sequence File. Record the classify information in storage and store its correlations as the the extended first class index;

Step 5. Classify all files which could not be classified at Step 2,3,4 by name order;

4.3 Experimental Results

4.3.1 Traditional Way

Puting 100,000 MP3 songs directly into Hadoop, the occupation of NameNode RAM is show in figure 2:

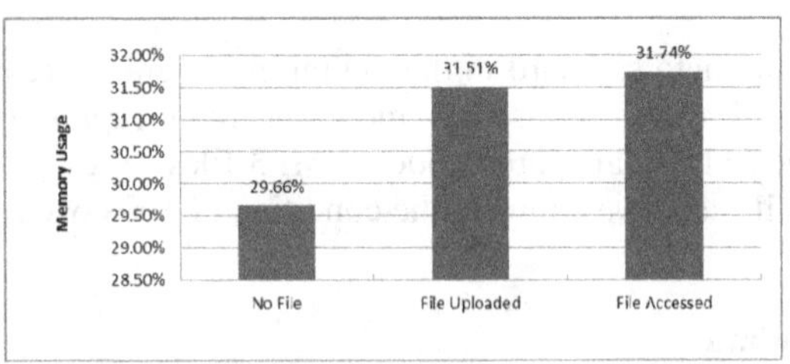

Fig. 2. The occupation of NameNode RAM

Reading the <Baby> and <Up> in the same album orderly, the time is needed shows in figure 3:

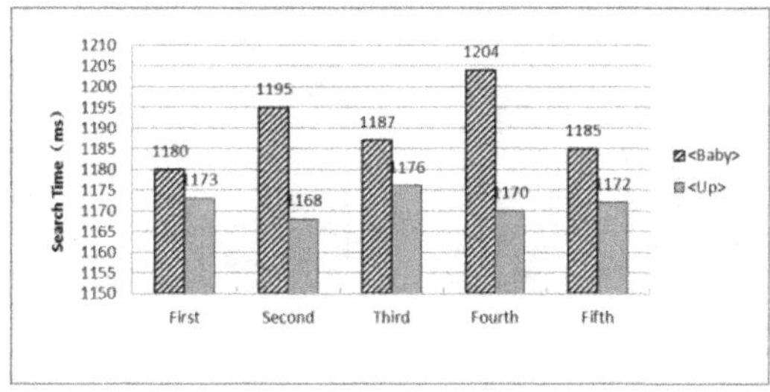

Fig. 3. The time cost in traditional way of file access

4.3.2 New Approach

Using the architecture this paper brings, The occupation of NamNode RAM compared with the traditional way is showed in figure 4:

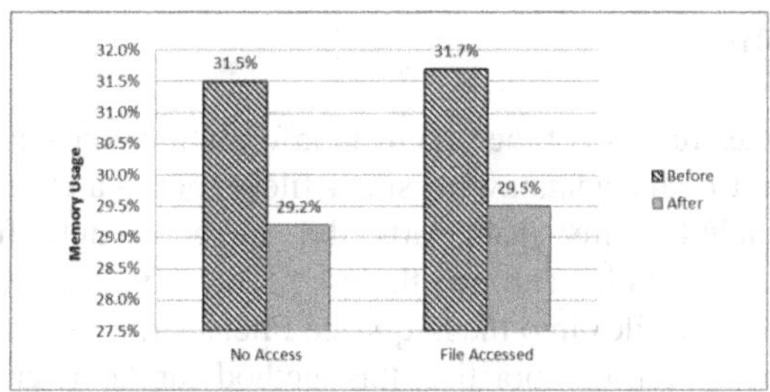

Fig. 4. The NameNode RAM occupation

As can be seen from figure 4, using the way this paper proposed, there is a notable decrease in NameNode RAM usage. This also means Hadoop can manage more files in the same hardware configuration.

Reading the <Baby> and <Up> in the same album orderly, these two song are merge to the same SequenceFile throgh preprocessing. The time of five times reading shows in figure 5:

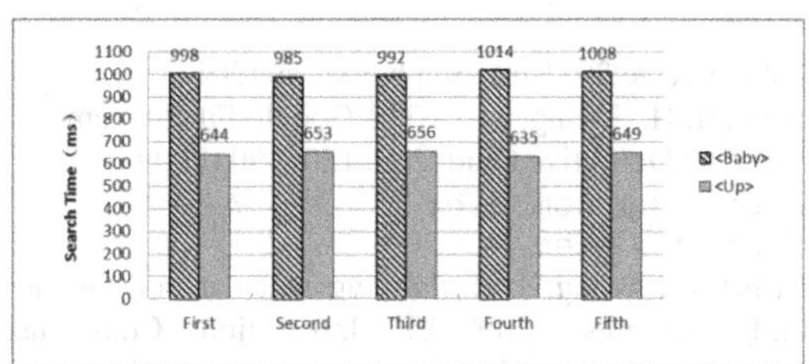

Fig. 5. The time cost of reading files after improvement

Comparing with the traditional way, the time cost is show in figure 6:

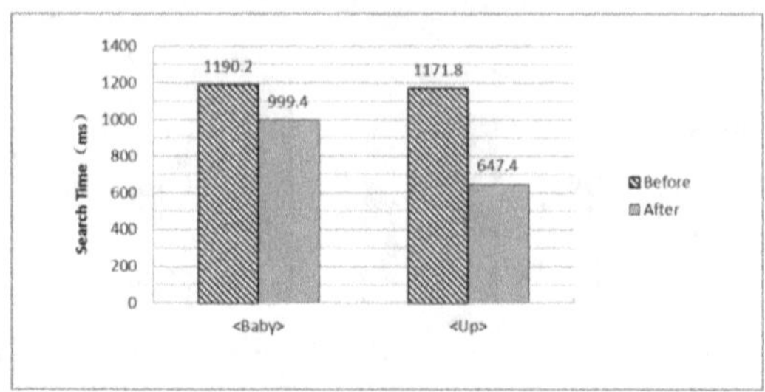

Fig. 6. The average time consuming contrast with the traditional way

As showed in the figure 6, used new approach, the performance will not be affected by including the the extended first class index. And belonging to the same SequenceFile, the operating system puts the file in cache. So when visiting it the second time, the performance will have a significant improvement.

5 Conclusion

In this paper, it makes full use of the rich meta information contained by the file itself and brings in a kind of metadata-aware small files storage architecture on Hadoop. This architecture includes three main parts, the merge module, first index module and metadata manage module; Through the use of classify algorithm in merge module, it merge the relevant files into the Sequence File, sharply reducing the number of files in HDFS. Verified through practice, this method can be a very good solution to the NameNode RAM bottleneck of dealing with small files on Hadoop. This paper mainly discusses the classify algorithm of MP3 files, the algorithms of other file type need further research.

Acknowledgement. This work is supported by National Natural Science Foundation of China (No.61070182, No. 61170209).

References

1. IDC EMC. Digital Universe 2011 Infographic Study (2011)
2. Ghemawat, S., Gobioff, H., Leung, S.-T.: The Google File System
3. White, T.: Hadoop: The Definitive Guide, pp. 150–190 (2009)
4. http://www.cloudera.com/blog/2009/02/the-small-files-problem/
5. Mackey, G., Sehrish, S., Wang, J.: Improving metadata management for small files in HDFS(C/OL). In: Proceedings of 2009 IEEE International Conference on Cluster Computing and Workshops (August 10, 2010), http://ieeexplore.ieee.org/stamp/stamp.jsp?tp=&arnumber=5289133
6. Liu, X., Han, J., Zhong, Y., Han, C., He, X.: Implementing WebGIS on Hadoop: A case study of improving small file I/O performance on HDFS. Cluster, 1–8 (2009)

7. Dong, B., Qiu, J., Zheng, Q., Zhong, X., Li, J., Li, Y.: A Novel Approach to Improving the Efficiency of Storing and Accessing Small Files on Hadoop: A Case Study by PowerPoint Files. In: Proceedings of IEEE SCC 2010, pp. 65–72 (2010)
8. Hadoop Sequence File, `http://hadoop.apache.org/common/docs/current/api/org/apache/hadoop/io/SequenceFile.html`
9. CouchDB(EB/OL) (2011), `http://couchdb.apache.org/docs/overview.html`
10. Memcached(EB/OL) (2011), `http://memcached.org/`
11. `http://top100.cn/`
12. MP3Format, `http://en.wikipedia.org/wiki/MP3`

Research on the Isomorphism of the Electronic-Government and Electronic-Commerce in Support System

Yunbiao Gu[1] and Jian Liu[2]

[1] School of Automation, Harbin Engineering University, Harbin 150001, China
[2] Tianjin Nankai District Science Technology and Information Committee
Guyunbiao@sina.com

Abstract. In this paper, the "isomorphism" of the e-government and e-commerce in the support system is elaborated and analysed .And a practiacal opinion is given for the harmonious development of the electronic government affairs and the electronic commerce.

Keywords: electronic government affairs, e-commerce, isomorphism, support system.

1 Introduction

1.1 Electronic Government Affairs

The Meaning of the Electronic Government Affairs

E-government is a procedure in which the government uses the advanced electronic information technology (computer, network, telephone, mobile phone, and digital television) to realize the government affairs information digitization, governmental affairs opening-up, office efficiency, service network, etc. The electronic government is that the government agencies use modern network communication and computer technology for internal and external management and service function by the optimized integration and reorganization. On the principle of possibility and efficiency this will provide the public and the government itself with high efficiency, high quality and honesty a set of management and service.

Components of Electronic Government Affairs

Generally speaking, the electronic government affairs consist of three parts: The first is the electronization and networking of the internal office functions of the government sector; The second is to obtain access to real-time communication of information sharing between different government sectors via network; The third is an outward manifestation of government department, which is based on the efficiency of the first and second part, and the validity conditions of information's exchange. Therefore, an efficient e-government system should be the organic combination of these three parts [1] .

F.L. Wang et al. (Eds.): WISM 2012, LNCS 7529, pp. 144–149, 2012.

Electronic-Government Applications

The application of electronic government affairs have the following five aspects: The first is the opening of government affairs. Governments will publish a mass of government information to the public through the government website. The second is to provide online services. The third is to realize the resources sharing. Governments at all levels, to the mass provide government owned public databases of information resources for the public, so as to realize the bonus use of public information resources. The forth is the internal office electronic. The meeting notice, information communication, policy propaganda, regulations promulgation, the survey and so on, Can all be handled by E-mail in order to accelerate the flow of information[2]. The fifth is to provide security.

1.2 Electronic Commerce

The Meaning of Electronic Commerce

The Electronic Commerce, which is abbreviated to EC, contains two aspects: one is the electronic platform, the other is business activities.

E-commerce refers to realizing the kinds of business activities using of simple, rapid, low cost of electronic communication. To give a simple example, you trade with clients by email or fax which seems to be regarded as electronic commerce; However, the electronic commerce is to finished mainly by EDI (electronic data interchange) and Internet. However with the growing sophistication of Internet technology, the essential development of electronic commerce is based on Internet technology[3]. So electronic commerce is also referred to as "IC (Internet Commerce).

The Components of Electronic Commerce's

E-commerce includes four parts: The first is the enterprise and the consumer (public) between the electronic between enterprises and t trading consumer (public) (B2C); The second is the electronic trading between enterprising (B2B); The third is the electronic trading between the enterprise and the government (B2G), for example the government procurement; The fourth is the electronic trading between consumers (C2C).

The Essence of Electronic Commerce

Therefore, three conditions are required which are: website construction, to realize the electronic commerce, the corresponding enterprise management and information work flow based, work flow.

2 Relationship between the Electronic Government Affairs and Electronic Commerce

First of all, e-government and e-commerce develop greatly based on computer network technology . So e-government and e-commerce development depends on the degree of the corresponding information technology directly, especially on the standards.

Second, the openness of the network will in inevitably bring about potential troubles of the theft of information. So it is the precondition to solve the relations between information sharing and confidentiality and integrity, between the openness and the protection of privacy, between interconnectivity and local segregation, which is to realize safe electronic government, electronic commerce[4].

2.1 The Close Relationship between the Electronic Government Affairs and the Electronic Commerce

Government informationalization is the premise of developing electronic government affairs and enterprise informationalization is the premise of developing electronic commerce. E-government is a electronized and networked management activities, in which the government supervises and serves its internal departments, related government agencies, and the public .Similarly, the electronic commerce is a electronized and networked business activities for the enterprises and its related ones, , government agencies and the average consumers (the public) ,etc.

Therefore, the intersection of e-government and e-commerce is the government and the enterprise information interface. The development of electronic government affairs cannot be proceeded smoothly without the support of the electronic commerce, and vice versa. So e-government and e-commerce are complementary with each other.

2.2 Relations of Electronic Government Affairs and Electronic Commerce

Isomorphism of electronic government affairs and electronic commerce in support system are reflected as follows:

The Isomorphism of Government Affairs Process Reengineering (GPR) and Business Process Reengineering (BPR)
In order to work better in the information society and the market economy conditions, enterprise will make the business process recombinated and updated, which is business process reengineering; Similarly, in order to use information technology to raise work efficiency and social service level , the government will also make the traditional government administrative management and service process recombinated and rehabilitated, which is government affairs process reengineering.

The Isomorphism of Government Resources Planning (GRP) and Enterprise Resource Planning (ERP)
In order to improve the market competitive power, enterprise is supposed to achieve the configuration optimization of all internal and external resources available, to speed up the reaction to the market, which is the enterprise resources planning; Similarly, in order to save the operation cost and improve its management efficiency, the government also needs to consider how to optimize the allocation of resources, which is the government resources planning.

The Isomorphism of Government Relationship Management (GRM) and Customer Relationship Management (CRM)

In order to attract customers, retain customers, continuously upgrade customers in market sales, and provide personalized service, enterprise is supposed to use information technology for the maintenance, management and renewal of relationship between the customers of enterprise, which is our customer relationship management; Although the government and the enterprise actually differ in many ways, the government's important function of society is to manage society and provide public services. The government is to establish a good government image, and use information technology to improve the ability to manage society and serve the public, which is the government relationship management.

The Isomorphism of Government Supply Chain Management (GSCM) and Enterprise Supply Chain Management (ESCM)

In order to effectively meet customers' demand, enterprise will unite the upstream and downstream firms as the partners, and at the same time, they will use information technology to complete the purchase of raw material and fuel, and the management of product supply. The process is the enterprise supply chain management. Though it is not necessary to act like enterprise, the government, as an entity unit, is bound to purchase materials to meet the requirements of the normal operation of the government department , such as office equipment, fuel, electricity, etc, and achieve informationalization management of the material purchasing , which is government supply chain management. In addition, if we take the government document flow as a special form of supply chain, means of enterprise supply chain management can be accommodated to the government of business process management.

The Isomorphism of Enterprise Information Portal (EIP) and the Government Information Portal (GIP)

In order to facilitate the interactive information exchange with the outside world, and quickly get outside information and improve the efficiency of the enterprises, the enterprises will establish the enterprise information web portal, which is the enterprise information portal; similarly, in order to facilitate the interactive information exchange with the outside world, and quickly get outside information and improve its efficiency, the government will build the government affairs portal, which is the government information portal.

The Isomorphism of Business Cooperation (CC) and Collaborative Government Affairs (CG)

In order to realize a win-win situation in market competition, the related enterprises will use information technology for interactive communication, collaborative development of electronic commerce, which is business cooperation. Because the enterprise or individual should deal with different government departments, traditionally they should have to deal with these departments one after another. Now since the government can use information technology to develop collaborative work, the enterprise or individual will only fill in an electronic form in a web portal, and then the rest will be done by relevant government departments. This can not only improve the work efficiency, but also can reduce the cost of the enterprise or individual, which is collaborative government affairs. So the collaborative commerce is the growing trend of electronic commerce. Similarly, collaborative government affairs is the growing trend of electronic government affairs.

3 The Promoting Function of Electronic Government Affairs to Electronic Commerce

It depends not only on the market economy, but also the promotion of the government. To get the electronic commerce into a higher level. And the important measure is developing electronic government, which is reflected as follows: the government online purchase will induce the enterprise informationalization, present the demonstration effect, and promote the development of the BtoB; The development of the electronic government affairs can promote the development of BtoG, BtoC; The formal network market can drive the electronic commerce development; Enterprise's supervisory function to the electronic government affairs; The enterprise service role in electronic government affairs; Guarantee and promotion of enterprise electronic commerce to the electronic government affairs .

4 The Coordinated Development of the Electronic Government Affairs and the Electronic Commerce

In the coordination development of e-government and e-commerce, the major contradiction is on the side of government and the various functional departments. The government and its departments should continuously study the new development of the electronic commerce and foresee the developing trend in the future, so as to make prompt improvement and update the services provide by electronic government affairs and related to enterprises . In this way electronic government affairs can provide faster, better and more convenient service for enterprises.

For example, when the merchants of electronic commerce have trouble in applying for a business license, in tax, and in other dispute with businessmen or customers, the government should make use of the electronic government affairs for electronic business enterprise to provide convenience to business license issuing, for tax department to collect tax online in a more reasonable, convenient and initiative way, and for the justice department to provide enterprise with easy legal assistance with the help of network platform, etc. These are all the important subjects the government probes into.

5 Conclusions

To sum up, from application perspectives, the government and the enterprise are two kinds of disparate organization, which have completely different application, The government management and enterprise management have essential difference. One is public management, while the other is business management. But they both have the commonness of certain significance in an abstract system, that is, a hierarchical organization structure, thus the "isomorphism" system relationship is brought about. This commonness provides theoretical possibility for ERP or other enterprise management software to translate into electronic government affairs. This is the latest trend of international electronic government affairs.

References

1. ITU Internet reports 2005: The Internet of Things (EB/OL) (2010)
2. Wilding, R.: Lean, leaner, leanest. International Journal of Physical Distribution & Logistics Management 25(3/4) (1996)
3. Holmes, D.: eGov: eBusiness strategies for government. McGraw-Hill (2002)
4. Relationship of E-government and E-commerce, http://www.ciia.org.cn

On the Deployment of Wireless Sensor Networks with Regular Topology Patterns

Wen-xiang Li and Ya-jie Ma

School of Information Science and Engineering
Wuhan University of Science and Technology, Wuhan, China
liwx2001@yahoo.com.cn, mayajie@wust.edu.cn

Abstract. In wireless sensor networks, such regular topology patterns as Square, Hexagon and Triangle can fully cover the area, provide accurate positioning for abnormal event, and achieve better network performance than random topology. In this paper, we survey related works to explore the deployment efficiency and network performance under such cases as full sensing coverage, *k*-coverage and *l*-connectivity. By summarizing the conclusions on the performance index of Area Per Node, we provide guidelines for network deployment, and propose several directions for further research.

Keywords: wireless sensor networks, regular topology, coverage, connectivity.

1 Introduction

The deployment pattern for wireless sensor networks is vital to network performance and lifetime. An optimal topology can cover the area to be monitored effectively, and achieve decreased latency and energy consumption, together with improved throughput and fault-tolerance. The topologies can be divided into such 2 categories as random topology and regular topology, and the later can be divided further into such patterns as Triangle, Square, Hexagon and Combined Pattern. The random topology is applicable in scenarios where there is no strict constraint on the coverage, and can be achieved by on-spot manual or robotic deployment and random airdropping deployment. On the other hand, by on-spot manual or robotic deployment, regular topology can provide full coverage, accurate positioning and better network connectivity.

Current works explore the deployment model and network performance for regular topologies from the perspectives of communication connectivity and sensing coverage. This paper gives a survey on the methods and conclusions in these works. We analyze the index of Area Per Node for typical regular topology patterns from the critical situation to the general situation, including such cases as full coverage, *k*-coverage and *l*-connectivity. We also propose new directions to be studied further.

2 Performance Indices of Network Deployment Efficiency

The indices of network deployment efficiency include *l*-connectivity, *k*-coverage, Sensing Strength Factor d_{mm} [1], and Area Per Node (*APN* for short) [2,3,4], together with some network performance indices such as average path length, packet delivery ratio, energy consumption and transmission rate. *l*-connectivity denotes the number of

F.L. Wang et al. (Eds.): WISM 2012, LNCS 7529, pp. 150–158, 2012.

direct communication links to adjacent nodes for a node, and is dominated by the shape of pattern cell and the communication radius r_c. With enough r_c, the connectivity for Square, Triangle and Hexagon are 4, 6 and 3 respectively. *k*-coverage denotes the number of sensor nodes that can monitor a specified point, and is dominated by the sensing radius r_s. Generally we assume all nodes are homogeneous with identical r_c and r_s.

Given the Euclidean distance $Ed(s_i,y)$ from the monitored point y to the sensor node s_i in area A, $\min_i Ed(s_i,y)$ means the distance from y to the shortest sensor node. d_{mm} denotes the distance at which each point in A can be covered by at least one sensor node, i.e.

$$d_{mm} = \max_{y \in A} \min_i Ed(s_i, y) \tag{1}$$

As a full-scale index for connectivity and coverage, *APN* has been studied extensively. The higher it is, the higher the deployment efficiency is. *APN* denotes the average contribution of a node to the network deployment, and is defined in (2).

$$APN^p = A_p N_n / N_p \tag{2}$$

In (2), A_p stands for the area of the pattern cell, N_p stands for the number of nodes in a pattern cell, and N_n stands for the number of pattern cells that a node is concerned with. For example, given the length a of a side for square, we have $A_p=a^2$. A Square consists of 4 nodes, so N_p=4. And a node is concerned with 4 cells, so N_n=4.

3 Deployment Efficiency of Regular Topologies for Full Coverage

3.1 Description of Area Per Node

The case of full coverage means every point in the area can be covered by at least one node. In Fig.1, the critical coverage state means r_s can barely reach the central point in the cell, i.e., the farthest point to the sensor node, and $(APN)_s$ denotes the Area Per Node for critical coverage state. In Fig.1, this area is enclosed by dashed lines. For Square, Triangle and Hexagon, their $(APN)_s$ are described in (3), (4) and (5) respectively [2,5].

$$(APN^s)_s = 2r_s^2 \tag{3}$$

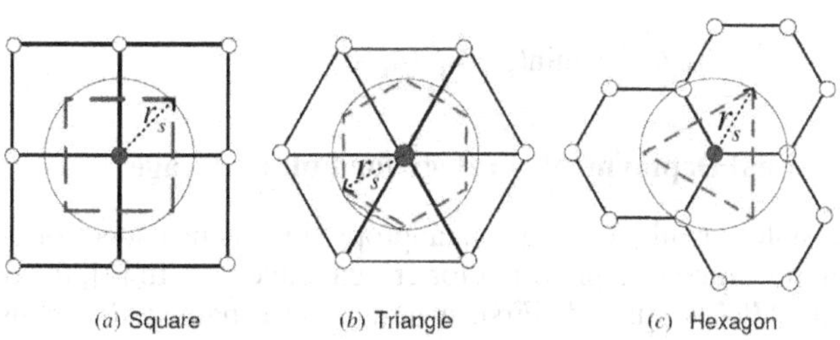

Fig. 1. Coverage state of critical sensing for regular topology

$$(APN^t)_s = 3\sqrt{3}r_s^2/2 \tag{4}$$

$$(APN^h)_s = 3\sqrt{3}r_s^2/4 \tag{5}$$

In Fig.2, the critical connectivity state means r_c can barely cover the adjacent sensor nodes in the cell, and $(APN)_c$ denotes the Area Per Node for critical connectivity state. For Square, Triangle and Hexagon, the $(APN)_c$ are described in (6), (7) and (8) respectively [2,5,6].

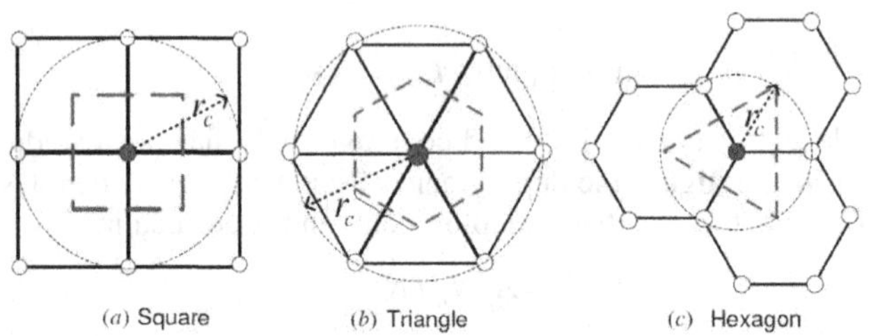

Fig. 2. Coverage state of critical communication for regular topology

$$(APN^s)_c = r_c^2 \tag{6}$$

$$(APN^t)_c = \sqrt{3}r_c^2/2 \tag{7}$$

$$(APN^h)_c = 3\sqrt{3}r_c^2/4 \tag{8}$$

With the same r_s, we have $(APN^t)_s > (APN^s)_s > (APN^h)_s$. And with the same r_c, we have $(APN^h)_c > (APN^s)_c > (APN^t)_c$. When the network is at the critical connectivity state and critical coverage state simultaneously, we have $r_c = r_s$ for Hexagon, $r_c = \sqrt{2}r_s$ for Square, and $r_c = \sqrt{3}r_s$ for Triangle.

At the general state, r_c and r_s are alterable. Combining the cases of coverage and connectivity, we get the *APN* for each pattern at the general state as follows [4]

$$APN^s = \min(2r_s^2, r_c^2) \tag{9}$$

$$APN^t = \min(3\sqrt{3}r_s^2/2, \sqrt{3}r_c^2/2) \tag{10}$$

$$APN^h = \min(3\sqrt{3}r_s^2/4, 3\sqrt{3}r_c^2/4) \tag{11}$$

3.2 The Optimal Deployment Strategy for Full Coverage

In the optimal state for full coverage, an appropriate r_c is provided for critical connectivity and an appropriate r_s is provided for critical coverage. In [4], the optimal pattern with maximum *APN* is studied. First, a new pattern named Rhombus is proposed. With identical r_c and $r_{s1}<r_{s2}<r_{s3}$, four sensor nodes A, B, C, and D are deployed at different positions for the 3 patterns in Fig.3. O is the critical sensing point and

$\theta = \angle$BAD = 2arccos(r_c/(2r_s)). At the left of Fig.3 is pattern Triangle with θ=π/3, at the right of Fig.3 is pattern Square with θ=π/2, and at the middle of Fig.3 is pattern Rhombus with π/3<θ<π/2. In Rhombus, there is no direct communication link between B and D, and O is closer to diagonal BD when $\sqrt{2} < r_c / r_s < \sqrt{3}$. So its *APN* is

$$APN^r = r_c^2 \sin\theta = r_c^3 \sqrt{4r_s^2 - r_c^2} / 2r_s^2 \tag{12}$$

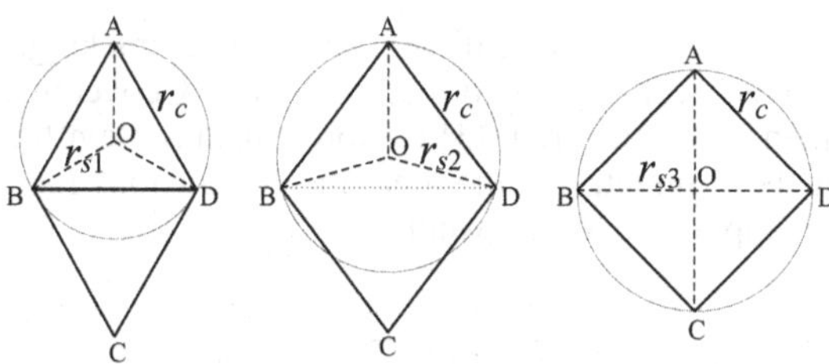

Fig. 3. Evolution from Triangle to Square with Rhombus

Let r_c/r_s be the variable ranging within 0 and 2, we plot the curves for (9), (10), (11) and (12) in Fig.4. It is observed that each pattern achieves the maximum *APN* among the 4 patterns at different range of r_c/r_s. When $APN^h = APN^s$, the first transition threshold for the pattern with maximum *APN* comes at Th1= $\sqrt{3\sqrt{3}/4} = 3^{0.75}/2$. When $APN^s = APN^r$, the second transition threshold comes at Th2= $\sqrt{2}$. When $APN^r = APN^t$, the third transition threshold comes at Th3= $\sqrt{3}$. Finally, the optimal deployment strategy is reduced to (13).

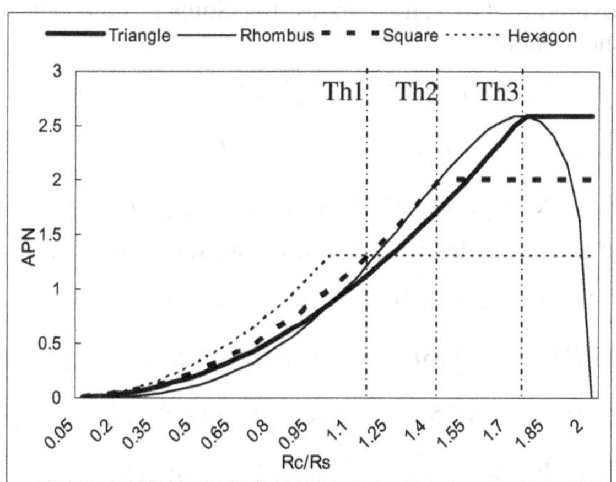

Fig. 4. The *APN* for several patterns with various r_c/r_s

$$APN_{\max} = \begin{cases} APN^h & 0 < r_c / r_s \le 3^{3/4}/2 \\ APN^s & 3^{3/4}/2 \le r_c / r_s \le \sqrt{2} \\ APN^r & \sqrt{2} \le r_c / r_s \le \sqrt{3} \\ APN^t & \sqrt{3} \le r_c / r_s \end{cases} \tag{13}$$

4 Deployment Efficiency of Regular Topologies for *k*-Coverage

In such application scenarios as multi-dimensional object tracking, the object is to be located by several sensors, this leads to the problem of *k*-coverage. If a point is covered by *k* nodes, then it is *k*-covered. [3] explores this problem for pattern Square in Fig. 5. Because of the symmetry of the area, the conclusions for the shadow region near node 5 are also applicable for the whole area.

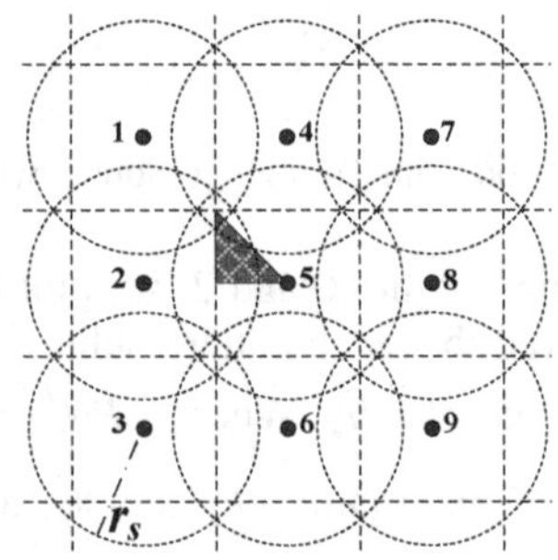

Fig. 5. The analytical model for *k*-coverage problem

Let the distance between adjacent nodes in horizontal or vertical direction be r_c. The method is to find out the range of r_c/r_s to achieve *k*-coverage, so that all points in the shadow region are covered by *k* sensor nodes, and some points in the shadow region are not covered by *k*+1 sensor nodes. For Square, the conclusions on 2, 3, and 4-coverage are shown below.

$$(APN^s)_{2s} = r_s^2, \quad 2/\sqrt{5} \le r_c / r_s \le 1 \tag{14}$$

$$(APN^s)_{3s} = 4r_s^2/5, \quad 3\sqrt{2}/5 \le r_c / r_s \le 2/\sqrt{5} \tag{15}$$

$$(APN^s)_{4s} = 18r_s^2/25, \quad 0 \le r_c / r_s \le 3\sqrt{2}/5 \tag{16}$$

And for Hexagon and Triangle, the conclusions are shown below.

$$(APN^h)_{2s} = 3\sqrt{3}r_s^2/4, \quad 2/\sqrt{7} \le r_c / r_s \le 1 \tag{17}$$

$$(APN^h)_{3s} = 3\sqrt{3}r_s^2/7, \quad 5/7 \le r_c / r_s \le 2/\sqrt{7} \tag{18}$$

$$(APN^h)_{4s} = 75\sqrt{3}r_s^2/196, \quad 0 \le r_c / r_s \le 5/7 \tag{19}$$

$$(APN^t)_{3s} = \sqrt{3}r_s^2/2, \quad \sqrt{3}/2 \le r_c / r_s \le 1 \tag{20}$$

$$(APN^t)_{4s} = 3\sqrt{3}r_s^2/8, \quad 0 \le r_c / r_s \le \sqrt{3}/2 \tag{21}$$

5 Deployment Efficiency of Regular Topologies for *l*-Connectivity

In some application scenarios not only full coverage is needed, but also *l*-connectivity is expected for better reliability and efficiency in communication. As the transient pattern from Hexagon (Fig. 6(a)) to Square (Fig. 6(c)), the Diamond [7] in Fig. 6(b) is a 4-connectivity quasi-Hexagon pattern with $r_{c1}>r_{c2}>r_{c3}$ and $r_{s1}>r_{s2}>r_{s3}$. From Fig. 6(a) to Fig. 6(c), the lengths of OA and OD do not change, while E, F, B, and C are approaching the horizontal line at O. So, in Fig. 6(b) O can connect to B, C, E and F directly, but not to A and D, for |OA| and |OD| are both longer than r_{c2}. The Diamond is superior to Square with larger Area Per Node (APN^d for short). And it is superior to Hexagon in that the connectivity drops from 6 in Hexagon to 4 to restrain communication interference.

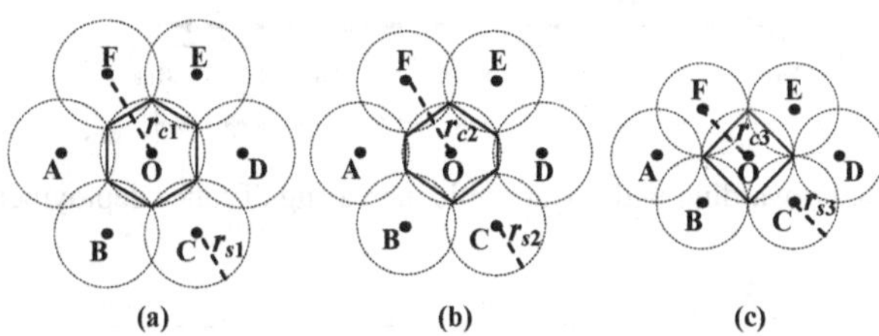

Fig. 6. Pattern evolution (a) Hexagon; (b) Diamond; (c) Square

The APN^d at O is the area enclosed by the solid lines. Let the distance from O to line EF be L, then $APN^d = L$*|FE|. Assume $\varphi=\angle EOF$, then $\cos\varphi/2=|FO|/(2|OG|)= r_c/(2r_s)$, so $|FE|=2r_c\sin(\varphi/2)$ and $L=r_c\cos(\varphi/2)$, further we have

$$APN^d = r_c^3\sqrt{4r_s^2 - r_c^2}\,/(2r_s^2), \quad \sqrt{2} \le r_c / r_s \le \sqrt{3} \tag{22}$$

More detailed works [8] design patterns for 3-connectivity and 5-connectivity at various r_c/r_s. Based on Hexagon, the authors set the link state of ON/OFF according to different strategies, and develop various combined patterns in Fig. 7.

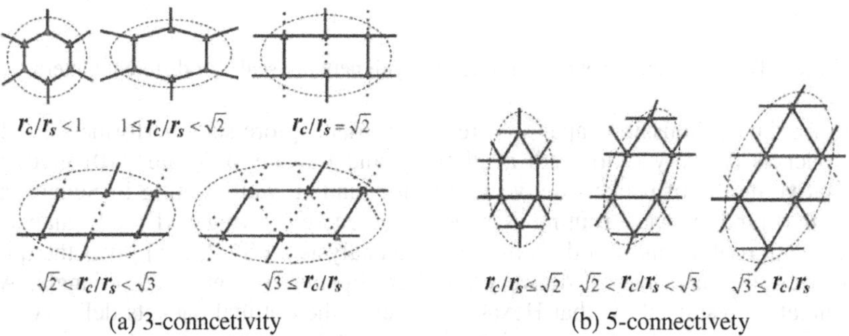

Fig. 7. Combined patterns for *l*-connectivity. (The solid lines are compulsory links, and the dash lines are optional links)

6 Network Performance Analyses on the Regular Topology Models

For prolonging network lifetime, [9] gives the quantitative description of energy consumption by nodes with different distance to Sink node, and makes performance comparison among 3 forwarding strategies, including hop by hop forwarding, forwarding with equal distance deployment and forwarding with logarithmic increasing distance deployment. The last strategy illustrated in Fig.8 can achieve longer network lifetime with the idea of "identical configuration and different distance". Another method for prolonging network lifetime [10] adopts the idea of "identical distance and different configuration". So the nearer to Sink node, the more backup nodes and energy are available.

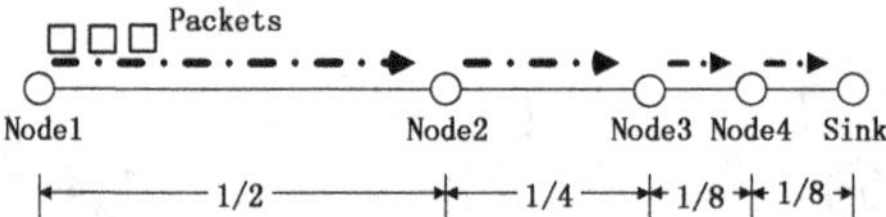

Fig. 8. Forwarding with logarithmic increasing distance deployment

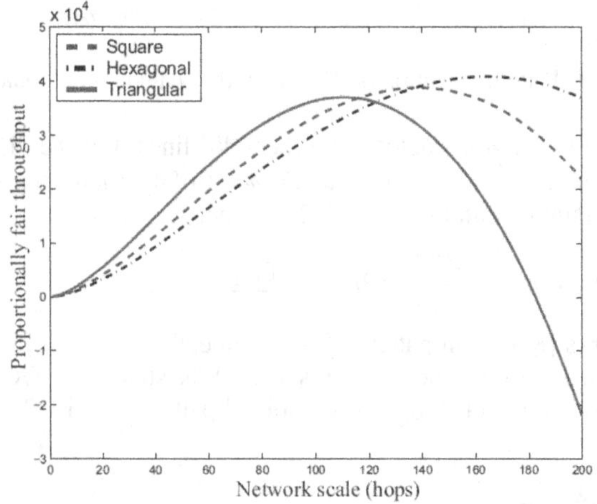

Fig. 9. The variation of network throughput with network scale for different pattern

As for the transmission capability, related works explore such performance indices as packets delivery ratio, path availability and transmission rate. [10] gives the quantitative model of packets delivery ratio for random walk routing in Square, and argues that random walk routing is superior in energy saving and load balance to shortest path routing and flood routing. By the analysis on SNR, [11] gives the quantitative description of packet delivery ratio for Square, Hexagon and Triangle. And the simulation results show that Hexagon achieves the optimal packets delivery ratio with the minimum communication interference, while the packets delivery ratio for Triangle is the lowest. Based on ALOHA protocol, [12] studies the quantitative description of path availability for Square, Hexagon and Triangle. The results from simulation experiments show that the nodes in Triangle have smaller probability of being isolated with 6-connectivity. There are less broken links because of shorter average path length in Triangle, so the average route lifetime is longer. However, the 3-connectivity in Hexagon leads to less communication interference, higher SNR and better connectivity. Aiming at maximizing network throughput, [13] describes the optimal transmission rate for Triangle, Square and Hexagon respectively. The simulation results in Fig. 9 show that when the networks scale, i.e. the number of hops, increases, Triangle, Square and Hexagon can provide the maximal throughput in turn.

7 Conclusions

The above research works have proposed the optimal deployment patterns for various r_c/r_s. And it is observed that node connectivity has obvious impact on network performance. There is less interference in Hexagon with smaller connectivity, and Triangle is more efficient in communication with higher connectivity.

More attention may be paid to such aspects as the features of new combined patterns, topology fault-tolerance and evolution, and such parameters as r_c and r_s should be taken into consideration for achieving optimal communication performance.

References

1. Tian, H., Shen, H., Matsuzawa, T.: Developing Energy-Efficient Topologies and Routing for Wireless Sensor Networks. In: Jin, H., Reed, D., Jiang, W. (eds.) NPC 2005. LNCS, vol. 3779, pp. 461–469. Springer, Heidelberg (2005)
2. Wang, Y., Agrawal, D.P.: Optimizing sensor networks for autonomous unmanned ground vehicles. In: Carapezza, E.M. (ed.) Unmanned/Unattended Sensors and Sensor Networks V. Proc. of SPIE, vol. 7112, p. 711215 (2008)
3. Miao, Z., Cui, L., Zhang, B., et al.: Deployment patterns for k-coverage and l-connectivity in wireless sensor networks. In: IET International Conference on Wireless Sensor Networks, Beijing, China, pp. 73–77 (2010)
4. Bai, X., Kumar, S., Xuan, D., et al.: Deploying wireless sensors to achieve both coverage and connectivity. In: 7th ACM International Symposium on Mobile Ad Hoc Networking and Computing, pp. 131–142. ACM Press, Florence (2006)
5. Biagioni, E.S., Sasaki, G.: Wireless sensor placement for reliable and efficient data collection. In: 36th Hawaii International Conference on System Sciences. IEEE CS Press (2003)
6. Rajagopal, I., Koushik, K., Suman, B.: Low-coordination topologies for redundancy in sensor networks. In: 6th ACM International Symposium on Mobile Ad Hoc Networking and Computing, pp. 332–342. ACM Press, Urbana-Champaign (2005)
7. Bai, X., Yun, Z., Xuan, D., et al.: Deploying four-connectivity and full-coverage wireless sensor networks. In: 27th IEEE INFOCOM, pp. 906–914. IEEE CS Press, Phonix (2008)
8. Yun, Z., Bai, X., Xuan, D., et al.: Complete optimal deployment patterns for full-coverage and k-connectivity (k < 6) wireless sensor networks. IEEE/ACM Transactions on Networking 18(3), 934–947 (2010)
9. Hui, T., Hong, S., Matthew, R.: Maximizing networking lifetime in wireless sensor networks with regular topologies. In: 9th International Conference on Parallel and Distributed Computing, Applications and Technologies, pp. 211–217. IEEE CS Press, Denudin (2008)

10. Hui, T., Hong, S., Teruo, M.: Random walk routing in WSNs with regular topologies. Journal of Computer Science and Technology 21(4), 496–502 (2006)
11. Liu, X., Haenggi, M.: Throughput analysis of fading sensor networks with regular and random topologies. Eurasip Journal on Wireless Communications and Networking 2005(4), 554–564 (2005)
12. Rajagopalan, R., Varshney Pramod, K.: Connectivity analysis of wireless sensor networks with regular topologies in the presence of channel fading. IEEE Transactions on Wireless Communications 8(7), 3475–3483 (2009)
13. Narayanan, S., Jun, J.H., Pandit, V., et al.: Proportionally fair rate allocation in regular wireless sensor networks. In: IEEE Conference on Computer Communications Workshops, pp. 549–554. IEEE Press, Shanghai (2011)

An Empirical Study on the Relationship among Trust and the Risky and Non-Risky Components of E-Commerce

Xibao Zhang

Department of International Economics & Trade, Qingdao University
7 East Hongkong Road, Qingdao 266071, China
xibao.zhang@qdu.edu.cn

Abstract. The influence of trust on the adoption of B2B international e-commerce by Chinese firms was empirically investigated by a questionnaire survey of Chinese international trade professionals. The major findings are that, first, for both the risky and non-risky components of e-commerce, trust only affects the adoption of some, but not all, of them. This contrasts with the current assumption that trust only impacts the adoption of the risky components of e-commerce. Secondly, there are also statistically significant relationships among the various components of e-commerce, both within each type and between them.

Keywords: Trust, International Electronic Commerce, International Trade, Risky Components of E-Commerce, Non-Risky Components of E-Commerce.

1 Introduction

The adoption and success of electronic commerce depend heavily on trust [e.g., 1, 2-4]. Low trust between trading partners in B2B e-commerce can result in a number of risk types including relational risks, performance risks, opportunistic behavior, uncertainty, and conflict, each of which impacts, to varying degrees, e-commerce adoption [5]. In addition, it is argued that not all aspects of e-commerce are affected by trust. This is because trust is relevant only in risky situations [6]. Therefore it should be expected that only the risky components of e-commerce is affected by trust. On the other hand, trust should not influence its non-risky components. This is, to some extent, supported by empirical evidence [7].

On the other hand, the assumption that the non-risky components of e-commerce are not related to trust needs further empirical testing. The purpose of this paper, therefore, is to study if and how trust influences the adoption of the non-risky components of e-commerce, and how they are related to its risky components. Specifically, this study extends the author's previous research work [7] by empirically investigating these relationships in the context of international e-commerce adoption by Chinese businesses.

F.L. Wang et al. (Eds.): WISM 2012, LNCS 7529, pp. 159–166, 2012.

2 Literature Review and Hypotheses

Not all aspects of international e-commerce are affected by trust [6]. And trust is relevant only in risky situations. As a result, international e-commerce should be decomposed into components based on whether or not risk is involved.

Electronic commerce is "the sharing of business information, maintaining business relationships, and conducting business transactions by means of telecommunications networks", including both inter-organizational relationships and transactions and intra-organizational processes that support these activities [8]. Therefore e-commerce comprises both inter-organizational and intra-organizational activities, and it includes not only online transactions but also other activities, such as information sharing and maintaining business relationships, that will eventually lead to online and/or offline transactions.

It is apparent that the risky and non-risky components of e-commerce are closely intertwined both theoretically and practically. It would not be possible, and it does not even make sense, to separate the risky components of e-commerce from their non-risky counterparts. Therefore it is the basic premise of this study that first, the risky and non-risky components of e-commerce are related, and second, risk influences the adoption of both types of components.

Empirically, [9] devised a five-component measurement framework for e-commerce in general, which includes e-mail, homepage, obtaining market research information, gathering customer information, and buying and selling online. Of these, only buying and selling online involves risk. In this study, buying and selling online is further broken into online contracting and online payment, because the types of trust that influence these two components are proposed to be different [6, 10]. It is self-evident that online contracting and online payment are the core components of international e-commerce. Therefore a complete international electronic commerce transaction includes both the peripheral (non-risky) components of e-mail, homepage, obtaining research information, gathering customer information, and the core (risky) components of online contracting and online payment.

Trust is generally defined as one party believing that the other party in a transaction is capable of and willing to carry out the promises it has made with regard to the transaction [11, 12]. It can be further broken into party trust and control trust. Party trust refers to believing in the other party of a transaction in its ability and willingness to carry out its obligations, while control trust is concerned with confidence in the controlling mechanism ensuring satisfactory performance of a transaction [6, 10].

There are objective and subjective reasons for both party trust and control trust. Objective reasons for trust are the relevant social indicators [6, 10], or cue-based factors [13]. They can create objective trust in e-commerce in general. In B2C e-commerce, Web assurance measures [14] such as third-party privacy seals, privacy statements, third-party security seals, and security features [15] also enhance user trust in e-commerce sites. Therefore it can be argued that this relationship should also hold for the risky components of B2B international e-commerce.

Hypothesis 1. The adoption of both risky and non-risky components of international e-commerce is directly related to the creation of objective trust.

Subjective trust reasons include personal experience, understanding, and communality [6, 10]. Personal experience refers to a party's past experience with a transaction partner or control procedure. Apparently it can also be argued that it influences the adoption of the risky components of international e-commerce. It can be argued that positive experience in international e-commerce with a certain partner will not only reinforce e-commerce relationship with this partner, but also encourage e-commerce adoption in trading with other partners in general. In other words, the influence of this type of experience can be hypothesized at the general level of international e-commerce adoption.

Hypothesis 2. The adoption of both the risky and non-risky components of international e-commerce is directly related to positive past experience in international e-commerce.

Understanding trust refers to one trusting a transaction partner because he or she understands the goals, capabilities, etc., of this partner, or one trusting a control procedures because he or she understands how it works. Communality trust, on the other hand, is created when one trusts a transaction party or a control procedure because other members of the community also trust it [6, 10]. Apparently these two types of trust should influence the adoption of e-commerce

Hypothesis 3. The adoption of both risky and non-risky components of international e-commerce is directly related to the creation of both communality trust and understanding trust.

It should be self-evident that the adoption of the non-risky components of e-commerce should lead to that of their risky counterparts. This is because the former lay the foundation for the adoption of the latter. For example, gathering customer information can logically lead to conducting e-commerce with this customer.

Hypothesis 4. The adoption of the risky components of international e-commerce is directly related to that of their non-risky counterparts.

3 Methodology

To test the hypotheses, a questionnaire was designed and sent out to professionals who are directly engaged in international trade in Chinese companies. The questionnaire consists of four parts. The first two parts are intended to gather demographic information of the businesses the respondents represent, where they were asked to indicate company size (as measured by the number of employees) and business type (international trade only vs. international trade + manufacturing). Respondents were asked to indicate their company size on an ordinal scale of 1 through 5: 1 (employing 1-5 people), 2 (6-20), 3 (21-50), 4 (51-200), and 5 (>200).

The third part of the questionnaire measures the respondents' e-commerce adoption by asking them to indicate the components of international e-commerce that their companies use in international trade. Here the measures were adapted from [9]. As discussed in the previous section, these include the non-risky components of e-mail, homepage, obtaining research information, gathering customer information (changed to 'finding trade leads' after pilot testing), and the risky components of online contracting and online payment. These measures, which transcend the corporate interface with the external environment, match [8]'s theoretical definition well. This part is a multiple-response question where the respondent could check any combination of the six items, depending on his or her firm's actual situation. These items are treated as dichotomous, i.e., the answer to which is either "yes (coded '2')" or "no (coded '1')".

The last part includes questions designed to gauge the level of various types of trust that the respondents have with regard to conducting international e-commerce. This is accomplished by measuring respondents' rating of antecedents of trust on a Likert scale of 1-5, with 1 representing 'strongly disagree' and 5 'strong agree'. Respondents were asked to rate statements designed to measure antecedents of the various types of trust.

The questionnaire was pilot tested, based on which two revisions were made. First, "gathering customer information" [9] was changed to "finding trade leads". Second, a note was added to part 4 that 'international e-commerce' in the statements refers to accomplishing a complete international e-commerce transaction, i.e., from researching customer information and finding trade leads online to contracting and paying online.

The revised questionnaire was then emailed to international trade professionals who are alumni of three major Chinese universities. They were asked either to print out the questionnaire, answer the questions, and send it back via mail or fax, or to complete the questionnaire electronically, save it, and email it back. A total of 349 copies of the questionnaire was emailed out, from which 140 completed copies were received, with 5 in hard copy, and the rest via email. Of these 135 are usable. The collected data were then coded and entered into a Microsoft Excel file and analyzed with SPSS.

For the sake of parsimony in displaying the statistical information in the next section, the following variable names are used: Homepage (homepage), Email (email), Trade Leads (finding trade leads), Research Info (obtaining research information), Online Contracting (international online contracting), Online Pay (international online payment), Understanding (understanding trust), Communality (communality trust), Objective (objective trust), Experience (trust based on experience in international e-commerce).

4 Results and Discussion

4.1 Characteristics of the Sample

As Table 1 shows, the respondent firms are relatively well distributed in terms of size and business type. It should be noted that international-trade-only firms are generally smaller in size than their international-trade + manufacturing cohorts, as is reflected in the fact that the median size for the former is in the 6-20 employees range, while that for the latter 51-200. This pattern reflects reality well in China, because for the same volume of trade, a trade-only firm would need to hire fewer people than its trading + manufacturing counterpart.

Table 1. Demographic Information of Respondent Firms

	Company Size (Number of Employees)											
	1-5		6-20		21-50		51-200		> 200		Subtotal	
Business Type	Freq.	%	Freq.	%	Freq.	%	Freq.	%	Freq.	%	Freq.	%
Int'l Trade	10	7.4%	16	11.9%	9	6.7%	13	9.6%	3	2.2%	51	37.8%
Int'l Trade + Mfg	3	2.2%	10	7.4%	12	8.9%	21	15.6%	38	28.1%	84	62.2%
Subtotal	13	9.6%	26	19.3%	21	15.6%	34	25.2%	41	30.4%	135	100%

4.2 Hypothesis Testing

Hypotheses 1-3 are concerned with the relationship between the different types of trust and international e-commerce components, while Hypothesis 4 states the relationship between its risky and non-risky components. For Hypotheses 1-3 independent-samples t tests were run on Objective, Experience, Communality, and Understanding together, with the various components of IEC each as a grouping variable in turn.

As Table 2 shows, the four types of antecedents of trust, i.e., Objective, Experience-IEC, Communality, and Understanding, all significantly influence group membership for Trade Leads and Online Contracting, with two-tailed t tests being significant ($p < 0.05$) for all. In addition, objective trust and experience trust affect Email, but this relationship is only significant at $p < 0.10$.

The different t test results for the various e-commerce components with regard to trust types suggest that there may be important relationships among the components themselves. Testing Hypothesis 4 answers this question. For this purpose, crosstab was run for pairs within the various e-commerce components. The results are shown in Table 3.

It can be seen that among the non-risky components, Trade Leads is significantly related to the other three components. In addition, only Trade Leads is significantly related to Online Contract, which is risky, while none of the other three are. The two risky components are also significantly related. Therefore, the relationship among the components of e-commerce can be depicted as Figure 1.

4.3 Discussion

Overall, there is strong empirical evidence that trust plays an important role in e-commerce adoption [16]. This study adds to knowledge by specifically focusing on the impact of trust on international electronic commerce adoption by Chinese international trade firms. Results indicate that trust not only affects the adoption of one of the risky components of e-commerce, it also influences that of the non-risky ones. Therefore the relationship between trust and e-commerce adoption is not as simple as asserted by [6, 10].

Table 2. Independent Samples T Test Results

	Levene's Test for Equality of Var.		t-test for Equality of Means						
	F	Sig.	t	df	Sig. (2-tailed)	Mean Diff.	Std. Error Diff.	95% Confidence Interval of Diff. Lower	Upper
Email[(a)]									
Objective	0.826	0.365	-1.678	133	0.096	-0.907	0.541	-1.976	0.162
Experience	1.481	0.225	-1.736	133	0.085	-0.880	0.507	-1.882	0.123
Communality	0.114	0.736	-1.492	133	0.138	-0.868	0.582	-2.019	0.283
Understanding	1.012	0.316	-1.292	133	0.198	-0.822	0.636	-2.079	0.436
Homepage[(a)]									
Objective	2.242	0.137	1.128	133	0.262	0.414	0.368	-0.313	1.141
Experience	2.821	0.095	-0.638	133	0.524	-0.221	0.346	-0.905	0.463
Communality	1.009	0.317	-1.270	133	0.206	-0.501	0.394	-1.280	0.279
Understanding	0.019	0.890	-1.611	133	0.109	-0.690	0.428	-1.537	0.157
Research Info[(a)]									
Objective	0.572	0.451	-1.678	133	0.296	-0.260	0.247	-0.749	0.230
Experience	0.068	0.795	-1.736	133	0.177	-0.314	0.231	-0.772	0.144
Communality	0.047	0.829	-1.492	133	0.301	-0.276	0.266	-0.801	0.250
Understanding	3.010	0.085	-1.292	133	0.505	-0.194	0.290	-0.768	0.380
Trade Leads									
Objective[(b)]	4.924	0.028	-3.032	61.763	0.004	-0.782	0.258	-1.297	-0.266
Experience[(b)]	7.331	0.008	-3.453	59.753	0.001	-0.839	0.243	-1.326	-0.353
Communality[(a)]	0.830	0.364	-2.689	133	0.008	-0.693	0.258	-1.203	-0.183
Understanding[(a)]	1.628	0.204	-3.433	133	0.001	-0.950	0.277	-1.497	-0.403
Online Contracting[(a)]									
Objective[(a)]	1.988	0.161	3.461	133	0.001	0.833	0.241	0.357	1.310
Experience[(a)]	6.983	0.009	5.775	117.501	0.000	1.352	0.234	0.888	1.815
Communality[(b)]	4.041	0.046	5.447	116.471	0.000	1.035	0.190	0.659	1.411
Understanding[(b)]	6.658	0.011	4.397	121.474	0.000	0.903	0.205	0.497	1.310
Online Payment[(a)]									
Objective	0.494	0.483	-0.974	133	0.332	-0.235	0.241	-0.711	0.242
Experience	1.553	0.215	0.908	133	0.365	0.239	0.263	-0.281	0.759
Communality	0.202	0.654	0.868	133	0.387	0.183	0.211	-0.234	0.600
Understanding	0.526	0.469	0.211	133	0.834	0.047	0.225	-0.398	0.493

a. Because Levene's test is not significant, only t statistics when assuming equal variances between the groups are shown.

b. Because Levene's test is significant, only t statistics when not assuming equal variances between the groups are shown.

Table 3. Crosstab Results (Chi-Square Test)

	Homepage	Research Info	Trade Leads	Online Contract	Online Pay
Email	0.605	0.499	0.063**	0.306	0.985
Homepage		0.978	0.003*	0.222	0.977
Research Info			0.006*	0.129	0.166
Trade Leads				0.003*	0.282
Online Contract					0.002*

* Significant at the 0.05 level.

** Significant at the 0.10 level.

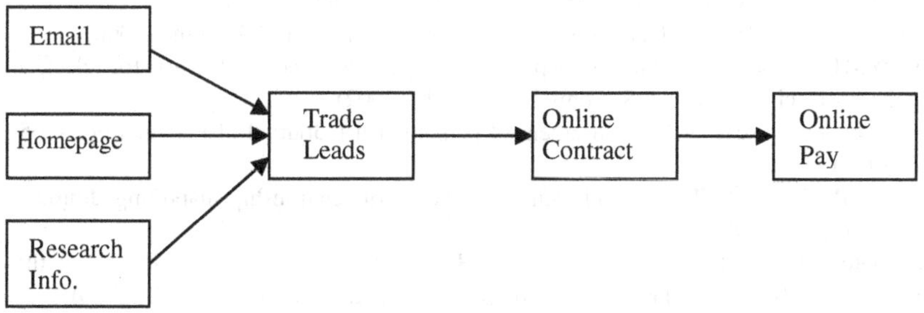

Fig. 1. Proposed Relationships among E-Commerce Components

5 Conclusion

The issue of trust in influencing the adoption of e-commerce has been primarily focused on its B2C type [e.g., 17, 18-22]. There has also been research on trust in C2C e-commerce [e.g., 23, 24]. Less attention has been given to B2B e-commerce [e.g., 25], especially international B2B e-commerce. However, it is important to study trust and international e-commerce adoption because the Internet does stimulate international trade [26, 27].

This study represents an attempt to empirically investigate the trust issue in this area in the context of Chinese firms engaged in international trade. As such, its theoretical contributions are primarily as follows. First of all, trust affects the adoption of some, but not all, of both the risky and non-risky components of e-commerce. Secondly, there are significant relationships among the various components of e-commerce, both between, and within, the two groups of components.

References

1. Ratnasingham, P.: The importance of trust in electronic commerce. Internet Research 8, 313–324 (1998)
2. Ratnasingham, P.: Trust in inter-organizational exchanges: A case study in business to business electronic commerce. Decision Support Systems 39, 525–544 (2005)
3. Slemp, C.: Electronic commerce success is a matter of trust. Network World 15, 42 (1998)
4. Urban, G.L., Amyx, C., Lorenzon, A.: Online Trust: State of the Art, New Frontiers, and Research Potential. Journal of Interactive Marketing 23, 179–190 (2009)
5. Ratnasingham, P.: Risks in low trust among trading partners in electronic commerce. Computers and Security 18, 587–592 (1999)
6. Tan, Y.-H., Thoen, W.: Toward a generic model of trust for electronic commerce. International Journal of Electronic Commerce 5, 61–74 (2000)
7. Zhang, X.: Trust in International Electronic Commerce: A Study of Chinese Companies Engaged in International Trade. In: WICOM 2008. IEEE Explore, Dalian (2008)
8. Zwass, V.: Electronic commerce: structures and issues. International Journal of Electronic Commerce 1, 3–23 (1996)

9. Moussi, C., Davey, B.: Internet-based electronic commerce: Perceived benefits and inhibitors. In: ACIS 2000. Queensland University of Technology, Brisbane (2000)
10. Tan, Y.-H., Thoen, W.: Formal aspects of a generic model of trust for electronic commerce. Decision Support Systems 33, 233–246 (2002)
11. Rotter, J.: A new scale for measurement of personal trust. Journal of Personality 35, 651–665 (1967)
12. Morgan, R., Hunt, S.: The commitment-trust theory of relationship marketing. Journal of Marketing 58, 20–38 (1994)
13. Warrington, T.B., Abgrab, N.J., Caldwell, H.M.: Building trust to develop competitive advantage in e-business relationships. Competitiveness Review 10, 160–168 (2000)
14. Kaplan, S.E., Nieschwietz, R.J.: A Web assurance services model of trust for B2C e-commerce. International Journal of Accounting Information Systems 4, 95–114 (2003)
15. Belanger, F., Hiller, J.S., Smith, W.J.: Trustworthiness in electronic commerce: The role of privacy, security, and site attributes. Journal of Strategic Information Systems 11, 245–270 (2002)
16. Wu, K., Zhao, Y., Zhu, Q., Tan, X., Zheng, H.: A meta-analysis of the impact of trust on technology acceptance model: Investigation of moderating influence of subject and context type. International Journal of Information Management (2011) (in Press, Corrected Proof)
17. Liu, C., Marchewka, J.T., Lu, J., Yu, C.-S.: Beyond concern–a privacy-trust-behavioral intention model of electronic commerce. Information and Management 42, 289–304 (2005)
18. Bahmanziari, T., Odom, M.D., Ugrin, J.C.: An experimental evaluation of the effects of internal and external e-Assurance on initial trust formation in B2C e-commerce. International Journal of Accounting Information Systems 10, 152–170 (2009)
19. Kim, C., Tao, W., Shin, N., Kim, K.-S.: An empirical study of customers' perceptions of security and trust in e-payment systems. Electronic Commerce Research and Applications 9, 84–95 (2010)
20. Kim, J., Jin, B., Swinney, J.L.: The role of etail quality, e-satisfaction and e-trust in online loyalty development process. Journal of Retailing and Consumer Services 16, 239–247 (2009)
21. Kim, M.-J., Chung, N., Lee, C.-K.: The effect of perceived trust on electronic commerce: Shopping online for tourism products and services in South Korea. Tourism Management 32, 256–265 (2011)
22. Palvia, P.: The role of trust in e-commerce relational exchange: A unified model. Information & Management 46, 213–220 (2009)
23. Chiu, C.-M., Huang, H.-Y., Yen, C.-H.: Antecedents of trust in online auctions. Electronic Commerce Research and Applications 9, 148–159 (2010)
24. Jones, K., Leonard, L.N.K.: Trust in consumer-to-consumer electronic commerce. Information & Management 45, 88–95 (2008)
25. Chien, S.-H., Chen, Y.-H., Hsu, C.-Y.: Exploring the impact of trust and relational embeddedness in e-marketplaces: An empirical study in Taiwan. Industrial Marketing Management (2011) (in Press, Corrected Proof)
26. Freund, C.L., Weinhold, D.: The effect of the Internet on international trade. Journal of International Economics 62, 171–189 (2004)
27. Freund, C.L., Weinhold, D.: The Internet and international trade in services. American Economic Review 92, 236–240 (2002)

Process Modeling and Reengineering in the Integration Stage of Electronic Government

Ning Zhang and Haifeng Li

School of Information, Central University of Finance and Economics,
Beijing, PRC, 100081
zhangning75@gmail.com, mydlhf@126.com

Abstract. Government Process Reengineering (GPR) is the kernel of electronic government, but it has high risk. A framework for Government Process Management (GPM) in the integration stage of electronic government is presented in this paper. According to the characteristics of the integration stage, role and activity combined process model are built based on Petri nets. Then the correctness, performance and optimization of Petri net based model are analyzed. For applying the GPM framework and the process modeling and analysis method, a real example of statistics process modeling and reengineering is illustrated finally.

Keywords: government process management, integration stage, electronic government, process modeling and analysis, Petri net.

1 Introduction

Technological changes have always catalyzed organizational changes. Electronic government (e-gov) is the use of information technology (especially Internet) to support government operations, engage citizens, and provide government services [1]. Government is the largest information holder and information technology user in the society. Since the 1990's, electronic government has been developing quickly in developed counties and been paid attention to more and more by all over the world.
Electronic government can be divided into four stages: initiation, interaction, transaction and integration [2]. In general, Chinese electronic government is in the interaction stage and preparing for the transaction stage. In the interaction stage, publics can have simple interaction with government such as sending e-mails or participating in discussion boards. In the transaction stage, publics can be provided full benefits from transactions over the Internet, such as purchasing licenses and permits, paying for taxes and fees.

Some advanced cities (such as Beijing and Shanghai) and departments (such as Ministry of Commerce) are now entering integration stage. In the integration stage, publics can be provided with one-stop services, i.e., services across the agencies and departments of government are integrated. It can be further divided into vertical and horizontal integration phases. Vertical integration is integration of government departments and organizations that operate the same or related systems at different

F.L. Wang et al. (Eds.): WISM 2012, LNCS 7529, pp. 167–174, 2012.

organizational levels. Horizontal integration is integration of government services for different functions.

Traditional government processes, which lead to low efficiency and coordination difficulty, cannot adapt to electronic government environment, especially in the integration stage. Government Process Reengineering (GPR), the thought of which comes from Business Process Reengineering (BPR), is the kernel of electronic government [3]. BPR has to be revolutionary to business, and that changes have to be radical in order to produce any desirable results. However, business history indicates that 70 percent of BPR projects failed. Due to powerful inertia and particularity of government, GPR has even more risks than BPR.

Based on the notion of Business Process Management (BPM) [4], this paper presents Government Process Management (GPM) to substitute for GPR, which can not only improve government performance and satisfy public's growing needs through end-to-end process excellence, but also reduce the risk greatly. A theoretical framework for GPM in the integration stage of e-gov is first presented. The process modeling and analysis method based on Petri nets is then detailed. Finally, a real example of process modeling and reengineering illustrated by Petri net is given.

2 Theoretical Framework for GPM in the Integration Stage

2.1 Hierarchy of GPM

Processes are designed normatively according to their current status in GPM. Only processes with major defects need to be reengineered. Hence we should first define different hierarchies of GPM to handle with processes in different situations. Here three hierarchies are defined.

- Process monitoring (the lowest level): monitoring processes to make them more efficient.
- Process improvement (the middle level): improving processes by finding out hidden problems to reduce losses and costs.
- Process reengineering (the highest level): reengineering processes which can no longer adapt to practical needs to make significant and radical changes.

Process reengineering has a great impact on government and takes much more time and resources than process improvement and monitoring. Therefore it should be considered carefully and implemented cautiously to reduce the risk.

2.2 Framework for GPM in the Integration Stage of E-gov

In the integration stage, model for both internal and external processes should be built to analyze and evaluate performance of current processes in detail to decide which hierarchy each core process should go into. Organizational change is often necessary at the same time with process reengineering. Figure 1 shows the framework for GPM in the integration stage of e-gov.

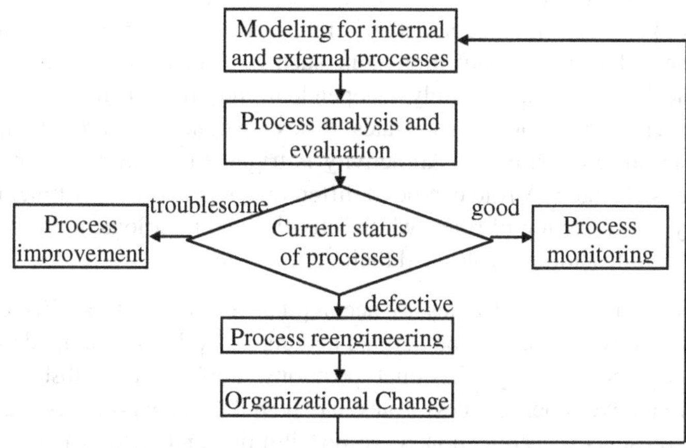

Fig. 1. Framework for GPM in the integration stage of e-gov

3 Process Modeling and Analysis in the Integration Stage

3.1 Comparative Analysis of Process Modeling Methods

The common process modeling methods and techniques include flow chart (FC), role activity diagram (RAD), IDEF suite, event-driven process chain (EPC), Petri nets, etc. Table 1 compares these different methods and techniques.

Table 1. Performance of different process modeling methods

	FC	RAD	IDEF	EPC	Petri
Process characteristics	functional	functional	functional	cross-functional	cross-functional
Abstract mechanism	no	no	yes	no	yes
Understand-ability	good	ok	ok	ok	ok
Computerized capacity	good	good	good	weak	good
Process Reengineering	weak	weak	weak	potential	potential
Dynamics	yes	no	no	yes	yes

According to the comparative analysis, Petri net would be the best solution for the government process modeling and analysis. It can be used for both activity-based model and role-based model. As to the integration stage of electronic government, combined process model should be built due to the different organizations and roles involved.

3.2 Role and Activity Combined Process Model Elements

Traditional process modeling methods and techniques are primarily designed around activities lack of considering different actors involved in the process. The concept of role is best defined by Biddle and Thomas in 1979 as "those behaviors characteristic of one or more persons within a context". From the view of organizational structure, a role is represented by an organization unit, such as a team or a department. In a process, a role is an active, relatively independent and abstract unit which includes elements like activities, resources and states. Roles are connected through information which is the premise of an event. An activity is triggered by an event, and then the state of a role is changed. A role can be in different status at different time. Processes are essentially the collection of roles, while activities and their logical relationship are only external manifestation of roles and their coordination.

- Role includes a set of activities to accomplish a specific task. Driven by the modeling objectives, roles can be defined and classified by functional departments, administrative positions, professional positions, persons, and abstract working concepts. In a government process, roles may be publics, departments, etc.
- Activity represents the collection of responsibilities and rights of each role. An activity describes a step pushing a process forward. The staring point, the ending point and the reason of triggering should be understood before an activity is defined.
- Coordination happens when a set of activities are created between two or more roles. Coordination is a necessary element constituting a process. Information is transmitted between roles directly or indirectly through coordination. It changes the status of roles.

3.3 Process Modeling Steps Based on Petri Nets

Petri nets are a class of modeling tools, which were originated by Petri. They have a well-defined mathematical foundation and an easy-to-understand graphical feature. Moreover, variation of the model can be carried out using Petri net analysis techniques. Petri nets are widely studied [5] and successfully applied, especially for modeling and analyzing of discrete event dynamic systems whose behavior is characterized by parallelism and synchronization.

A Petri net is a directed graph consisting of three structural components: places, transitions, and arcs. Places, which are drawn as circles, represent possible states or conditions of the system. Transitions, which are shown by bars or boxes, describe events that may modify system states. The relationship between places and transitions are represented by a set of arcs, which are the only connectors between a place and a transition in either direction. The dynamic behavior of a system can be represented using tokens, which graphically appear as black dots in places.

The role and activity combined process modeling steps based on Petri nets are as follows:

- Analyzing the different roles in the process and the activities in each role and the coordination between roles.
- Defining the state collection P, the event collection T and the relationship between them for each role.

- Matching the states of roles with places, the events of roles with transitions in Petri nets.
- Defining the coordination between different roles and connecting correlative transitions according to the sequence of information transmission.
- Determining the initial state of the model and the number of tokens in places.
- Simplifying the model for further analysis.

3.4 Process Model Analysis and Optimization

Process model analysis includes qualitative analysis and quantitative analysis.

- Qualitative analysis: checking the correctness of process and proving the model is valid. Qualitative analysis concentrates on properties such as deadlock freeness, absence of overflow, and the presence of certain mutual exclusion of wherever shared resources are used. The analysis can be conducted by using the business process modeling software tools, such as ExSpect.
- Quantitative analysis: calculating performance, responsiveness and resource utilization indices. Time and cost analyses are very important to find the current status of process to decide whether it should be improved or reengineered. Some extensions of Petri nets in which the notion of time is considered, for example, generalized stochastic Petri net (GSPN), are suitable for such analysis.
- Optimization: improving or reengineering process according to optimization rules, such as splitting, deleting or merging some elements of the model.

4 Statistics Process Modeling and Reengineering

Beijing Municipal Bureau of Statistics is now in the transaction stage and entering the vertical integration stage [6]. In order to provide publics with full benefits from transactions (especially statistics data collection) over the Internet, Beijing Municipal Bureau of Statics has been implementing statistics process reengineering in several recent years.

4.1 Process Modeling Based on Petri Nets

Beijing Municipal Bureau of Statistics has city-level institutions and subordinate district/county-level institutions and street/town-level institutions. In the institutions for each level, departments are divided based on statistics by national economy industry (such as manufacturing industry, construction industry, wholesale and retail industry) or by profession (such as investment, price, labor wage). Each department was relatively independent and accomplished its own tasks of data collection, data organization, data analysis, and data publication. Each survey object might face different departments or even institutions of different levels. The primary statistics process model before reengineering is shown in figure 2.

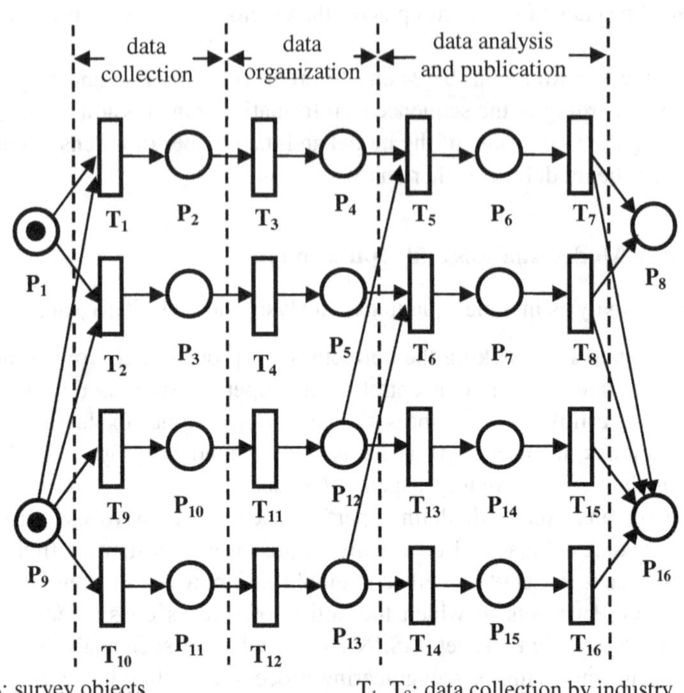

P_1, P_9: survey objects
P_2, P_{10}: basic data by industry
P_3, P_{11}: basic data by profession
P_4, P_{12}: summarized data by industry
P_5, P_{13}: summarized data by profession
P_6, P_{14}: statistics product by industry
P_7, P_{15}: statistics product by profession
P_8, P_{16}: publics

T_1, T_9: data collection by industry
T_2, T_{10}: data collection by profession
T_3, T_{11}: data organization by industry
T_4, T_{12}: data organization by profession
T_5, T_{13}: data analysis by industry
T_6, T_{14}: data analysis by profession
T_7, T_{15}: data publication by industry
T_8, T_{16}: data publication by profession

Fig. 2. Primary statistics process model before reengineering

The figure is divided into two different levels: the higher level and the lower level. If the higher level is city-level, the lower level is district/county-level, while if the higher level is district/county-level, the lower level is street/town-level. The coordination between the two levels include: the higher level collects data from survey objects of lower level; the lower level submits summarized data to the higher level; the higher level publishes statistics product to publics of lower level.

4.2 Process Analysis and Evaluation

Although each level and each department had their own autonomy in collecting and processing data, this kind of diverse way of primary statistics process had brought following problems:

- Survey objects had great difficulties in reporting data to different departments or even different institutions of each level since their reporting ways, reporting periods and data formats were usually different. This often increased unnecessary and extra burden to the survey objects.

- It was hard for different departments to exchange and summarize statistics data since these data had different structures and format standards. Although statistics institutions had large amount of data resources, these resources could not be utilized fully and shared widely.
- Diverse way often leads to management confusion. It was nearly impossible to unify behaviors of workers from different departments and levels. The phenomena of data missing and data inconsistency occurred frequently, which reduces data quality greatly.

The Petri net-based model is proved to be valid and the time and cost efficiency is relatively low.

4.3 Process Reengineering

In order to solve the above problems, the primary statistics process has been reengineered (as shown in figure 3).

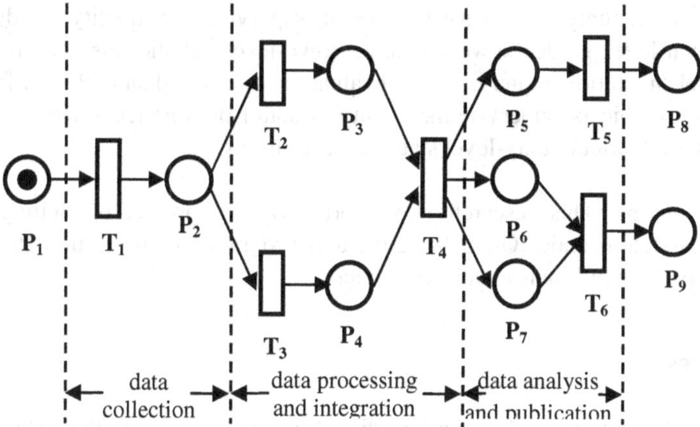

P_1: survey objects of all levels	T_1: data collection
P_2: basic data	T_2: data processing by industry
P_3: summarized data by industry	T_3: data processing by profession
P_4: summarized data by profession	T_4: data integration
P_5: statistics product data	T_5: data publication
P_6: integrated database	T_6: data analysis
P_7: data warehouse	

Fig. 3. Primary statistics process model after reengineering

A unified working platform has been built in Beijing Municipal Bureau of Statistics under electronic government environment. Survey objects report all statistics data through this platform directly. The data are then processed by industry or by profession in the process of data processing. The platform also provides functions of data integration, data analysis and data publication based on powerful data resources.

Statistics institutions of different levels and different departments work on this platform by the limit of different operational functions. Meta data standard is designed and utilized in the platform, which unifies data format for each level and each department.

4.4 Organizational Change

Instead of accomplishing their own tasks of data collection, data organization, data analysis and data publication, city-level, district/county-level, and street/town-level institutions perform the function of supervision in the new process. They take the responsibility of ensuring the timeliness, accuracy and completeness of statistics data from the corresponding level.

- Street/town-level institutions: query, verify, evaluate and accept data collected from street/town-level survey objects.
- District/county-level institutions: query, verify, evaluate and accept data collected from district/county-level survey objects; supervise the quality of data from street/town-level; guide the work of street/town-level statistics institutions.
- City-level institutions: query, verify, evaluate and accept data collected from city-level survey objects; supervise the quality of data from district/county-level; guide the work of district/county-level statistics institutions.

Acknowledgment. This research is supported by the project of Beijing Natural Science Foundation under Grant 9092014 and Program for Innovation Research in Central University of Finance and Economics.

References

1. Chandler, S., Emanuels, S.: Transformation Not Automation. In: Proceedings of 2nd European Conference on EGovernment, pp. 91–102. St. Catherine's College Oxford, UK (2002)
2. Layne, K., Lee, J.: Developing fully functional E-government: a four stage model. Government Information Quarterly 18, 122–136 (2001)
3. Ye, Y.: Government Process Reengineering: The Theoretical and Practical Researches. PhD dissertation at Jilin University, Changchun, China (2007)
4. van der Aalst, W.M.P., ter Hofstede, A.H.M., Weske, M.: Business Process Management: A Survey. In: van der Aalst, W.M.P., ter Hofstede, A.H.M., Weske, M. (eds.) BPM 2003. LNCS, vol. 2678, pp. 1–12. Springer, Heidelberg (2003)
5. Murata, T.: Petri nets: Properties, analysis, and applications. Proceedings of IEEE 77, 541–580 (1989)
6. Beijing Municipal Bureau of Statistics (2012), http://www.bjstats.gov.cn

Development of Vertical Industrial B2B in China: Based on Cases Study

Jinghuai She and Xiaotong Xing

College of Business Administration, Capital University of Economics and Business, China
shejinghuai@cueb.edu.cn

Abstract. Vertical Industrial B2B transaction market, especially raw materials vertical B2B transaction market, takes an important role in the Chinese circulation market. Based on economic theories and the current situation of Chinese vertical B2B transaction market, this essay firstly selects 3 representative market operators as case samples, secondly carries out multi-case analysis and comparison, finally concludes the common development laws of Chinese vertical B2B transaction market, which form information service to transaction service, then to resources integration of other value-adding service.

Keywords: vertical industry, B2B transaction market, development, case analysis.

1 Introduction

B2B Electronic Marketplaces is a typical electronic platform intermediary agent based on internet technology, which allows many buyers and sellers exchange goods or service information, trading on line and get value-adding service (Hadaya p, 2006) [1]. Applied to B2B Electronic Marketplace, which is also called electronic intermediary, electronic market maker, electronic business hub, etc., it shows an increasing trend (Hazra. J., 2004) [2]

Because of the lack of external environment and internal regulation for vertical industrial B2B transaction market in China, several severe problems occurred in the development of it recently, such as, manipulating price, misusing of clients' funds, gambling between clients and exchanges. These problems make the society worry about the development of spot EMarketplace. These problems are because of the lopsided development caused by the limited profit model.

This essay firstly lists relevant theories of vertical (raw materials) industrial B2B transaction market by literature review; secondly, analysis the vertical industrial B2B transaction market in China, especially businesses characteristics and profit models of raw material industry by using multi-case analysis method; finally, concluding the development trend of vertical industrial B2B transaction market in China by comparing analysis method so that give suggestions for vertical industrial B2B transaction market's healthy development.

F.L. Wang et al. (Eds.): WISM 2012, LNCS 7529, pp. 175–182, 2012.

2 Theories and Methods

Any economic phenomenon's appearance and rising will stimulate the application of existing economic theories to explain and understand it; this is also the driving force for the development of the economic theory. We try to analysis vertical industrial B2B transaction market by economics and electronic business platform theory.

2.1 Economics Theory

Economic theory especially market circulation and trade theory is the theoretical basis of analyzing vertical industrial B2B transaction market.

Transaction cost. It is the expenses that are needed to get accurate market information, the negotiation and regular contract costs. In other words, the transaction costs includes information searching costs, negotiation costs, contracting costs, monitoring of compliance costs and potential breach of contract costs （Ronald H. coase, 1937 ） [3]Vertical industrial B2B transaction market's establishment, development and formation of the market coordinate system are because it saves costs.

George Joseph Stigler (1961)[4]pointed out that, under incomplete information, the price will be affected by the price distribution and customers' searching costs, as a result, the actual price of the same goods may be different. While under the complete information, the prices deviation of the same goods will be nearly zero. By releasing all sorts of trading information by the various media, vertical industrial B2B transaction market is tiring to provide complete information and reduce trade organizations' losses caused by the information asymmetry.

Value chain theory. Poter(1985)[5]pointed out the concept of value chain and thought that the profitability of a company depends on whether it can get the value it created for the buyers or protect the value from being stolen. Hoh B, Eriksson K, Johansson J(1999) [6]the business net, which depends on each other, formed a chain of causation, this is important for improving the creating of value.

2.2 EMarketplace Theory

Vertical industrial B2B transaction market is based on internet technology, which allows many buyers and sellers exchange goods or service information, trading on line and get value-adding service. （Hadaya,p, 2006）[1]Applied to vertical industrial B2B transaction market, which are so called electronic intermediary, electronic market maker, electronic business hub, etc., it shows an increasing trend （Hazra. J.,2004）[2]The details of relevant theories can be explained from many aspects, such as classification, intermediary system, trade system and risk control, etc.

1 Intermediary system. Bakos （1998)[9] treated electronic business platform as an agent which can finds price information for market, facilitates trading, provide regulatory framework and provide virtual trading spaces on the internet through electronic methods.

2 Trading system. Electronic business platform create values in two ways: gathered mechanism and matched mechanism. The gather function brings many traders to a virtual platform and reduces cost by offering one-stop services. The matched mechanism works well in following conditions: (1) goods can be traded without check; (2) trading volume is much greater than trading costs;(3) traders can use the dynamic pricing mechanism smoothly; (4) companies eliminate the effect of crest and trough of demand by purchasing; (5) logistics can be achieved by third party; (6) the demand and the price keep changing.

3 Risk control. Saeed and Leith(2003) describe risks in three dimensions: trading risks, safety risks and privacy risks. Safety risks and privacy risks are characteristics of electronic market. Safety risks are caused by the lose of data and certification. The privacy risks includes improper data collection and unauthorized information publication.

4 Electronic trading market price discovery. Liu Xiaoxue and Wang Shennan(2009)'s research showed that electronic market and forward market have relationships of competition and complementation, the connected price signals formed by these two markets formed a long-term balance with spot price, and the balance affects and explains spot price a lot. This can be done by neither future price nor forward price[10].

To sum up, the appearance and development of vertical industrial B2B transaction market in China has theoretical basis and primary empirical analysis support; in fact, problems like risk control and trading mechanism of vertical industrial B2B transaction market in China also proved theories mentioned above.

2.3 Hypothesis

In 2008, Chinese E-commerce B2B electronic business operators have a revenue scale of 5.54 billion yuan, increased by 34.1%.Iresearch consulting company forecast that this number will increase to 15.23 billion yuan [11](figure 1). YIGUAN International (2008) published the "Chinese vertical B2B market special series report 2008 - raw materials industry" shows that, vertical B2B of raw material occupied 12.4% of civil vertical B2B. However, when look from income scale, vertical B2B of raw material have total revenue of 190 million, occupied 51% of whole vertical B2B market. Vertical B2B of raw material market is the most developed and active market. Vertical B2B of raw material will have a market scale of 1.05 billion, occupied 50.4% in 2011[12].

Based on theories and empirical researches above, considering the current conditions and development trend of Chinese vertical industry B2B market, we proposed following hyphenise:

Hypothesis 1: vertical industrial B2B transaction market shows the development trend of B2B electronic transactions. Every market's survival and development must be in a certain market conditions, having connection or extension with market operator's original business, therefore, vertical industrial B2B transaction market has different business models and profit models at this stage.

Hypothesis 2: the development of profit model of vertical industrial B2B transaction market is from provide information mainly to provide transition services, to the final applications of e-commerce integration.

3 Researches Design

3.1 Research Sample

In order to study vertical industrial B2B transaction market, we choose typical raw material industry enterprise: my steel, Lange and Treasure Island. They represent the most influential Chinese steel information service and transaction service providers, typical representatives of steel traders and steel warehouse transaction.

3.2 Data

Evidences of case study are from documents, files, interviews, observations, participated observation and physical evidences. □ Yin R K.1994□ [14]□ We collected data from various ways. Firstly, search internet and companies' websites. Secondly, quote scholars papers and third party's investigation results. Thirdly, communicate with manager and get data provided by the company.

4 Case Study

4.1 Lange Group Company (www.lgmi.com)

Lange started and benefited from professional steel market information provider. It has lots of steel manufactures and merchant partners, which provide base for its further development.

4.1.1 Case Description

4.1.1.1 Information Service and Its Profit Model

(1) Information service model

Lange steel net provides steel market information, economic news, market analysis forecast, and network technology for steel industry enterprises, In Beijing and other cities, the prices from Lange steel net is widely used as a unit price guideline for the settlement of the steel's traders. Market research. Lange steel net's information research center issues “the monthly report ", "research report", "the market forecast report", "China's steel industry's annual report", "the market research and advisory", "Lange steel index" and "Lange steel circulation industry the purchasing managers' index (LGSC-PMI)" ,etc.Value-added service. Lange steel net uses the text messages and daily mail business, which can send steel information to the customers in time and make customers master market information.

(2) Profit model

Network service income includes advertising, searching and web service fees, etc. according to the location and size of webs, advertisements fees are calculated. Lgsou provides services like recommendation of purchasers, searching the tops and instantly negotiation. It also offers web construction service and charge service fees. Lange provides virtual space for customer and charge rent fees.

4.1.1.2 Transaction Service and Profit Model

(1) Transaction service

Interim and forward transaction based on spot, uses Internet and E-commerce technology; takes advantage of electronic trading software based on highly specialized and normalized electronic trading platform, contracting parties trade on the electronic contracts of standardized steel products within contract time, the bank supervise the funds transference and the delivery is accomplished by the storage-logistics system. Spot transaction. Lange improved the marketing efficiency, lowered the cost and created values for manufacturers; dealers and customers by using computers and internet technology to connect every stage begin from accomplishment of producing until steel arrived at customers' hands. It includes JingMai sales, invite bidding and shopping.

(2) Profit model

Transaction fees. This includes fees caused by the signing, transfer, and delivery of electronic contracts, etc. Dealers pay the charges according to the trading volume and the charging standard released by market. Financial service charge. Lange steel net cooperates with domestic commercial banks, services for enterprise financing, connects bank capital and enterprise logistics, provides financing service for the general trading enterprise and solves the problems of financing and guarantee.

4.1.2 Case Interpretation

After 2008, lots of speculations occurred in many steel electronic markets, which led to shrinking of the market and investors' quit, most market daily turnover declined to 10 million tons. 95% traders of lange are professional traders without any speculations; price movements are stable and reflect the spot market trends, market management are standardized, however, Lange showed an increase. Until Dec 2009, average daily trading volume has reached 16.8 million tons. Since interim forward transaction started in 2006, its market share increased from 1.1% Dec 2006 to 10% .

4.2 Treasure Island (Beijing) Network Technology Co, Ltd (www.315.com.cn)

Comparing with Lange's strong information services, Treasure island (Beijing) network technology Co, Ltd provides a new trading model named "hard credit" which provides guarantee for credit mechanism of bulk stock EMarketplace.

4.2.1 Case Description

(1) Warehouse receipt trading model

Except for providing products information for dealers, Treasure Island's main trading service feature is warehouse receipt trading model. Sellers stay in the system and input present sales price, the buyer choose what they want. After buyers chose their goods, they transfer money to Treasure Island as commission and monitor the money through trading funds bank supervision system. Then, sellers send goods, logistics change the goods into warehouse receipt and buyer receive the goods.

(2)Profit model

The Treasure Island provides a pyramid services and inverted pyramid profit model: the lower level is the membership fees for providing industry information; the middle level is the commission from trading; the upper level is the services fees for providing all the services mentioned above.

4.2.2 Case Interpretation

Treasure Island's income includes information fees, advertising fees, commissions and financing fees for supply chain. Commissions and financing fees for supply chain occupied 80% of total income. Treasure Island provides steady base for its further development by offering "hard credit". Creation of trading mechanism propels expansion of trading scale. It perfectly explained relevant economic theories and electronic trading platforms theory.

4.3 My Steel Net (www.mysteel.com)

My steel net (short for steel net) is a Chinese vertical industrial B2B e-platform, which have the largest number of visitors of China iron and relative industries' business information, most web site users, and longest site visits time.

4.3.1 Case Description

Steel net was originally based on a third-party trading platform for the steel industry, but in 2000 when the Internet bubble burst, it transferred to the service information. Then, it sat up a research group consisting of professionals, to analysis the price of information which was simply collected and then formed research reports. Its specific business and profit model are as follows:

(1) membership. Steel net members are divided into grand A, grand B, and column the three tapes of membership, membership fee are 5000 Yuan / year, 3,000 Yuan/ year and 1,500 Yuan / year respectively. Besides information services, its e-commerce also takes membership, primarily provides business service of supply, purchase, search, and so on to clients.

(2) "Soso steel". In 2007 steel net uses information service network and its network platform, launched the steel industry search engine - "Soso steel", which then upgraded to an e-commerce network which used the network membership only for sellers, while needed to pay 1500 Yuan per year to become a member and then in order to release their own product information.

(3)The value-added information. Steel net also has incomes from e-commerce, research reports, publications and exhibition and other aspects. Form research and consulting business, steel net charges customers in consulting fees per year. steel net takes the advantages of their members to hold various professional conferences and exhibition using time to time, then to obtain value-added services.

4.3.2 Case Interpretation

Steel net is from the initial fail idea of electronic trading platform to provide information services, and then to be further developed into a comprehensive e-commerce trading platform, its development process reflects our B2B e-commerce platform, in particular, the development of vertical B2B e-trading platform's course and stages. Overall, although the steel net's service charge commissions grows faster, but its relatively simple revenue model, "member + advertising" , continues to occupy more than 80% of its revenue.

5 The Results and Discussion

Based on economic theories and electronic trading platforms theory, considering lange, Treasure Island and my steel net. By analyzing Lange, Treasure Island, and my steel net the three cases, we draw the following conclusions:

(1) The survival and development of vertical industry B2B transaction market must be based on the existing advantage of the market operator For example, steel net and Lange have the advantage of information service, and Treasure Island with an innovative "hard credit" trading system advantages, then the electronic trading market has a certain latecomer advantage. Therefore, in the future, vertical industry B2B transaction market's competition is mainly bound to market segments, and not only to realize the depth of their service development, but also will reflect the diversification of business model and profit model . This also confirms the hypothesis 1.

(2) Vertical industry B2B transaction market follows a developing model which is form information services to transaction services and to financial and other value-added services. For example, steel net and Lange develop from information to transactions, and Treasure Island develops from transaction to information and financial services, both ends of the extension. If the first-generation electronic market mainly based on information services or transaction services, then the second-generation electronic market will emphasize the gradual integration of e-commerce applications, and it will gradually integrate information flow, logistics and cash flow into a service platform. This supports prosthesis 2.

Therefore, integration resources of raw materials electronic trading market are the mature development stage of B2B e-market vertical industry, which can provide customers a full range of e-commerce solutions and integrate resource in every aspect of trade. Integration resources of raw materials electronic trading market is the future tendency of vertical industry B2B.

References

1. Hadaya, P.: Determinants of the Future Level of Use of Electronic Marketplaces: The Case of Canadian Firms. Electronic Commerce Research, 173–185 (2006)
2. Hazra, J., Mahadevan, B., Seshadri, S.: Capacity Allocation among Multiple Suppliers in an Electronic Market. Production and Operations Management, 161–170 (2004)
3. Coase, R.H.: The Nature of Firm Economic (1937)

4. Stigler, G.J.: The Economics of Information. Journal of Political Economy (1961)
5. Porter, M.E.: Competitive Advantage: Creating and Sustaining Superior Performance, New York (1985)
6. Hoh, B., Eriksson, K., Johansson, J.: Creating Value Through Mutual Commitment to Business Network Relationships. Strategic Management Journal 467–486 (1999)
7. Shi, X., Feng, G.: Recognition and research of main risks for bulk stock EMarketplace, management discuss (2010)
8. Liu, X., et al.: Competition and complementation: analysis of relationships between bulk stock market and forward market—empirical analysis based on evolution of sugar spots, forward and forward price, Beijing wholesale forum, memoir (2009)
9. Iresearch consulting company, report of Chinese B2B electronic business development 2008-2009 (2010)
10. Analysis International, series reports on chines vertical B2B market 2008—raw material (2009)
11. Yin, R.K.: Applications of Case Study Research. Sage, Beverly Hills (2003)

Personalized Recommendation System on Massive Content Processing Using Improved MFNN

Jianyang Li[1,2] and Xiaoping Liu[1]

[1] School of Computer and Information, Hefei University of Technology, 230009 Hefei, China

[2] Department of Information Engineering, Anhui Communications Technical College, 230051 Hefei, China

lijianyang@sina.com, xiaoping12@hotmail.com

Abstract. Though the research in personalized recommendation systems has become widespread for recent years, IEEE Internet Computing points out that current system can not meet the real large-scale e-commerce demands, and has some weakness such as low precision and slow reaction. We have proposed a structure of personalized recommendation system based on case intelligence, which originates from human experience learning, and can facilitate to integrate various artificial intelligence components. Addressing on user case retrieval problem, the paper uses constructive and understandable multi-layer feed-forward neural networks (MFNN), and employs covering algorithm to decrease the complexity of ANN algorithm. Testing from the two different domains, our experimental results indicate that the integrated method is feasible for the processing of vast and high dimensional data, and can improve the recommendation quality and support the users effectively. The paper finally signifies that the better performance mainly comes from the reliable constructing MFNN

Keywords: MFNN, recommendation system, massive content, case intelligence, retrieval efficiency.

1 Introduction

Websites have shown amazing growth in geometric progression and become an important means to obtain information [1].How to perform such efficient access to the information they need is difficult issues they must face, therefore various network applications are emerging in today's society to deal with so many problems of which the search engine is the most common tool for supporting people search, such as Yahoo, Google, and Baidu etc. Intelligent web tools have been used to help users search, locate and manage web documents. High level of web intelligent tools to generate user model learning, and can use the model of reasoning.

The biggest challenge most web sites face is lack of content, people are obliged in searching for their need information from the huge data-ocean. The statistics report from ACM symposium for e-commerce [2], indicates that the current e-commerce

F.L. Wang et al. (Eds.): WISM 2012, LNCS 7529, pp. 183–190, 2012.

recommendation system can not meet the large-scale e-commerce applications, and it has poor real-time problem accompany with the problem of weak quality in accuracy, e-commerce technologies and recommendation systems have gained a relatively big progress [3], but how to acquire the information in time people need from the mass merchandise has become increasingly difficult.

Reviewing the development of personalized recommendation system may be helpful to understand these problems [4]. Collaborative filtering is the most successful technology in recommendation systems at present, and there are many large sites using this technology for the user to provide more intelligent content from the entire world [5]. Both the User-based Collaborative Filtering and Item-based Collaborative Filtering are based on this basic principle: the relevance of user behavior and choice. The user's behavior choices here refers to download, purchase, evaluation, and so can explicitly or implicitly reflect the behavior of user preferences.

Many Artificial Intelligence tools are developed to construct recommendation system [6,7]. For example, the system uses data mining and other artificial intelligence technology, can analyze the collected data, obtain the behavior in e-commerce, and generate interests in the products for consumers to lead the consumer's purchase. Thus, how to use data effectively to improve the quality of recommendation and the performance of e-commerce recommendation system are central and difficult problems of the recommendation system.

In particular, the key problem is that the processing of massive user data [8]. Because of the recommended system is a data priority, the more accumulation of data, and the higher accuracy of recommendation. While the accumulation of the real user behavior data up to millions or even billions, how to suggest a valid recommend within a reasonable time is the greatest technical challenge for the recommendation system [9]. As we know that the user's behavior data must be changed in the websites, so the user data should be dynamic, that is exactly the same with case intelligence as we have described in the paper[10,11]. Recommendation system uses low level analogy reasoning, which is an important cognitive model of human sense. The "low level" means simple analogy, which is not inferring in different domain data. Focusing on such problems, the paper suggests a MFNN model in case-intelligent recommender to deal with user's data bases, which are huge and dynamic.

2 MFNN and System Model

Artificial neural network (ANN) is massively parallel distributed neurons which are connected for the calculation components, and it has the natural relationship with CBR. Recently, several successful theories have been put forward to integrate ANN into the CBR system and covering all the application aspects of it with the ANN components.

2.1 MFNN

Learning ability is one of the most important and most impressive features of ANN [12]. With the development of ANN, learning algorithm of ANN has a very important position. Each ANN model proposed are corresponding with its learning algorithms

[13], though some models can have a variety of algorithms, and some algorithms may be used for a variety of models. Current ANN model is widely used with the numerous algorithms.

In MFNN, each neuron in every layers is a processor which is used to complete a simple information processing; a set of weights are used to connect every layer, each connection weights are used to store a certain Information, also providing information channel. It has proved that 3-layer networks can realize any given function for approximate accuracy, thus it can be used to solve the nonlinear classification problem.

MFNN can be a digraph as input a vector x, then through the networks to get an output vector y, it can be used as a feedforward networks to complete the mapping x to y of a converter. MFNN has many models and algorithms, such as back-propagation (BP) network, radial basis function network (RBF), simulated annealing algorithm and their ameliorated algorithms.

MFNN has a lot of various improved algorithms, and has achieved many significant results as we have seen. After investigating the behavior of MFNN for case retrieval, the weaknesses such as having lower speed and local extreme value, are inherent in those algorithms, thus cannot be conquered thoroughly. For example, RBF recognizes and classifies samples dependence on their non-linear distance through projecting them to a multi dimensional space. Although RBF is a good similarity detector, it can hardly deal with huge user data in a reasonable time.

However, the current algorithm can not achieve satisfactory level. All of these changes are the networks become more complex (or performance function is more complex, such as linear becomes nonlinear, etc.; or structures become more complex), hopes to increase the complexity of the network to improve network Learning to improve learning speed. But so many changes cannot give the network design problems before finding a satisfactory outcome.

2.2 Multi-Layer Covering Algorithms

Multi-Layer covering algorithm is a constructive method for ANN, which is based on the geometrical representation McCulloch-Pitts neural model to build a three-layer model of weight and threshold neural networks [14]. This belongs to MFNN. The networks construct its layer-structure by means of input training data for itself. Firstly, Covering Algorithm assumes that each input vectors x of an n-dimension can be projected on a bounded set of a certain hyper-sphere S^n of a (n+l)-dimensional space (define the sphere radii is R), It is no doubt that the transformation must be achieved to the aim through widening the vector's dimension.

By this way of dimension expansion and space projection, the Domain Covering for the similar users in the user library will be well achieved and together with the selective information filtering. Besides, the acquired Covering Domains can be used as the input of the MFNN for case matching and the retrieval efficiency will been significantly enhanced.

The design stage of neural networks can be described as follows: each weight, threshold marked as $W(i), \theta(i)(i=1,2,3)$.

The first layer: assume total amounts of P neurons $A^1, A^2, \cdots, A^p, A^i$ represents the neuron of covering $C(i)$. $W(1) = (a^1), \theta(1) = (\theta_1)$

The second layer: select the same amounts of neurons as the first step, they are $B^1, B^2, \cdots, B^p$. $W_j^i(2) = \begin{cases} -1, & j < i \\ 1, & j = i \\ 0, & j > i \end{cases}, \theta_i(2) = i - 1, \begin{matrix}(i = 1,2,\cdots,p) \\ (j = 1,2,\cdots,p)\end{matrix}$.

The last layer: select total amounts of T neurons $C^1, C^2, \cdots, C^T$, T is the total classification type of samples. $W_j^i(3) = \begin{cases} 1, (j-i) \bmod T = 0 \\ 0, others \end{cases}, \theta_i(3) = i - 1,$

$$(i = 1, 2, \cdots, T), (j = 1, 2, \cdots, p)$$

The network adds a hidden layer to make the neuron weights increase linearly and decrease the complexity of the algorithm, in contrast, the original three-layer neural networks makes the neuron weights increase with index.

The personalized recommendation system that we proposed based on case intelligence mainly construct by three parts: input module, recommendation methods and output module, as shown in paper [5], and the MFNN module as we have described before is added as retrieve process.

The details of the components in our personalized recommendation system can be also seen in paper [5], which is not described. We care only about the MFNN component prepared for case matching-the similar user group, and the implementation process of personalized recommendation for the same with common recommendation system also not mentioned.

3 Experiments and Review

The next two experiments are designed for validating our system from the two directions, the one is from real world data, and the other is simulation for huge data.

3.1 Validation

The experimental data is downloaded from DELL technical support center, the total is 1,800 Emails from the service forum. Each Email is treated as a user-case, and we adapt the existing classification in our experiment as criteria from the forum, that is 12 classes according to their hardware involving:

Firstly, each Email is pretreated for stop-word processing and low-frequency-word processing, removing stop-words and low-frequency-words including in title and content, 3,420 words are extracted as the characteristic attributes through statistics to the training sets.

Secondly, each Email is expressed to text-feature vector, and becomes original user-case.

Finally, calculating the distinguished degrees of the word $Di(w)$, as the follows:

$$Di(w) = \begin{cases} \dfrac{Ni}{\sum\limits_{j \neq i} Nj} & \sum\limits_{j \neq i} Nj \neq 0 \\ 2 * Ni & = 0 \end{cases}$$

The largest 50 words sorted by $Di(w)$ are extracted as feature attributes for class i, and each user-case are projected in feature attribute. Macro-averaging is used to calculate all classes' means $F-score$:

$$F-score = \frac{2 * recall * precision}{recall + precision}$$

Table 1. Experimental results in web-text

Item	Aaverage F-score(%)	Aaverage retrieval Time(s)
KNN	62.3	42
OURS	86.3	0.3

Compared with classic KNN algorithm, our MFNN for case retrieval performs perfect, as it is shown in table 1.

3.2 Experiment in Vast Data

The first experiment shows the improved MFNN and its algorithm is reliable for its recommendation precision also with outlook speed, thus we decide to design the subsequent experiment to validate it for large-scale data. The data of "forest cover type" is downloaded from UCI repository and its main information describes as follow:

The actual forest cover type for a given observation (30 x 30 meter cell) is determined from US Forest Service (USFS) Region 2 Resource Information System (RIS) data. Independent variables were derived from data originally obtained from US Geological Survey (USGS) and USFS data. Data is in raw form (not scaled) and contains binary (0 or 1) columns of data for qualitative independent variables (wilderness areas and soil types). Number of instances (observations) 581012; Number of Attribute: 54 (12 measures, 10 quantitative variables, 4 binary wilderness areas and 40 binary soil type variables); Number of Class: 7; Missing Attribute: none.

We can see that the forest data are spare matrix with huge records. In the experiment, each record is regarded as a user case, to decrease the interference in the pretreatment of the real world. The simulation starts from 2,000 to 100,000 user cases, and the user cases are randomly selected. On account of the huge along with the high dimensional data, the KNN algorithm is shut down for directly dealing with the massive records, the results are only recorded for our MFNN.

In order to simulate the real E-commerce demands, we design the various amounts of samples, which are ranged from 2,000 to 100,000, and adapt 10-fold cross-validation methods to decrease the inferring of samples. Some are omitted in table 2 because of the limitation of the table length, but they are both described in fig 1.

Table 2. Main experimental results in Forest Covertype

Records	5000	10000	20000	40000	50000	100000
F-score(%)	63.1	79.2	82.9	82.4	81.6	83.2
Time(s)	9.8	40.7	125.7	259.5	319.8	698.3

From the table 2, we can find out that the F-score in each group is gradually increasing, which is exactly according with human's understanding of the real world. For example, while the amount of user cases is 2,000, the F-score is 68.3%; when the amounts of user cases reaches 100,000, the F-score is 82.9; when it reaches 100,000, F-score becomes 83.2%.

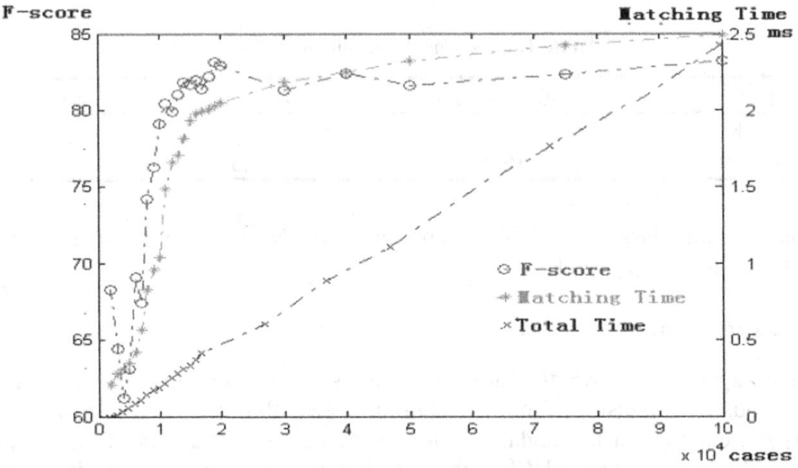

Fig. 1. The improved system performance

As the Fig 1 indicates, we can find that the running time is unacceptable, if we only mention the total time for this system, such in 10,000 user cases, the system needs to run 40.7s, but 698.3s (more than 11 minutes)in 100,000. In our MFNN model, the retrieval time includes 2 stages, the first spends on training the user cases, and the second spends on recommending the proper case. The first costs too many time (which can be run in the backstage), and few time in the second. Thus, the actual time user waiting for is from 0.21ms to 2.49ms corresponding with the training set. For another words, gaining the similar user case costs only 0.2ms to 2.5ms according with the total records from 2,000 to 100,000.

3.3 Analysis and Outlook

Many researches have found that learning from the huge amounts of data is more difficult while the amounts increasing, thus many papers validate their methods only in a limited range, as we do in the first experiment. From the second experiment, we know that user's behavior is gathered into a library, and can be trained for pretreated step. This means the retrieval time the user costs is too few, when a person use the recommendation system, the system pulls out the recommendation in time, as if it suggests online result.

Any algorithm shall engage with certain characters of the neural network, which is selected as one of the purpose and should be taken care of. After investigating the current artificial neural networks together with their learning methods, and mastering the learning course from overall viewpoint, covering algorithm is constructive neural networks. Our improved algorithm can construct the networks based on knowledge of all the samples given, including the design of the networks' structure and all the parameters.

It is a significant leap to study constructive neural networks, which is capable of large-scale problem solving. The study begins with the MP neuron model from the view of geometry, and traces into covering algorithm from forward propagation algorithm (compared with BP), the former algorithm of constructive neural networks. Because the weights and thresholds of each neuron is computing by the "distance" between the samples selected optimization, they are able to ensure that the network executes under "optimal" fault tolerance in some sense.

Suppose that n represents the scale of the samples, m for the dimension of the samples, and then the complexity of the covering algorithm is $O(n^2m)$: the maximum of total calculation is n^2m, if each covering circle covers k records, then the total calculation is n^2m/k.

No doubt local minimum does not exist in the algorithm, and the algorithm execution does not require multiple scans of the samples or repeated calculation iterations, thus it has the advantages of short learning time. Just like the two experiments shown that the improved method can perform well in various conditions. The success rate is increased along with the growing of the user library, though some F-scores are not very satisfied, which caused by the limitation of amount of user cases. As we know that a fewer user cases may lead to the generalized ability problem, and our future work is about how to reduce the difference in our experiment.

4 Conclusion

How to deal with the problems of the huge data and weak reaction in personalized recommendation system while it is running in real internet conditions is different question to researches. User behaviors are important for recommendation system, which are stored in the user data library; they are increasing while the recommender is running. Huge amounts of user behaviors with high dimensions are hardly dealt with in common ANN algorithms, moreover they are dynamic. So the paper proposes the improved method based on Multi-layer Feedforward Neural Network to solve these problems. The sequent investigations indicate that it has clear system structure, feasible component combination, easy integration and construction. Our experimental results suggests that it is suitable for the large-scale and high dimensional data processing, can overcome the problems of the excessive iteration as well as the low performing rate, so the MFNN approaches' practical application in the personalized recommendation system has been guaranteed.

Acknowledgment. This work is supported by the China Postdoctoral Fund of HFUT, and the National Science Project of Anhui province under grants KJ2011B054.

References

1. Yu, L., Liu, L.: Comparison and Analysis on E-Commence Recommendation Method in China. Systems Engineering & Practise 24, 96–101 (2004)
2. Al-Kofahi, K., Jackson, P., et al.: A Document Recommendation System Blending Retrieval and Categorization Technologies. In: AAAI Workshop, pp. 9–18 (2007)
3. Ben Schafer, J., Konstan, J., Riedl, J.: Recommender Systems in E-Commerce. In: E-COMMERCR 1999. ACM, Denver (1999)
4. Stormer, H.: Improving E-Commerce Recommender Systems by the Identification of Seasonal Products. In: AAAI Workshop, pp. 92–99 (2007)
5. Li, J., Li, R., Zheng, J., Zeng, Z.: A Personalized Recommendation System Based on Case Intelligence. In: International Conference on Educational and Information Technology (ICEIT 2010), pp. 162–166 (2010)
6. Kim, M.W., Kim, E.-J., Ryu, J.W.: A Collaborative Recommendation Based on Neural Networks. In: Lee, Y., Li, J., Whang, K.-Y., Lee, D. (eds.) DASFAA 2004. LNCS, vol. 2973, pp. 425–430. Springer, Heidelberg (2004)
7. Chu, W., Park, S.-T.: Personalized Recommendation on Dynamic Content Using Predictive Bilinear Models. In: 18th International World Wide Web Conference (WWW 2009), pp. 691–700 (2009)
8. Zhao, P.: Research on Complex Networks & Personalized Information Service on Internet. The University of Science and Technology, Hefei (2006)
9. Das, A., Datar, M., Garg, A., Rajaram, S.: Google news personalization: scalable online collaborative filtering. In: Proceedings of the International World Wide Web Conference (2007)
10. Li, J., Ni, Z., et al.: Case-based Reasonor Based on Multi-Layered Feedforward Neural Network. Computer Engineering 7, 188–190 (2006)
11. Ni, Z., Li, J., et al.: Survey of Case Decsion Techniques and Case Decision Support System. Computer Science 36, 18–23 (2009)
12. Patterson, D.W., Galushka, M., Rooney, N.: Efficient Real Time Maintenance of Retrieval Knowledge in Case-Based Reasoning. In: Ashley, K.D., Bridge, D.G. (eds.) ICCBR 2003. LNCS, vol. 2689, pp. 407–421. Springer, Heidelberg (2003)
13. Zhang, L., Zhang, B.: A Forward propagation learning algorithm of multilayered neural networks with feedback connections. Journal of Software 8, 252–258 (1997)
14. Gu, Y.-S., et al.: Case-base Maintenance Based on Representative Selection for 1-NN Algorithm. In: Proceedings of the Second ICMLC, pp. 2421–2425 (2003)

A Speaker Recognition Based Approach for Identifying Voice Spammer

Fei Wang, KeXing Yan, and Min Feng

Department of Electronics and Information Engineering, Huazhong University of Science and Technology, 430074 Wuhan, China
wangfei@hust.edu.cn, {yankexing118,sarah0304}@163.com

Abstract. With the blooming of personal communication, more and more people have to process irksome spam calls. However, conventional approaches can hardly gain satisfactory effect when spammers change their phone number or ID randomly. In this paper, we proposed a text-independent speaker recognition algorithm to identify spammer for large population scenario. Firstly, we discover that the codebooks from same person's different speeches are similar. Secondly, we design a vector quantization based CodeBook Filtering algorithm. The experiment results show that our algorithm could achieve a speed-up factor of 30.8:1 with an identification accuracy of 94.98%.

Keywords: Voice spam, speaker identification, vector quantization (VQ), text-independent.

1 Introduction

Recent years, personal communication continues to blossom in an exciting way. However, when it brings us convenience, it also provides simpler way to spammer to broadcast voice spam. Refer to our investigation, there are about 95.1% of the people had received spam calls and nearly 50% of the people would receive at least one spam call each day. It's believed that voice spam presents blooming tendency.

To against voice spam, there are a lot of approaches like VSD[1], PGM[2], SEAL[3], COD[4], Tetsuya[5], Bai[6], etc. have been developed. Nevertheless, all of these techniques are constructed over an assumption that the communication system could uniquely identify each end user. Hence, they can hardly gain satisfactory effect when spammers change their phone number or ID constantly. To avoid such disaster, we provide a novel approach to identify spammer through text-independent speaker identification in this paper.

According to the previous studying of voice spam [7], it is real-time and its call length is shorter, generally. Further, maybe there are a large number of spammers. Against voice spam by text-independent speaker identification, there are still some problems need to resolve: 1) use shorter speech to train and test; 2) gain higher speed-up factor in large scale population.

Text-independent speaker identification is a traditional hot topic. At premier stage, researchers paid more attention to decrease the error rate [8-11]. However, they could

F.L. Wang et al. (Eds.): WISM 2012, LNCS 7529, pp. 191–198, 2012.

hardly be applied directly more or less because these methods would consume unbearable computational resources. In the next stage, many methods have been proposed for speeding up the identification process. Generally, there are THREE main approaches:

1) *Simpler likelihood calculation*: The main idea is to reduce the complexity of each likelihood calculation. McLaughlin [12] improved the time complexity through removing the redundancy of GMM model and compressing the model size. Pellom [13] reorder the feature vectors so that they could prune the worst scoring speakers using a *beam search* technique. Auckenthaler [14] proposed GMM hash model, in which there is a shortlist of indices for each Gaussian component to speed up identification. The larger shortlist wins higher accuracy, and smaller shortlist receives higher speed-up factor.

2) *Hierarchical model*: This method provides hierarchical speaker model to accelerate computation. Beigi [15] proposes a hierarchical structuring of the speaker by merging closest GMMs. When identifying, it uses KNN-liked algorithm to search K best speakers and then compares detailed in them. Sun [16] using the ISODATA algorithm to cluster all speakers. When identification, the first step is to determine the speaker belong to which group, and then do conventional method to complete identification. Pan [17] provides another hierarchical method which computes two models in different size. They pruned out worst candidates by smaller CodeBook and do fine matching by larger CodeBook.

3) *Speaker pruning*: Kinnunen [18] proposes Static Pruning, Hierarchical Pruning, Adaptive Pruning and Confidence-Based Pruning to prune out K worst speakers. Aronowitz [19] provides a Top-N pruning exploits the property of GMM using approximated cross entropy (ACE).

In this paper, we focus on designing an efficient algorithm to index the special speaker. Think about GMM based algorithms need more speech data to train and test, in this paper, we study VQ based speaker recognition and provide follows contribution: 1) We discover models from different speech of same speaker are closed enough even speech length is very short. Hence, we can avoid large computation of the distance between the unknown speaker's feature vectors and the models of the speakers stored in the database. 2) We provide a codebook model based multistage filtering algorithm to reduce the scale of candidates within a large number of spammers. 3) We design a ratio nearest filtering algorithm to adapt each speaker.

This paper is organized as follows. In Section 2, we describe the original idea and presents detailed description of the approach. The experimental setup and test results are given in section 3. Finally, we draw a conclusion in section 4.

2 CodeBook Filtering

2.1 Basic Idea

According to the conventional VQ based speaker identification, the core elemetns are MFCC [20] or LPCC [21] feature vaectors, CB (CodeBook), and distance metric.

Currently, lots of research uses distance to denote the distortion between known speakers' CB C={c_1, ..., c_M} and unknown speaker's feature vectors X={x_1, ..., x_T} which is computed as the average quantization distortion [8]

$$D_{avg}(X,C)=\frac{1}{T}\sum_{i=1}^{T} min_{c_j\in C}\left\|x_i-c_j\right\|^2 \tag{1}$$

where M is dimension of CB. This method causes higher time complexity.

Propositions 1 The codebooks from same person's different speeches are similar.

No matter which piece of a speaker's speech is selected as training speech, it always gains minimal distortion between the CB and feature vectors. Therefore, we believe that the CBs which from a speaker's different speech are closed enough in distance. That is to say, we can compute the distance between two CBs to judge whether they are from the same person. However, it's difficult to sort vectors in a CB so that we cannot compute the distance between two CBs simply row by row. It seems that we have to evaluate the distortion of two CBs in conventional method. While, it's possible to simplify such evaluation because the distance bwtween two vectors is determined by vectors' *Euclid* norm and its' included angle.

Fig.1 shows that Euclid norms of 50 same speaker's different CBs are very closed. Refer to above analysis, we propose a Euclid norm based speaker filtering algorithm. Further, we utilize the nearest neighbor search method about high dimensional vector to obtain fewer candidate speakers with smaller distance which computed by (1). The block diagram of our approach is shown in Fig. 2.

The following notations will be used.

C_i	CB of speaker *i*
C_t	CB of test speech
N	size of the speaker database
CD	indices of the candidate speakers
R_N	ratio of Norm Filtering
R_D	ratio of Distance Filtering
CN_i	CB norm of speaker *i*

2.2 Norm Filtering (NF)

The idea is to filter most unlikely speakers by CB norm in a simple way. Think about maybe there are a large number of speakers in DB, we create a T-Tree for speakers' CB norm. R_N is the dominated parameter of NF, and it determines the accuracy and speed-up factor of NF. To protect accuracy of identification, we initialize R_N refer to the distribution of all speakers' CB norm.

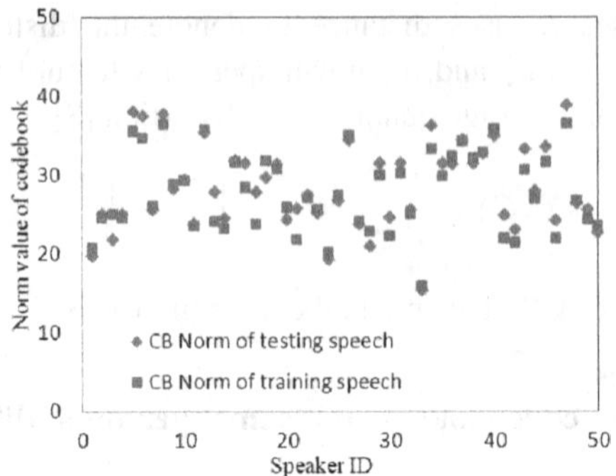

Fig. 1. Norm of training and testing speeches' codebook

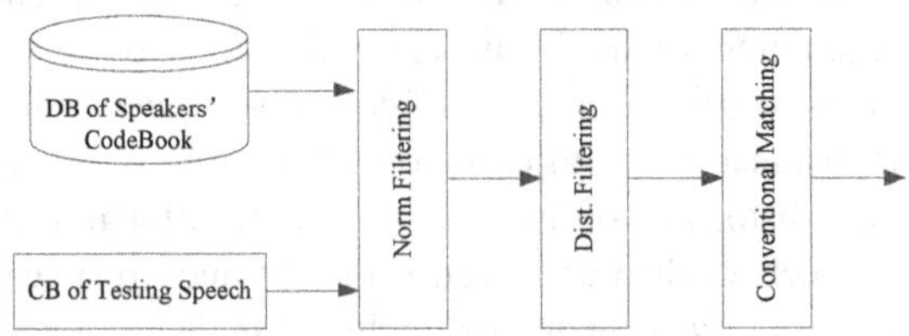

Fig. 2. Block diagram of codebook filtering

Algorithm 1. Norm Filtering (NF)

```
Initialization:
    CD := {1, 2, …, N};
    Initialize R_N;
    Calculate CN_i for all C_i and create T-Tree for all
CN_i;
Filtering:
    Calculate C_t according to test speech and it's CN_t;
    Search candidate speakers by T-Tree with the scope of
[CN_t *(1-R_N), CN_t *(1+R_N)];
    update CD according to search result;
Decision:
    target := min_i{D(X, C_i)|i•CD}.
```

2.3 Distance Filtering (DF)

Besides the norms of CBs from same speaker's different speech are closed, their average included angle is very small. Think about the time complexity of included angle and distance between 2 vectors are similar, we use its distance to do further filtering. Based the result of NF, we make further efforts to filter unlikely speaker by average distance between two CBs. R_D is also the control parameter of *DF* like R_N.

Algorithm 2. Distance Filtering (DF)

```
Initialization:
    CD := {1, 2, …, N};
    Initialize R_D;
Filtering:
    Calculate C_t according to test speech;
    Compute D_avg(C_t, C_i) for all i•CD, and min D_avg;
    FOR all i•CD
        IF D_avg(C_t, C_i) > R_D*min D_avg
            Filter out speaker i;
        END;
    END;
Decision:
    target := min_i{D(X, C_i)|i•CD}.
```

3 Experiments and Results

3.1 Experiment Environment

1) Speech Database

Our speech database is a collection of address from 499 speakers. The speech is recorded from the Internet video which mostly are the Chinese CCTV's "100 forum". Each person records nine pieces of speech from the same address, one speech is about 12s (silence removed) for training and others are about 1s~8s (silence removed) for testing.

2) Feature Modeling

We use the standard MFCC as the feature. Before computing MFCC vectors, a speech need pass a pre-emphasize filter, enframe and multiply with Hamming window. The 0th cepstral coefficient is abort and the codebook is generated by the LBG clustering algorithm [22].

We report two performance results: *accuracy of identification* and *speed-up factor*. The average identification time counts speech processing, MFCC, LBG clustering, filtering algorithm and decision among all speakers.

Table 1. Parameters of feature model

Parameter	Value
re-sampling	8 kHz, 16bit
pre-emphasis	0.9375
length of frame	32 ms
shift of frame	16 ms
window	Hamming
triangular filter	25
MFCC dimension	13MFCC + 13△MFCC
size of codebook	32

3) Experiment Environment

Our experiments are based on a simulation testbed we have developed. MFCC is computed by Matlab VoiceBox[23], and LBG and filtering algorithm are developed by C++ on Visual Studio 2008 express. All experiments were carried out on a computer of DELL Dimension E521 which has a 2.0GHz CPU and 2.0GB memory. The operating system is Windows XP sp3. Each experiment was run in single thread.

3.2 Test Results

The speaker identification accuracy for the baseline system was found to be 96.8%. Refer to the result from Fig. 3 (a), we can find that the accuracy of identification does not decline significantly when CodeBook Filtering has been implemented. At the best result, the ID rate of our approach still achieves 94.98%. Fig. 3 (a) also shows that our method can obtain satisfactory EER even if the test speech is 3~5s.

Fig. 3 (b) presents the relationship of accuracy and speed-up factor of our proposal when test speech is 5~8s. According to the result, at the best accuracy, speed-up factor is about 10~20. On the contrary, speed-up factor can reach 65 as the highest value. The accuracy still remains 55.9%. It shows that CodeBook Filtering can work well and be competent at voice spam filtering.

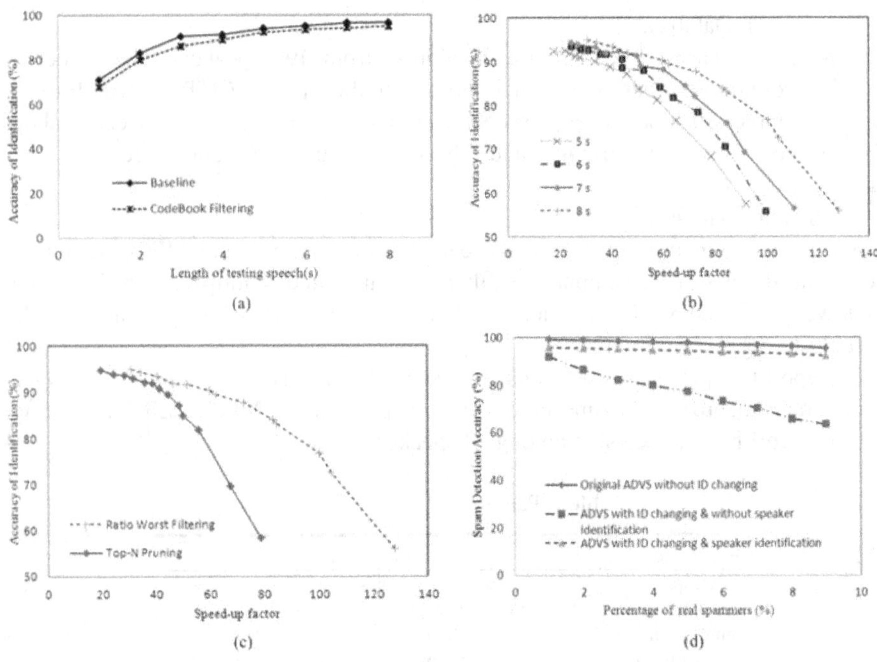

Fig. 3. (a)Accuracy of baseline system versus CodeBook Filtering's; (b) speed-up factor versus accuracy; (c) Ratio worst versus Top-N using testing speech when the length of test speech is 8s. (d)Accuracy of spam detection with vary percentage of real spammers when test speech is 5s length.

By contrast to Top-N pruning which implemented in Kinnunen [20] and Aronowitz [21], we provide ratio worst filtering. The contrast results contained in Fig. 3 (c) show that ratio worst filtering is more suitable for CodeBook Filtering because R_N and R_A are adapted to application environment.

Based the former research about spam filtering, we applied the algorithm into ADVS [24] and examined the effect of CodeBook Filtering. In this experiment, the basic testbed is referred to ADVS. We think about total number of real users is 499 which contained in speaker DB. We assume that each real spammer owns TWENTY different IDs or phone numbers. Furthermore, spammers would change their ID or phone number randomly. In addition, we think about all users (including spammer) are contained in a same domain. The rest parameters are referred to ADVS.

According to the results contained in Fig. 3 (d), the accuracy of spam detection will decrease sharply when the percentage of real spammers grows and the spammers change their ID or phone number randomly. However, the filter could work normally when our algorithm applied. Of course, the accuracy of spam detection meets a few decreasing because the accuracy of speaker identification cannot reach 100%.

4 Conclusion

To assure that ADVS could detect voice spam accuracy even if spammers own several IDs or phone numbers, we have presented an algorithm for efficient and accurate speaker recognition, in this paper. The algorithm is based on VQ and the basic theory that the codebooks from same person's different speeches are similar. By using our corpus, a speed-up factor of 30.8:1 was achieved with little degradation in the identification accuracy and absolute identification time is about 1.81s. In addition, we applied our schema within ADVS and the simulation results show that it could help ADVS to recognize spammer who owns multi dummy identities. The experiment results prove that our approach could be applied for large population scenario.

Acknowledgment. This work was supported by the National Natural Science Foundation of China (NSFC) under grant No. 61001070.

References

1. Dantu, R., Kolan, P.: Detecting Spam in VoIP Networks. In: USENIX. SRUTI 2005 Workshop, Cambridge, MA, pp. 31–37 (July 2005)
2. Shin, D., Shim, C.: Voice Spam Control with Gray Leveling. In: Proc. of the 2nd VoIP Security Workshop, Washington, DC (June 2005)
3. Quittek, J., Niccolini, S., Tartarelli, S., et al.: Detecting SPIT Calls by Checking Human Communication Patterns. In: Proc. of ICC, pp. 1979–1984. IEEE (2007)
4. Zhang, R., Gurtov, A.: Collaborative Reputation-based Voice Spam Filtering. In: Proc. of the 20th International Workshop on Database and Expert Systems Application. IEEE (2009)
5. Kusumoto, T., Chen, E.Y., Itoh, M.: Using Call Patterns to Detect Unwanted Communication Callers. In: Proceedings of the 9th Annual International Symposium on Applications and the Internet. IEEE (2009)

6. Bai, Y., Su, X., Bhargava, B.: Adaptive Voice Spam Control with User Behavior Analysis. In: Proceedings of 11th International Conference on High Performance Computing and Communications. IEEE (2009)
7. Wang, F., Mo, Y.J., Huang, B.X.: P2P-AVS: P2P based cooperative VoIP spam filtering. In: Proceedings of WCNC, pp. 3547–3552. IEEE (2007)
8. Soong, F.K., Rosenberg, A.E., Juang, B.-H., Rabiner, L.R.: A vector quantization approach to speaker recognition. AT&T Tech. J. 66(2), 14–26 (1987)
9. Reynolds, D.A., Rose, R.C.: Robust text-independent speaker identification using Gaussian mixture speaker models. IEEE Trans. Speech Audio Process. 3(1), 72–83 (1995)
10. Aronowitz, H., Burshtein, D., Amir, A.: Text independent speaker recognition using speaker dependent word spotting. In: Proc. Inter-speech, pp. 1789–1792 (2004)
11. Campbell, W.M., Campbell, J.P., Reynolds, D.A., Singer, E., Torres-Carrasquillo, P.A.: Support vector machines for speaker and language recognition. Comput. Speech Lang. 20(2-3), 210–229 (2006)
12. McLaughlin, J., Reynolds, D.A., Gleason, T.: A study of computation speed-ups of the GMM-UBM speaker recognition system. In: Proc. 6th European Conf. Speech Communication and Technology, Budapest, Hungary, pp. 1215–1218 (1999)
13. Pellom, B.L., Hansen, J.H.L.: An efficient scoring algorithm for gaussian mixture model based speaker identification. IEEE Signal Process. Letter 5(11), 281–284 (1998)
14. Auckenthaler, R., Mason, J.S.: Gaussian selection applied to text-in-dependent speaker verification. In: Proc. Speaker Odyssey: The Speaker Recognition Workshop, Crete, Greece, pp. 83–88 (2001)
15. Beigi, H.S.M., Maes, S.H., Sorensen, J.S., Chaudhari, U.V.: A hierarchical approach to large-scale speaker recognition. In: Proc. 6th Eur. Conf. Speech Communication and Technology, Budapest, Hungary, pp. 2203–2206 (1999)
16. Sun, B., Liu, W., Zhong, Q.: Hierarchical speaker identification using speaker clustering. In: Proc. Int. Conf. Natural Language Processing and Knowledge Engineering, Beijing, China, pp. 299–304 (2003)
17. Pan, Z., Kotani, K., Ohmi, T.: An on-line hierarchical method of speaker identification for large population. In: Proc. NORSIG 2000, Kolmården, Sweden (2000)
18. Kinnunen, T., Karpov, E., Fränti, P.: Real-Time Speaker Identification and Verification. IEEE Trans. Speech Audio Process. 14(1), 277–288 (2006)
19. Aronowitz, H., Burshtein, D.: Efficient Speaker Recognition Using Approximated Cross Entropy (ACE). IEEE Trans. Speech Audio Process. 15(7), 2033–2043 (2007)
20. Deller Jr., J.R., Hansen, J.H.L., Proakis, J.G.: Discrete-Time Processing of Speech Signals, 2nd edn. IEEE Press, New York (2000)
21. Huang, X., Acero, A., Hon, H.-W.: Spoken Language Processing: A Guide to Theory, Algorithm, and System Development. Prentice-Hall, Englewood Cliffs (2001)
22. Linde, Y., Buzo, A., Gray, R.M.: An algorithm for vector quantizer design. IEEE Trans. Commun. 28(1), 84–95 (1980)
23. http://www.ee.ic.ac.uk/hp/staff/dmb/voicebox/doc/voicebox/index.html
24. Wang, F., Wang, F.R., Huang, B.X., Yang, L.T.: ADVS: a reputation-based model on filtering SPIT over P2P-VoIP networks. Journal of Supercomputing, http://www.springerlink.com/content/19726tkkk76305t6/

Security Access Authentication System for IPv4/IPv6 Dual-Stack Campus Network Based on IpoE

Jinshan Zhu, Jiankang Guo, and Yueguang Ruan

Ningbo Institute of Technology, Zhejiang University,
315100 Ningbo, P.R. China
zjs@nit.net.cn

Abstract. By the analysis of network security access authentication technology in universities, this paper proposes the solution to security access authentication based on IPoE in the face of the reality of the outrunning of IPv4 addresses and general popularizing of IPv6 in universities. Having been implemented in Ningbo Institute of Technology, Zhejiang University, this scheme can meet the security access requirements for IPv4 /IPv6 dual-stack campus network of highly-dense users. Finally it discusses some extended application of the scheme in the practical situations.

Keywords: IPv6, CNGI, 802.1x, PPPoE, IPoE, DHCPv6.

1 Introduction

Since it was developed, Internet has been completely blended into people's life, work and study. Educational informatization in universities cannot do without the support of the Internet. Along with the outrunning of the IPv4 addresses, many Chinese universities have used IPv6 addresses and have one after another switched onto Cernet2, the demonstrating network of CNGI. On the other hand, the government requires more for the protection level of information security recently, which demands more for the network security for universities since its main users are the most creative undergraduates.

2 Traditional Security Access Authentication Technology and Its Problems

The network security access authentication for universities at present are mainly the following three: PPPoE, Web + Portal and 802.1 x. PPPoE (Point to Point Protocol over Ethernet) appeared early with its distinctive applications. Its technical standards, RFC2516, were defined in 1999. Since then it was widely used by telecom operators, as the currently popular broadband access way ADSL, takes this agreement[1]. This agreement requires clients to install the client software. Web Portal is used as one of the authentication mechanism of Web pages, which limits the users to pages by portal

F.L. Wang et al. (Eds.): WISM 2012, LNCS 7529, pp. 199–205, 2012.

server, as the users' visit to any page will be switched to a specific portal page, which can only be solved by the users as they enter their user name and password before they can log onto the page. Its advantage is the users do not need to install a special client authentication before they login[2]. And 802.1x protocol is a bi-layer protocol, which gets through its user authentication by the control of the open/close to the port of the switch. It has its advantage as it distributes its user authentication into the switches, which eases the pressure of dense authentication, whereas it is disadvantageous as each manufacturer adds a lot of private properties in the realization of their 802.1x agreement, which makes each manufacturer rely on their respective switches for their authentication. In addition, the users also need to install the client software. Below are characteristics of the three ways of authentication:

Table 1. Traditional Security Access Authentications

Authentication way	PPPOE	WEB/PORTAL	802.1x
Level of standardization	RFC2516	Private	Private
Cost of packaging	high	Low	Low
Mode of access control	By users	By ports	By users
Support of access to terminal	poor	powerful	poor
Dependency to equipment	low	Low	high
Connectivity to users	powerful	poor	powerful
Support to multi-users	poor	powerful	powerful
Isolation to virus	powerful	Poor	poor

From table 1, it can be seen that these three access authentications have their own characteristics. However, with the increasing development of the Internet, on one hand IPv4 addresses have been used up, and on the other hand, the next generation of Internet based on IPv6 grows popularly. Thus, a lot of universities and colleges have realized IPv4 /IPv6 dual-stack campus network. More generally, along with the advancement of fusion of three nets, mobile network and content network develop quickly as laptops, smart phones, tablet computers and IP TV enter into people's life. All requires that they have convenient access to the Internet. Obviously, the three traditional authentications can't meet the needs of Internet access as the ways based on the client dial-up cannot adapt to the various mobile terminal or sub-terminals while the traditional Web Portal cannot meet the increasingly demanding information security requirement of cell and flexibility in control of security accounts. This leads to the authentication way combined with PPPoE and Web Portal with their respective merits.

3 IPoE Working Principle

IPoE technology is a new way of access authentication recommended by the DSL BBS named WT-146. Based on DHCP protocol, it switches into RADIUS authentication messages to realize authentication and control of access for the users[3]. For the access to the user s' MAC address and other information of devices and ports, one can move forward the mechanism of the requiring access information by inserting DHCP Option 82 in the access equipments to replace the embedded PPPoE dial-up software in the user's terminals, thus the user terminals continue to maintain its original generality and flexibility to provide necessary guarantee of carrying on the evolution of multi-business of IP network. Its working principles as follows:

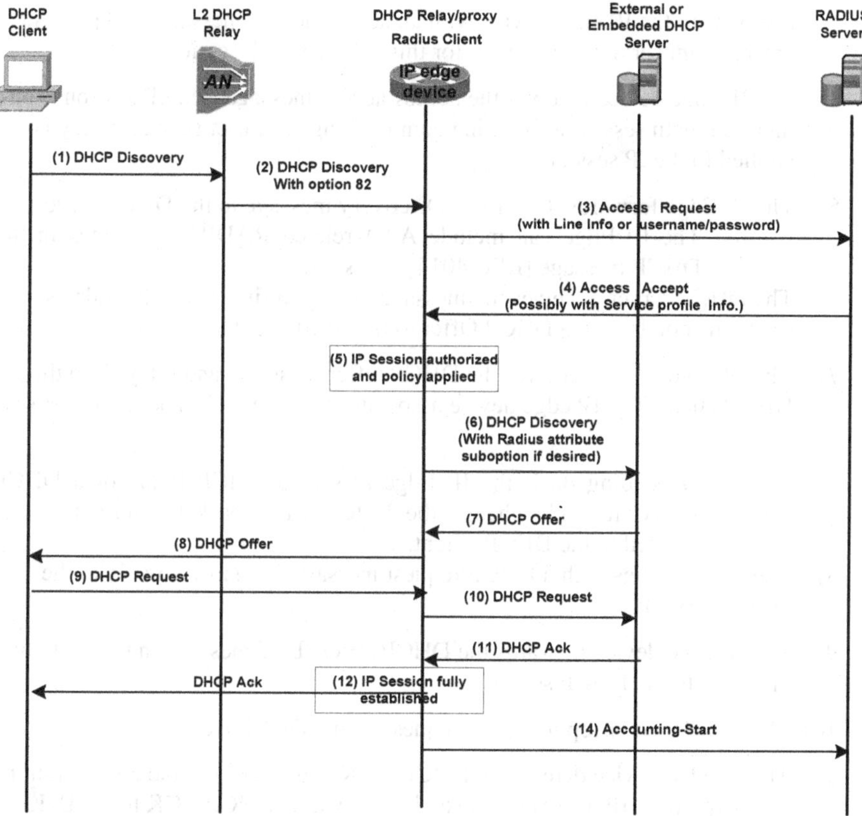

Fig. 1. Dynamic DHCP based IP session creation and authorization

1) The User PC sends a DHCP Discover message
2) The Access-Node forwards the message in the upstream direction to the IP edge, the available network supports DHCPv4 expansion option-option 60,

61 and 82 ,which IPv4 authentication needs ,while IPv6oE supports DHCPv4 expansion option-option 16,1 and 18[4]. The IP edge device detects that the DHCP Discover represents a new user sessions and creates an IP session context for this user (without knowing the IP address of the User). The IP edge device sends

a. The RADIUS access-request message could contain in the username field the information derived from DHCPv4 option 82 or DHCPv6 option 18 (for example, agent-circuit-id, or agent-remote-id, or source MAC address, or a combination there-of). The IP Edge may use a local rule to derive a Radius username/password dynamically (e.g. in case the RADIUS server doesn't support authenticating via Line information).

3) Following a successful authorization on a Radius database, the Radius server responds to the IP edge device with an access-accept message, and passes any relevant service parameters for this subscriber IP session.

4) The IP edge device receives the access-accept message. The IP session is authorized, with session accounting commencing. The user service policy is applied to the IP session.

5) The IP Edge forwards the DHCP-Discovery message to the DHCP server.

   b. The IP Edge can include AAA-related RADIUS attributes in the DHCP message [RFC 4014] if desired.

6) The DHCP server, receives the message and upon finding the IP address scope, responds with a DHCP Offer to the IP edge device.

7) The IP edge device receives the DHCP Offer message, and relays it to the DHCP client. The IP edge device associates the offered IP address to the session

   c. Depending on if the IP Edge acts as a DHCP Relay or a DHCP proxy, it might change the DHCP Offer packet before it is forwarded to the DHCP Client.

8) The client replies with a DHCP request message (possibly directly to the DHCP server)

9) The IP edge device forwards the DHCP REQUEST message in the upstream direction to the DHCP server.

10) The DHCP server replies to the request with a DHCP Ack

11) The IP edge device detects the DHCP ACK and associates the configuration parameter (e.g. IP, lease) to the session, forwards DHCP ACK to the DHCP client and starts keep-alive.

12) The IP edge device sends Accounting Start to RADIUS Server for session start.

13) The accounting start.

4 The Solution

Ningbo Institute of University, Zhejiang University is a full-time university of more than 10000 undergraduates. Due to the socialized logistics, the network for the students' apartment has been fully operating by telecom through PPPoE, which first of all has been separated in the long run with the administrative and didactical network as they are not connected. And secondly, as one needs to dial up to log onto the internet by PPPoE, he needs to install dialing software at the terminal, it also which severely limits the more and more wireless mobile terminal access needs. Thirdly, the coexisting bi-webs also results in not only chaotic accounts of users but also safety hazard of information. Under such circumstances, in March, 2012, the apartments were purchased back by the university under the guidance of Ningbo municipal government, hoping that the networks in the living zone and the teaching area be unified, thus serving teaching and research better for all the faculties and learners. In order to solve the above problems, the security authentication of IPoE was mainly adopted, whose principles as follows:

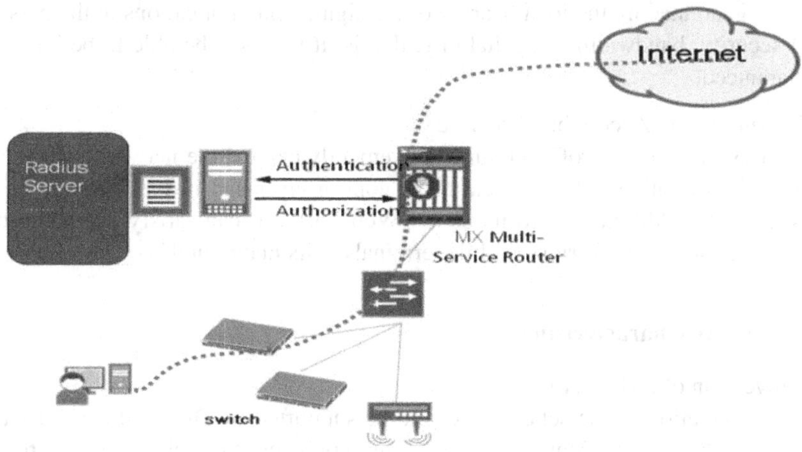

Fig. 2. Security authentication flow chart

4.1 Design Approaches

1) Simple and Flexible Flat Frame

The traditional network is made of three layers: the core, the assembly and the access. This implies an application needs to be realized by supporting sub-applications more than one. To be worse, as it demands too much of the functions of the equipment at the edge; it does not perform both with the users and equipment. In addition, it leads to the decrease of stability and reliability of the campus network equipment (especially the edged ones), thus, there is more and more pressure for the management and maintenance.

The flat frame simply divides the network into the core and the access layers, which is advantageous as it can be controlled centrally and extended conveniently. This is achieved as the most powerful and multi-functioned core equipment offers the service of control and management However, the gathering/access devices is functioned with their basic capabilities. This design can not only reduce the investment of the project as its operating cost greatly decreases, but also makes it easier to extend and manage with the network framework.

2) Refined Management
IPoE security authentication model can realize the effective separation between not only uses but also sessions. To start with, it helps to avoid the interference and influence between each other as it can segment and isolate. Secondly, all of the information concerning the users, including the user account, the MAC address, the IP address, the access location, the access method, the access time and the recognition and records of the serving behavior can be tracked and audited. In the third place, based on the identity of the users by controlling their accessible resources and occupying network bandwidth, it can be segmented and controlled. Finally, the refined management of the application of the network helps to perfect recognition and control of the flow, thus guaranteeing the load bearing of the significant applications in the perspective of security, bandwidth and reliability, that is, it makes it be able to be identified and guaranteed.

3) Expanded and Accessible Structure
Based on its flat frame, IPoE can quite conveniently expand the network or its applications as it is relatively independent of its control layer and it does not rely on the access layer. In addition, due to its use of Web Portal, it can satisfy all the need of access to the Internet by various mobile terminals at its maximum[5].

4.2 Security Characteristics

1) Prevention of DHCP Attack
The implementation of the scheme needs to be supported by DHCPv4 and DHCPv6. Although IPv6 supports Stateless DHCP, this approach can achieve better refined control of user access, whereas the traditional DHCP is vulnerable to be attacked by flood and fake DHCP server. The solution is to first limit the MAC address for port learning through binding of the MAC addresses and the Port to control the possible flood, and then manually set the normal port connected by DHCP server as the trustable server to prevent the deceiving DHCP server in into the network[6].

2) Prevention of ARP Attack
ARP deception is a common virus attack to the Internet as it uses the inherent designing vulnerability, namely the host actively send ARP response to refresh ARP tables of other hosts. the host which receives free ARP attacks without checking whether it is the security hole from its legitimate IP address . Such attacks can also use IP/MAC address table set by DHCP Snooping to dynamically detect ARP data packets; or it can use the binding information of DHCP Snooping to control and discard

the unmatched IP/MAC address , at the same time it can effectively prevent the attack similar with ARP deception by vlan isolation technology of QinQ[7] .

3) Terminal Management of IPoE Session

The characteristics of DHCP agreement implies a DHCP server will not timely release and recycle the IP address of the user as the session ends. To solve the security problem, WT-146 of DSL BBS requires that as IPOE Session ends, it must support Logout, Force, Arp Ping or BFD to realize the release of IP address in time.

5 Conclusion

IPoE security access authentication technology can effectively solve the security access authentication for universities which face IPv4/IPv6 dual-stack network. What's more, it can flexibly expand its network and its applications at the demands of its users, thus conveniently meeting the demands of logging at all kinds of terminals at the maximum. In practice, in order to unify the various operators in the living zone in Ningbo Institute of Technology, Zhejiang University, the technical frame provides a unified, open and fair operating interface , thus it on one hand helps to unify the teaching activities, and on the other hand ends the chaotic situation of managing by many operators. Last but not the least, in the implementation of the scheme based on IPv6oE, it takes Dibbler to fulfill its perfect support of Windows XP to DHCPv6 as the former doesn't know the current situation of DHCPv6.

References

1. Bian, Y.: Application of PPPoE for Campus Network. J. Computer and Information Technology 8, 20 (2008)
2. Congdon, S., et al.: RFC 3580 IEEE 802.1X Remote Authentication Dial In User Service (RADIUS). The Internet Society (2003)
3. Wojciech, D., Gilles, B., et al.: Working Text WT-146 Draft DSL Forum. Architecture and Transport Working Group (2007)
4. Wei, Z.: Normative Analysis of IPv6 Access Authentication. Telecommunication Network Technology 6, 26 (2011)
5. Ma, S.: Empirical Research on Smart Wireless Campus Network. Journal of Guangdong Polytechnic Normal University (Science) 3, 19 (2011)
6. Chen, Q.: Application of DHCP Snooping in Network. World of Network Manager 2, 42 (2011)
7. Guo, Z., Wu, X.: Active Protection Scheme against ARP Attack to Campus Network. Computer Project 5, 181–183 (2011)

Information Encryption Based on Virtual Optical Imaging System and Chen's Chaos

Wei Zhu[1,2], Geng Yang[2], Jian Xu[2], and Xiao-ling Yang[2]

[1] College of Science Nanjing University of Posts and Telecommunications
Nanjing 210003 Jiangsu, P.R.China
[2] Key Lab of Broadband Wireless Communication & Sensor Network Technology of Ministry of Education Nanjing University of Posts and Telecommunications
Nanjing 210003 Jiangsu, P.R.China
{zhuwei,yangg,xuj}@njupt.edu.cn, yxl_at_njupt@126.com

Abstract. A new scheme of information encryption is presented in this paper, which employs Chen's chaos to construct the random mask of the virtual optical imaging system instead of randomly generated by computer. The generation of random mask, the procedures of encryption and decryption are described in detail. Numerical simulations prove that the constructed random mask is uniformly distributed and the encryption algorithm is effective. The experimental results also demonstrate that the proposed encryption algorithm has advantages of large key space and high key sensitivities.

Keywords: Chaos, Virtual optical imaging System, Encryption/decryption.

1 Introduction

As an important supplement and Promotion to traditional security techniques, optical information encryption technique is progressing rapidly in recent years. Various kinds of optical encryption schemes have been proposed and implemented during the past decade [1-4], which involves many applications such as intellectual properties protection; secrete military communication, data/image encryption, etc. There are good reasons to consider using optical information processing to protecting data and images. For example, every pixel of a two-dimensional image can be both relayed and processed at the same time by the optical encrypting system. So, when there is a large volume of information to be processed, it can be transmitted rapidly. Besides the advantage of parallel processing, optical encrypting system has multiple-keys such as wavelength, and spatial frequency, to increase security level. Any unauthorized party would be very difficult to decode the information without having the correct multiple-keys. In [3], an enhanced scheme to virtual optical imaging system (EVOIS) was proposed by introducing a random mask. With the random mask, we can obviously improve the security of the system. However, the random mask and the original information should be the same size, therefore, the key size and the overhead of passing keys greatly increased.

F.L. Wang et al. (Eds.): WISM 2012, LNCS 7529, pp. 206–213, 2012.

In the last decades, Chaos based encryption algorithms have attracted interests of many researchers [5-8]. Most of the given image encryption algorithms adopt the methods of shuffling the positions and changing the grey values of image pixel based on the random sequences from all kinds of chaos system [6,7]. These work all have been motivated by the chaotic properties, i.e. highly sensitive to initial values, pseudorandom property and unpredictability. These Features are very consistent with the random mask of the optical information encryption system.

In this paper, a novel image encryption scheme is suggested. In order to reduce the volume of the keys, we use chen's chaos system [5] to generate the random mask of EVOIS. We simulate the proposed algorithm and analyze the sensitivity of the keys. The rest of this paper is organized as follows. Section 2 presents the proposed algorithm. Section 3 describes some simulation outcomes, some security analysis are given in Section 4.Finally, Section 5 concludes the paper.

2 The Proposed Encryption Algorithm

2.1 An Enhanced Scheme to Virtual Optical Imaging Technique (EVOIS) [3]

In EVOIS, which is schematically shown in Fig. 2, we suppose that both information plane and random mask are "illuminated" with the same coherent light wave with virtual wavelength λ. Under the Fresnel approximation, the propagation of virtual light wave between the information plane and front plane of the lens can be described with a Fresnel diffraction transformation. The Fresnel diffraction patterns of information plane and random mask will interfere each other at front plane of imaging lens to form an interference gram. The interference gram is further converted by lens transmission function to back plane of the lens. The complex amplitude distribution at back plane of imaging lens is taken as encrypted data, and it will be sent as an ordinary digital file through a communication channel safely.

The complex amplitude distribution on the information plane, front plane of the lens, back plane of the lens and the image plane are denoted by $U_0(x_0, y_0)$, $U_{L1}(\xi,\eta)$, $U_{L2}(\xi,\eta)$ and $U_i(x_i, y_i)$ respectively. Under a virtual environment, the imaging process mentioned earlier can be simulated with a computer in a digital domain. In the case of propagating from information plane to front plane of the imaging lens, $U_0(x_0, y_0)$ are sampled to $N \times N$ discrete points with steps $\Delta x_0, \Delta y_0$ along the coordinates. x_0 and y_0 are replaced by $k\Delta x_0$ and $l\Delta y_0$ .Similarly $\Delta\xi,\Delta\eta$ is the sample steps of the front plane of the lens, and $m\Delta\xi$, $n\Delta\eta$ are used to replace ξ and η . Here, k,l,m,n are integers from 0 to N-1.

We can get the discrete Fresnel diffraction (DFD) transformation and express it as following equation:

$$U_{L1}(m,n) = DFD\left[U_0(k,l),\lambda,d_0\right] \tag{1}$$

where

$$
\begin{aligned}
&DFD\left[U_0(k,l),\lambda,d_0\right] \\
&= \frac{1}{j\lambda d_0}\exp\left(j\frac{2\pi d_0}{\lambda}\right)\exp\left[j\frac{\pi}{\lambda d_0}\left(m^2\Delta\xi^2 + n^2\Delta\eta^2\right)\right] \\
&\times\sum_{k=0}^{N-1}\sum_{l=0}^{N-1}U_0(k,l)\exp\left[j\frac{\pi}{\lambda d_0}(k^2\Delta x^2 + l^2\Delta y^2)\right] \\
&\times\exp\left[-j2\pi\left(\frac{mk}{N}+\frac{nl}{N}\right)\right]
\end{aligned} \tag{2}
$$

From the numerical formulation point of view, we use discrete Fourier transformation to represent discrete Fresnel diffraction transformation in Eq. (2) .This means that the standard algorithms for a fast Fourier transformation (FFT) can be applied to simplify the calculation.Furthermore, the transmission function of imaging lens should also be rewritten in its discrete form in order to carry out the numerical simulation.

$$
t(m,n,f) = \exp\left[-j\frac{k}{2f}\left(m^2\Delta\xi^2 + n^2\Delta\eta^2\right)\right] \tag{3}
$$

where $m,n = 0,1,...,N-1$. k is the wave number, and f is the focal length of the lens.

2.2 Chen's Chaotic System

Chen's system can be described by the following non-linear differential equations [5]:

$$
\begin{cases}
\dot{x} = a(y-x) \\
\dot{y} = (c-a)x - xz + cy \\
\dot{z} = xy - bz
\end{cases} \tag{4}
$$

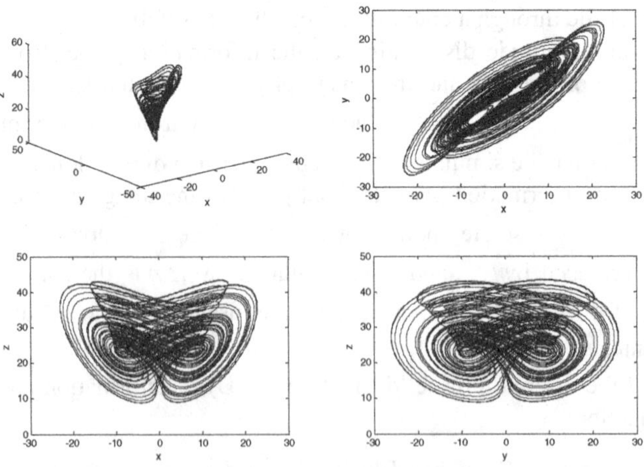

Fig. 1. Phase portraits of chen' s chaotic system for $(a,b,c) = (35,3,28)$, $(x_0, y_0, z_0) = (-10,0,37)$

Where, x, y, z are the state variables, and a,b,c are three system parameters. When $a = 35, b = 3, c \in [20, 28.4]$, the system is chaotic. The parameter □ in front of the state variable y leads to abundant dynamic characters of the new system. Therefore, the dynamical property of the Chen's chaotic system is more complicated than the famous Lorenz chaotic system. This feature is very useful in secure communications. The chaos attractors are shown in Fig.1

In our scheme, three discrete variables of the Chen's chaotic system are adopted to form the random mask.

Step 1: Iterate the Chen's chaotic system for N_0 times by fourth order Rounge-Kutta algorithm with step length $h = 0.001$.

Step 2: For each iteration, we can get three values x_i, y_i and z_i. These decimal values are preprocessed first as follows

$$x_i = \mathrm{mod}((abs(x_i) - floor(abs(x_i))) \times 10^{14}, 256) \tag{5}$$

Where $abs(x)$ returns the absolute value of x. $floor(x)$ rounds the elements of x to the nearest integers less than or equal to x. $\mathrm{mod}(x, y)$ returns the remainder after division. Obviously, $x_i \in [0, 255]$, and we preprocess y_i and z_i as the same way.

Step 3: For each iteration, generate $\overline{x}_i$ by the following formula:

$$\overline{x}_i = \mathrm{mod}(x_i, 3)\,. \tag{6}$$

According to $\overline{x}_i$, we select the corresponding group to form the random mask from Table 1.

Table 1. Different combination of state variables of chen's chaotic system

$\overline{x}_i$	Combination of state variables $C_{\overline{x}_i}$
0	(x_i, y_i)
1	(y_i, z_i)
2	(z_i, x_i)

Without loss of generality, we assume the dimension of the plain-image is $M \times N$, and then we need a random mask of the same dimension. The element $U_i (i = 1, 2, ..., M \times N)$ of the random can be defined according to the following formula

$$\begin{cases} U_{2\times(i-1)+1} = C_{\overline{x}_i}{}^1 \\ U_{2\times(i-1)+2} = C_{\overline{x}_i}{}^2 \end{cases} \tag{7}$$

Where $i = 1, 2, ...$ represent the i th iteration of the chen's chaotic system, and $C_{\overline{x}_i}{}^1, C_{\overline{x}_i}{}^2$ represents the fist and second state variables of combination $C_{\overline{x}_i}$ .The process will not end until the set $\{U_1, U_2, ... U_{M\times N}\}$ is all formed. At last, we get the random mask U_{rm} , which is a matrix of M rows and N columns, by reshape the set $\{U_1, U_2, ... U_{M\times N}\}$.

2.3 Information Encryption with EVOIS Based on Chen's Chaos

In our algorithm, we use the above matrix U_{rm} as the random mask of EVOIS . By utilizing the special characteristics of chaotic systems, we can both increase the security and effectively reduce the size of the keys. Fig.2 is the block diagram of the proposed image encryption algorithm.

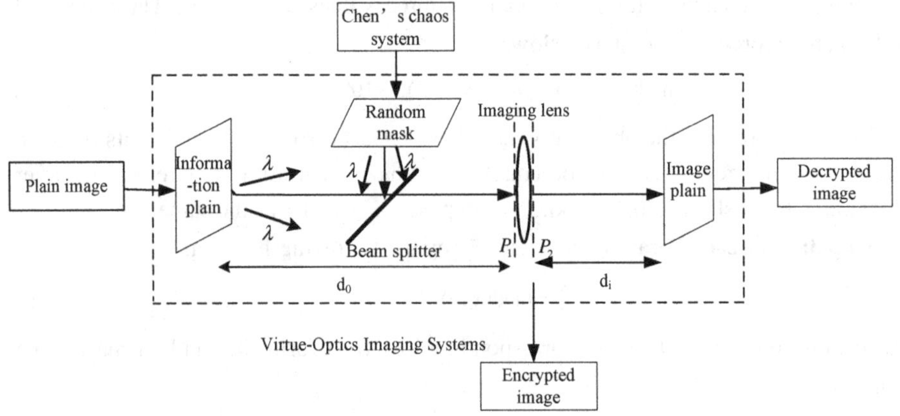

Fig.2. Block diagram of the image encryption

1. **The encryption algorithm**

The encryption process can be described as the following equation:

$$U_{L2}(m,n) = \{DFD[U_0(k,l);\lambda,d_0] + DFD[U_{rm}(k,l);\lambda,d]\} \times t(m,n;f) \tag{8}$$

2. **Decryption algorithm**

The decryption process is to compute the complex amplitude distribution on the image plane (U_i) based on the enciphered data and correct keys. The lens law, $1/d_0 + 1/d = 1/f$, reveals that distance between the image plane and back plane of the lens, labeled with d_i , is not a independent parameter and determined by key parameters d_0 and f . The definite steps of the decryption algorithm are as follows:

$$U_i(m,n) = DFD[U'_0(k,l), \lambda, d_i(d_0, f)] \tag{9}$$

Where

$$U'_0(m,n) = U_{L2}(m,n) - DFD[U_{rm}(k,l), \lambda, d] \times t(m,n,f) \tag{10}$$

At last, we can recover the original information U_0 by reversing U_i .

3 Numerical Simulation

Experimental analysis of the proposed encryption algorithm in this paper has been done in Matlab2010. We choose the gray "Lena" with size 512×512 as original image and the multiple keys is set up as follows: $c = 28, x_0 = 0, y_0 = 1, z_0 = 0, d_0 = 0.1,$ $d = 0.01, f = 0.025, \lambda = 623e-9$. Effectiveness of encryption and decryption algorithms are shown in Fig.3 and Fig.4.

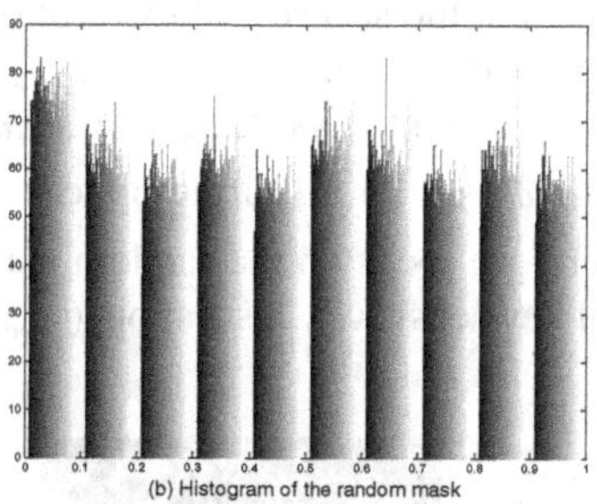

Fig. 2. Random mask generated from Chen's chaos systems

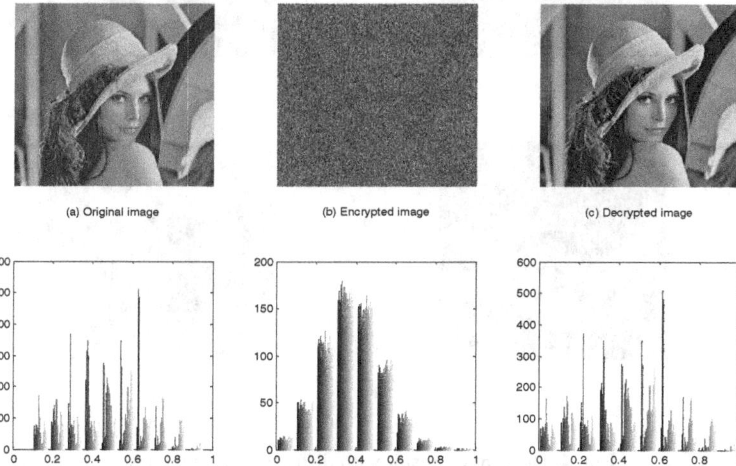

Fig. 3. Result of image encryption and decryption

Fig.3.shows Chen's chaos systems can generate a random mask, which is like the uniformly distributed noise. The original image 'Lena', the encrypted image of it and the decrypted image with correct keys are shown in Fig.4 (a), Fig.4 (b) and Fig4(c) respectively. The corresponding histograms of them are shown in Fig.4 (d), (e) and (f). We can see from Fig.4 (b) that encrypted image has completely hidden the original image and could not see the outline of it. Histogram of the decrypted image, shown in Fig.4 (d), is more flat than original image, which mean that the algorithm can prevent statistical attacks. Fig.4 (c) gives a perfect retrieve of the original image. Overall, experimental results demonstrate that the proposed encryption algorithm is effective.

4 Security Analysis

We carry out the security analysis of our proposed algorithm from key sensitivity and the size of key space.

There are eight keys $(c, x_0, y_0, z_0, d_0, d, \lambda, f)$ in our algorithm. The sensitivity about the keys (d_0, d, λ, f) of virtue optical imaging systems has been analyzed detailed in [3]. Here, we focus on the random mask we proposed based on Chen's chaos system. Several key sensitivity tests about (c, x_0, y_0, z_0) are performed in Fig.5.

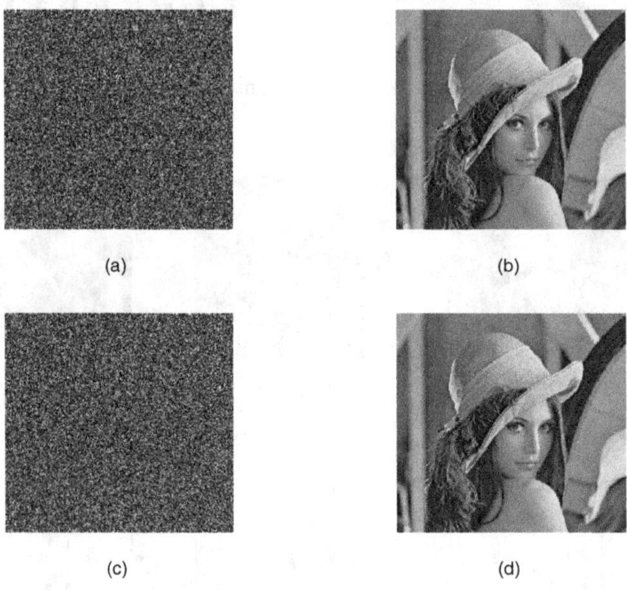

Fig. 4. Decryption result with different initial values

Fig.5 (a) is the decrypted image when key x_0 is with a discrepancy of 10^{-15}, at the condition of other seven keys are correct. Fig.5 (b) is the decrypted image with a discrepancy of 10^{-16} on x_0 .The other two initial value y_0 and z_0 have the similar precision with x_0 . Fig.5(c) and Fig.5 (d) are decrypted image with the same keys that used in encryption algorithm except parameter c of Chen's chaos system, which is with discrepancy of 10^{-14} and 10^{-15} respective. So, it is easy to see the encryption algorithm is very sensitive to the keys of random mask, and a very small change of the key will lead to the wrong decryption result.

In our algorithm, the parameter c and initial values of Chen's chaos system are used as secret keys. According to the above sensitivity test, the system parameter c has the precision of 10^{-14}, and the three initial values have the precision of 10^{-15} each. Therefore the key space size reach $10^{14}\times10^{15}\times10^{15}\times10^{14}\approx2^{192}$, without counting the keys (d_0,d,λ,f) .This is sufficient to resist all kinds of brute-force attacks.

5 Conclusions

In this paper, a novel encryption algorithm with virtual optical imaging system is proposed, which uses Chen's chaos system to generate the random mask. Some security analyses are given to certificate the sensitivities of parameters. We found the multiple keys are sensitive and security level is significantly increased when Chen's chaos system is introduced to produce the random mask. Our scheme may have some potential application in image encryption and information transmission based on Internet.

Acknowledgments. This work was supported by the National Basic Research Program of China (973 Program) under Grant No.2011CB302903, the National Science Foundation of China under Grant No.60873231, the Natural Science Foundation of Jiangsu Province under Grant No.BK2009426, the Natural Science Research Project of Jiangsu Education Department under Grant No.09KJD510008, the Natural Science Foundation of the Jiangsu Higher Education Institutions of China under Grant No.11KJA520002, Research Fund for the Doctoral Program of Higher Education of China No.20113223110003.

References

1. Kim, H., Lee, Y.: Encryption of a volume hologram by complementary input image and binary amplitude mask. Optics Communications 258(1), 9–17 (2006)
2. Nishchal, N.K., Naughton, T.J.: Flexible optical encryption with multiple users and multiple security levels. Optics Communications 284(4), 735–739 (2011)
3. Peng, X., Cui, Z.Y., Tan, T.N.: Information encryption with virtual-optics imaging system. Optics Communications 212(4-6), 235–245 (2002)
4. Guan, Z.H., Huang, F., Guan, W.: Chaos-based image encryption algorithm. Physics Letters A 346, 143–147 (2005)
5. Chen, G., Ueta, T.: Yet another chaotic attractor. Int. J. Bifurcation and Chaos 9(7), 1465–1466 (1999)
6. Ge, X., Liu, F.L., Lu, B., Wang, W., Chen, J.: An image encryption algorithm based on spatiotemporal chaos in DCT domain. In: 2010 The 2nd IEEE International Conference on Information Management and Engineering (ICIME), pp. 267–270 (2010)
7. Seyed, M., Sattar, M.: A fast color image encryption algorithm based on coupled two-dimensional piecewise chaotic map. Signal Processing 92(5), 1202–1215 (2012)
8. Chen, J.Y., Zhou, J.W., Wong, K.W.: A Modified Chaos-Based Joint Compression and Encryption Scheme. IEEE Transactions on Circuits and Systems-II 58(2), 110–114 (2011)

A New Scheme with Secure Cookie against SSLStrip Attack

Sendong Zhao[1,*], Wu Yang[1], Ding Wang[2], and Wenzhen Qiu[3]

[1] Information Security Research Center, Harbin Engineering University, Harbin City 150001, China
[2] College of Computer Science and Technology, Harbin Engineering University, Harbin City 150001, China
[3] China Telecom Co., Ltd. Fuzhou Branch, Fuzhou City 350005, China
zhaosendong@hrbeu.edu.cn

Abstract. In 2009 Moxie Marlinspike proposed a new Man-in-the-Middle (MitM) attack on secure socket layer (SSL) called SSLStrip attack at Black Hat DC, which is a serious threat to Web users. Some solutions have been proposed in literature. However, until now there is no practical countermeasure to resist on such attack. In this paper, we propose a new scheme to defend against SSLStrip attack by improving the previous secure cookie protocols and using proxy pattern and reverse proxy pattern. It implements a secure LAN guaranteed proxy in client-side, a secure server guaranteed proxy in server-side and a cookie authentication mechanism to provide the following security services: source authentication, integrity control and defending SSLStrip attack.

Keywords: Secure Cookie, Integrity, HMAC, MitM, SSLStrip, Proxy Pattern.

1 Introduction

As one of the most important and wildly used network protocols, HTTP has attracted much attention of computer science researchers. HTTP itself is not a dependable and secure protocol, so it needs security measures of some auxiliary. Secure Socket Layer is used as a countermeasure to protect the security and dependability of http connection currently. However, a new MitM attack–SSLStrip attack launched a new challenge to it. This attack makes SSL vulnerable. Nick Nikiforakis *et al.* [1] presented a method to avoid a SSLStrip attack while using browser's history information. A scheme with cue information is proposed by Shin *et al.* [2], which rely on web user's active exploration. Nevertheless, those suggestions only indicate ways to avoid a SSLStrip attack and not how to actively stop it. Cookie is a technical means to keep the HTTP connection state, and it also can be used to save user's information, such as identification number ID, password, visited pages, residence time, ways of shopping on the Web and

* Corresponding author.

F.L. Wang et al. (Eds.): WISM 2012, LNCS 7529, pp. 214–221, 2012.

the times of accessing to the Web. When users re-link the Web server, browser will read the cookie information and send it to the Web site. Cookie can even check whether the HTTP connection is secure or not. So cookie can be utilized to defend against SSLStrip attack.

In this paper, we point out the drawbacks of the state-of-the-art cookie protocols. Also, we present a new secure cookie protocol and a new topology structure between client and server to solve the drawbacks of previous secure cookie protocols especially to defend SSLStrip.

The rest of the paper is organized as follows: In section 2, the SSLStrip attack [3] is explained and analyzed in detail. We track the previous important secure cookie protocols, then point out their main weaknesses briefly in section 3. In section 4, we propose a new scheme, containing two aspects. One is our secure cookie protocol, and the other is secure LAN guaranteed proxy and secure server guaranteed proxy. Section 5 gives an experiment and a performance analysis. In section 6, we analyze the security of our new scheme. Conclusion and future work are discussed in section 7.

2 SSLStrip Attack

SSLStrip attack was first proposed by Moxie Marlinspike [3] at Black Hat DC in 2009 and eventually culminated in the release of a tool known as SSLStrip. At a high level, SSLStrip allows an adversaries to insert themselves in the middle of a valid SSL connection [3]. The user believes that they have a true SSL connection established, while the adversary has the ability to view the user's Web traffic in clear-text. SSLStrip attack is based on the behaviors of Internet users. Most Internet users enter a URL directly in address box, rarely adding protocol types in front of the URL. Although the connection should be on SSL, they don't type 'https://' in front of the URL. This behavior results in the establishment of secure connection needs jumping from unsecure connection (port 80) to secure connection (port 443). The jumping has two forms: user's active jumping and user's passive jumping. When user type 'www.google.com.hk', a Web page will be returned. From this page's html source code, we find that the 'Sign in' button corresponds to a HTTPS hyperlink. It will jump to a secure connection (port 443) when users click this button. When users enter 'mail.google.com' in the browser's address box, the server will return a redirect packet which could jump to secure connection.

The whole process can be summarized into two stages. In the first stage, SSLStrip records and stores URLs whose protocol type is changed from HTTPS to HTTP, as shown in sequence 1-4 of Fig.1. In the second stage, SSLStrip decides to use clear-text communication or cipher-text communication depending on the result of matching client request URL and SSLStrip stored URL. If matched, SSLStrip uses cipher-text communication; otherwise, it uses clear-text communication, as shown in sequence 5-8 of Fig.1.

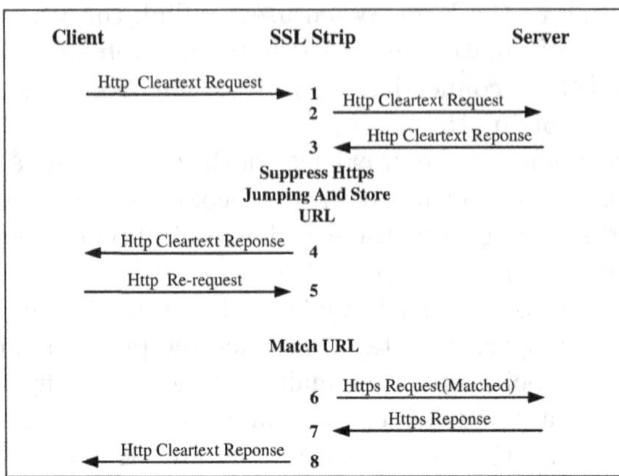

Fig. 1. The attack process of SSLStrip as MitM between client and server

3 Previous Secure Cookie Mechanism

In this section, we track the previous important secure cookie protocols, then point out their main weaknesses briefly.

Fu *et al.* [4] proposed a Web authentication mechanism that mainly uses secure cookie as authentication token. While Web server makes use of HMAC [5] to encrypt cookie and transmit it to client. They used cookie to store the authentication token with client. This mechanism can defend some kinds of attacks. However, it has the following three weaknesses: it does not have a mechanism for providing high-level confidentiality; it is vulnerable to cookie re-play attacks; and its mechanism for defending against volume attacks is inefficient and non-scalable. To improve these weaknesses, Liu *et al.* [6] proposed a secure cookie protocol. The process for verifying a cookie created using Alex's protocol is shown as follow. Note that if FALSE is returned, the cookie is deemed invalid and the client must login in again using his username and password.

Step 1. Compare the cookie's expiration time and the server's current time. If the cookie has expired, then return FALSE.

Step 2. Compute the encryption key: k=$HMAC_{sk}$(username||expires).

Step 3. Decrypt the encrypted data using k.

Step 4. Compute $HMAC_k$(username||expires||data||session-key), and compare it with the keyed-hash message authentication code of the cookie. If they match, then return TRUE; otherwise return FALSE.

Pujolle *et al.* [7] proposed a secure cookie mechanism that implements an intermediary reverse Proxy patterns. They modified the grammatical definition of the cookie header based on EBNF form. In Guy's mechanism, in order to guarantee the integrity of cookie sent to server, the server must store every cookie since cookie has clear-text cookie-av used only to store data field. Furthermore,

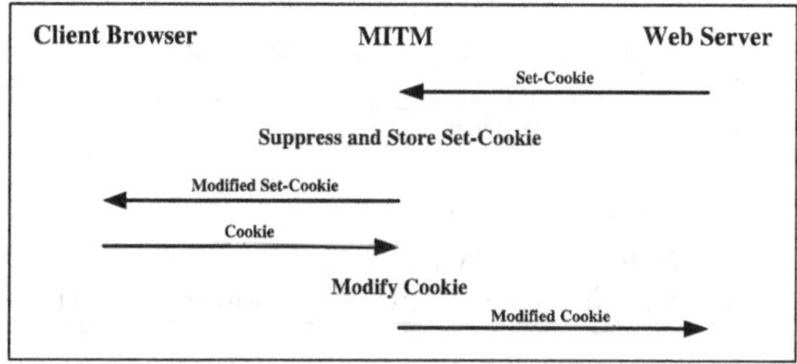

Fig. 2. MitM hijack cookie procedure

because of complete clear-text cookie information, this mechanism has no confidentiality about cookie. Again, the above secure cookie protocols have to use both secure cookie and SSL to ensure secure communication between client and server. However, there is no SSL upon most occasions. In section 2, we have discussed this in detail. In Guy's mechanism, server must store every cookie. Obviously, it has no confidentiality about cookie data and also it cannot defend SSLStrip attack. Although Alex's secure cookie protocol has improved Fu's cookie protocol and dealt with those three weaknesses, it also has weaknesses. Its cookie integrity is relying on using SSL. Also, it is vulnerable to SSLStrip attack. Furthermore, if this secure protocol is not implemented on SSL, the integrity of Cookie and Set-Cookie is lost. An adversary between client and server as a MitM, changes the Set-Cookie from server and meanwhile records the original Set-Cookie [3]. And then the adversary sends it to client. When client sends Cookie to the server, the MitM modifies the cookie according to original Set-Cookie which is recorded. This process is shown in Fig.2.

4 Our Scheme

Our scheme contains a secure cookie protocol and a new topology structure using proxy pattern an reverse proxy pattern. The secure cookie protocol is implemented between client and server. The secure cookie protocol is shown in Table 2. Recall that $HMAC_k$=$HMAC_k$(username||expires||data||CID||SN), sk denotes the secret key server proxy, k=$HMAC_{sk}$(username||expires), ks denotes the private key of SSGP for signature, SN denotes secure note of cookie. SN is either $Secure$ or $Unsecure$. ESN=$Sig_{ks}(h(SN \parallel CID))$. CID is a random digit generated for each new cookie. If the value of SN is $Secure$, the connection is on SSL protocol and vice versa. The client browser will not upload cookie to the server as a request if the value is $Secure$ and the protocol type of request URL is not HTTPS. The server will not return a login web page. So, the client must send a request again. It is worth mentioning that SN and ESN are mandatory fields in our secure cookie protocol.

The topology structure mainly contains secure LAN guaranteed proxy (SLGP) and secure server guaranteed proxy (SSGP), as shown in Fig.3. It is clear to see that SLGP and SSGP are implemented on the gateway of client LAN and server

Table 1. Notation

Symbol	Description
$\|\|$	String concatenation operation
$H(\bullet)$	Hash function
$E_{key}(M)$	Encrypt M using key *key*
$Sig_{key}(M)$	Sign M using key *key*
$HMAC_{key}(M)$	Keyed-hash message authentication code of M with key *key*

Table 2. Our secure cookie protocol

Set-cookie Header						
username	expires	$E_k(data)$	SN	CID	ESN	$HMAC_k$
Cookie Header						
username	expires	$E_k(data)$	SN			$HMAC_k$

Ethernet, respectively. To facilitate research, we assume that the LAN of clients is a secure LAN. It means that there are no such attacks as arp spoofing attack. That can guarantee the absolute security of the client LAN. On client-side, the data from outside of the Secure LAN is sent to SLGP. After processing, the data is sent to the gateway of the LAN again. On server-side, the data from inside and outside of server Ethernet is sent to SSGP. After processing, the data is sent to the gateway of the Ethernet again. The duty of the SLGP is to check the integrity of essential attributes of Set-Cookie from server. The duty of the SSGP is to reconstruct Set-Cookie from Web servers and check the integrity and validity of cookie from client. If the data itself is cipher-text, SLGP will do nothing. From Figure 4, the operating mechanism of our scheme can be summarized as follows.

Step 1. Client sends request to server.
Step 2. SSGP receives cookie and checks its integrity and validity. Then, send it to server.

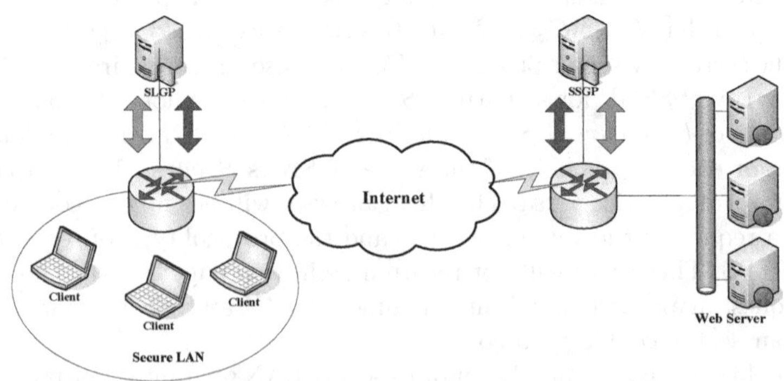

Fig. 3. The topology of our scheme

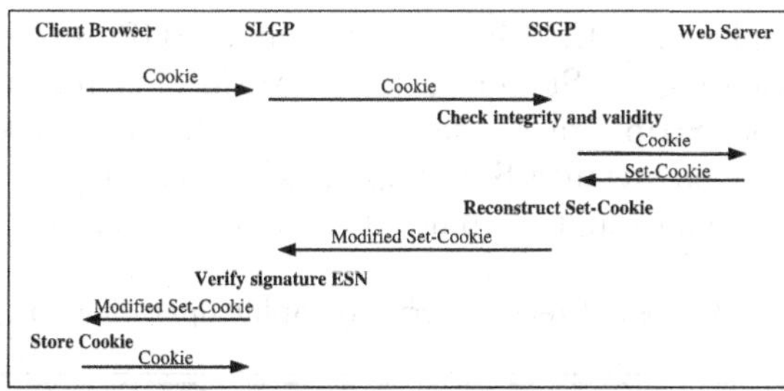

Fig. 4. The mechanism of our scheme

Step 3. Server receives the request data and generates a Set-Cookie. Then send it to SSGP.
Step 4. SSGP reconstructs the Set-Cookie. Then send the modified Set-Cookie to client-side again.
Step 5. SLGP receives Set-Cookie and verifies the signature ESN. If the signature is not valid, drop the packet.
Step 6. SLGP send the Set-Cookie to client. The client receives the Set-Cookie and stores it in disk or RAM.
Step 7. The client sends request again with a new cookie.
Step 8. Goto Step 2.

In the above mechanism, SLGP should check whether Set-Cookie has SN or not. If not, SLGP drops this response packet. SLGP apply public key of SSGP to SN as a signature verification key verifies ESN as the message signed with the corresponding private key ks if the value of field SN is Unsecure. If the signature is not valid, drop the response packet.

SSGP reconstruct the Set-Cookie when servicing a request. The process of reconstruction is like this: calculate k=$HMAC_{sk}$(username||expires) using the secret key sk of the proxy. Then calculate HMACk using key k. Compute ESN using key ks. Encrypt data using key k. Once those are done, add ESN, $HMAC_k$, SN to Set-Cookie. Replace data with $E_k(data)$ in Set-Cookie. SSGP should check the integrity and validity of cookie when receiving a request with cookie. A series of options of SSGP is to compute $HMAC_{sk}$(username||expires), and consider it as key k to compute $HMAC_k$. Then compare it with the keyed-hash message authentication code of the cookie. If they match, send clear cookie to servers; otherwise, drop this request. If the data is encrypted, then decrypt it. The cookie sent to server includes only clear-text of username, expires and data.

Our scheme is composed of a secure cookie protocol, a SLGP and a SSGP. In order to achieve enough security, all the components of our secure scheme are inseparable.

5 Experiment and Performance Analysis

Using our scheme, five groups of experiment are conducted separately. In client-side, we use IE, FireFox, Oprea, Safari and Chrome as Web browser successively.

In addition, we implement SLGP in client-side and SSGP in testing server-side meanwhile implementing SSLStrip attack as a MitM. The result of our scheme defending against SSLStrip attack is shown in Table 4. It can be seen from Table 4 that our scheme defend against SSLStrip attack successfully in the browser of IE, FireFox, Oprea, Safari and Chrome. Our scheme adds two times $HAMC$ computing and one time verifying signature option, compared with traditional cookie protocol. HAMC computing is a hash computation and verifying signature option is a modular exponentiation operation. Cost of relevant cryptographic operations is shown in table 3. The extra time of our scheme is two times hash consuming and one time modular exponentiation operation consuming. In the interaction of protocol process, our scheme add none extra time data transmission compared with traditional cookie protocol. So, total extra cost of our scheme is $0.325ms$ according to our experiment. Although our scheme spent some extra time, our scheme improve the security of network interaction which uses HTTPS.

Table 3. Cost of relevant cryptographic operations(ms)

Cryptographic operation	Modular exponentiation($\|n\|$=1024 ,$\|e\|$=32)	Hash(SHA-1)
Time	0.273	0.026

Table 4. Evaluation of our scheme against SSLStrip attack

Browser	IE	FireFox	Opera	Safari	Chrome
Success	YES	YES	YES	YES	YES

6 Security Analysis

In this section we present an informal analysis of the security properties of our new scheme, including the following five services: authentication, integrity, and anti-SSLStrip attack in detail.

Authentication: A secure cookie protocol should allow the server to verify that the client has been authenticated within a certain time period [6]. Moreover, any client should not be able to forge a valid cookie. In our scheme, the server can get the cookie from client, in which every field is the initial value by using detection of SSGP. So the server can authenticate the client exactly. Obviously, the Set-Cookie reconstructed by SSGP is hardly to forge. Above all, the server can verify the validity of the client easily and accurately.

Integrity: Our scheme uses secure cookie protocol, SLGP and SSGP to ensure the integrity of both essential part of Set-Cookie and Cookie. If adversaries modify the Cookie from client, the SSGP will encrypt the Cookie domains with HMAC, and compare it with keyed-hash message authentication code of Cookie. If they are matched, it states that the Cookie is modified. Then the server will not service the request. If adversaries modify the field SN of Set-Cookie from server or delete this Set-Cookie fields, SLGP will firstly check the integrity of domain items, verify signature ESN. If there are any abnormal evidences, SLGP will drop whole response packet.

Anti-SSLStrip: The core of SSLStrip attack is to make users cannot recognize what the connection should be. It should be HTTP or HTTPS. Nevertheless, our secure cookie protocol has mandatory fields: SN and ESN. In our scheme, SLGP will check these fields. So the client will know what the connection should be originally. For this reason, our scheme can defend SSLStrip attack.

7 Conclusions

Our contributions in this paper are threefold. First, we discover the drawbacks of state-of-the-art cookie protocols. Second, we optimize the secure cookie protocol of Alex and present a new secure cookie protocol. Third, we present a new topology structure between client and server combined with the new secure cookie protocol to solve the drawbacks of previous secure cookie protocols especially to defend SSLStrip attack. However, due to the serious harm of SSLStrip attack, the cost of resisting this attack is inevitably high. It is not very easy to deploy reflected in our scheme. Our scheme also requires the security of client LAN. Our further work will focus on improving the deploy-ability of our scheme and reducing security requirement of client LAN.

References

1. Nikiforakis, N., Younan, Y., Joosen, W.: HProxy: Client-Side Detection of SSL Stripping Attacks. In: Kreibich, C., Jahnke, M. (eds.) DIMVA 2010. LNCS, vol. 6201, pp. 200–218. Springer, Heidelberg (2010)
2. Shin, D., Lopes, R.: An empirical study of visual security cues to prevent the sslstripping attack. In: Proceedings of the 27th Annual Computer Security Applications Conference, ACSAC 2011, pp. 287–296. ACM, New York (2011)
3. Callegati, F., Cerroni, W., Ramilli, M.: Man-in-the-middle attack to the https protocol. IEEE Security Privacy 7(1), 78–81 (2009)
4. Fu, K., Sit, E., Smith, K., Feamster, N.: Dos and don'ts of client authentication on the web. In: Proceedings of the 10th Conference on USENIX Security Symposium, SSYM 2001, vol. 10, pp. 19–35. USENIX Association, Berkeley (2001)
5. Bellare, M., Canetti, R., Krawczyk, H.: Keying Hash Functions for Message Authentication. In: Koblitz, N. (ed.) CRYPTO 1996. LNCS, vol. 1109, pp. 1–15. Springer, Heidelberg (1996)
6. Liu, A., Kovacs, J., Huang, C.T., Gouda, M.: A secure cookie protocol. In: Proceeding of 14th International Conference on Computer Communications and Networks, ICCCN 2005, pp. 333–338 (October 2005)
7. Pujolle, G., Serhrouchni, A., Ayadi, I.: Secure session management with cookies. In: Processing of 7th International Conference on Information, Communications and Signal, ICICS 2009, pp. 1–6 (December 2009)

ID-Based Signatures from Lattices in the Random Oracle Model*

Chunxiang Gu[1], Li Chen[2], and Yonghui Zheng[1]

[1] Zhengzhou Information Science and Technology Institute, Zhengzhou 450002, China
[2] Henan University of Economics and Law, Zhengzhou 450002, China

Abstract. Cryptosystems based on the hardness of lattice problems are becoming more and more popular in the research community. These kind of crypto schemes have many potential advantages: their resistance so far to cryptanalysis by quantum algorithms, their asymptotic efficiency and conceptual simplicity, and the guarantee that their random instances are as hard as the hardness of lattice problems in worst case. In this paper, we propose an ID-based signature scheme using lattices and then make some discussion on its extensions. The constructions use lattice basis delegation with fixed-dimension technique for extracting users' secret key, and the security can be reduced to the small integer solution problem (SIS) problem in the random oracle model.

Keywords: Lattices, ID-Based signature, Blind signature, provable secure.

1 Introduction

ID-based public key cryptography (ID-PKC) was first proposed by Shamir[1] to simplify key management procedure of traditional certificate-based PKI. In ID-PKC, an entity's public key is directly derived from certain aspects of its identity, such as an IP address or an e-mail address associated with a user. Private keys are generated for entities by a trusted third party called a private key generator (PKG). The direct derivation of public keys in ID-PKC eliminates the need for certificates and some of the problems associated with them.

Due to the contribution of Boneh and Franklin[2], many constructions[3] of ID-PKC has been proposed based on groups with a bilinear pairing. Recently, lattices have emerged as a powerful mathematical platform on which to build a rich variety of cryptographic primitives. Starting from the work of Ajtai[4], many new constructions with lattices have been proposed, such as one-way functions and hash functions[4], trapdoor functions[5], public-key encryption schemes[6], fully homomorphic encryption schemes[7], and even an array of ID-based encryption schemes[8-10]. Lattice-based schemes are attractive due to many potential advantages: their asymptotic efficiency and conceptual simplicity; their resistance so far to

* Research supported by the Foundation for the Key Sci & Tech Research Project of He Nan (No.112102210007).

F.L. Wang et al. (Eds.): WISM 2012, LNCS 7529, pp. 222–230, 2012.

cryptanalysis by quantum algorithms (as opposed to schemes based on factoring or discrete log); and the guarantee that their random instances are as hard as the hardness of lattice problems in worst case. X. Boyen[11] said that "*all those factors conspire to make lattices a prime choice, if not the primary one yet, for mathematical crypto design looking out into the future.*"

In this paper, we propose an ID-based signature scheme using lattices and then make some discussion on its extensions. The schemes use lattice basis delegation with fixed-dimension technique for extracting users' secret key, and hence achieve short secret-keys and short signatures. The schemes are unforgeable and hold blindness (for blind version) based on the hardness of small integer solution (SIS) problem of lattices in the random oracle model. Using the similar technique in [10,14] , it can also be modified to obtain a version secure in the standard model.

The rest of this paper is organized as follows: In Section 2, we recall some preliminary works. In Section 3 and 4, we describe the details of our new schemes and their security proofs. Finally, we conclude in Section 5.

2 Preliminaries

2.1 Identity Based Signatures

Recall that an ID-based signature scheme consists of four polynomial-time algorithms[12]: *Setup, Extract, Sign*, and *Verify*. The *Setup* algorithm generates public system parameters and a secret master kye. The *Extract* algorithm uses the master key to extract a private key corresponding to a given identity. The *Sign* algorithm signs messages with a user's private key and the *Verify* algorithm verifies signatures using the identity.

The general notion of security of an ID-based signature scheme is existential unforgeability on adaptively chosen message and chosen identity attacks (EUF-ID-ACMA). We say that an ID-based signature scheme is EUF-ID-ACMA secure if no polynomial time adversary $\mathcal{A}$ has a non-negligible advantage against a challenger $\mathcal{C}$ in the following game:

1. $\mathcal{C}$ runs *Setup* algorithm of the scheme and gives the system parameters to $\mathcal{A}$.
2. $\mathcal{A}$ issues the following query as he wants:
   (a) *Hash* function query. $\mathcal{C}$ computers the value of the hash function for the requested input and sends the value to $\mathcal{A}$.
   (b) *Extract* query. Given an identity id, $\mathcal{C}$ returns the private key corresponding to id which is obtained by running *Extract* algorithm.
   (c) *Sign* query. Given an identity id and a message m, $\mathcal{C}$ returns a signature which is obtained by running *Sign* algorithm.
3. $\mathcal{A}$ outputs (id^*, m^*, σ^*), where σ^* is a signature of identity id^* for message m^*, such that id^* and (id^*, m^*) are not equal to the inputs of any query to *Extract* and *Sign*, respectively. $\mathcal{A}$ wins the game if σ^* is a valid signature of m^* for id^*.

With ID-based signature, a lot of extent primitives have also proposed, such as ID-based blind signatures, ID-based ring signatures, and so on. For example, ID-based blind signature scheme[13] can be considered as the combination of a general blind signature scheme and an ID-based one. It uses an interactive protocol between signer and user instead of the *Sign* algorithm. It allows the user to get a signature of a message in a way that the signer learns neither the message nor the resulting signature. Besides the unforgeability, the security of an ID-based blind signature scheme also requires the blindness property.

2.2 Lattices and Sampling

Definition 1. Given n linearly independent vectors $b_1, b_2, ..., b_n \in R^m$, the lattice Λ generated by them is denoted $L(b_1, b_2, ..., b_n)$ and define as:

$$L(b_1, b_2, ..., b_n) = \{\sum_{i=1}^{m} x_i b_i \mid x_i \in Z\}$$

The vectors $b_1, b_2, ..., b_n$ are called the basic of the lattice. Let $B = \{b_1, b_2, ..., b_n\}$, we let $\widetilde{B}$ denote its Gram-Scahmidt orthogonalization of the vectors $b_1, b_2, ..., b_n$ taken in that order, and we let $\| S \|$ denote the length of the longest vector in B for the Euclidean norm.

Definition 2. For q prime and $A \in Z_q^{n \times m}$ and $u \in Z_q^n$, define:

$$\Lambda_q^{\perp}(A) = \{e \in Z^m \mid A \cdot e = 0 \bmod q\}$$

$$\Lambda_q^{u}(A) = \{e \in Z^m \mid A \cdot e = u \bmod q\}$$

Ajtai [4] and later Alwen and Peikert[14] showed how to sample an essentially uniform matrix $A \in Z_q^{n \times m}$ with an associated basis T_A of $\Lambda_q^{\perp}(A)$ with low Gram-Scahmidt norm.

Propositon 1 [15]. For any prime $q \geq 2$ and $m \geq 5n \log q$, there exists a probabilistic polynomial-time algorithm ***TrapGen***(q, n) that outputs a pair $A \in Z_q^{n \times m}$ and T_A such that A is statistically close to uniform and T_A is a basis for $\Lambda_q^{\perp}(A)$ with length $L = \| \widetilde{T}_A \| \leq m\omega(\sqrt{\log m})$ with all but $n^{-\omega(1)}$ probability.

We further we review Gaussian functions used in lattice based cryptographic constructions.

Definition 3. Let m be a positive integer and $\Lambda \in R^m$ be a m dimensional lattice. For any vector $c \in R^m$ and any positive parameter $\sigma \in R_{>0}$, we define:

$$\rho_{\sigma,c}(x) = \exp(-\pi \| x - c \|^2 / \sigma^2) \text{ and } \rho_{\sigma,c}(\Lambda) = \sum_{x \in \Lambda} \rho_{\sigma,c}(x)$$

The discrete Gaussian distribution over Λ with center c and parameter σ is

$$\forall y \in \Lambda \quad D_{\Lambda,\sigma,c}(y) = \frac{\rho_{\sigma,c}(y)}{\rho_{\sigma,c}(\Lambda)}$$

For notational convenience, $\rho_{\Lambda,\sigma,0}$ and $D_{\Lambda,\sigma,0}$ are abbreviated as ρ_σ and $D_{\Lambda,\sigma}$.

Gentry *et.al.* [5] showed an poly-time algorithm ***SampleDom***(A,σ) to sample an x from distribution $D_{Z^m,\sigma}$ for which the distribution of $A \cdot x \bmod q$ is uniform over Z_q^n. They also construct the following algorithm for sampling from the didcrete Gaussian $D_{\Lambda,\sigma,c}$, given a basis T_A for the m-dimensional lattice $\Lambda_q^\perp(A)$ with $\sigma \geq \| \tilde{T}_A \| \omega(\sqrt{\log m})$.

SamplePre(A,T_A,u,σ): On input a matrix $A \in Z_q^{n\times m}$ with 'short' trapdoor basis T_A for $\Lambda_q^\perp(A)$, a target image $u \in Z_q^n$ and a Gaussian parameter $\sigma \geq \| \tilde{T}_A \| \omega(\sqrt{\log m})$, outputs a sample $e \in Z_q^m$ from a distribution that is within negligible statistical distance of $D_{\Lambda_q^u(A),\sigma}$.

Agrawal *et.al* [10] proposed a simple delegation mechanism that does not increase the dimension. A matrix $R \in Z^{m\times m}$ is said Z_q-invertible if $R \bmod q$ is invertible as a matrix in $Z_q^{m\times m}$. The mechanism makes use of Z_q-invertible matrices with all the columns are "small" or "low norm". Let $\sigma_R = \sqrt{n \log q} \cdot \omega(\sqrt{\log m})$, we let $D_{m\times m}$ denote the distribution on matrices in $Z^{m\times m}$ defined as $(D_{Z^m,\sigma_R})^m$ conditioned on the resulting matrix being Z_q-invertible. Agrawal *et.al* show an algorithm:

BasisDel (A,R,T_A,σ): On input a matrix $A \in Z_q^{n\times m}$ with 'short' trapdoor basis T_A for $\Lambda_q^\perp(A)$, a Z_q-invertible matrices $R \in Z^{m\times m}$ sampled from $D_{m\times m}$, a parameter $\sigma \in R_{>0}$, outputs a random basis T_B of $\Lambda_q^\perp(B)$ where $B = A \cdot R^{-1} \in Z_q^{n\times m}$, and with overwhelming probability $\| \tilde{T}_B \| / \| \tilde{T}_A \| \leq \sigma_R m \omega(\log^{3/2} m)$.

For the security proofs of our scheme, we also introduce some more poly-time algorithms:

SampleR(1^n)[10]: On input a security parameter 1^n, the algorithm samples a matrice $R \in Z^{m\times m}$ from a distribution that is statistically close to $D_{m\times m}$.

SampleRwithBasis(A)[10]: On input a random rank *n* matrix $A \in Z_q^{n\times m}$, the algorithm outputs a matrix $R \in Z^{m\times m}$ sampled from $D_{m\times m}$ along with a short basis for $\Lambda_q^\perp(AR^{-1})$.

Security of our constructions reduces to the *small integer solution problem* (SIS) problem. We describe it as follows.

Definition 4. The small integer solution SIS problem (in the Euclidean norm) is as follows: given an integer q, a matrix $A \in Z_q^{n \times m}$, and a real β, find a nonzero integer vector $e \in Z^m$ such that $A \cdot e = 0 \bmod q$ and $\| e \|_2 \leq \beta$.

For security parameter n, we let $SIS_{q,m,\beta}$ denote the ensemble over instances $q(n), A, \beta(n))$, where $A \in Z_q^{n \times m}$ is uniformly random. Micciancio and Regev [16] showed that for any poly-bounded m, β and for any prime $q \geq \beta \cdot \omega(\sqrt{n \log n})$, the average-case problem $SIS_{q,m,\beta}$ is as hard as approximating the SIVP problem in the worst case within a factor $\tilde{O}(\beta \cdot \sqrt{n})$.

3 The ID-Based Signature Scheme from Lattices

3.1 Construction

Let n, m be positive integers, q be prime, with $q \geq 2$ and $m \geq 5n \log q$.

- *Setup*(1^n): On input a security parameter n, invoke *TrapGen(q,n)* to generate a uniformly random matrix $A \in Z_q^{n \times m}$ along with a short basis T_A for $\Lambda_q^{\perp}(A)$, and output the public parameters $PP = A$ and master key $MK = T_A$.
- *Extract*(PP, id, T_A): To extract a decryption key corresponding to the identity $id \in \{0,1\}^*$ using the master secret T_A.
  1. Compute $R_{id} = H_1(id) \in Z^{m \times m}$, $F_{id} = A \cdot R_{id}^{-1}$.
  2. Generate $T_{id} \leftarrow BasisDel(A, R, T_A, s)$ to obtain a short random basis for $\Lambda_q^{\perp}(F_{id})$, where s is a parameters for Gaussian sampling.
  3. Output the identity based private key T_{id}
- *Sign*(T_{id}, m) : For input a message m, compute $h \leftarrow H_2(m) \in R_n$, $\delta \leftarrow SamplePre(T_{id}, h)$, and output δ as the signature.
- *Verify*$(PP, id, (m, \delta))$: Let $R_{id} = H_1(id)$, $F_{id} = A \cdot R_{id}^{-1} \in Z_q^{n \times m}$. If $\delta \in D_n$ and $F_{id}\delta \bmod q = hash(m)$, accept. Else, reject.

The scheme's correctness is obvious. It can be seen as an identity-based Full-Domain Hash (FDH) scheme. And it is easy to translate into a probabilistic identity-based FDH scheme by appending "salt" for each signature.

3.2 Proof of Security

Proposition 2. In the random oracle mode, suppose there is an polynomial-time adversary $\mathcal{F}$ that can break the existential unforgeability of the signature scheme with probability ε, then we can construct a polynomial-time algorithm $\mathcal{S}$ that can solve the $ISIS_{q,m,s\sqrt{m}}$ with probability $\varepsilon(Q_{H_1}-1)/Q_{H_1}{}^2$, where Q_{H_1} denotes the number of queries that $\mathcal{F}$ can query to the random oracle $H_1(.)$.

Proof: With the adversary $\mathcal{F}$, we construct the algorithm $\mathcal{S}$ as follows:

1. For input an instance $(q, A \in Z_q^{n \times m}, 2s\sqrt{m})$, $\mathcal{S}$ samples a random matrices $R^* \sim D_{m \times m}$ by running $R^* \leftarrow SampleR(1^m)$, and gives the system parameters $A_0 = A \cdot R^*$ to $\mathcal{F}$.

2. $\mathcal{S}$ selects a random integer $Q^* \in [1, Q_{H_1}]$, and answers queries of $\mathcal{F}$ as follows:

-- $H_1(.)$ query: For the Q_th query input $id \in \{0,1\}^*$, if this is query number Q^*, $\mathcal{S}$ defines $H_1(id) = R^*$ and return $H_1(id)$. Otherwise, $\mathcal{S}$ runs $SampleRwithBasis(A)$ to obtain a random $R \sim D_{m \times m}$ and a short basis T_B for $B = A_0 R^{-1} \bmod q$, saves the tuple (id, R, B, T_B) in H_1_list for future use, and returns $H_1(id) = R$.

-- $H_2(.)$ query: For input $m \in \{0,1\}^*$, $\mathcal{S}$ chooses a random $e_m \leftarrow D_{Z^m,s}$ by running the algorithm $SampleDom(1^n)$, returns $y_m = Ae_j \bmod q \in Z_q^n$ and stores (m, e_m, y_m) in the H_2_list for future use.

-- *Extract*(.) query: For input $id \in \{0,1\}^*$, if $H_1(id) = R^*$, $\mathcal{S}$ aborts and fails. Otherwise, $\mathcal{S}$ retrieves the saved tuple (id, R, B, T_B) from the H_1_list (w.l.o.g., we can assume that an extraction query on id is preceded by a $H_1(.)$ query with id). By construction $B = A_0 R^{-1} \bmod q$, and T_B is a short Basis for $\Lambda_q^{\perp}(B)$. Return T_B as the secret key to the adversary.

-- *Sign*(.) query: For input identity $id \in \{0,1\}^*$ and message $m \in \{0,1\}^*$, it can be assumed, without loss of generality, that $\mathcal{F}$ has made $H_2(.)$ query on m and $H_1(.)$ query on id. If $H_1(id) = R^*$, $\mathcal{S}$ retrieves the tuple (m, e_m, y_m) in the H_2_list and returns e_m. Otherwise, $\mathcal{S}$ retrieves the saved tuple (id, R, B, T_B) from the H_1_list, runs the algorithm $e_m' \leftarrow SamplePre(T_B, B, y_m)$ and returns e_m' as the reply.

3. When $\mathcal{F}$ outputs a valid forgery (id^*, m^*, σ^*) such that id^* and (id^*, m^*) are not equal to the inputs of any query to *Extract*(.) and *Sign*(.). If $H_1(id^*) = R^*$, $\mathcal{S}$ looks up (m^*, e^*) in H_2_list. Then $A \cdot \sigma^* = H(m^*) = A \cdot e^* \bmod q$, and $A(\sigma^* - e^*) = 0 \bmod q$. With the min-entropy property, $\sigma^* \neq e^*$ with overwhelming probability expect with negligible $2^{-\omega(\log n)}$, hence $\mathcal{S}$ can output $\sigma^* - e^*$ as the solution for the $ISIS_{q,m,s\sqrt{m}}$ instance.

Because Q^* is chosen randomly in 1 and Q_{H_1}, $\mathcal{S}$ does not aborts in the simulation of *Extract*(.) with probability $1-1/Q_{H_1}$. At setp3, the probability of $H_1(id^*)=R^*$ is $1/Q_{H_1}$. So we conclude the probability that $\mathcal{S}$ solves the $ISIS_{q,m,s\sqrt{m}}$ instance is $\frac{\varepsilon(Q_{H_1}-1)}{{Q_{H_1}}^2}$.

4 Extension

With ID-based signature, we can construct some extension schemes. In this section, we propose an ID-based blind signature scheme. Blind signature plays an important role in many applications such as electronic payment and electronic voting. The *Setup*, *Extract*, *Verify* algorithms are the same as that in the above scheme.

Blind signature issuing protocol:
Suppose that m is the message to be signed. The protocol is shown in Fig. 1.

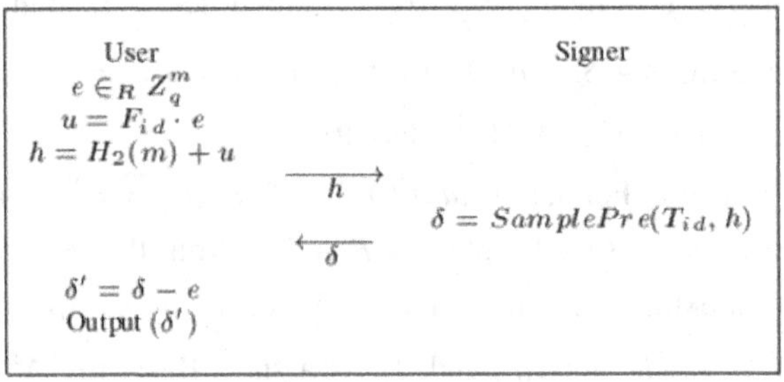

User		Signer
$e \in_R Z_q^m$ $u = F_{id} \cdot e$ $h = H_2(m) + u$	$\xrightarrow{h}$	
		$\delta = SamplePre(T_{id}, h)$
	$\xleftarrow{\delta}$	
$\delta' = \delta - e$ Output (δ')		

Fig. 1. The blind signature issuing protocol

--(Blinding) The User randomly chooses $e \in Z_q^m$, compute $u = F_{id} \cdot e$, $h = H_2(m) + u$, and sends h to the signer.

--(Signing) The signer sends back $\delta = Sample\Pr e(T_{id}, h)$.

--(Unblinding) The user computes $\delta' = \delta - e$, and outputs δ' as the blind signature of the message m.

The verification of the signature is justified by the following equations:

$$F_{id}\delta' \bmod q = F_{id}(\delta - e) \bmod q = h - u = H_2(m)$$

The above scheme holds unforgeability with the proof almost identitcal the prior one. The blindness can be proved in the full version of this paper.

Proposition 3. The proposed scheme holds blindness.

5 Conclusion

Lattice-based cryptosystems are becoming an increasingly popular in the research community. In this paper, we propose an ID-based signature scheme and its blind version using lattices. The schemes use lattice basis delegation with fixed-dimension technique for extracting users' secret key, and hence achieve short secret-keys and short signatures. The security of our constructions can be reduced to the *small integer solution problem* (SIS) problem in the random oracle model.

References

1. Shamir, A.: Identity-Based Cryptosystems and Signature Schemes. In: Blakely, G.R., Chaum, D. (eds.) CRYPTO 1984. LNCS, vol. 196, pp. 47–53. Springer, Heidelberg (1985)
2. Boneh, D., Franklin, M.: Identity-Based Encryption from the Weil Pairing. In: Kilian, J. (ed.) CRYPTO 2001. LNCS, vol. 2139, pp. 213–229. Springer, Heidelberg (2001)
3. Gorantla, M.C., Gangishetti, R., Saxena, A.: A Survey on ID-Based Cryptographic Primitives. Cryptology ePrint Archive, Report 2005/094, `http://eprint.iacr.org/2005/094.pdf`
4. Ajtai, M.: Generating hard instances of lattice problems (extended abstract). In: STOC 1996: Proceedings of the Twenty-Eighth Annual ACM Symposium on Theory of Computing, pp. 99–108. ACM, New York (1996)
5. Gentry, C., Peikert, C., Vaikuntanathan, V.: Trapdoors for hard lattices and new cryptographic constructions. In: Ladner, R.E., Dwork, C. (eds.) STOC, pp. 197–206. ACM (2008)
6. Ajtai, M., Dwork, C.: A public-key cryptosystem with worst-case/average-case equivalence. In: STOC, pp. 284–293 (1997)
7. Gentry, C.: Fully homomorphic encryption using ideal lattices. In: STOC, pp. 169–178 (2009)
8. Gentry, C., Peikert, C., Vaikuntanathan, V.: Trapdoors for hard lattices and new cryptographic constructions. In: Ladner, R.E., Dwork, C. (eds.) STOC, pp. 197–206. ACM (2008)
9. Agrawal, S., Boneh, D., Boyen, X.: Efficient Lattice (H)IBE in the Standard Model. In: Gilbert, H. (ed.) EUROCRYPT 2010. LNCS, vol. 6110, pp. 553–572. Springer, Heidelberg (2010)
10. Agrawal, S., Boneh, D., Boyen, X.: Lattice Basis Delegation in Fixed Dimension and Shorter-Ciphertext Hierarchical IBE. In: Rabin, T. (ed.) CRYPTO 2010. LNCS, vol. 6223, pp. 98–115. Springer, Heidelberg (2010)
11. Boyen, X.: Expressive Encryption Systems from Lattices. In: Lin, D., Tsudik, G., Wang, X. (eds.) CANS 2011. LNCS, vol. 7092, pp. 1–12. Springer, Heidelberg (2011)
12. Cha, J.C., Cheon, J.H.: An Identity-Based Signature from Gap Diffie-Hellman Groups. In: Desmedt, Y.G. (ed.) PKC 2003. LNCS, vol. 2567, pp. 18–30. Springer, Heidelberg (2002)

13. Zhang, F., Kim, K.: ID-Based Blind Signature and Ring Signature from Pairings. In: Zheng, Y. (ed.) ASIACRYPT 2002. LNCS, vol. 2501, pp. 533–547. Springer, Heidelberg (2002)
14. Boyen, X.: Lattice Mixing and Vanishing Trapdoors: A Framework for Fully Secure Short Signatures and More. In: Nguyen, P.Q., Pointcheval, D. (eds.) PKC 2010. LNCS, vol. 6056, pp. 499–517. Springer, Heidelberg (2010)
15. Alwen, J., Peikert, C.: Generating shorter bases for hard random lattices. In: STACS, pp. 75–86 (2009)
16. Micciancio, D., Regev, O.: Worst-case to average-case reductions based on Gaussian measures. SIAM J. Comput. 37(1), 267–302 (2007)

Strongly Secure Attribute-Based Authenticated Key Exchange with Traceability

Fugeng Zeng[1,2], Chunxiang Xu[1], Xiujie Zhang[1], and Jian Liu[1]

[1] School of Computer Science and Engineering, University of Electronic Science and Technology of China, Chengdu, 611731, China
[2] School of Science and Engineering, Qiongzhou University, Hainan 572000, China
zengfugeng@gmail.com, chxxu@uestc.edu.cn, {2008xiujie,cxljian}@163.com

Abstract. Attribute-Based Encryption (ABE) scheme is an epidemic and emerging computing paradigm currently and has a broad application in the area of fine-grained access control. Combining with *NAXOS* transformation, we proposed a strongly secure Attribute-based Authenticated key exchange protocol, whose security proof is under the gap Bilinear Diffie-Hellman assumption in the Attribute-Based eCK(ABeCK) model. We also use group signature to achieve traceability of the user.

Keywords: ABE, NAXOS, ABeCK, traceability.

1 Introduction

Authenticated Key Exchange (AKE) plays an important role in the field of secure communications. In an open network environment, AKE protocol enables two parties, Alice (A) and Bob (B), to establish a shared session key over an insecure channel. Later, the session key can be used to ensure data confidentiality and integrity between A and B using efficient symmetric encryptions and message authentication codes. To achieve authentication, traditional public key AKE protocol must know two parties' identities, however, in a distributed environment such as an epidemic and emerging cloud computing, the reveal of identities will affect users' privacy. Attribute-Based AKE (ABAKE) can hide the identity information of the individual, which allows users to achieve mutual authentication and establish a session secret key by their attributes and some fine-grained access control policy.

As a prototype of ABAKE, Ateniese et al.[1] proposed a secret handshakes scheme with dynamic and fuzzy matching, which allows two members of the same group to secretly authenticate to each other and agree on a shared key for further communication. Wang et al. [2] proposed simple variants of ABAKE. In their protocols, attributes are regarded as identification strings, and their protocols are in fact a kind of ID-based AKE rather than ABAKE. Gorantla et al. [3] proposed attribute-based group AKE protocol, whose protocol is generically constructed with a attribute-based KEM, and allows users access control based on the users' attributes

F.L. Wang et al. (Eds.): WISM 2012, LNCS 7529, pp. 231–238, 2012.

and security is proved in a security model based on the Bellare Rogaway (BR) model [4],which is not allowed the reveal of EphemeralKey. Birkett and Stebila [5] demonstrated how to realize a secure predicate-based key exchange protocol by combining any secure predicate-based signature scheme with the basic Diffie-Hellman key exchange protocol. They prove security of their protocol without random oracles in a security model based on the BR model.

The BR model does not capture this property to resistant to leakage of ephemeral secret keys. Leakage of ephemeral secret keys will occur due to various factors, for example, a pseudorandom generator implemented in the system is poor, and the ephemeral secret key itself is leaked by physical attacks such as side channel attacks. Thus, if we consider such cases, resistance to leakage of ephemeral secret keys is important for AKE protocols and the extended Canetti-Krawzcyk (eCK) security model [6], which captures leakage of ephemeral secret keys, is used for security proofs of some recent AKE protocols. To prove the security in the eCK model, it is known that the NAXOS technique [6] is effective. The NAXOS technique involves that each user applies a hash function to the static and ephemeral secret keys and computes an ephemeral public key by using the output of the hash function as the exponent of the ephemeral public key. An adversary cannot know the output of the function as long as the adversary cannot obtain the static secret key even if the ephemeral secret key is leaked.

Yoneyama[7] presented a two-party attribute-based authenticated key exchange scheme secure in the the eCK model, and proved the security of their scheme under the gap Bilinear Diffie-Hellman assumption. The recent research was completed by Fujioka et al [8]. They proposed two strongly secure Predicate-Based Authenticated Key Exchange protocols secure in the proposed predicate-based eCK security model.

However, all of the protocols above do not consider traceability. In our work, we use group signature to achieve traceability of the user. Assuming under the environment of University of Electronic Science and Technology (UESTC), there are Mathematics(Math) workgroup and Computer Science (CS) workgroup. Alice associated with attributes γ_A from Math want to establish a session key between someone from CS with access control policy δ_A without exposing her own identity. At the same time someone from CS associated with attributes γ_B (maybe Bob) can also develop an access control policy δ_B at the premise of that Bob does not know the identity of Alice. When there is a dispute, such as Alice would like to know the identity who establishes the session key with her, she can seek the help of the CS Workgroup administrator, and then trace the identity of Bob.

Related Work

Attribute-Based Encryption: ABE was first introduced by Sahai and Waters [9] in the context of a generalization of IBE called Fuzzy IBE, which is an ABE that allows only single threshold access structures. Later, two flavors of ABE were proposed: key-policy ABE, where a ciphertext is associated to a list of attributes, and a secret key is associated to a policy for decryption; and ciphertext-policy ABE, where secret keys are associated to a list of attributes The first construction of KP-ABE which allows any monotone access structures was provided by Goyal et al. [10]. After then, Bethencourt, Sahai, and Waters [11] proposed the first CP-ABE scheme, but the security of their scheme was only proved in the generic group model. Ostrovsky, Sahai, and Waters subsequently extended both schemes to support any non monotone structures[12]. Cheung and Newport [13] proposed a CP-ABE scheme that allows only AND gate which could be proved in the standard model, while Goyal et al. [14] proposed a bounded CP-ABE scheme which allows only a priori bounded size of gates. Waters [15] proposed the first fully expressive CP-ABE in the standard model.

Group Signature: Group signatures, introduced by Chaum and Heyst [16], provide anonymity for signers. Any member of the group can sign messages, but the resulting signature keeps the identity of the signer secret. Often there is a third party that can undo the signature anonymity (trace) using a special trapdoor. Boneh et al.[17] construct a short group signature scheme. Signatures in their scheme are approximately the size of a standard RSA signature with the same security. Security of their group signature is based on the strong Diffie-Hellman assumption and Decision Linear assumption.

Organization of the paper*:* In section 2, we review the concepts of preliminaries. We introduce the definition and secure model of ABAKE in section 3. Then we introduce our ABAKE scheme in section 4 and give its security analysis in section 5. Finally, we conclude our work.

2 Preliminaries

We give a brief review on the property of pairings and some candidate hard problems from pairings that will be used later.

2.1 Bilinear Map

Let G_1 and G_2 be cyclic groups of prime order p with the multiplicative group action and g is a generator of G_1. Let $e: G_1 \times G_1 \rightarrow G_2$ be a map with the following properties:

1.Bilinearity: $e(g^a, g^b) = e(g, g)^{ab}$ for $\forall$ a,b□Z_p^*

2. Non-degeneracy: $e(g, g) \neq 1$.

3.Computability: There is an efficient algorithm to compute $e(u, v)$ for all u, v□G_1.

2.2 Gap Bilinear Diffie-Hellman Assumption

Let κ be the security parameter and G be a cyclic group of a prime order p with a generator g and G_T be a cyclic group of the prime order p with a generator g_T. Let e

$: G \times G \rightarrow G_T$ be a bilinear map. We say that G, G_T are bilinear groups with the pairing e.The gap Bilinear Diffie-Hellman problem is as follows. We define the BDH function BDH $: G^3 \rightarrow G_T$ as $\text{BDH}\left(g^a, g^b, g^c\right) = e(g, g, g)^{abc}$, and the DBDH predicate DBDH: $G^4 \rightarrow \{0,1\}$ as a function which takes an input $\left(g^a, g^b, g^c, e(g,g)^d\right)$ and returns 1 if $abc = d \bmod p$ and 0 otherwise. An adversary A is given input $\alpha = g^a, \beta = g^b, \gamma = g^c \in G$ selected uniformly random and oracle access to DBDH $(\cdot,\cdot,\cdot,\cdot)$ oracle, and tries to compute BDH (α, β, γ) .For adversary A, we define advantage $Adv^{GBDH}(A) = \Pr\left[(\alpha, \beta, \gamma) \in G, A^{GBDH(\cdot,\cdot,\cdot,\cdot)}(g, \alpha, \beta, \gamma)\right] = BDH(\alpha, \beta, \gamma)$, where the probability is taken over the choices of g^a, g^b, g^c and the random tape of A.

Definition 1 (Gap Bilinear Diffie-Hellman Assumption). We say that the GBDH assumption in G holds if for all polynomial-time adversary A, the advantage $Adv^{GBDH}(A)$ is negligible in security parameter κ.

3 Security Model

In this section, we referred an eCK security model for the ABAKE. We use the eCK model from [7].

4 Proposed ABAKE Protocol

In this section, we describe the proposed ABAKE protocol.

Setup: Let k be the security parameter and G, G_T be bilinear groups with pairing $e: G \times G \rightarrow G_T$ of order k-bit prime p with generators g, g_T respectively. Let $H: \{0,1\}^* \rightarrow \{0,1\}^k$ and $H': \{0,1\}^* \rightarrow Z_p$ be cryptographic hash functions modeled as random oracles. And let $\Delta_{i,s}$ for $i \in Z_p$, a set S of elements in Z_p be the Lagrange coefficient such that $\Delta_{i,s} = \prod_{j \in S, j \neq i} \frac{x-j}{i-j}$.

Key Generation: The key generator (algorithm) randomly selects a master secret key $z \in Z_p$ and $\{t_i \in Z_p\}_{i \in U}$ for each attribute. Also, the key generator publishes the master public key $Z = g_T^z \in G_T$ and $\{T_i = g^{t_i}\}_{i\in U}$. At the same time, we use group signature scheme to the workgroup referred to Section 1. A group signature private key will be assigned to each member from different workgroup as well as group public key is given in accordance with literature [17]. For example user Alice from the Math group, we will set a private key sk_{Alice}, Bob from CS group will obtain a private key sk_{Bob} for signature. We can make use of group signature to achieve authentication and traceability.

Key Extraction: For a given access tree γ_A of user U_A, the key extractor (algorithm) computes the static secret key $\{D_u\}_{u\in L(\gamma_A)}$ by choosing a polynomial q_u for each node u in A as follows:

First, the key extractor lets the degree of the polynomial q_u to be $d_u = k_u - 1$. For the root node u_r, $q_{u_r}(0) = z$ and other d_{u_r} points of q_{u_r} are randomly chosen from Z_p. Thus, q_{u_r} is fixed and other $c_{u_r} - d_{u_r}$ points are determined. For any other node u, set $q_u(0) = q_u(indes(u'))$, where u' is the parent node of u and other d_u points of q_u are randomly chosen from Z_p. Polynomials of all nodes are recursively determined with this procedure. Next, the key extractor computes a secret value for each leaf node u as $D_u = g^{\frac{q_u(0)}{t_i}}$, where $i{=}att(u)$. Finally, the key extractor returns the set $\{D_u\}_{u\in L(\gamma_A)}$ of the above secret values as the static secret key. The static secret key $\{D_u\}_{u\in L(\gamma_B)}$ for an access tree γ_B of user U_B is derived from the same procedure.

Key Exchange: In the following description, user U_A is the session initiator and user U_B is the session responder.

1. First, U_A determines a set $\delta_A \subset$U of attributes in which he hopes δ_A satisfies the access tree γ_B of U_B. U_A randomly chooses an ephemeral private key $\tilde{x} \in Z_p$ .Then, U_A computes $x = H'(\{D_u\}_{u\in L(\gamma_A)}, \tilde{x})$, and the ephemeral public key $X = g^x$ and $\{T_i^x\}_{i\in\delta_A}$ .U_A sends $M_1 = \{X, \{T_i^x\}_{i\in\delta_A}, \delta_A\}$ and $Sign_{sk_A}(M_1)$ to U_B.

2. Upon receiving $M_1 = \{X, \{T_i^x\}_{i\in\delta_A}, \delta_A\}$ and $Sign_{sk_A}(M_1)$, firstly U_B use the group public key from U_A to verify whether the signature is correct or not. If it's not correct, it will abort. Otherwise, U_B determines a set of attributes δ_B in which he hopes δ_B satisfies the access tree γ_A of U_A. U_B randomly chooses an ephemeral private key $\tilde{y} \in Z_p$. Then, U_B computes $y = H'(\{D_u\}_{u\in L(\gamma_B)}, \tilde{y})$, and the ephemeral

public key $Y = g^y$ and $\{T_i^y\}_{i\in\delta_B}$ .U_B sends M_2={Y, $\{T_i^y\}_{i\in\delta_B}$, δ_B }and $Sign_{sk_B}(M_2)$ to U_A.

U_B computes the shared secrets as follows: First, for ach leaf node u in γ_B, U_B computes $e(D_u, T_i^x) = e(g^{\frac{q_u(0)}{t_i}} g^{xt_i},) = e(g,g)^{xq_u(0)}$ where $i = att(u)$, if $att(u) \in \delta_A$. Next, for each non-leaf node u in γ_B, set S_u={u_c|u_c is a child node of u s.t. $e(g,g)^{xq_{u_c}(0)}$ obtained.}.

If $|S_u^{'}| \geq k_u$,U_B lets $S_u \subset S^{'}$,s.t. $|S_u| = k_u$ and computes $\prod_{u_c \in S_u} (e(g,g)^{xq_{u_c}(0)})^{\Delta_{i,S_u}(0)} = e(g,g)^{xq_u(0)}$,where $i = index(u_c)$. This computation validly works by using polynomial interpolation. On the output of the root node u_r of γ_B, U_B obtains $e(g,g)^{xq_r(0)} = e(g,g)^{xz}$ if δ_A satisfies the access tree γ_B.

Then, U_B sets the shared secrets $\sigma_1 = e(g,g)^{xz}, \sigma_2 = Z^y, \sigma_3 = X^y$ and the session key $K = H(\sigma_1, \sigma_2, \sigma_3, M_1, M_2)$. U_B completes the session with session key K.

3. Upon receiving M_2={ Y ,$\{T_i^y\}_{i\in\delta_B}$,δ_B }and $Sign_{sk_B}(M_2)$, firstly U_A use the group public key from U_B to verify whether the signature is correct or not. If it's not correct, it will abort. Otherwise, if δ_B satisfies the access tree γ_A, U_A obtains $e(g,g)^{yq_u(0)}$ and computes the shared secrets similarly as the step 2. U_A computes $\sigma_1 = Z^x, \sigma_2 = e(g,g)^{yz}, \sigma_3 = Y^x$, U_A completes the session with session key $K = H(\sigma_1, \sigma_2, \sigma_3, M_1, M_2)$.

The protocol runs between U_A and U_B. Its description is given in Figure 1.

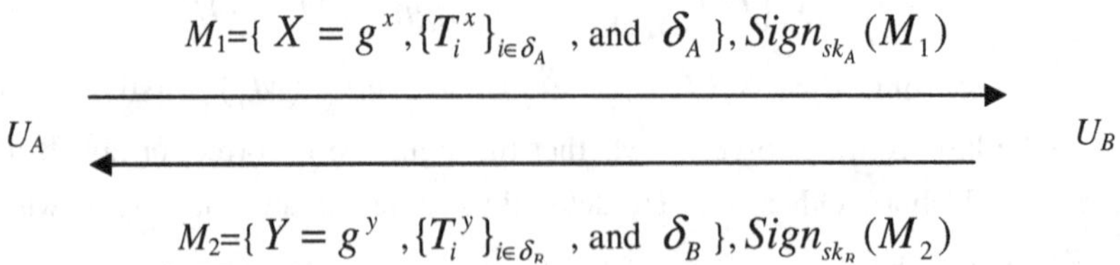

Fig. 1. Strongly Secure Attribute-based Authenticated Key Exchange with traceability

5 Security Analysis

In this section, we discuss the security proof of the protocol. We obtain Fig.2 by making some modification of the Fig.1. We can find that the first two pass messages can be viewed as an ABAKE with non traceability. The last two pass messages can be viewed as a group signature. So we obtain the following theorem:

Theorem 1: If G is a group, where the *gap BDH* assumption holds and H and H' are random oracles, the proposed attribute-based AKE protocol is selective-condition secure in the Attribute-based eCK model described in Sect. 3.

The proof is similar as the literature [8], if necessary, please refer to literature [8]

Theorem 2: Suppose the construction of group signature scheme [17] is secure and traceable, then we have provided a secure Attribute-based Authenticated Key Exchange with traceability.

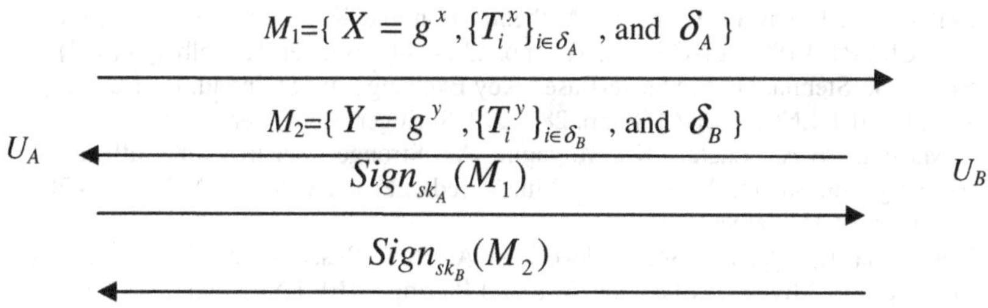

Fig. 2. Some modification of the Fig.1

6 Comparison and Conclusion

Comparison of the proposed protocols with existing Attribute-based AKE protocols is summarized in the Table 1. We compared the round complexity, security model, assumption and traceability among the protocols.

Table 1. Comparison with the existing Attribute-based AKE protocols

Protocol	Round complexity	Security model	assumption	traceability
BS[5]	Three pass	BR	PB-signature	no
FSY[8]	Two pass	eCK	Gap BDH	no
Our protocol	Two pass	eCK	Gap BDH	yes

Combining with NAXOS transformation, we proposed a strongly secure Attribute-based Authenticated key exchange protocol, whose security proof is under the gap Bilinear Diffie-Hellman assumption in the Attribute-Based eCK(ABeCK) model. We also use group signature to achieve traceability of the user.

Acknowledgments. This work is supported by Science and Technology on Communication Security Laboratory Foundation(NO.9140C110301110C1103) and The Weaponry Equipment Pre-Research Foundation, the PLA General Armament Department (NO.9140A04020311DZ02) and Scientific Research Foundation of Qiongzhou University(QYQN201241).

References

1. Ateniese, G., Kirsch, J., Blanton, M.: Secret handshakes with dynamic and fuzzy matching. In: Proceedings of the Network and Distributed System Security Symposium, pp. 159–177 (2007)
2. Wang, H., Xu, Q., Ban, T.: A provably secure two- party attribute-based key agreement protocol. In: Proceedings of Intelligent Information Hiding and Multimedia Signal Processing, pp. 1042–1045. IEEE Computer Society (2009)
3. Gorantla, M.C., Boyd, C., González Nieto, J.M.: Attribute-Based Authenticated Key Exchange. In: Steinfeld, R., Hawkes, P. (eds.) ACISP 2010. LNCS, vol. 6168, pp. 300–317. Springer, Heidelberg (2010)
4. Bellare, M., Rogaway, P.: Entity Authentication and Key Distribution. In: Stinson, D.R. (ed.) CRYPTO 1993. LNCS, vol. 773, pp. 232–249. Springer, Heidelberg (1994)
5. Birkett, J., Stebila, D.: Predicate-Based Key Exchange. In: Steinfeld, R., Hawkes, P. (eds.) ACISP 2010. LNCS, vol. 6168, pp. 282–299. Springer, Heidelberg (2010)
6. LaMacchia, B.A., Lauter, K., Mityagin, A.: Stronger Security of Authenticated Key Exchange. In: Susilo, W., Liu, J.K., Mu, Y. (eds.) ProvSec 2007. LNCS, vol. 4784, pp. 1–16. Springer, Heidelberg (2007)
7. Yoneyama, K.: Strongly Secure Two-Pass Attribute-Based Authenticated Key Exchange. In: Joye, M., Miyaji, A., Otsuka, A. (eds.) Pairing 2010. LNCS, vol. 6487, pp. 147–166. Springer, Heidelberg (2010)
8. Fujioka, A., Suzuki, K., Yoneyama, K.: Strongly Secure Predicate-Based Authenticated Key Exchange: Definition and Constructions. IEICE Transactions on Fundamentals of Electronics, Communications and Computer Sciences (1), 40–56 (2012)
9. Sahai, A., Waters, B.: Fuzzy Identity-Based Encryption. In: Cramer, R. (ed.) EUROCRYPT 2005. LNCS, vol. 3494, pp. 457–473. Springer, Heidelberg (2005)
10. Goyal, V., Pandey, O., Sahai, A., Waters, B.: Attribute-based encryption for fine-grained access control of encrypted data. In: ACM CCS 2006, pp. 89–98 (2006)
11. Bethencourt, J., Sahai, A., Waters, B.: Ciphertext-Policy Attribute-Based Encryption. In: IEEE Symposium on Security and Privacy (S&P), pp. 321–334 (2007)
12. Ostrovsky, R., Sahai, A., Waters, B.: Attribute-based encryption with non-monotonic access structures. In: ACM CCS 2007, pp. 195–203 (2007)
13. Cheung, L., Newport, C.: Provably secure ciphertext policy ABE. In: ACM CCS 2007, pp. 456–465 (2007)
14. Goyal, V., Jain, A., Pandey, O., Sahai, A.: Bounded Ciphertext Policy Attribute Based Encryption. In: Aceto, L., Damgård, I., Goldberg, L.A., Halldórsson, M.M., Ingólfsdóttir, A., Walukiewicz, I. (eds.) ICALP 2008, Part II. LNCS, vol. 5126, pp. 579–591. Springer, Heidelberg (2008)
15. Waters, B.: Ciphertext-Policy Attribute-Based Encryption: An Expressive, Efficient, and Provably Secure Realization. In: Catalano, D., Fazio, N., Gennaro, R., Nicolosi, A. (eds.) PKC 2011. LNCS, vol. 6571, pp. 53–70. Springer, Heidelberg (2011)
16. Chaum, D., van Heyst, E.: Group Signatures. In: Davies, D.W. (ed.) EUROCRYPT 1991. LNCS, vol. 547, pp. 257–265. Springer, Heidelberg (1991)
17. Boneh, D., Boyen, X., Shacham, H.: Short Group Signatures. In: Franklin, M. (ed.) CRYPTO 2004. LNCS, vol. 3152, pp. 41–55. Springer, Heidelberg (2004)

A New Public Key Signature Scheme Based on Multivariate Polynomials

Feng Yuan, Shangwei Zhao, Haiwen Ou, and Shengwei Xu

Beijing Electronic Science and Technology Institute, Beijing, China
fyuan1234@yahoo.cn,
{zsw,ouhw,xusw}@besti.edu.cn

Abstract. This paper proposes a new public key signature scheme based on multivariate polynomials over a finite field with characteristic 2. This scheme has a very simple internal transformation, allowing for efficient signature generation and verification. The security of the scheme is analyzed in detail. The result indicates that the new signature scheme can withstand all known attacks effectively.

Keywords: public key cryptography, polynomials, mapping, attack, finite field.

1 Introduction

Multivariate public key cryptosystems (MPKC) are asymmetric primitives based on hard computational problems involving multivariate polynomials. In the last twenty years, multivariate public key cryptosystems have increasingly been seen by some as a possible alternative to number theoretic-based cryptosystems such as RSA and elliptic curve schemes, and are considered viable options for post-quantum cryptography. This constructions rely on the proven theorem that solving a set of multivariate polynomial equations over a finite field is, in general, an NP-hard problem [1].

In 1988, Matsumoto and Imai [2] presented the C^* scheme with multivariate polynomials. This scheme was based on the idea of "hiding" a monomial by two invertible affine maps. However, Patarin [3] indicated the C^* scheme was insecure. Then Patarin [4] presented the new scheme HFE. The Hidden Field Equation (HFE) cryptosystem is a family of cryptosystem based on a more complicated hidden function over an extension field. The practical cryptanalysis of HFE scheme was provided by Faugère and Joux [5]. In 1998, Patarin et al. [6] proposed the C^{*-} scheme with the minus method. The SFLASH scheme [7] is the C^{*-} scheme with specific parameters, which was selected in 2003 by the NESSIE European Consortium as one of the three recommended public key signature schemes [8]. In 2007, Dubois et al. [9, 10] developed the differential attack and applied the method to attack the SFLASH scheme practically. In 2008, Ding et al. [11] proposed the projected C^{*-} scheme. Although the projected C^{*-} scheme is a simple modification of the C^{*-} scheme, it can resist the differential attack efficiently.

F.L. Wang et al. (Eds.): WISM 2012, LNCS 7529, pp. 239–245, 2012.

In this paper, we presents a new public key signature scheme based on multivariate polynomials over a finite field with characteristic 2. The internal mapping is very simple over an extension field. We analyze the security of the new signature scheme in detail. The result shows that the signature scheme can not only withstand the Linearization Equations attack [3], Rank attack [12], XL&Gröbner basis attacks [13, 14] and Kipnis-Shamir attack [15], but also the differential attack [9, 10].□

In section 2, we propose the new multivariate public key signature scheme. In section 3, we discuss the basic cryptanalytic property of the internal transformation. In section 4, we introduce the security analysis of the signature scheme in detail. Section 5 is the conclusion of this paper.

2 The New Signature Scheme

Let k be a small finite field of 2 elements. Let K be an n-degree extension of the finite field k with characteristic two, and $K = k[x]/(f(x))$, where $f(x)$ is any irreducible polynomial of degree n over finite field k, i.e., $k = \mathrm{GF}(2)$, $K = \mathrm{GF}(2^n)$.

Let $\phi: K \to k^n$ be the standard k-linear isomorphism between K and k^n given by

$$\phi(a_0 + a_1 x + ... + a_{n-1} x^{n-1}) = (a_0, a_1, ..., a_{n-1}) .$$

Define the internal transformation $F: K \to K$ by

$$F(X) = X^{2^\theta + 3}$$

where $n=2\theta+1$, $(17, 2^\theta + 3) = 1$. We can know that F is an invertible mapping, if t is an integer such that $t(2^\theta + 3) \equiv 1 \bmod (2^n - 1)$, then F^{-1} is simply $F^{-1}(X) = X^t$.

Let r and s be two integers between 0 and n. Let $T^-: k^n \to k^{n-r}$ be the projection of invertible affine mapping $T: k^n \to k^n$ on the last r coordinates and $U^-: k^{n-s} \to k^n$ be the restriction of invertible affine mapping $U: k^n \to k^n$ on the last s coordinates. Then we will have a new map $P: k^{n-s} \to k^{n-r}$ defined by

$$\begin{aligned} P(x_1, ..., x_{n-s}) &= T^- \circ F \circ U^-(x_1, ..., x_{n-s}) \\ &= T^- \circ \phi \circ F \circ \phi^{-1} \circ U^-(x_1, ..., x_{n-s}) \\ &= (f_1, ..., f_{n-r}) . \end{aligned}$$

2.1 Key Generation

The Public Keys

1) The field k including its additive and multiplicative structure;
2) The $n-r$ polynomials: $f_1, ..., f_{n-r} \in k[x_1, ..., x_{n-s}]$.

The Private Keys

The private keys include two invertible affine mappings U and T.

2.2 Signature Generation

The document is $Y^- = (y_1,\ldots,y_{n-r})$, a vector in k^{n-r}. A user first chooses r random elements $y_{n-r+1},\ldots,y_n$, which are appended to Y^- to produce $Y = (y_1,\ldots,y_n)$ in k^n. Then the user recovers $m = (x_1,\ldots,x_n)$ by inverting T, F and U. If the m has its last s coordinates to 0, then its first $n-s$ coordinates are a valid signature for Y^-. Otherwise, the user discards the m and tries with another random $y_{n-r+1},\ldots,y_n$. When $r>s$, the process ends with probability 1 and costs on average 2^s inversions of F.

2.3 Signature Verification

A signature (Y^-, m) can be checked by computing $Y^- = P(m)$, which is extremely fast since it only involves the evaluation of a small number of cubic equations over the small finite field $k = \mathrm{GF}(2)$.

In order to the security and efficiency, the new signature scheme with the following parameter choices will be viable: $\theta = 55, n = 111, r = 33$ and $s = 3$.

3 The Cryptanalytic Property of the Internal Transformation

We first recall the definition of APN function [16]. The APN function opposes then an optimum resistance to differential cryptanalysis.

For any $0 \neq A \in k^n$, $k = \mathrm{GF}(2)$, the differential of the function $F{:}k^n \to k^n$ is as follow:

$$D_A F(X) = F(X + A) - F(X).$$

Definition 3.1. Let F be a function from k^n into k^n. For any A and B in k^n, we denote

$$\delta(A, B) = \#\{X \in k^n, F(X + A) - F(X) = B\}$$

where $\#E$ is the cardinality of any set E. Then, we have

$$\delta_F = \max_{A,B\in k^n,\, A\neq 0} \#\{X \in k^n, F(X + A) - F(X) = B\} = 2$$

and the functions for which equality holds are said to be almost perfect nonlinear (APN).

The property of APN function can be easily found as follows.

Proposition 3.2. Let F be any function over k^n. Then, F is APN if and only if, for any nonzero $A \in k^n$, the set $\{D_A F(X) \mid X \in k^n\}$ has cardinality 2^{n-1}.

Then we review the Lemma 3.3 as follows.

Lemma 3.3. [17] Let $K = \mathrm{GF}(2^n)$, where $n=2\theta+1$, then

$$q(X) = X^{2^{\theta+1}+1} + X^3 + X$$

is a permutation over K.

From the Lemma 3.3, the Theorem 3.4 can be found as follows.

Theorem 3.4. The function $F(X)=X^{2^\theta+3}$ is a APN function, where $n=2\theta+1$, $(17, 2^\theta+3)=1$.

Proof. Let $l=2^\theta+3$

$$D_A F(X)=F(X+A)-F(X)=A^l[(\frac{X}{A}+1)^l-(\frac{X}{A})^l]$$

Define $Y=X/A$, then

$$D_A F(X)=A^l[(Y+1)^l-Y^l]=A^l[(X^{2^\theta}+X)(X^2+X+1)+1]$$

we denote

$$D_A F(X)=A^l[q(X^{2^\theta}+X)+1]$$

Since $X^{2^\theta}+X$ is a 2-1 mapping, and $q(X)$ is a permutation polynomial, then $D_A F(X)$ is a 2-1 mapping over *K*. Hence $F(X)=X^{2^\theta+3}$ is a APN function.

From the Theorem 3.4, The internal transformation $F(X)=X^{2^\theta+3}$ has the APN property.

4 Security Analysis

In this section, we analyze the security of the signature scheme in detail.

4.1 Linearization Equations Attack

Linearization Equations attack was proposed by Patarin [3] in 1995. In the new scheme, it can be noted that Patarin's bilinear equations also exist between $(x_1,x_2,\dots,x_{n-s})$ and $(f_1,f_2,\dots,f_{n-r})$. Since

$$Y=F(X)=X^{2^\theta+3}$$

If we multiply both sides by *XY*, we then have

$$XY^2=X^{2^\theta+4}Y$$

$$XY^2=X^{2^\theta+2^2}Y$$

or equivalently,

$$XY^2-X^{2^\theta+2^2}Y=0$$

Then we can find the Linearization Equations from the public key polynomials as follows.

$$\sum a_{ijk}x_ix_jf_k+\sum b_{ij}x_ix_j+\sum c_{ij}x_if_j+\sum d_ix_i+\sum e_if_i+g=0$$

where $a_{ijk},b_{ij},c_{ij},d_i,e_i,g\in k=\mathrm{GF}(2)$. It is easy to see that the Linearization Equations is quadratic. We can not solve the a system of Linearization Equations. Therefore, the new scheme can resist the Linearization Equations attack effectively.

4.2 Rank Attack

The Low Rank attack is to find a large kernel shared by a small subset of the space spanned by the public key matrices. The converse, to find a small kernel shared by a many linear combinations of the public key matrices, may be called a High Rank attack. It happens when a variable appears in too few internal transformation equations.

The High Rank attack has a complexity of $O(q^u(un^2 + n^3/6))$ [12], where u is every variable appears in at least u internal transformation equations, n is the variable number of internal transformation equations, q is the number of elements of finite field K. When $u = 1$, $q = 2^{111}$ and $n = 1$, the complexity of the High Rank attack is at least $O(2^{111})$, so the new scheme can resist the High Rank attack. The Low Rank attack has a complexity of $O(q^{\lceil \frac{m}{n} \rceil r} m^3)$ [12], where m is the number of internal transformation equations, n is the variable number, r is minimum rank, q is the number of elements of finite field K. When $q=2^{111}$, $m = n = 1$ and $r = 1$ or 2, if $r = 1$, the complexity of the Low Rank attack is at least $O(2^{111})$; if $r = 2$, the complexity of the Low Rank attack is at least $O(2^{222})$. Therefore, the new scheme can also resist the Low Rank attack.

4.3 XL&Gröbner Attacks

This attack uses the general method of solving a system of multivariate polynomial equations, for example using the Grobner basis method, including the Buchberger algorithm, its improvements (such as F_4 [14] and F_5 [18]), the XL algorithm [13], and so on. Currently XL algorithm has also been widely discussed but F_4 is seen to be more efficient [19]. When $k = \mathrm{GF}(2)$ is a two element field, the complexity of the attack is $O(2^{0.87n})$ [20], where n is the number of the variable in the multivariate equations. When $n = 111 - 3 = 108$, the complexity is at least $O(2^{94})$, hence the new scheme can also resist the XL&Gröbner attacks effectively.

4.4 Kipnis-Shamir Attack

The key idea of Kipnis-Shamir attack [15] is to transform the map P into a map $\phi^{-1} \circ P \circ \phi$, in order to use the underlying algebraic structure over the extension field. Then we can see a large kernel shared by a many linear combinations of the public key matrices. In fact, it is a Low Rank attack problem. The key to success of the Kipnis-Shamir attack mainly depends on

$$(\phi^{-1} \circ T^{--1} \circ \phi) \circ (\phi^{-1} \circ P \circ \phi) = F \circ (\phi^{-1} \circ U^{-} \circ \phi)$$

But according to the form of T^{-} in the new scheme, T^{-1} does not exist, so the new scheme can resist the Kipnis-Shamir attack.

4.5 Differential Attack

The quadratic differential of the internal transformation F is as follow:

$$\begin{aligned} DF_{B,A}(X) &= DF_B(F(X+A)-F(X)) \\ &= F(X+A+B)-F(X+B)-F(X+A)+F(X) \\ &= (X+A+B)^{2^\theta+3}-(X+B)^{2^\theta+3}-(X+A)^{2^\theta+3}+X^{2^\theta+3} \end{aligned}$$

Since

$$\begin{aligned} &F(X+A)-F(X)) \\ &= X^{2+1}A^{2^\theta}+A^2X^{2^\theta+1}+XA^{2^\theta+2}+AX^{2^\theta+2}+X^2A^{2^\theta+1}+A^{2+1}X^{2^\theta}+A^{2^\theta+2+1} \end{aligned}$$

we can see that

$$\begin{aligned} &DF_{B,A}(X) \\ &= X(A^{2^\theta}B^2+A^2B^{2^\theta})+X^2(A^{2^\theta}B+AB^{2^\theta})+(A+B)^{2^\theta+2+1}+A^{2^\theta+2+1}+B^{2^\theta+2+1} \end{aligned}$$

In 2008, Ding et al. [11] found that the internal large field K was restricted to a purely linear subspace and the symmetry exploited by the differential attack did no longer exist. Based on this important result, they proposed the projected C^{*-} scheme. In the new scheme, we also adopt the project method. By the analysis above, we can know that the new scheme can resist the differential attack.

5 Conclusion

In this paper, we applies the minus method and project method to present a multivariate public key signature scheme over a finite field with characteristic 2. We discuss the basic cryptanalytic property of the internal mapping. The security of the new signature scheme is investigated in detail. The result shows that the signature scheme can withstand the known attacks efficiently.

Acknowledgments. This work is supported by the National Natural Science Foundation of China under Grants No.61103210, the Fundamental Research Funds for the Central Universities: YZDJ 1102 and YZDJ 1103, and the Fund of BESTI Information Security Key Laboratory: YQNJ 1005.

References

1. Ding, J., Gower, J.E., Schmidt, D.S.: Multivariate public key cryptosystems. Springer, New York (2006)
2. Matsumoto, T., Imai, H.: Public Quadratic Polynomial-Tuples for Efficient Signature-Verification and Message-Encryption. In: Günther, C.G. (ed.) EUROCRYPT 1988. LNCS, vol. 330, pp. 419–453. Springer, Heidelberg (1988)

3. Patarin, J.: Cryptanalysis of the Matsumoto and Imai Public Key Scheme of Eurocrypt '88. In: Coppersmith, D. (ed.) CRYPTO 1995. LNCS, vol. 963, pp. 248–261. Springer, Heidelberg (1995)
4. Patarin, J.: Hidden Fields Equations (HFE) and Isomorphisms of Polynomials (IP): Two New Families of Asymmetric Algorithms. In: Maurer, U.M. (ed.) EUROCRYPT 1996. LNCS, vol. 1070, pp. 33–48. Springer, Heidelberg (1996)
5. Faugère, J.C., Joux, A.: Algebraic Cryptanalysis of Hidden Field Equation (HFE) Cryptosystems Using Gröbner Bases. In: Boneh, D. (ed.) CRYPTO 2003. LNCS, vol. 2729, pp. 44–60. Springer, Heidelberg (2003)
6. Patarin, J., Goubin, L., Courtois, N.: C^*_{-+} and HM: Variations around Two Schemes of T. Matsumoto and H. Imai. In: Ohta, K., Pei, D. (eds.) ASIACRYPT 1998. LNCS, vol. 1514, pp. 35–50. Springer, Heidelberg (1998)
7. Patarin, J., Courtois, N., Goubin, L.: FLASH, a Fast Multivariate Signature Algorithm. In: Naccache, D. (ed.) CT-RSA 2001. LNCS, vol. 2020, pp. 298–307. Springer, Heidelberg (2001)
8. NESSIE, New European Schemes for Signatures, Integrity, and Encryption. Portfolio of Recommended Cryptographic Primitives, `http://www.nessie.eu.org`
9. Dubois, V., Fouque, P.-A., Stern, J.: Cryptanalysis of SFLASH with Slightly Modified Parameters. In: Naor, M. (ed.) EUROCRYPT 2007. LNCS, vol. 4515, pp. 264–275. Springer, Heidelberg (2007)
10. Dubois, V., Fouque, P.A., Shamir, A., Stern, J.: Practical Cryptanalysis of SFLASH. In: Menezes, A. (ed.) CRYPTO 2007. LNCS, vol. 4622, pp. 1–12. Springer, Heidelberg (2007)
11. Ding, J., Dubois, V., Yang, B.-Y., Chen, O.C.-H., Cheng, C.-M.: Could SFLASH be Repaired? In: Aceto, L., Damgård, I., Goldberg, L.A., Halldórsson, M.M., Ingólfsdóttir, A., Walukiewicz, I. (eds.) ICALP 2008, Part II. LNCS, vol. 5126, pp. 691–701. Springer, Heidelberg (2008)
12. Yang, B., Chen, J.: Building Secure Tame-like Multivariate Public-Key Cryptosystems: The New TTS. In: Boyd, C., González Nieto, J.M. (eds.) ACISP 2005. LNCS, vol. 3574, pp. 518–531. Springer, Heidelberg (2005)
13. Courtois, N.T., Klimov, A., Patarin, J., Shamir, A.: Efficient Algorithms for Solving Overdefined Systems of Multivariate Polynomial Equations. In: Preneel, B. (ed.) EUROCRYPT 2000. LNCS, vol. 1807, pp. 392–407. Springer, Heidelberg (2000)
14. Faugère, J.C.: A new efficient algorithm for computing Gröbner bases (F_4). Journal of Pure and Applied Algebra 139, 61–88 (1999)
15. Kipnis, A., Shamir, A.: Cryptanalysis of the HFE Public Key Cryptosystem by Relinearization. In: Wiener, M. (ed.) CRYPTO 1999. LNCS, vol. 1666, pp. 19–30. Springer, Heidelberg (1999)
16. Berger, T.P., Canteaut, A., Charpin, P., Laigle, C.Y.: On almost perfect nonlinear functions over $F^n{}_2$. IEEE Transactions on Information Theory 52(9), 4160–4170 (2006)
17. Dobbertin, H.: Almost perfect nonlinear power functions over GF(2^n): the Welch case. IEEE Transactions on Information Theory 45(4), 1271–1275 (1999)
18. Faugère, J.C.: A new efficient algorithm for computing Gröbner bases without reduction to zero (F_5). In: International Symposium on Symbolic and Algebraic Computation, ISSAC 2002, pp. 75–83. ACM Press (2002)
19. Ars, G., Faugère, J.-C., Imai, H., Kawazoe, M., Sugita, M.: Comparison Between XL and Gröbner Basis Algorithms. In: Lee, P.J. (ed.) ASIACRYPT 2004. LNCS, vol. 3329, pp. 338–353. Springer, Heidelberg (2004)
20. Yang, B., Chen, J., Courtois, N.: On Asymptotic Security Estimates in XL and Gröbner Bases-Related Algebraic Cryptanalysis. In: López, J., Qing, S., Okamoto, E. (eds.) ICICS 2004. LNCS, vol. 3269, pp. 401–413. Springer, Heidelberg (2004)

Comments on an Advanced Dynamic ID-Based Authentication Scheme for Cloud Computing

Ding Wang[1,2], Ying Mei[1], Chun-guang Ma[2], and Zhen-shan Cui[2]

[1] Department of Training, Automobile Sergeant Institute of PLA, Bengbu, China
[2] Harbin Engineering University, Harbin City 150001, China
wangdingg@mail.nankai.edu.cn

Abstract. The design of secure remote user authentication schemes for mobile devices in Cloud Computing is still an open and quite challenging problem, though many such schemes have been published lately. Recently, Chen et al. pointed out that Yang and Chang's ID-based authentication scheme based on elliptic curve cryptography (ECC) is vulnerable to various attacks, and then presented an improved password based authentication scheme using ECC to overcome the drawbacks. Based on heuristic security analysis, Chen et al. claimed that their scheme is more secure and can withstand all related attacks. In this paper, however, we show that Chen et al.'s scheme cannot achieve the claimed security goals and report its flaws: (1) It is vulnerable to offline password guessing attack; (2) It fails to preserve user anonymity; (3) It is prone to key compromise impersonation attack; (4) It suffers from the clock synchronization problem. The cryptanalysis demonstrates that the scheme under study is unfit for practical use in Cloud Computing environment.

Keywords: Dynamic ID, Authentication protocol, Elliptic curve cryptography, Cryptanalysis, Cloud Computing.

1 Introduction

With the advent of the Cloud Computing era, various services are provided on cloud to allow mobile users to manage their businesses, store data and use cloud services without investing in new infrastructure. Generally, the cloud infrastructure provides various kinds of services in a distributed environment, while sharing resources and the storage of data in a remote data center. Without knowledge of, expertise in, or control over the cloud infrastructure, mobile users can access data or request services anytime and anywhere, and thus it is of great concern to protect the users and the systems' privacy and security from malicious adversaries [16]. Accordingly, user authentication becomes an important security mechanism for remote systems to assure the legitimacy of the communication participants by acquisition of corroborative evidence.

In 2009, Yang and Chang [15] presented an identity-based remote user authentication scheme for mobile users based on ECC. Although the Yang-Chang's scheme preserves advantages of both the elliptic curve and identity-based cryptosystems, and is efficient than most of the previous schemes, in 2011, Islam et

F.L. Wang et al. (Eds.): WISM 2012, LNCS 7529, pp. 246–253, 2012.

al. [6] showed that the Yang-Chang's scheme suffers from clock synchronization problem, known session-specific temporary information attack, no provision of user anonymity and forward secrecy. Almost at the same time, Chen et al. [3] also identified two other security flaws, i.e. insider attack and impersonation attack, in Yang-Chang's scheme. To remedy these security flaws, Chen et al. further proposed an advanced password based authentication scheme using ECC. The authors claimed that their improved scheme provides mutual authentication and is free from all known cryptographic attacks, such as replay attack, impersonation attack and known session-specific temporary information attack, and is suitable for Cloud Computing environment.

In this paper, however, we will show that, although Chen et al.'s scheme is superior to the previous solutions for implementation on mobile devices, we find their scheme cannot achieve the claimed security: their scheme is vulnerable to offline password guessing attack, and key compromise impersonation attack, and suffers from clock synchronization problem. In addition, their scheme fails to preserve user anonymity, which is an important objective in their scheme.

The remainder of this paper is organized as follows: in Section 2, we review Chen et al.'s scheme. Section 3 describes the weaknesses of Chen et al.'s scheme. Section 4 concludes the paper.

2 Review of Chen et al.'s Scheme

In this section, we examine the identity-based remote user authentication scheme for mobile users based on ECC proposed by Chen et al. [3] in 2011. Chen et al.'s scheme, summarized in Fig.1, consists of three phases: the system initialization phase, the registration phase, the authentication and session key agreement phase. For ease of presentation, we employ some intuitive abbreviations and notations listed in Table 1. We will follow the original notations in Chen et al.'s scheme as closely as possible.

Table 1. Notations

Symbol	Description
U_A	the user A
S	remote server
ID_A	identity of user U_A
$\oplus$	the bitwise XOR operation
$\|\|$	the string concatenation operation
$h(\cdot)$	collision free one-way hash function
q_s	secret key of remote server S
$\mathcal{O}$	the point at infinity
$\mathcal{P}$	base point of the elliptic curve group of order n such that $n \cdot \mathcal{P} = \mathcal{O}$
Q_s	public key of remote server S, where $Q_s = q_s \cdot \mathcal{P}$
$A \rightarrow B : M$	message M is transferred through a common channel from A to B
$A \Rightarrow B : M$	message M is transferred through a secure channel from A to B

2.1 The System Initialization Phase

Before the system begins, server S performs as follows:

Step S1. S chooses an elliptic curve equation $E_P(a, b)$ with order n.
Step S2. Selects a base point P with the order n over $E_P(a, b)$, where n is a large number for the security considerations. And then, S computes its private/public key pair (q_s, Q_s) where $Q_S = q_s \cdot P$
Step S3. Chooses three one-way hash functions $H_1(\cdot)$, $H_2(\cdot)$ and $H_3(\cdot)$, where $H_1(\cdot) : \{0, 1\} \rightarrow G_p$, $H_2(\cdot) : \{0, 1\} \rightarrow Z_P^*$ and $H_3(\cdot) : \{0, 1\} \rightarrow Z_P^*$, where G_p denotes a cyclic addition group of P.
Step S4. Stores q_s as a private key and publishes message $\{E_p(a, b), P, H_1(\cdot), H_2(\cdot), H_3(\cdot), Q_s\}$.

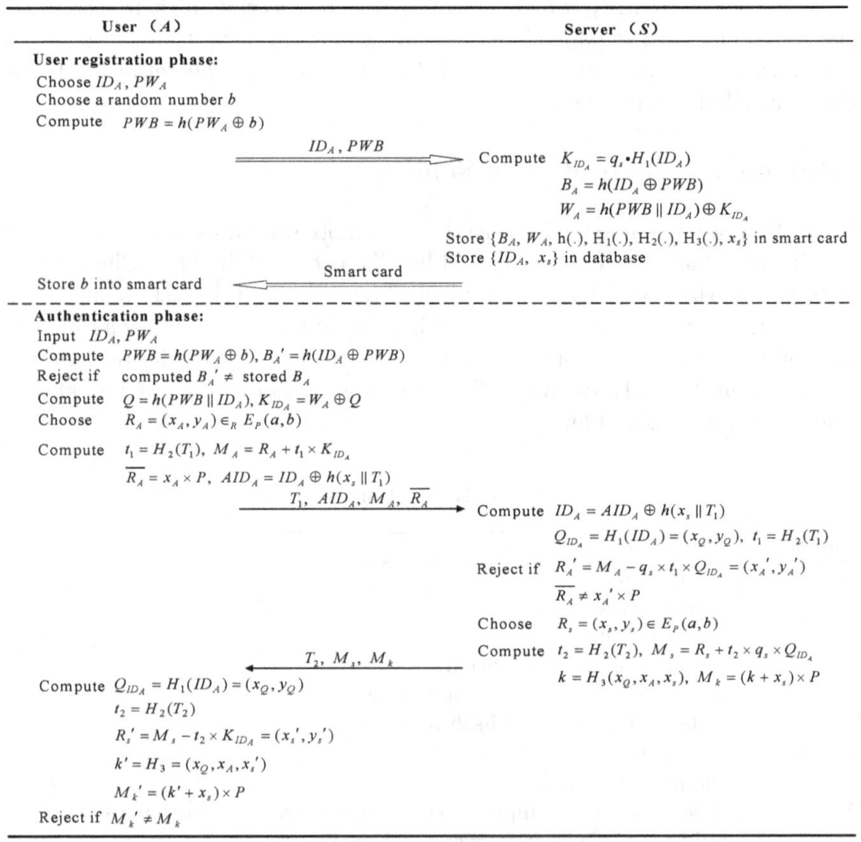

Fig. 1. Chen et al.'s remote authentication scheme for Cloud Computing

2.2 The Registration Phase

The registration phase involves the following operations:

Step R1. U_A chooses his/her ID_A and password PW_A and generates a random number b for calculating $PWB = h(PW_A \oplus b)$. Then, U_A submits ID_A and PWB to the server S.

Step R2. S computes $K_{IDA} = q_s \cdot H_1(ID_A) \in G_p$, where K_{ID_A} is U_A's authentication key.

Step R3. S computes $B_A = h(ID_A \oplus PWB)$ and $W_A = h(PWB \| ID_A) \oplus K_{IDA}$.

Step R4. S stores $\langle B_A, W_A, h(\cdot), H_1(.), H_2(.), H_3(.), x_s \rangle$ on a smart and sends the smart card to U_A over a secure channel (Here x_s is a secret key shared with users).

Step R5. Upon U_A receiving the smart card, U_A stores the random number b in the smart card. Note that now the smart card contains $\{B_A, W_A, h(\cdot), b, H_1(.), H_2(.), H_3(.), x_s\}$.

Step R6. U_A enters his/her ID_A and PW_A to verify whether $B_A = h(ID_A \oplus h(PW_A \oplus b))$. If it is hold, U_A accepts the smart card.

2.3 Mutual Authentication with Key Agreement Phase

When U_A wants to login to S, the following operations will be performed:

Step V1. U_A inserts the smart card into the card reader and enters his/her ID_A and PW_A.

Step V2. Smart card calculates $PWB = h(pw_A \oplus b)$ and $B'_A = h(ID_A \oplus PWB)$ and checks whether $B'_A = B_A$. If it holds, smart card calculates $Q = h(PWB \| ID_A)$ and $K_{IDA} = W_A \oplus Q$.

Step V3. After U_A obtaining his/her authentication key K_{IDA}, U_A (actually the smart cardchooses a random point $R_A = (x_A, y_A) \in E_{P(a,b)}$, where x_A and y_A are x and y coordinating point of R_A.

Step V4. U_A computes $t_1 = H_2(T_1)$, $M_A = R_A + t_1 \times K_{ID_A}$ and $\overline{R}_A = x_A \times P$ at the timestamp T_1.

Step V5. U_A computes $AID_A = ID_A \oplus h(x_s \| T_1)$ to obtain an anonymous ID.

Step V6. U_A sends message $m_1 = \{T_1, AID_A\ , M_A, \overline{R}_A\}$ to S.

Step V7. After receiving m_1, S performs the following operations to obtain $Q_{IDA} = (x_Q, y_Q)$ and $R'_A = (x'_A, y'_A)$ of U_A as follows:

$$ID_A = AID_A \oplus h(x_s \| T_1);$$
$$Q_{IDA} = H_1(ID_A);$$
$$t_1 = H_2(T_1);$$
$$R'_A = M_A - q_s \times t_1 \times Q_{IDA}.$$

Step V8. S verifies whether $\overline{R}_A = x'_A \times P$. If it holds, U_A is authenticated by S.

Step V9. S chooses a random point $R_S = (x_S, y_S) \in E_p(a, b)$.

Step V10. S computes $t_2 = H_2(T_2)$,$M_S = R_S + t_2 \times q_S \times Q_{ID_A}$, session key $k = H_3(x_Q, x_A, x_S)$ and $M_k = (k + x_S) \times P$ at the timestamp T_2.

Step V11. S sends message $m_2 = \langle T_2, M_S, M_k \rangle$ to U_A.

Step V12. After receiving m_2,U_A performs the following computations to obtain $Q_{IDA} = (x_Q, y_Q)$ and $R'_s = (x'_s, y'_s)$ of S:
$Q_{IDA} = H_1(ID_A)$;
$t_2 = H_2(T_2)$;
$R'_S = M_S - t_2 \times K_{IDA}$.

Step V13. U_A computes $k' = H_3(x_Q, x_A, x'_S)$ and $M'_k = (k' + x'_s) \cdot P$ to verify whether $M'_k = M_k$. If it holds, S is authenticated by U_A.

Finally, U_A and S empoly k as a session key for securing subsequent data communications.

3 Cryptanalysis of Chen et al.'s Scheme

With superior properties over other related schemes and a long list of arguments of security features that their scheme possesses presented, Chen et al.'s scheme seems desirable at first glance. However, their security arguments are still specific-attack-scenario-based and without some degree of rigorousness, and thus it is not fully convincing. We find that Chen et al.'s scheme still fails to serve its purposes and demonstrate its security flaws in the following.

3.1 Offline Password Guessing Attack

Although tamper resistant smart card is assumed in many authentication schemes, but such an assumption is difficult and undesirable in practice. Many researchers have pointed out that the secret information stored in a smartcard can be breached by analyzing the leaked information [9] or by monitoring the power consumption [7,10]. In 2006, Yang et al. [14] pointed out that, previous schemes based on the tamper resistance assumption of the smart card are vulnerable to various types of attacks, such as user impersonation attacks, server masquerading attacks, and offline password guessing attacks, etc., once an adversary has obtained the secret information stored in a user's smart card and/or just some intermediate computational results in the smart card. Since then, researchers in this area began to pay attention to this issue and admired schemes based on non-tamper resistance assumption of the smart card is proposed, typical examples include [8, 12, 13]. Hence, in the following, we assume that the secret data stored in the smart card could be extracted out once the smart card is somehow obtained (stolen or picked up) by the adversary $\mathcal{A}$. Note that this assumption is widely assumed in the security analysis of such schemes [8, 11, 12, 14].

What's more, in Chen et al.'s scheme, a user is allowed to choose her own password at will during the registration phase; the user usually tends to select a password, *e.g.*, his home phone number or birthday, which can be easily remembered for his convenience. Hence, these easy-to-remember passwords, called weak passwords, have low entropy and thus are potentially vulnerable to offline password guessing attack.

Let us consider the following scenarios. In case the legitimate user U_A's smart card is in the possession of an adversary $\mathcal{A}$ and the parameters stored in it, like B_A, W_A, b and x_s, is revealed. Once the login request message $m_1 = \{T_1, AID_A, M_A, \overline{R}_A\}$ during any authentication process is intercepted by $\mathcal{A}$, an offline password guessing attack can be launched as follows:

Step 1. Computes the identity of U_A as $ID_A = AID_A \oplus h(x_s \parallel T_1)$, where AID_A, T_1 is intercepted and x_s is revealed from the smart card;
Step 2. Guesses the value of PW_A to be PW_A^* from a dictionary space $\mathcal{D}$;
Step 3. Computes $PWB^* = h(PW_A^* \oplus b)$, as b is revealed from the smart card;
Step 4. Computes $B_A^* = h(ID_A \oplus PWB^*)$;
Step 5. Verifies the correctness of PW_A^* by checking if the computed B_A^* is equal to the revealed B_A;
Step 6. Repeats Step 1, 2, 3, 4 and 5 of this procedure until the correct value of PW_A is found.

As the size of the password dictionary, i.e. $|\mathcal{D}|$, is very limited in practice, the above attack procedure can be completed in polynomial time. After guessing the correct value of PW_A, $\mathcal{A}$ can compute the valid authentication key $K_{ID_A} = W_A \oplus h(PWB \parallel ID_A) = W_A \oplus h(h(PW_A \oplus b) \parallel ID_A)$. With the correct K_{ID_A}, $\mathcal{A}$ can impersonate U_A to send a valid login request message $\{T_1, AID_A, M_A, \overline{R}_A\}$ to the service provider server S, and successfully masquerade as a legitimate user U_i to server S.

3.2 No Provision of User Anonymity

Let us consider the following scenarios. A malicious privileged user $\mathcal{A}$ having his own smart card or stolen card can gather information x_s from the obtained smart card as stated in Section 3.1, while x_s is shared among all the users. Then, $\mathcal{A}$ can compute ID_A corresponding to any user U_A as follows:

Step 1. Eavesdrops and intercepts a login request message $\{T_1, AID_C, M_C, \overline{R}_C\}$ of any user, without loss of generality, assume it is U_C, from the public communication channel;
Step 2. Computes $ID_C = AID_C \oplus h(x_s \parallel T_1)$, where x_s is revealed from the smart card and AID_C, T_1 is intercepted from the public channel as stated in Step 1;

It is obvious to see that user anonymity will be breached once the parameter x_s is extracted out. Hence, Chen et al.'s scheme fails to preserve user anonymity, which is the most essential security feature a dynamic identity-based authentication scheme is designed to support.

3.3 Key Compromise Impersonation Attack

Suppose the long-term secret key q_s of the server S is leaked out by accident or intentionally stolen by the adversary $\mathcal{A}$. Without loss of generality, we assume one of U_A's previous login requests, say $\{T_1, AID_A\ , M_A, \overline{R}_A\}$, is intercepted by $\mathcal{A}$. Once the value of q_s is obtained, with the intercepted $\{T_1, AID_A\ , M_A, \overline{R}_A\}$, $\mathcal{A}$ can impersonate the legitimate user U_i since then through the following method:

Step 1. Guesses the value of ID_A to be ID_A^* from a dictionary space $\mathcal{D}_{ID}$;
Step 2. Computes $k_{ID_A}^* = q_s \cdot H_1(ID_A^*)$;
Step 3. Computes $t_1 = H_2(T_1)$;
Step 4. Computes $R_A^* = M_A - t_1 \times k_{ID_A}^* = (x_A^*, y_A^*)$;
Step 5. Computes $\overline{R}_A^* = x_A^* \times P$;
Step 6. Verifies the correctness of ID_A^* by checking if the computed $\overline{R}_A^*$ is equal to the intercepted $\overline{R}_A$;
Step 7. Repeats Step 1, 2, 3, 4, 5 and 6 of this procedure until the correct value of ID_A is found;
Step 8. Computes $K_{IDA} = q_s \cdot H_1(ID_A)$.

As the size of the identity dictionary, i.e. $|\mathcal{D}_{ID}|$, is often more limited than the password dictionary size $|\mathcal{D}|$ in practice, the above attack procedure can be completed in polynomial time. After guessing the correct value of K_{ID_A}, $\mathcal{A}$ can impersonate U_A since then. Hence, Chen et al.'s scheme cannot withstand key compromise impersonation attack.

3.4 Clock Synchronization Problem

It is widely accepted that, remote user authentication schemes employing timestamp may still suffer from replay attacks as the transmission delay is unpredictable in real networks [5]. In addition, clock synchronization is difficult and expensive in existing network environment, especially in wireless networks [4] and distributed networks [1]. Hence, these schemes employing timestamp mechanism to resist replay attacks is not suitable for mobile applications [2, 6]. In Chen et al.'s scheme, this principle is violated.

4 Conclusion

In this paper, we have shown that Chen et al.'s scheme suffers from the offline password guessing, no provision of user anonymity and key compromise impersonation attack. In addition, their scheme suffers from the clock synchronization problem. In conclusion, Although Chen et al.'s scheme possesses many attractive features, it, in fact, does not provide all of the security properties that they claimed and only radical revisions of the protocol can possibly eliminate the identified pitfalls.

Acknowledgments. This research was in part supported by the National Natural Science Foundation of China (NSFC No. 61170241 and No. 61073042).

References

1. Baldoni, R., Corsaro, A., Querzoni, L., Scipioni, S., Piergiovanni, S.: Coupling-based internal clock synchronization for large-scale dynamic distributed systems. IEEE Transactions on Parallel and Distributed Systems 21(5), 607–619 (2010)
2. Chang, C.C., Lee, C.Y.: A secure single sign-on mechanism for distributed computer networks. IEEE Transactions on Industrial Electronics 59(1), 629–637 (2012)
3. Chen, T., Yeh, H., Shih, W.: An advanced ecc dynamic id-based remote mutual authentication scheme for cloud computing. In: Proceedings of the 2011 Fifth FTRA International Conference on Multimedia and Ubiquitous Engineering, pp. 155–159. IEEE Computer Society (2011)
4. Giridhar, A., Kumar, P.: Distributed clock synchronization over wireless networks: Algorithms and analysis. In: 2006 45th IEEE Conference on Decision and Control, pp. 4915–4920. IEEE (2006)
5. Gong, L.: A security risk of depending on synchronized clocks. ACM SIGOPS Operating Systems Review 26(1), 49–53 (1992)
6. Islam, S.H., Biswas, G.: A more efficient and secure id-based remote mutual authentication with key agreement scheme for mobile devices on elliptic curve cryptosystem. Journal of Systems and Software 84(11), 1892–1898 (2011)
7. Kasper, T., Oswald, D., Paar, C.: Side-Channel Analysis of Cryptographic RFIDs with Analog Demodulation. In: Juels, A., Paar, C. (eds.) RFIDSec 2011. LNCS, vol. 7055, pp. 61–77. Springer, Heidelberg (2012)
8. Ma, C.-G., Wang, D., Zhang, Q.-M.: Cryptanalysis and Improvement of Sood et al.'s Dynamic ID-Based Authentication Scheme. In: Ramanujam, R., Ramaswamy, S. (eds.) ICDCIT 2012. LNCS, vol. 7154, pp. 141–152. Springer, Heidelberg (2012)
9. Mangard, S., Oswald, E., Standaert, F.X.: One for all-all for one: unifying standard differential power analysis attacks. IET Information Security 5(2), 100–110 (2011)
10. Messerges, T.S., Dabbish, E.A., Sloan, R.H.: Examining smart-card security under the threat of power analysis attacks. IEEE Transactions on Computers 51(5), 541–552 (2002)
11. Wang, D., Ma, C.G.: On the security of an improved password authentication scheme based on ecc. Cryptology ePrint Archive, Report 2012/190 (2012), `http://eprint.iacr.org/2012/190.pdf`
12. Wang, D., Ma, C.-G., Wu, P.: Secure Password-Based Remote User Authentication Scheme with Non-tamper Resistant Smart Cards. In: Cuppens-Boulahia, N., Cuppens, F., Garcia-Alfaro, J. (eds.) DBSec 2012. LNCS, vol. 7371, pp. 114–121. Springer, Heidelberg (2012)
13. Xu, J., Zhu, W., Feng, D.: An improved smart card based password authentication scheme with provable security. Computer Standards & Interfaces 31(4), 723–728 (2009)
14. Yang, G., Wong, D.S., Wang, H., Deng, X.: Formal Analysis and Systematic Construction of Two-Factor Authentication Scheme (Short Paper). In: Ning, P., Qing, S., Li, N. (eds.) ICICS 2006. LNCS, vol. 4307, pp. 82–91. Springer, Heidelberg (2006)
15. Yang, J., Chang, C.: An id-based remote mutual authentication with key agreement scheme for mobile devices on elliptic curve cryptosystem. Computers & security 28(3-4), 138–143 (2009)
16. Yu, H., Powell, N., Stembridge, D., Yuan, X.: Cloud computing and security challenges. In: Proceedings of the 50th Annual Southeast Regional Conference, pp. 298–302. ACM (2012)

Research on Security Management in Active Network Node Operating Systems

Yongchun Cao*, Yabin Shao*, and Zhengqi Cai

School of Mathematics and Computer Science, Northwest University for Nationalities, Lanzhou 730030, China

{cych33908,yb-shao}@163.com, caizhengqi@126.com

Abstract. This paper analyzes the security requirements and problems with which active nodes are confronted, and proposes a general security management subsystem in Active Network NodeOS. The subsystem implements through four functions: resource management, hop-by-hop authentication, credential management and security policy management. Resource management avoids excessive use of resources by constraining the maximum available resource quantity of each application. Hop-by-hop authentication is realized by adding hop-by-hop integrity option to ANEP header, which accomplishes the previous hop authentication and the hop-by-hop integrity checking. The function of credential management is to obtain the credential, authenticate the validity of the credential, and make origin authentication and end-to-end integrity checking by using the principal's public key carried in the credential. Security policy management is realized by embedding a reformed KeyNote Trust Management system into NodeOS kernel to complete access control to node resource.

Keywords: Active network, NodeOS, Security management, Trust management.

1 Introduction

Active network appears as a new type network architecture. It is first advanced by the DARPA active networks research community. An active network consists of a set of nodes (not all of which need to be "active"), each active node is a router or switch, with the ability of dealing with active packets. Active packets are packets carrying codes in active networks, the codes are executed when active packets arrive at an active node. The execution may change the forwarding path of active packets, cause new packets to be sent or change the state of the node's environment. Active network is the ultimate method to solve the restrictive problems of network development because of its ability to introduce customized new services flexibly and quickly.

The active packets coming from network can change the state of an active node, which makes the security problems of active network very serious; obviously, if the security is not guaranteed, the network is of no practical use. Many researchers did a lot of studies on this. At present, the relatively important studies on active networks

* Corresponding authors.

F.L. Wang et al. (Eds.): WISM 2012, LNCS 7529, pp. 254–263, 2012.

are as the following: The DARPA active networks research community advanced a security architecture for active network [2], and accordingly to this, implemented an active network secure prototype SANTS [3]. SANTS paid particular emphasis on the introduction of traditional technology of authentication, authorization and encryption, and provided protection to active network on EE. Janos [4] is also a Java-orientated active network prototype to provide protection to active network on EE, mainly supported by the security system of Java 2 to provide protection. The common defect of the two prototypes is NodeOS itself lacks precaution mechanism, which requires higher security of EE; and moreover, once the attack rounds EE, the defense line of the system will be completely destroyed. Actually, the security system of a hierarchical structure should be built up from the bottom to the top. So, SANE [5] proposed a method to build up a reliable system, which guarantees a node starts running in a safe state, and the node runs in the safe state all along by the dynamic integrity checking during the operation of the system. SANE ensures the security of the system by safe programming language, and a type safe language PLAN is specifically defined, but the defect of the system is of no generality. AMP [6] is another NodeOS providing protection on operation system level; it implements the interface of NodeOS needed to be protected by individual trust server, and calls the protected operation by PRC across the address space. The defect of AMP is PRC calling may result in frequent switch of the context, and greatly influences the system performance.

The main task of this paper is to analyze the security requirements of active nodes to design and realize a general security management system of active network. The system is to establish security system of active networks on NodeOS layer, this system provides self-protection on operation system layer, and provides powerful security support for EE and AA. This system has no special security requirements for EE that runs in NodeOS and it possesses generality because this system does not rely on some special technology (e.g. the language of type safe and namespaces restriction or hardware).

2 Security Requirements Analysis of Active Networks

According to the active network architecture[1] advanced by DARPA , the functionality of the active network node is divided into the Node Operating System (NodeOS), the Execution Environments (EEs), and the Active Applications (AAs). The general organization of these components is shown in figure 1. The NodeOS is responsible for the allocation and scheduling of resources, one or more of the EE running on it, each EE defines and realizes an active application programming model, and AA programming implements customized service of users.

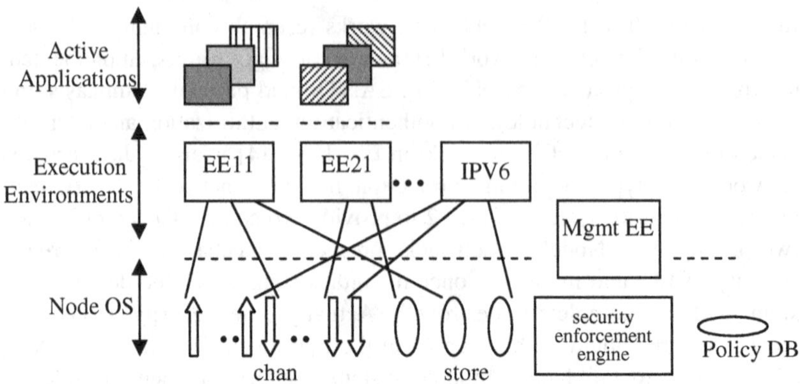

Fig. 1. Active network architecture

The active network provides a platform on which network services can be experimented with, developed, and deployed; at the same time provides a new way for networks, Quality of Service (QoS) and reliable multicast. But so far active networks have not been widely implemented and applied; the main reason is, in addition to encounter resistance and difficulties while reconstructing traditional intermediate nodes, its safety has not yet been solved. Because active networks allow users programming intermediate nodes of networks, and the execution of the active code which arrived at the active node needs to access the resource of the node, it makes the security facing a greater threat.

The entities which are related to the security in active networks include NodeOS, EE, AA and the end user, with the first three running at the active node. NodeOS, EE and AA are mainly concerned with the Legitimacy of the resource usage, the availability of the service and the integrity of the state in their own field, while the end user cares for the authenticity, integrity and confidentiality [2]. As far as each entity is concerned, security threat could probably come from all other entities, therefore every entity needs to take some measures to protect its own security.

For each active node, the main security threat comes from the foreign codes. The malicious code can make the node fail to provide the normal service by using the resource of the node excessively or modifying the state of the node without authorization, and the access of the malicious code to the confidential data in the node can probably result in data disclosure. Attackers may directly produce active packets carrying malicious code, or add malicious code after intercepting the active packet, even may also replay attacks with “old” active packets.

In general, the security of active networks includes two aspects, one is the security of the active nodes, and the other is the security of the active packets. The security of the active nodes is how to ensure that the resource of the nodes and the service will not be accessed without authorization or will not be damaged; the security of the active packets is how to ensure that the active packets will not be illegitimately changed while transmitting.

3 Design of Security Management System

The effective methods to deal with the security threat mentioned above are resource usage limit, access control and validity checking of packets. By prescribing the maximum available resource quantity of each application and suspending running application when the resource usage is over-limited, resource usage limit can avoid excessive use of resources. Access control mechanism requires granting certain authority to each visitor in advance, checking its authority when a visitor makes access requests, and making decision whether or not to allow this access. This avoids the illegitimate access of the resources and data. Through source authentication and integrity checking, validity checking of packets ensures that only legitimate packets can be processed by nodes, which can exclude some malicious packets to some degree.

Resource access needs to get the authorization in advance. A principal in security architecture is an entity (e.g. the owner of the active application) that can be authorized; the authority to access resource is granted to the acceptable principal by the owner of the resource. It is impracticable to grant authority to the principal directly when there are a large number of principals or the exact principle was not known in advance. In such a case, we need to introduce an intermedium between the principal and the authority; generally the intermedium is a set of security attributes. The owner only grants authority to the attributes, and the principal inherits the authority from the attributes. Here, attributes set and mapping the attributes to the authority make up the access policy of application. The policy is established by the principals of application, but the system must provide the necessary policy management mechanism.

When a method is called, the principle should declare its attributes because the authority is bound up with the attributes. In order to prove the attributes that the principal declared are true, the principal needs carrying a certificate that was issued by the certificate authority of the third party; this certificate is also called the credential in networks. The credential binds the principal, the principal's attributes and the principal's public keys together, attaching the signature issued by the certificate authority. Active packets usually carry more than one credentials and use different credential in different secure domain, because active packets will traverse multiple secure domains during the transmission and each secure domain requires different secure attributes. Generally the active packet only carries index or location of the credentials on the purpose of decreasing the cost of the active packet; and the credentials are saved in the distributed, global accessible Certificate Authority Center, which can be got back by the relevant nodes when needed. Therefore, the system must provide the needed management functionalities for the credentials, which must include automatic acquisition, buffering, authentication, and so on.

The frame structure of NodeOS's security management system that we designed is shown in Figure 2. It is made up of four modules: resource management, hop-by-hop authentication of packets, credential management and security policy management. Resource management can solve the security problem of resource using in NodeOS by resource admission control, resource allocation and adaptation, resource usage monitoring and policing. Hop-by-hop authentication delivers the unmodified packets that come from the legitimate previous hop to credential management. Credential management gets the credential back from the local buffer or distributed certificate system, authenticates the validity of the credential, and then checks the end-to-end integrity by using the public key carried in the credential. Security policy management is responsible for management of policy database, and computes the access authority of the principal according to the system's security policy, the principal's credential and the action request.

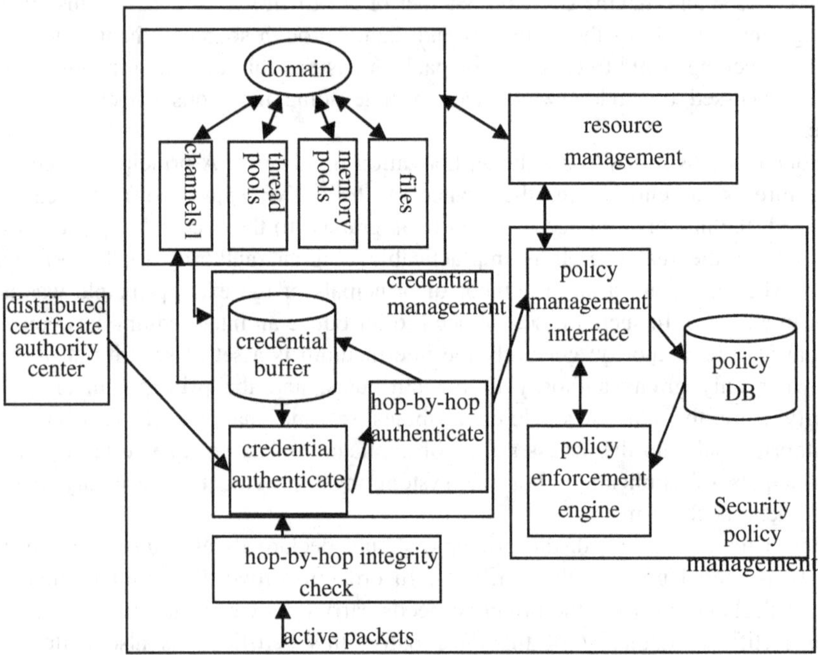

Fig. 2. Active network NodeOS security management system

4 Implementation of Security Management System

In the following part, we are going to discuss the implementation methods of the above four modules in NodeOS's security management system.

4.1 Resource Management in NodeOS

The main duty of active nodes is to store, compute and forward packets, and this need consume multiple resources such as CPU, memory and bandwidth. Five resource abstractions, which are defined in NodeOS interface specification, are used to encapsulate resources of nodes — they are thread pools, memory pools, channels, files and domains. The first four respectively encapsulate the resources of computing, memory, communication and permanent storage of the system; and domains congregate and allocate these resources. Domain, a resource container, which includes all needed resources for dealing with a specific packet stream, is a main abstract structure of the system used to account, control admission and allocation. The concept of domain best illustrates resource management ideas which take the packets stream as the center, and this is the key difference between NodeOS and traditional OS.

The minimal granularity of resource allocation is domain, and in order to manage resources on a higher level, domains are generally organized as cluster. Cluster has its hierarchical structure; a higher level management implements its macro-control of resource through controlling the hierarchical structure of cluster and resources allocation proportion of each cluster. The basic principle of resource allocation is "fairness", that is to say, resource allocation must be in conformity with the distributing characteristics of the node business flow. Hierarchical structure and resources allocation rate of cluster are problems to be studied in resource allocation strategy.

For each AA executed on a node, NodeOS establishes a domain and allocates needed resources, which are mainly CPU, memory and bandwidth. After confirmed needed resources for AA, whether AA can obtain these resources or not, it is determined by admission control checking. Resource distribution strategies prescribe resources that each cluster can obtain, and NodeOS must guarantee the proper implementation of these strategies. Therefore, NodeOS needs to control the admission of the request of each establishing domain, viz., to check whether the establishment of the domain will violate resource allocation strategies for resource allocation rules of each cluster, only the domains which do not violate the resource allocation strategies can be established.

As the resource requirements can not be accurately evaluated and some resource requirements of application may vary according to the march of time, the adjustment of the resource allocation limit to the domains should be allowed. For this, NodeOS must track and monitor each domain's resource usage, assess the tendency of its variation, and determine whether and how to make adjustment. To ensure the implementation of the resource allocation strategies, resource usage control is also needed, viz., to limit the application's excessive use of resources. The security of NodeOS's resource usage is realized by the strategies and methods which are used to monitor, adjust and control the resource usage. The implementation of resource management in the security management system won't be discussed in details in this paper.

4.2 Packets Validity Checking

Packets validity checking includes two parts, source authentication and integrity checking. Because, while traveling through a number of nodes from source to destination, the active packets may be modified legitimately in these nodes, or be modified illegitimately by the attackers, this makes the packets integrity checking complicated. On the one hand, permission of the active packets modification by the intermediate nodes means the nodes need to ensure the source authentication and integrity checking hop-by-hop, which guarantees the previous node modification is completely delivered to the next node. On the other hand, not all of the content of the packet is permitted to be modified, which means there is the need of end-to-end integrity checking to guarantee the impermissible modification (namely the invariable parts of the active packet, such as static code and data) unable to be modified. Besides, attackers can probably falsify an active packet by a counterfeit identification; so, the identification of the active packets' sender needs to be identified by the nodes. It is now clear that the packets source authentication must include origin authentication and previous hop authentication; the former ensures whether the packets originated from one declared principal, the latter makes sure whether the packets come from the declared previous node. Integrity checking includes two parts as well — end-to-end integrity checking and hop-by-hop integrity checking, the former checks whether the invariable parts of the active packet was modified after sent from the origin source, the latter checks whether the packet was modified after sent from the previous node(mainly in relation to the variable parts of the active packet).

1) Hop-by-hop Authentication

Hop-by-hop authentication is responsible for checking the validity of active packets which come from the previous active node(probably passing some common nodes); it includes the previous hop authentication and the hop-by-hop integrity checking. Because the digital signature has dual functions of making certain of sender's identity and ensuring the integrity of messages, both source authentication and integrity checking of packets can be accomplished by digital signature. Taking the operating efficiency into account, we introduce the digital signature based on symmetric key.

IETF Active Networks Group proposed an Active Network Encapsulation Protocol (ANEP) [7] to ensure the interoperability and the effective use of the existing network. This Protocol defined the encapsulation method of active packets while transmitting in the network, mainly defined the form of ANEP header. The option field in ANEP header is the critical point that ANEP supports multiprotocol. One can introduce a new protocol easily through defining a new option type and adding it to ANEP header. Hop-by-hop authentication is realized just by adding hop-by-hop integrity option to ANEP header; this option is the last option in ANEP header, the format is shown in figure 3.

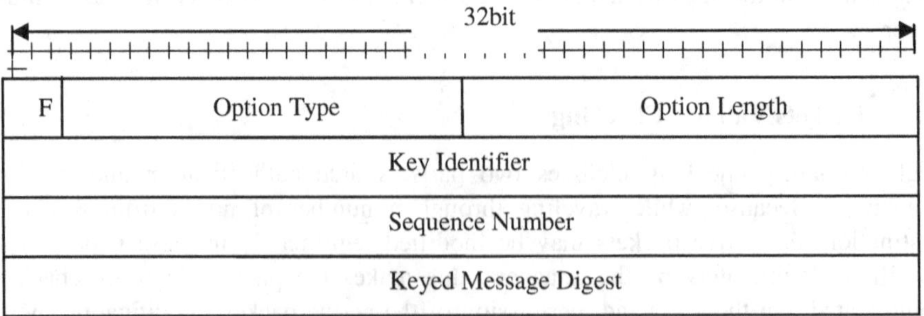

Fig. 3. Hop-by-hop integrity option

Keyed Message Digest is the digital signature of packets, which is produced by the previous active node that signs the parts of the packet except the hop-by-hop integrity option; the receiver judges whether the integrity is damaged through signature certification. We choose HMAC as the signature algorithm. HMAC is a signature algorithm based on the key, which computes a secure hash representation of packets according to the specific key and the hash algorithm. In this system, we have each pair of neighboring nodes use one symmetric key; and we can choose one of the following hash algorithms — SHA1, MD2, MD4, MD5 or RIPEMD-160.

The utmost problem of using symmetric key is generation, maintenance and transportation, but these tasks are usually done by the application layer, and NodeOS does not participate in. So we only maintain a key table which consisting of neighboring nodes, and the content of the key table is generated and maintained by the upper- layer protocol (e.g. key exchange protocol). Key table takes the neighboring node's ID as the index, and the node's ID, which is an unsigned number generated by node's address and other useful information, is used to uniquely identify a node, and Key Identifier in hop-by-hop integrity option corresponds to the node's ID. Each item in the key table stores the key and hash algorithms which are used to signature while communicating the node and the corresponding neighboring node; the table also stores the largest Sequence Number used by the node and the neighboring node while communicating.

When one node sends the active packet to the next hop: (1) consult the key table by using the next hop node's ID; (2) compute the digital signature of the packet by using the key and the hash algorithms listed in the table, and then add 1 to the latest Sequence Number; (3) insert the node's ID, the Sequence Number of the packet and the digital signature to the corresponding domain of the hop-by-hop integrity option, and send it to the next hop.

When one node gets the active packet from the previous hop: (1) use Key Identifier in the hop-by-hop integrity option as the index to consult the key table; (2) examine the Sequence Number of the packet; and if the Sequence Number of the packet is less than the Sequence Number listed in the table, discard the packet (because it means this packet is obsolete); (3) compute the digital signature of the packet by using the key and the hash algorithms listed in the table, and then compare it with the HMAC identification number carried in the packet; discard the packet if it does not conform to the identification number; (4)update the Sequence Number in the table as the received Sequence Number of the packet; (5) remove the hop-by-hop integrity option and deliver the active packet to the credential management.

2) Credential Management

Credential management obtains the credential according to the credential index or address carried by the active packets, authenticates the validity of the credential, then makes origin authentication and end-to-end integrity checking by using the principal's public key carried in the credential.

The credential must carry enough information besides the security attributes to identify the principal, so we select X.509v3 [8] certificate as the credential. X.509 is a general certificate standard; the extended domain it provided can be used to store the extended information such as the security attributes. The most important information carried in the certificate is the public key of the principal and the algorithm needed by the principal to signature. The certificate is usually stored in a distributed certificate directory for obtaining. We select DNSSEC（DNS Security Extension）as the distributed certificate directory, taking security and interoperability of the existing network into consideration. DNSSEC provides secure DNS services, and the certificate is stored in DNS as a kind of resource. Credential management implements the interface with DNSSEC, gets the certificate from DNSSEC, and authenticates it.

The active packet may carry more than one credentials, and each credential needs to be authenticate individually; in addition, the end-to-end signature may cover different areas of the active packets when using different credentials, so we need to check the end-to-end integrity again for each credential. Because the end-to-end signature is only used for the invariable area of the active packet, so we need to divide the payload of the active packet into variable area and invariable area; moreover, the active packet may also carry online security policy. A typical active packet includes Static Payload, Dynamic Payload, Credential-Authenticator Tuples, Online Policy and Hop-by-Hop Integrity, and they all appear as options of ANEP header. Credential-Authenticator Tuples option consists of a list of Credential-Authenticator Tuples; each tuple includes three parts: the information of the credential (including the credential ID, storage position and obtaining methods), coverage areas of the end-to-end signature and Authenticator.

When the active packets are passed to credential management after the hop-by-hop integrity check, credential management checks the validity of the credential carried in the packets one by one. To each Credential-Authenticator Tuples:(1) Search the appropriate credential according to credential ID from the local buffers, and if it is not existing in the local buffers, call DNSSEC Client Application Program to get it back from the distributed certificate directory according to the position of the credential and the obtaining method, then certificate its signature, discard the credentials that failed to be certificated; (2) Extract the public key and the signature algorithms of the principals from the credentials, authenticate the end-to-end signature validity according to the areas that the signature covers, and discard the credentials that failed to be certificated; (3) Extract the security attributes carried in the credentials and online security policy carried in the active packets; (4) Store it at local if the credential is obtained by remote access. Because the end-to-end signature has dual functions of identity authentication and integrity checking, origin authentication and end-to-end integrity checking are actually accomplished by the above process.

5 Conclusion

This paper analyzes the security requirements and problems with which active nodes are confronted, discusses the thoughts to solve these problems based on authorization; and proposes a security management system of active networks which includes security policy management, credential management and hop-by-hop authentication of packets, providing functions of policy database management and policy computation, acquisition and authentication of credential,. hop-by-hop validity checking of packets and so on; thus provides authorization management and access control services for the nodes' security.

The security system provides protection on NodeOS layer, which provides the most effective protection to the nodes, simplifies the security design of the upper layer, improves the operational efficiency of system, and permits EE of any trusts running on NodeOS because there are no specific requirements of EE security in such NodeOS. In addition, The security system possesses generality because it does not rely on some special technology.

Acknowledgement. This work is supported by the National Scientific Fund of China (No.11161041), and Fundamental Research Funds for the Central Universities(No. zyz2012081).

References

1. DARPA AN Architecture Working Group. Architectural Framework for Active Networks (2000)
2. DARPA An Security Working Group. Security Architecture for Active Nets (2001)
3. Murphy, S., et al.: Strong Security for Active Networks. In: IEEE OPENARCH (2001)
4. Tullmann, P., et al.: Janos: A Java-oriented OS for Active Network Nodes. IEEE Journal on Selected Areas in Communications 19(3), 501–510 (2001)
5. Alexander, D., et al.: A Secure Active Network Environment Architecture: Realization in SwitchWare. IEEE Network 12(3), 37–45 (1998)
6. Dandekar, H., et al.: AMP: Experiences with Building an Exokernel-Based Platform for Active Networking. In: Proceedings of the DARPA Active Networks Conference and Exposition (2002)
7. IETF Active Networks Group. Active Network Encapsulation Protocol (ANEP). RFC Draft (1997)
8. IETF Network Working Group. Internet X.509 Public Key Infrastructure Certificate and Certificate Revocation List (CRL) Profile. RFC 3280 (2002)
9. Blaze, M., et al.: Decentralized Trust Management. In: Proceedings of the 17th IEEE Symposium on Security and Privacy, Oakland, USA, pp. 164–173 (1996)
10. IETF Network Working Group. The KeyNote Trust-Management System Version 2. RFC 2704 (1999)
11. http://www.cis.upenn.edu/~keynote/code/keynote.tar.gz

An Integrity Verification Scheme for Multiple Replicas in Clouds*

La Zhang, Qingzhong Li**, Yuliang Shi, Lin Li, and Wenxiao He

School of Computer Science and Technology, Shandong University, Shandong Provincial Key Laboratory of Software Engineering
zhangla168@163.com, {lqz,liangyus}@sdu.edu.cn, ducklilin@126.com, zhangruihua1012@163.com

Abstract. Recently, cloud computing is the fundamental change happening in the field of Information Technology. In SaaS (software as a service) model, both applications software and databases will be deployed to the centralized large data centers, where the management of the data and services may not be fully trustworthy. Many storage systems rely on replicas to increase the availability and durability of data, but secure replicas storage brings about many new security challenges. In this paper, based on multi-tenancy data-sharing storage model, we propose an integrity verification scheme which allows a third party auditor (TPA) to verify the integrity of multiple replicas stored in clouds through random sampling and periodic verification. In particular, via the double-layer authenticating construction, we achieve the isolation of different tenants' replicas and dynamic data operations. Extensive performance analysis about sampling in different conditions shows correctness of the proposed scheme in this paper.

Keywords: Cloud Computing, SaaS, Integrity Verification, Multiple Replicas, Storage Security.

1 Introduction

Cloud computing is a promising service platform that provides the services and resources for users. SaaS, as an application of cloud computing, is a new software delivery model. Because its economy, high expansibility can release the heavy pressure of IT infrastructure management and maintenance, more and more enterprises are attracted by its advantages. In order to ensure the availability and durability of data, replication becomes a fundamental principle in cloud storage environments, so users can access other replicas to meet availability goals when a replica fails.

* The research is supported by the Natural Science Foundation of Shandong Province of China under Grant No.ZR2010FQ026, No.2009ZRB019YT, the Key Technology R&D Program of Shandong Province under Grant No.2010GGX10105, the Independent Innovation Foundation of Shandong University under Grant No.2012TS074.

** Corresponding author.

F.L. Wang et al. (Eds.): WISM 2012, LNCS 7529, pp. 264–274, 2012.

However, storing multiple replicas in clouds brings about many new security challenges. One of the biggest concerns with cloud data storage is that of replicas integrity verification at untrusted servers. For example, some service providers might neglect to keep or deliberately delete rarely accessed replica data which belong to an ordinary tenant in order to save storage space for profits. In addition, the storage service provider, which experiences Byzantine failures occasionally, may decide to hide the errors from the tenants instead of maintaining data timely for the benefit of their own. Although users can't discover the data errors obviously during accessing the replica, it has seriously broken the service level agreement (SLA) and affected the availability of data. Tenants have the rights to verify the integrity of their replicas to maintain their own interests. What is more serious is that the disrupted data could spread to other replicas when consistency updating between replica data.

Considering the above situation, the core of the problem can be generalized as how to find an efficient way to perform integrity verifications. In order to solve this problem, many schemes are proposed under different systems and security models [1, 2, 3, 4, 5, 6, 7]. Nowadays, in Database-as-a-Service model, these methods focus on the signature-based, challenge-response methods and probability-based approaches [8, 9, 10, 11, 12, 13]. Ateniese et al. [10] introduced the notion called Proof of Data Possession (PDP) which allowed a cloud client to verify the integrity of its data outsourced to the cloud in a very efficient way. Juels and Kaliski [9] introduced the notion called Proof of Retrievability (POR) which offered an extra property that the client can actually recover the data outsourced to the cloud. Halevi et al. [7] proposed the notion called Proof of Ownership (POW) as well as concrete constructions. Reference [8] defined Multiple-File Remote Data Checking (MF-RDC), an RDC model suitable for the special data update model of cloud storage, and MF-RDC checks the intactness of a dynamic file group consisting of a growing number of static files.

Based on multi-tenancy data-sharing storage model, it is well believed that the isolation of different tenants' replicas and supporting of data dynamics can be of vital importance to the integrity verification of multiple replicas but the articles in this field are rare. In view of these features of cloud data storage, in this paper we propose a periodic sampling approach to verify the replicas. In contract with whole checking, random sampling checking greatly reduces the workload of verification while still achieve an effective detection of misbehavior. In this way, the verification activities are efficiently scheduled in a verification period. Therefore, this method can detect the exceptions in time, and reduce the sampling numbers in each checking. For improving the efficiency of Integrity verification, we propose the double-layer authenticating construction which can support the isolation of different tenants during verifying replicas and data error localization.

Our contribution can be summarized as follows: (1) We propose the double-layer authenticating construction which can achieve the data share and the isolation of different tenants with the help of effective constructions such as MHT (Merkle Hash Tree) and MABTree (the masked authenticating B+-tree); (2) We propose an efficient approach based on random sampling and periodic verification for improving the performance of verifying the replicas; (3) We support data error localization and dynamic data operations; (4) Extensive security and performance analysis based on a proof-of-concept prototype about sampling in different conditions evaluate the feasibility and viability of our proposed scheme in this paper.

The rest of the paper is organized as follows. Section 2 overviews related works. In section 3, we describe the system model. Then, we provide the detailed description of our integrity verification scheme and process in section 4. In section 5 we propose performance analysis and show the correctness of our scheme. Finally, we conclude in section 6.

2 Related Works

Recently, many approaches were proposed to assure cloud data storage integrity [1, 2, 3, 4, 5, 6, 7]. Reference [1] constructed an elegant verification scheme for public verifiability and dynamic data operations, and improved the POR model by manipulating the classic Merkle Hash Tree (MHT) construction for block tag authentication. Reference [2] proposed an effective and flexible distributed schema which achieved the integration of storage correctness insurance and data error localization and supported efficient dynamic operations on data blocks by utilizing the homomorphic token with distributed verification of erasure-coded data. Reference [3] introduced HAIL, a high-availability and integrity layer, which was a distributed cryptographic system that allows a set of servers to prove to a client that a stored file is intact and retrievable. Reference [4] proposed an efficient approach based on probabilistic query and periodic verification for improving the performance of audit services. Reference [7] introduced the notion of proofs-of-ownership, which lets a client efficiently prove to a server that that the client holds a file, rather than just some short information about it. However, these approaches just consider the data integrity for file storage, which is not inefficient for SaaS model where software and databases are both deployed at service provider's platform.

Reference [11] proposed a provably-secure scheme that allowed a client that stores t replicas of a file in a storage system to verify through multiple-replica provable data possession (MR-PDP). The scheme used a constant amount of metadata for any number of replicas and new replicas may be created dynamically without pre-processing the file again. Reference [8] proposed constructions of two efficient MF-RDC schemes which checked the intactness of a dynamic file group consisting of a growing number of static files. However, all these methods are not available directly for SaaS model.

Reference [12] proposed a new integrity checking scheme using the masked authenticating B+-tree (MABTree) as the underlying data structure, which reduced both the computational overhead in the integrity check process and the verification time of the query results. This scheme could be used in our paper. Reference [5, 6] defined data integrity concept for SaaS data storage security. Basing on the meta-data driven data storage model and data chunking technology, SaaS data integrity verification can be realized based on the integrity verification of data chunks. It can be seen that while existing schemes are proposed to aiming at providing integrity verification under different data storage systems, based on multi-tenancy data-sharing storage model, multiple replica integrity verification has not been fully addressed. How to achieve a secure and efficient design to solve this problem remains an open challenging task in clouds.

3 System Model

As shown in Fig.1, our architecture for cloud data storage defines three different entities as follows: Tenant□ an entity, which has large data to be stored in clouds and relies on the cloud for data maintenance and computation, owns multiple users; Cloud Server Provider (CSP): an entity, which provides data storage service and has significant storage space and computation resources to maintain tenants' data; Third Party Auditor (TPA): a TPA, which has expertise and capabilities that tenants do not have, is trusted to manage or monitor stored data on behalf of the tenants upon request. In this paper we assume that the service platform (SP) is trustworthy.

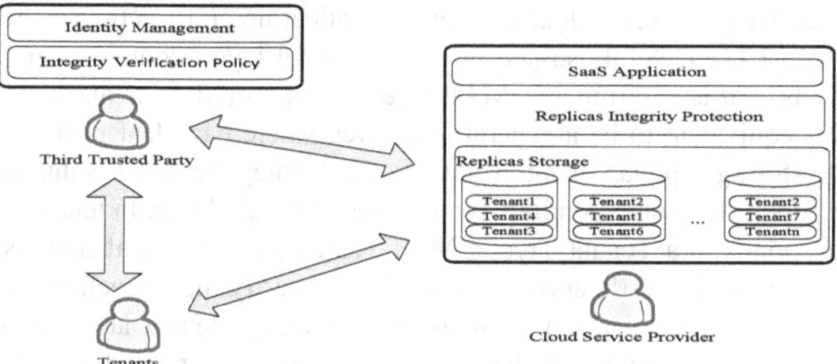

Fig. 1. Cloud data storage architecture

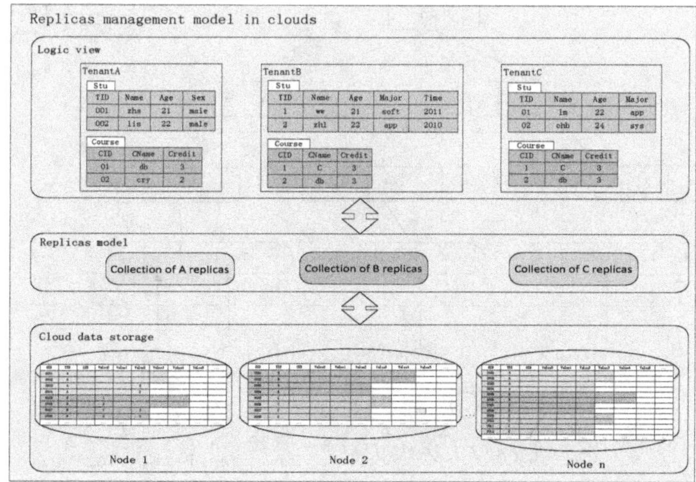

Fig. 2. Replicas management model in clouds

Based on multi-tenancy data-sharing storage model, multi tenants share the common data structure as illustrated in Fig.2. First, the data is partitioned horizontally according to tenants' ID. All of the application data of a tenant is concentrated in the same partition and generated a collection of replicas. Then, based on replica placement strategy, replicas are placed into different nodes. We propose the hypothesis: A tenant's total amount of data is no greater than the capacity of a single data node, any of replicas cannot be divided and stored into different data nodes, and a data node can hold replicas of hundreds of tenants. We don't consider the problem about how to place replicas for the present.

4 Our Integrity Verification Scheme

4.1 Double-Layer Authenticating Construction

As an improvement of the existing masked authenticating B+-tree (MABTree) methods [12], we proposed a double-layer authenticating construction including total layer and tenant layer. As illustrated in Fig.3, in total layer, common computational information is extracted from the MABTree and is stored in a mask vector. The MABTree is equivalent to be a general index tree where each leaf node represents a server node storing replicas of multi tenants. Each integrity check value of the leaf node is the root of the corresponding Tenant Tree of tenant layer. In tenant layer, each tenant corresponds to a Tenant Tree, a MHT [13], where each leaf node is the hash value of a tuple. In the verification process, we first verify the correctness of integrity check value of the special tenant in total layer. Then, in tenant layer we verify the correctness of the sampling tuples. The following are related definitions and descriptions about the double-layer authenticating construction.

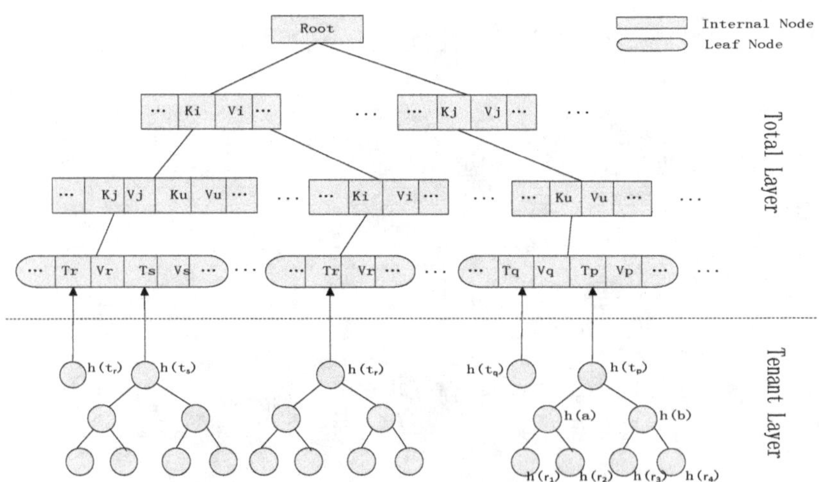

Fig. 3. Double-layer authenticating construction

Definition 1: The masked authenticating B+-tree. Let MABTree = (Root, INode, LNode, Nmask). Root denotes the root node, INode is a set of internal nodes, LNode is a set of leaf nodes, Nmask is the node mask used to calculate the node check values.

Definition 2: Let LNode = (T_1, V_1, T_2, V_2, …, T_n, V_n), T_i is a pointer to the Tenant Tree, V_i is the check value of the Tenant Tree which T_i is pointing to. Let INode = (K_1, V_1, K_2, V_2, …, K_n, V_n), K_i is the pointer to the i-th child node, V_i is the check value of the i-th child node which K_i is pointing to. N is the degree of MABTree. Let Nmask = (A_1, A_2, …, A_n), it is a n-dimensional prime vector, A_i is a randomly selected prime.

Definition 3: Compute the check value V_r of leaf nodes via the following equation (1), $h(t_r)$ is the root of the Tenant Tree which T_r is pointing to in tenant layer, where $1 \leq r \leq n$.

$$V_r = h(t_r) \tag{1}$$

Definition 4: Compute the check value of internal nodes. Let $N = \prod_{i=1}^{n} A_i$, $N_i = N/A_i = \prod_{j=1, j \neq i}^{n} A_j$, $N_i x_i \equiv 1 \bmod A_i$, x_i is the multiplicative inverse of modulo A_i. N_i and A_i are coprime numbers, so inverse x_i must exist. The check value V_d is computed by the following equation (2).

$$V_d \equiv \sum_{i=1}^{n} h(D.V_i) \times N_i \times y_i \ mod\ N \tag{2}$$

According to the Chinese Remainder Theorem, V_d is the unique solution of congruence equation (2). $D.V_i$ is the i-th check value of node D where Vu is stored, where $1 \leq d \leq n$.

$$V_d \equiv h(D.V_i) \ mod\ A_i \tag{3}$$

Definition 5: The Tenant Tree. A Merkle Hash Tree (MHT), a well-studied authentication structure, corresponds to a tenant Ti. It is constructed as a binary tree where the leaves in the MHT are the hashes of tuples. Each non-leaf node stores a hash value, which is generated by its child nodes values after being connected and transformed by a one-way hash function. With this method of calculating the non-leaf nodes, the root node R_i is generated as a check value V_i of the leaf node of MABTree.

We hypothesize that parameters r_1, r_2, r_3, r_4 denote four tuples of tenant T_p, h is a one-way hash function, $h(t_p)$ is the root node of MHT, h(a), h(b), $h(r_1)$, $h(r_2)$, $h(r_3)$, $h(r_4)$ denotes hash values of other nodes, as shown in Fig.3, a= $h(r_1) \parallel h(r_2)$, b= $h(r_3) \parallel h(r_4)$, t_p= $h(a) \parallel h(b)$.

In our double-layer authenticating construction, by avoiding mass exponential computation, the scheme reduces both the computational overhead in the integrity verification process and the verification time of the query results. More importantly, we don't need access other check values in the same node in the process of checking or updating a certain check value, so that we achieve the isolation of different tenants during verifying replicas. Also, we can locate error tuple through the MHT construction in tenant layer.

4.2 Integrity Verification Process

There are n tenants, T_1, T_2, …, T_n, tenant T_i generates and stores multi replicas, on t servers, S_1, S_2, …, S_t. N_i denotes the replica number of tenant T_i, F_r denotes the replica stored in server S_r, where $1 \leq i \leq n$, $1 \leq r \leq t$, $1 \leq N_i \leq t$.

4.2.1 Notations and Algorithms

--*KeyGen(1^k):* This probabilistic algorithm is run by the tenant. It takes a security parameter *k* as input, and returns public key *pk* and private key *sk*.
--*SigGen(sk):* This algorithm is run by the third party auditor in SP. It takes as input private key *sk* and the root R of MABTree, and outputs the signature $sig_{sk}(H(R))$.
--*ProofGen(chal):* TPA generates a random challenge *chal* and challenges CSP. It takes as input *chal* and outputs a data integrity proof *P* for the tuples specified by *chal.*
--*ProofVerify(P, chal, pk):* This algorithm can be run by either the tenant or the third party auditor upon receipt of the proof *P*. It takes as input the public key *pk*, the challenge *chal*, and the proof *P*, and outputs *true* if the integrity of replica is verified as correct, or *false* otherwise.
--*Update(update):* It takes as input a data operation request *update* from the tenant and outputs an updated tuple, updated Tenant Tree and MABTree and a new proof *P'*.
--*UpdateVerify (pk, update, P')*: This algorithm is run by the TPA. It takes as input public key *pk*, an operation request *update*, and the proof *P'* from CSP. If the verification successes, it outputs a signature $sig_{sk}(H(R'))$ for the new root R', or *false* otherwise.

4.2.2 Setup Phase

Firstly, the tenant's public key and private key are generated by invoking *KeyGen(1^k)*. The tenant T_i chooses a random $\alpha \leftarrow \mathbb{Z}_p$ and computes $\gamma \leftarrow g^{\alpha}$, the secret key is $sk = (\alpha)$ and the public key is $pk = (\gamma)$. Then, based on replica placement strategy, tenant Ti sends its N_i replicas to N_i serves and deletes them from its local storage, while a root R based on the double-layer authenticating construction of MABTree and Tenant Tree is generated. The specific calculation method depends on the above definitions. Next, TAP signs the root R under the private key α: $(H(R))^{\alpha} \rightarrow sig_{sk}(H(R))$ and sends the signature into clouds.

4.2.3 Integrity Verification Phase

As illustrated in Fig.4(a), the verifier (TPA or the tenant T_i) can verify the integrity of multiple replicas stored in clouds by challenging a serve S_r or all serves which stores replicas in parallel. This is an interactive proof protocol between CSP and TPA. TPA generates the message *chal* for S_r includes *c* (the number of challenged tuples), *k* (a set which determines the positions of the challenged tuples), and sends the *chal (c,k)* to the server S_r. Upon receiving the responses from the prover, TPA first authenticates the root R and special check value V_i of leaf node of MABTree that is the root R_i of corresponding Tenant Tree, and then checks the correctness of sampling tuples.

In the total layer, the verifier authenticates the root R of MABTree by checking $e(sig_{sk}(H(R),\ g)) = e(H(R)g^{\alpha})$. If the authentication fails, the verifier returns *false*. Otherwise the verifier authenticates special check value V_i of leaf node of MABTree. The verification process is shown from the bottom to top in MABTree.

(1) Find the check value V_i of leaf nodes and verify the equation (1).

(2) Let D be the leaf node which stores check value V_i, let F be the parent node of D, the check value V_d of F corresponds to node D. Verify the equation (3).

(3) If node F where the check value V_d stores is the root node, the authentication is successful. Otherwise, verify the check values until the root node. If any equation fails, then return *false*.

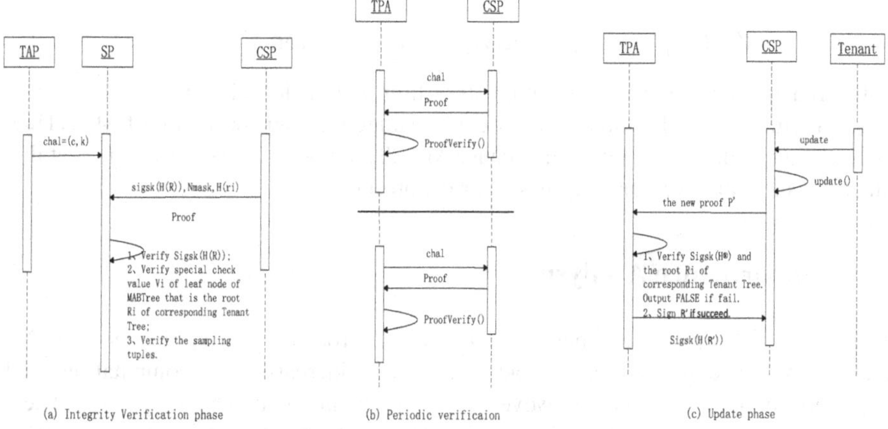

Fig. 4. Integrity verification process

If the signature verification or any equation fails, then returns *false*. Otherwise, check the correctness of sampling tuples in tenant layer.

In tenant layer, the verification process is also shown from the bottom to top in MHT. Verify the hash value of sampling tuples until the root node of the Tenant Tree. If the hash value of the root is equal to the verified $h(t_i)$ in total layer, the verification successes and return *true*, otherwise *false*.

4.2.4 Update Phase

As illustrated in Fig.4(c), suppose the tenant T_i wants to modify the i-th tuple *r* to *r'*. The tenant T_i generates an update request message $updata = (M, i, r')$ and sends to cloud servers, where *M* denotes the modification operation. Upon receiving the request, CSP replaces the tuple *r* with *r'* and *H(r)* with *H(r')* in the Tenant Tree construction, and generates the new root R' in MABTree and a new proof *P'* by running *Update(update)*. After receiving the new proof *P'* for this operation, TAP first authenticates R by checking $e(sig_{sk}(H(R),\ g)) = e(H(R)g^{\alpha})$, and special check value V_i of leaf node of MABTree that is the root R_i of corresponding Tenant Tree. If it is not true, output *false*, otherwise output *true*. Then TAP signs the new root R' by $sig_{sk}(H(R'))$ and sends it into clouds.

In update MABTree phase, update check value V_r of leaf node D via the following equation (4). h(t_r') is new root of the corresponding Tenant Tree. Let F be the father node of D, updating check value V_d via the following equation (5). Update the check value until the root of MABTree. After finishing the update, the new check value still meet the verification equation. Node mask has been computed in constructing the MABTree, so don't need to compute it in update phase. More importantly, don't need to access other tenants' data when update a tenant' date, that is to complete the update of the corresponding check value of the parent node only via new value and old value, which achieves the isolation of different tenants.

$$V_r' = h(t_r') \tag{4}$$

$$V_d' \equiv V_d + \big(h(V_r') - h(V_r)\big) \times N_r \times x_r \bmod N \tag{5}$$

Compared to tuple modification, tuple insertion and tuple deletion can change the logic structure of the Tenant Tree, but can't change the logic structure of MABTree. Only adding or deleting a tenant can cause MABTree nodes to split or merge, but this situation is less and we don't discuss it for the present.

5 Performance Analysis

In this paper, we propose a periodic sampling approach to verify the replicas. No doubt too frequent verification activities will increase the computation and communication overheads. However, less frequent activities may not detect abnormality timely. Therefore, in order to detect the exceptions in time and also reduce overhead, we have to achieve a balance between random sampling verification and periodic verification and find an efficient strategy to improve the performance.

5.1 Sampling Verification

We propose a random sampling approach in order to reduce the workload. The detection probability P of disrupted tuples is an important parameter to guarantee that these tuples can be detected in time. Assume on a server there are e disrupted tuples out of the n tuples which belong to the same tenant. The probability of disrupted blocks is ρ=e/n. Let c be the number of sampling tuples for a challenge in the protocol proof. We have detection probability P=1-((n-e)/n)^c=1-(1-ρ)^c. Hence, the number of sampling tuples is c=log(1-P)/log(1-ρ) ≈(P·n)/e, sampling ratio (the ratio of sampling tuples in total tuples) is w=c/n=log(1-P)/(n·log(1-ρ))≈P/e. When n is infinitely great, w tends to be a constant value. As shown in Fig 5, the results of sampling ratio w under different detection probabilities P (from 0.5 to 0.99), different number of total tuples n (from 100 to 10000), and constant number of disrupted tuples e(10, 100). To achieve P of the least 99%, sampling ratio w is close to 4.5% and 45% of n where e is 100 and 10.

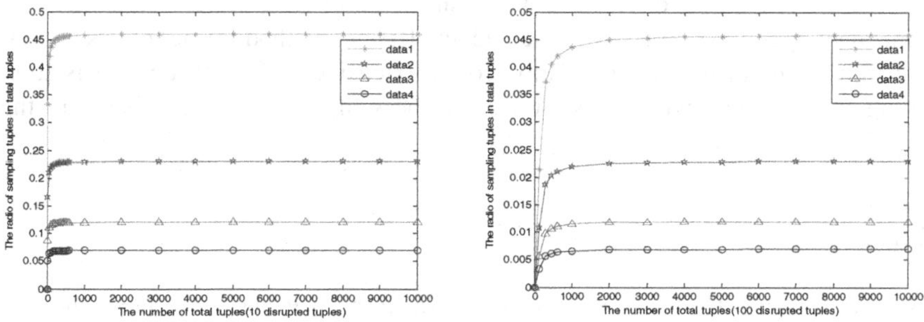

Fig. 5. The sampling ratio under different detection probabilities

5.2 Periodic Verification

We assume that each tenant has a verification period T, which depends on how important it is for the tenant. For example, some tenants assign verification period as one week, some assign it as one month. Assume we make use of the verification frequency f to denote the number of occurrences of a verification event per unit time. The detection probability in unit time $P_e = 1 - (1 - P)^f$. In terms of $P = 1 - (1 - \rho)^c = 1 - (1 - \rho)^{n \cdot w}$, we have the equation $P_e = 1 - (1 - P)^f = 1 - (1 - \rho)^{n \cdot w \cdot f}$ and the equation $f = \frac{\log(1 - P_e)}{w \cdot n \cdot \log(1 - \rho)}$. This means that the verification frequency f is inversely proportional to sampling ratio w. In Fig 6, we show the relationship between f and w under 10 and 100 disrupted tuples for 10,000 tuples. Hence, appropriate verification frequency would greatly reduce sampling numbers, as well as computation and communication overheads.

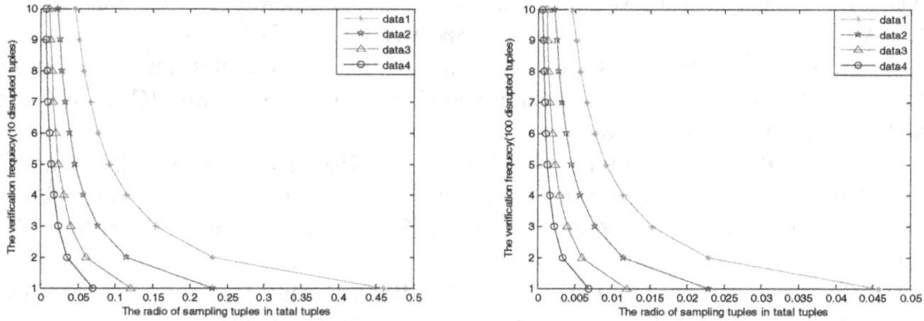

Fig. 6. The verification frequency under different sampling ratios

6 Conclusions

In this paper, based on multi-tenancy data-sharing storage model, we present the double-layer authenticating construction which allows a third party auditor (TPA) to verify the integrity of multiple replicas stored in clouds. Our construction is designed to meet the isolation of different tenants and support data error localization and dynamic data operations. We also presented an efficient method for periodic sampling verification to minimize the computation costs. Extensive performance analysis about sampling in different conditions shows correctness of the proposed scheme in this paper.

References

1. Wang, Q., Wang, C., Li, J., Ren, K., Lou, W.: Enabling Public Verifiability and Data Dynamics for Storage Security in Cloud Computing. In: Backes, M., Ning, P. (eds.) ESORICS 2009. LNCS, vol. 5789, pp. 355–370. Springer, Heidelberg (2009)
2. Wang, C., Wang, Q., Ren, K., Lou, W.: Ensuring data storage security in cloud computing. In: 17th IEEE International Workshop on Quality of Service (IWQoS 2009), pp. 1–9. IEEE Press, New York (2009)
3. Bowers, K., Juels, A., Oprea, A.: HAIL: a high-availability and integrity layer for cloud storage. In: Proceedings of the 2009 ACM Conference on Computer and Communications Security (CCS 2009), pp. 187–198. ACM, New York (2009)
4. Zhu, Y., Wang, H., Hu, Z., Ahn, G., Hu, H., Yau, S.: Dynamic Audit Services for Integrity Verification of Outsourced Storages in Clouds. In: The 26th Symposium On Applied Computing, pp. 1550–1556. ACMSAC, Taiwan (2011)
5. Shi, Y., Zhang, K., Li, Q.: A New Data Integrity Verification Mechanism for SaaS. In: Wang, F.L., Gong, Z., Luo, X., Lei, J. (eds.) WISM 2010. LNCS, vol. 6318, pp. 236–243. Springer, Heidelberg (2010)
6. Shi, Y., Zhang, K., Li, Q.: Meta-data Driven Data Chunk Based Secure Data Storage for SaaS. International Journal of Digital Content Technology and its Applications 5, 173–185 (2011)
7. Pinkas, B., Shulman-Peleg, A., Halevi, S., Harnik, D.: Proofs of ownership in remote storage systems. Cryptology ePrint Archive, Report 2011/207 (2011)
8. Xiao, D., Yang, Y., Yao, W., Wu, C., Liu, J., Yang, Y.: Multiple-File Remote Data Checking for cloud storage. Computers and Security 31(2), 192–205 (2012)
9. Juels, A., Kaliski Jr., B.: PORs: proofs of retrievability for large files. In: Proceedings of the 2007 ACM Conference on Computer and Communications Security (CCS 2007), pp. 584–597. ACM, New York (2007)
10. Ateniese, G., Burns, R., Curtmola, R., Herring, J., Kissner, L., Peterson, Z., Song, D.: Provable data possession at untrusted stores. In: Proceedings of the 2007 ACM Conference on Computer and Communications Security (CSS 2007), pp. 598–609. ACM, New York (2007)
11. Reza, C., Osama, K., Randal, B., Giuseppe, A.: MR-PDP: Multiple-Replica Provable Data Possession. In: The 28th International Conference on Distributed Computing Systems, pp. 411–420. IEEE Press, Beijing (2008)
12. Xian, H., Feng, D.: An Integrity Checking Scheme in Outsourced Database Model. Journal of Computer Research and Development 47(6), 1107–1115 (2010)
13. Merkle, R.C.: Protocols for public key cryptosystems. In: Proc. of IEEE Symposium on Security and Privacy, pp. 122–134. IEEE Computer Society, Washington (1980)

Multi-stage Attack Detection Algorithm Based on Hidden Markov Model

Nurbol Luktarhan[1], Xue Jia[1], Liang Hu[2], and Nannan Xie[2,*]

[1] Information Science and Engineering College, Xinjiang University, Urumqi, China
[2] Computer Science and Technology College, Jilin University, Changchun, China
jienn10@jlu.edu.cn

Abstract. With the growing amount and kinds of network intrusion, multi-stage attack is becoming the one of the main methos of the network security threaten. The hidden Markov model is a kind of probabilistic model, which is widely used in speech recognition, text and image processing. In this paper, a Multi-stage Attack Detection Algorithm Based on Hidden Markov Model is proposed.And inorder to improve the performance of this algorithm,aother algorithm aims at false positive filter is also put forward. Experiments show that the algorithm has good perfomance in multi-stage attack detection.

Keywords: Hidden Markov Model, Multi-stage attack, False positive Filter.

1 Introduction

Nowadays, with more and more network intrusion events, the importance of network security is getting more and more attention. Symantic's reports in July 2010[1] and April 2011[2] show that the type of network intrusion has transformed from the original "pure" Trojan horses and viruses into complex attacks of the combination of the "Internet", "Teamwork" and "Commercial Interests", especially the multi-stage attacks, to achieve a network intrusion, which not only threats networks but also difficult to predict. It is such a huge challenge to the existing equipments of security detection. In this case, the detection of multi-stage attack has become a hot research topic in network security field.

Commonly the process of multi-stage attack is composed of a number of interrelated attack steps, the previous one being the occurring condition of the latter one which made it more difficult to detect. To solve the problem that it is difficult to detect multi-stage attack, researchers put forward intrusion alert correlation and attack scenario recognition [3], which is aimed to the effective identification of multi-stage attack scenario. There are some representative research of attack intention recognition technology, such as the event correlation method based on attack premise and result[4], Hidden Colored Petri-Net model[5] and similarity analysis. These methods have generally achieved better research results, but the costs of establishment of the prior knowledge base and the selection of large training samples are large[6].

Hidden Markov models(HMM) is one of the methods which achieve a better effect in the field of speech recognition, and HMM can modeling the complex attacks and

* Corresponding author.

F.L. Wang et al. (Eds.): WISM 2012, LNCS 7529, pp. 275–282, 2012.

has a better classification ability of attack sequence[7]. In this paper, the application of HMM in the multi-stage attack detection is analysed, and an multi-step attack detection algorithm based on HMM is proposed. Experiments show that this method can effectively recognize the multi-stage attack.

2 Related Works

2.1 Typical Methods of Multi-stage Attack Detection

A number of relational analysis methods have been put forward in the research of security event correlation analysis, but there are few typical and influential methods. Wang Li introduces several typical methods of security event correlation [3]:

A Correlation polymerization method based probability similarity

It is a quantitative method, which makes correlation aggregation on security events, can be used for the analysis of real-time security event aggregation correlation. The method was proposed by Valdes etc. in U.S. scientific research EMERALD[8]. Since it uses quantitative calculation, the main advantage of this method is avoiding the complexity of define formulation correlation, and it performs well in aggregating some alarms with high similarity.

B Cross-correlation method based on machine learning

Researchers from IBM believe that in large number of false positives, the majorities are resulted from quite a few causes. Classify the alarms resulted from the same cause and then find the cause, large quantities of false positives can be removed, and the number of alarms can be reduced. Based on this, Pietraszek etc. proposed a IDS false alarms removal method based on machine learning[9] . This method has good practical applicability, has been applied to actual products and can precisely analyses false alarms while it needs a lot of manual intervention and demands professional quality of the security analysts.

C Attack scenario construction method based on causes and effects

This method was first proposed in 2010 by Templeton et al.[10]. The basic idea is to associate two security incidents by comparing two incidents, the consequence of one incident that happened earlier, and the prerequisite of the incident that happened later. The advantage of this method is that new attacks formed by the combination of different attacks can be identified without knowing the whole attack. The deficiency is that those attacks cannot be handled without knowing the causes and effects of the attack, in addition, the consumption of computing resources is larger.

2.2 HMM

HMM is a statistical model which is widely used in the fields of text, image, voice processing and so on. It is developed from the Markov chain[11]. HMM is a double random process, which contains a Markov chain with a hidden state transition Q= { q1, q2, ... , qt, ... ,qT} (qt is the state of system at time t), and a random observation sequence O= { O1 , O2, ... , Ot, ... ,OT}(Ot is the observed values of the system at time t) . The state transition process of Q can only be inferred through the observation sequence O.

The complete HMM consists of five features[6]:N, number of state, Corresponds to a finite state set S = { S1, S2, ... , SN};M, total number of observations, corresponds to finite set of observations O={V1, V2, …, VM};π, the initial state distribution;A, State transition probability;B, the probability of the observed values generate.

This Markov model consists of two stochastic processes: One is the Markov process, also known as the basic random process, describing the transitions between the states. The other random process describe the statistical relationship between the state and the observed values.

Complex network attacks always can be divided into several stages, each stage depends on the results of the previous stage, the conversion of the attack stages meet the characteristics of Markov. Each stage of the attack can achieve its goal through a variety kinds of ways, which lead to the multiple alarm events, and some alarm events may appear in more than one attack step. There is no mapping relationship between attack steps and alarming time. In a certain moment, which kind of attack, which step of a attack event can not be know exactly, it needs to be inferred from more than one alarm time. There is good correspondence between attack steps, alarm time and the HMM state.

3 Multi-stage Attack Detection Algorithm Based on HMM

3.1 The Basic Definition

Definition 1: A hidden Markov model is a triplet (π, A, B) . $\Pi = (\pi_i)$: Initialized probability vector ; $A = (a_{ij})$: State transition matrix ; $P_r(x_{i_t} \mid x_{j_{t-1}})$; $B = (b_{ij})$: Confusion matrix ; $P_r(y_i \mid x_j)$;

Every probability in the state transition matrix and confusion matrix are independent of time, that is, when the system evolution, the matrix does not change with time, and this is the model that used in proposed model in the next part.

Definition 2: Attack scenario model set $\Lambda = \{\lambda_1, \lambda_2, ..., \lambda_n\}$;

Definition 3: $P(\vec{O}|\lambda)$, The model expresses the probability of an observed sequence appear in the system.

3.2 The Algorithm Description

Input：Alarm sequence $\vec{O} = o_1, o_2, ..., o_T$
Output：Probability of occurrence of intrusion scenarios P
Parameter：Hidden Markov model $\lambda = (\Pi, A, B)$
Step1. Training a hidden Markov model for each attack scenario $\lambda = (\Pi, A, B)$
Step2. Get a collection of an attack scenario model $\Lambda = \{\lambda_1, \lambda_2, ..., \lambda_n\}$
Step3. Let i = 1,if i<n，calculate $P(\vec{O}|\lambda_i)$

Step4. Find the largest $P(\vec{O}|\lambda_i)$ corresponding to the attack scene , that is the scene of the attack happened currently.
Step5. End.

Training Hidden Markov Models for each attack, we need to collect attack scene data, which can be used to train two attack scenarios. In the determination of hidden Markov model parameters, since each step based on the previous step attack, according to the means of attackers and the familiarity of the target network, some attack steps can be omitted. The initial probability $\Pi = \{\pi_i\}$ is set as a descending sequence accord to the attack steps, so the initial probability of a previous attack step is large than the follow one.

After define the parameters Π, A, B , make the alarm sequence as the observation sequence, use the above algorithm, can calculate the probability that an alarm sequence appearing in this model, as the probability of occurrence of this intrusion.

3.3 Performance Optimizing with False Positive

Intrusion detection technology has some limitations, this is an important reason to cause false positives.The reason for the high rate of intrusion detection system false positive can be summarized as follows:caused by the specific topology of the network itself,caused by the specific application protocol,caused by the special equipment,caused by network intrusion detection system's vulnerability.The introduction of intrusion detection system false alarm filter is the method to improve the effection of multi-step attack to identify.

Definition 4: By listing all possible hidden state sequences and calculating corresponding probabilities of each combination to find the most possible hidden state sequence. The most likely hidden state sequence is a combination which makes the below probability biggest : Pr (observed sequence I hidden state combination).

For example, the alarm sequence O = {"alert category ID1","alert category ID2 "," alert category ID3 "} shows the observed sequence, The most likely hidden state sequence is the one corresponds to the maximum probability of the following probabilities,which expressed as: Pr (alert category ID1 "," alert category ID2 "," alert category ID3 "I false alarm, false, alarm, false alarm), Pr (alarm category ID1", "alert category ID2","alert category ID3 "I false, alarm, false, alarm, attack), ..., Pr ("alert category ID1","alert categoryID2", "alert category ID3" lattack, attack, attack).

The Algorithm Description:

Inpute: alert file that caused by intrusion detection system.

Outpute: Alarm sequence Q （Q1, Q2, ...Qn) ,the alert category that output by intrusion detection system.

Parameter: Hidden Markov model HMM{M, N, A, B, pi}.The total time value T; alarm feature observation sequence O (O1, O2,, ... On), the name of the initial alert ID value."

Step1. Declares a local probability delta, so that for each i (1iN) make the delta [1] [i] = pi[i]*B[i][O[1]], through the product of the initial probability of hidden state and the corresponding observation probability, computing the local probability when t = 1.

Step2.t increase 1,make j=1,make j cycles N times, in every cycle, make i=1 cycles N

Times, computing delta[t-1][i]*(phmm->A[i][j]),and reached the maximum value Qt = MAXtj;

Step3. If t is less than or equal to T, Goto Step2:

Step4. According to the most likely state path backtracking in the entire grid. when backtracking finishied, will generate the observed sequence that most likely alert to hidden state sequence Q.

Step5. In the algorithm,step2 determine the most likely path to reach the next state, And make the record about how to reach the next state. Specifically, First, by examining the produnct of all of the transition probability and the maximum partial probability obtained in the prebious step, And then record the biggest one. Also includes trigger this probability state in the previous step. Step 3 Make sure the t=T the most likely hidden state when the system complete.

4 Experiments and Evaluation

Experiments were conducted using the LLDOS1.0 and LLDOS2.0 datasets in order to check the performance of the proposed approach for Multi-stage attack detection. The LLDOS1.0 and LLDOS2.0 dataset are extension of the DARPA2000 dataset, which is prepared and managed by MIT Lincoln Labs. The evaluation criteria Correct Rate is defined as "the number of correlation alerts detected/ the total number of correlation alerts".

4.1 Experiment Result

(1) Attack detection test

It can be seen from the figure 1 that the probability of multi-step attack scenario for LLDOS1.0 are increasing,and the probability of occur LLDOS2.0.1 multi-step attack scenario is maintained at the above of a steady data.

(2) The algorithm performance with and without false positive filter

In order to explore the effection of the false alarm to the result of the test,based on the preliminary work, in the following experiment,false positive filter module to join or remove from the algorithm ,under the two cases,compare the effection of multi-step attack scenario recognition algorithm.

Using the Intrusion scenario analysis module to analyze Snort alerts.we use the data to training and testing that the data used in the simulates attack scenarios in the previous section.These attack scenarios are belong to the same type,so it should be regarded as the training in an attack scene.The results predicted value of intrusion scene recognition using the testing data are close to an upward trend.the scene recognition results without using the false alarm filter module shown in Figure 2. Because there are a large number of false positive information, so we could not analyze the occurrence was what kind of attack scenarios.

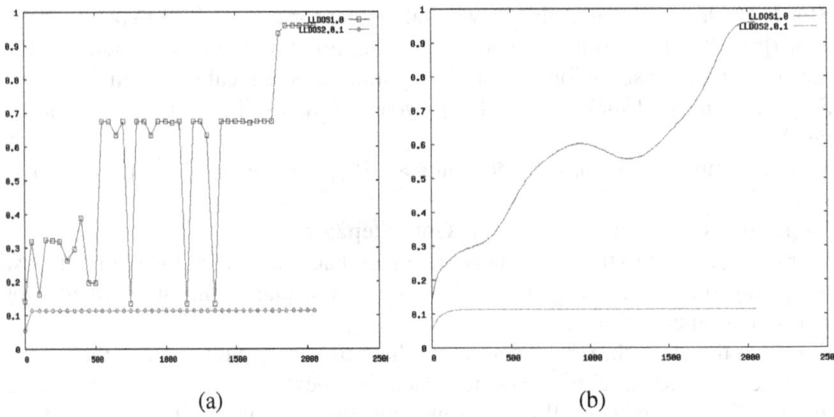

Fig. 1. Result of multi-stage attack detection(b is Bezier Curve)

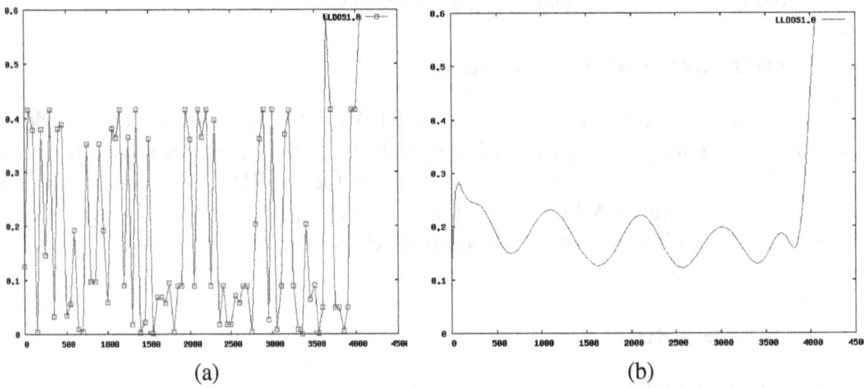

Fig. 2. Result of multi-stage attack detection without false positive filter(b is Bezier Curve)

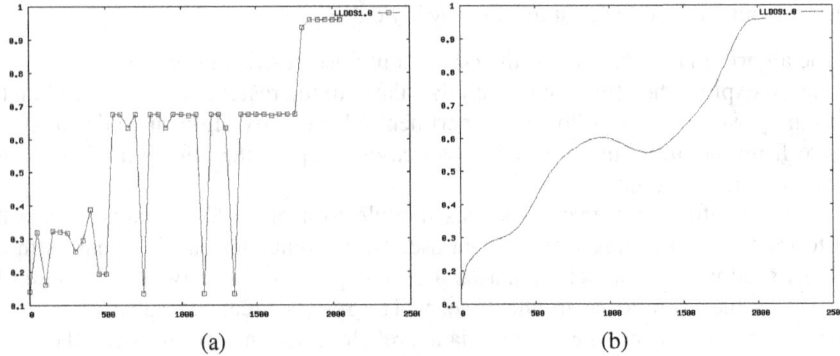

Fig. 3. Result of multi-stage attack detection with false positive filter(b is Bezier Curve)

Figure 3 is the scene recognition results using the false alarm filter module and the corresponding Bezier curves.From these two graphs can be seen,In-depth with the attack in the process of Scene Recognition with the false alarm filter module,The calculated probability value of corresponding attack scenario is also increasing.When the entire attack scene is completed, the corresponding probability of occurrence of the attack scene is close to 1.

The tests of mark the correct rate of attack steps also using the cross-comparison approach, Table 1 is ten times the cross on the results .It can be seen that,in most,the false alarm filter module to improve the correct rate of attack steps marked intrusion scenario analysis.

Table 1. Comparison of correct rate

No.	Correct rate without false positive filter	Correct rate with false positive filter
1	86.08%	99.98%
2	88.89%	99.96%
3	92.29%	95.00%
4	96.27%	99.76%
5	98.39%	96.81%
6	92.73%	95.00%
7	94.63%	94.96%
8	95.67%	94.49%
9	91.10%	99.70%
10	93.05%	95.06%

4.2 The Basic Definition Result Analyse

From the above experiments, we come to the conclusions:
(1)The multi-stage attack detection algorithm based on HMM has been proposed in this paper can effectively identify the multi-step attack scenarios.
(2)Comparing the algorithm with adding the false positive filter function, we can see that the detection rate is different.

5 Conclusion

In this paper, we use Hidden Markov Model to network intrusion, and propose an algorithm to achieve multi-stage attack detection. The method can detect the defined multi-step attack. It reduce the computing amout relatively and get high accuracy. Because we consider the entire attack sequence other than single step attack, the algorithm we proposed have good validity in multi-stage attack detection.

In future work, we will try this method in the online intrusion detection systems which developed by our team, and evaluate its performance in a real network environment, and given the demo of the practical application of this approach, and provide the online experimental data of the intrusion alert correlation.

Acknowledgment. This work is supported by the National Grand Fundamental Research 973 Program of China under Grant No. 2009CB320706, the National High Technology Research and Development Program(863) of China under Grant No. 2011AA010101, the National Natural Science Foundation of China under Grant No. 61073009 and 61163052, Program of XinJiang University Doctorial Start Foundation No BS110126.

References

1. Symantec Global Internet Security Threat Report trends, `http://eval.symantec.com/mktginfo/enterprise/white_papers/bwhitepaper_internet_security_threat_report_xiv_04-2009.en-us.pdf`
2. Symantec Internet Security Threat Reaport, `http://www.symantec.com/zh/cn/theme.jsp?themeid=istr`
3. Li, W.: Study on Method of network Multi-stage Attack Plan Recognition, Wu Han (2007)
4. Lindqvist, U., Porras, P.A.: Detecting computer and network misuse throughproduction-based exstemsy system toolset (P-BEST). In: Proceedings of the IEEE Comuter Society Symposium on Research in Security and Privacy, pp. 146–161. IEEE Press, Washington, D.C. (1999)
5. Liang, Y., Zhou, J., Yan, P.: Network Intrusion Detection System Based on CPN and Mobile Agent. J. Computer Engineering 16, 106–108 (2003)
6. Sun, Y., Zhong, Q., Su, J.: Research on Intention Recognition Based on HMM. J. Computer Engineering & Science 29(8), 19–22 (2007)
7. Balthrop, J., Esponda, F., Forrest, S., et al.: Coverage and Generalization in an Artificial Immune System. In: Proceedings of the Genetic and Evolutionary Computation Conference, pp. 3–10. Morgan Kaufmann, New York (2002)
8. Valdes, A., Skinner, K.: Adaptive, Model-Based Monitoring for Cyber Attack Detection. In: Debar, H., Mé, L., Wu, S.F. (eds.) RAID 2000. LNCS, vol. 1907, pp. 80–92. Springer, Heidelberg (2000)
9. Pietraszek, T., Tanner, A.: Data mining and machine learning-Towards reducing false positives in intrusion detection. Information Security Technical Report 10, 169–183 (2005)
10. Templeton, S.J., Levitt, K.: A requires/provides model for computer attacks. In: Proceedings of the New Security Paradigms Workshop 2000, Cork Ireland, pp. 31–38 (2000)
11. Zhong, A.M., Jia, C.F.: Study on the application of hidden Markov models to computer intrusion detection. In: Proceedings of the 5th World Congress on Intelligent Control and Automation, pp. 4352–4357 (2004)
12. Chen, X., Wen, Z.: The Research on Network Intrusion Detection Method Based on RBF-HMM. J. Network Security 1, 9–11 (2011)

Security Analysis of a Secure and Practical Dynamic Identity-Based Remote User Authentication Scheme

Mo-han Zhang[1], Chen-guang Yang[1], and Ding Wang[2,*]

[1] College of Software Engineering, Sichuan University, Chengdu City 610225, China
[2] Automobile Sergeant Institute of PLA, Bengbu City 233011, China
zmh3169@126.com, wangdingg@yeah.net

Abstract. In 2005, Lee et al. proposed a secure smart card based remote user authentication scheme to improve the security of Chien et al.'s scheme. More recently, Sood et al. pointed out that Lee et al.'s scheme is still vulnerable to the reflection attack, off-line password guessing attack, user impersonation attack and fails to preserve user anonymity. Consequently, Sood et al. proposed a more secure remote user authentication scheme, which is an improvement over Lee et al.'s scheme to overcome their security drawbacks. In this study, however, we find that Sood et al.'s scheme still cannot achieve the claimed security and report its following flaws: (1) It fails to preserve user anonymity under their non-tamper resistance assumption of the smart card; (2) It cannot withstand stolen-verifier attack. The proposed cryptanalysis discourages any use of the scheme for practical applications.

Keywords: Cryptanalysis, Authentication protocol, User anonymity, Smart card, Stolen-verifier attack.

1 Introduction

Since Lamport [1] proposed the first password authentication scheme in 1981, many password-based authentication schemes have been presented to improve the efficiency and security of remote authentication. In 2002, Chien et al. [2] found that a previous scheme presented by Sun [3] only achieves unilateral user authentication, more precisely, only authentication server authenticates the legitimacy of the remote user, the user cannot make sure the legitimacy of the corresponding server, which is an insecure factor. Hence, Chien et al. further developed another remote user authentication scheme and claimed that their scheme provides mutual authentication, requires no verification table, supports local password update and uses only a few hash operations. However, as pointed out by Hsu [4] and Ku et al. [5], Chien et al.'s scheme is susceptible to the insider attack and parallel session attack. Later, Lee et al. [6] improved Chien et al.'s scheme to cope with this drawback. In 2009, Xu et al. [7] demonstrated that

* Corresponding author.

F.L. Wang et al. (Eds.): WISM 2012, LNCS 7529, pp. 283–290, 2012.

Lee et al. scheme is prone to offline password guessing attack provided that the sensitive information stored in the smart card is disclosed by the adversary.

More recently, Sood et al. [8]] further found that Lee et al.'s [6] scheme is also vulnerable to malicious user attack, impersonation attack and reflection attack, and fails to protect the user's anonymity in insecure communication channel. Then they further proposed an enhanced version and claimed that their scheme is efficient and can overcome all the identified security drawbacks of Lee et al.'s scheme even if the smart card is non-tamper resistant. Unfortunately, in this paper, we will demonstrate that Sood et al.'s scheme still cannot withstand stolen verifier attack. And to our surprise, user anonymity, which is the most essential security feature a dynamic ID-based authentication scheme is designed to support, cannot be preserved either.

The remainder of this paper is organized as follows: in Section 2, we review Sood et al.'s authentication scheme. Section 3 describes the weaknesses of Sood et al.'s scheme. Section 4 concludes the paper.

2 Review of Sood et al.'s Scheme

In this section, we examine the dynamic identity-based authentication scheme using smart cards proposed by Sood et al. [8] in 2010. Their scheme, summarized in Fig.1, consists of four phases: the registration phase, the login phase, the verification and session key agreement phase and the password change phase. For simplicity of presentation, we employ some intuitive abbreviations and notations listed in Table 1.

Table 1. Notations

Symbol	Description
U_i	i^{th} user
S	remote server
$\mathcal{M}$	malicious attacker
ID_i	identity of user U_i
PW_i	password of user U_i
x	the secret key of remote server S
$H(\cdot)$	collision free one-way hash function
$\oplus$	the bitwise XOR operation
$\|\|$	the string concatenation operation
$A \rightarrow B : C$	message C is transferred through a common channel from A to B
$A \Rightarrow B : C$	message $C$is transferred through a secure channel from A to B

2.1 Registration Phase

The registration phase involves the following operations:

1) U_i chooses his/her identity ID_i, password PW_i, and a random number b, to compute $A_i = H(ID_i||b)$.

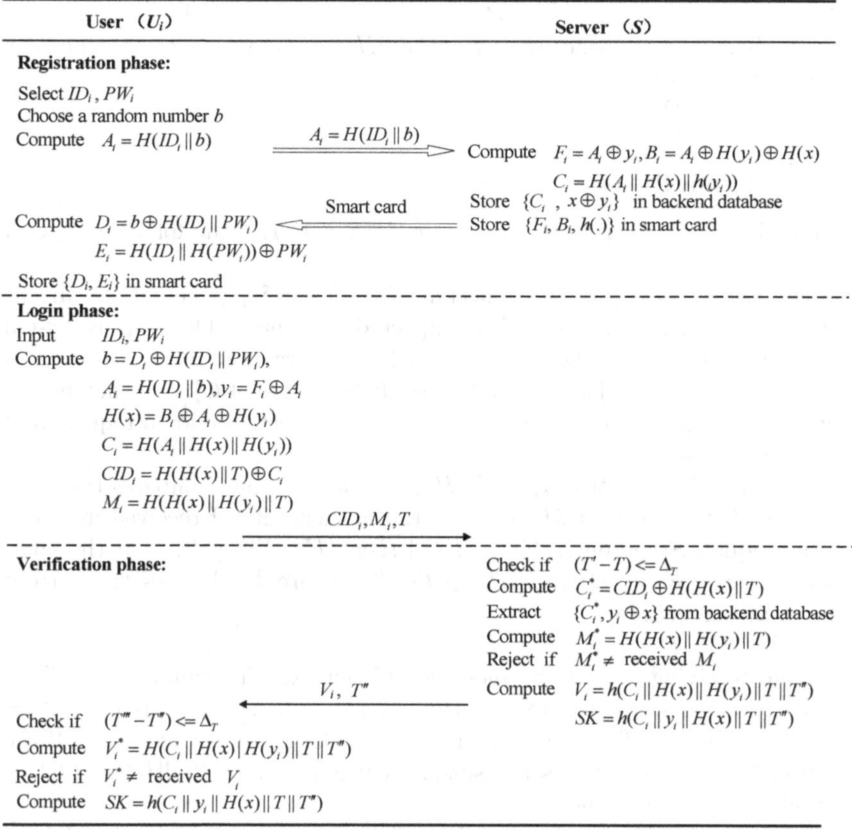

Fig. 1. Sood et al.'s remote user authentication scheme

2) $U_i \Rightarrow S: \{A_i\}$.

3) On receiving the registration message from U_i, the server S computes $F_i = A_i \oplus y_i$, $B_i = A_i \oplus H(y_i) \oplus H(x)$, and $C_i = H(A_i||H(x)||H(y_i)$, where x is the secret key of remote server S. S chooses the value of y_i corresponding to each user in such a way that the value of C_i must be unique for each user. Then, S stores $y_i \oplus x$ corresponding to C_i in its database and issues the smart card containing security parameters $(F_i, B_i, H(\cdot))$ to the user U_i.

4) $S \Rightarrow U_i$: A smart card containing security parameters $\{F_i, B_i, H(\cdot)\}$.

5) U_i computes security parameters $D_i = H(ID_i||PW_i) \oplus b$, $E_i = H(ID_i||H(PW_i)) \oplus PW_i$ and enters the values of D_i and E_i in his smart card.

2.2 Login Phase

When U_i wants to login to S, the following operations will be performed:

1) U_i inserts her smart card into the card reader, and inputs ID_i^* and PW_i^*.

2) The smart card computes $E_i^* = H(ID_i^*||H(PW_i^*)) \oplus PW_i^*$, and checks $E_i^* \stackrel{?}{=} E_i$. After the validation, smart card computes $b = D_i \oplus H(ID_i||PW_i)$, $A_i =$

$H(ID_i||b)$, $y_i = F_i \oplus A_i$, $H(x) = B_i \oplus A_i \oplus H(y_i)$, $C_i = H(A_i||H(x)||H(y_i))$, $CID_i = H(H(x)||T) \oplus C_i$ and $M_i = H(H(x)||H(y_i)||T)$, where T is the current timestamp on user side.

3) $U_i \rightarrow S : \{CID_i, M_i, T\}$.

2.3 Verification Phase

After receiving the login request from user U_i, S performs the following operations:

1) S checks the validity of T by checking $T' - T <= \nabla_T$, where T' is current date and time of the server S and ∇_T is expected time interval for a transmission delay. S computes $C_i^* = CID_i \oplus H(H(x)||T)$ and extracts y_i from $y_i \oplus x$ corresponding to C_i^* from the backend database. If the value of C_i^* does not match with any value of C_i in the database of server S, S rejects the login request and terminates this session.

2) The server S computes $M_i^* = H(H(x)||H(y_i)||T)$ and compares the computed M_i^* with the received M_i to check the authenticity of received message. If they are equal, S computes $V_i = H(C_i||H(x)||H(y_i)||T||T'')$, and the common session key $SK = H(C_i||y_i||H(x)||T||T'')$, where T'' denotes the current timestamp on server side.

3) $S \rightarrow U_i : \{V_i, T''\}$.

4) Upon receiving the reply message, U_i checks the validity of T''. If the verification fails, U_i terminates the session. Then U_i computes $V_i^* = H(C_i||H(x)||H(y_i)||T||T'')$ and compares it with the received value of V_i. If the verification holds, U_i computes the session key $SK = H(C_i||y_i||H(x)||T||T'')$. Otherwise, U_i terminates the session.

5) After authenticating each other, U_i and S use the same session key SK to secure ensuing data communications.

2.4 Password Change Phase

When U_i wants to change the old password PW_i to the new password PW_i^{new}, this phase will be involved. Since this phase has little relevance with our discussion, it is omitted here to save space.

3 Cryptanalysis of Sood et al.'s Scheme

There are three assumptions explicitly or implicitly made by Sood et al.'s scheme [8]:

i. An attacker $\mathcal{M}$ has total control over the communication channel between the user U and the remote server S. That is, the attacker can insert, alter, delete, or intercept any messages transmitted in the channel.
ii. The attacker can get access to the user's smart card to extract the secret parameters stored in the smart card.
iii. The passwords are weak.

Note that the above three assumptions, which are also made in the latest works [7,9–12], are indeed reasonable: (1) Assumption *i* is accordant with the standard Dolev-Yao adversary model [13]; (2) Assumption *ii* is practical in consideration of the state-of-art side-channel attack techniques [14–16]; and (3) Assumption *iii* reveals the reality that a user is allowed to choose her own password PW at will during the registration phase and password change phase, usually the user is apt to select a password which is easily remembered for her convenience and thus the human-memorable password tends to be "weak password" [17,18].

In the following investigation into the security flaws of Sood et al.'s remote user authentication scheme, based on the above assumptions, we assume that an attacker can reveal the secret values $\{B_i, D_i, E_i, V_i\}$ stored in the legitimate user's smart card, and the attacker can also intercept or block the login request $\{CID_i, M_i, T\}$ from the user U_i and the reply message $\{V_i, T''\}$ from S.

3.1 No Provision of User Anonymity

A protocol with user anonymity prevents an attacker from acquiring sensitive personal information about an individual's preferences, social circle, lifestyles, shopping patterns, etc. by analyzing the login information, the services, or the communications being accessed [19]. What's more, the leakage of the user specific information may make an unauthorized entity to track the user's current location and login history [20]. Therefore, assuring anonymity does not only protect user privacy but also make remote user authentication protocols more robust.

To preserve user anonymity, a straight forward and effective way is to employ the "dynamic ID technique": user's real identity is concealed in session-variant pseudo-identities. The authentication schemes employing this technique are so-called "dynamic-ID" schemes, and Sood et al.'s scheme falls into this category. However, Sood et al.'s scheme fails to preserve user anonymity, which is the crucial aim a dynamic-ID scheme is designed to achieve.

In Sood et al.'s scheme, a malicious privileged user U_i having his own smart card can gather information D_i, F_i and B_i from his own smart card. Then he can compute $b = D_i \oplus H(ID_i||PW_i)$ and $A_i = H(ID_i||b)$ because the malicious user U_i knows his own identity ID_i and password PW_i corresponding to his smart card. With the values of A_i, B_i and F_i, U_i can find out the value of $H(x)$ as $H(x) = B_i \oplus A_i \oplus H(F_i \oplus A_i)$. Then the attacker can successfully learn some sensitive user-specific information about any legitimate client U_k through the following steps:

Step 1. Eavesdrops and intercepts a login request message (CID_k, M_k, T) of user U_k from the public communication channel;
Step 2. Computes $L_1 = H(H(x)||T)$, where $H(x)$ and T are known;
Step 3. Computes $L_2 = CID_k \oplus L_1$.

It is obvious that L_2 is unconditionally equal to C_k, while the value of $C_k = H(A_k||H(x)||H(y_k))$ is kept the same for all the login requests of user U_k and is specific to user U_k. This value C_k can be seen as user U_k's identification, and an attacker can, therefore, use this information to trace and identify the user U_k's login requests.

From the above attack, any legal user who logins to the remote server would be exposed to attacker U_i, and thus the scheme fails to achieve user anonymity, which is the most essential security feature a dynamic identity-based authentication scheme is designed to support.

3.2 Stolen-Verifier Attack

Let us consider the following scenarios. A malicious privileged user U_i having his own smart card can gather information B_i, D_i and F_i from his own smart card. Then he can find out the value of y_i as follows:

Step 1. Computes $b = D_i \oplus H(ID_i \oplus PW_i)$, because the malicious user U_i knows his own identity ID_i and password PW_i corresponding to his smart card;
Step 2. Computes $A_i = H(ID_i||b)$;
Step 3. Computes $y_i = F_i \oplus A_i$, where F_i is extracted from the smart card.

In case the verifier table in the database of the server S is leaked out or stolen by this malicious user, he can compute y_k corresponding to any user U_k as follows:

Step 1′. Computes the private key x of the server S as $x = (y_i \oplus x) \oplus y_i$, where the value of y_i is obtained from the above Step 1 to Step 3;
Step 2′. Computes the sensitive parameter y_k of U_k as $y_k = (x \oplus y_k) \oplus x$, where $(x \oplus y_k)$ is U_k's verifier.

With the values of x and y_k, the malicious user U_i can launch user impersonation attacks successfully as follows:

Step 1″. Computes C_k as described in Setion 3.1;
Step 2″. Computes $CID_k = H(H(x)||T) \oplus C_k$ and $M_k = H(H(x)||H(y_k)||T)$, where T is the current timestamp, and sends $\{CID_k, M_k, T\}$ to the server S;
Step 3″. Computes session key $SK = H(C_i||y_i||H(x)||\ T||T'')$ on receiving the response from S.

Since the timestamp T is valid and the values of CID_k and M_k are directly computed with the correct x and y_k, S will find no abnormality in the login request sent by U_i. Hence, the malicious user U_i can impersonate U_k at his will. In a similar way, U_i can masquerade server S with no difficulty. As a result, the entire system will be compromised.

4 Conclusion

In this paper, we have demonstrated that Sood et al.'s scheme suffers from the stolen-verifier attack. In addition, their scheme fails to provide the property of user anonymity, thereby contradicting the claims made in [8]. In conclusion, although Sood et al.'s scheme has many attractive features, it, in fact, does not provide all of the security properties that they claimed and only radical revisions of the protocol can possibly eliminate the identified defects. Therefore, the scheme under study is not recommended for any practical application.

References

1. Lamport, L.: Password authentication with insecure communication. Communications of the ACM 24(11), 770–772 (1981)
2. Chien, H.Y., Jan, J.K., Tseng, Y.M.: An efficient and practical solution to remote authentication: smart card. Computers & Security 21(4), 372–375 (2002)
3. Sun, H.M.: An Efficient Remote User Authentication Scheme using Smart Cards. IEEE Transactions on Consumer Electronics 46(4), 958–961 (2000)
4. Hsu, C.L.: Security of two remote user authentication schemes using smart cards. IEEE Transactions on Consumer Electronics 49(4), 1196–1198 (2003)
5. Ku, W.C., Chen, S.M.: Weaknesses and improvements of an efficient password based remote user authentication scheme using smart cards. IEEE Transactions on Consumer Electronics 50(1), 204–207 (2004)
6. Lee, S.W., Kim, H.S., Yoo, K.Y.: Improvement of chien et al.'s remote user authentication scheme using smart cards. Computer Standards & Interfaces 27(2), 181–183 (2005)
7. Xu, J., Zhu, W., Feng, D.: An improved smart card based password authentication scheme with provable security. Computer Standards & Interfaces 31(4), 723–728 (2009)
8. Sood, S.K., Sarje, A.K., Singh, K.: Secure Dynamic Identity-Based Remote User Authentication Scheme. In: Janowski, T., Mohanty, H. (eds.) ICDCIT 2010. LNCS, vol. 5966, pp. 224–235. Springer, Heidelberg (2010)
9. Khan, M., Kim, S., Alghathbar, K.: Cryptanalysis and security enhancement of 'a more efficient and secure dynamic id-based remote user authentication scheme'. Computer Communications 34(3), 305–309 (2011)
10. Wang, D., Ma, C.-G., Wu, P.: Secure Password-Based Remote User Authentication Scheme with Non-tamper Resistant Smart Cards. In: Cuppens-Boulahia, N., Cuppens, F., Garcia-Alfaro, J. (eds.) DBSec 2012. LNCS, vol. 7371, pp. 114–121. Springer, Heidelberg (2012)
11. Wang, D., Ma, C.G.: On the security of an improved password authentication scheme based on ecc. Cryptology ePrint Archive, Report 2012/190 (2012), `http://eprint.iacr.org/2012/190.pdf`
12. Ma, C.G., Wang, D., Zhang, Q.M.: Cryptanalysis and Improvement of Sood et al.'s Dynamic ID-Based Authentication Scheme. In: Ramanujam, R., Ramaswamy, S. (eds.) ICDCIT 2012. LNCS, vol. 7154, pp. 141–152. Springer, Heidelberg (2012)
13. Dolev, D., Yao, A.C.: On the security of public key protocols. IEEE Transactions on Information Theory 29(2), 198–208 (1983)
14. Messerges, T.S., Dabbish, E.A., Sloan, R.H.: Examining smart-card security under the threat of power analysis attacks. IEEE Transactions on Computers 51(5), 541–552 (2002)
15. Mangard, S., Oswald, E., Standaert, F.X.: One for all-all for one: unifying standard differential power analysis attacks. IET Information Security 5(2), 100–110 (2011)

16. Gu, K., Wu, L.J., Li, X.Y., Zhang, X.M.: Design and implementation of an electromagnetic analysis system for smart cards. In: 2011 Seventh International Conference on Computational Intelligence and Security, pp. 653–656. IEEE Press, New York (2011)
17. Spaford, E.H.: Opus: Preventing weak password choices. Computers & Security 11(3), 273–278 (1992)
18. Campbell, J., Ma, W., Kleeman, D.: Impact of restrictive composition policy on user password choices. Behaviour & Information Technology 30(3), 379–388 (2011)
19. Bao, F., Deng, R.: Privacy Protection for Transactions of Digital Goods. In: Qing, S., Okamoto, T., Zhou, J. (eds.) ICICS 2001. LNCS, vol. 2229, pp. 202–213. Springer, Heidelberg (2001)
20. Tang, C., Wu, D.: Mobile privacy in wireless networks-revisited. IEEE Transactions on Wireless Communications 7(3), 1035–1042 (2008)

Formal Construction of Secure Information Transmission in Office Automation System

Xiaole Li[1], Lianggang Nie[1], Jianrong Yin[1], and Yuanlu Lu[2]

[1] Experimental teaching center, Guangxi University of Finance and Economics, Nanning 530003, China

[2] Guangxi Polytechnic College, Nanning 530026, China

28224384@qq.com

Abstract. This paper mainly studies formal construction of secure information transmission in Office Automation system. With a comprehensive analysis on security requirements in Office Automation system, the processes of information transmission can be divided into two types: user-to-server and user-to-user. Based on this, the formal processes of secure information transmission are constructed with composition method in symmetric key cryptosystem. The formal analysis shows that secrecy, integrity, availability and identifiability of information during transmission could be insured by this construction, which could be used as a direction for development of other various application systems in digital campus, from the viewpoint of information security.

Keywords: Office Automation, information transmission, formal construction, symmetric key cryptosystem.

1 Introduction

As an important part of Digital Campus, Office Automation system (OA system for short) refers to the varied computer machinery and software used to digitally create, collect, store, manipulate, and relay office information. Raw data storage, electronic transfer, and the management of electronic business information comprise the basic activities of OA system, which helps in optimizing or automating existing office procedures.

As the spread of OA system construction, the security of information transmission is faced to more and more threats[1], [2]- how to avoid illegal stealing, altering or destroying of important information during transmission, such as user name, password, personal information, or records about office affairs, scientific research, or finance, is very important.

With a comprehensive analysis on security requirements of information transmission in OA system, an information transmission model is built from the viewpoint of information security- the secrecy, integrity, availability, and identifiability of information sender during transmission should be insured in the process of information transmission, which could be precisely specified by formal methods. And based on the successful research of formal methods in design of

F.L. Wang et al. (Eds.): WISM 2012, LNCS 7529, pp. 291–298, 2012.

security protocols and application (including APG, Authentication Test, Composition, and so on) [3], [4], [5], [6], [7], a formal construction of secure information transmission of OA system in symmetric key cryptosystem, is given. The formal analysis shows that secrecy, integrity, availability, and identifiability of information during transmission could be insured by this construction, which could be used as a direction for development of various application systems in digital campus, from the viewpoint of information security.

2 Security Requirements of Information Transmission in OA System

The construction of OA system is related to many fields about colleges and universities, such as internal communication platform, information release platform, daily office, file management, meeting management, organization management, system management, auxiliary management, and so on, as shown in Fig. 1. These are important parts of management system in school, supplying information service of network resource.

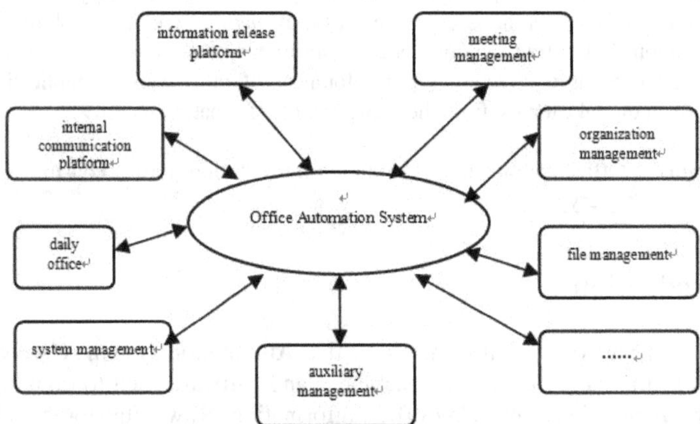

Fig. 1. The model for OA system

In order to avoid the illegal stealing, altering or destroying of important information during transmission in OA system, such as user name, password, personal information, and records about office affairs, scientific research, or finance, the following security requirements of information transmission should be satisfied:

- Information Secrecy- Through the form of $\{| x |\}_k$ in symmetric key cryptosystem, important information x, such as user name, password, personal information, and records about office affairs, scientific research, or finance, should be encrypted by symmetric key cryptosystem or asymmetric key cryptosystem during transmission, in order to avoid the illegal stealing, altering or destroying.

- Information Integrity- Through the form of $auth_{k(x)}$ in symmetric key cryptosystem, important information should be ensured from being altered; or perceived if it is altered.
- Information Availability- Using private key k in symmetric key cryptosystem, the $\{| x |\}_k$ should be decrypted by authorized user. So that the information x could be available for authorized users, such as general user, department manage, system administrator or server.
- Identifiability for senders- The identity of sender should be obtained from decrypted term by symmetric key cryptosystem, so that it could ensure that the receiver could distinguish whether the information is sent by correct sender with effective measures: the server recognizes the identity of receiver before sending information, to ensure the validity of transmission; on the other hand, the receiver recognizes the identity of sender after receiving information, to ensure the authenticity of transmission.

3 Formal Construction of Secure Information Transmission in OA System

Based on the security requirements, a formal construction of secure information transmission in OA system is given. During the transmission process, first, the certificate is transmitted for the verification of identity; then, the important information and operations are encrypted in symmetric key cryptosystem. As a result, the information could only be decrypted by authorized users; on the other hand, the secrecy, integrity, availability, and identifiability of the sender must be ensured.

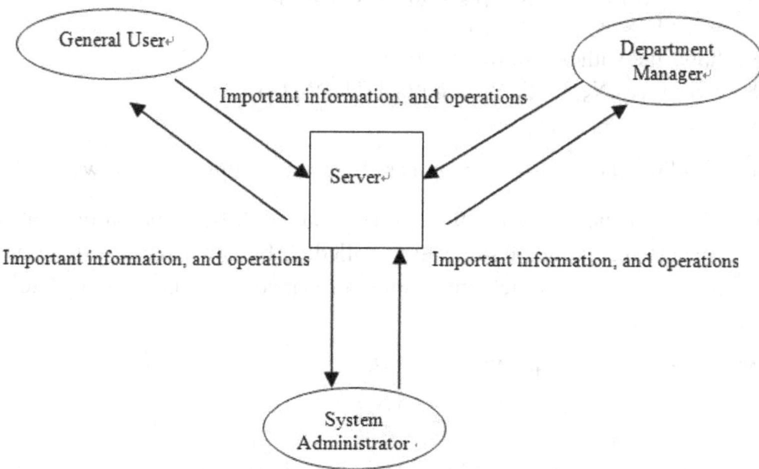

Fig. 2. The model for mutual information transmission in OA system from the viewpoint of information security

3.1 The Processes of Information Transmission between User and Server, Named ITUS(Information Transmission between User and Server) for Short

According to Fig. 2, there are three similar types of transmission processes between user and server: general user-to-server, department manager-to-server, and system administrator-to-server. Based on this model, a formal construction of secure information transmission between user and server in OA system is given by composition.

The formal construction is the mutual authentication and transmission of secrets without an authentication center in symmetric key cryptosystem, if two distinctive keys are used by each participant. It is assumed that the server is honest and competent, sharing a secret key before communication with the general user. There are two participants: user U and server S, in symmetric key cryptosystem. U uses k_{US} and S uses k_{SU} to communicate with the other, respectively; N_U and N_S are the nonce of U and S respectively that can be used only once; x_1 and y_1 are the secrets exchanged between U and S, while x_1 includes the information submitted to S from U and y_1 includes the information returned to U from S; in the $\beta_x = \beta_y = (U, S)$, β_x and β_y are the binding information of x_1 and y_1; the $auth_{kUS}(*)$ or $auth_{kSU}(*)$ is used to represent a message originating from someone who possesses the key k_{US} or k_{SU}. The goal of ITUS is to ensure the mutual authentication between U and S as well as the exchange of secret x_1 and y_1.

Two information security primitives for authentication are used in the process of formal construction:

The p1 shows the authentication of U by S:

M1 $U \rightarrow S : \beta_x, \{|N_U, x_1, \beta_x|\}_{kUS}, auth_{kUS}(N_U, x_1, \beta_x)$
M2 $S \rightarrow U : U, < N_U, x_1, \beta_x >_{kSU}$
The p2 shows the authentication of S by U:
N1 $S \rightarrow U : \beta_y, \{| N_S, y_1, \beta_y |\}_{kSU}, auth_{kSU}(N_S, y_1, \beta_y)$
N2 $U \rightarrow S : S, < N_S, y_1, \beta_y >_{kUS}$

Based on p1 and p2, the construction process by composition is as follows:

a. The e+ (*) means sending of some term called *, by some participant, while the e- (*) means receiving of some term called *, by some participant. Based on this, the sequence of events which are actions performed by agents, can be determined as follows:

e+ (M_1) < e- (M_1) < e+ (M_2) < e- (M_2)
e+ (N_1) < e- (N_1) < e+ (N_2) < e- (N_2)
e+ (M_1) < e+ (N_1)
$\beta = \beta_x = \beta_y = \{U, S\}$, so p= p1⊗p2 is obtained by composition:
p : M1 $U \rightarrow S : \beta, \{| N_U, x_1, \beta|\}_{kUS}, auth_{kUS}(N_U, x_1, \beta)$
MN $S \rightarrow U : U, < N_U, x_1, \beta>_{kSU}, \beta, \{| N_S, y_1, \beta |\}_{kSU}, auth_{kSU} (N_S, y_1, \beta)$
N2 $U \rightarrow S : S, < N_S\ y_1, \beta>_{kUS}$

b. It should be ensured that two authentication goals in the same mutual transmission run, to distinguish different runs of the same transmission from each other, avoiding replay attacks in multiple runs. So term binding c= (x_1, y_1, β) is added:

p : M1 $U \rightarrow S$: β, $\{|N_U, x_1, \beta|\}_{kUS}$, $auth_{kUS}(N_U, x_1, \beta)$

M2 $S \rightarrow U$: $<c>_{kSU}$

MN $S \rightarrow U$: U, $<N_U, x_1, \beta>_{kSU}$, β, $\{|N_S, y_1, \beta|\}_{kSU}$, $auth_{kSU}(N_S, y_1, \beta)$

N2 $U \rightarrow S$: S, $<N_S\ y_1, \beta>_{kUS}$

c. To simplify the primitive p, as the authentication replay could be replaced by c:

p : M1 $U \rightarrow S$: β, $\{|N_U, x_1, \beta|\}_{kUS}$, $auth_{kUS}(N_U, x_1, \beta)$

MN $S \rightarrow U$: β, $\{|N_S, y_1, \beta|\}_{kSU}$, $<c>_{kSU}$

N2 $U \rightarrow S$: S, $<N_S, y_1, \beta>_{kUS}$

d. The sub formal construction ITUS is obtained by composition:

ITUS $U \rightarrow S$: β, $\{|N_U, x_1, \beta|\}_{kUS}$, $auth_{kUS}(N_U, x_1, \beta)$

$S \rightarrow U$: β, $\{|N_S, y_1, \beta|\}_{kSU}$, $<c>_{kSU}$

$U \rightarrow S$: S, $<N_S, y_1, \beta>_{kUS}$

The mutual authentication between U and S as well as the exchange of secret x_1 and y_1, is ensured by ITUS. The formal construction for information transmission between user and server is obtained.

3.2 The Processes of Information Transmission between User and User, Named ITUU(Information Transmission between User and User) for Short

The formal construction of information transmission between user and user is structurally very similar to that in Chapter 3.1. However, due to the assumption that users are participants not to be trusted, some features are distinctive. In symmetric key cryptosystem, there is a disadvantage in key management since every pair of participants should share a secret key before communication with each other. To overcome this disadvantage, the authentication center, such as the server in Chapter 3.1, is usually introduced so that each participant shares a secret key with the servers instead, and the servers intervene in communication between participants.

According to Fig. 2, there are six similar types of transmission processes between user and user: general user-to-general user, general user-to-department manager, general user-to-system administrator, department manager-to-department manager, department manager-to-system administrator, and system administrator-to-system administrator. Based on this model, a formal construction of secure information transmission between user and user in OA system is given by composition.

It is assumed that the authentication server is honest and competent. There are three participants: user U_1, user U_2, and server S. The authentication server plays many different roles. It delivers a message coming from one participant to the other, and it generates a fresh key and distributes it to a group of participants. Moreover, the authentication server is the only entity which can see all parameters used in a single run. Assume that S is an authentication server with faithful behavior and each participant shares a secret key with S. U_1 uses k_{US1} and S uses k_{SU1} to communicate with the other respectively; U_2 uses k_{US2} and S uses k_{SU2} to communicate with the other respectively; N_{U1}, N_{U2} and NS are the nonce of U_1, U_2, and S respectively that can be used only once; $c'_{SU1}= (N_{U1}, U_1, U_2, S)$, and $c'_{SU2}= (N_{U2}, U_1, U_2, S)$; the $auth_{kUS1}(*)$ or $auth_{kSU2}(*)$ is used to represent a message originating from someone who possesses the key k_{US1} or k_{US2}. The goal of ITUU is to ensure the mutual authentication between U and S as well as the exchange of secret x_2 and y_2.

Two information security primitives for authentication are used in the process of formal construction:

The p1 shows the authentication of U_1 by S:

$M1\ U_1 \rightarrow S : U_1, U_2, S, \{| x_2, c'_{SU1} |\}_{kUS1}, auth_{kUS1}(c'_{SU1})$
$M2\ S \rightarrow U2 : U2, < y2, c'_{SU2}>_{kSU2}$
The p2 shows the authentication of U_2 by S:
$N1\ U_2 \rightarrow S : U_1, U_2, S, \{| y_2, c'_{SU2} |\}_{kUS2}, auth_{kUS2}(c'_{SU2})$
$N2\ S \rightarrow U_1 : U_1, < x2, c'_{SU1}>_{kSU1}$

Based on p1 and p2, the construction process by composition is as follows:

a. The e+ (*) means sending of some term called *, by some participant, while the e- (*) means receiving of some term called *, by some participant. Based on this, the sequence of events which are actions performed by agents, can be determined as follows:

$e+(M_1) < e-(M_1) < e+(M_2) < e-(M_2)$
$e+(N_1) < e-(N_1) < e+(N_2) < e-(N_2)$
$e+(N_1) < e+(M_2)$

So p= p1⊗p2 is obtained by composition:

$p : M1\ U_1 \rightarrow S : U_1, U_2, S, \{| x_2, c'_{SU1} |\}_{kUS1}, auth_{kUS1}(c'_{SU1})$
$N1\ U_2 \rightarrow S : U_1, U_2, S, \{| y_2, c'_{SU2} |\}_{kUS2}, auth_{kUS2}(c'_{SU2})$
$M2\ S \rightarrow U_2 : U_2, < y_2, c'_{SU2}>_{kSU2}$
$N2\ S \rightarrow U_1 : U_1, < x_2, c'_{SU1}>_{kSU1}$

b. It should be ensured that two authentication goals in the same mutual transmission run, to distinguish different runs of the same transmission from each other, avoiding replay attacks in multiple runs. So term binding $c= (x_1, y_1)$ is added:

$p : M_1\ U_1 \rightarrow S : U_1, U_2, S, \{| x_2, c'_{SU1} |\}_{kUS1}, auth_{kUS1}(c'_{SU1})$
$N_1\ U_2 \rightarrow S : U_1, U_2, S, \{| y_2, c'_{SU2} |\}_{kUS2}, auth_{kUS2}(c'_{SU2})$
$M_2\ S \rightarrow U_2 : U_2, < y_2, c'_{SU2}>_{kSU2}, <c>_{kSU2}$
$N_2\ S \rightarrow U_1 : U_1, < x_2, c'_{SU1}>_{kSU1}, <c>_{kSU1}$

c. To simplify the primitive p, as the authentication replay and c could be combined:

$p : M_1\ U_1 \rightarrow S : U_1, U_2, S, \{| x_2, c'_{SU1} |\}_{kUS1}, auth_{kUS1}(c'_{SU1})$
$N1\ U_2 \rightarrow S : U_1, U_2, S, \{| y_2, c'_{SU2} |\}_{kUS2}, auth_{kUS2}(c'_{SU2})$
$M2\ S \rightarrow U_2 : U_2, < c, c'_{SU2}>_{kSU2}$

N2 $S \rightarrow U_1 : U_1, < c, c'_{SU1} >_{kSU1}$

d. The sub formal construction ITUU is obtained by composition:

ITUU $U_1 \rightarrow S : U_1, U_2, S, \{| x_2, c'_{SU1} |\}_{kUS1}, auth_{kUS1}(c'_{SU1})$

$U_2 \rightarrow S : U_1, U_2, S, \{| y_2, c'_{SU2} |\}_{kUS2}, auth_{kUS2}(c'_{SU2})$

$S \rightarrow U_2 : U_2, < c, c'_{SU2} >_{kSU2}$

$S \rightarrow U_1 : U_1, < c, c'_{SU1} >_{kSU1}$

The mutual authentication between U_1 and U_2 as well as the exchange of secret x_2 and y_2, is ensured by ITUU.

4 Verification of Formal Construction

The analysis shows that secrecy, integrity, availability, and identifiability of sender during transmission can be insured by this formal construction. Take the ITUS for example:

4.1 Information Secrecy

The information including x_1 and y_1 are secret during transmission by $\{| N_U, x_1, \beta |\}_{kUS}$ and $\{| N_S, y_1, \beta |\}_{kSU}$.

4.2 Information Integrity

The information $auth_{kUS}(N_U, x_1, \beta)$ can only be generated by the participant owning k_{US}, while the x_1 and binding information β can be used as verification for information integrity.

4.3 Information Availability

The information including secret $\{| N_U, x_1, \beta |\}_{kUS}$ or $\{| N_S, y_1, \beta |\}_{kSU}$ could be decrypted by authorized user(general user, department manager, or system administrator) using private keys.

4.4 Identifiability of Information Sender

The authentication for identity of general user by server: the user name and password could be obtained from the secret x_1 decrypted from $\{| N_U, x_1, \beta |\}_{kUS}$, by which the identity of general user could be authenticated by server.

The authentication for identity of server by general user: the information sent by server could be obtained from the secret y_1 decrypted from $\{| N_S, y_1, \beta |\}_{kSU}$, by which the identity of server could be authenticated by general user.

5 Conclusions

As an important goal of informatization in colleges and universities, the security of information transmission in OA system is paid more and more attention, the processes of information transmission can be divided into two types: user-to-server and user-to-user., according to the mutual model during transmission. Based on this, the security requirements are obtained.

The composition method in protocol design is based on "challenge and response" exchange, constructing complex protocol by using primitives, especially adaptive for the specification of two-part communication [5]. Based on this, a formal construction for information transmission is given. The analysis shows that secrecy, integrity, availability, and identifiability of sender during transmission can be insured by this construction.

From the viewpoint of information transmission security, there are security requirements during information transmission in the constituent parts of digital campus, such as digital library, educational administration system, scientific research administration system, management system, and so on. In order to avoid the illegal stealing, altering or destroying of important information during transmission, such as user name, password, personal information, or records about office affairs, scientific research, or finance, the formal construction could be used as a direction for development of various application systems, or common rules for information transmission in digital campus.

References

1. Hummel, S.: Network Security Model-Defining an Enterprise Security Strategy, http://computerdetailreviews.com/2010/07/network-security-model-defining-an-enterprise-security-strategy/
2. Ramaraj, E., Karthikeyan, S., Hemalatha, M.: A Design of Security Protocol using Hybrid Encryption Technique (AES- Rijndael and RSA). International Journal of The Computer 17(1), 78–86 (2009)
3. Guttman, J.D.: Security protocol design via authentication tests. In: Proc. the 15th IEEE Computer Security Foundations Workshop, pp. 92–103 (2002)
4. Song, D.: An automatic approach for building secure systems. PhD thesis. University of California at Berkeley, California (2002)
5. Choi, H.-J.: Security protocol design by composition. PhD thesis. University of Cambridge, Cambridge (2006)
6. Yang, J., Cao, T.: A Verifier-based Password-Authenticated Key Exchange Protocol via Elliptic Curves. Journal of Computational Information Systems 7(2), 548–553 (2011)
7. Gu, X., Zhang, Z., Qiu, J.: Research on Wireless Security Protocol Design. Computer Science 38(9), 103–107 (2011)

A Novel CBCD Scheme Based on Local Features Category

Jie Hou, Baolong Guo, and Jinfu Wu

Institute of Intelligent Control and Image Engineering (ICIE),
Xidian University, Xian, China
jaymarquis007@gmail.com,
blguo@xidian.edu.cn,
wu.nihewo@163.com

Abstract. Global and local features have been applied extensively to improve the performance of Content-Based Copy Detection (CBCD) systems. A novel CBCD scheme which uses global features to category local features is proposed in this paper, containing frame prerecession, database establishment and copy detection. The scheme aims to reach a fast speed and a high accuracy. A proper combination of global and local features in our scheme breeds higher efficiency and accuracy than taking use of any single feature alone. Experimental results indicate that the scheme achieves a rapid scene index and an accuracy frame-frame match.

Keywords: Video Copy Detection, Local Feature Category.

1 Introduction

With the explosive growth of the video service on the Internet, the necessity of protecting the copy right of online media is more obvious than ever. Thus effective methods of detecting video copies turn out to be dramatically valuable under current circumstance.

A variety of algorithms have been developed in this area, they can be roughly divided into two main streams——Digital Watermarking and Content Based Copy Detection (CBCD). Digital Watermarking mainly focus on using the embedded information to detect illegal copies. CBCD algorithms use invariant features to analyze the similarity between two videos. Due to its high efficiency and independence of extra information, CBCD has attracted more attention from researchers.

A novel CBCD scheme based on the separation of high dynamic objects and static information is proposed in this paper and is proved to have features including rapid scene indexing and precise frame-frame matching.

The rest of this paper is organized as follows. In Section2, important previous work on CBCD is reviewed, and a brief introduction of the new scheme is presented. Then a clear illustration of the scheme is given in Section 3. Experiment of a system based on the scheme is carried out in Section 4 along with the analysis on the result. Finally, drawbacks, merits are concluded in Section 5.

F.L. Wang et al. (Eds.): WISM 2012, LNCS 7529, pp. 299–305, 2012.

2 Previous Work

Vast amounts of work have been done in the area of CBCD, which breeds massive algorithms mainly focusing on analyzing global features, local features and coarse features [1].

The most common method of using global features is dividing each frame in to severe blocks and ranking these blocks according their average luminance [2]. Interest points in different frames are detected and used to construct different descriptors in many local feature based methods [3] [4]. Other features like human face information [5] can be used in copyright detection as well.

Methods based on global features obtain reasonable computational cost but tend to be vulnerable to spatial attacks. On the contrary, algorithms based local features are proved to have high robustness but much higher computational complexity. A variety of algorithms are developed, however, few investigations have been done to use global features to category local features in order to accelerate the matching process and reduce the computational complexity. We aim to test the feasibility of a novel scheme, which packets local features based on global information. It is hoped that a balance between accuracy and computational cost can be reached under our structure.

3 Our Work

3.1 Brief Introduction of the Scheme

Our original idea in designing the scheme is to extract the most instrumental features from massive information contained by video crops, in order to save the size of database and to enhance the efficiency of the scheme at the same time. The scheme is designed to be efficient enough to use complex descriptors like SIFT as the local feature to reach high detection accuracy at a reasonable calculation cost.

When human beings search a specific item in video, static information like background or luminance are used first to make coarse search for the scene where the object might appear. Then highly dynamic objects are paid more attention to in order to find the specific temporal and spatial location of the item. Inspired by the scheme described above, local features in a video crop are extracted and organized as shown in Fig. 1.

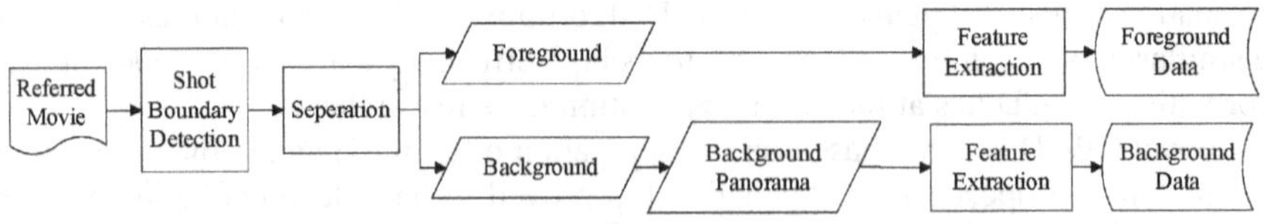

Fig. 1. Extraction and organization processes of local features

First of all, dynamic areas of each frame in the video are marked, then foreground and background information are separated. Local features are extracted from two separated groups, packaged with other information and saved into the database. When we attempt to find the original video of a copy, the procedure is quite similar to the human searching activity described above.

The extraction and organization of information turn out to be challenging tasks if we want to adopt this scheme on CBCD. In the following part of this section, details about the extraction and organization of foreground and background information will be illustrated.

3.2 Separation and Preprocesses

We define the highly dynamic area and temporal static area of a frame as the *foreground* and the *background* respectively.

Effective algorithms are introduced in this part to distinguish the foreground area from the background one. Then preprocesses including background rebuilding, deleted selection and panorama mosaic are carried out to reduce the redundant information.

3.2.1 Separation of Foreground Information and Rebuilding of Background

An effective method was proposed by Lu Peizhong[6] to generate a panorama from a series of pictures. A Change Detection Measurement (CDM) value was calculated in the algorithm to distinguish the foreground area from the background area (**1**).

$$CMD = \begin{cases} 1 & , foreground \\ 0 & , background \end{cases} \tag{1}$$

A virtual pure background can be obtained by replacing the foreground area with the background information contained in the latter frame.

i. Establish separation moldboards of every frame based on (**2**) according to the CDM values.

$$Front(x,y) = \begin{cases} i'_1(x,y), CMD(x,y) = 1 \\ 0 \qquad , CMD(x,y) = 0 \end{cases} \tag{2}$$

i'_1 is obtained by transforming the original first frame i_1 with the affine transform matrix between two frames. i'_2 is the second frame.

ii. Replacing foreground pixels in the latter frame with the corresponding pixels in the i'_1 according to (3).

$$Back(x,y) = \begin{cases} i'_1(x,y), Front(x,y) = 0 \\ i'_2(x,y), Front(x,y) > 0 \end{cases} \tag{3}$$

3.2.2 Deleted Selection of Background Image and Panorama Mosaic

Background Image Deleted Selection: In order to reduce the repeated information contained in the series of background frames and enhance the computational efficiency of the panorama mosaic, a delete selection of them is carried out. The selected images contain more valuable interest points and few repeated information.

Panorama Mosaic: A Shot Boundary Detection was carried out to divide the entire video into different scenes. Then every scene was marked with a unique serial number. Preprocessed background images are documented according to different scene numbers. We use open source platform *Hugin*[7] to mosaic the documented background pictures into a panorama.

3.3 Establishment of Database

Extraction and Selection of SIFT Points: We use the classical SIFT extraction method[9], and deleted selected these points one-step further.

The Structure of Database: A two-level database structure containing the Scene Level and the Frame Level is presented in this section. The background information packaged in Background Descriptors are involved in the Scene Level, and the foreground information packaged in Foreground Descriptors was involved in the Frame Level.

A *Foreground Descriptor* containing SIFT descriptors and position information of a single dynamic frame are defined as (4).

$$F(FNm) = (Def, FNs, SN) \tag{4}$$

FNm is the frame serial number in the video crop; Def is the SIFT descriptors extracted from the foreground image; FNs is the frame serial number in the scene; SN is the scene serial number.
A *Background Descriptor* containing SIFT descriptors of a scene's panorama together with the Scene Number are defined as (5).

$$B(SN) = (Deb, Lob) \tag{5}$$

SN is the serial number of the current scene; Deb contains the SIFT descriptors extracted from the background panorama; Lob contains the locations of SIFT interest points detected in the panorama.

3.4 Copy Detection

The extraction of different information and panorama mosaic is effective in reducing the redundancy of data, but these procedures will prolong the preprocessing time which are not supposed to happen when we tackling with suspected video copies. Thus we directly extract the descriptors from the suspected video copies. Main steps of the match procedure can be observed in Fig. 2.

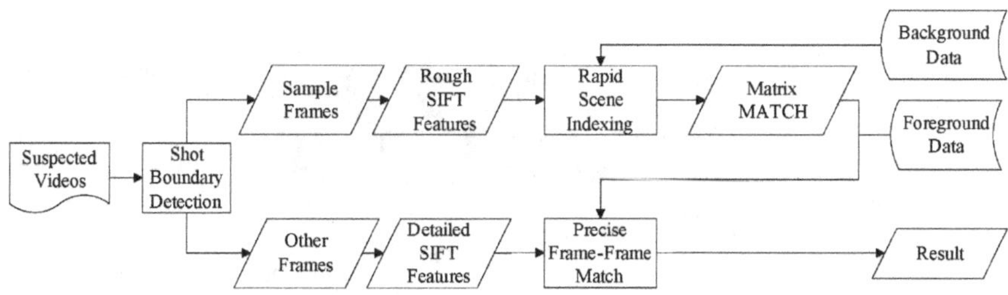

Fig. 2. Copy right detection of suspected videos

To determine whether an input video crop is a copy or not, shot boundary detection was carried out to separate the crop into different scenes. SIFT descriptors are extracted from the sample frame of each scene and used for the match with background descriptors in the database. A matrix *MATCH* (**6**) is generated to record the similarity level between scene i in the suspected video and scene j in the database:

$$MATCH(i,j)=\begin{cases}1, & \textit{sufficient pairs matched}\\0, & \textit{else}\end{cases} \tag{6}$$

Through inquiring the matrix, scenes with high suspension of copyright infringement can be acknowledged. Then the similarity between frames in copies and dynamic data are calculated under these scenes. If a frame in the suspected copy contains same dynamic descriptors from the database, we determine the frame as a copied frame. While majority of frames from a suspected copy are judged as copied frames, there're enough proof to assert a copyright infringement.

4 Experimentals

A Video database has been established to sever our experiment.400-minute-long video including comics, movies, sports and TV series are imported. We believe it is sufficient to measure the efficiency, accuracy and robustness of the proposed scheme.

4.1 Efficiency and Accuracy

4.1.1 Performance of Rapid Indexing

A test of relationship between the numbers of copy scenes, scenes in the database and totally indexing time are carried out.

In order to achieve a widely applicable result, database of diverse scales and suspected videos with varying number of scenes are applied in the experiment. We randomly selected 100, 200, 500, 700, 1000 scenes respectively from the database to be samples of multiple scale databases. Relatively, videos containing 10, 20, 30, 40, 50 scenes are cited as suspected videos. The relationship between the scales of videos and time cost of indexing is presented in Fig. 3.

Less than six seconds are consumed on indexing 10 scenes from 100 scenes. Additionally, two and half minutes are spent in indexing 50 from 1000. The result gives strong evidence that the scheme has a high efficiency on scene indexing.

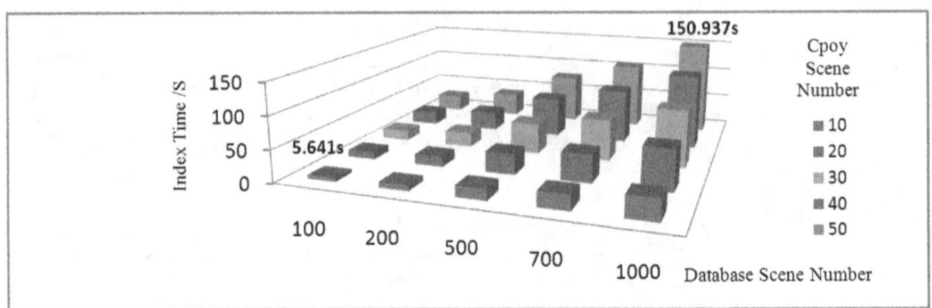

Fig. 3. Time spent on scene indexing

4.1.2 Frame-Frame Match Accuracy

We judged the performance of Frame-Frame match by analyzing its precision as well as recall rate. Scenes containing diverse numbers of frames are referred here in order to find out a convincing result which is widely applicable in most cases. Precisions and Recall Rates of the match are plotted in Fig. 4 with diverse frame numbers in the scene.

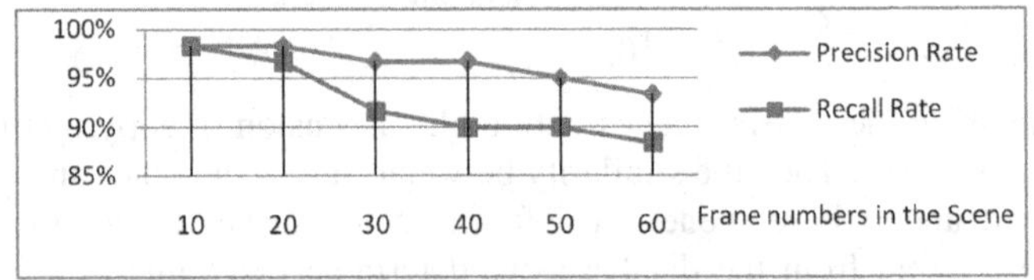

Fig. 4. Precision rate and recall rate of frame-frame match

From the chart we can clearly find that short scenes with 10 sample frames reached a high precision and recall rate. Minimum precision rate and recall rate are around 90% although slight drops of them can be witnessed with the increasing of numbers of sample frames in a scene.

4.2 Handling of Attacks

Four common used attacks are applied on the copies including Change of contrast, blurring, cropping and window embedding. Due to the diverse degrees of attacks might happen to the suspected copies, we adopted these attacks on different lengths on the suspected videos. Again, precision and accuracy of the match is tested here to measure the scheme's robustness against different kinds of attacks.

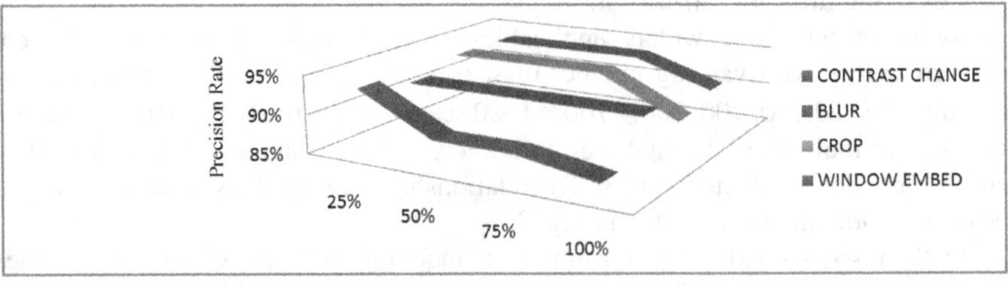

Fig. 5. Precision rate of frame-frame match under different levels of attacks

In our experiment, 25%, 50%, 75% and 100% of the frames in suspected video copies are attacked and precision rates are collected on different situations. Precision rate under different levels of attacks are plotted in Fig. 5. Precision rate remains relatively stable against blurring, cropping and picture embedding. A drop of precision appears when the contrast of a video changes but the rate is still around 85%. We believe improved algorithms which take the distribution of the interest points into consideration might help enhancing the robustness. Based on the data and analyses above, the proposed scheme has a terrific performance of scene indexing and a high matching accuracy which is robust to some common attacks.

5 Conclusion

The rapid development of HD video and online video sharing makes it challenging to achieve a CBCD scheme both fast and accurate. We proposed a novel scheme based on the separation of dynamic and static items in this paper, and reduced the redundant information. The scheme proved to be highly effective and computational inexpensive.

It should be noted that this study has examined only the preprocess and organization of the local features, a limitation of this study is a lack of intelligent algorithm for the matching from frame to frame. Notwithstanding its limitation, this study does suggest that a proper organization of different information breeds a high efficient scheme. Further studies on application of other features on the scheme and intelligent methods of indexing and matching will be summarized in our next study.

Acknowledgement. This work is supported by National Natural Science Foundation of China under Grant No.60802077.

References

1. Lian, S., Nikolaidis, N., Sencar, H.T.: Content-Based Video Copy Detection – A Survey. In: Sencar, H.T., Velastin, S., Nikolaidis, N., Lian, S. (eds.) Intelligent Multimedia Analysis for Security Applications. SCI, vol. 282, pp. 253–273. Springer, Heidelberg (2010)
2. Law-To, J., Chen, L., Joly, A., Laptev, I., Buisson, O., Gouet Brunet, V., Boujemaa, N., Stentiford, F.: Video copy detection: a comparative study. In: Proceedings of the 6th ACM International Conference on Image and Video Retrieval (CIVR 2007), pp. 371–378 (2007)
3. Roopalakshmi, R., Ram Mohana Reddy, G.: A Novel CBCD Approach Using MPEG-7 Motion Activity Descriptors. In: IEEE International Symposium on Multimedia Multimedia (ISM), pp. 179–184 (2011)
4. Joly, A., Buisson, O., Frelicot, C.: Content-based copy detection using distortion-based probabilistic similarity search. IEEE Transactions on Multimedia 9(2), 293–306 (2007)
5. Cotsaces, C., Nikolaidis, N., Pitas, I.: Semantic video fingerprinting and retrieval using face information. Signal Processing: Image Communication 24(7), 598–613 (2009)
6. Lu, P., Wu, L.: Double Hierarchical Algorithm for Video Mosaics. In: Shum, H.-Y., Liao, M., Chang, S.-F. (eds.) PCM 2001. LNCS, vol. 2195, pp. 206–213. Springer, Heidelberg (2001)
7. `http://hugin.sourceforge.net/` (accessed April 11, 2012)
8. Lowe, D.G.: Distinctive Image Features from Scale-Invariant Keypoints. International Journal of Computer Vision 60(2), 91–110 (2004)

Research and Improvement on Association Rule Algorithm Based on FP-Growth

Jingbo Yuan[1] and Shunli Ding[2]

[1] Institute of Information Management Technology and Application, Northeastern University at Qinhuangdao, Qinhuangdao, China

[2] Dept. of Information Engineering, Environmental Management College of China, Qinhuangdao, China

jingboyuan@hotmail.com, dingsl@163.com

Abstract. Association rules mining (ARM) is one of the most useful techniques in the field of knowledge discovery and data mining and so on. Frequent item sets mining plays an important role in association rules mining. Apriori algorithm and FP-growth algorithm are famous algorithms to find frequent item sets. Based on analyzing on an association rule mining algorithm, a new association rule mining algorithm, called HSP-growth algorithm, is presented to generate the simplest frequent item sets and mine association rules from the sets. HSP-growth algorithm uses Huffman tree to describe frequent item sets. The basic idea and process of the algorithm are described and how to affects association rule mining is discussed. The performance study and the experimental results show that the HSP-growth algorithm has higher mining efficiency in execution time and is more efficient than Apriori algorithm and FP-growth algorithm.

Keywords: association rule mining, frequent item sets, Apriori algorithm, FP-growth algorithm.

1 Introduction

Association rules mining is an important research topic in the data mining and is widely applied in intrusion detection and other fields. An association rule is defined as the relation between the item sets, since its introduction in 1993[1] the process of finding the association rules has received a great deal of attention. In fact, a broad variety of algorithms for mining association rules have been developed [2-4].

Given a transaction set, association rules mining is a procedure to generate association rules that the support and credibility respectively is greater than the user given minimum supports and minimum reliability. Generally, association rules mining can be divided two phases: the first phase is to mine all frequent item sets based on user given minimum support count (*min_support*), i.e. each of these item sets meets the support is not less than the *min_support.* In fact, these frequent item sets may have contain relations. In general, only concerning about the Frequent Large Item Set, which do not included in other frequent item sets. The second phase is to

F.L. Wang et al. (Eds.): WISM 2012, LNCS 7529, pp. 306–313, 2012.

produce strong association rules from the frequent item sets; these rules must assure minimum support and minimum confidence.

The performance of discovering association rules is largely determined by the first phase[5]. Finding frequent item sets is the most expensive step in association rule mining. The paper mainly study the method to mine frequent item sets from a huge set of items, consequently to generate association rules.

2 Correlation Algorithms Analysis

No matter how many association rule mining algorithms there are, they can be classified into two categories: one category produces frequent item candidate set and the other doesn't produce frequent item candidate set. Apriori algorithm[6] and FP-Growth algorithm[7] is the most representative respectively.

2.1 Apriori Algorithm Analysis

The Apriori algorithm adopts candidates' generations-and-testing methodology to produce the frequent item sets. That is, Apriori uses an iterative method of layer-by-layer search, which *kth* item set is used to search *(k+1)th* item set. First find first frequent item set, written *L1*. *L1* is used to search second item set *L2*, in this way until cannot find frequent item set. Each search requires a database scan.

In the case of the massive data the operational process of Apriori algorithm mainly has two issues. One is that it scan the database many times, resulting in time cost is very high. Apriori algorithm scans repeatedly the database to calculate the Support of candidate sets and then generates a new candidate set that its length plus 1. Especially in the case of the long item sets the Apriori approach suffer from the lack of the scalability, due to the exponential increasing of the algorithm's complexity. For example for mining frequent set $X=\{x_1, x_2, ..., x_{1000}\}$, Apriori need to scan the database 1000 times. The other is that operation process produces a lot of candidate sets to lead to space cost is also very high. In large transaction databases there may be a large number of candidates set to deal with and this operation is very time consuming.

2.2 FP-Growth Algorithm Analysis

FP-growth approach mines frequent item sets without candidate generation and has been proposed as an alternative to the Apriori-based approach. FP-growth adopts the divide-and-conquer methodology to decompose mining tasks into smaller ones in order to produce the frequent item sets.

After first time scan, it compresses frequent item sets in the database into a frequent pattern tree (FP-tree), meanwhile still keep associated information. The FP-tree is divided into a number of condition databases and each database relates a frequent item set, and then mines these condition databases respectively. The algorithm scans the database only twice. It directly compresses the database into a frequent pattern tree and uses the tree to generate association rules. FP-growth has good adaptability for different length rules, and does not need to generate candidate sets, so greatly enhances the efficiency of time and space compared with Apriori. But this algorithm has also several deficiencies:

(a) To require scanning the database two times and uses sorting in the first, which is feasible for smaller transactions, but for massive transactions the performance will inevitably reduced.

(b) The process of building tree is complex. When insert an item, it requires to compare the prefix with tree other branches resulting in increasing time.

(c) The node structure of building tree is not clear enough. The defined node structure only meets the number of required minimum successor, which lead to waste storage space.

(d) The process of finding frequent largest item set complicates. To find frequent largest item set need to get every frequent item set and then do compare, so increase overhead of storage space and time.

(e) Don't make full preparation for calculation confidence level. Generated maximum frequent item set don't contain all the frequent item sets, so had to traverse the tree to find the frequent item sets in the next confidence level calculation.

3 HSP-Growth Algorithm

In view of a few deficiencies of above algorithm the paper proposes an algorithm called HSP-growth (Huffman Sequence Pattern-growth) based on Huffman transform algorithm and the characteristics of the item sequence sets and its operation. The algorithm uses Huffman coding form to descript information and uses binary tree structure to store information. A database scan can efficiently generate association rules.

3.1 Algorithm Ideas

HSP-growth algorithm uses Huffman tree to describe frequent item sets. The path from non-leaf nodes to left children or right children is represented with 1 or 0 respectively. Each item set want to add to the transaction database will correspond to a one-dimensional sequence. The corresponding value on sequence is 1 if an item is in the item set, or the value is 0.

In accordance with Huffman coding an item set is inserted into the binary tree. Insert process first scans the value of first node on the sequence:

(a) When the value is 1 look up whether the left children pointer of current node is null. If null, creates a node with support count 1 and the left children pointer of current node points to the node, and then sets the current node as its left subtree; if no null, the count of its left subtree plus 1.

(b) When the value is 0 look up whether the right children pointer of current node is null. If null, creates a node with count 0 and the right children pointer of current node points at the node, and then sets the current node as its right subtree; if no null, the count of its right subtree plus 1.

Next scans the second value on the sequence according to same method until the last node with nonzero value is deal with.

Final to find maximal frequent item set, *Max-items*. Firstly, find out Huffman codes of each path traversing binary tree from the head node to all the leaf nodes. Each node corresponding code value is the count value of the node. Next, traverse all Huffman codes with nonzero count value and let the total count of these codes as the count value of the last node, and then on processed sequence find out items that count value is not less than the minimum support. The collection of these items is frequent item set and store to *Max-items*. Next, in order process countdown second item on the sequence, the count value of the last item is as 0. Then find out a frequent item set. If the frequent item set don't contain in *Max-items*, or contain in but its support isn't greater than the original support, then store the frequent item set to *Max-items*, else discard. In this way until first item on the sequence is processed. Now the *Max-items* is the maximal frequent item set that contains all frequent item sets and but is the simplest.

3.2 The Algorithm Implementation

For a detailed description of HSP-growth algorithm, defined as follows:

Definition 1 HSP-tree. HSP-tree is a weighted binary tree formed by such nodes: each node includes four fields: *layers, count, Lchild and Rchild*, where *layers* registers node's layer, *count* registers the number of transactions represented by the portion of the path reaching this node. *Lchild* is left child node pointer and *Rchild* is right child node pointer. The layers of the head node set as 0 in order to distinguish the other nodes.

```
The node structure of HSP_Tree is defined as following.
Struct HSP_Tree
   { int    layers;
   int   count;
   Tree  *Lchild;
   Tree  *Rchild;
   };
```

Definition 2 Entire item set. The entire item set is entire collection of each item in the transaction database. Each item set includes in entire item set.

Definition 3 One dimension sequence. The one dimension sequence forms as $(S_1, S_2, ..., S_n)$, where $Si \in (0,1)$ and $i=1,2,..., n$. n is the dimension of entire item set. *Si* corresponds to the *ith* item of entire item set. If the *ith* item exists in an item set, then *Si* of one dimension sequence corresponding the item set is 1, else is 0.

Definition 4 The simplest frequent item set library. The library consists of the least frequent item sets but contains all frequent item sets in the database.
HSP-growth algorithm divides into two procedures.

Step 1 Construct HSP-tree. The algorithm is as follows.
Input: all item sets *Pitems*
Output: HSP-tree
Method: Call *HSP_tree Creat_tree(DB Pitems)*, which is implemented as follows.

```
Procedure HSP_tree Creat_tree( DB Pitems)
  { HSP_tree *head=new( HSP_tree)
   while(Pitems++ no null)
   { list=turn( *Pitems);
     while( list++ no null)
      { if( list=1)
         Update(Lchild);
         else if( list=0)
         Update( Rchild); }
      }
     return head;
  }
```

Step 2 Generate the simplest frequent item set. The algorithm is as follows.
Input: the constructed HSP-tree and Minimum Support
Output: the simplest frequent item set, *Maxitems*
Method: Call *Maxitems Find_Maxitems(HSP_tree Head, Support Sup)*, which is implemented as follows.

```
Procedure Maxitems Find_Maxitems( HSP_tree Head, Support
Sup)
    { Path=turn( Head);
     while(the ith dimension of path)
     { while (the sequence no end)
         if(ith dimensional •0)
            add weight on the path to SumPath;
         if( SumPath >=sup)  Maxitems+=path2item(
         SumPath);
         simplest frequent item set
     }
    }
```

4 Algorithm Analysis and Experiment

4.1 Algorithm Analysis

Algorithm complexity is a measure for algorithms efficiency and is an important basis for the evaluating an algorithm performance. The algorithm complexity of generating frequent item sets depends mainly on the complexity of generating candidate item sets and computing support of generated candidate item sets.

Apriori algorithm needs to scan the database many times. FP-growth algorithm needs to scan the database twice. However, HSP-growth algorithm don't need to access the original database in generating association rules and only access to the generated one-dimension sequence set, therefore greatly reduces the time complexity of the algorithm. The Comparison of time complexity of three algorithms is as shown in the following table1. Where m is the dimension of entire item set and n is the number of transaction sets.

Table 1. Comparison of time complexity in HSP-growth, Apriori and FP-growth

Algorithm		The 1st time	The 2nd time	The 3rd time	The 4th time	The *m*th time	Total times
Apriori	Times	$C_n^1 \times n$	$C_n^2 \times n$	$C_n^3 \times n$	$C_n^4 \times n$	$C_n^m \times n$	$2^m \times n$
	Function	get one dimensional frequent sets	get two dimensional frequent sets	get three dimensional frequent sets	get four dimensional frequent sets	get m dimensional frequent sets	
FP-growth	Times	$C_m^1 \times n$	m^2	m×n	$(m+1)2^m$		$m(2n+m+2^m)+2^m$
	Function	get one dimensional frequent sets	sort	construct tree	find the maximum frequent set		
HSP-growth	Times	m×n	2^m	$2^m \times m$			$2^m \times (1+m)+mn$
	Function	construct tree	get one dimensional frequent sets	find the simplest frequent set			

In order to have a clear comparison of three algorithms, consider the following two circumstances.

(a) The dimension of a transaction set database unchanged. With the increasing of the number of transaction sets, the time complexity of three algorithms is shown in Figure 1, where the dimension of item set m equals 5.

(b) The number of transaction sets unchanged. With the increasing of the dimension of a transaction set database, the time complexity of HSP-growth and FP-growth algorithm is shown in Figure 2, where the number of transaction sets n equals 1000.

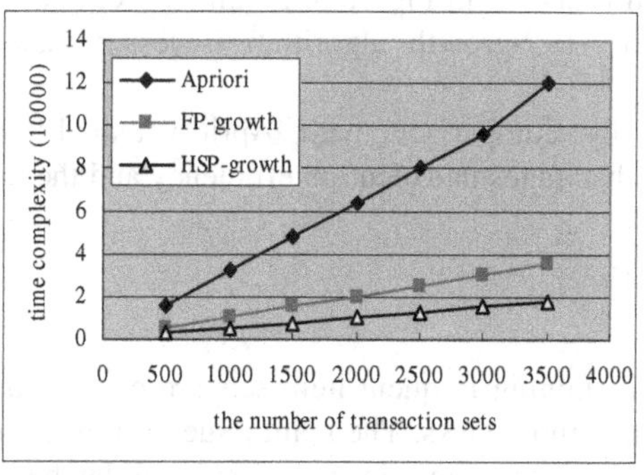

Fig. 1. The comparison of the time complexity as m=5

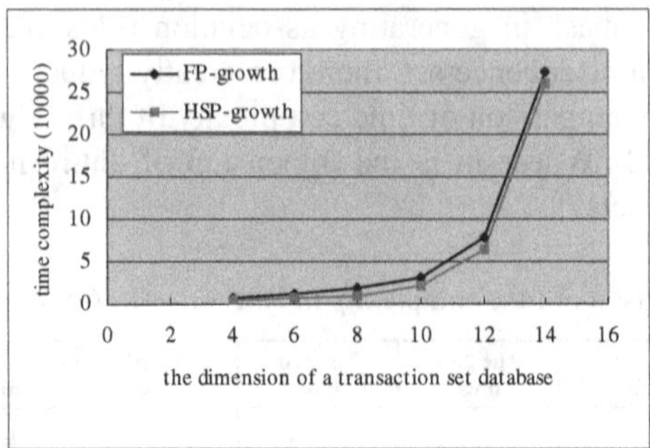

Fig. 2. The comparison of the time complexity as n=1000

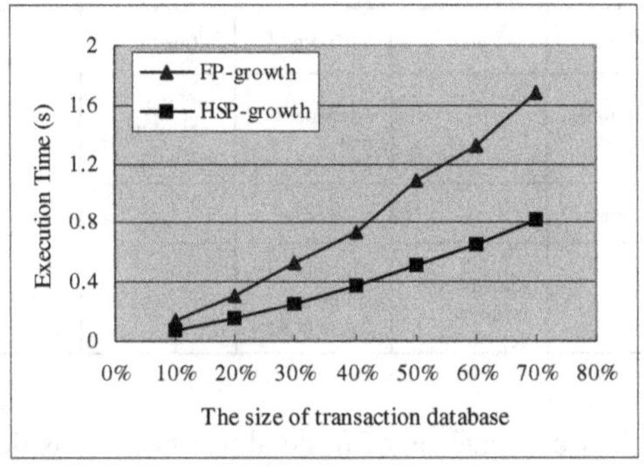

Fig. 3. The comparison of the execution time

4.2 Experiment Result

In order to better explain execution efficiency of HSP-growth algorithm, the paper performed test for HSP-growth and FP-growth in the same experiment environment. Experimental data set uses KDDCup99[8]. The size of transaction set database are respectively 10%-70% of KDDCup99. The comparison of execution time of HSP-growth and FP-growth is shown in Figure 3. It can be seen that under the same scale the execution time of HSP-growth algorithm is lesser than that of FP-growth algorithm.

As stated above, by comparison with Apriori and FP-growth HSP-growth algorithm is obvious advantages in execution efficiency and the time complexity.

5 Conclusion

Efficient algorithms for mining frequent item sets are crucial for mining association rules and for other data mining tasks. The main issue in frequent patterns mining is to reduce the number of database scans since transactional database is usually too large to fit into the main memory. In this paper, an improved FP-growth-based algorithm,

HSP-growth, is proposed. HSP-growth algorithm only scan database once. We analyzed and tested the algorithm performance based on both runtime experiments and time complexity analysis. When generating frequent item sets and association rules in the database, the algorithm can effectively reduce the execution time. HSP-growth performs better than Apriori and FP-growth in the same conditions.

References

1. Agrawal, R., Imieliński, T., Swami, A.: Mining association rules between sets of items in large databases. In: Proceedings of the 1993 ACM SIGMOD International Conference on Management of Data, pp. 207–216 (1993)
2. Jitendra, A., Jain, R.C.: An Efficient Algorithm for Mining Frequent Itemsets. In: International Conference on Advances in Computing, Control, & Telecommunication Technologies, pp. 179–183 (2009)
3. Zhang, W., Liao, H., Zhao, N.: Research on the FP Growth Algorithm about Association Rule Mining. In: 2008 International Seminar on Business and Information Management, vol. 1, pp. 315–318 (2008)
4. Vivek, T., Vipin, T., Shailendra, G., Renu, T.: Association rule mining: A graph based approach for mining frequent itemsets. In: 2010 International Conference on Networking and Information Technology (ICNIT), pp. 309–313 (2010)
5. Han, J., Kamber, M.: Data Mining:concepts and techniques, 2nd edn. The Morgan Kaufmann Series in Data Management Systems (2006)
6. Agrawal, R., Srikant, R.: Fast algorithms for mining association rules in large databases. In: Proceedings of the 20th International Conference on Very Large Data Bases, pp. 487–499 (1994)
7. Han, J., Pei, H., Yin, Y.: Mining Frequent Patterns without Candidate Generation. In: Proc. Conf. on the Management of Data (SIGMOD 2000), Dallas, TX, pp. 1–12. ACM Press (2000)
8. http://kdd.ics.uci.edu/databases/kddcup99/kddcup99.html

Encrypted Remote User Authentication Scheme by Using Smart Card

Ali A.Yassin, Hai Jin, Ayad Ibrahim, and Deqing Zou

Services Computing Technology and System Lab
Cluster and Grid Computing Lab
School of Computer Science and Technology
Huazhong University of Science and Technology
Wuhan, 430074, China
aliadel2005alamre@yahoo.com, hjin@hust.edu.cn

Abstract. Smart card-based authentication is considered as one of the most excessively used and applied solutions for remote user authentication. In this paper, we display Wang et al.'s scheme and indicate many shortcomings in their scheme. Password guessing, masquerade, *Denial-Of-Service* (DOS) and insider attacks could be effective. To outfight the drawbacks, we propose a strong, more secure and practical scheme, which is aimed to withstand well-known attacks. In addition, our proposed scheme provides many pivotal merits: more functions for security and effectiveness, mutual authentication, key agreement, freely chosen password, secure password change, and user anonymity. Moreover, our proposed scheme is shown to be secure against replay attack, password guessing attack, DOS attack, insider attack, and impersonate attack. Furthermore, the security analysis of our work gains it to appear in applications with high-security requirements.

Keywords: smart card, mutual authentication, password, key agreement.

1 Introduction

Password-based authentication has been widely used in several e-business applications and internet security protocols due to their low cost, ability to be carried, efficiency, flexibility and cryptographic properties. Classical password-based remote authentication scheme [1, 2] is depended on a password table saved on the server. This scheme cannot resist malicious attacks such as offline password dictionary attacks, replay attacks, and password table tampering. Moreover, it also has several weaknesses such as the cost of protecting and preserving the password table. After that, many other schemes [3, 4] have been presented to enhance the security and performance. However, these solutions are not immune to resist some of sophisticated attacks such as dictionary attacks, offline attacks, replay attack, and observing power consumption. In order to offer good solutions and enhance the network security, there are many authentication schemes that proposed [5, 6] to use the smart card as a second authentication factor. Das et al. [7] proposed a scheme which is secure against replay attack, password guessing attack, forgery attack, dictionary attack, and identity

F.L. Wang et al. (Eds.): WISM 2012, LNCS 7529, pp. 314–323, 2012.

theft. The research in [8] denoted drawbacks of Das et al.'s scheme, which suffers from disclosing the identity of user's authentication messages. Wang et al. [9] presented a more secure dynamic ID-based remote user authentication scheme and demonstrated the weakness of Das et al.'s scheme such as impersonate attack and lack mutual authentication. However, Wang et al.'s scheme suffers from malicious attacks and has some feasible security risks.

In this paper, we review a security analysis of Wang et al.'s scheme which is not immune to password guessing attack, DOS attack and server impersonate attack. We also propose a new efficient secure smart card based on a remote password authentication scheme that overcomes not only the weaknesses of Wang et al.'s scheme but also enjoys in many features like efficiency, flexible password-based remote mutual authentication, user anonymity. Furthermore, users can freely select and update their passwords, a server and a user can construct authenticated sessions' keys and our scheme generates once a key for each user's login request in authentication phase. Moreover, our scheme can resist many attacks such as replay attack, insider attack, off-line attack, parallel-session attack and DOS attack.

The remainder of the paper is organized as follows. Section 2 reviews Wang et al.'s scheme. Security analysis of Wang et al.'s scheme is presented in section 3. Our proposed scheme is presented in section 4. Section 5 discusses the security analysis of our proposed scheme. Section 6 compares the functionality of our proposed scheme with the related authentication schemes. Section 7 gives our conclusions.

2 Review of Wang et al.'s Scheme

Wang et al.'s scheme consists of registration phase, login phase, verification phase, and password change phase.

2.1 Registration Phase

In this phase, everybody who wishes to register at the authenticated server must gain a smart card. U_i selects his ID_i ‹ then sends it to the remote server S over a secure channel. After S receives $U_i's$ request, S executes following operations:

1. S selects a password PW_i for U_i and calculates $N_i = h(PW_i) \oplus h(x) \oplus ID_i$, where x is the secret key preserved by S privately.
2. S saves the data $(h(.), N_i, y)$ into a new smart card, where y is the remote server's secret key which is saved in each new smart card.
3. S sends the smart card and PW_i to U_i through a secure channel.

$$S \Rightarrow U_i : Smart\,Card\,and\,PW_i .$$

2.2 Login Phase

Once the user wishes to login to the remote server, he must first insert his smart card to the terminal machine, and enters his password PW_i . Then, the smart card performs the following computations:

1. The smart card computes $CID_i = h(PW_i) \oplus h(N_i \oplus y \oplus T) \oplus ID_i$, where T is the current time stamp of this phase.
2. The smart card sends the message $M_i = \{ID_i, CID_i, N_i, T\}$ to S. $Smart\,Card \rightarrow S: ID_i, CID_i, N_i, T$.

2.3 Verification Phase

Upon receiving the login request from the user at time T', S performs the following operations to verify the validity of U_i:

1. S checks the legality of time stamp by testing $T' - T \leq \Delta T$. If so, S directly accepts the user's login message M_i, otherwise terminates the login phase.
2. S calculates $h'(PW_i) = CID_i \oplus h(N_i \oplus y \oplus T) \oplus ID_i$, $ID_i' = N_i \oplus h(x) \oplus h'(PW_i)$, and checks whether ID_i' is equal to ID_i. If so, S accepts the user's login request; otherwise, S rejects the user's login request.
3. Finally, S computes $a' = h(h'(PW_i) \oplus y \oplus T')$ and sends the message $\{a', T'\}$ to U_i.

Upon receiving the responding message $\{a', T'\}$ from S at time T'', U_i ensures from the identity of the server S by performing the following steps:

1. U_i checks whether T'' is a fresh time. If $T'' - T \leq \Delta T$, U_i performs next step. Otherwise U_i terminates this phase.
2. U_i calculates $a = h(h(PW_i) \oplus y \oplus T')$ and ensures whether a is equal to a'. If so, U_i verifies that S is valid.

2.4 Password Change Phase

This phase will be entreated if the user wants to change his password from PW_i to PW_i'. U_i inserts his smart card into the card reader and pushes both PW_i and PW_i'. Then, the smart card calculates $N_i^n = N_i \oplus h(PW_i) \oplus h(PW_i')$ and stores N_i^n instead of N_i.

3 Cryptanalysis of Wang et al.'s Scheme

In this section, we display weaknesses of Wang et al.'s scheme which suffers from impersonation attack, offline password guessing attack, DOS attack, and insider user's attack.

Impersonation Attack. The attacker can apply this type attack when intercepts user's previous login message $M_i = \{ID_i, CID_i, N_i, T\}$. Then, he tries to impersonate legal user U_i to log in the remote server S at time $T^* > T$. The attacker performs the following computations to complete this attack:

1. The attacker calculates $\Delta T = T^* \oplus T$, $ID_i^* = ID_i \oplus \Delta T$, $CID_i^* = CID_i \oplus \Delta T$, and $N^* = N_i \oplus \Delta T$.
2. Now, an attacker has the ability to send forged login message to the destination server $M_i^* = \{ID_i^*, CID_i^*, N_i^*, T_i^*\}$.
3. The server S verifies whether the current time T_i^* is genuine. If so, computes $h^*(PW_i)$ as follows.

$$\begin{aligned} h^*(PW_i) &= CID_i^* \oplus h(N_i^* \oplus y \oplus T^*) \oplus ID_i^* \\ &= (CID_i \oplus \Delta T) \oplus h((N_i \oplus \Delta T) \oplus y \oplus (T \oplus \Delta T)) \oplus (ID_i \oplus \Delta T) \\ &= CID_i \oplus h(N_i \oplus y \oplus T) \oplus ID_i \\ &= h(PW_i) \end{aligned}$$

Then, S computes $ID_i^{'*}$ as follow:

$$\begin{aligned} ID_i^{'*} &= N_i^* \oplus h(x) \oplus h^*(PW_i) \\ &= (N_i \oplus \Delta T) \oplus h(x) \oplus h(PW_i) \\ &= (h(PW_i) \oplus h(x) \oplus ID_i \oplus \Delta T) \oplus h(x) \oplus h(PW_i) \\ &= ID_i \oplus \Delta T \\ &= ID_i^* \end{aligned}$$

Finally, S verifies the attacker while $ID_i^{'*}$ is equal to ID_i^* in the user's login request. Therefore, the attacker could impersonate the legitimate user to login the remote server.

Offline Password Guessing Attack. When the user loses his smart card, the attacker can obtain the server's secret key y saved in user's smart card. The attacker eavesdrops the common channel between user and server and then, can access to important information $\{ID_i, CID_i, N_i, T\}$. Lastly, this attack is implemented with the following computations:

1. The attacker calculates $X_i^* = h(N_i \oplus y \oplus T)$ by depending on N_i, y, T.
2. The attacker calculates $y_i^* = CID_i \oplus X_i \oplus ID_i = h(PW_i)$.
3. The attacker selects a random password PW_i^*, calculates $h(PW_i^*)$ and compares it with $h(PW_i)$. If so, the attacker infers that PW_i^* is user's password. Otherwise, the attacker selects another password nominee and performs the same processes, until he locates the valid password.

Denial-Of-Service Attack. The attacker can apply this type of malicious attacks in the password change phase. So, Wang et al.'s scheme is insecure in this phase. When an attacker stoles the smart card, he can change user's password in short time. The attacker performs this attack as follows.

1. He inserts user's smart card in card reader and submits a random string as new password PW_i^n without knowing the real password PW_i.
2. User's smart card computes $N_i^n = N_i \oplus h(PW_i^n) \oplus h(PW_i)$.
3. User's smart card stores in its memory N_i^n instead of N_i.

Thus, DOS attack can achieve user's smart card.

Insider User's Attack. The insider user has registered in the remote server, considers as an attacker when attempts to know some secret keys of its server and the other legitimate users. An attacker applies the following steps to complete this attack.

1. The attacker computes $h(x) = N_i^* \oplus ID_i^* \oplus h(PW_i^*)$.
2. The attacker computes $h(PW_i) = N_i \oplus N_i^* \oplus ID_i \oplus ID_i^* \oplus h(PW_i^*)$.

The $h(x)$ is the server's secret information which used to build a registered user's authentication information, and $h(PW_i)$ is directly associated with the password of user. The revelation of this information means that an attacker can detect the secret x and PW_i by performing brute-force guessing attack, password dictionary attack, and even the brute-force guessing attack.

4 Our Proposed Scheme

There are four phases in our scheme: registration phase, login phase, mutual authentication with key agreement phase, and change password phase.

4.1 Registration Phase

In this phase, everybody to be registered on the remote server is supplied a smart card. To initialize, the user U_i submits his selected identity ID_i and hashed password PW_i to the remote server S over a secure channel. Upon receiving the user's registration request, the server S performs the following operations:

1. S computes $N_i = h(PW_i) \,||\, h(X_s) \oplus h(ID_i)^{X_s}$, $M_i = h(PW_i) \oplus h(X_s)$, where X_s is a secret key kept by S in private.
2. S saves the data $\{h(.), N_i, M_i\}$ on a new smart card and sends a smart card to U_i over a secure channel. $S \Rightarrow U_i : Smart\,card.$

4.2 Login Phase

When a user U_i wishes to login S, U_i attaches his smart card in the card reader and inputs his password PW_i. The smart card fulfills the following steps:

1. Compute $h(X_s) = M_i \oplus h(PW_i)$ and $Z' = h(PW_i) || h(X_s)$.
2. Generate a random number r_i and perform the following steps:
   - Compute $K_i = h(r_i \oplus Z')$, $C_i = K_i \oplus (Z' \oplus N_i)^{r_i}$, and $f_i = h(ID_i)^{r_i}$.
   - Calculate $CID_i = Z' \oplus h(T \oplus r_i)$, where T is the current time stamp of the input device.
   - Encrypt (r_i, T, N_i, CID_i) by using K_i.
3. User's smart card sends login request message M to the remote server. $SmartCard \rightarrow S : M = (C_i, f_i, E_{K_i}(r_i, T, N_i, CID_i))$

4.3 Authentication Phase

Upon receiving user's login request message at time T', S performs the following computations:

1. S computes $K_i = C_i \oplus f_i^{X_s}$, and decrypts $E_{K_i}(r_i, T, N_i, CID_i)$.
2. S checks the legitimacy of the timestamp T. If $T' - T \leq \Delta T$ contains, S persists the next step. Otherwise, S terminates this phase.
3. S computes $Z'' = CID_i \oplus h(T \oplus r_i) = h(PW_i) || h(X_s)$, checks whether $(N_i \oplus Z'')^{r_i}$ is equal to $f_i^{X_s}$. If so, S accepts the user's login request.
4. S computes $a' = h(Z'' || r_i || T')$ and sends message $M' = E_{K_i}(a', T')$ to U_i. $S \rightarrow U_i : M'$.
5. When U_i receives the message $M' = E_{K_i}(a', T')$ at time T'', U_i executes the following steps:
   - Check whether $T'' - T' \leq \Delta T$. If not hold, U_i overthrows the message M' and terminates this phase. Otherwise, U_i continues the next step.
   - U_i decrypts message M' by using K_i, computes $a = h(Z' || r_i || T')$, and compares a with a'. If so, U_i decides that the remote server S is authenticated.

4.4 Password Change Phase

When U_i wants to change his password from PW_i to PW_i^n, U_i implores this phase. The password change phase needs to pass the following steps:

1. U_i must execute the above phases login and mutual authentication. The server S authenticates his old password PW_i.

2. After the successful mutual authentication, U_i enters his new password PW_i^n. Then, smart card computes $h(X_s) = M_i \oplus h(PW_i)$, $N_i^n = N_i \oplus h(PW_i) \,||\, h(X_s) \oplus h(PW_i^n) \,||\, h(X_s)$ and replaces the old N_i with the new N_i^n.

5 Security Analysis of Our Proposed Scheme

In this section, we analyze our proposed scheme and display that our work can withstand several famous attacks and enjoys many security properties.

Theorem1. Our proposed scheme can support mutual authentication.

Proof. A mutual authentication feature gains both the server and the user to authenticate each other. In our work, authentication of U_i to S represents by $M = (C_i, f_i, E_{K_i}(r_i, T, N_i, CID_i))$. Also, authentication of U_i to S by $E_{K_i}(a', T')$. An adversary is not able to generate $(K_i, h(x_s), h(PW_i), r_i)$. In addition, U_i and S securely exchange $C_i = K_i \oplus (Z' \oplus N_i)^{r_i}$ and $K_i = C_i \oplus f_i^{X_s}$ in the login and authentication phases, respectively. The authenticated session key is demonstrated as follows:

$$C_i = K_i \oplus (Z' \oplus N_i)^{r_i} = K_i \oplus (h(ID_i)^{X_s})^{r_i}, \; K_i = C_i \oplus f_i^{X_s} = C_i \oplus (h(ID_i)^{r_i})^{X_s}.$$

Thus, our proposed scheme provides mutual authentication.

Theorem 2. Our proposed scheme can support user anonymity.

Proof. If an attacker eavesdrops on user's login request message, he fails to infer the user's identity from encrypting message $E_{K_i}(r_i, T, N_i, CID_i)$ since it is encrypted with K_i, which is anonymous to the attacker. Also, the ciphertext does not possess the real user's identity ID_i where the server verifies user's identity in an anonymous manner between $(N_i \oplus Z'')^{r_i}$ and $f_i^{X_s}$. Additionally, we use the timestamp in the login phase, user's login request message is changed each login time where it's parameters $\{T, r_i, K_i, CID_i, f_i\}$ change in each login session. Therefore, it is impossible for the attacker to reveal a user's identity. Obviously, our proposed scheme can support user anonymity.

Theorem 3. Our proposed scheme can support security of the stored data and resists a password guessing attack.

Proof. In our proposed scheme, the remote server S stores only secret information $\{h(.), N_i, M_i\}$ on the smart card. The secret information $\{h(.), N_i, M_i\}$ derived from the user's smart card do not assist an attacker without the user's password $h(PW_i)$ and server's secret key $h(X_s)$ to retrieve the user's secret key K_i, since $h(X_s) = M_i \oplus h(PW_i)$, $Z' = h(PW_i) \,||\, h(X_s)$, $K_i = h(r_i \oplus Z')$, and

$C_i = K_i \oplus (Z' \oplus N_i)^{r_i}$. If an attacker is attempting to retrieve K_i by combining dictionary attacks with the recover secret information $\{h(.), N_i, M_i\}$, he is required to locate both $h(X_s)$ and $h(PW_i)$ to compute $K_i = h(r_i \oplus Z')$. On the other hand, the attacker can gain (C_i, f_i) by eavesdropping the insecure channel between U_i and S. The attacker cannot get useful information about the user's password from these values, because other information is encrypted by user's secret key K_i. Thus, our proposed scheme provides security of the stored data and resists a password guessing attack.

Theorem 4. Our proposed scheme can resist the server impersonate attack.

Proof. User's smart card contains two values $N_i = h(PW_i) || h(X_s) \oplus h(ID_i)^{X_s}$ and $M_i = h(PW_i) \oplus h(X_s)$. Since the user knows his password PW_i, he can obtain the value of $h(ID_i)^{X_s}$. However, this value is based on user's identity, it is not the same for all users. The attacker cannot play a role of the server by this value and he fails to get the values $\{X_s, K_i\}$. They are used to decrypt the ciphertext $E_{K_i}(r_i, T, N_i, CID_i)$ sent by U_i where K_i is computed by $K_i = C_i \oplus f_i^{X_s}$. Therefore, the proposed scheme can resist the server impersonate attack.

Theorem 5. Our proposed scheme can withstand an insider attack and user impersonate attack.

Proof. In our proposed scheme, when U_i wishes to register with a remote server, he sends $ID_i, h(PW_i)$ instead of ID_i, PW_i. Due to the utilization of the one-way hash function $h()$, it is difficult for the server to extract the password of user from the hashed value. In addition, when the attacker wants to impersonate the valid user, he requires to forge a legal login request message $(C_i, f_i, E_{K_i}(r_i, T, N_i, CID_i))$, in which $K_i = h(r_i \oplus Z')$, $C_i = K_i \oplus (Z' \oplus N_i)^{r_i}$, $f_i = h(ID_i)^{r_i}$, $CID_i = Z' \oplus h(T \oplus r_i)$. However, the attacker cannot obtain the server's secret key $h(X_s)$ and fails to forge such a message.

Theorem 6. Our proposed scheme can resist DOS attack.

Proof. This attack means that an attacker changes password verification information of user's smart card to another information. As a result, a legal user cannot complete his login to the server request successfully. In our proposed scheme, user's smart card checks the legitimacy of user identity ID_i and password PW_i before password change phase. If we assume that the attacker inserts the user's smart card into the terminated machine, he must guess the values of user identity and password. These values are not stored directly in the smart card but they are combined with other values, e.g. $h(ID_i)^{X_s}, h(PW_i) || h(X_S)$ where X_S is not stored in smart card. Therefore, the attacker cannot obtain X_S, PW_i, ID_i to apply DOS attack.

Theorem 7. Our proposed scheme can withstand the parallel session attack.

Proof. In our work, an attacker cannot impersonate a valid user by constructing a legal login message in another continuous execution from the authentic execution since the server's submitted message $M'=E_{K_i}(a',T')$ is encrypted by K_i, which is anonymous to the attacker and a' generates once for each mutual authentication phase. Hence, our proposed scheme can withstand the parallel session attack.

6 Comparison with Related Works

We compare security properties and computational cost of our proposed scheme with five authentication schemes, including Das et al. [7], Liao et al. [8], Wang et al. [9], Yoon and Yoo [10], and Khan et al. [11]. Table 1 describes comparison of security properties.

Table 1. Comparison of authentication schemes

	Our Scheme	*Das et al. [7]*	*Liao et al. [8]*	*Wang et al. [9]*	*Yoon and Yoo [10]*	*Khan et al. [11]*
C1	*Yes*	*Yes*	*Yes*	*No*	*Yes*	*Yes*
C2	*Yes*	*Yes*	*Yes*	*No*	*Yes*	*No*
C3	*Yes*	*No*	*Yes*	*No*	*Yes*	*Yes*
C4	*Yes*	*No*	*No*	*No*	*No*	*No*
C5	*Yes*	*No*	*Yes*	*Yes*	*Yes*	*Yes*
C6	*Yes*	*No*	*Yes*	*No*	*Yes*	*Yes*

C1: Freely chosen password; C2: User anonymity; C3: Secure password change; C4: session key agreement; C5: Mutual authentication, C6: No password reveal.

7 Conclusion

In this paper, we illustrate that Wang et al.'s scheme is vulnerable to impersonation attacks, off-line guessing attack, DOS attack, insider attack and does not provide anonymity for the users. Furthermore, under the supposition of smart card security infringe, this scheme will be totally failed by the insider attacker. We propose a good improvement for Wang et al.'s scheme. Moreover, we present a robust smartcard based password authentication scheme. The security analysis explains that our proposed scheme is immune against the attacks mentioned in this paper. The comparisons of previous works display the advantages of our scheme in details.

Acknowledgements. The work is supported by National High-tech R&D Program (863 Program) under grant No.2012AA012600, National Science and Technology Major Project of Ministry of Science and Technology of China under grant No.2010ZX-03004-001-03, and Wuhan City Programs for Science and Technology Development under grant No.201010621211.

References

1. Lamport, L.: Password authentication with insecure communication. Communications of the ACM 24(11), 770–772 (1981)
2. Haller, N.: The s/key one-time password system. In: Proceedings of Internet Society Symposium on Network and Distributed Systems, California, pp. 151–157 (1994)
3. Tan, K., Zhu, H.: Remote password authentication scheme based on cross-product. Computer Communications 22(4), 390–393 (1999)
4. Hwang, M.-S., Li, L.-H.: A new remote user authentication scheme using smart cards. IEEE Transactions on Consumer Electronics 46, 28–30 (2000)
5. Sun, H.-M.: An efficient remote use authentication scheme using smart cards. IEEE Transactionson on Consumer Electronics 46, 958–961 (2000)
6. Fan, C., Chan, Y., Zhang, Z.: Robust remote authentication scheme with smart cards. Computers and Security 24(8), 619–628 (2005)
7. Das, M.L., Saxena, A., Gulati, V.P.: A dynamic ID-based remote user authentication scheme. IEEE Transactions on Consumer Electronics 50(2), 629–631 (2004)
8. Liao, I.E., Lee, C.C., Hwang, M.S.: Security enhancement for a dynamic ID-based remote user authentication scheme. In: Proceeding of International Conference on Next Generation Web Services Practice, Seoul, Korea, pp. 437–440 (2005)
9. Wang, Y.Y., Liu, J.Y., Xiao, F.X., Dan, J.: A more efficient and secure dynamic ID-based remote user authentication scheme. Computer Communications 4(32), 583–585 (2009)
10. Yoon, E.J., Yoo, K.Y.: Improving the Dynamic ID-Based Remote Mutual Authentication Scheme. In: Meersman, R., Tari, Z., Herrero, P. (eds.) OTM 2006 Workshops, Part I. LNCS, vol. 4277, pp. 499–507. Springer, Heidelberg (2006)
11. Khan, M.K., Kim, S.K., Alghathbar, K.: Cryptanalysis and security enhancement of a more efficient & secure dynamic ID-based remote user authentication scheme. Computer Communications 34, 305–309 (2011)

A Web Visualization System of Cyberinfrastructure Resources

Yuwei Wang[1,2], Kaichao Wu[1], Danhuai Guo[1], Wenting Xiang[1,2], and Baoping Yan[1]

[1] Computer Network Information Center, Chinese Academy of Sciences
Beijing 100190, China
[2] Graduate University of Chinese Academy of Sciences
Beijing 100049, China
{ahwangyuwei,kaichao,guodanhuai,xiangwenting,ybp}@cnic.cn

Abstract. With the development of cyberinfrastructure (CI), the massive heterogeneous resources emerge gradually. The resources are multi-source, heterogeneous and spatiotemporally distributed. Traditional structural management of resources has been unable to meet the needs of global resource sharing because of the lack of efficiency in expressing spatiotemporal features of CI. Since the CI resources leap over time and space, it is essential to establish a unified spatiotemporal metadata model for CI services across different disciplines. The adoption of a system with intuitive resource visualization and management would aid interoperability and collaboration of scientific research activities among organizations and domains. In this paper, we propose a spatiotemporal metadata model which characterizes the resources as spatial entities with time and space dimensions. Furthermore, a web-based GIS platform is implemented for spatiotemporal visualization. The well-known data, computing and network resources are used as the cases in the model and the visualization system. This allows the easy integration of emerging CI services into the system. Our work here would provide the scientists with great benefits for the integration, management, spatiotemporal analysis of CI resources and auxiliary optimization of resource distribution.

Keywords: cyberinfrastructure, spatiotemporal visualization, GIS.

1 Introduction

The development of CI resources has largely facilitated breakthrough in the process of the data-intensive science and engineering research. These resources include the large-scale scientific data sets generated by numerous scientific researches and applications, the computing power owned by abundant supercomputing institutions through-out the world and high-speed network under rapid development. As massive heterogeneous resources emerge, the lack of efficiency of traditional structural management of resources is exposed gradually. These existing approaches do not take into account the spatial and temporal characteristics of dynamic CI resources adequately and are unable to meet the needs of global resource sharing.

F.L. Wang et al. (Eds.): WISM 2012, LNCS 7529, pp. 324–332, 2012.

Because of the diversity and distribution characteristics in space, time and discipline dimensions, the complexity of management, monitoring and analysis of CI resources increases significantly. Traditionally, the access has been restricted to a small number of organizations and individuals Thus, the comprehensive and timely sharing, and the exchange of the resources, is obstructed largely. At present, people can search relevant information online with keyword-based search engines, but this approach does not guarantee the accuracy, integrity, timeliness and validity of CI resources. It is difficult to obtain an overall view of the spatiotemporal distribution, status and efficiency of the resources. Moreover, scientists may search the data resources through the data portal system or search the computing resources through the Grid platforms.

In this paper, we summarize some characteristics of the CI resources, especially in temporal and spatial dimensions, and propose a spatiotemporal schema for modeling various CI resources. Data, computing and network resources are used as the cases in the schema. Furthermore, the resource information is integrated in a unified framework which enables resource registration, storage, release, analysis and discovery. A GIS-based system is implemented to allow the user to interact with the spatiotemporal information in a visual environment. Our system allows users to search for a variety of resources in both space and time dimensions, and provides users with contextual spatiotemporal understanding.

2 Background

The research and development of global CI aims to construct a common high performance network environment for scientific research. The CI has facilitated large-scale scientific activities, applications, collaboration and sharing across disciplines, organizations, and policy barriers [1]. Many current studies, such as [1], [2], [3] [4], focus on efficient integration of complex software, networks, computational platforms, middleware, visualization tools and virtual organizations for future progress across the frontier of large-scale scientific and engineering tasks. According to the current general statement, the CI mainly includes data, networking, computing resources, application services and tools.

Traditionally, the access to research resources has been restricted to a small number of organizations and individuals participating in the projects. There are many studies concentrating on providing a metadata model for scientific data resources of specific domains or a generic model for multidisciplinary scientific data resources. For example, GEOSS [6] is built for data interoperability of global earth observations. The iPlant [3] is proposed to facilitate the collaboration among plant biology community. Metadata models also are implemented in some scientific data portal systems [7]. The data resources are accessed through the data portal system which is located on many data servers. Moreover, a metadata model [9] is proposed for integration of a large range of differing types of data resources across a wide number of scientific domains.

There are a series of significant spatial characteristics in the CI resources, including the locations of resources, users and the spatial relationships. For example, the dataset of Asian climate falls within the global dataset on climate. Due to the fact that the resource-sharing and transmission of massive data rely on the high-speed world-wide network, it is required highly for supercomputing resources to meet their huge computational needs owing to tremendous computational complexity. Hence, interdependence and cooperation among different types of resources jointly helps the achievement of the current highly-sophisticated and large-scale computing tasks.

The CI resources are multi-source, heterogeneous and spatiotemporally distributed. They are supported and managed by a range of organizations in different spatial locations. The metadata of CI resources must be designed in a coordinated collection, storage mechanism. We propose a spatiotemporal metadata model which characterizes the CI resources as spatial entities with time and space dimensions. A web-based GIS platform is implemented for spatiotemporal visualization of CI resource service. Emerging CI services can be integrated easily into the system. By collecting, modeling, storing and releasing the spatiotemporal and other information, users can explore the CI environment interactively, and acquire intuitively the overall understanding about various types of resources.

3 Spatiotemporal Metadata of CI Resources

In order to establish a common visualization mechanism over various types of resources, a simple comprehensive metadata is required. Hence, a unified spatiotemporal modeling is built for abstracting spatial and temporal characteristics of CI resources. The generic model described with UML diagrams is shown in Fig. 1, in which well-known data, supercomputing and network resources are using as typical cases. In Fig. 1, the attributes written in uppercase represent the time or spatial characteristics. For reasons of space, we select some features of particular interest and do not make a complete description of all features of CI resources.

The model has a hierarchical structure of CI, data resources, supercomputing resources and network resources. The data resources consist of databases, datasets and files. There exist a hierarchical structure of supercomputing centers and supercomputing systems. Some public properties among all types of resources are shown in Fig. 1, such as name (unique identifier), resource type (data, computing, network, storage space, etc.), and the contact information which is associated with spatial location and is often consistent with management organization. The creators and holders of data and computing resources are determined by the storage nodes and the supercomputing center. Considering the data resources, they are typically stored in many storage systems physically distributed throughout cooperative organizations. Interconnection among network nodes forms a spatial topology structure. Each CI has one or more associated organizations that describes who are undertaking the CI project.

The data resource may be collected from many countries and research institutions. Time and space extents are usually associated with data themselves. For example, the MODIS global land cover product MOD12Q1 data sets are collected from NASA, U.S. with the time span of 2001-2004, bi-annual time-steps and global coverage. The data resource would have the access URI, subject, description, amount and creation time information associated with it. The subject field contains some domain specific keywords which can describe discipline that the resource belongs to. In addition, data resources may belong to different hierarchies depending on the classification. For example, a scientific data repository may be a subject database, thematic database, reference database or professional database [5]. A data repository may contain several sub-databases, and each sub-database also contains many datasets. Each of them can be a physical or logical database, file, or file collection. Specific time and space characteristics (creation date, update date, storage location, etc.) exist in each level. For every hierarchic data resources, access logs are collected to establish a logical foundation for statistical analysis of its running state, and the records about the visitor location, access time, the number of web pages and size, and enable OLAP operations.

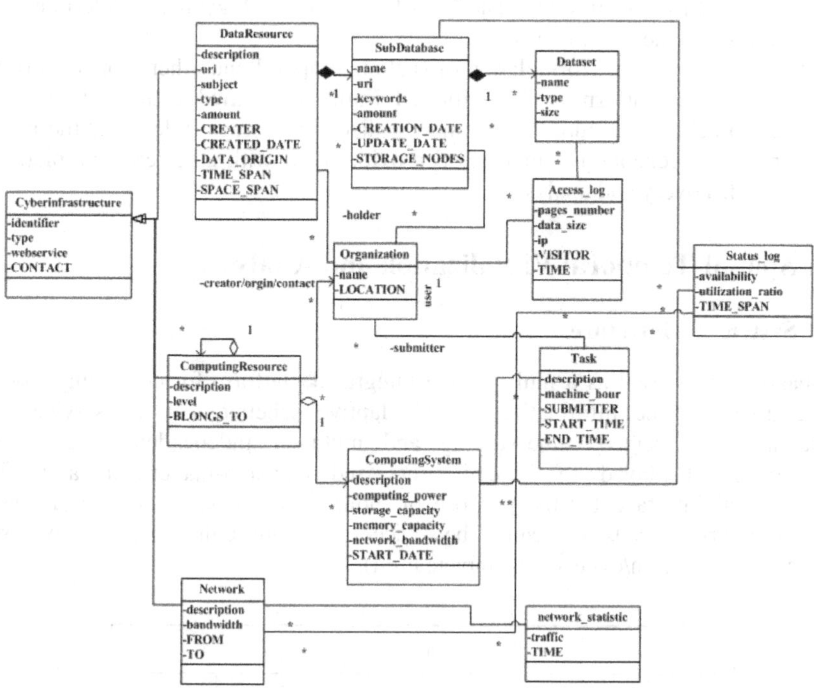

Fig. 1. A spatiotemporal modeling of CI resources

There is a similar level in computing resources. For example, Grid system is set up by many co-organizations. The world-class high performance computing environment is made up of many supercomputing centers. The supercomputing center is classified as main, branch, or institute-level center in possession of a number of supercomputing systems. Basic properties of supercomputing systems include unique identifier, creation time, computing power, memory capacity, storage capacity and network bandwidth. The computing tasks assigned to the system are associated with submitters, start time, end time and other spatiotemporal attributes. The interconnection topology and network traffic statistics in arbitrary periods are peculiar characteristics of network resources. The records of overall running state of the resource are equipped obviously with time dimension, for the simple reason that availability and utilization rate of service vary over time. The status log records the state of some aspect of the environment over a series of ranges in time and space.

The organization object in Fig. 1 can represent an individual, research institute, city, region, country, or even the world depending on the context and the information granularity. Spatial information is expressed through the organization object associated with the resource. The measures at different granularity enable the roll-up and drill-down operations in the statistical analysis of resource usage in any spatial extent, and enable spatiotemporal visualization of CI. Since there are multiple storage nodes for a dataset and its backups, and a computing system can encompass multiple sub-systems as well, spatial locations of data and computational resources can be described as multi-point structure. The spatial network topology also is determined by the locations of the network nodes.

On the basis of abstract formulation of spatial, temporal and other properties of CI resources, the basic uniform model is formed. Further, a unified technical framework is required to achieve resource registration, collection, storage, release of the multi-source and heterogeneous resources, and provide users with an interactive platform for intuitive discovery and analysis.

4 Spatial-Temporal Visualization and Analysis

4.1 System Architecture

Web-based GIS is used as a multi-source integrated platform for collecting, storing and releasing service information in an adaptive schema. The CI services are geographically dispersed, time-varying and mutually independent. Various CI resources are displayed to general users and decision-makers in a unified spatiotemporal interface for the purpose of monitoring, management and analysis. The prototype system can be accessed through the web site (http://cybergis.cnic.cn/cyberviz/mainpage.html).

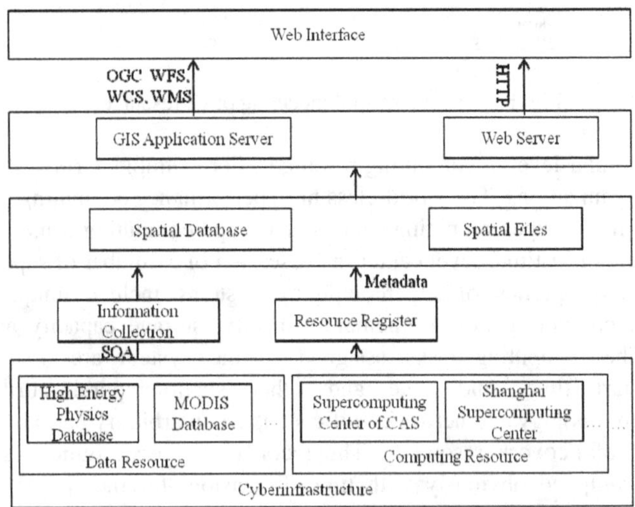

Fig. 2. System architecture of cyberinfrastructure visualization

The system is scalable for meeting the need of the rapidly growing CI and multi-source heterogeneous data collection and integration. The metadata must be designed in a coordinated collection, storage mechanism. The overall system architecture is shown in Fig. 2. It is similar to most of the GIS systems, in which dynamic resource registration mechanism is critical for collecting spatial information and statistics, and achieving uniform resource integration. Under this architecture, an independent access interface is provided by each CI application service, so that the spatiotemporal statistics can be integrated by using open standards such as service-oriented architecture. The resources can be discovered through a metadata catalog in the system.

The fundamental geographic data environment is composed of spatiotemporal database and spatial files. The database is established by registering the service information and integrating the historical service records. The database is constituted with metadata which indicate the resource properties, and statistical spatiotemporal data and other data. The spatial files contains vector data and raster data, such as the base maps from National Fundamental Geographic Database provided by National Geomatics Center of China, and the maps free accessed from OpenStreetMap. The GIS application server, which is independent of the web application server, is responsible for publishing GIS data and responding to the client's geographic operations. In the web browser, the use of open source GIS library is advantageous to execute the data inquiry and parse, filter and display the response on compliance with protocol Open Geospatial Consortium (OGC) WFS, WCS, WMS [8] protocols.

4.2 Resource Registration and Information Collection

Simple and consistent access to real-time state and historical data of dispersed CI resources is a prerequisite for the system. The integrated nature of the system determines that the consistent metadata is essential for automatically access of the statistics data, sharing of resource services, and on-the-fly search in resource catalogs. Resource entries constitute the entire metadata, providing sufficient description of resource information. They are acquired in the dynamic resource registration mechanism form the foundation for catalogs, web portals, and other discovery services [10]. Using the standard service architecture SOA, the coupling between our system and the distributed resource service system can be greatly reduced. The architecture is designed to combine cross-organizational services rapidly and build our application quickly.

In order to seamlessly integrate multi-source data, the URIs of the WSDL files are provided in metadata catalog. All metadata are collected in resource registration module. Various CI applications release their WSDL files as service providers. The URLs of these files are recorded and our system acts as a service registry in resource registration step. We follow the address and fetch the service records into our database. The system accesses self-governed resource information as a service consumer via SOAP-based interoperation in data collection step.

4.3 Storage and Release

The data collected in previous steps must be organized effectively for subsequent release and visualization. In our system, open source spatial database is chosen to provide standard access interface and operating mode. The spatiotemporal data in vector form are stored in the database, and the base maps in spatial files are transformed and then are stored as well.

The resource status is real-time accessed via their service interfaces for real-time monitoring, and the statistics are accessed periodically for enabling the historical statistical analysis. Historical statistics then is aggregated according to the time intervals and the spatial regions for supporting multi-granularity OLAP. An objective of the data release is to make data available to users using web services and to push data into larger federated systems. An open-source GIS application server is adopted to publish the spatial data under OGC protocols.

4.4 Visualization and Analysis

The geospatial data is open and accessible under a series of geospatial web services specifications (i.e., WFS, WMS and WCS). Some open source JavaScript libraries are used to request, display and render maps on the web browser. The interaction with the system, including zoom, pan, marker and feature selection operations, are implemented without server-side dependencies. For example, when a resource is selected, a WFS request in an XML file is sent to the GIS server. The response in GML format containing the resource information is parsed and then displayed. The timeline and region controls are also used to search and to locate the relevant resources by filtering resources in time and space dimensions. The time series animation of resource running statistics can help domain experts discover potential spatiotemporal patterns.

Consider data resources, it is difficult for users to search efficiently and monitor comprehensively these resources managed by multiple organizations in different disciplines. Through the previously mentioned modeling, scientific data resources, which describe different spatial regions with different time ranges, are integrated in the spatiotemporal visualization platform. Users are allowed to query directly resources from the registered metadata catalog, and the relevant results are organized by resource type. The left side of Fig. 3 shows the list of all current data resources relevant with “植物” (plant).

Users can get a general understanding of all available resources from the visual web interface and make efficient selection by combining both historical statistics and their underlying knowledge. For example, users can assess and compare intuitively the available massive data resources and their backups based on utilization rates, network load and transmission distances. Furthermore, users want to search for supercomputing resources to carry out the large-scale computational task. At this time, users should consider the tradeoff between the cost caused by data transmission from the different data sources to the supercomputer centers and the waiting time in busy supercomputer centers.

The overall understanding provides more reliable decision support in site selection and layout optimization of the CI through the participation of decision makers. Fig. 3 shows spatial statistics of Scientific Database of China Plant Species [5] in June 2010. The red bubbles denote a number of storage nodes which the resource is located on. The lines in the figure indicate usage statistics through the data portal system from the

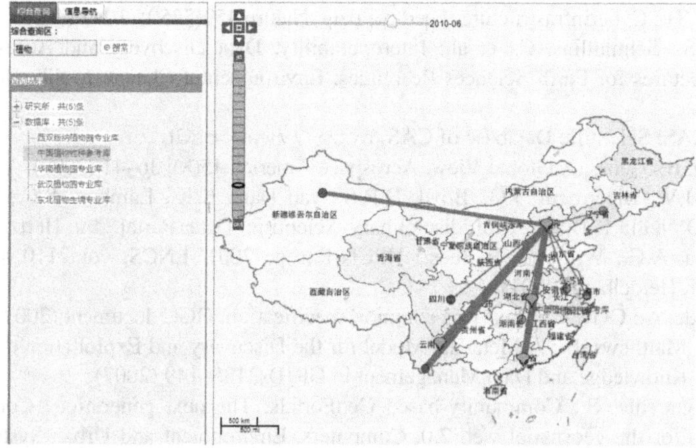

Fig. 3. Spatial-temporal visualization of the CI resources

major users. The line width in the flow map indicates the data size. It is obvious that the use of the database concentrated on organizations in southern China. Therefore, if the database or backup is deployed on storage nodes in southern China, the transmission costs during service running can be reduced. At this time, usage of computational resources in the branch of Supercomputing Center of Chinese Academy of Sciences which locate in Kunming, China, about 20%, is significantly lower than other supercomputing centers. This is significant that users may prefer the plant database's backup and relatively idle branch in the same place to get the data and perform their computational tasks.

5 Conclusions

A spatiotemporal schema is designed in this paper which models CI resources geographically. The time and space characteristics of resource distribution and usage can be effectively presented with such mechanism. A web-based GIS platform is also implemented for spatiotemporal visualization of CI services. The unified resource discovery, monitoring and assessment are implemented through the process of resource registration, data collection, storage, release, display and analysis. Moreover, some examples about data, computing and network resources has shown the system's ability to enable decision support for resource searching, planning and service running. Our current work is focusing on establishing an integrated spatiotemporal visualization platform. Spatial analysis tools and network flow visualization tools remains to be investigated on the basis of the system in future.

References

1. NSF Cyberinfrastructure Council: NSF's cyberinfrastructure vision for 21st century discovery, `http://www.nsf.gov/od/oci/ci-v7.pdf` (retrieved)
2. Carter, P.E., Green, G.: Networks of contextualized data: a framework for cyberinfrastructure data management. Communications of the ACM 52(2), 105–109 (2009)

3. Ledford, H.: Cyberinfrastructure: Feed me data. Nature 459(7250), 1047 (2009)
4. Nativi, S., Schmullius, C., et al.: Interoperability, Data Discovery and Access: The e-Infrastructures for Earth Sciences Resources. Environmental Change in Siberia, 213–231 (2010)
5. CNIC,CAS: Scientific Database of CAS, `http://www.csdb.cn`
6. Iannotta, B.: Geoss: A Global View. Aerospace America 45(8), 36–41 (2007)
7. Ashby, J.V., Bicarregui, J.C., Boyd, D.R.S., van Dam, K.K., Lambert, S.C., Matthews, B.M., O'Neill, K.D.: A Multidisciplinary Scientific Data Portal. In: Hertzberger, B., Hoekstra, A.G., Williams, R. (eds.) HPCN-Europe 2001. LNCS, vol. 2110, pp. 13–22. Springer, Heidelberg (2001)
8. Whiteside, A.: OGC Web services common specification. OGC document (2007)
9. Sufi, S., Matthews, B.: A Metadata Model for the Discovery and Exploitation of Scientific Studies. Knowledge and Data Management in GRIDs, 135–149 (2007)
10. De Longueville, B.: Community-based GeoPortals: The next generation? Concepts and methods for the geospatial web 2.0. Computers, Environment and Urban Systems 34(4), 299–308 (2010)

A Novel Clustering Mechanism Based on Image-Oriented Correlation Coefficient for Wireless Multimedia Sensor Networks

Rong Wu[1,2], Ruchuan Wang[1,2,3], Chao Sha[1,2], and Liyang Qian[1]

[1] College of Computer, Nanjing University of Posts and Telecommunications, Nanjing, 210003, China

[2] Jiangsu High Technology Research Key Laboratory for Wireless Sensor Networks, Nanjing, 210003, China

[3] Key Lab of Broadband Wireless Communication and Sensor Network Technology of Ministry of Education, Nanjing University of Posts and Telecommunications, Nanjing 210003, China

Abstract. Wireless multimedia sensor networks consist of unattended multimedia nodes with limited storage, energy and computation and communication capabilities. And energy is the most crucial resource for nodes. Clustering technique has been proven to be an effective approach for data-gathering in wireless multimedia sensor networks. Therefore, this paper presents a novel clustering mechanism based on image-oriented correlation coefficient to reduce redundancy transmission so as to conserve network energy. And the aim is achieved through taking advantage of the correlation coefficient of captured images and RSSI value of nodes to avoid excessive computation and transmission. Moreover, a new cluster header election method is also presented to make the network's load balance on energy. The simulation results show that the correlation coefficient threshold greatly affects the number and average size of clusters. And compared with another clustering algorithm based on overlapping FoVs, our clustering mechanism is more effective in prolonging network lifetime.

Keywords: Clustering Mechanism, Wireless Multimedia Sensor Networks, Correlation Coefficient, RSSI, Network Lifetime.

1 Introduction

In recent years, Wireless Multimedia Sensor Networks (WMSN)[1][2] have attracted a plethora of researchers' attention due to their vast potential applications and features of low cost. The multimedia nodes can be deployed at the interested areas to sensor various kinds of multimedia data such as image, audio and video. However, multimedia nodes have limited resources and usually cost much more energy than traditional nodes when transmitting multimedia data, it is crucial to design a wireless multimedia sensor network that can save nodes' energy and maximize network lifetime.

F.L. Wang et al. (Eds.): WISM 2012, LNCS 7529, pp. 333–341, 2012.

So far, there are some researchers have done a great deal of studies on the problem of saving nodes' energy from different aspects. Literature [3-11] have suggest several approaches for reducing energy consumption in wireless sensor networks, such as increasing the density of sensor nodes to reduce transmission range, reducing standby energy consumption via suitable protocol design, and advanced hardware implementation methods. And Clustering algorithms have also been proven to be a feasible solution. However, S. Soro et al.[12] pointed out that existing clustering algorithms of traditional wireless sensor networks are not suitable in the wireless multimedia sensor networks for the sensing model of multimedia nodes is quite different from that of common scalar nodes. Hence, K. Obraczka et al.[13] suggested that the sensing area of multimedia nodes should be an important factor to be considered. Paper [14] presents a clustering algorithm for wireless multimedia sensor networks which based on the overlapped Field of View (FoV) areas. It compute the overlapping sensing area of two nodes through their coordinates and orientation information. However, this algorithm only works well when there are little obstructions in the field of view and all nodes are in such a situation that similar to the 2D-plane. In addition, it is difficult to be accurately informed of all nodes' locations and directions in the real environment.

Here, we proposed a clustering mechanism based on image-oriented correlation coefficient (CMICC). In this mechanism, correlation coefficient of collected images is used to calculate the redundancy of sensed data by different nodes and then put certain nodes into the same cluster. And, a RSSI-assisted method is used to reduce the computational complexity of correlation coefficient. Besides, we propose a new cluster header election method to balance energy load in each node so as to prolong the whole network's lifetime more effectively.

The remainder of the paper is organized as follows: Section 2 illustrates the relevant definitions and network model. Section 3 describes the detail of CMICC. Simulation results are shown in Section 4. Finally, concluding remarks are made in Section 5.

2 Definitions and Network Model

2.1 Relevant Definitions

Definition 1 (Correlation Coefficient): the correlation coefficient of two multimedia nodes illustrates the relativity degree of images collected by the two nodes. That is the coverage redundancy of the two nodes.

In [15], it proposed a correlation coefficient computational algorithm which is image-oriented and independent with the node model. It defines the correlation coefficient of node i and node j as ρ_{ij}, and $\rho_{ij} = O_{ij} / T_{ij}$. Where O_{ij} denotes the overlapping area of two images captured by node i and node j, and T_{ij} is the total area of the two images. Also, the image's area is expressed by the number of pixels in the image.

For example, if there is overlap between the left parts of image *i* (captured by node *i*) and the right parts of image *j* (captured by node *j*), then

$$\rho_{ij} = \frac{O_{ij}}{T_{ij}} = \frac{\sum_{l=1}^{R} L_{il}}{R \times C \times 2 - \sum_{l=1}^{R} L_{il}} \tag{1}$$

Where *R* and *C* denote the row and column of a completed image, respectively. L_{il} denotes the coincident point's pixel location made by line *l* of image *i* and the right parts of image *j*.

Definition 2 (RSSI): RSSI indicates the signal strength of wireless communication of one node receive from the other. And we can use it to judge the distance and obstruction situation between the sending and receiving nodes.

From [16], we know that the value of RSSI obey the log-normal distribution, and the general form of the probability model of signal propagation is given as follows:

$$RSSI(d) = RSSI(d_r) - 10n_p \lg(\frac{d}{d_r}) + X_s \tag{2}$$

Where d_r is the reference range, n_p is the path loss exponent which depends on the RF environment. And X_s are the random variables that obey Gaussian distribution with mean 0 and standard deviation *S*. It is used to reduce the error of signal strength.

According to formula (2), we can derive distance *d* between the sending and receiving nodes.

Moreover, we use a simplified model shown in [17] for the wireless channel model of the network as follows.

$$RSSI(d) = \begin{cases} P_t - 40.2 - 10 * 2.0 * \lg d, (d \le 8m) \\ P_t - 58.5 - 10 * 3.3 * \lg d, (d > 8m) \end{cases} \tag{3}$$

Where P_t is the transmitted signal power of the sending node.

Hence, from formula (3), we can obtain the calculating formula of P_t.

2.2 Network Model

Before studying the details of the proposed clustering mechanism, we make some assumptions about the nodes and the underlying network model:

(1) The wireless multimedia sensor network is a high-density static network.
(2) Sink node is located far away from the square sensing field and it has powerful storage, energy, computation and communication capabilities.
(3) The initial energy of each multimedia node is the same and can not be replenished.
(4) Each multimedia node can use power control to vary the amount of transmission power, which depends on the distance to the receiver.
(5) Every multimedia node is equipped with GPS unit.
(6) The initial transmitted power of all multimedia nodes is the same.

3 A Novel Clustering Mechanism Based on Image-Oriented Correlation Coefficient (CMICC)

Here we propose a novel clustering mechanism based on image-oriented correlation coefficient (CMICC). Since correlation coefficient of nodes reflect the redundancy degree of the images captured by nodes, we can take advantage of it to reduce redundant data transmission and only send data which is not duplicated with others.

Moreover, since RSSI can indicate the distance of two nodes, and it is not necessary to calculate the correlation coefficient of nodes which are far away or have big obstacles between them for there is little possibility that they have overlapping sensor area. So we can use RSSI to decrease the calculation times of correlation coefficient.

Also, in order to prolong the network lifetime further, we also present a new cluster header election method based on cost. The main idea is that we define a cluster header competition parameter which contains two elements. One is the node's residual energy, the other is the total amount of energy consumption that cost by other nodes when they transmit their images to it. Hence, a node with certain residual energy and least total amount of energy cost will be selected as the header of the cluster.

We consider a monitor area with N wireless multimedia nodes, represented by the set $S = (S_1, S_2,..., S_N)$. In order to avoid high energy assumption of nodes for long distance transmission within the cluster, we set the maximum number of multimedia nodes within a cluster to m ($m \leq N$).

The specific clustering mechanism is described as the following three phases: Correlation Coefficient Matrix Building Phase, Clustering Phase and Cluster Header Election Phase.

1) Correlation Coefficient Matrix Building Phase:

①All nodes $\{S_1,S_2,\ldots,S_N\}$ start initializing. Each of them broadcasts a message which only contains its own ID, and receives ID messages from other multimedia nodes;

② Each node records the corresponding RSSI value receives from others. We use r_{ij} to denote the signal strength of node i receives from node j. If node i didn't receive node j's message, set r_{ij} to 0. And send the RSSI value results to Sink node;

③When having received all the results, Sink node constructs a signal strength matrix R of $N \times N$ as follows and creates an assisted set P, $P = \{P_1,P_2,\ldots,P_N\}$;

$$R = \begin{bmatrix} 0 & r_{12} & r_{13} & \cdots & r_{1N} \\ r_{21} & 0 & r_{23} & \cdots & r_{2N} \\ r_{31} & r_{32} & 0 & \cdots & r_{3N} \\ \cdots & \cdots & \cdots & \cdots & \cdots \\ r_{N1} & r_{N2} & r_{N3} & \cdots & 0 \end{bmatrix}$$

④ For the i_{th} (i=1,2,…,N) column, traverse the matrix R from the first row to look for node j(j=1,2,…,N) which satisfies the condition $r_{ij} \geq \beta$ (β is a predefined threshold constant). If there is node j (j=1, 2, …, N) meets the above condition which means node i and node j have overlapping sensed field, node j will be put into the cluster P_i. Otherwise, continue to search the next row;

⑤ i+1, if $i \le N+1$, then go to ④. Otherwise, go to ⑥;

⑥ Sink node broadcasts the images collecting request to all of the multimedia nodes in the network;

⑦ Each node begins to capture its first image after receiving the request message and transmits the image to Sink node;

⑧ After receiving all the images, Sink node starts to calculate the correlation coefficient ρ_{ij} of any two nodes within the cluster P_i(i=1,2,…,N) and set ρ_{ij}=0 when node j is not in the cluster P_i. Therefore, a simplified correlation coefficient matrix M is established as follows.

$$M = \begin{bmatrix} 1 & \rho_{12} & \rho_{13} & \cdots & \rho_{1N} \\ \rho_{21} & 1 & \rho_{23} & \cdots & \rho_{2N} \\ \rho_{31} & \rho_{32} & 1 & \cdots & \rho_{3N} \\ \cdots & \cdots & \cdots & \cdots & \cdots \\ \rho_{N1} & \rho_{N2} & \rho_{N3} & \cdots & 1 \end{bmatrix}$$

2) Clustering Phase:

① Sink node creates a new empty cluster set T, $T=\{T_1,T_2,\ldots,T_N\}$ and select the node with minimal ID which assumed to be S_i(i=1,2,…N) from nodes set S, and put it into the empty cluster T_i, then delete node S_i from set S. Then traverse the correlation coefficient matrix M in ascending order of the ID of remaining nodes. Put all of the nodes j (j=1,2,…,N) which satisfy the condition $\rho_{ij} \ge \alpha$ (where α is a correlation coefficient threshold which is depended on the actual needs) into T_i;

② If the number of nodes in T_i is more than m, calculate the sum of the correlation coefficient of each pair of nodes within the cluster, recorded as $\sum \rho_{st}$ (s, t is the ID of the multimedia nodes in the cluster). And sort the results in descending order, keep the top-ranked m nodes and remove the rest from T_i;

③ If the number of nodes in T_i is no more than m, calculate the average value of the correlation coefficient of each pair of nodes within the cluster. If the result is more than α, keep the node, otherwise, remove it from T_i;

④ Judge whether set S is empty. If it is, the clustering process is considered to be finished. Otherwise, repeat above clustering phase until S is empty.

3) Cluster Header Election Phase

① A Node whose residual energy is more than the energy threshold ω_0 is able to be the candidate header;

② Every candidate header calculate the total energy cost of the network in accordance with formula (4) and (5);

③ Elect the node which has the least total cost to be the header.

$$Cost(i,j) = P_{tj} * (1-\rho_{ij}) * k, (i, j \in 1,2,...,n, i \neq j) \quad (4)$$

Where $Cost(i,j)$ denotes energy cost of node j when it sends data to node i, k is the amount of bytes of a completed image, k=2×R×C, and n is the number of nodes of a cluster.

$$Cost_{Total} = \sum_{j=1, j\neq i}^{n} Cost(i,j), (i = 1,2,...,n) \quad (5)$$

In addition, we can achieve more energy-saving by regulating the nodes' transmission power to minimal in accordance with formula (6).

$$Pt = \begin{cases} RSSI_{\min} + 40.2 + 10 * 2.0 * \lg d, (d \leq 8m) \\ RSSI_{\min} + 58.5 + 10 * 3.3 * \lg d, (d > 8m) \end{cases} \quad (6)$$

Where $RSSI_{min}$ denotes the minimal signal strength which means any two nodes within the cluster can communicate with each other. And d can be obtained from formula (2).

4 Simulation and Analysis

In this section, we evaluate the performance of our proposed algorithm (CMICC) via MATLAB. For simplicity, an ideal MAC layer and error-free communication links are assumed. The simulation parameters are given in Table 1.

Table 1. Simulation Parameters

Parameter	Value
Network size	100*m*×100*m*
Sink location	(50,150)*m*
Number of multimedia nodes	50
Initial Energy	2*J*
Data Packet Size	4000*bits*
Initial RSSI	-70*dBm*
d_r	87*m*

A. Number of Clusters in CMICC

In our setup, 50 multimedia nodes are uniformly distributed in a region of a 100m×100m square. Because the value of RSSI represents the communication distance between two nodes, so it greatly affects the final results of clustering. Here, we set each node's value of RSSI $\beta = -70dBm$, so that the node's communication radius can be maximal.

Since the correlation coefficient threshold α is a judgment which determines whether the nodes can join the cluster, by varying α, we can control the number of clusters of the network. Fig 1(a), (b), (c) show the results of clustering when $\alpha = 0.25$, $\alpha = 0.30$, $\alpha = 0.35$, respectively.

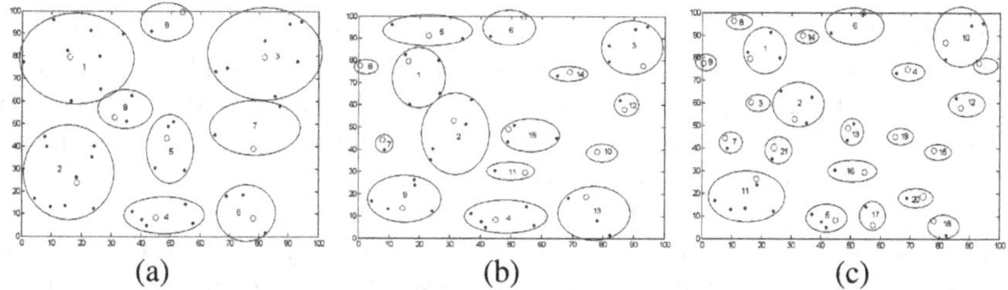

(a) (b) (c)

Fig. 1. The graph result of the clustering

Table 2. Result of Experiment A

α	average number of clusters	Average size of a cluster
$\alpha = 0.25$	9.21	5.56
$\alpha = 0.30$	15.58	3.33
$\alpha = 0.35$	22.46	2.22

From above experiment result, we can find that the number of clusters increases as the value of α increases, while the average size of a cluster decreases as α increases.

B. Average Size of Clusters

In this set of experiments, we study the results of clustering when the value of correlation coefficient α ranges from 0 to 1. We conducted the experiments under the condition of 50, 100 and 150 multimedia nodes in a network, respectively. Each case we carried out 10 experiments and recorded the number and size of clusters for every experiment. Fig 2 depicts the average cluster size under different correlation coefficient threshold.

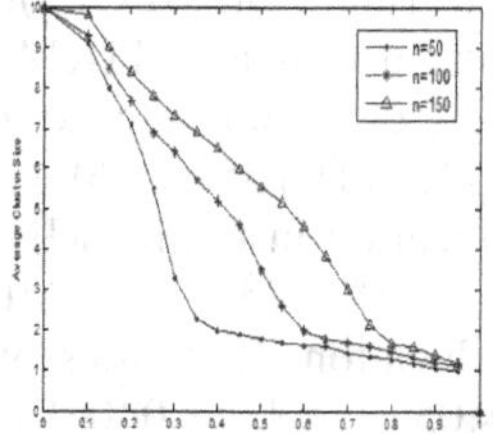

Fig. 2. Impact of correlation coefficient on average size of clusters

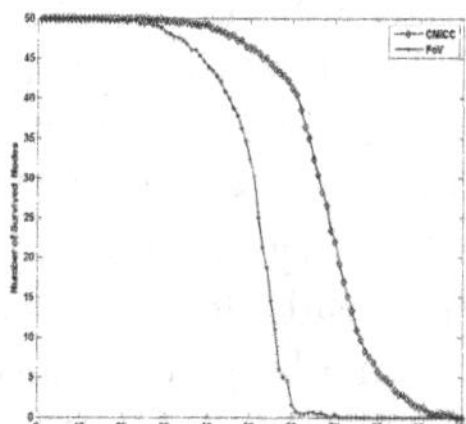

Fig. 3. The number of survived nodes over time

From Fig 2 we can observe that the more the nodes, the larger the number of nodes in a cluster, and the average size of a cluster decreases as α increases. Besides, there are parts of each curve drop quickly during such intervals. This is due to the number of nodes in a cluster is more than other case when the correlation coefficient threshold α falls within these intervals.

C. Network Lifetime
Finally, we examine the network's lifetime of two clustering algorithm by evaluating of the time until the first node dies and the time until the last node dies. We compare our clustering mechanism with FoV[14] (a clustering algorithm based on overlapping FoVs for WMSN), the results is shown as Fig 3.

It is clear that CMICC has better performance in increasing network lifetime compared with FoV. Although, FoV achieves energy conservation through coordination of nodes belonging to the same cluster to avoid redundant sensing and processing, but it is impossible to keep away from sensing repeated information completely. And it is well know that data transmission cost more energy consumption than data gathering. Contrastly, our algorithm eliminates redundant data before nodes start to transmit their captured images, and will cost less transmission power. Therefore, it prolongs the network lifetime effectively.

5 Concluding Remarks

In this paper, we propose CMICC, a clustering mechanism based on the image-oriented correlation coefficient. This mechanism makes the multimedia nodes which capture duplicated data join the same cluster, so that it will avoid redundant data transmission since each node only needs to send parts of the image instead of a completed image to the cluster header. Besides, we make use of RSSI to reduce computation times of correlation coefficient. And, a new cluster header election method is presented to balance energy load in every node. Thus, it prolongs the network lifetime effectively.

Acknowledgement. The subject is sponsored by the National Natural Science Foundation of P. R. China (No. 60973139, 61170065, 61171053, 61003039, 61003236, 61103195), the Natural Science Foundation of Jiangsu Province(BK2011755), Scientific & Technological Support Project (Industry) of Jiangsu Province(No.BE2010197, BE2010198, BE2011844, BE2011189), Natural Science Key Fund for Colleges and Universities in Jiangsu Province (11KJA520001), Project sponsored by Jiangsu provincial research scheme of natural science for higher education institutions (10KJB520013, 11KJB520014, 11KJB520016), Scientific Research & Industry Promotion Project for Higher Education Institutions(JH2010-14, JHB2011-9), Postdoctoral Foundation (20100480048, 2012M511753, 1101011B), Science & Technology Innovation Fund for higher education institutions of Jiangsu Province(CX10B-196Z, CX10B-200Z, CXZZ11-0405, CXZZ11-0406), Doctoral Fund of Ministry of Education of China(20103223120007, 20113223110002)and key Laboratory Foundation of Information Technology processing of Jiangsu Province(KJS1022), Project Funded by the Priority Academic Program Development of Jiangsu Higher Education Institutions(yx002001).

References

1. Akyildiz, F., Melodia, T., Chowdhury, K.R.: A survey on wireless multimedia sensor networks. Computer Networks 51, 921–960 (2007)
2. Akyildiz, I.F., Melodia, T., Chowdhury, K.R.: Wireless multimedia sensor networks: applications and testbeds. Proceedings of the IEEE 96(10), 1588–1605 (2008)
3. Amis, A.D., Prakash, R., Vuong, T.H.P., Huynh, D.T.: Max-Min D-Cluster Formation in Wireless Ad Hoc Networks. In: Nineteenth Annual Joint Conference of the IEEE Computer and Communications Societies, vol. 1, pp. 32–41 (March 2000)
4. Bandyopadhyay, S., Coyle, E.J.: An energy efficient hierarchical clustering algorithm for wireless sensor networks. In: Twenty-Second Annual Joint Conference of the IEEE Computer and Communications, vol. 3, pp. 1713–1723 (April 2003)
5. Chatterjee, M., Das, S.K., Turgut, D.: WCA: A Weighted Clustering Algorithm for Mobile Ad hoc Networking. Cluster Computing 5, 193–204 (2002)
6. Chen, B., Jamieson, K., Balakrishnan, H., Morris, R.: Span: an energy-efficient coordination algorithm for topology maintenance in ad hoc wireless networks. In: Proc. of the 7th ACM International Conf. on Mobile Computing and Networking, pp. 85–96 (2001)
7. Choi, W., Shah, P., Das, S.K.: A framework for energy-saving data gathering using two-phase clustering in wireless sensor networks. In: Proceedings of Mobile and Ubiquitous Systems (Mobiquitous): Networking and Services, Boston (2004)
8. Frontini, M., Sormani, E.: Some variant of Newton's method with third-order convergence. Applied Mathematics and Computation 140, 419–426 (2003)
9. Kirousis, L.M., Kranakis, E., Krizanc, D., Pelc, A.: Power consumption in packet radio networks. Theoretical Computer Science 243, 289–305 (2000)
10. Krishnan, R., Starobinski, D.: Efficient clustering algorithms for self-organizing wireless sensor networks. Ad Hoc Networks 4, 36–59 (2006)
11. Liu, J.S., Hung, C., Lin, R.: Energy-efficiency clustering protocol in wireless sensor networks. Ad Hoc Networks 3, 371–388 (2005)
12. Soro, S., Heinzelman, W.: On the coverage problem in video-based wireless sensor networks. In: Proceeding of IEEE International Conference of Broadband Communications and Systems (BroadNets), Boston (2005)
13. Obraczka, K., Manduchi, R., Garcia-Luna-Aveces, J.J.: Managing the Information Flow in Visual Sensor Networks. In: Proceedings of the 5th International Symposium on Wireless Personal Multimedia Communications (2002)
14. Alaei, M., Barcelo-Ordinas, J.M.: Node clustering based on overlapping FoVs for wireless multimedia sensor networks. In: Wireless Communications and Networking Conference (WCNC), pp. 1–6 (April 2010)
15. Ma, Z., Wang, R., Ping, H., Sha, C.: A redundancy eliminating scheme based on the suture technology for Image Sensor networks. Journal of Southeast University (Natural) 2, 280–284 (2011)
16. Che, Y., Xu, W.: Researches on the RSSI-based positioning technology in Wireless Sensor Networks. Sensors and Instrumentation 26(4), 82–84 (2010)
17. Hu, K.: IEEE 802.15.4 simulation and performance analysis. Huazhong University of Science and Technology, Wuhan (2006)

Design of Underground Miner Positioning System Based on ZigBee Technology

Xin Xu, Ping Zheng, Lingling Li, Haisheng Chen, Jianping Ye, and Jun Wang

School of Electrical and Information Engineering, Xihua University, 610039 Chengdu, Sichuan, China
{xxtoday,zp5757}@126.com

Abstract. This article begins with analysis of the particularity of coal mine structure, for the lack of mine wired network in some respects, such as freshness, flexibility, coverage and so on, a scenario of Wireless Sensor Network (WSN) based on ZigBee technology is presented. And it can be used for miner positioning and related information collection in the coal mine. The tree network topology is selected to set up the WSN.The result of this experimental indicate that the scenario can meet the demands of underground communication, and overcome the existing problems of single function and wiring complexity. The scenario provides a better solution for the underground miner positioning and related information collection.

Keywords: Wireless Sensor Network (WSN), ZigBee, miner positioning, information collection.

1 Introduction

China is a country that mainly depends on coal as the energy, both the coal production and consumption rank first in the world. In the future for a long period of time, the coal-dominated energy structure will not change. Due to the complex structure and bad environment of the coal mine, some accidents have occurred from time to time, more attention needs to pay to the coal mine safety. Therefore, it is important to monitor the environment parameters when mining coal; Moreover, when the accidents happening, it is of great significance for the rescue to locate the position exactly of trapped miners.

Currently, the RFID (Radio Frequency Identification) technology is widely used in the mine for the miner positioning. The concrete realization of the method is: the RFID tags on miners communicate with the RFID readers in the key position [1], RFID readers record the time that miners passed the readers, then infer the miners' approximate location via the time. There are some disadvantages of this approach. On one hand, the RFID reader needs a larger space to deploy, whereas the mine space is relatively small. On the other hand, the approach is not conducive to the expansion of the network. But ZigBee is a rising wireless network technology which is of short space, low complicacy, low power consumption and low cost. It is the characteristics

F.L. Wang et al. (Eds.): WISM 2012, LNCS 7529, pp. 342–349, 2012.

to make up for these deficiencies. In the past decade, ZigBee technology has been successfully applied in the field of smart home, intelligent transportation, modern agriculture, medical care etc [2]. Therefore, basing on the complex structure of the mine, in this paper, a scenario of WSN based on ZigBee technology is presented. This scenario not only can be used for the miner positioning, but also for the detection and transmission of gas concentration, temperature, air pressure and other parameters.

This paper is organized as follows. In Section 2, we present the mine structure analysis and the choice of ZigBee network topology. And the whole structure of system is proposed in Section 3. In the end, we draw the conclusions.

2 Analysis of Mine Structure and ZigBee Technology

2.1 Analysis of Mine Structure

It is meaningful to get miners' position in the main tunnel, underground work locations and other important places; the choice for the appropriate network topology has important significance for the completion of position task and the aid mechanism. Tunnel will generally appear in four forms, as shown in Fig.1. No matter what the tunnel structure may be, as far as ZigBee is in the visual range, it will follow the same rules for communication. For the curve and cross shaped structure, adding a node to provide a certain degree of redundancy, which will be beneficial to the data transmission.

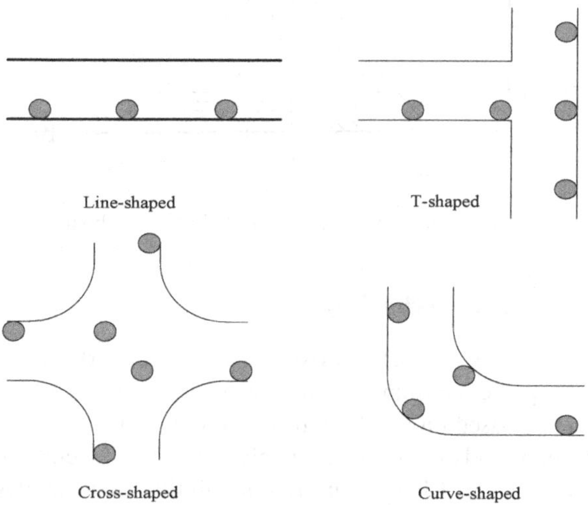

Fig. 1. Four forms of tunnel

Although the tunnel forms have little impact on communications, there are strict requirements for the adoption of machinery and equipment. In order to meet the demands of machine equipment and temporary double track train, the width of tunnel can be calculated the following method [3]:

$$B = a + 2A_1 + c \tag{1}$$

Where B is the tunnel width, a is the width of non-sideway, A_1 is the max width of transportation facilities, and c is the sideway width. In accordance with relevant provisions, take the sideway width c=840mm, non-sideway width a=400mm, and A_1 =1180mm. In accordance with (1), the pure width of tunnel B=3600mm=3.6m, therefore, the tunnel width should be about 2m to 4m. In this design, take the max width of the tunnel B=4m, as shown in Fig.2, the communication range is proposed. Assume that a ZigBee node is at the point L=e, it can communicate with the farthest node at the point L=g, thus the node has communication capability of area is S1=2×4× (g-e). But another node is at L=f, it can as far as communicate with the node at L=h, and its communication area is S2=2×4×(h-f). The overlap S1 and S2 can be educed, that is S3=4×(g-f). The S3 region can be covered by the adjacent ZigBee nodes. In this way, this section of the tunnel is divided into the common area S3 and the other two separate areas. By the communication between ZigBee nodes carried by miners and nodes fixed in the tunnel, the location of the miners can be morally fixed.

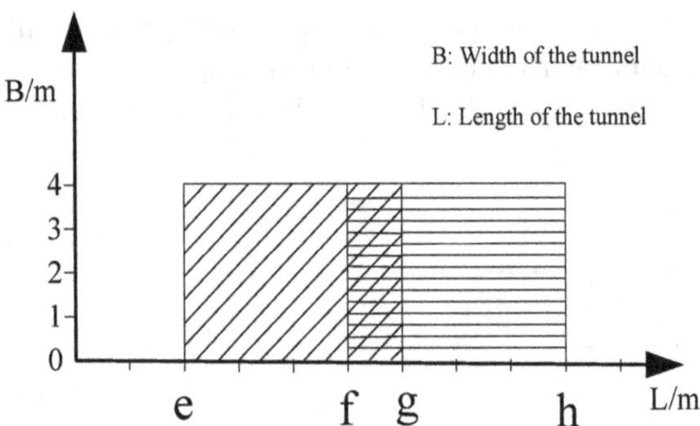

Fig. 2. Example of ZigBee communication in the tunnel

2.2 Analysis of ZigBee Technology

ZigBee (IEEE 802.15.4 standard) is a rising wireless network technology which is a short distance, low power consumption, low data rate and low cost [4]. In the network with IEEE 802.15.4, based on the discrepancy of communication ability, equipments are divided into two kinds of devices, namely FFD (full function device) and RFD (reduced function device). RFD has no routing ability, transmits less data quantity, and usually is adopted as an end device in the network. If not at work, it will be in the dormancy state. Whereas FFD is in the state of working at all times, and holds the ability of control and management in the network. Due to the tasks of devices, ZigBee alliance calls them coordinator, router and end device [5]. Coordinator is a special type of router, and is responsible for forming the network, setting up an addressing scheme, and maintaining all routing tables. Router extends network area coverage, and connects to the coordinator and other routers, and also support child devices.

The end device can transmit or receive messages; it must be connected to both the coordinator and a router, and do not support child devices.

ZigBee defines the network layer specification for star, tree and mesh network topologies and provides a framework for application programming in the application layer. As shown in Fig.3.

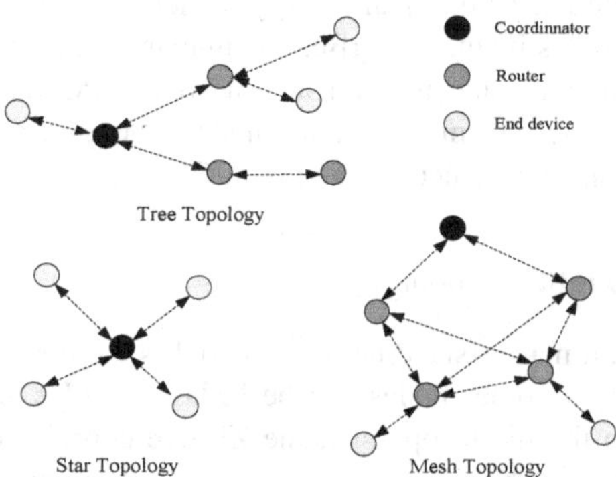

Fig. 3. ZigBee network topology

3 Proposed the Whole Structure of System

The proposed system is to build up a miner positioning and information collection system. As shown in Fig.4, the system mainly consists of ground control centre and undermine ZigBee network. The personnel computer (PC), the industrial personal

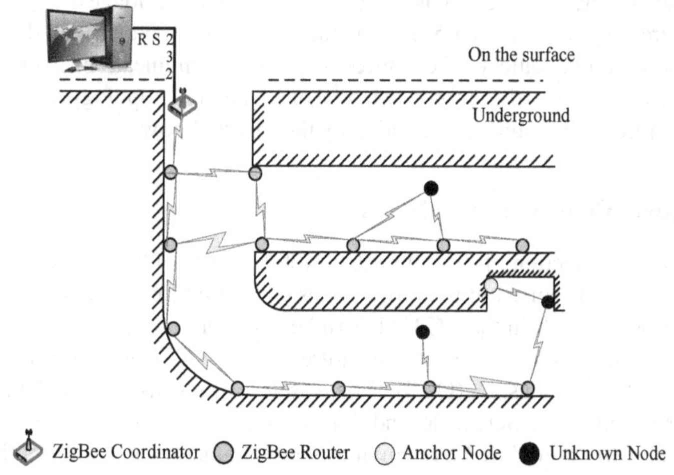

Fig. 4. System of Whole Structure

computer (IPC), and related equipments constitute the ground control center, whose role is to store, compute the collection, and operate corresponding equipments in the mine [6]. Coordinator, the core of ZigBee network, connected to the computer through the RS232 bus, forwards the mine information data to the PC. In the tunnel of coal mine, router nodes are placed at intervals along the tunnel. In order to obtain energy conveniently, these nodes can be set in the neighborhood of the mine lamp. However, in the some places that gain energy difficultly, some anchor nodes are set. Miners' wireless modules included ZigBee module and sensors, are used as unknown node, collect information data and transmit it to the PC through routers and the coordinator [7]. Then the system can acquire undermine environment parameters and reach the miners' position by calculation of the PC.

3.1 Selection of ZigBee Topology

In the star topology, a master-slave network model is adopted. A FFD plays the role of PAN coordinator, the other nodes can be FFDs or RFDs, communicate with the PAN coordinator. In the mesh topology, the ZigBee coordinator is in the charge of starting the network and choosing certain key network parameters, whereas ZigBee routers are responsible for network extension [8]. But the tree topology, in a sense, is deemed as a complicated, expanding star topology. In tree networks, routers move data and control messages through the network using a hierarchical routing strategy [9]. Compared to the mesh topology, it is more straightforward to establish a tree topology. In addition, the tree network has the wider or more comprehensive application.

It is precisely for this reason that the system chooses the tree topology. In addition, another important reason is the restrictions of router energy consumption. Many articles adopt mesh topology as the network structure, and use #5 alkaline batteries to supply to the router; however, it is experimentally that the longest supply time of battery-powered is no more than 5 days. If these routers were configured undermine, for the reason of the battery, the entire network communications would become unstable. It is based on the above two reasons, the tree topology is used as the network structure, and routers are supplied by the external power.

3.2 Design of Communication Nodes

In this design, the selection of the chip is CC2430/CC2431 of TI Company. As shown in Fig.5, the CC2431 is used in unknown nodes carried on miners, and the CC2430 is used in other nodes [10]. In the CC2431 peripherals, there are four sensors, which are gas concentration sensors (GCS), temperature sensors (TS), pressure sensors (PS), oxygen concentration sensors (OCS). The wireless transmission of data is in the 2.4GHZ ISM (Industrial Scientific and Medical) opening band, and the option of antenna is Sub-Miniature-A (SMA). Environment parameters are collected by sensors and sent to the nearest router, and then are transmitted to the ground center with the help of WSN.

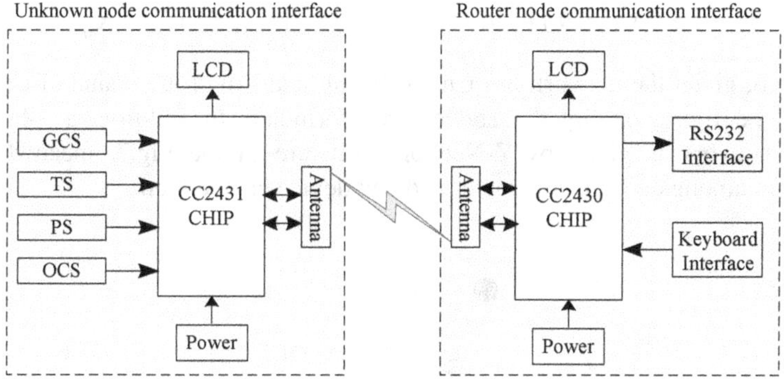

Fig. 5. Design of communication nodes

3.3 Analysis of Routing Algorithm

In the ZigBee network, network addresses are assigned using a distributed addressing scheme that is designed to provide every potential parent with a finite sub-block of network addresses. These addresses are unique and are given by a parent to its children. Figure 6 shows the parent node can be allocated the number of addresses (Cskip) and each node address (Addr). When the coordinator establishes a new network, it has a depth of 0, while its children have a depth of 1. If the node (i) wants to join the network, and connects to the node (k), then the node (k) is the parent node of node (i). According to the Addr and network depth (Depth k), the node (k) shall assign node (i) Addr and network depth (Depth i =Depth k +1).

Assumed values for the maximum number of children a parent may have, nwkMaxChildren (Cm), the maximum number of routers a parent may have as children, nwkMaxRouters (Rm), and the maximum depth in the network, nwkMaxDepth (Lm), then the Cskip (d) can be calculated as follows, given network depth, d [11].

$$Cskip(d) = \begin{cases} 1 + C_m(L_m - d - 1) \ , & if \quad R_m = 1 \\ \dfrac{1 + C_m - R_m - C_m \bullet R_m{}^{L_m - d - 1}}{1 - R_m} \ , & otherwise \end{cases} \tag{2}$$

If an end device, without routing capabilities, connects to the parent node as the nth node. Based on the parent address, A_k and the depth of the child node, d, the child node Addr: A_i is given by the following equation [12]:

$$A_i = A_k + Cskip(d) \bullet R_m + n, \qquad 1 \le n \le (C_m - R_m) \tag{3}$$

If a router joins the network, it will get the Addr:

$$A_i = A_k + 1 + Cskip(d) \bullet (n-1), \qquad 1 \le n \le (C_m - R_m) \tag{4}$$

In the Fig.6, given the network has Cm=4, Rm=4 and Lm=4, the value of Cskip and Addr can be figure out by the above three formulas. In addition, a 3-layer tree topology network is given by Z-Network software in the Fig.7, meanwhile, the relationship among coordinator, router and end device can be seen.

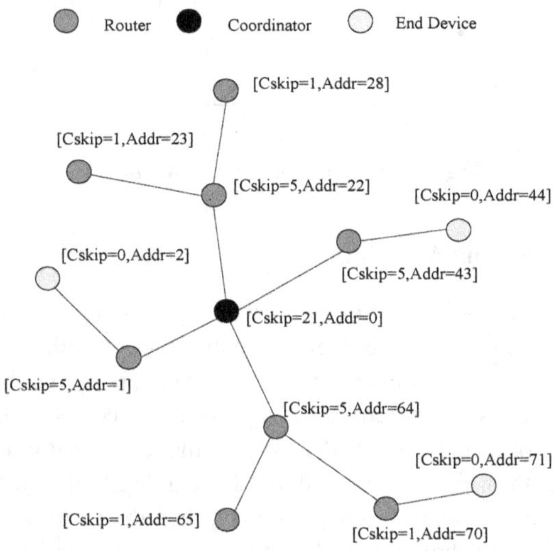

Fig. 6. An example of the tree network address assignment [13]

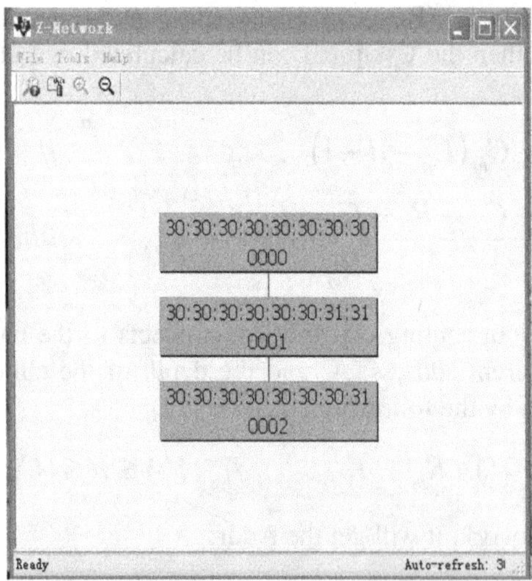

Fig. 7. Tree topology displayed in Z-Network

4 Conclusion

Based on the analysis of mine structure and ZigBee technology, a scenario has been proposed in this paper. By comparing the characteristics of the three topologies, the tree topology is used to establish WSN in the mine. The experimental studies show that the scenario can be better to locate miners and acquire the relevant parameters in the pit. The whole system is easy to install and expand, and the practical value is worth popularizing.

Acknowledgments. This work is supported by the Intelligent Electrical Equipment Laboratory Project (No. XZD0819), and the Innovation Fund of Postgraduate of XIHUA University of China (No. Ycjj201293).

References

1. Worek, C., Szczurkowski, M., Krzak, L., Warzecha, M.: RFID Reader Designed for Underground Operation in Mines. Przeglad Elektrotechniczny 83, 94–97 (2007)
2. Han, D.-M., Lim, J.-H.: Smart Home Energy Management System using IEEE 802.15.4 and ZigBee. IEEE Transactions on Consumer Electronics 56, 1403–1410 (2010)
3. Sun, Y., Tao, Z.: Design of Personnel Management System for Coal Miners Based on ZigBee (2009)
4. Wang, C.G., Sohraby, K., Jana, R.: Voice Communications over ZigBee Networks. IEEE Communications Magazine 46, 121–127 (2006)
5. ZigBee Alliance, ZigBee Specification. version1.1 (November 2006)
6. Terada, M.: Application of ZigBee Sensor Network to Data Acquisition and Monitoring. Measurement Science Review 9, 183–186 (2009)
7. Yang, D.: A Coal Mine Environmental Monitor System with Localization Function Based on ZigBee-Compliant Platform. In: 7th International Conference on Wireless Communications, Networking and Mobile Computing, pp. 23–25. IEEE Press, Beijing (2011)
8. Chan, H.K.: Agent-Based Factory Level Wireless Local Positioning System With ZigBee Technology. IEEE Systems Journal 4, 179–185 (2010)
9. Nadimi, E.S., Sogaard, H.T., Bak, T.: ZigBee-based Wireless Sensor Networks for Classifying the Behaviour of a Aerd of Animals Using Classification Trees. Biosystems Engineering 100, 167–176 (2008)
10. Ding, G., Sahinoglu, Z., Orlik, P.: Tree-based Data Broadcast in IEEE 802.15.4 and ZigBee Networks. IEEE Transactions on Mobile Computing 5, 1561–1574 (2006)
11. Kim, H.S., Yoon, J.: Hybrid Distributed Stochastic Addressing Scheme for ZigBee/IEEE 802.15.4 Wireless Sensor Networks. Etri Journal 33, 704–711 (2011)
12. Shiuan, P.M., Hung, T.C., Chee, T.Y.: The Orphan Problem in ZigBee Wireless Networks. IEEE Transactions on Mobile Computing 8, 1573–1584 (2009)
13. ZigBee Alliance, ZigBee Specification. Version r13 (December 2006)

Metadata Management of Context Resources in Context-Aware Middleware System

Rong Tan, Junzhong Gu, Zhou Zhong, and Peng Chen

Department of Computer Science and Technology,
East China Normal University, P.R. China
{rtan,jzgu,zhzhong,pchen}@ica.stc.sh.cn

Abstract. Metadata is the data about data. With the development of information technologies, it is realized that metadata is of great importance in discovery, identification, localization and access of resources. For the context-aware middleware, it is necessary to provide the metadata of context resources connected to it for the context consumers. However, the complexity of context-aware computing results in the diversity of context resources which makes it difficult in explicitly describing the metadata. In this paper, an XML-based markup language called VirtualSensorML is proposed for the purpose of providing model and schema to describe the metadata of context resources in an understandable format. It defines fundamental elements and various sub elements to describe the common and special properties of different context resources. The metadata management of context resources is currently implemented in a multi-agent based context-aware middleware. The evaluation demonstrates that it is efficient to improve the discovery of context resources.

Keywords: Metadata management, Context-aware computing, Context resource management, Context-aware middleware.

1 Introduction

Metadata is the "data about data" which can be considered as specification of data structures or data about data content. With the rapid development of information technologies, it is imperative for people to be able to exchange, share and manage information. The proposal of metadata is to better enable them to discover and access the resources required[8]. Metadata used to improve data management of large-scale data storage related systems such as electronic libraries and data warehouses[2][3]. As people gradually realized the importance of metadata in discovering, locating, identifying and assessing resources, more and more applications and systems, especially intelligent ones[4,7], place metadata management into one of the priorities.

For a context-aware application developed with a context-aware middleware, it should be able to know what context the middleware can provide, where is the context from, how to acquire the context, and what are the special properties of the context required. Above related information are understood as the metadata of the context resources. Therefore, it is necessary for a context-aware middleware to

F.L. Wang et al. (Eds.): WISM 2012, LNCS 7529, pp. 350–357, 2012.

provide certain metadata management of context resources. However, the complexity of context-aware computing leads to the difficulties in description of the metadata of context resources, which reflects in the following two aspects:

- The types of context resources are various. The diversity of contextual information results in the diversity of context resources as well. Context can be collected from a physical sensor or a logical one. While logical resources are usually in the form of software, it is very difficult to make an accurate and standard description of them.
- Interactions between context resources and middleware are various. Communications and interactions between context resources and middleware can through either the external networks or local APIs. While some context resources are encapsulated into standard services and the detailed interactive information are described by standard profiles such as WSDL documents[12], most context resources directly interact with context-aware middleware without any additional descriptions.

Former researches such as [5,6,9,11] have provided certain frameworks for creation of metadata profiles describing the capabilities of context resources. However, most of them focus on physical resources and pay little attention on logical ones. In this paper, we aim to provide a well-defined metadata description framework for the context resources in context-aware middleware. An XML-based model and schema called VirtualSensorML is proposed. It defines some fundamental elements of the metadata such as identification, capabilities, location and outputs and related sub elements for describing the detailed properties. The creation and modification of metadata profile can be done by VSML editor which is an editing tool for VirtualSensorML. In addition, the GaCAM(Generalized Adapt Context Aware Middleware)[1], a multi-agent context-aware middleware, is introduced. We will discuss the metadata management of context resources including how to deal with the metadata profiles and how context consumers can discover and access the resources required by these profiles in GaCAM.

The remainder of this paper is structured as follows. In Section 2, the generic definition of context resource and its category are described. The VirtualSensorML is introduced in Section 3. Section 4 illustrates the metadata management of context resources in GaCAM and an evaluation is provided. Finally, Section 5 concludes the paper.

2 Context Resources in Context-Aware Computing

A context resource is a data source that collects information which is used to characterize the situation of an entity. The collected contextual information ranges from data that can specify, determine or clarify a phenomenon in real world to information in the profile which describes an entity in cyber world.

According to the implementation of context resource, it can be categorized into physical context resource and logical one. Physical resource is a physical device that captures or detects the environmental elements in real world. For example, GPS is a frequently used physical resource which returns the coordinate pairs (longitude and latitude) of an entity. In contrast, logical resource is a data source that indirectly interacts with physical layer, and it often has no computational ability. What's more important, a logical resource may make use of different physical resources and other logical resources to form a higher-level context which usually used during context preprocessing and context reasoning to derive new contextual information.

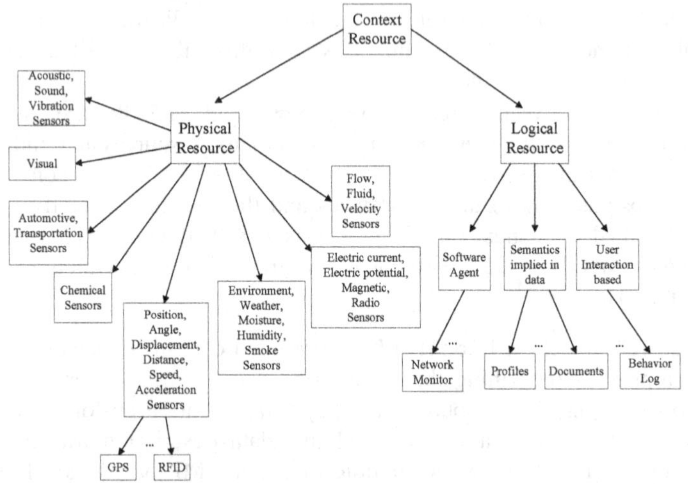

Fig. 1. Context resource category in context-aware computing

Fig. 1 shows a classified general context resource category. Context resource is divided into physical resource and logical resource. And physical resources are categorized into eight sub-classes and each sub-class has more detailed resource types. For example, the GPS sensor is an instance of location resources which is the fifth sub-class of physical resources. The classification of physical resources is mainly based on the sensor classification on Wikipedia. On the other hand, logical resources are classified into three main sub-classes: Software Agent, Semantics implied in data and User interaction based. Software Agent resources are in forms of software to collect contextual information. They can be the software which monitors the conditions of cyber world such as a network monitor or a device condition monitor which reports the current condition of CPU, memory, battery et al. in the device. Semantics implied in data records the contexts of semantics such as user profiles, group profiles, calendars and documents. User interaction based resources collect the contexts of user interactions such as the operation history and usage logs.

3 VirtualSensorML

VirutalSensorML is an XML-based markup language with the purpose of providing a proper model and schema to explicitly describe the common and special properties of various context resources in an understandable format. It presents the concept of resource group and resource field which is used to describe a set of context resources such as a small sensor network and defines metadata elements to overcome the deficiencies of current solutions which are not capable of describing the logical context resources.

3.1 Resource Group and Resource Field

Context-aware middleware usually interacts with more than one context resources. And sometimes a number of context resources will interact with middleware as a whole. For example, a small sensor network contains a number of various sensors and communicates with middleware through one of the node. Obviously, in this scenario to provide each context resource a metadata profile will definitely leads to a lot of redundant information. Therefore, in order to reduce the redundancies, it is desirable to describe the set of resources which are located nearby or grouped together for a same purpose as a whole and abstract their common properties. The definitions of Resource Group and Resource Field element are described below:

- *Resource Group* is composed by a collection of context resources which are usually located nearby or grouped together for a same purpose.
- *Resource Field* is composed by a set of resource groups. For example, a huge sensor network which comprises a number of small internal sensor networks is a sensor field.

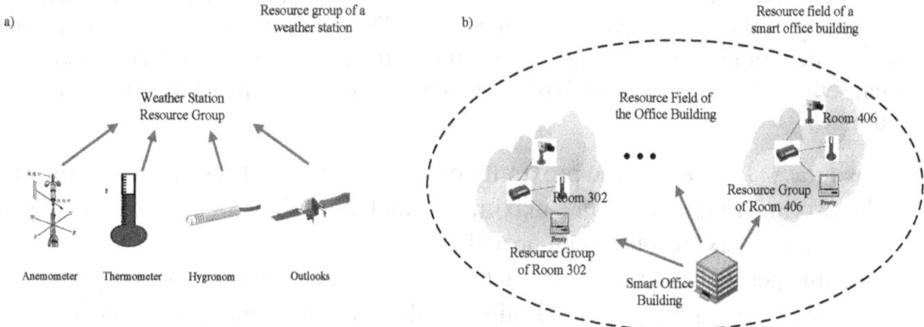

Fig. 2. a) A resource group presents a weather station and b) Resource groups and resource field of a smart office building

Fig. 2a shows a resource group example. The resource group which presents a weather station is composed of a thermometer, a hygronom, an anemometer and an agent which describes the outlooks of the weather. Fig. 2b illustrates another resource group and resource field example. In a smart office building, there is a small sensor network deployed in every room to detect the change of environment according to which intelligent apparatus such as air conditioner can adjust their operations. The left cloud encloses three sensors and a proxy is a resource group of room 302 and the right cloud represents the resource group of room 406. All the sensor groups make up a resource field (the dotted ellipse) of the smart office building.

3.2 Location of Context Resource

Location information also plays an important role in metadata description of context resource. Accurate location information not only makes context consumers locate the resource required easily, but also helps establish the connections among contexts.

However, to describe location is also a challenging job since location of context resource is various. For example, location can be identified by either a geospatial position or a logical address such as IP address. VirtualSensorML introduces four ways to describe the location information of context resources. They are:

- Common address. The most general way to represent a location with elements such as country, street, road, etc.
- Geospatial position. A coordinate pair containing the longitude and latitude.
- Network address. An IP address or a URL link indicates the accessible location of the resources.
- Descriptive location. Using descriptions and keywords to describe the location. In this case, the context resource is usually under mobile and dynamic environment which cannot use a fixed position to describe the location.

3.3 Properties of Context Resource

The quality of contextual information is closely associated with the properties of context resource. For example, for a GPS, its accuracy and sample period determine the accuracy of the coordinate pair it generated. Therefore, to describe the properties in details is beneficial for the context consumers to deal with context correspondingly. VirtualSensorML defines four basic elements to describe the properties of context resources.

- Accuracy. It indicates the proximity of the measurement results to the true value.
- Measuring range. It is the maximum and minimum of measurement results which a context resource can provide.
- Sample period. The frequency that a context resource updates its outputs.
- Certainty. It implies the credibility of the measurement results generated by a context resource.

Besides the above four properties, with respect to different context resource types, VirtualSensorML also defines some elements to describe the special properties. For instance, a video camera which provides the image contextual information may has the properties such as focal length range, optical zoom, digital zoom and resolution. On the other hand, VirtualSensorML also defines some operational environment requirement related elements such as operational temperature, operational humidity, supply voltage, supply current and power level.

3.4 Outputs of Context Resource

It is necessary to describe the outputs since different unit and data type of measurement results may have different meanings. A general example is the Fahrenheit and Celsius thermometers. What's more important, as the outputs of logical resources are always not standard, to list the detailed output contents what the context resource can provide is especially essential. Hence, metadata description of context resource written in VirtualSensorML is required to list output contents and their related units and data types.

3.5 VSML Editor

VSML Editor is a GUI tool to create and modify metadata profile of context resources according to VirtualSensorML model and schema. In VSML Editor, left column is a list of different kinds of context resources. When creating a new metadata profile, user can select a context resource type from the list, and VSML Editor will create a basic profile according to the selected type. The middle area is the main editing area, user can directly text profile elements in it. Below the editing area is a check box for syntax check according to the VirtualSensorML schema. The right side will show panels containing components such as text box, list box, radio, etc. of different parts of metadata profile such as Location and Properties for users to edit the detailed elements. See Fig. 3.

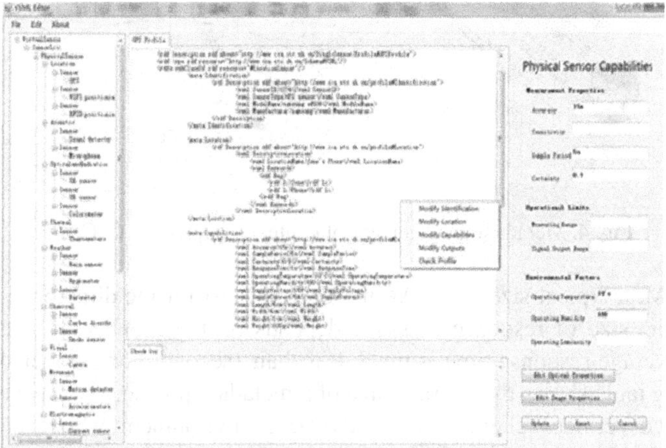

Fig. 3. GUI of VirtualSenserML Editor

4 Metadata Management

GaCAM is a multi-agent context-aware middleware which aims to provide a flexible and domain-independent support for context-aware services. The core functionalities of GaCAM are composed of context collection, context modeling, context preprocessing, context reasoning, context storage and access management and services composition and dispatch. Agents are responsible for related functionalities. The interfaces to external context-aware applications are wrapped as web services which mean applications can request the middleware for the contexts by SOAP messages.

The metadata management of context resources is achieved in the sensing agent container. An agent container contains a number of agents and manages the life-cycle of them. The sensing agent container contains all the sensing agents and is capable of creating, suspending, resuming and killing them. Every sensing agent is considered as a mediator which is connected to an external context resource and collects the contextual information. Directory of sensing agents alive maintained by the sensor agent container will be provided for the context-aware application when it connects to GaCAM. If the application is interested in a particular context collected by a sensing agent, it can query the middleware through a DescribeResource method for the metadata profile of the related context resource connected to this sensing agent. And then the application can decide whether the context meets its need. The metadata profile is stored in related sensing agent and can be accessed through SOAP request. Fig. 4 illustrates the above descriptions.

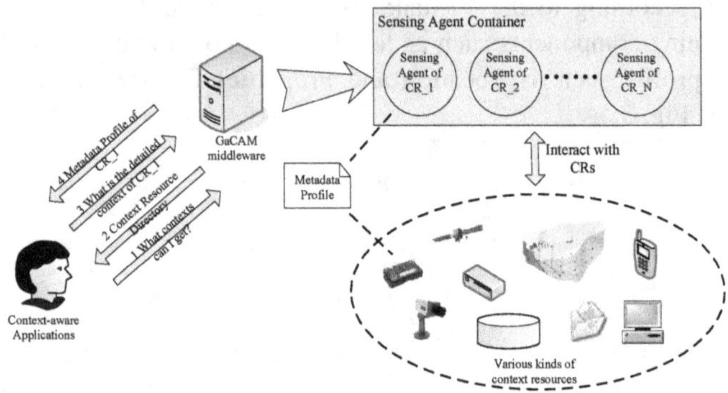

Fig. 4. Metadata management of context resources in GaCAM

In GaCAM, a typical size of SOAP message request for the directory is about 450 bytes, and the size of response message depends on the number of sensing agents alive. The average response time is much less than 1second. For the metadata profile, the following table shows the average size of a metadata profile. We have developed a location-based services client under a dynamic environment whose bandwidth is about 50k/s. Evaluation(See Table 1) shows that the response time of metadata profile arriving client is quite short. Therefore, it is reasonable to believe that metadata profile written in VirtualSensorML is efficient in context resource discovery, and improve the QoC(Quality of Context).

Table 1. Average Size of Metadata Profile and Its Response Time

Context resource type	Average size(bytes)	Average response time(ms)
Single Resource	3,679	746
Resource Group	11,745	1284
Resource Field	177,020	3209

5 Conclusions

In this paper, we present the VirtualSensorML which concentrates on providing the metadata description schema of context resources in a more general manner. And the characteristics of context resources used in context-aware computing are illustrated as well. Compared with the current solutions, VirtualSensorML is more suitable to be used to model the logical resources. In addition, the metadata management of context resources in a multi-agent context-aware middleware is described in details. The evaluation shows that metadata profiles written in VritualSensorML are efficient in resource discovery. Our future work is to explore the metadata of the services and their components in context-aware systems. We believe that a well-defined metadata management framework will certainly improve the quality of context-aware services.

Acknowledgments. This work is supported by the International Cooperation Foundation of Shanghai Science and Technology under Grant No. 09510703000.

References

1. Gu, J.Z.: Middleware for Physical and Logical Context Awareness. In: International Conference on Signal Processing, Image Processing and Pattern Recognition, pp. 366–378. SERSC Press, Jeju Island (2011)
2. Hüner, K., Otto, B., Österle, H.: Collaborative management of business metadata. International Journal of Information Management 31(4), 366–373 (2011)
3. Cathro, W.: Metadata: An Overview. Technical report, Standards Australia Seminar (1997)
4. Adel, A., Abbdellah, B.: CxQWS: Context-aware Quality Semantic Web Service. In: 2010 International Conference on Machine and Web Intelligence, pp. 69–76. IEEE Press, Algiers (2010)
5. Composite capability/preference profiles (cc/pp): Structure and vocabularies 1.0, `http://www.w3.org/TR/CCPP-struct-vocab`
6. UAProf Profile Repository, `http://w3development.de/rdf/uaprofrepository`
7. Bellavista, P., Corradi, A., Montanari, R.: Context-aware middleware for resource management in the wireless Internet. IEEE Trans. on Software Engineering 29, 1086–1099 (2003)
8. Hans, W.N., Manfred, A.J., Matthias, J., Georg, V.Z., Harald, H.: Managing multiple requirements perspectives with metamodels. IEEE Software 13(2), 37–48 (1996)
9. Sensor Model Language (SensorML), `http://www.opengeospatial.org/standards/sensorml/`
10. Ota, N., Kramer, W.T.C.: TinyML: Meta-data for Wireless Networks. Technical Report, UCB (2003)
11. IEEE 1451 Homepage, `http://ieee1451.nist.gov/`
12. Web Services Description Language (WSDL) 1.1 Homepage, `http://www.w3.org/TR/wsdl`

Research on Scientific Data Sharing Platform of Hydrology and Water Resources Based on Service Composition

Dingsheng Wan[1], Juan Tang[1], Dazheng Yu[2], Yufeng Yu[1], and Xingzhong Guan[3]

[1]School of Computer & Information, Hohai University, Nanjing 210098, China
[2] Bureau of Hydrology, Ministry of Water Resources, Beijing 100053, China
[3] Jiangxi Provincial Hydrology Bureau, Nanchang 330002, China
dshwan@hhu.edu.cn

Abstract. According to the demand for scientific data sharing of hydrology and water resources, through the introduction of Web Service and service composition development technology, we carried out a detailed design on the sharing platform from the aspects of service packaging, service registration, service publishing and service composition based on WSBPEL process. In this paper, we achieved the scientific data sharing platform of hydrology and water resources based on service composition in the private network environment of water system. The platform breaks the structure of traditional data sharing platform, and it can reassemble services to generate more complex business services, providing a basis for implementing sharing platform's development from data sharing to service sharing.

Keywords: Service Composition, Data Sharing, Web Service, WSBPEL, Hydrology and Water Resources.

1 Introduction

The scientific data sharing of hydrology and water resources in our country has achieved preliminary results since launched, and it provides query and download capabilities for the main scientific data of hydrology and water resources to the outside. The literature[1] proposes a scientific data sharing system of hydrology and water resources, the sub-node of which stores the main data regarding province and river basin as a unit, and the central node only stores part of the data; The literature[2] designs a distributed data exchange system based on metadata and data sets. In this paper we consider that the research of data sharing for water basin not only needs data, but also needs mathematical methods to process and analyze data, thus to share more rich and intuitive analysis results, and to make the scientific data sharing platform developing from data sharing to service sharing. The implementation of service sharing requires a mix of many public services working together to complete the complex application, which is unable to be achieved by the existing scientific data sharing platform of hydrology and water resources.

F.L. Wang et al. (Eds.): WISM 2012, LNCS 7529, pp. 358–365, 2012.

This article will deeply study the scientific data sharing platform of hydrology and water resources based on service composition, by regarding each node as a service provider, and the central node as a service manager and publisher, and using service to solve existing problems instead of original application. The main issues need to be solved are as following: ①The creation and packaging of functional services. Each sub-node has an independent site of hydrology and water resources data, and provides a lot of rich application functions, so we need to use different methods to package and publish the existing functions into services, and use a unified J2EE technology to create, package and publish the application functions, providing service support for scientific data sharing platform of hydrology and water resources. ②The registration and discovery of services. With the feature-rich application services, how to use them in the platform has become the main problem. This paper adopts a UDDI-based service registration and discovery mechanism, and displays a variety of services to the user for them to choose and use. ③The use of service composition. Using WSBPEL (Web Services Business Process Execution Language) to combine the services users need to generate new services, in order to accomplish specific business requirements.

2 System Requirements Analysis

The scientific data sharing platform of hydrology and water resources is based on hydrology and water resources scientific data, and it uses networks, databases, and information processing technologies to provide users with data sharing services. The existing sharing platform provides users with inquiry and retrieval functions of hydrological data based on data inquiry, but it is lack of data reprocessing function. The service composition-based scientific data sharing platform of hydrology and water resources proposed in this paper will introduce the concept of service and service composition, in order to build a set of service-based sharing platform. The sharing platform mechanism will abandon the disadvantage of developing corresponding portal sites for each node, and it just need to package business logic function provided by each node into Web Service and then register and publish it to registration center. This mechanism can reduce the development effort, but also can take full advantages of functions in the existing information systems the sub-nodes have, and can provide information services for users in water basin, without providing specific hydrology and water resources data.

The platform needs to be built as a distributed sharing platform in which platform managers, service providers and service users are active agents. This will relieve the burden on central node, but also make sub-node service useful without logging sub-node information system, to achieve single sign-on service. The service manager is responsible for the operation and management of the entire platform, and provides functional service for users and provides service registration and publishing platform for providers. The service provider registers and publishes services through the sharing platform, and provides service to users. The service user obtains services through the platform, and can use service composition according to specific needs.

3 System Design for Sharing Platform

3.1 System Architecture

The architecture of the sharing platform has two types of nodes in general: the central node responsible for the management of the entire platform, and the sub-node providing functional services. The sub-node uses the loose coupling mechanism of Web Service to build the interaction with other nodes, thus ensuring that sub-nodes can operate independently and provide services outside.

The central node has a central database which is used to store service information and user information provided by each node; the central node as the total node of the entire platform, is responsible for the management and monitoring work, to uniformly coordinate and organize the interaction between various services.

A sub-node regards river basin, provinces and cities as a unit, and exercises their respective responsibilities by providing services to the central node. Registering the various functions of sub-nodes to service registration center of central node in the way of service, and manipulating data through the user selecting services they need, can improve user's operability, and lower data interaction amount.

3.2 System Logical Structure

The sharing platform of hydrology and water resources can be divided into four logic layers: presentation layer, web service layer, business logic layer and database layer.

Presentation layer: The customer get access to the sharing platform through a browser, select different menus to enter corresponding functions, enter query conditions, and select services to conduct a detailed inquiry. All the results will be displayed to the user through page, and it also provides image showing effect.

Web service layer: Web service layer is the core layer of the sharing platform, and its main function is to accomplish Web Service with specific functions through service registration center. It contains an atomic service with single function, but also provides function of combining services produced by atomic services combination.

Business logic layer: Business logic layer is an important part of the whole sharing platform to complete business functions, and it is the interaction link of Web service layer, presentation layer and database layer. The business logic layer establishes corresponding relationship with database through data models or business rules, and provides the specific capabilities to process underlying business logic.

Database layer: The central node database is used to store service information, users-related information and other information data while the sub-node database stores hydrology and water resources data of each river basin and province and city.

3.3 System Functional Setup

According to the requirement analysis of scientific data sharing platform of hydrology and water resources, the sharing platform is divided into two parts: foreground sharing portal and background management system, as shown in figure 1. The foreground portal site provides users with sharing services, and the main function modules include query function, graph function, service registration, service composition and service execution module. The background management system is oriented to the platform administrator, providing administrator with various functions including user management function and service management function.

The sharing platform is a service-based sharing system, therefore query function, service execution and service composition module are all carried out on the basis of service registration module, and users can find and use all the registered services from service registration center, ultimately achieving the purpose of integrated sharing.

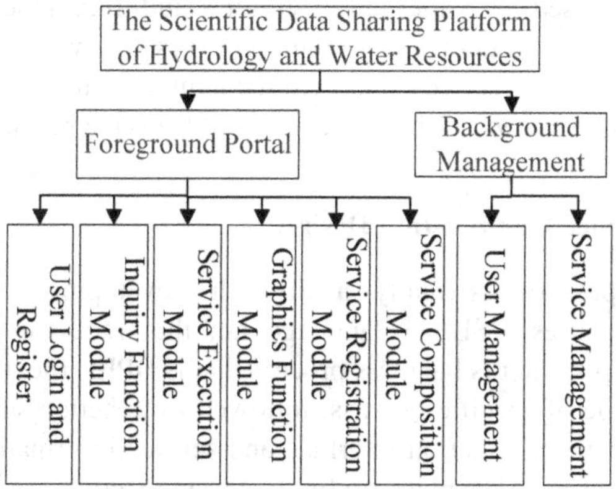

Fig.1. System Functions

3.4 Registration Center Design

The scientific data sharing platform of hydrology and water resources based on service composition needs to uniformly manage and use services provided by sub-nodes through the service registration center, then users can directly query, modify and register services from the foreground portal of service registration center; the platform administrator can audit and manage services for registration through the background of service registration center, and has the delete function on the service.

The service registration center uses UDDI mechanism, open source project JUDDI as registry software, and objects and methods UDDI4J API provides to access the registration center. JUDDI[3] is a realization based on Java and in line with UDDI specification, and it is also an open source project. The UDDI4J[4] API provides developers with two logical parts including query API and publish API. The query API searches registered service information through Find_X method, and gets information through Get_X method. The publish API publishes service information to registration center database through Save_X method.

The foreground of service registration center has two main functions: to provide sub-nodes with service registration and modification function, and to provide users with service inquiry function. The service registration page provides information which need to be filled out when registering service, including service name, service category, service binding information and service reference information, which corresponding to information required to be completed in the four data structures UDDI contains. Service query page provides function to query services according to different classifications, such as their names and types. Service modification page provides users with the function of modifying service information, such as when service name of sub-node changes or service deployment address changes.

The background of the service registration is administrator-oriented, and by returning the registered service list, administrators carry out the management work of service. It provides the review and deletion function of services. When the sub-nodes register service to the service registration center, the service will not immediately be used, but it will be used after the administrator's auditing. When the administrator enters the service management background, the system will prompt all the latest registered but unapproved services, and then the administrator can review and publish them and can delete the service which does not meet service requirements.

3.5 System Service Composition Design

The service composition is really a difficult development technique. Service composition usually uses WSBPEL language, and the process in WSBPEL language contains two concepts: First is business process that is BPEL file, the file describes the implementation logic of specific process; followed by function service that is WSDL file, the file describes the function interface and service information when process is called to be a Web service. The platform firstly gets registered services from the service registration center through the service selection module, and after selecting the needed services, the core module of the service composition combines and generates services to corresponding process file and service description file, and deploys to the process execution engine for users to call, as shown in figure 2.

The design of service selection module: The service selection module provides the selection of registered service, and when the user need service composition, they query the registered service from service registration through service selection module, and select services they need to be combined, and the service selection module provides a graphical display function.

The design of service composition core module: The core module of service composition is the core to complete the functions of service composition, and its main function is to combine and generate services selected by users to process documentation according to WSBPEL specification, and at the same time generate to corresponding service description document. Process documentation will be deployed to the process execution engine to perform, and the service description document will be deployed to Web server. The functions user use will be obtained from the service description document. This paper studies the service composition approach according to the idea of literature[5], and the ideas are shown in figure 3.

The entire combination function is divided into two parts: service definition and process definition, and they are defined at the same time.

Process definition is a developing and editing process. Firstly after the user enter service composition module, it will define the name of process automatically; In the platform this paper studies, the input of service composition is scientific data of hydrology and water resources, therefore we treat data query process as the start-up activity<receive>of process, define <variables> according to the data structure service definition provides, and introduce the service that user select as a <partnerLink> into process, and to be <invoke> object of basic activities; Finally, after arranging activities according to a certain logic relationship, through <reply> activity it forms a complete process execution, and publish it to the process execution engine.

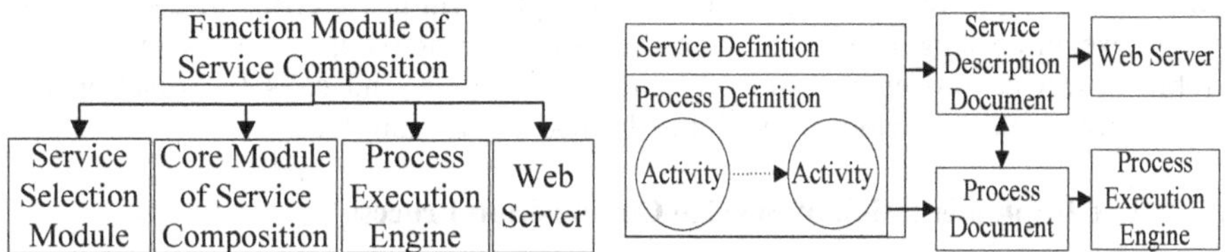

Fig. 2. Service Composition

Fig. 3. Core Module of Service Composition

Service definition is defined at the same time with process definition, and the specific feature is to package defined processes to services and release them. First, <import> selected service to process service and define it as a <partnerLinkType>, for process to use; then define the data structures<types> and <message> needed in the process service, providing types for the definition of service variables; finally, bind process as a service and define port, providing call interface for the user, and deploy the service to the Web server at the end of the service definition.

4 The Implementation and Application of Sharing Platform

4.1 System Implementation

The function of the sharing platform can be achieved through service and service composition, and from the point of system's versatility, it mainly needs to implement service packaging, service publishing, service registration, service composition and other functions.

Service Packaging: After the system overall analysis and the analysis of the existing system of sub-nodes, extract and package the existing function of sub-nodes, and create the lacking services. In accordance with the requirement analysis and design, the key services: metadata query service, basic data query service, real-time data query service, regression analysis service, neural network service, multivariate statistical analysis service, time series analysis service, spectrum analysis service, wavelet analysis service, charts service and so on.

Service Publishing: After the service packaging we need to publish the service to the network for users or developers to use. The water system uses a separate network; therefore we only need to deploy service to the hydraulic private network. All the services packaged by J2EE technology in scientific data sharing platform can be deployed to Web server just as Apache Tomcat, and can be deployed to Microsoft IIS server using .NET platform and DCOM / DLL packaged services.

Service Registration: Service registration is to store the service name, service source and service address URL information as a service description to a database, and users view and select the service they need to use through the page. This article uses the open source service registration center JUDDI and Oracle database to complete the registration and storage of services, and uses the services UDDI4J API provides to carry out service registration, query, delete and other operations.

Service Composition: The function of service composition is to allow users to complete corresponding function by selecting registered service and combining them. The service composition features of the sharing platform uses WSBPEL 2.0 standard process execution language to combine the atomic service, deploys and executes the combined service processes using Apache ODE server, and calls the function the composition service provides through the defined service interface.

4.2 The Implementation of Service Composition Process

The specific process of service composition is shown in figure 4, as follows: ①Users select the service queried from the service registration center when enter the service composition function. ②After selecting the desired service, the backend system will automatically enter the process editing and service editing functions: define process file, including the process name, namespace and so on, this function is done automatically in the background; define service description document, including the service name, namespace, and so on, this function is done automatically in the background. ③Define elements for processes and services: define the link of process, variables and activities; define data structure, messages and actions of the services. ④Arrange activities in a certain order, to produce process documentation; Package the operation offered by process documents, and define the service call interface. ⑤Deploy process documentation (BPLE document) to the process execution engine

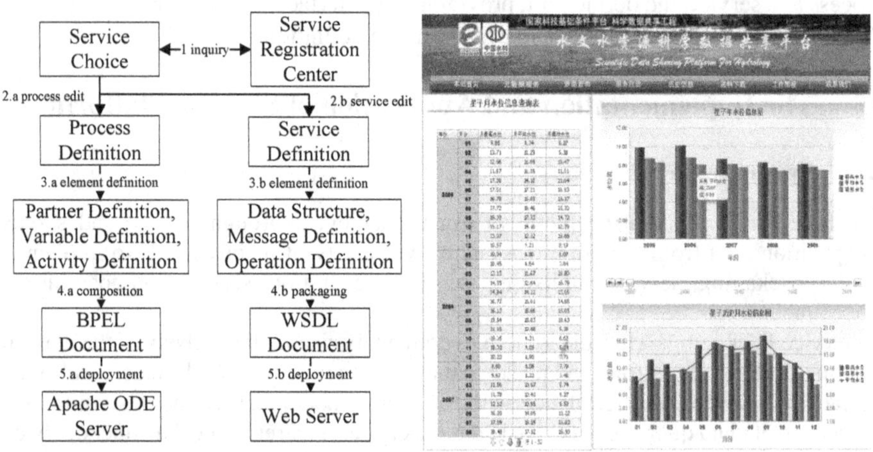

Fig. 4. Service Composition process

Fig. 5. Display of Executive Result

Apache ODE server; Deploy service description documents (WSDL document) to the Web server (Tomcat).

4.3 System Application Examples

The system defines the inquiry service of monthly water level characteristic value, the statistical chart service of yearly water level information and hydrological characteristics over the years and months of the hydrological stations. First through service registration center, we register the service, and display it to users; then users select and combine the service. Users first select the inquiry service of monthly water level characteristic value, which provides data query function and provides data source for chart service; then users choose the chart service of yearly water level information and hydrological characteristics over the years and months to analyze and show query results. Figure 5 shows the query statistics of 2005-2007 years and months' water level data of Poyang Lake Stars Station with this system.

5 Conclusion

In recent years there are more and more studies on service composition, and the developed composition methods and techniques are endless, especially in the aspect of service dynamic combination and semantic combination. The scientific data sharing platform of hydrology and water resources based on service composition in this paper only combines existing Web Service to generate more complex business logic functions, and there are many aspects need to be further studied.

Acknowledgement. This work is supported by National Natural Science Foundation of China (No.51079040), National Science and Technology Infrastructure Construction Project（2005DKA32000） and '948 Projects' of MWR (Ministry of Water Resources P. R. China)(No.201016).

References

1. Zhang, S., Cheng, Y., Wu, L., Zhu, X.: Hydrology and Water Resources Science Data Sharing Practice. In: Proceedings of Chinese Hydraulic Engineering Society 2006 Academic Annual Conference and 2006 Hydrology Academic Symposium (2006)
2. Geng, Q., Miao, G., Xu, H., Zhu, X.: Framework and Realization of Water Scientific Data Sharing Network. Water Conservancy and Hydropower Technology 39(5), 85–87 (2008)
3. Apache juddi(EB/OL), http://juddi.apache.org
4. UDDI4J(EB/OL), http://uddi4j.sourceforge.net
5. Zhu, Y.: Research on Key Technology of Earth System Scientific Data Sharing. Science Press, Beijing (2009)

Study on the Scheme of Tianjin Area E-commerce Platform Construction

Yunbiao Gu

School of Automation, Harbin Engineering University, Harbin 150001, China
guyunbiao@sina.com

Abstract. The paper puts forward the concept of e-commerce platform construction in Tianjin for the first time and make an analysis and a prediction of its development trend on the basis of analyzing the characteristics of the development of Tianjin Area. It presents a mode of "3+1" construction. This paper also conducts an analysis of the potential economic benefits, and gives some advice on designing its operation and management mode.

Keywords: electronic commerce, regional e-commerce platform, city platform, industrial chain area.

1 Macro Environment Analysis of the Development of E-commerce

E-commerce, which was emerge lately, undergoes a profound change in the traditional business model, such as the Internet, cloud computing, Internet of things. With the growing development of IT technology, E-commerce, marked with the network shopping, develops with an average speed of more than 50%[1].

The internet has removed the regional limits and will expand market into the world. Thus, the previous customers divided by regions are integrated, which put the Internet to the ideal of space integration. The advantage of city location declines rapidly, global operation becomes easier, and mobile office can be seen everywhere. The Internet opens the never-ending market for enterprises—7×24 hours business and home office, infinite operation time. The Internet powerful information processing capacity, low communication cost makes it easier to find out the hidden customers in small groups. Therefore, many of the small customers are pulled into the market, causing the market capacity to expand.

The electronic commerce makes resources flow freely and smoothly in the world. The only geographical barrier has been broken through by "siphon" principle. Regional economic "siphon resources" phenomenon began to appear, arising from virtual gathering. As long as there is a pressure difference of price, service, quality, resources will flow to aggregation area, and the capital, assets, personnel resources will be globally relocated. Entity competitiveness and virtual industry chain appear. Through

F.L. Wang et al. (Eds.): WISM 2012, LNCS 7529, pp. 366–371, 2012.

the "siphon resources", small businesses picks up the power to compete with big enterprises, and remote areas obtain the direct opportunity to compete with the developed area.

Cloud computing and three nets fusion have been included in speeding up cultivating and developing strategic new industry by the State Council. Beijing, Shanghai, Shenzhen, Hangzhou, Wuxi have become a cloud computing services pilot city. This marks the launch of the cloud computing in china. Three nets fusion will also further facilitate shopping on the Internet, and promote the rapid development of electronic commerce[2].

2 Developing Electronic Business Is the Best Choice for the Economic Transformation of Tianjin Central Area

From the perspective of entity economy, although Tianjin manufacturing industry is relatively complete and of many categories, yet because land resources are relatively scarce, the industrial structure is not rational, independent development ability is weak, etc. There is a restriction to the development of entity economy. The main economic form is to develop modern high-end service. It is imperative that we should develop e-commerce. Through the development of the regional industry and the electronic commerce, the national advantage resources can be attracted to Tianjin via the network and in turn special industry in Tianjin will be introduced to the whole China and even the world. The electronic commerce will greatly change the downtown Tianjin industrial structure and drive the development of all the industries to go hand in hand. Therefore, the development of electronic commerce is the best choice for the economic transformation of Tianjin central area.

3 The Content of Tianjin Area E-commerce Platform Construction

Tianjin e-commerce platform content can be summarized as "3+1"mode, namely the three kind of platform—city platform, the platform of characteristics and advantage industries; a area—the e-commerce industry chain area.

3.1 Construction of Three Electronic Commercial Platform

City Platform

City e-commerce platform is a comprehensive portal website which integrates local network shopping, classification of information, social business network, the micro median and so on, serving the area.

First of all, city e-commerce platform will provide an entry for all industries and trades, display their information platform and transaction navigation; Second, the city platform provides commercial information services, such as information trading, local shopping, social business relationship for small and medium enterprises. Third, the city platform provides personal services, such as entertainment, life consumption, making friends, communication and knowledge consulting, etc for end-consumers.

Tianjin city e-commerce platform, part of the digital city, lays a solid foundation for constructing an intelligent Tianjin. The construction of Tianjin electronic commercial platform will provide local enterprises with B2B, trade and trade service information B2C; enterprise and individual, social business network, micro media, cloud technology application service; the city, a general electronic business service.

Industry Platform

Industry platform gathers the advantage industries of the whole city and the enterprise resources of all levels in the city and around the country. It extends transaction collaboration to downstream distributor and upstream suppliers and provides services for industries and enterprises like network trade, online payment, market quotation, brand alliance, online logistics, centralized purchasing, online bidding, business management, business in coordination, instant messaging and mobile business services. It is the leading domestic Internet industry market with advanced concepts and technology.

Tianjin has some advantage industries like electronics, metallurgy, petrochemicals, auto and textile, some of which are highly competitive in the nation. By selecting some prominent ones with characteristics and potential capacity to carry out e-commerce, it is possible to make targeted industry type electronic commercial platform, provide basic services such as trading, management and coordination, online payment, mobile business and instant communications, and other basic services, and provide value-added services like finance, enterprise group buying, marketing, integrity certification, talent recruitment, data, etc.

The development adopts incremental, gradually-expanding strategies, based on local, province, gradually to the nation, thus advancing into the world. Consequently it will accelerate the virtual gathering of the industry and build vertical clustering electronic business platform.

Personalized Platform

Building several personalized platforms (such as chain type platform: chemicals, clocks, cables, etc) refer to online wholesale (B2B), online retail (B2C), online purchase, and Vendor Managed Inventory (VMI), Distributors Managed Inventory (DMI) so that new economy mode integrated with virtual gathering and entity gathering is achieved.

The whole chain integration electronic commerce is based on supply chain industry, centers on the leading enterprises of the supply chain, and gathers each section of the supply chain (including upstream supplier, downstream dealers and end-customers). It provides them with the "one-stop" work style order in coordination, settlement in coordination, logistics synergy, sales, inventory coordination, collaborative purchasing services. The whole chain integration electronic commerce platform for both the wholesale and retail trade pattern is the new B2B and B2C e-commerce platform[3].

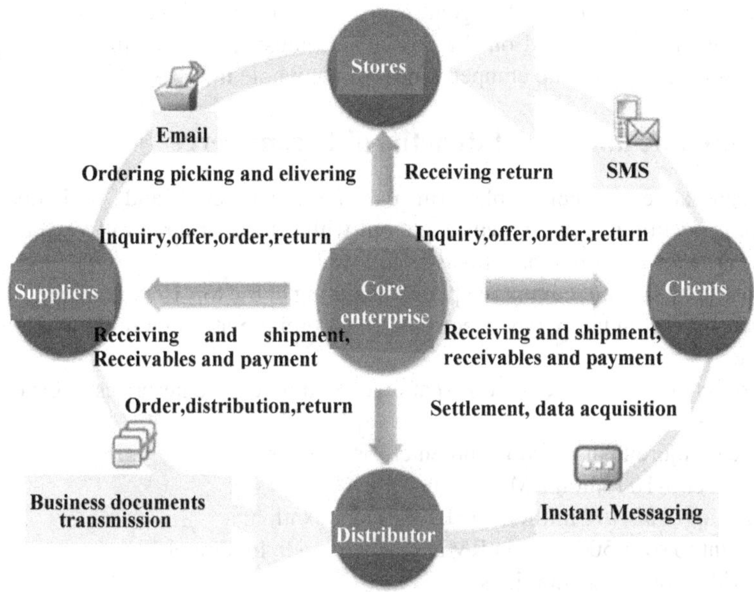

Fig. 1. The Scheme of the Use of Supply Chain

3.2 Construct Integration Electronic Commerce Chain Area

Once the time arrives several years later, it is possible to build a integration electronic commerce chain area to rally enterprises related to the electronic commerce, industry, government, association, cloud computing and financial supporting service resources. It consists of four parts: to build characteristic regional headquarters and enterprise base; to construct integrated electronic commerce association community so as to gather e-commerce industry interest-related resources; to construct "SOHO For IEC" community to bring in high-end pioneering talents; to build "three-dimensional shop" to gather high-end products and business popularity[4].

3.3 Construct E-commerce Industry Service System

Through the e-commerce industry platform, national industry resources can be effectively integrated and cooperation between enterprises of all levels of supply chain can be achieved so as to transform the product competition form from "enterprise and enterprise" competition to "supply chain and supply chain" competition. Meanwhile, high-level cooperation between chain members will lead to simplification of procedure and improve the long-term competitiveness of the supply chain.

From the perspective of global supply network, constructing industry e-commerce platform for small and medium-sized enterprises in Tianjin, namely global supply chain collaboration platform, can not only meet the demands of present enterprises but also help those enterprises to seek new business opportunities and business partners in the global range, to participate in global trade and to promote the high-speed development of enterprises. Our goal is to construct e-commerce platform with industry characteristics, based on the advantage industries of Tianjin, and integrate industry resources to build up competiveness of the whole industry.

4 Economic and Social Benefits of E-commerce

If the regional e-commerce platform can be constructed and well operated, considerable economic benefits will come out with efforts of three to five years. The overall forecast of economic benefits is as follows:

To construct Tianjin e-commerce platform aiming at the top 15 nationwide.

To build several industry type electronic commercial platform aiming at the top 10 nationwide.

To support and cultivate several national e-commerce enterprises registered in Tianjin.

To develop more than 20,000 valid enterprise clients.

To turn over 5-10 billion RMB yearly.

To reach up to 50-100 million RMB yearly in taxation.

To amount to over 500 million RMB yearly in trade fund pool.

To provide more than 5000 jobs.

At the same time, considerable social benefits can be brought out with the platform construction and operation and the area construction and investment invitation: The first is to develop three-dimensional economy and "siphon" advantage resources, thus keeping the existing resources and bringing in resources in other provinces and cities (such as products, capital, taxation, talents, etc) ; The second is to improve business environment and boost the development of enterprises. Small and medium-sized enterprises get development with e-commerce, large enterprises get industrial upgrading, local economy structure gets optimization and growth mode gets changed; The third is to rely on e-commerce to promote the regional science and technology level. E-commerce will drive the network, cloud computing, enterprises applications, etc so as to enhance Tianjin's science and technology.

5 Conclusions

In the thesis, it is put forward for the first time that Tianjin area e-commerce platform construction scheme will realize the integrating development of city platform, characteristic industry platform, the advantage industry platform and whole e-commerce industrial chain area. Through the reasonable management model and operation mode, it is bound to bring considerable economic and social benefits and promote sound and fast economic and social development of TianjinConclusion

In the essay, it is put forward for the first time that Tianjin area e-commerce platform construction scheme will realize the integrating development of city platform, characteristic industry platform, the advantage industry platform and whole e-commerce industrial chain area. Through the reasonable management model and operation mode, it is bound to bring considerable economic and social benefits and promote sound and fast economic and social development of Tianjin.

References

1. Marshall, L.F.: What is the Right Supply Chain for Your Product? HBR (Match-April 1997)
2. Koller: Discriminative Probabilistic Models for Relational Data. In: Proceedings of the 18th Conference on Uncertainty in Artificial Intelligence, pp. 485–492 (2002)
3. Karray, F., Alemzadeh, M., Saleh, J.A., et al.: Human-Computer Interaction: Overview on state of the Art. Inter. J. Smart Sensing and Intelligent Systems, 129–136 (2008)
4. Shi, Y., Zhang, Y.: Practice of Electronic Business Management. China's Water Power Press (2009)
5. Discussion of Government Resource Plan, http://www.echinagov.com

Research on Chinese Hydrological Data Quality Management

Yufeng Yu[1], Yuelong Zhu[1], Jianxin Zhang [2], and Jingjin Jiang [3]

[1] College of Computer & Information, Hohai University, Nanjing 210098, China
[2] Bureau of Hydrology, Ministry of Water Resources, Beijing 100053, China
[3] College of Jin Cheng, Nanjing University of Aeronautics & Astronautics, Nanjing 211156, China
yfyu@hhu.edu.cn

Abstract. Data quality has become increasingly important in information constructions and low data quality will influence the decision-making process related to design, operation, and management of hydrology application. Although many researches could be found that discuss data quality in many areas, few literature exist that particularly focuses on data quality in the field of hydrology. In this paper, we first analyze the key dimensions such as completeness, consistency and accuracy of hydrology date quality, and then propose an efficient date quality management framework based on those dimensions. Moreover, a general date quality assessment model to assess the data quality in these dimensions is also provided. At the end of paper, we proposed a series of methods and techniques to improve the data quality in hydrology database, and carried out in practice to prove it.

Keywords: Data Quality, Assessment, Improvement, Dimension.

1 Introduction

With the widely use of large-scale database systems and auto-acquisition instruments, huge amounts of data is generated and stored in hydrological departments. Such tremendous growth of hydrological data, however, has exacerbated two existing problems. On the one hand, the fast-growing amount of data has far exceeded our human ability for comprehension without powerful tools. Consequently, more effective techniques and methods should be developed to transform such data into useful information and knowledge, which is known as the area of data mining and knowledge discovery. On the other hand, researchers oftentimes face with the lack of high quality data that fully satisfies their needs. These two problems are actually closely related: of high quality data is indispensable for data mining techniques to extract reliable and useful knowledge.

The topic of data quality is covered in many literatures. Data quality is the extent of data to meet specific user expectations [1]. In Literature [2], the quality of the data mainly refers to the extent to which an information system achieves the coherence

F.L. Wang et al. (Eds.): WISM 2012, LNCS 7529, pp. 372–379, 2012.

between models and data instances, and to which both its models and instances achieve the accuracy, consistency, completeness and minimal nature. Moreover, there are a number of literatures which research spatial data quality, statistical data quality, manufacturing aspects of data quality, and so on [3, 4].However, as is indicated by Kulikowski [5], there is no universal set of data quality descriptors as well as authoritative standard model and algorithm of data quality management exists.

Among various application fields, hydrology is an area where data quality bears particular importance. As the basis of water resource management decision making, hydrological data is useful only when it is of good quality. For flood forecasting and control, drought relief, or other forms of water conservation and utilization activities, any decision making based on low-quality data might result in some unwanted, sometimes even disastrous consequences.

In the past few years, there are a large number of researches discussing data quality in hydrological informatics [6,7,8]. However, existing works in this aspect basically discuss the data quality issues in the context of hydrologic survey. To our knowledge, few data quality researches are published in the background of hydrological database quality assessment and improvement. This paper is an attempt to introduce the experience and methods accumulated in years' practice, which might be beneficial to hydrological workers and researches with regard to data quality issue.

2 Key Dimensions of Hydrological Data Quality

The scopes of data quality dimensions to consider are varying according to applications. With the years' experience of data collection, integration and analysis in hydrological field, we regard the following data quality dimensions are worth more attention: completeness, consistency, and accuracy. It should be noticed that the discussion of these dimensions here does not mean their superior importance than others. They are highlighted because most data quality problems in hydrological application domains belong to these dimensions. Consequently, we focus on these three dimensions in this section and introduce them one by one as below.

2.1 Completeness

Completeness is defined as the degree to which a given data collection includes data describing the corresponding set of real-world objects. Completeness often includes three types: schema completeness, column completeness and population completeness.

In relational data model, the definitions of completeness can be provided by considering the granularity of the model elements, i.e., value, tuple, attribute and relations.

1) A value completeness, to capture the presence of null values for some fields of a tuple;

2) A tuple completeness, to characterize the completeness of a tuple with respect to the values of all its fields;

3) An attribute completeness, to measure the number of null values of a specific attribute in a relation;

4) A relation completeness, to capture the presence of null values in a whole relation.

In the research area of hydrological data quality, completeness is often related to missing values processing.

2.2 Consistency

Consistency refers to the violation of semantic rules defined over a set of data items. With reference to the relational theory, integrity constraints are a type of such semantic rules. In the statistical field, data edits are typical semantic rules that allow for consistency checks.

In the issues of hydrological data quality, consistency often refers to data dependency, which can be divided into three categories: key dependencies, inclusion dependencies and functional dependencies.

Key dependency describes the uniqueness of a table attributes, which can be represented with a set of attributes directly. Given a relationship instance R with a set of properties $A=\{A_1, A_2 \ldots A_n\}$. If K, a subset of A, is subject to key dependencies, then all the items containing K data tuple must not be repeated. Key dependency can help us identify and eliminate duplicate data.

An inclusion dependency shows an attribute A contains in another attribute B of relation R_1, or contains in attribute C of relation R_2. Inclusion dependency can be described formally as *A.value* IN *B.value*.

Functional dependency represents attribute A of relation R has function relationship with another attribute sets $\{B, C, \ldots\}$. Functional dependency can divide into explicit function dependency and implicit function dependency, the former indicates an attribute has clear function relationship with one or more other attribute(s), while the latter reveals the attribute has logical associations with other attribute(s).

2.3 Accuracy

Accuracy is fundamental of data quality and defined as the degree of correctness and precision to which a data value correctly represents the attributes of the real-world object. In general, accuracy should at least be reached to a point where the risk of misinterpretation as a result of the use of the data is minimized.

The major constraint of this minimization is that exact measurements are often prohibitively expensive, especially in light of the fact that the cost due to errors is very hard to quantify· Errors affecting the accuracy of data measured in the field of water resources can be isolated, which are local errors that may occur at regular or irregular intervals (i.e., outliers), or persistent, which are errors that are propagated over some intervals (i.e., shifts or trends). Both isolated and persistent errors can either be random, which means that they are not the result of identifiable structures, or systematic, which implies that they follow some known structure. Origins of errors include factors due to anything from malfunctioning measurement instruments (i.e., technical limitations, condition) to errors in the processing of the data measured (e.g., coding errors, typos). Therefore, the issues of data accuracy mainly focus on detecting anomaly with respect to distinct features or patterns.

3 Hydrological Date Quality Management Frameworks

The problems mentioned in the previous section are the main obstacles in hydrology on the way to high data quality. It is extremely important to manage the data quality for data analysis and knowledge discovery. Seminal works in the area of data quality have defined various extensive frameworks to review systems within organizations. Essentially, a data quality framework is a tool to define a model of its data environment, identify relevant data quality attributes, analyze data quality attributes in their current or future context, and provide guidance for data quality improvement.

However, the date quality issue often involves the knowledge dependence and application independence, it can't expect complete all the date quality issue at once. The date quality management framework should follow such a process: assessment - analysis - improvement - assessment.

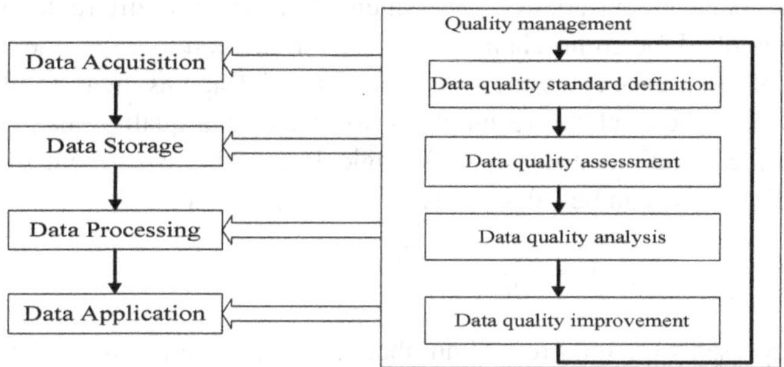

Fig. 1. Date Quality Management Framework

3.1 Data Quality Assessment

The main task of data quality assessment is to judge the level of data quality by choosing appropriate method to calculate the quantitative score of the key dimension. Therefore, data quality assessment is the foundation and necessary prerequisite to improve data quality. It can give a reasonable assessment to the whole or part data quality status, which can help data consumer to understand the data quality level of the application system, and then to take appropriate process to improve data quality.

Combined with structure of data quality assessment system and requirements of hydrology data quality assessment, it can establish assessment model of hydrology data quality as following:

$$M=\{P,D,R,W\} \tag{1}$$

P is the perspective of data quality assessment, which can be divided into two categories: new production data of the system and historical data in the database. D represents the assessment dimension of data quality, which includes a (accuracy), c

(completeness) and x (consistency). R is the assessment algorithm or rule for each dimension. W is attribute weight of different dimension and rules. Figure 2 shows an all data quality assessment model.

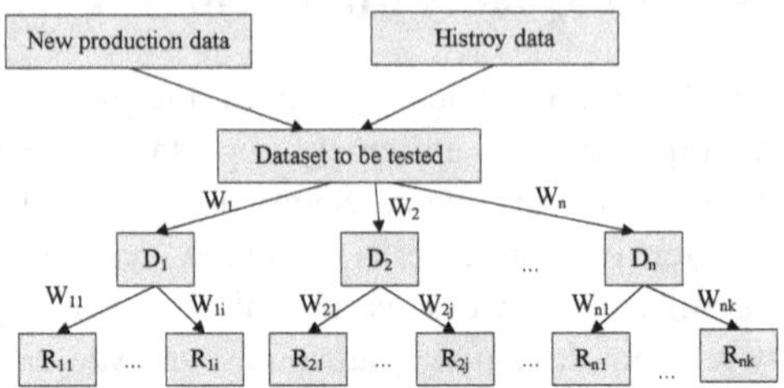

Fig. 2. Data Quality Assessment Model

The methods of data quality assessment generally include direct assessment, indirect assessment and comprehensive assessment [8]. Since the requirement of data quality and accuracy is different among the foreground, background and the department of management-control, assessment standards are different. Therefore, in this paper we applied the comprehensive assessment method to assess the data source.

Generally, it often uses assessment scores, which defined as the ratio of the expect records number to the total record numbers, to assess data quality quantificationally. According to the actual situation of hydrological data, the assessment score S_i of ith data quality dimension can be calculated as following formulas.

$$S_i = \sum_{j=1}^{n}\left(W_j * \frac{N_j}{M_j}\right) * 100 \tag{2}$$

Here N_j represents the total record numbers that meet the jth rule of that dimension and M_j stands for the total record numbers of given dataset. W_j is the weight for the jth rule of the ith dimension and its value be signed in (0, 1), furthermore, $\sum_{i=1}^{n} W_i = 1$. And n is the total rule numbers of the ith dimension.

Based on the results of the individual dimension data quality assessment, the overall score of data quality assessment can be calculated as follow:

$$S = \sum_{i=1}^{k} W_i * S_i \tag{3}$$

Here k is the total dimension numbers of the data to be assessed and W_j is the weight for the ith dimension, which's value also be signed in (0, 1),and $\sum_{i=1}^{n} W_i = 1$.

3.2 Data Quality Improvement

The technology for date quality improving mainly involves two aspects: schema level and instance level [9]. The former focus mainly on how to understand data schema and then to redesign data schema based on existing data instances, while the latter usually adopts data cleaning, which applies duplicate detection, anomaly detection, logical error detection, missing data processing and inconsistent data processing to achieve data quality improvement and enhancement.

1) Duplicate object detection
Duplicate detection tries to match inexact duplicate records that refer to the same real world entity but are not syntactically equivalent. The duplicate problem is tackled using a pairwise approximate record matching algorithm, which sorts the records in the database and then compares the neighboring records to detect whether the record-pairs to be identical or not.

2) Missing values processing
The actual hydrological data often has numbers of missing data, which makes effect to the analytic result. The main purpose of missing value imputation is to avoid analytical bias caused by large numbers of missing value in the data analyzing phases.

It adopts different methods to process hydrological missing value according to the missing location. Here uses the water level as an example to show different types of missing value processing:

DT (Date) missing: Fill the appropriate value based on the context record. For example, a certain station has a missing date value in the daily water level table (See Table 1). It can directly use the date value "1997-2-23" to fill the missing value.

AVZ (Water level) missing: Uses the average water level for many days to fill the missing value. For example, if the water level of the third record in table 1 is missing, it can use the value "2.83"- the average of 2 days before and after the missing records- to fill this missing value.

Record missing: If the entire record is missing, it fill missing DT and AVZ with appropriate value according to the above two cases respectively.

Table 1. Daily water level of a certain station

STCD	DT	AVZ
63201850	1959-9-3	2.82
63201850	1959-9-4	2.83
63201850		2.82
63201850	1959-9-6	2.84
63201850	1959-9-7	2.83

3) Outlier detection
Outliers are observations that appear to be inconsistent with the remainder of the data or deviate remarkably from the bulk of the data [10], for example, an extreme high flows in streamflow analyses. Simply discarding outliers could lead to underestimation of the estimated model but preservation of outliers normally complicates the statistical analyses.

We here use the mean value of a neighborhood of data points to determine whether a particular data point is an outlier in hydrological datasets. Given a water level time series $D=<d_1=(v_1,t_1),\ d_2=(v_2,t_2),\ldots d_n=(v_n,t_n)>$,where point $d_i=(v_i,\ t_i)$ stands for the observation v_i at the moment t_i; defines neighborhood point set $\eta_i^{(k)}=\{d_{i-k},\ldots d_{i-1}, d_{i+1},\ldots d_{i+k}\}$ for the point d_i,. In order to detect point d_i as an outlier or not, compute the mean of neighborhood point $m_i^{(k)}$.Then Calculate the absolute error $e_i^{(k)}$ between $m_i^{(k)}$ and v_i and compare it to a specified threshold τ, If $e_i^{(k)} <\tau$, keep v_i ,otherwise label x_i as an outlier and replace v_i with $m_i^{(k)}$ to obtain a clean time series.

4) Inconsistent data processing

An inconsistent data will occur when integrates data from several independent data source. Considered from the data itself and the relationship between data entity, inconsistent data processing can be divided into single-table data consistency check and multi-table consistency check.

Single-table consistency check use to improve the single table's data consistency. It recalculates the value to be assessed according to the relationship between certain elements of the same record in the table and then compares it with the corresponding value stored in the database. If these are inconsistency between them, audits and corrects stored value with calculated value artificially.

Multi-table consistency check use to improve hydrological elements between different tables. It recalculates month's (year's) statistics value according to the relationships between daily value and monthly (yearly) value, and then compares them to the data values stored in the database. If these are inconsistency between them, audit and correct stored value with calculated value artificially.

4 Result

The experimental result of data quality improvement in in a provinces' hydrological database is shown in figure 3.

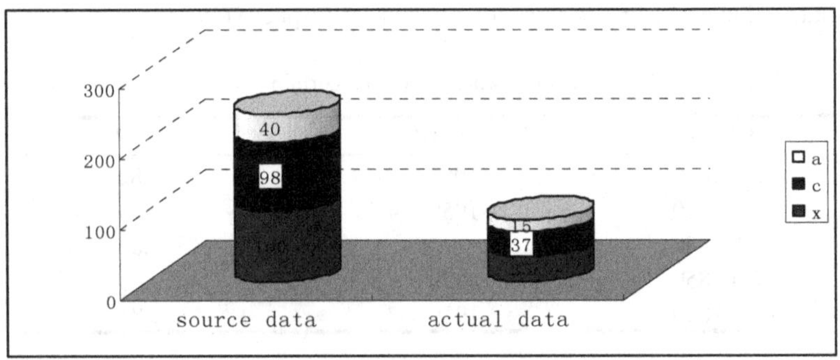

Fig. 3. Date Quality improvement result

In the figure, blue is on behalf of the consistency problem, black is on behalf of the completeness problem and white is on behalf of the accuracy problem. From this picture it can be seen that after the improvement process, the data quality problems were significantly improved. To analyze specifically, the frequency of problem data in the data table declined from 238 times to 87 times, and basically meet the user's requirement.

5 Conclusion

Data quality is a key issue in hydrology information. Due to the high domain-specificity of data quality, it is of great necessity to identify key dimensions of data quality in hydrological. In this paper, three data quality aspects are highlighted: consistency, completeness and accuracy. Moreover, methods and techniques to handle these data quality problems are also presented in this paper.

Although this paper is focused on these three dimensions, it does not mean their superior importance than other aspects. They are highlighted because most data quality problems found in hydrological practice in past years could fall into these dimensions. This paper is expected to help the researchers decide to acquire data with satisfactory quality for more reliable analysis and decision making.

In the future work, we will study some other assessment standards and algorithms which will enhance the data quality. And then the data quality in hydrology will be improved more and can gradually realize the goal of decision support.

Acknowledgement. This work is supported by the Natural Science Foundation of China (No.51079040) and National Science and Technology Infrastructure of China (No.2005DKA32000).

References

1. Cappiello, C., Francalanci, C., Pernici, B.: Data quality assessment from user's pespective. In: IQIS (2004)
2. Aebi, D., Perrochon, L.: Towards improving data quality. In: Proc. of the International Conference on Information Systems and Management of Data, pp. 273–281 (1993)
3. CaoRuichang, Chien-Mingwu: Information Quality and its evaluation index system. Information Research 84(4), 6–9 (2004)
4. Yu, M., Bo, P.: Information gap and its application to the evaluation of the product information quality. Chinese Mechanical Engineering 15(17), 1557–1561 (2004)
5. Kulikowski, J.L.: Data Quality Assessment, Encyclopedia of Database Technologies and Applications, pp. 116–120. Idea Group (2005)
6. Cole, R.A.J., Johnston, H.T., Robinson, D.: The use of flow duration curves as a data quality tool. Hydrol. Sci. J. 48, 939–951 (2003)
7. Petersen-Overleir, A., Soot, A., Reitan, T.: Bayesian Rating Curve Inference as a Streamflow Data Quality Assessment Tool. Water Resour. Manage. 23, 1835–1842 (2009)
8. Yang, X., Liu, Y.: Data mining based study on quality of water level data of Three Gorges Reservoir Automatic Dispatching System. Water Resour. Hydropower Eng. 42(11), 98–101 (2011)
9. Wang, C., Ma, K.L.: A statistical approach to volume data quality assessment. IEEE Trans. Vis Comput. Graph. 14(3), 590–602 (2008)
10. Knorr, E.M., Ng, R.T.: Algorithms for Mining Distance-Based Outliers in Large Datasets. In: Proc. of VLDB 1998, pp. 392–403 (1998)

Using IoT Technologies to Resolve the Food Safety Problem – An Analysis Based on Chinese Food Standards

Yun Gu[1], Weili Han[1,3], Lirong Zheng[3], and Bo Jin[2]

[1] Software School, Fudan University
{11212010011,wlhan}@fudan.edu.cn
[2] Key Lab of Information Network Security of Ministry of Public Security
jinbo@stars.org.cn
[3] Wuxi Institute of Fudan University
lrzheng@fudan.edu.cn

Abstract. Despite the complicated conditions in the food supply chain, we identify the risks in it based on the food related standards in China. We analyze the standards and summarize the types of the risks, then map them into the five main phases in the food supply chain, which are production, processing, transportation, storage and sale. The classification will help us to find the key risks in each procedure and take measures to mitigate them. We also investigate how to use the Internet of Things to mitigate the risks. Some of them can have been solved and some of them are solvable. However, we argue there are still some risks cannot be covered by the IoT technologies. We believe this research can help other researcher find how well the IoT technologies can resolve the food safety problem.

Keywords: Food Safety, Internet of Things, Risk, Food Supply Chain.

1 Introduction

Food safety received a lot of publicity after several food accidents occurred around the world, such as the "2008 Chinese Milk Scandal" [1]. How to keep the food safety in the complicated food industry is a headbreaking problem for both governments and common users. Fortunately, researchers around the world try to employ the Internet of Things (IoT for short) to solve the food safety problem.

A lot of technical work has been done in the IoT. New paradigm shift are introduced [2] and information sharing is applied [3]. Among all the IoT technologies, RFID (Radio Frequency IDentification) is one of the hottest topics. Researchers used it to trace the products and monitor them [4]. Besides, in order to suit the food industry, researchers invented smaller tags, which is more convenient and cheaper to be used [5]. Several kinds of management systems are good at the food safety solutions as well [6]. The systems help the food factories to manage the staff and the products. At the same time, the systems offer the useful information about the food to the customers and make warnings when food accident happened [6].

Though there are so many achievements in the field of IoT applications to keep the food safety [4, 5, 6, 8, 14]. To the best of our knowledge, rare research pays attention to

F.L. Wang et al. (Eds.): WISM 2012, LNCS 7529, pp. 380–392, 2012.

the whole risks identification in the food industry, and investigating how well the IoT technologies can resolve the food safety problem. Researchers often focus on only a part of the food safety in this industry, such as tracing and monitoring or apply the IoT technologies in one phase of the food supply chain. However, to identify all risks in the food supply chain is important to the implementation of the IoT for food safety. Only after we identify the risks, we can know what the aspects we should pay attention to. Only after we take all the risks into account, we can assure the food safety in the complex food industry. In this paper, we choose the Chinese food standards in public as the base to identify the risks.

A typical food supply chain consists of production, processing, transportation, storage and sale. By analyzing them, we identify the risks in the food industry. After investigating over 1,200 standards, we obtain 619 related standards and map the key risks in the standards into the five phases:

- In the production, the risks come from additive, package, label, production operation, factory hygiene, operator, fungi, microbe, nutritive fortifier, element limitation, chemicals residue, radiation food requirement, radioactive material and specific food requirement.
- In the processing, the risks come from additive, package, label processing operation, factory hygiene, operator, fungi, microbe, nutritive fortifier, element imitation, chemicals residue, radiation food requirement, radioactive material and specific food requirement.
- In the transportation, the risks come from package, label, operator, transportation operation, fungi, microbe, radiation food requirement and specific food requirement.
- In the storage, the risks come from package, label, storage operation, operator, fungi, microbe, chemicals residue, radiation food requirement and specific food requirement.
- In the sale, the risks come from package, label, acceptable daily intake, operator, fungi, microbe, radiation food requirement and specific food requirement.

After we identify these risks, we try to use the IoT technologies to mitigate them. Using the IoT technologies introduced, we can solve some of the food safety problems, including label, transportation operation and storage operation. They have already been solved, and realized in practical or in the researches. Besides, some of the risks listed are solvable by the development of the IoT technologies, including additive, package, fungi, microbe, nutritive fortifier, element limitation, chemical residue, radiation food requirement, radioactive material, specific food requirement, operator, Acceptable Daily Intake and processing operation. However, there are still some risks cannot be solved by IoT itself perfectly, including factory hygiene and production operation. Other technologies and labor are needed to mitigate these risks.

The rest of the paper is organized as follows: section II introduces the related work. Section III presents IoT technologies and their architecture. Section IV presents the standards we investigated, the classification of the risks and the meaning of the risks in the production. In section V, we map the IoT technologies to the risks introduced in the section three. In section VI, we conclude our work and discuss the future work.

2 Related Work

IoT technologies have already been widely applied to the pharmaceutical supply chain and improve the safety of drugs greatly [7]. Similarly, researchers try to apply the IoT technologies to the food supply chain to keep the food safety. However, the food industry is more complicated than the pharmaceutical industry.

In order to suit the condition that livestock is moving and the plants are small, smaller tags are invented and contactless identification is provided to enhance the system's usability [5]. Besides, the real-time, accurate information sharing will be helpful in the management of food, and it can do a lot in realizing the intelligent recognition, location, tracking, monitoring and management for food [3]. After sharing the information of food, customers can make sure the safety of the good more easily and the efficiency of the food supply chain will be improved. RFID is one of the hottest topics in the IoT. It has the potential to offer food retailers a wide range of benefits [9]. Using the RFID technology, tighter management and control can be realized in the food supply chain. The relevant system can reduce the labor cost as well as improve the customer service level. The major benefits of RFID are that it uniquely and massive recognizes each item accurately without touch, tracks items as they move through the supply chain, and shares information with business partners, allowing collaboration on inventory management, planning, forecasting and replenishment [11].

In addition, researchers developed a lot of implementations. Applications were invented to monitor of the fresh agricultural products, control the food security sources strictly and build information management system of fresh agricultural products based on IoT, which helps to increase supply chain integration level and reducing supply chain management costs to improve supply chain efficiency [7, 8]. Ambient Assisted Living encompasses technical systems were invented to support elderly people in their daily routine, which help the old to have an independent and safe lifestyle as long as possible [10]. A system and method for managing food production, inventory and delivery in a restaurant was also realized [12].

Decision making and self-regulating also play important roles in the food safety, relating systems have been developed [13, 14].

Although so many researches have been done so far, we still do not know what the whole risks in the food supply chain. These researches pay attention to mitigate the risks while none of them point out the whole risks in the chain. However, identify all the risks in the food industry is meaningful and helpful when we try to solve the food safety problem.

3 IoT Technologies

The IoT consists of many distributed resources which are provided and required by different users and organizations around the real world. The architecture of the IoT can be usually divided into three layers [20]: sensing layer, network layer and application layer.

Sensing Layer: This layer is responsible for data acquisition and collaboration. The event or state of "things" in the physical world such as temperature, concentration, and multi-media data has been perceived and acquired by sensing devices, such as sensors,

RFID tags, cameras and GPS terminals. The layer is in charge of short data transmission, context awareness, and massive information processing.

Network Layer: This layer is the most familiar layer to us in the IoT for we have used this system for many years and it has truly brought us great evolution and convenience. In this layer, there are three main problems to solve: address, network integration and resource management. (i) Addressing: it means each thing in the IoT needs to be mapped to one address. (ii) Network Integration: IoT is a complicated, real-time heterogeneous network with the large scale. Multiple terminals are deployed here and they need to be integrated well. (iii) Resource Management: massive, dynamic, dispersible data have to be stored, transmitted or processed in this layer. These requirements lead to both the access networks and the core network to design a new topological structure or interactive way to improve the network resource utilization efficiency and throughput.

Application Layer: This layer includes the Support sub-layer and the Service sub-layer. After storing, processing and analyzing data intelligently, the Support sub-layer delivers the results based on users' requests. The Service sub-layer mainly includes the applications designed for users.

With the help of the internet of things, problems in many fields can be come over. Following are many applications of the internet of things.

Global fresh food training

According to our knowledge, approximately 10% of the fresh fruits and vegetables coming from different parts of the world into European market are wasted during transportation, distribution, storage, and retail process. The main causes of fresh food damage during the handling process are microbial infections, biochemical changes due to biological processes, physical food injuries due to improper environmental conditions, and mechanical damage due to mishandling. Global fresh food training systems can help us to monitor in the fresh food supply chain and prevent the risks in it.

Health care and medical applications at home and in hospitals

Many elderly people must leave their homes to move into a nursing home when the risk for living alone in their own homes becomes too high. However, the IoT can help people to monitor the old so that they can enjoy themselves at home. Besides, it also can offer disabled people the assistance, and support them to achieve a good quality of life and allows them to participate in the social and economic life [18].

Access control and mobile tickets

Another application that has a huge potential is to turn mobile devices into access devices, which enables consumers to use mobile devices instead of traditional plastic access cards to verify identity in commercial and consumer settings.

IoT Household Appliances

In a smart home, a mobile terminal can be used to control all IoT appliances either locally or remotely, including TV, humidifier, air conditioner, electric curtain, etc. All the IoT household appliances can be inter-connected and can talk to each other, forming an IoT smart home. Those household appliances will work in a cooperative manner to greatly improve the user experience

In addition, the IoT technologies also play an important role in Identity Security, Smart Grid, Location-based Service, Mobile Contactless Payment, social networking, Robot Taxi, smart museum and gym, Mobile Interaction, Smart Healthcare, Real-world Objects Search, Participatory Sensing, Augmented Reality and Automation, monitoring and control of industrial production processes [19, 20].

4 Risk Identification in Food Supply

In this section, we present the standards we acquired and divide the risks in them into several categories according to the procedures of food industry.

We get the standards from the website of china food safety [15] and food mate [16]. From the former one, we get complete food-related standards from 2010 to 2011 in china. From the later one, we get the most part of food-related standards from 1985 to 2009. There are 1,289 standards in total. 526 standards in them are about how to measure the elements in food and the experiments, which will not be analyzed here. In the other 763 standards, 619 standards are valid since the rest standards are out of date.

The risks in the food industry can be separated into five categories according to the phases of food industry. The five phases are production, processing, transportation, storage and sale.

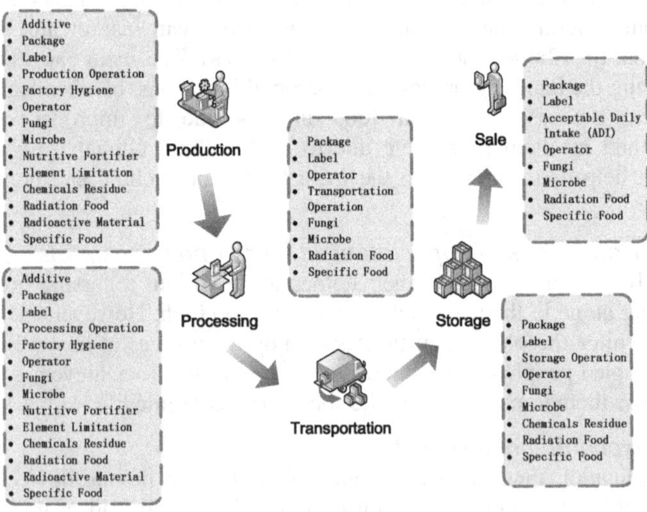

Fig. 1. Risks in the food industry

Production: For plants, such as vegetables, production includes the plantation, gathering, washing, packing and other operations making a seed grow to a plant, then to raw ingredient or a product. For livestock, such as pigs, production includes birth, feeding, slaughter, packing and other operations from the birth of the livestock to its grown-up, then to make it to raw ingredient or a product.

Processing: it means the methods and techniques used to transform raw ingredients into food for human consumption. Food processing takes clean, harvested or slaughtered and butchered components and uses them to produce marketable food products [17].

Transportation: it means move the food product from one place to another, which includes the receiving, shipping, land transport, air transport and ocean transport.

Storage: it includes the storage of harvested and processed plant and animal food products.

Sale: It relates to wholesaler, retailers and the end consumers. In this procedure, the storage of the products, the way selling products to customer and the return of goods need to be considered.

After classify and analyze the standards, we use figure 1 to show the risks in each phase of the food industry.

We choose the risks in the production to have more detailed explanation because the number of risks in this procedure is the most and these risks are typical.

Additive: the substances added to food to preserve flavor or enhance its taste and appearance. The additive in the product should below its limitation provided by correlative standard.

For example, GB 6227.1-2010 gives the limitation of Sunset Yellow like the table 1.

Table 1. Limitations of sunset yellow

Item	Index	method of calibration
Sunset Yellow, w% ⩾	87.0	appendix A.4
loss on drying, chlorid, sulfate in total, w% ⩽	13.0	appendix A.5
water-insoluble, w% ⩽	0.20	appendix A.6
sodium p-aminobenzenesulfonate, w% ⩽	0.20	appendix A.7
2-naphthol-6-sulfo, w%	0.30	appendix A.8
6,6'-OBI (naphthalenesulfonate) 2Na, w% ⩽	1.0	appendix A.9
4,4'-Two acid disodium salt, w% ⩽	0.10	appendix A.10
1-Phenyl azo-2-naphthol/(mg/kg) ⩽	10.0	appendix A.11
Sulfonated aromatic primary amine /% ⩽	0.01	appendix A.12
Vice-dye, w% ⩽	4.0	appendix A.13
AS/(mg/kg) ⩽	1.0	appendix A.14
PA/(mg/kg) ⩽	10.0	appendix A.15
Hg/(mg/kg) ⩽	1.0	appendix A.16

Table 2. GB 19303-2003

Factory design and facilities	site area and layout
	workshop
	sanitary fixture
	processing unit
	storage and transportation facilities
	hygienic quality facilities
Sanitary control	institution and staff
	equipment maintenance
	cleaning and disinfection
	garbage treatment
	environmental hygiene
	anti-pest
	dangerous cargo management
	staff hygiene
Process	raw material and accessories
	prevention of cross contamination
	Process
	Package
	Storage
	transportation
Hygienic quality control	organization and staff
	management system
	raw material
	hygienic quality control in process
	hygienic quality control for products
	other contents in hygienic quality control
Staff management	detail information

In order to keep the safety of food in the production, we should use the method offered in the standard to check the content of each item listed in the table.

Package: it means the packing for products. It requires protection, tampering resistance, and special physical, chemical, or biological needs.

Label: for different kinds of food, the label on the package should offer the enough information. For example, GB 7718-2011 gives the standard for the label of prepackaging food. It requires the food name, burden sheet, net contents, specification, producer, contact, address, date of manufacture, quality guarantee period, condition of storage, food producing license No., product No. and other required information.

Production Operation: there are some standards gives the requirements for the factory of specific food, which may include the requirements for environment, control, operators and other factors.

For example, GB 19303-2003 is a standard for hygienic practice of cooked meat and meat-product factory. Table II lists its requirements for hygienic practice of cooked meat and meat-product factory in five aspects, including factory design and facilities, sanitary control, process, hygienic quality control and staff management. Besides, the interests in each aspect are also listed.

Factory Hygiene: there are some standards for specific factory, which includes requirements for factory environment, operations, additives, packages and other aspects. For example, GB12696-90 is designed for winery. Besides the requirements set in other standards, it has some specific requirements for winery, such as the pesticide is forbidden 15 days before the harvest of grape.

Operator: operators' misoperation may also bring risks into the food industry, including doing intentionally or with premeditation

Fungi: with the passage of time, the content of fungi will increase, which means the food is out of date or goes bad. So the content of fungi should be lower than the limitation in the standards. For example, GB 2761-2011 lists six common fungi and their limitations in different kinds of food.

Microbe: it is similar to the fungi. The increasing content of microbe in food also means the food may be out of date or goes bad. So the content of microbe should be lower than the limitation in the corresponding standards.

Nutritive Fortifier: Nutritive fortifier must in its valid scope. GB 14880 gives the scopes of common nutritive fortifiers in different kinds of foods. Note that different from the fungi and microbe, the nutritive fortifier has not only upper bound but also lower limit.

Element Limitation: the contents of chemical elements, such as calcium, iron, selenium, must be lower than the requirement since over-englobement is bad. For example, GB 13106-1991 introduces the tolerance limit of zinc in foods.

Chemical Residue: in the planting or feeding, some chemical residue may be left on the product or the raw material, which will cause risks in production. For example, pesticide is a big problem which the contents of them must take care. GB 2763

presents maximum residue limits for pesticides in food, in which we can find the limitations for different kinds of pesticides.

Radiation Food: nowadays, some foods are grown with radiation in order to delays the fruit ripening or prevent the growth of microorganism. This kind of food has their own corresponding standards, which has different requirements from the ordinary ones. For example, GB 14891.5-1997 is a standard for radiation fruits and vegetables, which includes the limited amount of radiation.

Radioactive Material: radioactive materials will be bad for human beings, so the content of them in foods should also be paid attention. GB 14882-1994 gives the Limited concentrations of radioactive materials in foods. For more detailed requirements, there are other standards for specific material in different kind of foods.

Specific Food: In addition to the basic risks listed above, specific food has its specific requirements and limitations. For example, dairy products get much concern in china. There are a lot of standards for them. GB 13102-2010 is designed for evaporated milk, sweetened condensed milk and formulated condensed milk. In the standard, raw material has its tough limit and Fungi, Microbe and Nutritive Fortifier has specific requirement.

5 Apply the IoT Technologies to Food Safety

Facing the risks in the food industry, we try to employ the techniques in the Internet of things to the food safety. But we argue that some of the risks can be solved by IoT while others cannot. On the basis of this condition, we divide the risks into three classes as showed in the table III.

Table 3. Classes of the risks

Solved	Label, Transportation Operation, Storage Operation
Solvable	Additive, Package, Fungi, Microbe, Nutritive Fortifier, Element Limitation, Chemical Residue, Radiation Food, Radioactive Material, Specific Food, Processing Operation, Acceptable Daily Intake, Operator
Insoluble	Factory Hygiene, Production Operation

Solved: The risks cannot only be solved by the IoT technologies but also have already been realized in practice or in research.

Label: we can use RFID readers to check whether the information on the label is enough. It is very convenient and efficient. Besides, in the ordinary conditions, the information will be input into the information system, so the operations, using the reader to check the label won't bring any redundant operation or waste any time.

Transportation Operation: in the transportation, the IoT technologies can help to check the real-time temperature, humidity, accelerated speed and other interests. In addition, the IoT technique, especially RFID technologies can help us to record the information in shipping and receiving.

Storage Operation: during the storage, the IoT technologies can help to check the real-time temperature, humidity and other interests, which can prevent the existent risks. Besides, the IoT can records the information at intervals, which costs less than the labor force.

Solvable: The risks can be solved by IoT according its development.

Additive: the additives in the products can be ferreted out by the IoT. But using present IoT technologies, it may be little hard to check each product in the food industry. We can send sample drawn to the inspection center and take the result of the sample as the result for the whole. Besides, the IoT devices can alert when the additives over the limit.

Package: the package of products can be checked by the IoT. If there is any damage about the package of the products, the readers can help to find the problem and give us an alarm. Besides, in the process of packing, the data about the packing material, such as the hardness, can be recorded in the information system to ensure they are qualified to be the package of the products.

Fungi: the fungi in the products can be ferreted out by the IoT. Like the additive, it is little hard for us to check whole products. So we can take sample to the inspection center to check the fungi. Besides, we can check the fungi content in the environment to get the fungi condition of the products. When fungi over-range the limit, the IoT devices will give alarm.

Microbe: the microbe in the products can also be ferreted out by the IoT technique like the fungi. We can check the sample to get the microbe condition of the products or observe the content of microbe in the environment of the products. When fungi over-range the limit, the IoT devices will give alarm.

Nutritive Fortifier: it can also be ferreted out by the IoT. We can send the sample to the inspection center to have a check and record the information in the system. If the nutritive fortifier is over the limit, treatment should be taken to ensure the safety of the food.

Chemical Residue: the chemical residue mainly brought by the growing phase of the livestock or the plant. So in the production and processing procedure, we must use IoT technique to check the content of them. The method is similar to the microbe. We can send sample to the inspection center or check the content in the environment. Besides, alarms are necessary when the chemical residue over-range the limit.

Element Limitation: the chemical element in food will have bad effect for the health of people. Maintaining them in the proper range is important. We can send the sample to the inspection center and record the data in the information system. If there is any problem with the data, measures should be taken to ensure the safety of food.

Radiation Food: the requirement for the radiation food is a little different from the ordinary food. Beside the interests we should take care for the ordinary people, the radiation amount should also be paid attention. We can use the RFID to make sure the radiation amount is in the valid range.

Radioactive Material: we can use the RFID devices to check the content of radioactive materials in the environment. If the content is over-ranged, alarms should be made.

Specific Food: the standard for specific food mainly gives more detailed requirements in the fungi, microbe, additives and the interests like them. Because the objective limit like additives, fungi and microbes can be ferreted out by the IoT technique, we can also use the IoT to meet the requirements for specific food.

Processing Operation: in the processing of the food industry, the main risks are from the additives, fungi, microbes, machine hygiene and the raw materials. The former three has been discussed above. The machine hygiene can be checked by the devices or human being. The raw material can be maintained the safety like other food that we will check the interests such as fungi, microbe and other interests of it.

Acceptable Daily Intake: it is the information customer should know when they have the food. The information on the product or its tag should mark out the acceptable daily intake. Besides, if the customer wants to take several foods together, the IoT devices will count the total of the content of the elements in them and check whether the sum is below the acceptable daily intake. In addition, the information system can share the information online and send the acceptable daily intake when the IoT devices send requests.

Operator: the operator may mistake in the production, process, storage, transportation and sale. The management of operator is very difficult since we cannot predict what kind of mistakes he will make. But we can use video to record the events. Though the cost may be high, the problem is still solvable.

Insolvable: The risks are hard to be resolved in whole, although we can provide some solutions to mitigate part of risks.

Factory Hygiene: the risks in the factory hygiene focus on the environment of the factory, the operation regulation, the operator hygiene and the detailed requirement for a specific kind of factory. For the objective requirements such as package, fungi, temperature, the IoT can check and take record. However, it can do little in the subjective aspects, such as operation regulation and operator hygiene.

Production Operation: Production operation has many factors, including factory design and facilities, sanitary control, process, hygienic quality control and staff management. It is more complex than the operations in other procedures. We can use IoT to solve the risks in process, sanitary control and hygienic quality control. But the factory design and facilities and the staff management are not suitable for the devices to solve. For example, whether the factory design is logical is out of the range of the ability of IoT devices since the requirement is subjective instead of objective. We still need labors to help the IoT to mitigate these risks.

6 Conclusion

Our main contribution is the identification of the risks in the whole food supply chain by investigating the related standards in the food industry in China. We analyze the standards and find out the classification of the kinds of the standards, then divide the risks in the standards into five food supply phrases. So we can know what is the key risks we should pay attention to in each phase of the food industry.

We also try to use the IoT technologies to avoid the risks introduced in the paper. We show the present IoT applications and researches in the food supply chain at first. Then we analyze the risks and divide them into three classes: solved, solvable and insolvable. They are divided in this way according to whether the risk has been solved by the IoT technologies and whether the risk can be solved by the IoT technologies in the future.

Acknowledgement. This paper is supported by the 863 project (Grant NO: 2011AA100701), the Project-sponsored by SRF for ROCS, SEM, and the Opening Project of Key Lab of Information Network Security of Ministry of Public Security (The Third Research Institute of Ministry of Public Security). The corresponding author of this paper is Bo Jin.

References

1. Wikipedia, 2008 Chinese milk scandal (2011)
2. Zouganeli, E., Svinnset, I.E.: Connected objects and the Internet of things — A paradigm shift. Photonics in Switching (2009)
3. Yue, T.: Internet of Things Technology Application in the Food Supply Chain Management. In: E-Business and E-Government, ICEE (2011)
4. Vergaraa, A., Llobet, E.: An RFID reader with onboard sensing capability for monitoring fruit quality. Sensors and Actuators B: Chemical 127(1), 143–149 (2007)
5. Jones, P.: Networked RFID for use in the Food Chain. In: Emerging Technologies and Factory Automation (2006)
6. Song, B., Xing, Q.: On Security Detecting Architecture of Food Industry Based on Internet of Things. In: Automation and Logistics, ICAL (2011)
7. Gu, Y., Jing, T.: The IoT Research in Supply Chain Management of Fresh Agricultural Products. In: Artificial Intelligence, Management Science and Electronic Commerce, AIMSEC (2011)
8. Li, L.: Application of the Internet of Thing In Green Agricultural Products Supply Chain Management. In: Intelligent Computation Technology and Automation, ICICTA (2011)
9. Thompson, C.: Radio frequency tags for identifying legitimate drug products discussed by tech industry. American Journal of Health-System Pharmacy, 1430 (2004)
10. Dohr, A., Modre-Opsrian, R., Drobics, M., Hayn, D., Schreier, G.: The Internet of Things for Ambient Assisted Living. In: Information Technology: New Generations, ITNG (2010)
11. Vijayaraman, B.S., Osyk, B.A.: An empirical study of RFID implementation in the warehousing industry. International Journal of Logistics Management 17(1), 6–20
12. Schackmuth, G., Sus, G.A.: RFID Food Production, Inventory and Delivery Management System for a Restaurant, Pub. No: US 2007/0251521 A1
13. McMeekin, T.A., Baranyi, J.: Information systems in food safety management. International Journal of Food Microbiology 112(3), 181–194 (2006)

14. Clothier, B.L.: Temperature Self-regulating Food Delivery System, Patent No: US 6232585 B1
15. http://www.chinafoodsafety.net
16. http://down.foodmate.net/standard/
17. http://en.wikipedia.org/wiki/Food_production
18. Domingo, M.C.: An overview of the Internet of Things for people with disabilities. Journal of Network and Computer Applications 35(2), 584–596 (2012)
19. Atzori, L., Iera, A.: The Internet of Things: A survey. Computer Networks 54(15), 2787–2805 (2010)
20. Zheng, L., Zhang, H., Han, W., Zhou, X., He, J., Zhang, Z., Gu, Y., Wang, J.: Technologies, Applications, and Governance in the Internet of Things. In: Vermesan, O., Friess, P. (eds.) Internet of Things - Global Technological and Societal Trends. From Smart Environments and Spaces to Green ICT. River Publishers (2011)

Towards Better Cross-Cloud Data Integration: Using P2P and ETL Together

Jian Dai and Shuanzhu Du

4# South Fourth Street, Zhong Guan Cun, Institute of Software, Chinese Academy of Sciences, Beijing, China, 100190
daijian@nfs.iscas.ac.cn

Abstract. Cloud computing has been increasingly used in a variety of scenarios and environments to deal with huge amounts of data, accordingly, data integration among different clouds have also been extensively concerned by both academia and the business community. With the division of private and public clouds and different contents and functions of clouds, cross-cloud data integration becomes a grand challenge. Various methods are proposed to cope with the challenge; however, how to achieve better cross-cloud data integration still involves a lot of tradeoff analysis. In this paper, we propose weaving P2P technique into ETL process to integrate cross-cloud data. And we built a middleware prototype system to testify the feasibility of our approach.

Keywords: P2P, ETL, data integration.

1 Introduction

Cloud computing makes computing and storage located in a large number of distributed computers, rather than local computers or servers, the operation of the enterprise data center and the Internet are more similar, which enables enterprises to focus energy and resources to the desired application. Following this tendency, a lot of clouds come out.

However, clouds are different. Many large corporations like Amazon, Google and Microsoft have their own cloud technologies, such as Amazon AWS, Google APP Engine, IBM and Google the blue cloud plans. These technologies on the basis can be divided according to what kind of the cloud is (private cloud, public cloud or hybrid cloud) or which fields the cloud applications are used (education, industry or health). That's a major reason why massive and heterogeneous data are generated [1].

Although various clouds are producing various data, starting from the point of view of end-users and applications, many applications require more than one field or one data type and structure for collaboration, for example: a corporation may use Google's E-mail, but its ERP system is a application server deployed on the enterprise LAN systems, therefore, a question may arise: Can it use data integrated ERP system with E-mail system? Similarly, a city may have different domain clouds, such as

F.L. Wang et al. (Eds.): WISM 2012, LNCS 7529, pp. 393–401, 2012.

health cloud and Education cloud, but for the decision-maker of the city, a question may arise: Can different cloud data integrated together?

Generally speaking, how to integrate data is heavily depended upon how data are stored and managed. We introduce three more representative data generation ways below and give a brief analysis, which are Google Big Table, Amazon Dynamo and Microsoft Azure.

The fundamental idea of the big table is a sparse, distributed, persistent storage of multi-dimensional sorting Map for data management. If such kind of data format is used to integrate the cross-cloud data, obvious drawback is its data unit is too tiny to facilitate high-speed data collaboration.

The goal of Amazon Dynamo is to achieve the machine-level data management, but due to its tight information encapsulation, not be able to provide cross-cloud data integration.

Microsoft Azure, however, mainly focuses on the SQL Data Services cloud computing environment, and lacks of cross-cloud data integration support.

As we observed, there are three major features of cross-cloud data:

1. Cross-cloud data are on large scale. In cloud computing systems, there could be thousands or even millions of computers connected, and these computers continuously exchanging data with each other or application. The cloud computing systems need keep the latest versions of data, and in most cases, also need keep the historical versions of important data for some time period in order to support state tracking, statistical analysis, and data mining.

2. Cross-cloud data are usually heterogeneous. The same type of cloud computing system may contain multiple categories of data sources such as traffic, hydrological, geological, biomedical data sources, and so forth. Each category can be further divided into different kinds of low-level categories. For instance, traffic data sources can include GPS data, RFID data, video-based traffic-flow analysis data, traffic loop data, road condition data, and so forth. The data from different sources may have different data structures and different semantics, which greatly increases the difficulties in data integration.

3. Cross-cloud data are highly dynamic. Every time when cloud computing system receives or generates a new data record, it need insert the new version to the corresponding database and meanwhile delete old versions when they become obsolete. Since a cloud computing system may contain huge number of data sources and each data source generates new data frequently, the data to be integrated at the target are highly dynamic, which poses heavy update pressure on the system.

Admittedly, not matter Microsoft, Amazon or Google has made great efforts towards better data integration cross-cloud, however, to achieve better data integration, as far as we concerned, P2P, ETL and middleware techniques will be essential contributors [8] [9][10].

In this paper, our contributions can be summarized as follows:

(1) Analyzed why P2P is an essential technique to achieve better cross-cloud data integration;

(2) Analyzed why ETL is an essential technique to achieve better cross-cloud data integration;

(3) Proposed an approach to weave P2P into ETL process using middleware technique;

(4) Testified the feasibility of using P2P and ETL together in a prototype system;

The rest of the paper is organized as follows. The second section depicts why P2P is an essential technique with literature review; the third section depicts why ETL is an essential technique with literature review; section 4 gives an introduction on how we weave P2P into ETL process to achieve better cross-cloud data integration; section five reports experimental and performance results and the final section is our conclusion and discussion.

2 Using P2P in Cross-Cloud Data Integration

Peer-to-peer (abbreviated to P2P) refers to a computer network where each computer in the network can act as a client or server for the other computers in the network, allowing shared access to files and peripherals without the need for a central server. P2P network requires all computers in the network to use the same or a compatible program to connect to each other and access files and other resources found on the other computer [2]. As stated in section one, the cross-cloud data integration involves a large number of computers distributed in different clouds. To achieve better cross-cloud data integration, how to efficiently and seamlessly organize those computers is a major challenge. And from the conceptual level of P2P, it is natural and efficient to organize cross-cloud computers in this kind of network.

Moreover, the architecture of P2P is a distributed network that straightforwardly partitions tasks or workloads among peers. Another problem of cross-cloud data integration is that failure of cloud computers are common scenario. Since peers are equally privileged participants in the application, it avoids a bottleneck. And strategies can easily be made to cope with the failure of cloud computers, such as data duplication strategy in contrast to the traditional client-server model where the message exchanges are usually one-way only.

Meanwhile, as an essential technique, Distributed hash tables (DHTs) which are a class of decentralized distributed systems that provide a lookup service similar to a hash table: (key, value) pairs make sure that cross-cloud data integration can be done with global unique data resource key. DHTs offer responsibility for maintaining the mapping from keys to values is distributed among the nodes, in such a way that a change in the set of participants causes a minimal amount of disruption. This allows DHTs to scale to extremely large numbers of cross-cloud nodes and to handle continual node arrivals, departures, and failures [3].

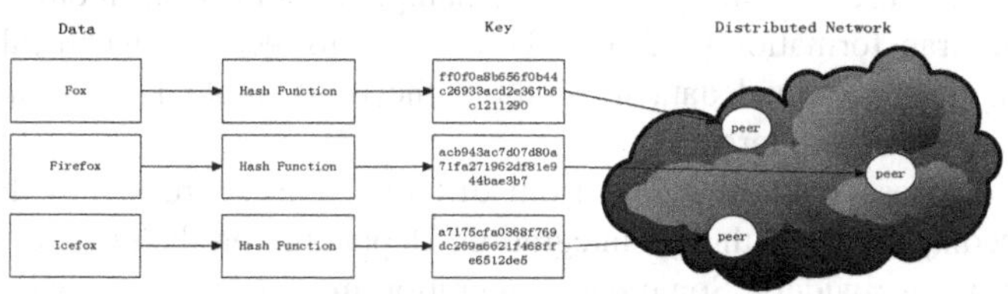

Fig. 1. Distributed hash tables

In cross-cloud P2P networks, clients share resources, which may include bandwidth, storage space, and computing power. This property is one of the major advantages of using P2P networks among clouds because it makes the setup and running costs very small for the end target of data integration. As nodes arrive and demand increases, the total capacity of the cross-cloud P2P network also increases, and with certain strategies the likelihood of failure decreases [7]. If one peer on the network fails to function properly during data integration, the whole network is not largely compromised. In contrast, in typical client–server architecture, clients share only their demands with the system, but not their resources. In this case, as more computers join the system (which can likely happen during cross-cloud data integration), fewer resources are available to serve each client, and if the central server fails, the entire network fails. The decentralized nature of P2P networks increases robustness of cross-cloud data integration process because it removes the single point of failure that can be inherent in a client-server based system.

3 Using ETL in Cross-Cloud Data Integration

Extract, transform and load (ETL) is a process in data integration usage and especially in distributed data usage that involves [5]:

1. Extracting data from multiple data sources
2. Transforming it to fit operational needs (which can include quality levels)
3. Loading it into the end data integration target

The first stage of an ETL process involves extracting the data from the different data sources. In many cases this is the most challenging aspect of ETL, as extracting data correctly and efficiently will set the stage for how subsequent processes will go. In cross-cloud data integration, data sources are distributed computers in different clouds. An intrinsic part of the extraction involves how to organize and communicate with data sources, if properly networked; the extraction efficiency could be greatly improved [4]. Considering each separate data source may also use a different data organization/format, the extraction sub-process should adopt a unified strategy to deal with different data sources. In general, the goal of the extraction phase is to convert the data into a single format which is appropriate for transformation processing.

The transform stage applies a series of rules or functions to the extracted data from the source to derive the data for loading or integrating into the end target. Some data sources will require very little or even no manipulation of data. In other cases, one or more of the transformation types may be required to meet the technical needs of the end target. In cross-cloud data integration, the rules applied are a major factor to satisfy the needs of end target.

The load phase loads the data from different data sources into the end target, usually accompanying with data integration. Depending on the goals of the ETL, this process may vary widely. Some data integration may overwrite existing information with cumulative information, frequently updating extract data is done on daily, weekly or monthly basis. Other data integration may add new data in a historicized form, for example, hourly [6]. In cross-cloud data integration process, often both time and space are two considerations to integrate data. Accordingly, different strategic design choices should be made. Data may be replaced or appended dependent on the time when it arrived and space where it arrived. And a history log can kept the trail of all changes during data integration.

4 Weaving P2P into ETL Process

We introduce middleware architecture to weave P2P into ETL process. This cross-cloud middleware is software deployed on each cloud computing machine and data integration target machine and some accessory machines which consist of a P2P network and run ETL process in an organized way. The middleware located in each machine composes a P2P network for the exchange of information which can provide effective interaction, shield a variety of different cloud environment implementation details.

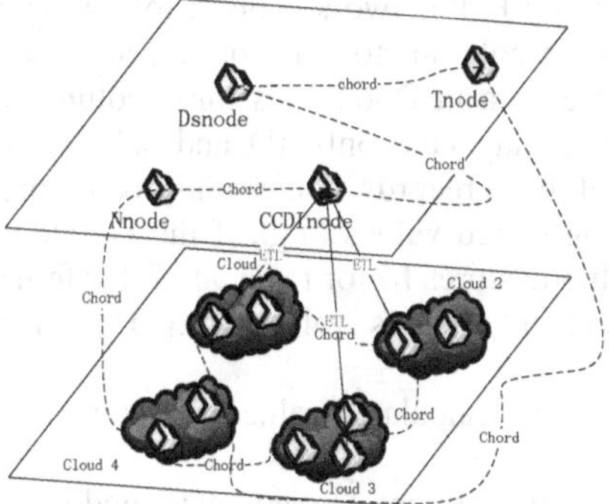

Fig. 2. Using P2P and ETL together in Cross-Cloud Data Integration

There are five types of nodes in P2P network:

Task node (Tnode for short): used to generate a variety of task message sent to the task execution node; Naming Service the node (Nnode): used to collect information of each node and the node in which cloud computing environment; Data Serving Node (DSnode): used to save the data resources index information, providing data management services. Node in the cloud computing environment (Dnode): used for cross-cloud applications, data sources, and may come from different cloud computing environments. Cross-cloud data integration target node (CCDInode): the need for the application of cross-cloud data node.

To weave P2P into ETL process, three major algorithms are proposed.

Algorithm 1: Join P2P Network Algorithm.

Step1: a JoinP2PMessage is sent to Nnode, containing the cloud and node description as a request message. Step2: if this node is accepted by Nnode, it sends its DataInfoMessage to the DMnode, containing the data information it can provide; Step3: Nnode sends NodeID to this node, meanwhile notify its adjacent nodes in the chord ring to update their finger table.

Algorithm 2: Leave P2P Network Algorithm

Step1: a LeaveP2PMessage is sent to Nnode as a request to leave the chord ring; Step2: if there are live connections between this node and the others, LeaveNotifyMessage will be sent to notify the others that the node is leaving, thus the others can preserve connection states in order to use when this node joins chord ring next time. Step3:LeaveMessages are sent to its adjacent nodes to notify they to update finger tables.

Algorithm 3: Data Integration Algorithm

Step1: a DataIntegrationMessage is sent to Tnode as a request to initialize a data integration process from a data integration target node, containing the ETL related arguments; Step2:Tnode schedules two task messages and distribute them to Nnode, DSnode accordingly. One task message is sent to Nnode, which requires ETL realted Dnodes information, the other task message is sent to DSnode, which requires ETL related data information. Step3: Sub-ETL processes are started in DSnodes, and data are sent forward to the data integration target node.

The ETL process built up P2P network support following operations:

Select: Selecting part of columns to load (or selecting null columns not to load). For example, if the source data in Dnode has three columns, such as: ID, age, and salary, then the extraction may take only ID and salary. Similarly, the extraction mechanism may ignore all those records where salary is not present (salary = null).

Translate: Translating coded values (e.g., if the Dnode stores 1 for male and 2 for female, but the CCDInode stores M for male and F for female).

Encode: Encoding heterogeneous values into a unified form (e.g., mapping "Male" to "1").

Calculate: Deriving a new calculated value from source data (e.g., sale_amount = month * unit_price)

Sort: Sorting data from different Dnode in ascending order or in descending order.

Deduplicate: Joining data from multiple Dnodes and deduplicating the data.

Aggregate: Aggregation (for example, rollup — summarizing multiple rows of data — total sales for each store, and for each region, etc.)

Surrogate: Generating surrogate-key values from data elicited from multiple Dnodes.(for example, using SHA-2 to hash the multiple major keys to a new major key)

Rotate: rotating multiple columns into multiple rows or vice versa.

Split: Splitting a column into multiple columns (e.g., putting a comma-separated list specified as a string in one column as individual values in different columns)

Merge: Disaggregation of repeating columns into a separate detail table (e.g., moving a series of addresses in one record into single addresses in a set of records in a linked address table)

V&V (verify and validate): Verifying data from multiple Dnodes in both quantified and qualitative way. (e.g., there should be only male or female in one software company).Validate data when verified. If validation fails, it may result in a full, partial or no rejection of the data, and thus none, some or all the data is handed over to the real use of CCDInode, depending on the rule design and exception handling.

5 Experimental Evaluation

This section shows experimental results tested on our prototype system, which proves the feasibility and effectiveness of using P2P and ETL together. The prototype system is developed by JAVA programming language, used an open-source chord software and an open-source Message Queue software. With the help of JAVA Virtual Machine, our prototype system can run on many different operating system platforms. Thus we can testify our cross-cloud data integration approach.

The topology we chose to do experiments is shown in Figure 3, and the machine configuration is shown in Table 1. In the P2P network, we use two machines as the cross-cloud data integration target nodes; three machines as data service node, named node and task service node respectively; eight machines as four cloud computing environments to simulate cloud computing environment.

Table 1. Configuration details of the experiment environment

Name	Configuration	Details
cross-cloud data integration target nodes (2 nodes)	hardware	CPU: AMD Anthon(tm) II X2 215 Processor Memory: 2.00GB Disk: WDC WD3200AAKS – 75S9A0 320G
	Operating system	MICROSOFT WINDOWS XP professional 2002 Service pack 3
DSnode, Nnode, Tnode(3 nodes)	hardware	CPU: AMD Anthon(tm) II X2 215 Processor Memory: 2.00GB Disk: WDC WD3200AAKS – 75S9A0 320G
	Operating system	Cent OS
Dnodes (8 nodes)	hardware	CPU: AMD Anthon(tm) II X2 215 Processor Memory: 2.00GB Disk: WDC WD3200AAKS – 75S9A0 320G
	Operating system	ubuntu-hadoop-release

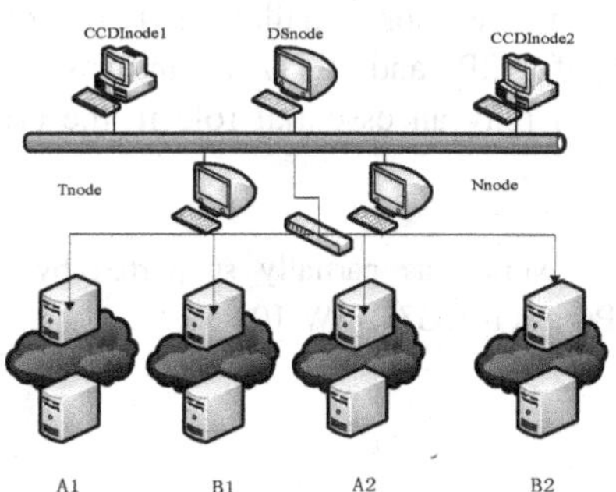

Fig. 3. Topology of our experiment environment

As shown in Fig.3, each computer running the middleware is connected in the P2P network. The experiments are designed as follows: WordCount program is running on cloud A1, cloud B1, cloud A2 and cloud B2 respectively to count the frequency of the word appearance times in specified text file. Cross-cloud data integration goal is to check out the results of the implementation of A1, B1, A2, B2 and adding the results together; another cross-cloud data integration goal is to calculate the sum of A1 and B1.

We designed one indicator to testify the feasibility of using P2P and ETL together in cross-cloud data integration process. DPS – the amount of data the middle can integration per second. As depicted in Fig. 4, the average dps varies depending on the size of the input data source (size of the word count file). Extracting 8M data from 4 clouds using P2P network and ETL process costs about 2sec. Thus, the average cross-cloud data integration speed is around (8*4)/2=16M/s, which testified the feasibility.

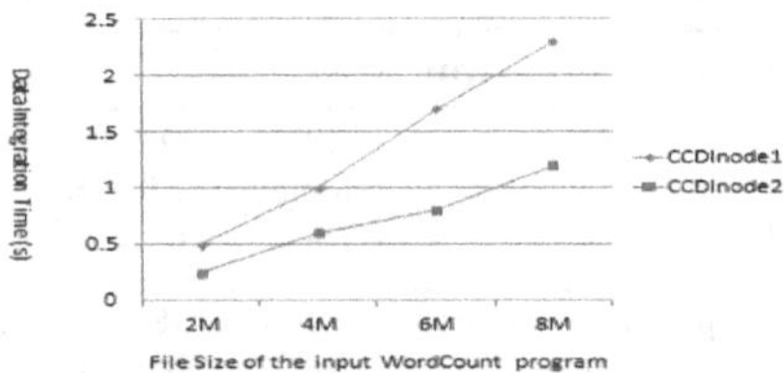

Fig. 4. Results of our experiment

6 Conclusion

In this paper, we analyzed why P2P and ETL are two important techniques in cross-cloud data integration process, proposed a middleware based framework to use P2P and ETL together, and finally designed experiments to illustrate how to use P2P and ETL together in cross-cloud data integration. We believe that with cloud computing is becoming increasingly popular, massive and heterogeneous data will be largely generated. And there may be more challenges in handling the cross-cloud data integration. However, if P2P and ETL techniques are properly refined and customized, they can still play an essential role in the cross-cloud data integration process.

Acknowledgment. This work was partially supported by NSFC under Grants NSFC 91124001 and CAS Key Project is KGZD-EW-102-3-3.

References

1. Halevy, A.Y., Ashish, N., Bitton, D., Carey, M., Draper, D., Pollock, J., Rosenthal, A., Sikka, V.: Enterprise Information Integration: Successes, Challenges and Controversies. In: SIGMOD 2005 (2005)

2. Fuxman, A., Kolaitis, P.G., Tan, W.-C.: Peer Data Exchange. ACM Transactions on Database Systems 31(4), 1454–1498 (2006)
3. De Giacomo, G., Lembo, D., Lenzerini, M., Rosati, R.: On Reconciling Data Exchange, Data Integration, and Peer Data Management. In: PODS 2007, June 11-13 (2007)
4. Sampaio, A., Mendonça, N.: Uni4Cloud: An Approach based on Open Standards for Deployment and Management of Multi-cloud Applications. In: SECLOUD 2011 (2011)
5. Vassiliadis, P., Simitsis, A., Skiadopoulos, S.: Conceptual Modeling for ETL Processes. In: DOLAP 2002 (2002)
6. Majchrzak, T.A., Jansen, T., Kuchen, H.: Efficiency evaluation of open source ETL tools. In: SAC 2011 (2011)
7. Zhuge, H., Liu, J.: A Novel Heterogeneous Data Integration Approach for P2P Semantic Link Network. In: WWW 2004 (2004)
8. Aggarwal, V., Feldmann, A.: Can ISPs and P2P Users Cooperate for Improved Performance? In: SIGCOMM 2007 (2007)
9. Bennett, T.A., Bayrak, C.: Bridging The Data Integration Gap: From Theory to Implementation. ACM SIGSOFT Software Engineering Notes 36(4) (July 2011)
10. Michael Maximilien, E., Ranabahu, A., Engehausen, R.: IBM Altocumulus: A Cross-Cloud Middleware and Platform. In: OOPSLA 2009 (2009)

Design of Intelligent Maintenance Decision-Making System for Fixed Equipment in Petrochemical Plants

Guangpei Cong [1], Jinji Gao [2], Bing Hou[3], and Guochen Wang[4]

[1] School of Chemical Engineering, Dalian University of Technology, Dalian 116024, China
[2] School of Mechantronic Engineering, Beijing University of Chemical Technology, Beijing 100029, China
[3] School of Petroleum Engineering, China University of Petroleum, Beijing, 102249, China;
[4]The 2nd Engineering Department, Petrochina Pipeline Bureau the 5th branch company , Cangzhou, 062552, China

Guangpeicong@yahoo.com.cn, Gaojinji@263.net, houbing9802@163.com, 8188719@qq.com

Abstract. The current study proposes a intelligent maintenance decision-making system is as a fixed equipment maintenance strategies assessment system. Because the maintenance decision-making is based on monitoring of equipment damage state and risk in the system, the system is established on the distributed database and internet technology. Moreover, different maintenance strategies, such as corrective, preventive, and predictive maintenance strategies after periodic and in-service inspections, are determined according to equipment failure consequences and risks. According to the data in the risk database and damage state database, the decision-making regards the "System" as a decision-making object and performs the risk-concentrated analysis, the operational cycle after maintenance and cost-effective maintenance assessment, where the multi-objective optimization of maintenance strategies on safety, availability, and economic efficiency is accomplished.

Keywords: Accident, Mechanical failure, Risk-based maintenance (RBM), Risk-based inspection (RBI), Risk-concentrated factor analysis, Maintenance effectiveness.

1 Introduction

Accidents seriously influence the long-cycle production and economic profits in the petrochemical industry. Statistically, mechanical failure accounts for 56% of major accidents caused by incorrect maintenance programs [1], whereas breakdown maintenance and loss of shutdown consume percent of the productive costs in the petrochemical industry. This proportion (30%–40%)has increased in recent years because of changes in operational and reservoir conditions[2]. Especially in recent years, since a large amount of different crude oils was introduced in China, refinery plants had to frequently face various categories of crude oil that even went beyond 120 categories in some plants. Thus, petrochemical plants performed reformations to

F.L. Wang et al. (Eds.): WISM 2012, LNCS 7529, pp. 402–413, 2012.

increase their oil output in general. On the other hand, safety management strategies for critical systems usually involve multiple dimensions, including design philosophy, maintenance policies, and procedures for personnel hiring, training, and evaluation [4], indicating that reasonable maintenance and redesign play important roles in safety management strategies. In addition, equipment tends to become large-scale, and thus, maintenance cost is not secondary anymore [5]. For example, statistically, maintenance costs tend to be very high, accounting for approximately 20%–50% of the total operating budget of process systems [6]. Moreover, a reduction of approximately 40%–60% in operating costs can be achieved through effective maintenance strategies [3]. Therefore, a highly effective maintenance strategy not only guarantees safety but also increases economic profits.

2 Intelligent Maintenance Decision-Making System

In current maintenance strategies, the most conservative approaches at one end of the spectrum rely on robust system design, frequent preventive maintenance, and early responses to warnings. At the other end of the spectrum, aggressive strategies are driven by demanding production schedules, single-string system designs, and minimal inspection and maintenance to obtain maximum production with minimum interruptions [7]. In maintenance activities, the considerations on reliability and availability of equipment are necessary to avoid excessive consumption and serious accidents [8], and thus, cost-effective maintenance strategies are widely sought. Risk-based maintenance (RBM) strategies have recently received increasing attention from researchers, in which the consequence of failure is a financial concern as risk-based maintenance (RBI) in ref [9]. Moreover, the probability of unexpected events is determined using the fault tree analysis, whereas the maximum failure probability of the key equipment under risk constraints is determined using the reverse fault tree analysis. According to reliability limits, the interval of periodic preventive maintenance (PM) for the key equipment is obtained under the condition that the default maintenance effect is perfect [10,13]. Moreover, the interval is modeled using the Weibull distribution [14,17], in which all kinds of mean time are regarded as modeling parameters.

2.1 Model of Intelligent Maintenance Decision-Making System

Modern maintenance decision-making models for either fixed or dynamic equipment often possess the following characteristics:

(1)Fault (dynamic equipment) or failure (fixed equipment) mechanism analyses may be regarded as kernel.
(2)The analytical results of the mechanisms are used to screen the status of a fault or failure and to locate the position where the fault or failure occurs.
(3)The status of the fault or failure should be monitored (dynamic equipment) or inspected (fixed equipment) at the position obtained in 2).

(4)Feedback on the status of the mechanism analytic system is used to re-analyze and then form a closed loop.

In addition, failure accidents caused by corrosion in the investigated object-fixed equipment in the petrochemical industry account for 77% of the total failure accidents in China.

In summary, shutdown inspection, in-service inspection, risk analysis, and corrosion analysis technologies are combined to constitute a closed-loop maintenance decision-making model based on risk and damage status inspection, as shown in Fig. 1. The corrosion mechanism analysis is regarded as a failure analysis, and the potential hazard caused by corrosion is quantified with risk in the model. Equipment is distinguished according to different ranks of consequence and risk to implement corrective, preventive, and predictive maintenance services after periodic and in-service inspections.

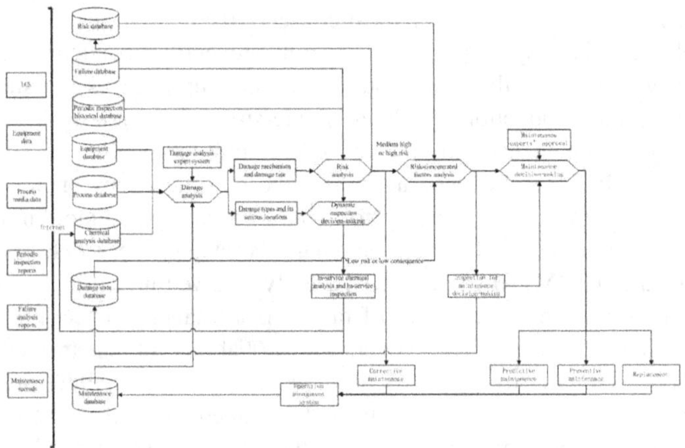

Fig. 1. Intelligent maintenance decision-making system

2.2 Composition of the Intelligent System

From the Fig.1, the intelligent system is consist of two subsystem, the equipment damage status and risk monitoring subsystem and maintenance decision-making system. The latter is performed based on the data accumulated through the former. Besides the inspection results will be feed back to the damage state database to amend the Omission of former.

3 Equipment Damage Status and Risk Monitoring Subsystem

The subsystem is a soft measure system from Fig.1. Hence for accomplishing the monitoring to equipment damage status, it needs experts' experiences except for equipment data, process data etc. Moreover, corrosion analysis professionally requires the abundant knowledge and analytical ability of associated fields, such as process, equipment design, and corrosion. Consequently, the formation of a panel of experts is a widely acknowledged method for encouraging communication among process, equipment, and corrosion experts. However, Chinese petrochemical plants have difficulties in training and organizing a panel of experts. After its formation, a panel cannot continuously go around petrochemical plants to solve all problems on corrosion. Thus, the model of equipment status inspection is designed to provide a net system framework. In the net system, petrochemical plants accept the guidance of the panel to complete the actions at the client terminal, such as data collection and inspection, as shown as Fig. 2.

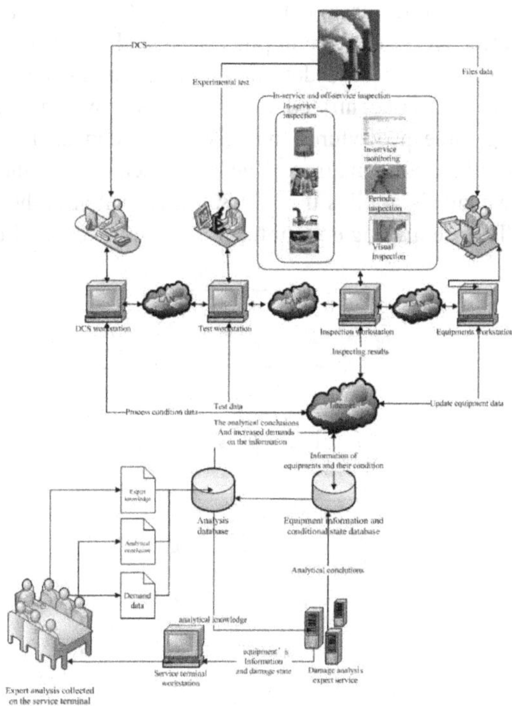

Fig. 2. Net system of the equipment status monitoring subsystem

Moreover, using some assistant systems (corrosion analytical expert system among others), experts are concentrated at service terminals to complete professional analyses, such as corrosion analysis and maintenance decision-making, among others. The distributed system for inspection and maintenance has the following two advantages:

(1) The collected data can be guided and amended at any moment to avoid insufficiency that is usually catastrophic to corrosion and risk analyses.

(2) Analysis and decision-making can be adjusted rapidly using the feedback from the updated data at the terminal.

4 Maintenance Decision-Making Subsystem

In this subsystem, high- and medium high-risk equipment are important for maintenance decision-making and for ensuring safety, long-term operation, and maintenance benefit, which are the root causes of high risks. Moreover, the operational cycle after maintenance and maintenance effectiveness are analyzed.

4.1 Risk-Concentrated Factor Analysis and Confirmation

In this procedure, the data in risk database will be utilized for the risk-concentrated factor analysis. Risk-concentrated factor analysis, which targets maintenance activities that can effectively reduce equipment risk and improve equipment safety, is a necessary procedure prior to the determination of maintenance strategies. Fig. 4 shows the model for risk-concentrated factor analysis, which searches for the root causes of risk using three procedures, namely, risk trend analysis, corrosion factor analysis, and inspection confirmation. The identification of the scope needed to perform risk factor analysis applies the "20/80" law, which is the theoretical base of risk management. The identifying equation can be described as follows:

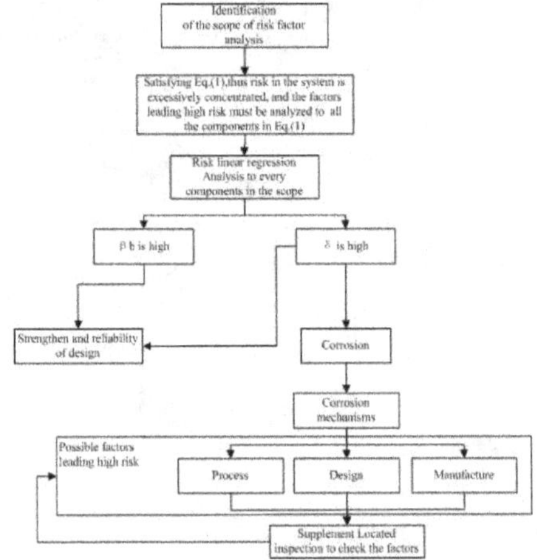

Fig. 3. Risk-concentrated factor analysis model

$$\frac{\sum_{i=1}^{N_H} \mathrm{Risk}_{ki} + \sum_{j=1}^{N_{MH}} \mathrm{Risk}_{kj}}{\sum_{l=1}^{N_k} \mathrm{Risk}_l} > 80\% \text{ and } \frac{\sum_{i=1}^{N_H} n_{ki} + \sum_{j=1}^{N_{MH}} n_{kj}}{\sum_{l=1}^{N_k} n_l} < 20\% \tag{1}$$

The detail in risk-concentrated factor analysis are described in Table 1.

Table 1. The detail of risk-concentrated factor analysis

Analytic content	Analytic objective	Analytic method	Data sources
The intercept of the linear regression of the component risk β_b; slope of the linear regression of the component riskδ	Distinguish the impacts of the factors that strengthen and improve design reliability and the corrosion factors on risk	linear regression analysis	Risk database
Design strength and reliability factors	Confirm design factors to impact component risk including stress concentration, material selection, wall thickness, welding material and process, and PWHT (Post weld heat treatment) etc.	Damage analysis expert system	Risk database, Equipment database
Process factors of corrosion	Confirm design factors to impact corrosion including corrosive media concentration, operation temperature, dew point corrosion, flow velocity, fluid state	Damage analysis expert system	Process database, Equipment database, Chemical analysis database
Design factors of corrosion	Confirm design factors to impact corrosion including dead leg, stress concentration, erosion/corrosion, and basic material selection	Damage analysis expert system	Equipment database
Manufacturing factors of corrosion	Confirm design factors to impact corrosion including wrong welding material selection, welding process, and PWHT	Damage analysis expert system	Equipment database, Maintenanc e database

Finally, supplemented inspection is performed to confirm all the factors obtained from the procedures above and the confirmed factors, which are all supplied to following maintenance decision-making procedures, will be saved into the damage state database, chemical analysis database etc.

4.2 Analysis of the Operational Cycle after Maintenance

Besides safety, prolonging the operational cycle is an important objective of maintenance decision-making. In the maintenance model that regards availability and reliability as optimal objectives, the maintenance decision-making function is not only concerned on the average availability of the equipment system, but also on the deviation of availabilities in the equipment system [18]. As far as fixed equipment is concerned, availability involves original reliability and its increment. Thus, the status of the planned equipment maintained after maintenance, as well as the current reliability and the increment of reliability of all the components around the equipment are considered to reduce time deviation, which causes components to reach reliability or risk limit and improves the availability of the whole unit. Based on this philosophy, the following equipment management structure was established:

Unit→System→Equipment→Component

where "System" technically refers to the group of equipment and pipes that can independently complete a certain process function. In a "System," the specific process function performed by the system would not be accomplished if a component fails, resulting in shutdown or huge economic loss. Thus, the "System," as the basic unit that performs maintenance decision-making, should be the objective of maintenance decision-making to adjust the deviation of reliabilities of the components in the system. "Component" refers to the minimized unit that implements specific equipment management jobs, such as repair, risk analysis, and corrosion analysis, among others.

In this procedure, based on the data in damage state database and risk database, the difference between the expected net production profit per year and the increment of the risk of the system per year is the optimal model after various factors, such as direct, indirect, and in-service maintenance costs, among others, which are considered synthetically. The advantage of the model includes greater susceptibility to rapidly increasing risk at the end of the equipment life, which could lead to more accurate decision-making. As far as the "System" is concerned, the optimal maintenance time and scope can be confirmed as follows:

$$\text{Min:} \quad \text{Rate per year}_{\text{real-profit}}(t) = E(t) - \Delta R(t) \tag{2}$$

$$\text{Subjected to:} \quad E(t) = B - M(t) \tag{3}$$

$$\Delta R(t) = \Delta R(t_k) = R(t_k) - R(t_{k-1}) \tag{4}$$

$$R(t) = \text{Max}\{g_i(t)\} \times \text{COF} \tag{5}$$

$$M(t) = \sum_{j=1}^{N(t)} m_j + Pr_{leak} \frac{m_0 + l_{low}}{t} \tag{6}$$

$$m_j = \frac{m_s + l_d \times D_s}{t} \tag{7}$$

$$N(t) = N_H(t) + N_{MH}(t) + N_{ME}(t) \tag{8}$$

$$\text{Risk}_{comi} / \text{Risk}_{sys} \geq \varepsilon \tag{9}$$

where $g_i(t)$ is the probability function of the failure of component i in a "System" at variable time t, whose default is subject to gamma stochastic process, B is the economic profits per year, m_j is the average maintenance cost of component j per year, Pr_{leak} is the probability of the leak during the interval for the next preventive maintenance, m_0 is the cost of the pressured seal, l_{low} is the loss due to low loading when the pressured seal is used, m_s is the maintenance cost, l_d is the loss per day during the maintenance, D_s refers to the maintenance days including maintenance delay, supply delay, access time, flaw inspection, replacement and repair, equipment revalidation [19], and restarting of the unit, $M(t)$ is the average maintenance cost per year of the "System", $N(t)$ is the amount of the planned maintenance components, which is the function at variable time t determined using Eq. (8), $N_H(t)$ is the amount of high-risk components, $N_{MH}(t)$ is the amount of medium high-risk components, $N_{ME}(t)$ is the amount of medium-risk components determined using Eq. (9), $Risk_{comi}$ is the risk of component *i*, $Risk_{sys}$ is the total risk of the "System", and ε is the threshold parameter for the ratio of the risk between the component and the system selected and adjusted by the user.

From the previous equipment management structure, it is known that a unit consists of some systems, so the most optimal operational cycle of a unit is decided by the shortest one of the most optimal operational cycles of systems in the unit. According to the viewpoint above, after the users in plants have chosen the most suitable unit continuous operational cycle ΔT_{op} with the unit operational plan, the most optimal maintenance plan can be determined through calculating the shortest one of the most optimal system continuous operational cycle , namely, the shortest one of the most optimal maintenance time $t_{op\text{-}shortest}$ with Eq.(2)~ (9) after the specific maintenance plan has been performed. In other words, the most optimal maintenance time of every maintained system after some maintenance plan has been performed, $t_{op\text{-}i}$, can be assessed through Eq.(2)~ (9), in which the shortest one $t_{op\text{-}shortest}$ will be utilized to determine the most optimal maintenance plan through comparing $t_{op\text{-}shortest}$ and ΔT_{op}. The maintenance plan with the least positive difference of ($t_{op\text{-}shortest}$ − ΔT_{op}) will be the most optimal one.

4.3 Assessment of Maintenance Effect

In the procedure, based on the maintenance database and risk database, the risk-effective cost of maintenance plan is assessed. The "System" in equipment management structure is the basic decision-making objective in this study. In theory, components may have the same or different corrosion mechanisms, although the former is mostly used in practice. Based on the concept of the "System" discussed in Section 1.3, the “Buckets effect” obviously exists in a "System", whether maintained or not. The risk of the "System" is dominant by the component with maximum risk, as shown in Eq. (10). Therefore, the change in "System" risk should be balanced by the risk of the “shortest board” in the maintained "System".

$$Q = \mathrm{Max}\{q_1, q_2, \ldots, q_M\} \tag{10}$$

where q_i is the probability of failure of component *i* in a system.

Fixed equipment maintenance includes many contents, such as recovery repairing, material updating, and reliability redesigning, among others. However, regardless of what the maintenance includes, its final impact on the equipment has the following aspects:

(1) Change in equipment risk
(2) Change in equipment risk rate

Thus, a maintenance assessment factor should include the following two aspects:

$$\mathrm{Factor}_{maintenace} = \frac{risk_{original} - risk_{after\text{-}maintenance}}{cost_{maintenance}} + \frac{B\Delta t}{\sum_{i=1}^{\Delta t} \Delta risk_i} \tag{11}$$

where $risk_{original}$ is the original risk prior to maintenance, $risk_{after\text{-}maintenance}$ is the remaining risk after maintenance, $cost_{maintenance}$ is the maintenance cost, Δt is the expected interval to the succeeding maintenance, and$\Delta risk_i$ is the increment of risks in year i after maintenance.

According to Eq. (10), let the probability of the failure of component i in a "System" be$\theta_i q_i$ after maintenance, and thus, the risk of the "System" after maintenance can be expressed as follows:

$$\text{risk}_{\text{after-maintenance}} = Q \times CoF = \text{Max}\{\theta_1 q_1, \theta_2 q_2, ..., \theta_M q_M\} \times CoF \qquad (12)$$

5 An Application Case

In this paper, Maoming petrochemical company is as a case, in which the specific application system is shown as Fig. 4.

In the simplified application system, corrosion survey, RBI, in-service located thickness reading, chemical analysis, and periodic inspection are combined together to predict the next planned shutdown maintenance and open ratio of the fixed equipment, among others. The specific flow is as follows:

(1) Equipment information, daily chemical data, inspection data, and operational information are standardized and managed in accordance to different units.
(2) The equipment status inspection plan is determined after corrosion analysis, corrosion distribution analysis, and RBI according to the data and information from the corrosion survey.
(3) The original maintenance decision-making will be determined according to all analyses and equipment statuses.
(4) Inspections for maintenance decision-making are performed to determine the risk-concentrated factors, and then the final work orders for maintenance will be scheduled after the assessment and amendment of experts.
(5) The work orders are inputted into the ERP(Enterprise Resource Planning system). The next open ratio of fixed equipment is then predicted in the next maintenance, and the equipment status inspection plan is performed prior to the following maintenance.
The practice of this simplified application system for many years with 121 categories of refined crude oil leads to the quite complicated corrosion conditions in Maoming company since 2000. The open ratio of fixed equipment and the inspection maintenance cost obviously decreased from 2002 to 2007, as shown in Tables 2 and 3.

On the other hand, unplanned shutdowns have been continually reduced and the continuous operational cycle of all units in the refinery plant prolonged because of effective damage status inspection and equipment predictive maintenance. For example, the operational cycle of Maoming prolonged from 3 years to 6 years in the lubricating oil unit and from 1.5 years to 3 years in other units.

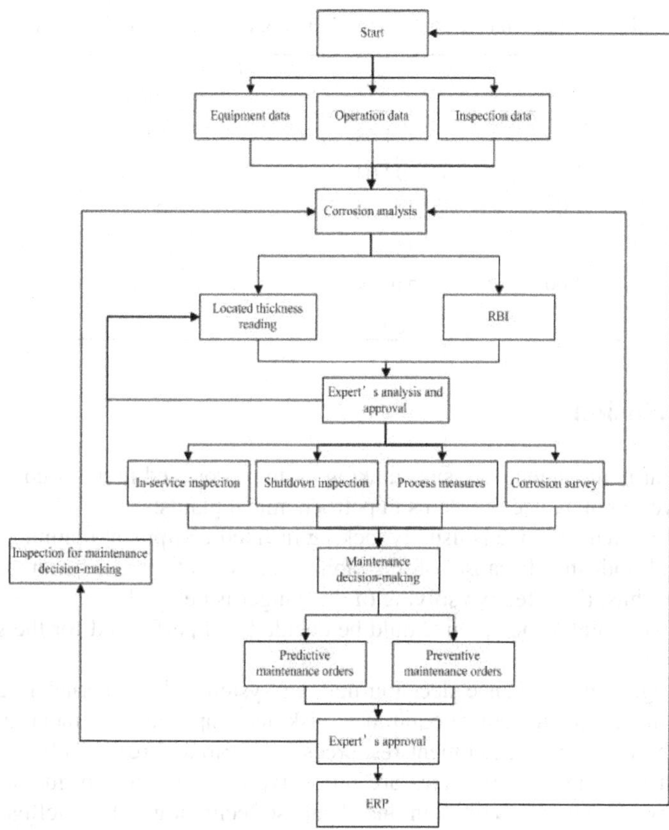

Fig. 4. Simplified application system in Maoming petrochemical company

Table 2. The open ratio of fixed equipment in different units in 2007

Order	Unit	Equipment open ratio
1	Residue hydrotreating	From 100% to 41.72%
2	Hydrocracking	From 100% to 71.3%
3	#3 Sulfur recovery and tail-gas treating	From 100% to 77.69%
4	#4 Sulfur recovery	From 100% to 77.69%
5	#1 Solvent recuperation	From 100% to 90%
6	#2 Solvent recuperation	From 100% to 90%
7	#3 Distilling	From 100% to 11.83%
8	Lubricating oil	

Table 3. Maintenance cost in the Maoming petrochemical company

Year	Cost (Million)	Cost and total fixed assets ratio (%)
2002	334.68	4.12
2003	293.04	3.9
2004	248.00	3.2
2005	256.79	2.96
2006	249.58	2.88
2007	200.00	2.06

6 Conclusion

The intelligent maintenance decision-making system proposed in this study solves the following two maintenance problems in petrochemical plants:
1)Pertinence of maintenance is usually lacking in fixed equipment maintenance in the petrochemical industry because of the ability constraint in corrosion and failure analyses, and thus, the effective solving of the danger is delayed.
2)A preventive maintenance plan should be decided and performed for the safety and long-cycle product.

The intelligent maintenance decision-making system whose kernel is equipment risk management aims to reduce equipment risk and improve equipment availability. Moreover, the ranking of equipment resources is distributed reasonably according to the risk, and maintenance strategies are purposively determined to guarantee safety and reliability. Through information and database technology, the intelligent system becomes a distributed maintenance decision-making net system that consists of two subsystem, namely, equipment damage and risk status monitoring subsystem and intelligent maintenance decision-making subsystem. The former is the base and different data are saved into corresponding database through the net system. In the latter, based on the data in database, Risk-concentrated factors in high- and medium high-risk components, the operational cycle of a unit after maintenance, Assessment of maintenance effect are performed step by step to satisfy the double demands on safety and economic effectiveness. Through the simplified application of this system proposed in this study in the Maoming petrochemical company in SINOPAC, with the intelligent system equipment open ratios have decreased in all units in the refinery plant, and maintenance cost has decreased to 50% from 2002 to 2007. However, the average continuous operational cycle per unit has increased to 100%, indicating that safety and economic effectiveness have obviously improved using this intelligent system proposed in this study.

Acknowledgements. The authors would like to acknowledge the support of the National Science & Technology Pillar Program during the Twelfth Five-Year Plan Period (Approved Grant No. 2011BAK06B02 and 2011Bak06B03) and the National Basic Research Program of China (973 Program) (Approved Grant No. 2012CB026000).

References

1. Kim, J., Lee, Y., Lim, W., Moo, I.: Application of TRIZ crativity intensification approach to chemical process safety. Journal of Loss Prevention in the Process Industries 22, 1039–1043 (2009)
2. Ramirez, P., Utne, I.B.: Challenges with ageing plants. Process Safety Progress 30, 196–199 (2011)
3. Dhillion, B.S.: Engineering maintenance, a modern approach. CRC Press, New York (2002)
4. Cowing, M.M., Cornell, M.E.P., Glynn, P.W.: Dynamic modeling of the trade off between productivity and safety in critical engineering systems. Reliability Engineering and System Safety 86, 269–284 (2004)
5. Yongli, H.: System Control and Equipment Management for Energy & Power Engineering. Huazhong University of Science and Technology Press, Wuhan (2006)
6. Tan, J.S., Kramer, M.A.: A general framework for preventive maintenance optimization in chemical process operations. Computers Chemical Engineering 21, 1451–1469 (1997)
7. Bertolinia, M., Bevilacqua, M., Ciarapica, F.E., Giacchetta, G.: Development of Risk-Based Inspection and Maintenance procedures for an oil refinery. Journal of Loss Prevention in the Process Industries 22, 244–253 (2009)
8. Hu, H., Cheng, G., Li, Y., Tang, Y.: Risk-based maintenance strategy and its applications in a petrochemical reforming reaction system. Journal of Loss Prevention in the Process Industries 22, 392–397 (2009)
9. American Petroleum Institute: API 581 Risk-Based Inspection Base Resource Document (2008)
10. Khan, F.I., Haddara, M.M.: Risk-based maintenance (RBM): a quantitative approach for maintenance/inspection scheduling and planning. Journal of Loss Prevention in the Process Industries 16, 561–573 (2003)
11. Khan, F.I., Haddara, M.R.: Risk-based maintenance (RBM): a new approach for process plant inspection and maintenance. Process Safety Progress 23, 252–265 (2004)
12. Khan, F.I., Haddara, M.: Risk-based maintenance of ethylene oxide production facilities. Journal of Hazardous Materials 108, 147–159 (2004)
13. Krishnasamy, L., Khan, F., Haddara, M.: Development of a risk-based maintenance (RBM) strategy for a power-generating plant. Journal of Loss Prevention in the Process Industries 18, 69–81 (2005)
14. Pham, H., Wang, H.: Imperfect maintenance. European Journal of Operational Research 94, 425–438 (1996)
15. Shin, I., Lim, T.J., Lie, C.H.: Estimating parameters of intensity function and maintenance effect for repairable unit. Reliability Engineering and System Safety 54, 1–10 (1996)
16. Martorell, S., Sanchez, A., Serradell, V.: Age-dependent reliability model considering effects of maintenance and working conditions. Reliability Engineering and System Safety 64, 19–31 (1999)
17. Jayabalan, V., Chaudhuri, D.: Cost optimization of maintenance scheduling for a system with assured reliability. IEEE Transaction on Reliability 41, 21–25 (1992)
18. Kančev, D., Gjorgiev, B., Čepin, M.: Optimization of test interval for ageing equipment: A multi-objective genetic algorithm approach. Journal of Loss Prevention in the Process Industries 24, 397–404 (2011)
19. Krishnasamy, L., Khan, F., Haddara, M.: Development of a risk-based maintenance (RBM) strategy for a power-generating plant. Journal of Loss Prevention in the Process Industries 18, 69–81 (2005)

Dimensional Modeling for Landslide Monitoring Data Warehouse

Hongbo Mei and Guangdao Hu

Faculty of Earth Resources, China University of Geosciences,
Wuhan 430074,China
hbmei@cug.edu.cn

Abstract. There selects starlike structure which is one of the dimensional modeling as the logical construction of landslide monitoring data warehouse, and establishes a bus matrix for the data warehouse. According to the characteristics and subject requirements of landslide monitoring data, this paper focuses on six fact tables such as daily rainfall, daily water level, fissure relative displacement and five shared dimensions such as location, time, monitoring point and so on. Using those fact tables and dimensions to establish corresponding muti-demensional models with building blocks approach which provides subject-oriented data support for landslide hazard prediction and waning command.

Keywords: landslide monitoring, data warehouse, dimensional modeling.

1 Introduction

Landslide is one of the main types of geological disasters, which has lots of characteristics such as wide geographical distribution, high occurrence frequency, fast movement, serious disaster loss.Its hazards and destructiveness is only after eatthquakes and volcanoes.For this reason, it has a great significance to do the research that how to monitor and forecast landslide which can improve the rapid response capability for sudden geological disaster and to realize the disaster prevention and mitigation effectively[1][2].

In order to achieve the forecasting and early warning command, there need for not just high-level methodology and model library for statistical analysing and fully utilizing those information, but also the tool for quickly finding and mining useful information in enormous amount of monitoring database to make accurate and effective decisions quickly. But for traditional operational database, that would be hard and unable to achieve. However, data warehouse can satisfy the need for quickly search and data mining beccause of the functional features of subject-oriented, integrativity, diachronism, so that to provide an important support for early warning and command decision-making[3].

F.L. Wang et al. (Eds.): WISM 2012, LNCS 7529, pp. 414–421, 2012.

2 The Logical Structure and Algorithm Process of Dimensional Modeling

The key to the success of data warehouse structure design lies in data organization, integrated storage and effective management and maintenance.That requires a good data warehouse model.

Logical modeling as the core of data warehouse is developed on topic model, and also is the physical basis for designing data warehouse. At present, the mainstay techniques for data warehouse modeling can be divided into two categories. One is entity-relationship modeling, the other is dimension modeling. The latter is widely accepted as the leading technology which can be expert in handling atomatic type data and fast querying data to demonstrate data warehouse. Dimentional modeling can not only overcome the shortages of query slowness and understand difficulties of entity-relationship modeling, but also achieve the design requirements of data warehouse. The simplictiy of design Implementation is the lifeblood of dimensional modeling, and its basic elements are fact tables and dimention tables[4].

2.1 The Typical Logic Structure of Dimentional Modeling

Dimensional modeling is an analysis model which contrapose relatively independent business. It has two typical logic structures: starlike and snowflake.

As shown in fig.1, the fact table is in the center of starlike structure, surrounded by dimension tables. Snowflake, a expansion of starlike, consists multi-layer structure of demension table. In order to avoid data redundancy, it uses multiple tables to describe a complex dimension table as the fig.2. Considering the demand of On-Line Analytical Processing (OLAP) user system and the processing efficiency of Extract Transform and Load (ETL), starlike structure is characterized by high redundancy, fast aggregation, high analysis efficiency etc.

While snowflake structure with clear hierarchy can easily interact with On-Line Transaction Processing (OLTP) user system[5]. Because of the large amount of landslide monitoring data, it is worth to achieve fast online retrieval by the sacrifice of storage space. So this paper adopts starlike structure for the construction of landslide monitoring data warehouse.

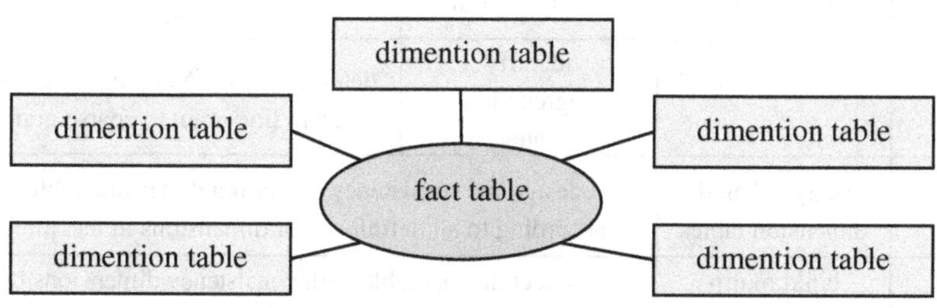

Fig. 1. Starlike structure

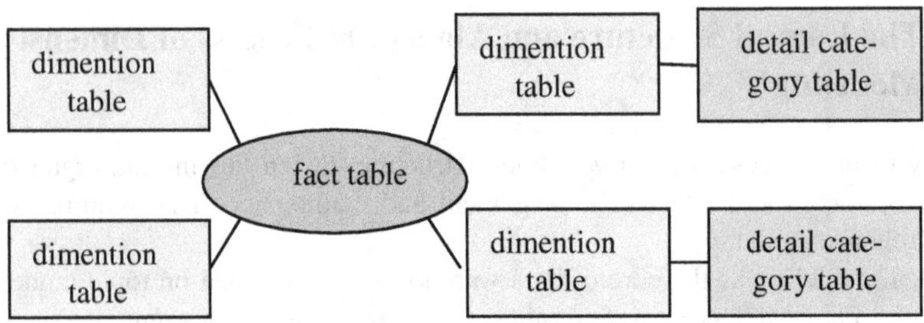

Fig. 2. Snowflake structure

2.2 Algorithm Process of Dimensional Modeling

Data warehouse must provide consistent information for similar inquires. A method to keep this consistency is to create consistent dimension tables and fact tables which are used and shared by all modules and subjects in data warehouse: establish bus structure of data warehouse. On this basis, constantly adjust or parallelly add some new subjects according to user's requirements, and gradually build an integrated data warehouse. Table 1 show the algorithm process of dimensional modeling[6].

Table 1. Program algorithm of dimensional modeling

No.	Task	Subtask	Task scheduler
1	create bus structure of data warehouse	list subjects	analyse the related subjects by row rank of matrix in the data warehouse
		list the various dimensions	analyse all dimensions by column of the matrix
		Identify the intersection	marking the correlation between subject row and dimension column
2	design all kinds of fact tables	select subject	choose some specific subject
		select dimension	choose all kinds of dimensions associated with the fact table
		identify measure	determine the contents in the fact table
		identify aggregation strategy	determine how to aggregate measure from fine grain to coarse grain
3	design shared dimension tables	design the consistency of sharing dimension tables according to all definitions of dimensions in fact tables	
4	build multi-dimensional datasets	connect the fact table with consistency dimensions by foreign-keys, and form the multi-dimensional dataset by starlike model	

3 The Dimensional Modeling Process of the Landslide Monitoring Data Warehouse

3.1 The Bus Matrix of Landslide Monitoring Data Warehouse

Through defined standard bus interface for data warehouse environment, each subject can be implemented by different teams at different time, and then can be integrated together like building blocks.

One row in the data warehouse bus matrix is corresponding to certain business subject.The column of matrix is bus bandwidth, corresponding with public dimentions.Table 2 gives a subset of the landslide monitoring data warehouse , and "√" indicates that this column is related to certain subject row[7].

Table 2. The bus matrix in landslide monitoring data warehouse (subset)

Dimension / Subject	Location	Time	Monitoring point	Monitoring category	Monitoring position
Daily precipitation	√	√			
Daily water Level	√	√			
Fissure relative displacement	√	√	√		
Surface drain flow	√	√	√		
Seepage flow by Weir method	√	√	√		
Monitoring and predicting workload	√	√		√	√

3.2 Determine Subjects of Decision-Making

Subject is a standard of classifying data in the higher level. Each subject is basically related to a macroscopical analysis area. Subject is based on the requirements of analysis to determine, which is different with organizing data by the requirements of data processing.

In data warehouse, each subject is realized by a group of relational tables, that is to say, subject is still based on the relational database. The subjects related to landsilde monitoring: daily precipitation, daily water level, fissure relative displacement, surface drain flow, seepage flow by Weir method, monitoring and predicting workload, etc.

3.3 Determine Measurements and Aggregation Strategy

Measurement is a practical significance data concerned by decision-maker. Different business subjects should select different measurements. Some measure examples related to landslide monitoring are listed below: precipitation, water level, water temperature, water level altitude, water level difference,etc..

According to different measurements, functional operation sets can be selected to summarize measurements from fine grain to coarse grain, such as maximum, average, and summation. For example, daily precipitaion can be cumulated together to form the rainfall overall monthly. As to water level, it can be collected by calculating average.

3.4 Design Fact Tables

Fact table based on dimensional table is a kind of data tables to count detail data.Every data warehouse contains one or more fact tables. For instance, the center of starlike structure is a fact table. The main feature of fact table is that it contains metric data which can provide information of operating history data by summary. Each fact table also includes one or several indexes which contain foreign keys of relevant dimensions.

Combined with the characteristics of landslide monitoring data and theme demand, the Composition of some fact tables is listed as follows.

1) Fact table of daily precipitation. Measurement is rainfall, and referenced dimensions are location and time.

2) Fact table of daily water level. Measurements are water level, water temperature, water level altitude, water level difference. Referenced dimensions are location and time.

3) Fact table of fissure relative displacement. Measurements are accumulated gripping displacement, observed value of opening and closing. Referenced dimensions are location,time and monitoring point.

4) Fact table of surface drain flow. Measurements are square of collecting rain and flow. Referenced dimensions are location,time and monitoring point.

5) Fact table of seepage flow by weir method. Measurements are upstream water level, tail water level, measured flow, water temperature, standard flow, transparency, air temperature, rainfall. Referenced dimensions are location,time and monitoring point.

6) Fact table of monitoring and predicting workload. Measurement is monitoring number. Referenced dimensions are location,time, monitoring category, monitoring position.

3.5 Design Shared Dimentions

Corresponding shared dimension in data warehouse means that every used dimension is a strict mathematical subset of the most subtle dimension. What can show the consistency are conforming key words, attribute name, definition and value [8].

The key of dimension modeling is the modeling to dimension level and the relationship of them. Some levels in landslide monitoring data warehouse lists as follows.

1) Dimension of location includes four levels: province, district, township and hazard body.

2) Dimension of time includes three levels: year, month and day.

3) Dimension of monitoring point includes two levels: monitoring method and monitoring point.

4) Dimension of monitoring category includes two levels: all categories and monitoring category.

5) Dimension of monitoring position includes two levels: all positions and important points.

3.6 Establish Multi-dimension Datasets

Starlike structure combines fact table with dimesion table. Fact table usually has two or more foreign key (FK). FK is a bridge to connect fact table and dimension table. Fig.3 and fig.4 below is two instances about establishing multi-dimension datasets through connecting by FK.

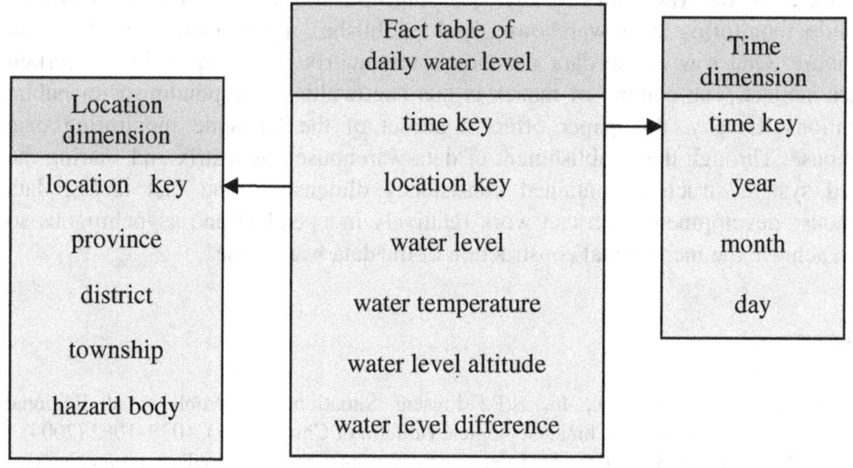

Fig. 3. Multi-dimension dataset of daily water level

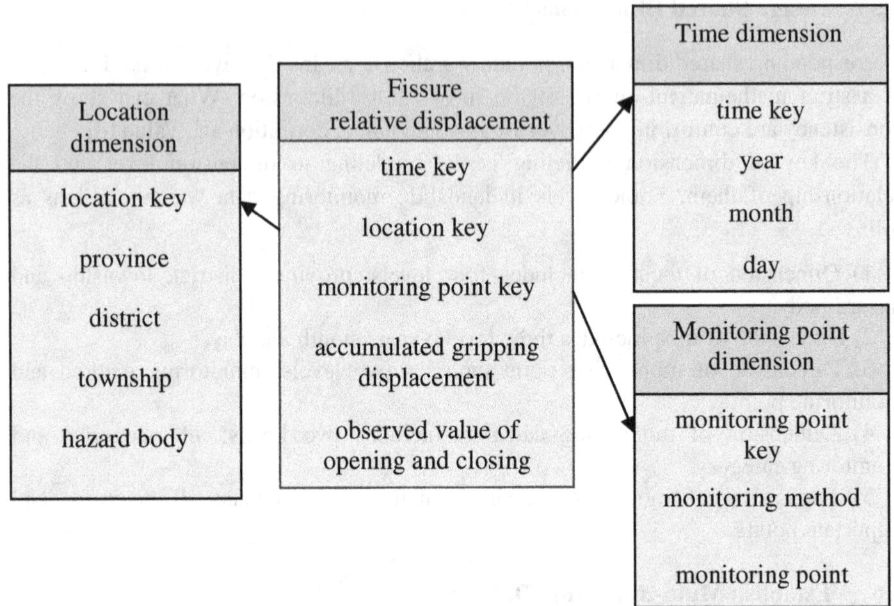

Fig. 4. Multi-dimension dataset of fissure relative displacement

4 Conclusion

This paper selects starlike structure of dimensional modeling which is expert in handling atomatic type data and fast querying data as the logical construction of landslide monitoring data warehouse, and establishes a bus matrix for the data warehouse. One row in the data warehouse bus matrix is corresponding to certain service subject. The column of matrix is bus bandwidth, corresponding with public dimentions. Hereby, this paper offers a subset of the landslide monitoring data warehouse. Through the establishment of data warehouse bus matrix and sharing the unified system structure contained consistency dimensions and fact tables, data warehouse development team can work relatively independent and asynchronous so that to achieve the incremental construction of the data warehouse.

References

1. Huang, R.Q., Xiang, X.Q., Ju, N.P.: Present Situation and Problems of Regional Geohazards Assessment in China. Geological Bulletin of China 23(11), 1078–1082 (2004)
2. Wang, N.Q., Wang, Y.F., Luo, D.H., et al.: Review of Landslide Prediction and Forecast Research in China. Geological Review 54(3), 355–361 (2008)
3. Hu, G.D., Li, Z.H., Mei, H.B.: Geohazards Data Warehouse Design and Development of the Three Gorges Reservoir Area. Earth Science-Journal of China University of Geosciences 36(2), 255–261 (2011)

4. Ralph, K., Margy, R.: The Data Warehouse Toolkit: the Complete Guide to Dimensional Modeling, 2nd edn. Wiley Computer Publishing (2002)
5. Bai, S.L.: Modeling of Data Warehouse Dimension and ETL Process. Liaoning Higher Vocational Technical Institute Journal on Technology and Application 10(10), 61–63 (2008)
6. Li, J., Zhao, Z.S.: Design Strategies of Data Warehouse Dimensional Modeling. Vocational and Technical Education Forum (6), 264–266 (2006)
7. Mei, H.B.: Study on Data Warehouse System of Monomer Landslide Hazard in Three Gorges Reservior Area. Doctoral Thesis of China University of Geosciences, Wuhan (2010)
8. Gu, H.T.: Dimensional Modeling and Data Processing in Data Warehouse for Regional Power Dispatch 26(10), 49–53 (2006)

A New Fuzzy Risk Evaluation Method for Uncertain Network Public Sentiment Emergency

Qiansheng Zhang[1], Xia Li[1], and Yirong Huang[2,*]

[1] School of Informatics, Guangdong University of Foreign Studies, Guangzhou 510420, China
[2] SunYat-sen Business School, Sun Yat-sen University, Guangzhou 510275, China
huangyr@mail.sysu.edu.cn, zhqiansh01@126.com

Abstract. This paper presents a fuzzy risk evaluating method for network public sentiment emergency in which the risk factors are easily assessed by fuzzy value and the risk grades of each risk factor are evaluated by fuzzy linguistic value. By using the fuzzy entropy formula and the fuzzy comprehensive judgment matrix, we calculate the occurring probability of the unknown network public sentiment emergency. According to the interval that the calculated fuzzy comprehensive risk probability lies in, the risk rate of the network public sentiment emergency can be determined, which can further facilitate the urgent emergency decision making.

Keywords: Network Sentiment Emergency, Entropy, Fuzzy Judgment Matrix, Fuzzy Risk Evaluation

1 Introduction

Network public sentiment is the set of public opinions of some event with some influence and strength. Recently, Network sentiment analysis and early warning become very important research issues. As is well known, the uncontrolled network sentiments easily incur the emergency. Simultaneously, the abrupt events affect network public sentiment. In the past decades, Zeng [14, 15] and Zhang [16] proposed the methods of selecting sentiment indexes and determining their weights for network sentiment emergency. Peng [6] discussed the close relationship between network public sentiment and emergency. Also some authors [9, 13] have proposed many early warning decision methods for network emergency. However, most of the existing early warning mechanisms and decision methods can only deal with the emergency under precise condition and certain environment. In fact, due to the increasing complexity of the socio-economic environment and the lack of knowledge about the problem domain, most of the real-world problems, such as network public sentiment emergency decision analysis and risk evaluation, are involved variety of fuzziness, like fuzzy value and fuzzy linguistic value.

Especially, in the risk evaluation process of network sentiment emergency it will inevitably involve some uncertain factors including the economy depression, the wicked public moral, the improper network real-time broadcasting, the network group

[*] Corresponding author.

F.L. Wang et al. (Eds.): WISM 2012, LNCS 7529, pp. 422–430, 2012.

attack, and the political assembly, as well as the religious conflict, etc. Also, the values of above risk factors are easily expressed by fuzzy value or fuzzy linguistic values. As is well known, the unexpected and uncontrolled risks easily incur the emergency in the public network environment. So, in order to decrease the possibility of network sentiment emergency, there is much need to analyze and control the risk of Network sentiment emergency [1]. And, it is a necessitous task for the government department decision-maker to adopt the corresponding strategy to avoid and reduce the occurrence of Network sentiment emergency in public according to the evaluated fuzzy comprehensive risk.

Recently, many researchers studied the risk analysis of emergency in [5, 8, 10, 11], but the fuzzy risk factor was not considered. Although Zhong[12] employed fuzzy entropy and comprehensive judgment method to calculate the occurring probability of risk, but it is based on traditional Shannon entropy. And Li [4] discussed the fuzzy comprehensive risk judgment of HR outsourcing, it is unable to determine the accurate risk grade. In fact, most of the existing risk analysis methods can only deal with the precise risk computing. Even most of the fuzzy risk computing methods still focus on the qualitative analysis, they can not effectively cope with the risk evaluating and risk grade determining involved fuzzy risk factors. Thus, in this paper we aim to propose a new fuzzy risk evaluation approach for network sentiment emergency in the uncertain environment.

2 Prelimilaries

Fuzzy set introduced by Zadeh[17] is a useful extension of the classic set, which has been proved to be more suitable way for dealing with vagueness and uncertainty. Particularly, the information entropy [7], similarity measure and distance measure [3] of fuzzy sets play very important roles in the extensive application areas, such as medical diagnosis, decision-making [2] and pattern recognition.

Definition 1[17]. A fuzzy set A in the universe $X=\{x_1,x_2,\cdots,x_n\}$ is defined as $A=\{(x_i,\ \mu_A(x_i))/x_i\in X\}$, and the condition $0\le\mu_A(x_i)\le 1$ must hold for any $x_i\in X$, where $\mu_A(x_i)$ is called the membership degree of element x_i to the fuzzy set A. We denote all the fuzzy sets in universe X by $F(X)$.

Definition 2. Let A,B be two fuzzy sets in the finite universe $X=\{x_1,x_2,\cdots,x_n\}$, the union, intersection and complement of fuzzy sets are defined as follows.

$$A\cup B=\{(x_i,\mu_A(x_i)\vee\mu_B(x_i))|x_i\in X\},$$

$$A\cap B=\{(x_i,\mu_A(x_i)\wedge\mu_B(x_i))|x_i\in X\},$$

$$A^c=\{(x_i,1-\mu_A(x_i))|x_i\in X\}.$$

Definition 3. A mapping from all the fuzzy sets in universe X to interval [0,1] is named as the fuzzy entropy, if it satisfies the following extension of DeLuca-Termini axioms:

(p1) $H(A)=0$ (minimum) , if A is a crisp set, i.e., $\mu_A(x_i)=0$ *or* 1 for all $x_i \in X$;

(p2) $H(A)=1$ (maximum), iff $\mu_A(x_i)=0.5$ for all $x_i \in X$;

(p3) $H(A^*) \le H(A)$, if $A^* \lll A$, i.e., A^* is a sharpened version of A defined as

$$\begin{cases} \mu_{A^*}(x_i) \le \mu_A(x_i) \ , & for \ \ \mu_A(x_i) \le 0.5 \\ \mu_{A^*}(x_i) \ge \mu_A(x_i) \ , & for \ \ \mu_A(x_i) \ge 0.5 \end{cases} .$$

(p4) $H(A) = H(A^c)$, where A^c is the complement set of fuzzy set A .

Notably, one can easily see that the following formula

$$H(A)=\frac{1}{n}\sum_{i=1}^{n}\frac{\mu_A(x_i)\wedge\mu_{A^c}(x_i)}{\mu_A(x_i)\vee\mu_{A^c}(x_i)}, \tag{1}$$

is an information entropy measure of fuzzy set A , since it fulfills all the properties (p1)-(p4).

3 A Fuzzy Risk Analysis Method for Network Sentiment Emergency

As is well known, the Network sentiment emergency will possibly face to many kinds of risk factors with different probabilities. Especially in the uncertain environment, the accurate value of risk factor is difficult to measure. However, it can be easily evaluated by fuzzy set in the real-life world. By fuzzy comprehensive judgment matrix, we can conveniently judge the possibility grade of each risk factor. And by employing the proposed fuzzy risk analysis method, we can effectively evaluate the corresponding risk grade of Network sentiment emergency.

Through questionnaire survey and statistic analysis, we can easily get some important risk factors of the Network sentiment emergency. As we know, the risk factors that incur the Network sentiment emergency mainly include the economy risk, the political and moral system risk, the network group attack risk, and the religious conflict risk, as well as the public assembly risk, etc. Suppose the set of risk factors in Network sentiment emergency is denoted by $U=\{u_1,u_2,\cdots,u_m\}$. Generally, the risk factor is always fuzzy concept, for example, undesirable economic state, adverse political environment, wicked network broadcasting, bad public moral, and so on.

As we know, the accurate value of the severity of loss and the occurring probability of each risk factor is difficult to measure in real-life complex environment. On the contrary, people and experts can easily evaluate the above probability of each risk factors of Network sentiment emergency by using fuzzy language terms like P = {Very Low, Low, Medium, High, Very High} rather than by using accurate real numbers.

In order to evaluate the risk grade of Network sentiment emergency and provide early warning for the related government department, we should set the different risk grades firstly. For convenience and for sake of effectively early warning for the Network sentiment emergency, here the five risk grades of Network sentiment emergency are pre-established and characterized by the following fuzzy linguistic terms as below.

Table 1. Linguistic terms for rating the risk grades of network public sentiment emergency

Linguistic terms	Interval degrees
Very likely occur (VL)	[0.8, 1.0]
likely occur (L)	[0.6,0.8]
Medium (M)	[0.4,0.6]
Unlikely occur (U)	[0.2,0.4]
Very Unlikely occur (VU)	[0.0, 0.2]

For simplicity, we denote all the five risk grades by the set $G=\{\ G_1$ (VL), G_2 (L), G_3 (M), G_4 (U), G_5 (VU)}. Based on the above risk analysis and the previous formulae, here we give the fuzzy risk comprehensive evaluation process for the Network sentiment emergency involved with fuzzy risk evaluation value under the uncertain environment.

Step 1. Let U be the set of all fuzzy risk factors $\{u_1,u_2,\cdots,u_m\}$, and given the judgment set $V=\{v_1,v_2,\cdots,v_n\}$ for risk occurring probability, $v_j\ (j=1,2,\cdots,n)$ denotes the different probability grade of the occurrence of each risk factor. Assume the fuzzy judgment matrix is $R=(r_{ij})_{m\times n}$, r_{ij} is the fuzzy membership value of risk factor u_i with respect to the judgment criteria v_j, which can be given by the knowledge of related field experts.

Step 2. By using entropy formula (**1**), we compute the entropy of each fuzzy value in the fuzzy comprehensive judgment matrix and get the entropy matrix of this judgment matrix as $D=(h_{ij})_{m\times n}$, where h_{ij} is the entropy value of r_{ij}.

Step 3. Normalize the entropy values in the decision matrix by using the equation.

$$\bar{h}_{ij}=\frac{h_{ij}}{\max_j h_{ij}},\qquad j=1,2,\cdots,n\,,\ i=1,2,\cdots,m\,, \tag{2}$$

And the normalized entropy matrix is expressed as $D=(\bar{h}_{ij})_{m\times n}$.

Step 4. Calculate the occurring probability (weight) of fuzzy risk factor by applying the following transformer.

$$w_i=\frac{1-\sum_{j=1}^{n}\bar{h}_{ij}}{m-\sum_{i=1}^{m}\sum_{j=1}^{n}\bar{h}_{ij}},\qquad i=1,2,\cdots,m. \tag{3}$$

Step 5. Calculate the comprehensive risk value according to the above probability and the jugement matrix of all the risk factors in Network sentiment emergency by the following

$$P = W_U R W_V', \tag{4}$$

where W_U denotes the weight vector of all the risk factors; W_V is the weight vector of all the judgement criteria v_j, W_V' is the transpose of W_V.

Step 6. Calculate the similarity measure $S(R,G_j)$ between the fuzzy comprehensive risk value P and each pre-established risk grade G_j,

Step 7. Determine the risk rate of the unknown Network sentiment emergency. From the calculated risk value we can determine the risk grade of the Network sentiment emergency. If $P \in G_k$, where G_k is the k-th risk grade in the pre-established grade set $G = \{ G_1(\text{VL}), G_2(\text{L}), G_3(\text{M}), G_4(\text{U}), G_5(\text{VU}) \}$, then the risk rate of this Network sentiment emergency should belong to the given grade G_k.

According to the risk grade, the related government department decision-maker can make effective decision to raise the corresponding alarm for public. Moreover, the fuzzy risk evaluation grade of this Network sentiment emergency can help related managers adopt the corresponding urgent decision strategy to avoid and reduce the risk of emergency, or decrease the occurring probability of unexpected emergency in Network environment.

4 Application Example

Recently, fuzzy sets, as an useful tool to deal with imperfect data and imprecise knowledge, have drawn the attention of many researchers in order to perform pattern recognition, medical diagnosis and decision making. In this section, we give a numeric example to illustrate the application of the proposed fuzzy risk evaluating method for Network sentiment emergency in uncertain environment.

Suppose the government department will implement a policy, there is much need to evaluate the risk of Network sentiment emergency regarding this policy. Assume the Network sentiment emergency are subject to many fuzzy risk factors $\{ u_1, u_2, \cdots, u_m \}$, which can be evaluated by n fuzzy judgment criteria $\{v_1, v_2, \cdots, v_n\}$. Assume the fuzzy comprehensive judgment matrix is expressed by $R = (r_{ij})_{m \times n}$, C denotes the set of n judgment criteria, and $r_{ij} = \mu_i(v_j)$ denotes the membership degree that the i th risk factor belongs to the j th judgment criteria. The fuzzy risk evaluating problem is to decide which risk grade the network sentiment emergency should belong to. And the main task is to determine the occurring probability of each fuzzy risk factor and calculate the fuzzy comprehensive risk value of Network sentiment emergency.

Example 1. Suppose a government department will carry out one policy, it needs to monitor the Network sentiments and evaluate the risk of possible emergency. Assume the set of fuzzy risk factors U ={ u_1 (poor policy and economy environment), u_2 (disloyal network broadcasting), u_3 (wicked network group attack), u_4 (severe religious conflict)} must be taken into account in the Network sentiment emergency. And the occurring probability of each risk factor is unknown, but may be evaluated by a fuzzy judgment criteria in V ={ v_1 (Very Low), v_2 (Low), v_3 (Medium), v_4 (High), v_5 (Very High)}. The evaluating results are expressed by a fuzzy comprehensive judgment matrix, where $v_j (1 \le j \le 5)$ denotes the different occurring possibility value of each risk factor of Network sentiment emergency, and r_{ij} is the membership degree of i-th fuzzy risk factor belongs to the j-th judgement criteria as in Table 2. Also, the five risk grades of Network sentiment emergency expressed by fuzzy linguistic terms are pre-established and correspond to the five interval numbers in Table 1.

Table 2. The fuzzy judgment matrix of risk factors regarding the judgment criteria

Risk factor	v_1	v_2	v_3	v_4	v_5
u_1	0.45	0.7	0.4	0.6	0.8
u_2	0.7	0.4	0.35	0.85	0.65
u_3	0.15	0.3	0.6	0.2	0.25
u_4	0.4	0.6	0.8	0.3	0.55

Our main task is to determine the risk grade of the Network sentiment emergency involved with fuzzy values. And then the related government department can adopt the corresponding decision strategy to avoid or decrease the risk of Network sentiment emergency. According to the above mentioned fuzzy risk evaluating method, in what follows we will employ the fuzzy entropy measure to calculate the probability of each fuzzy risk factor in Network sentiment emergency.

First, from the fuzzy value Table 2 we express it as the following fuzzy comprehensive judgment matrix of risk factors with respect to all judgment criteria.

$$R = (r_{ij})_{4\times 5} = \begin{pmatrix} 0.45 & 0.7 & 0.4 & 0.6 & 0.8 \\ 0.7 & 0.4 & 0.35 & 0.85 & 0.65 \\ 0.15 & 0.3 & 0.6 & 0.2 & 0.25 \\ 0.4 & 0.6 & 0.8 & 0.3 & 0.55 \end{pmatrix},$$

where r_{ij} represents the membership value of risk factor u_i with respect to probability grade v_j, for example, $r_{21} = 0.7$ represents the true membership of risk factor u_2 belong to the occurring probability grade v_1 is 0.7.

By using the fuzzy entropy formula (**1**), we can compute the information entropy of each fuzzy value in the above fuzzy judgment matrix and get the following entropy matrix.

$$D = (h_{ij})_{4\times 5} = \begin{pmatrix} 0.8182 & 0.4286 & 0.6667 & 0.6667 & 0.25 \\ 0.4286 & 0.6667 & 0.5385 & 0.1765 & 0.5385 \\ 0.1765 & 0.4286 & 0.6667 & 0.250 & 0.3333 \\ 0.6667 & 0.6667, & 0.250 & 0.4286 & 0.8182 \end{pmatrix}.$$

With formula (**2**), we transform the above entropy matrix to the normalized entropy matrix below.

$$\overline{D} = (\overline{h}_{ij})_{4\times 5} = \begin{pmatrix} 1.000 & 0.5238 & 0.8148 & 0.8148 & 0.3055 \\ 0.6429 & 1.000 & 0.8077 & 0.2647 & 0.8077 \\ 0.2647 & 0.6429 & 1.0000 & 0.3750 & 0.4999 \\ 0.8148 & 0.8148, & 0.3055 & 0.5238 & 1.000 \end{pmatrix}.$$

Then, with the formula (**3**), we can obtain the occurring probability of each risk factor by the following formula $w_i = \dfrac{1-\sum_{j=1}^{5} \overline{h}_{ij}}{4-\sum_{i=1}^{4}\sum_{j=1}^{5} \overline{h}_{ij}}$, $i = 1,2,\cdots,4$. So, we can get the weight vector of all the fuzzy risk factors as W_U =(0.2666, 0.2735, 0.1933, 0.2666). In fact, it represents the occurring probability vector of all the risk factors in the Network sentiment emergency.

Next, assume the weight vector of all the judgment criteria in set V is $W_V = (0.2,0.1,0.3,0.1,0.3)$.

From previous formula (**4**), we calculate the fuzzy comprehensive risk value below.

$$P = W_U R W_V'$$

$$=(0.2666, 0.2735, 0.1933, 0.2666)\begin{pmatrix} 0.45 & 0.7 & 0.40 & 0.60 & 0.80 \\ 0.70 & 0.4 & 0.35 & 0.85 & 0.65 \\ 0.15 & 0.3 & 0.60 & 0.20 & 0.25 \\ 0.40 & 0.6 & 0.80 & 0.30 & 0.55 \end{pmatrix}\begin{pmatrix} 0.2 \\ 0.1 \\ 0.3 \\ 0.1 \\ 0.3 \end{pmatrix}$$

$$= 0.5272.$$

Since $0.5272 \in [0.4, 0.6]$, then the fuzzy comprehensive risk value of Network sentiment emergency should belong to G_3 grade, and the risk of this emergency may be medium. That is to say, the network sentiment emergency will induce some risk of grade G_3. The related government department decision-maker must select the corresponding strategy to decrease the likely occurring risk of the Network sentiment emergency before implementing some public policy.

Acknowledgments. This work is supported by the Humanities and Social Sciences Youth Foundation of Ministry of Education of China (Nos. 12YJCZH281, 11YJCZH086, 10YJC790104), the Guangzhou Social Science Planning Project "The study of early warning index selection and urgent decision mechanism for city significant emergency in uncertain environment" (No. 2012GJ31), the National Natural Science Foundation (Nos. 60974019, 61070061, 60964005), the Fundamental Research Funds for the Central Universities in China, and the Guangdong Province High-level Talents Project.

References

1. Han, L., Lai, L., Yang, S., Li, F.: The risk evaluation for the city water supply emergency in Beijing by using AHP model. Beijing Water Utilities 6, 36–38 (2009)
2. Hong, D.H.: Multicriteria fuzzy decision-making problems based on vague set theory. Fuzzy Sets and Systems 114(1), 103–113 (2000)
3. Liu, X.C.: Entropy, distance measure and similarity measure of fuzzy sets and their relations. Fuzzy Sets and Systems 52(3), 305–318 (1992)
4. Li, L.: The Risk Analysis of HR Epiboly Based on the Fuzzy Integrated Evaluation. Journal of Zhongnan University of Economics and Law 4, 35–40 (2007)
5. Liu, J.K., Yuan, G.L.: The risk management investigation of critical emergency. Journal of College of Railway Police 1, 35–39 (2006)
6. Peng, Z.H.: Discussion on population emergency and network public sentiment. Journal of Shanghai Public Security College 1, 46–50 (2008)
7. Shang, X.G., Jiang, W.S.: A note on fuzzy information measures. Pattern Recognition Letters 18, 425–432 (1997)
8. Wang, P., Lin, J.P.: The evaluation system investigation of emergency in the main railroad. Railway Computer Applications 18(9), 1–3 (2009)

9. Wu, S.Z., Li, S.H.: The study of mechanism for network public sentiment early warning. Journal of Chinese People's Public Security University 3, 38–42 (2008)
10. Xiao, K.H.: The risk early warning model study of emergency in supply chain. Journal of Henan University of Technology 8(1), 54–57 (2012)
11. Ye, J.Z., She, L.: Risk analysis and management of network emergency: a suggested framework. Soft Science 25(12), 59–67 (2011)
12. Zhong, R.Q., Xie, Y.J.: Application of fuzzy risk assessment and entropy coefficient in the software out-sourcing. Science Technology and Engineering 11, 2625–2628 (2011)
13. Zhang, L.L.: The investigation on the model of emergency network public sentiment early warning. Library and Information Service 11, 135–138 (2010)
14. Zeng, R.X., Xu, X.L.: The early warning system, index and mechanism of network public sentiment emergency. Journal of Informatics 11, 52–55 (2009)
15. Zeng, R.X.: The construction of early warning index of network public sentiment. Informatics Theory and Practice 33, 77–80 (2010)
16. Zhang, W.P.: Study on the index selecting and weight evaluation for emergency early warning. Journal of Chinese People's Public Security University 6, 80–89 (2008)
17. Zadeh, L.A.: Fuzzy sets. Information and Control 8(3), 338–353 (1965)

Service Lifecycle Management in Distributed JBI Environment

Ben Wang, Xingshe Zhou, Gang Yang, and Yunfeng Lou

School of Computer, Northwestern Polytechnical University, 710072, Xi'an, China
{wben,zhouxs,yeungg}nwpu.edu.cn, mf_870507@163.com

Abstract. Enterprise Service Bus is widely-used, flexible, SOA-based infrastructure. Lifecycle management is a necessity for component reusing, maintenance, monitoring, as well as service governance. But existing approaches of service lifecycle model and management tools are not suitable for ESB environment, especially for distributed ESB environment. In this paper, we present a lifecycle model for ESB services with the consideration of both JBI components and service assemblies. A web based tool is also introduced to assist administrators to manage and monitor each phase of ESB services lifecycle.

Keywords: Service Lifecycle, Java Business Integration, Register, Repository.

1 Introduction

ESB (Enterprise Service Bus) is a kind of SOA-based infrastructure for business application, especially legacy application, integration, which provides the underpinning for loosely coupled, highly distributed, events driven, service-based applications. This infrastructure allows third-party software or application systems to be "plugged in" as standard components by kinds of adapters, and allows those components to interoperate in a predictable and reliable manner by message router despite being produced by separate providers [1].

Service lifecycle management is an important item in SOA environment for component reusing, product development, maintenance, functional and non-functional monitoring. R. Weinreich and A. Wiesauer in [4] present a service lifecycle model for SOA with essential elements: product-based development of services, support for service reuse, controlled service and system evolution as well as support for process monitoring during development and operating. Ref. [5] presents a generic business and software service lifecycle and aligns it with the common management layers in organizations. To manage services throughout the lifecycle is also a necessity for service governance [5], [6].

But existing approaches can not make adjustment to ESB services. In this paper, we will present a service lifecycle model for ESB environment, especially for distributed ESB environment, to manage both the JBI components and service assemblies. The analysis is based on a commercial ESB production called

F.L. Wang et al. (Eds.): WISM 2012, LNCS 7529, pp. 431–438, 2012.

SynchroESB which will be introduced in section 2. A web based tool named Synchro Registry and repository is designed and implemented to configure, manage and monitor each phases of ESB services.

This paper is organized as follows. In section 2 we will introduce the basic information and architecture of SynchroESB. In section 3 we will present our lifecycle module of JBI components and service assemblies, followed by the design and implementation of Synchro Registry and Repository in section 4. The conclusions are given in section 5.

2 Introduction of SynchroESB

SynchroESB [3] is a commercial software product on the basis of NpuESB which is a distributed ESB system based on JBI (Java Business Integration) environment. Compared with centralized ESB, SynchroESB has the superior capabilities of load balancing, single node failure tolerating, persistent messaging, and meanwhile avoiding performance bottleneck [2].

The Architecture of SynchroESB as shown in Figure 1 could be divided into three layers, communication layer, ESB layer and tools layer.

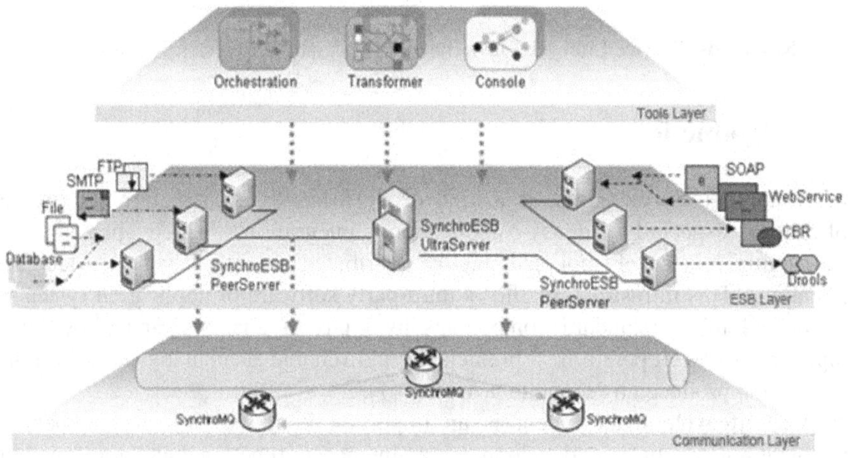

Fig. 1. Architecture of SynchroESB

ESB layer is composed of a set of PeerServers and an UltraServer. Each PeerServer is a standalone JBI environment capable of hosting specification-compliant JBI components. External software and application systems, such as Database system, FTP server and Web Services, can be adapted as Service Engine (SE) or Binding Component (BC) for intercommunication with standardized messages, and those components can be installed in one or more PeerServers. We employ UltraServer as the manager of each JBI nodes, including PeerServer management module, JBI components management module, business process management module, log module and security module.

Communication layer is based on distributed SynchroMQs which are implementations of Java Message Service (JMS). JMS enables distributed communication loosely coupled, reliable, asynchronous and provides the ability of persistent message transmission [2].

Tools layer consists of three GUI based software including Orchestration, Transformer and Console which are employed to interact between the users and SynchroESB. Orchestration is a visual tool for Service Assembly design, deploy, management and monitor. Transformer is used for normalized message schema unified between separate components with different message format. Console is a web based tool for users to configure and monitor the entire SynchroESB environment.

3 JBI Services' Lifecycle Model

Based on the holistic view of SynchroESB, we present a JBI services' lifecycle model (see Figure 2) in the following.

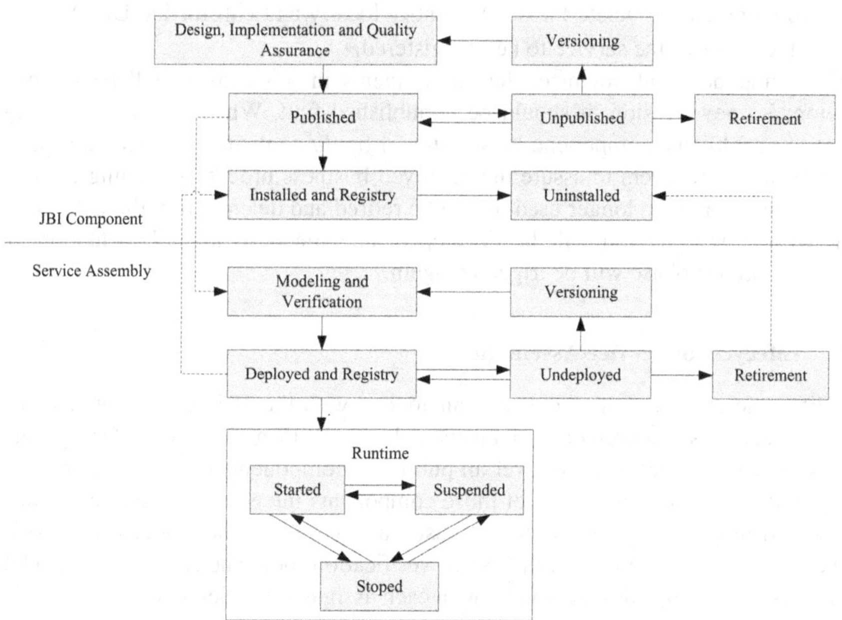

Fig. 2. Lifecycle of JBI components and service assemblies

The lifecycle model is separated into two parts: the lifecycle of JBI components and the lifecycle of service assemblies.

3.1 Lifecycle of JBI Components

The lifecycle of JBI components starts with the creation of a component by design, implementation and quality assurance. Based on the required functionality of business logic, system developers should design and implement a new JBI component which must comply with the JBI specification. Another important factor is the non-functional requirement in terms of reliability, security, interoperability, etc. Quality assurance could ensure all of those meet the requirements.

When the quality assurance is done, the component is ready to be used, we call it published. The published component is archived as a zip file and will be uploaded to the UltraServer. Users who design the service assemblies by the Orchestration will be find it through UltraServer and could use it in the new service assemblies. Once the component is employed in a service assembly, when the service assembly is deployed, the component will be downloaded from the UltraServer to the assigned PeerServer. The PeerServer running the hot deploy program to install the component locally and registry it to the UltraServer, thus, other users can be search and use it without duplicate install. In this statement, we call the component is installed and registry.

When a component is not used anymore by its service consumer in the PeerServer, the service uninstall phase will be triggered. In this phase, the PeerServer do the uninstall command to delete the service entity locally and call for the UltraServer to change the status of the service to be unregistered.

If a component did not meet the requirements in practical, it will be retired or developed a new version, it should be unpublished first. When the system manager wants to unpublish a component, he/she should firstly confirm it has been uninstalled in all of the PeerServers to assure the deployed business processes running correctly. If the component is no longer used, it will be retired and deleted from the UltraServer. Otherwise a new version will be developed, and the design, implementation, and quality assurance phase will be triggered again.

3.2 Lifecycle of Service Assemblies

The lifecycle of service assemblies is analogous with the JBI components'. It starts with the analysis requirements of business logic. With assistance of Orchestration tool, system developers can discover all published components in the UltraServer, and then, by drag and drop the icons of those components integrated as service assembly, they could model the business process easily and visually. The modeled process can be verified by Pi calculus with the use of verification tool. The result of the modeled process is a set of zip files associated with each assigned PeerServers.

After the phase of model and verification, the service assembly can be deployed to the PeerServers and registered in the UltraServer. The deployment program will copy the zip files to each assigned PeerServers and JBI components which are used in the process will be downloaded and installed into the corresponding PeerServer according to the modeled zip files.

In our lifecycle model, service assembly deployment and registry is followed by the phase of runtime. In this activity, service assemblies have three states: started, stopped and suspended. These three states can be changed between each other by the reason of users' commands or the program's controls.

While the service assembly will be retired or a new version has to be developed, it should be undeployed firstly, because only non-running service assemblies can be changed the state to undeployed. For the data safety purpose, it should make sure there is no user is running this service. Meanwhile, the registry program will be triggered in the UltraServer to remove the service records from the registry center.

The retirement stage will delete all associated deployment files and data files from the PeerServers. If the JBI components called by the retired service assembly have gone out of use by other services, the JBI components' uninstall activity will also be triggered.

4 Synchro Registry and Repository

Previously, the management of JBI components' and service assemblies' lifecycle is based on the UltraServer, as we analyzed in previous section. But this mode has the following disadvantages.

Firstly, the UltraServer did not be designed for lifecycle management of JBI services; therefore, the UltraServer has no cognition of the model of JBI services' lifecycle and has not store and index all of the information needed.

What is more, the UltraServer is a RMI-based application, there is no GUI console provided. Users can not query and locate their information quickly and conveniently.

For the requirement of lifecycle management for both JBI components and service assemblies, we design and implement a web-based service registry and repository software named Synchro Registry and Repository Center, short as SynchroRR.

4.1 Architecture of SynchroRR

Structurally, the SynchroRR includes three parts: Registry Center, Repository Center and web-based management console, as illustrated in Figure 3. The Registry Center is capable of store and location both logical and physical addresses of JBI services for SynchroESB, including the functions of service publication, service discovery and service categorization. The Repository Center is employed for storage of service entities and their backups and transmission of service entities to PeerServers when required. Users can access SynchroRR by web-based management console to query and manage the JBI services through web explorer over a network.

The Registry Center is build based on the application services. Each deployed services will registry themselves into the Registry Center, including the logical and physical addresses, the description file based on XML, and the properties stored in HSQL database. In addition, the Registry Center provides the function to categorize the services based on different layers to improve the query efficiency and simplify the information management. The Registry Center is also implemented the capability of recording the QoS information while the service is running. Through all of these efforts, PeerServers and users could query and locate the services conveniently and quickly.

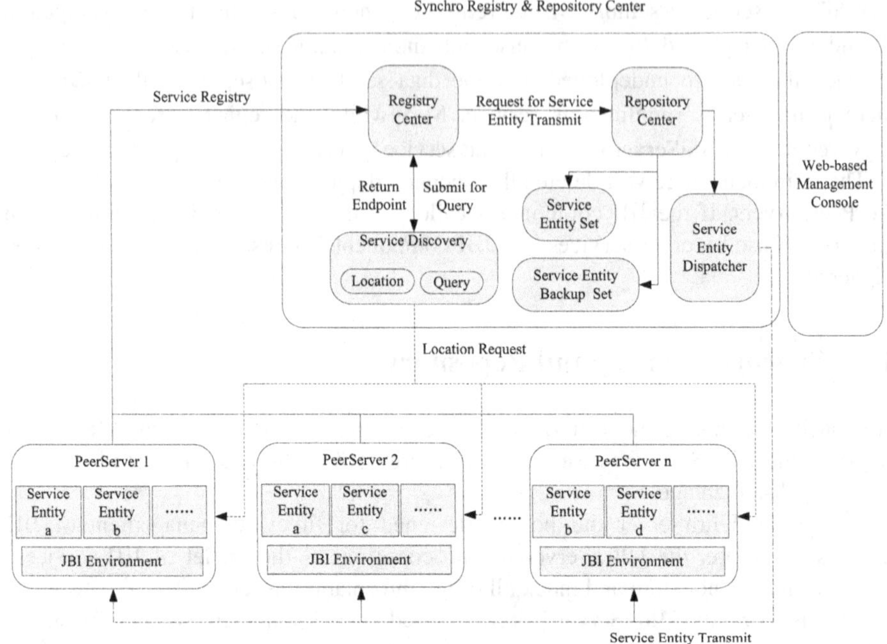

Fig. 3. Architecture of SynchroRR

The Repository Center is used for storing services' entities, including jar files, configure files, documents, such as WSDL, XDS, WS-Policy, SCDL, etc. and other supported files. Some fine grain sized information, for instance, the *port number* and *port type* derived from WSDL file, are also involved in the Repository Center in some cases. A service entity backup contains whatever the service needed at run time; it will be transmitted to PeerServers requested and installed or deployed dynamically to be ready for using. Another important effort is version control. As we discussed, the services version can be changed in the lifecycle, for the sake of service assembly running correctly, the Repository Center must select the JBI components in concern adaptively before the transmission.

Users could access the system through web-based management console including the capabilities of user management, activity monitor, service query and resource management. For each service registered or uploaded into the repository, users could list the information such as creation time, last update time, type , URL, etc., rank it from 1 to 5 classes, modify the properties and description, label it with key words, and write a comment on the service.

4.2 Information Management of SynchroRR

For information management is very important for the SynchroRR configuration, we designed an XML file named ServiceRepositoryConfig.xml to configure and store repository and services information.

The following form shows an example and illustrates the basic grammar of the XML file.

```
<?xml version="1.0" encoding="UTF-8" ?>
<ServiceRepositoryConfig>
<ServiceRepositoryHeader launchable='true'
          isErrorHandlingSupported='true'>
<Name>ServiceRepository</Name>
<CreationDate>07/27/2011</CreationDate>
<Version>1.0</Version>
<Label>Production</Label>
<Category>US</Category>
<ShortDescription>This is a ServiceRepository Module
         Description.</ShortDescription>
</ServiceRepositoryHeader>
<ServiceRepositoryFileInfo>
<Service type=″UnregisteredService″ autorun="false">
<parameter name=″Location″
         value=″%SynchroESB%/ServiceRepository/Unregist
         eredServices/″/>
<parameter name=″Service_Name″ value=″XXXService″/>
<parameter name=″IconResource_Name″
         value=″XXXServiceIcon.gif″/>
<parameter name=″ConfigResource_Name″ value=″
         ServiceDescriptor.xml″/>
</Service>
<Service Type=″RegisteredService″ autorun="false">
<parameter name=″Location″
         value=″ %SynchroESB%/ServiceRepository/Registe
         redServices/″/>
<parameter name=″Service_Name″ value=″YYYService″/>
<parameter name=″IconResource_Name″
         value=″YYYServiceIcon.gif″/>
<parameter name=″ConfigResource_Name″ value=″
         ServiceDescriptor.xml″/>
<parameter name=″ServiceJarResource_Name″
         value=″YYYService.jar″/>
<parameter name=″UIJarResource_Name″
         value=″YYYServiceui.jar″/>
</Service>
</ServiceRepositoryFileInfo>
</ServiceRepositoryConfig>
```

The element *ServiceRepositoryHeader* and its sub-elements contain the basic information of repository including name, creation time, version, description, etc.

The element *ServiceRepositoryFileInfo* and its sub-elements record storage information of unregistered and registered services, such as service name, the type of the service, storage location, and resource files associated with the service.

5 Conclusion

The successful implementation of an ESB system requires the supports of specific requirements like service reuse, service evolution, and quality assurance during the whole service lifecycle. The ESB service lifecycle models presented in this article is on the basis of SynchroESB. However, the presented concepts tools are general enough to be used in other JBI-based systems.

Acknowledgements. This work is supported by "HEGAOJI" Funding of China (grant code 2009ZX01043-002-001-NPU).

References

1. Chappell, D.A.: Enterprise Service Bus. O'Reilly Publishing (2004)
2. Ning, F.: Distributed Enterprise Service Bus Based on JBI. In: The 3rd International Conference on Grid and Pervasive Computing, pp. 292–297 (2008)
3. Synsoft, `http://www.synsoft.com.cn`
4. Weinreich, R., Wiesauer, A.: A Service Lifecycle and Information Model for Service-Oriented Architectures. In: 2009 Computation World: Future Computing, Service Computation, Cognitive, Adaptive, Content, Patterns, pp. 346–352 (2009)
5. Kohlborn, T., Korthaus, A., Rosemann, M.: Business and Software Service Lifecycle Management. In: 2009 IEEE International Enterprise Distributed Object Computing Conference, pp. 87–96 (2009)
6. Niemann, M., Janiesch, C., Repp, N., Steinmetz, R.: Challenges of Governance Approaches for Service-Oriented Architectures. In: 2009 3rd IEEE International Conference on Digital Ecosystems and Technologies, pp. 600–605 (2009)

Graded BDI Models for Agent Architectures Based on Łukasiewicz Logic and Propositional Dynamic Logic

Xiaojun Zhang[1], Min Jiang[2], Changle Zhou[2], and Yijiang Hao[3]

[1] Department of Philosophy, Xiamen University, Xiamen 361005, China
[2] Department of Cognitive Science, Xiamen University, Xiamen 361005, China
Fujian Provincial Key Laboratory of Brain-Like Intelligent Systems, Xiamen University
[3] Institute of Philosophy, Chinese Academy of Social Sciences, Beijing 100732, China
{zhangxj666,haoyijiang}@yahoo.cn, {dozero,minjiang}@xmu.edu.cn

Abstract. In the recent past, the most influential theories with respect to agent technology are the Belief-Desire-Intention (BDI) model. This paper not only expands past achievements but also makes novel proposals. It gives a very elegant way within a uniform and flexible logical framework for the graded BDI model, that is, for reasoning about probabilities and other uncertainty models, such as degrees of desires and intentions. We choose to blend the infinite-valued Łukasiewicz Logic and Propositional Dynamic Logic to formulize the graded BDI agent model. After presenting the language and semantics for this model, we propose axioms and rules for the graded BDI logic and prove a soundness and completeness result. On the basis of dealing with composite action, we illustrate interrelations between contexts for the model. It is hoped that the present study will make contributions to represent and reason under uncertainty as well as providing a formal support for engineer complex distributed systems.

Keywords: Graded Belief-Desire-Intention (BDI) Agents, Uncertain Reasoning, Contexts, Actions, Plans.

1 Introduction

In the recent past years, several works have proposed theories and architectures to give multi-agent systems a formal support. Among them, one of the most common agent theories is based on an intentional stance [1]. And the most widely used intentional formal system is the BDI agent architecture presented by Rao and Georgeff [2, 3]. BDI agents are agents whose behavior is guided by beliefs (B), desires (D) (or goals) and intentions (I). This conceptual model was developed by Bratman [4]. In order to help the designer to reason about the expected behavior of the systems, BDI architecture proposes a formal representation of the properties that agents should have. This architecture has been applied in several of the most significant multi-agent applications developed up to now.

In order to incorporate a model to represent and reason under uncertainty, Casali et al. [5] introduce a general model for BDI agents, and an architecture able to model these graded mental attitudes. The architecture presented in our paper is inspired by the work of them. Modeling different intentional notions by means of three modalities (B, D, I) is only in one logical framework in our paper. And our model is not only based on

F.L. Wang et al. (Eds.): WISM 2012, LNCS 7529, pp. 439–450, 2012.

Łukasiewicz Logic but also based on Propositional Dynamic Logic. These are different from the work of Casali et al. [5].

The BDI architectures proposed so far mostly deal with two-valued information. But quantified information about how possible a particular world is to be actual one should be expressed. More specifically, quantifying degrees in beliefs, desires and intentions should be done. It is very evident that taking into consideration this graded information could improve the agent's performance. There are a few works that partially address this issue [6, 7, 8]. We present in this paper an efficient graded BDI architectures based on fuzzy logic for agents that interact in an uncertain environment in terms of dynamics, specifying an architecture able to deal with the environment uncertainty and with graded mental attitudes in a general model. In our paper, belief degrees mean what extent the agent believes a formula is true; desire degrees allow the agent to set different levels of preference; intention degrees give a preference measure. On the basis of the representation and interaction of these three attitudes, one can model agents having different kinds of behavior.

This paper is organized as follows: after this introduction we survey in Section 2 the graded BDI agent model. Section 3 and 4 present the language and semantics for this model. In Section 5 and 6, we propose axioms and rules for the graded BDI logic and prove a soundness and completeness result. Section 7 deals with composite action. We illustrate interrelations between contexts for the graded BDI agent model in Section 8. An example of a graded BDI agent for buying houses is given in Section 9. Finally, Section 10 contains some concluding remarks and future work.

2 Graded BDI Agent Model

The architecture presented in our paper is inspired by the work of Casali et al. [5] about graded BDI models for agent architecture. Their work is based on the work of Parsons et al. [7] about multi-context BDI agent. The multi-context system specification of an agent contains units or contexts, logics and bridge rules channeling the propagation of consequences among theories [9].

Uncertainty reasoning can be dealt with by defining suitable modal theories over suitable many-valued logics. We decide to use the modal many-valued approach developed by Hájek et al. [10, 11] to represent and reason about graded notions of beliefs, desires and intentions. The basic idea is as follows: belief degrees are to be modeled as probabilities. A modal formula $B\varphi$ is interpreted as "φ is probable", then $B\varphi$ is also a fuzzy formula which may be more or less true depending on the probability of φ, and which faithfully respects the uncertainty model chosen to represent the degrees of belief. In the case of desire and intention degrees, the basic idea is similar. Intention degrees are to model a measure of the benefit relation involved in the actions toward a goal.

The infinite-valued Łukasiewicz Logic whose main connectives are based on the arithmetic addition in the unit interval [0, 1] can define a logic to reason about degrees of necessity and possibility, so we choose Łukasiewicz Logic to deal with additive measures like probabilities. In Łukasiewicz Logic, truth values are real number from the unit interval [0, 1], and truth functions are: $\neg x = 1 - x$, $x \rightarrow y = \min(1, 1 - x + y)$.

Casali et al. have mental contexts to represent beliefs (*BC*), desires (*DC*) and intentions (*IC*), and consider two functional contexts: for planning (*PC*) and communication (CC), then an agent is defined as a group of interconnected units: ({BC, DC, IC, PC, CC}, Δ_{br}) [5]. Belief context is to model the agent's belief about environment; regardless of environment and of the cost involved in actually achieving them, desire context is to model the agent's preferences. In order to set intentions, desires are used as pro-active. Intentions depend on the agent's knowledge about the world, which may allow –or not- the agent to set plan to change the world into a desired one. Each plan allows the agent to move from its current world to another, where a given formula is satisfied, so it must be feasible and make true its corresponding desire that the plan is built for, that is, the current state of the world must satisfy the preconditions. Communication context is the agent's door to external worlds, receiving and sending messages to and from other agents in the multi-agent society where graded BDI agents live.

3 The Graded BDI Language

To reason about the uncertainty of beliefs (B), desires (D) and intensions (I), we define a language for BDI representation based on Łukasiewicz Logic [10] and Propositional Dynamic Logic which has been used to model agents' action in [12] and [13]. A classical propositional language L is defined upon a countable set of propositional variables and connectives $\neg$ and $\rightarrow$, we extend the language L to represent actions which must be part of the agent's BDI set.

Now we define the language L_{BDI}. In this paper we introduce a language for BDI L_{BDI} by adding action modalities of the form $[\alpha]$ where α is an action, and three fuzzy modal operator B, D and I, to the classical propositional language L. $B\varphi$ means that "φ is believable". $D\varphi$ means that "φ is desired" and its truth degree refers to the agent's level of satisfaction would φ become true, and "the degree of $I\varphi$ is δ" means that the truth degree of the expression "φ is intended" is δ. The language L_{BDI} has expressions of two sorts: propositions or formulae φ, ψ, ... and actions (or programs) α, β, ... There are countably many atomic symbols of each sort. The set of all atomic propositions is denoted as Φ_0, and the set of all propositions denoted as Φ. The set of all atomic actions is denoted as Π_0, and the set of all actions (including atomic actions and plans which are in fact composite actions) denoted as Π. Actions and propositions are built inductively from the atomic ones using classical propositional operators $\neg$ (negativity) and $\rightarrow$ (implication), action operators; (composition), $\cup$ (nondeterministic choice) and $*$ (iteration), and mixed operators [] (necessity) and ? (test). More formally, the set of plans Π and formulae L_{BDI} are defined as follow:

(1) $\Phi_0 \subseteq \Phi$
(2) $\Pi_0 \subseteq \Pi$
(3) if $\varphi, \psi \in \Phi$, then $\varphi \rightarrow \psi \in \Phi$ and $\neg\varphi \in \Phi$
(4) if $\alpha, \beta \in \Pi$, then $\alpha; \beta$, $\alpha \cup \beta$, $\alpha^* \in \Pi$
(5) if $\alpha \in \Pi$ and $\varphi \in \Phi$, then $[\alpha]\varphi \in \Phi$
(6) if $\varphi \in \Phi$, then $\varphi? \in \Pi$

In clause (5), the intuitive meaning of $[\alpha]\varphi$ is that "it is necessary that after executing α, φ is true."

In order to define a modal context language for *BC*, we also need to use the connectives of Łukasiewicz many-valued Logic to build *B-Modal* from elementary modal formulae and truth constants r_c, for each rational $r\in[0,1]$:

(7) if $\varphi\in L_{BDI}$, then φ, $B\varphi\in BC$
(8) if $r\in Q\cap[0,1]$, then $r_c\in BC$
(9) if $B\varphi$, $B\psi\in BC$, then $B\varphi \rightarrow_L B\psi\in BC$ and $B\varphi\&B\psi\in BC$

Other logic connectives for modal formulae can be from $\rightarrow_L$, & and $\bar{0}$ as in Łukasiewicz logic. In clause (9), $\rightarrow_L$ and & correspond to the implication and conjunction of Łukasiewicz Logic. A formula $B\varphi \rightarrow_L B\psi$ is 1-ture if and only if the truth value of $B\varphi$ is greater or equal to that of $B\psi$. And modal formulae of the type $r_c \rightarrow_L B\varphi$ mean that the probability of φ is at least r. For simplicity's sake, formulae of the type $r_c\rightarrow_L B\varphi$ is denoted as $(B\varphi, r_c)$.

Now we define the modal context language for *DC*, we also choose to use Łukasiewicz Logic to formulize graded desires by extended with a new connective Δ (known as Baaz's connective). For any modal *M*, if *M* has value < (smaller than) 1, then ΔM gets value 0; otherwise, if *M* has value 1, then ΔM gets value 1 as well. Thus, ΔM becomes a two-valued Boolean formula. And the type $r_c\rightarrow_L D\varphi$ is denoted as $(D\varphi, r_c)$.

(10) if $\varphi\in L_{BDI}$, then $D\varphi\in DC$
(11) if $r\in Q\cap[0,1]$, then $r_c\in DC$
(12) if $D\varphi$, $D\psi\in DC$, then $D\varphi \rightarrow_L D\psi\in DC$ and $D\varphi\&D\psi\in DC$

In this context, qualitative expressions like $D\varphi\rightarrow_L D\psi$ express that φ is at least as preferred as ψ. And expressions $(D\varphi, 1)$ means that the agent has maximum preference in φ and is fully satisfied if it is true.

The definition of the modal context language for *IC* is similar. The intention to make φ can be thought as the consequence of finding a feasible plan α that permits to achieve a state of the world where φ holds.

4 The Graded BDI Semantics

As in usual in modal logics, we define the semantics of the context language for BDI using a Kripke structure. In order to represent the world transitions caused by actions, we add to such structure a τ function, a probability measure ρ and a preference distribution θ over possible worlds, as well as a possibility distribution μ_w. Thus, we can define a BDI Kripke structure as a 6-tuple $K=\langle W, \upsilon, \rho, \tau, \theta, \{\mu_w\}_{w\in W}\rangle$ where:

(13) W is a non-empty set of possible worlds.

(14) υ: $\Phi\times W \rightarrow \{0, 1\}$ assigns to each propositional variable $\varphi\in\Phi$ and each possible world $w\in W$ a two-valued Boolean evaluation. That is, $\upsilon(\varphi, w)\in\{0, 1\}$.

(15) ρ: $2^W \rightarrow [0, 1]$ is a finitely additive probability measure on a Boolean algebra of subsets of W such that for $\varphi \in \Phi_0$, the set $\{w | \upsilon(\varphi, w)=1\}$ is measurable.

(16) τ: $\Pi_0 \rightarrow 2^{W \times W}$ provides a set of pairs of worlds denoting world transitions for each atomic action.

(17) θ: $W \rightarrow [0, 1]$ is a distribution of preferences over possible worlds. And $\theta(w) < \theta(w')$ means that w′ is more preferred than w.

(18) μ_w: $W \rightarrow [0, 1]$ is a possibility distribution, for each $w \in W$. Where $\mu_w(w') \in [0, 1]$ is the degree on which the agent may try to reach the possible world w′ from the possible world w.

L_{BDI} is defined by extending L using classical connectives and action modalities as follows:

(19) $\tau(\alpha; \beta) = \tau(\alpha) \circ \tau(\beta)$
(20) $\tau(\alpha \cup \beta) = \tau(\alpha) \cup \tau(\beta)$
(21) $\tau(\alpha^*) = \tau(\alpha)$
(22) $\tau(\varphi?) = \{(w, w) | \upsilon(\varphi, w)=1\}$
(23) $\upsilon([\alpha]\varphi, w) = \min\{\upsilon(\varphi, w_i) | (w, w_i) \in \tau(\alpha)\}$
(24) $\upsilon([\alpha]\varphi, w) = 1$ if only if $\upsilon(\varphi) = 1$ in all possible worlds w′ that may be reached through the action α from the world w.

By means of Łukasiewicz Logic truth-functions and the probabilistic interpretation of belief, υ is extended to B-formulae as follows:

(25) $\upsilon(B\varphi, w) = \rho(\{w' \in W | \upsilon(\varphi, w')=1\}$, for each φ
(26) $\upsilon(r_c, w) = r$, for all $r \in Q \cap [0,1]$
(27) $\upsilon(B\varphi \& B\psi, w) = \max(\upsilon(B\varphi)+\upsilon(B\psi)-1, 0)$
(28) $\upsilon(B\varphi \rightarrow_L B\psi, w) = \min(1-\upsilon(B\varphi)+\upsilon(B\psi), 1)$
(29) $||B\varphi||^K = td_{w \in W}\, \upsilon(B\varphi, w)$, where $||B\varphi||^K$ is the truth degree of a formula $B\varphi$ in the Kripke structure $K=\langle W, \upsilon, \rho, \tau, \theta, \{\mu_w\}_{w \in W}\rangle$.

As in B-formulae, by means of Łukasiewicz connectives, the preference distributions $\theta(w)$ for formulae $D\varphi$ and the unary connective Δ, υ is extended to D-formulae as follows:

(30) $\upsilon(D\varphi, w) = td\{\theta(w') | \upsilon(\varphi, w')= 1\}$
(31) $\upsilon(\Delta D\varphi, w) = 1$, if $\upsilon(D\varphi, w) = 1$
(32) $\upsilon(\Delta D\varphi, w) = 0$, if $\upsilon(D\varphi, w) \neq 1$
(33) $td\varnothing = 1$
(34) $\upsilon(D\bot, w) = 1$, for all $w \in W$

In clauses (30) and (33) *td* refers to the truth degree of a formula $D\varphi$ in the Kripke structure $K=\langle W, \upsilon, \rho, \tau, \theta, \{\mu_w\}_{w \in W}\rangle$. The evaluation of D-formulae only depends on the formula itself-represented in the preference measure over possible worlds where the agent is situated.

As in B- and D-formulae, by means of Łukasiewicz Logic and the possibility distributions μ_w for formulae $I\varphi$, υ is extended to I-formulae as follows:

(35) $N_w(S) = td\{1-\mu_w(s) \mid s \notin S\}$, where N_w refers to the necessity measure associated to the possibility distributions μ_w.

(36) $\upsilon(I\varphi, w) = N_w\{w' \mid \upsilon(\varphi, w') = 1\}$

5 Axioms and Rules for the Graded BDI Logic

On the basis of Propositional Dynamic Logic (see. e. g. [14]) and Łukasiewicz Logic (see. e. g. [11]), according to the graded BDI semantics, axioms for graded BDI logic are given as follows (37-56):

(37) Axioms of classical Propositional Logic for the non-modal formulae

(38) Axioms of the Łukasiewicz Logic for modal formulae, for example, axioms of Hájek basic logic (see. e.g. [11]) plus the three axioms: $\neg\neg B\varphi \to B\varphi$, $\neg\neg D\varphi \to D\varphi$ and $\neg\neg I\varphi \to I\varphi$.

Axioms for B over Łukasiewicz Logic are as follows [5]:

(39) $B(\varphi \to \psi) \to_L (B\varphi \to B\psi)$.

(40) $\neg_L B(\varphi) \equiv B(\neg\varphi)$

(41) $B(\varphi) \equiv \neg_L B(\varphi \wedge \neg\psi) \to_L B(\varphi \wedge \psi)$

Axioms for B over Propositional Dynamic Logic are the following:

(42) $[\alpha]B(\varphi \to \psi) \to ([\alpha]B\varphi \to [\alpha]B\psi)$

(43) $[\alpha]B(\varphi \wedge \psi) \leftrightarrow [\alpha]B\varphi \wedge [\alpha]B\psi$

(44) $[\alpha \cup \beta]B\varphi \leftrightarrow [\alpha]B\varphi \wedge [\beta]B\varphi$

(45) $[\alpha; \beta]B\varphi \leftrightarrow [\alpha][\beta]B\varphi$

(46) $[B\psi?]B\varphi \leftrightarrow (B\psi \to B\varphi)$

(47) $B\varphi \wedge [\alpha][\alpha^*]B\varphi \leftrightarrow [\alpha^*]B\varphi$

(48) $B\varphi \wedge [\alpha^*](B\varphi \to [\alpha]B\varphi) \leftrightarrow [\alpha^*]B\varphi$ (induction axiom)

Axioms for D and I over Propositional Dynamic Logic are similar, such as

(49) $[\alpha]D(\varphi \to \psi) \to ([\alpha]D\varphi \to [\alpha]D\psi)$

(50) $[\alpha]I(\varphi \to \psi) \to ([\alpha]I\varphi \to [\alpha]I\psi)$

Axioms for D and I over Łukasiewicz Logic are as follows:

(51) $D(\varphi \to \psi) \to_L (D\varphi \to D\psi)$

(52) $D(\varphi \wedge \psi) \equiv D\varphi \wedge D\psi$

(53) $\neg D(\bot)$

(54) $I(\varphi \to \psi) \to_L (I\varphi \to I\psi)$

(55) $\neg I(\bot)$

(56) $I(\varphi \wedge \psi) \equiv I\varphi \wedge I\psi$

Deduction rules for graded BDI logic are given as follows (57-59):

(57) Modus Ponens (MP): if φ and $\varphi \to \psi$ hold, then ψ holds.

(58) Necessitation Rule (NR): $[\alpha]\varphi$, $B\varphi$, $D\varphi$ and $I\varphi$ can be derived from φ.

(59) Implication Rule (IR): $D\varphi \to_L D\psi$ can be derived from $\varphi \to \psi$.

6 Soundness and Completeness for the Graded BDI Logic

Our Graded BDI Logic (GBDIL) is based on Propositional Dynamic Logic (PDL) and Łukasiewicz Logic (LL), which are sound and complete (cf. [12] and [11], respectively). As we know, PDL $\subset$ GBDIL and LL$\subset$ GBDIL. Thus, we can prove soundness and completeness for GBDIL by embedding of GBDIL into PDL or LL. Embedding operations are very useful at least in two respects. Firstly, sometimes they make it possible to interpret logical connectives in one logic language in terms of those in another logic language. Secondly, embeddings may preserve various properties of logics [15]; for example, if L_1 is embeddable into a sound and complete logic L_2, then L_1 is also sound and complete.

Now we need to find an effective translation function Tr such that, for all formulae $\varphi\in$ GBDIL if and only if $Tr(\varphi)\in$ PDL or $Tr(\varphi)\in$ LL. Indeed, let Tr be a map from GBDIL to PDL or LL defined as follows:

Definition 1

(60)Tr(p) = p, if p is a PDL or LL formula

(61) $Tr(\neg\varphi) = \neg Tr(\varphi)$, if φ is a GBDIL formula

(62) $Tr(\varphi\rightarrow\psi) = Tr(\varphi)\rightarrow Tr(\psi)$,for all GBDIL formulae φ and ψ

(63) $Tr(B\varphi) = Tr(\varphi)$, $Tr(D\varphi) = Tr(\varphi)$, $Tr(I\varphi) = Tr(\varphi)$, if φ is a GBDIL formula

In terms of the translation function Tr, the following theorem can be proved:

Theorem 1. The map Tr defined by definition 1 is an embedding of GBDIL to PDL or LL.

Proof: One can easily show that: all formulae $\varphi\in$ GBDIL if and only if $Tr(\varphi)\in$ PDL or $Tr(\varphi)\in$ LL. For instance, by clause (39), $B(\varphi\rightarrow\psi) \rightarrow_L (B\varphi\rightarrow B\psi)\in$ GBDIL, by definition 1, $Tr(B(\varphi\rightarrow\psi) \rightarrow_L (B\varphi\rightarrow B\psi)) = Tr(B(\varphi\rightarrow\psi)) \rightarrow_L \mathrm{Tr}\,(B\varphi\rightarrow B\psi) = Tr(\varphi\rightarrow\psi) \rightarrow_L (Tr(B\varphi)\rightarrow Tr(B\psi)) = (Tr(\varphi)\rightarrow Tr(\psi)) \rightarrow_L (Tr(\varphi)\rightarrow Tr(\psi)) = (\varphi\rightarrow\psi) \rightarrow_L (\varphi\rightarrow\psi)\in$ LL. Other clauses are similar to (39). Hence, the translation function Tr is an embedding of GBDIL to PDL or LL.

Propositional Dynamic Logic (PDL) and Łukasiewicz Logic (LL) are sound and complete, then the graded BDI logic (GBDIL) is sound and complete by the properties of embedding operations. In other words, if *GBDIL* is a finite theory over *BC*, *DC* and *IC* and ϕ is a modal formula, then $GBDIL\vdash\phi$ if and only if $\|\phi\|^K=1$ in each BDI Kripke structure K model of GBDIL.

7 Composite Actions of the Graded BDI Agent Model

As we known, plans are in fact composite actions which allow the agent to move from its current world to another, where a given formula is satisfied. There is indeed an associated cost according to the actions involved [2]. We choose to use a first-order language restricted to Horn Clauses (HL), where a theory of planning including the special predicates is the following:

(64) *action*(α, *Pre-*, $Cost_\alpha$), where $\alpha \in \Pi_0$ is an elementary action, *Pre-*$\subset$HL are the preconditions of the action α, and $Cost_\alpha \in [0,1]$ is the cost of α.

(65) *plan*(φ, *action*(α, *Pre-*, $Cost_\alpha$), *r*), where $\alpha \in \Pi$ is an elementary action or a composite action representing the plan to achieve $\varphi \in$ *Pre-*, and *r* is the belief degree (> 0) of actually achieving φ by performing α. It is assumed that only one instance of this predicate is generated per formula.

(66) *bestplan*(φ, *action*(α, *Pre-*, $Cost_\alpha$), *r*) is similar to (65), and only one instance with the best plan is generated.

We assume that the current state of the world must satisfy the preconditions and the plan must make true the desire that the plan is built for.

8 Interrelations between Contexts for the Graded BDI Agent Model

In this section, we try to show the interrelations between contexts. From the desires and beliefs of the agent, and the possible transformations using actions, the planner can build plans generated from actions to fulfill desires [5]. Relationships among *D*, *B*, and *P* contexts are the following:

(67) if *D*: $\neg\Delta\neg(D\varphi)$, *P*: *action*(α, *Pre*, $Cost_\alpha$), and *B*: ($B([\alpha]\varphi$, *r*), then *P*: *plan*(φ, *action*(α, *Pre-*, $Cost_\alpha$), *r*)

The intention degree trades off the benefit and the cost of achieving a goal, then the degree of $I\varphi$ for each composite action α that allows to reach the goal is deduced from the degree of $D\varphi$ and the cost of a plan that satisfies the desire φ. That is, the degree of $I\varphi$ is calculated by a function *f* as follows:

(68) if *D*: ($D\varphi$, *d*), and *P*: *plan*(φ, *action*(α, *Pre-*, $Cost_\alpha$), *r*), then *I*: ($I\varphi$, $f(d, b, r)$)

Different functions $f(d, b, r)$ may model different individual behaviors. Given full belief in achieving φ after performing α, the degree of the intention to bring about φ mainly depend on the satisfaction that it bring the agent and in the cost, for example, the complement to 1 of the normalized cost. Therefore, the function $f(d, b, r)$ might be defined as:

(69) $f(d, b, r) = r\,(b+(1-Cost_\alpha))/2$

The agent's interactions with the environment means that if the agent intends φ at i_{max}, then the agent will focus on the best plan that allows the agent to achieve the most intended goal, where i_{max} refers to the maximum degree of all the intentions. That is:

(70) if *I*: ($I\varphi$, i_{max}), and *P*: *bestplan*(φ, *action*(α, *Pre-*, $Cost_\alpha$), *r*), then *C*: *C*(*does*(α))

The agent perceives all the changes in the environment through the communication unit, thus, the interrelations between *B* and *C* contexts are as follows:

(71) if *C*: β, then *B*: $B\beta$.

9 Example of a Graded BDI Agent for Buying Houses

Suppose we choose to instruct our investment agent to look for a buying-houses package. We instruct the agent with two desires, first and more important, we hope the house near a school, and second we hope the house near a subway. We restrict its buying range as we hope to buy no more than 10 kilometers from Guomao. To decide to buy a house the agent will have to take into account the benefit (with respect to near a school and a subway) and the cost of the buying. In this case, *BC*, *DC*, *PC* contexts are as follows:

Desire Contexts (*DC*): The agent has the following desires:

(72) (*D*(*near a school*), *d*=0.82)

(73) (*D*(*near a subway*), *d*=0.78)

(74) (*D*(*near a school* $\wedge$ *near a subway*), *d*=0.96)

(75) (*D*(*distance* $\leq$ 10km), *d*=0.85)

Belief Contexts (*BC*): The BDI theory contains knowledge about the interrelations between possible actions that the agent can take and formulae made true by their execution. For this example, actions would be buying different houses. In this scenario, we consider five houses:

(76) Π_0 ={*Blusky*, *Grenpark*, *Whitfish*, *Yellostar*, *Redbord*}

The degree of *B*([α] *near a subway*) is the probability of *near a subway* after buying α. We choose to give the buying agent the following belief:

(77) *B*([*Grenpark*]*near a subway*, r_1=1)

(78) *B*([*Blusky*]*near a subway*, r_1=0.45)

(79) *B*([*Whitfish*]*near a subway*, r_1=0.66)

(80) *B*([*Yellostar*]*near a subway*, r_1=0.58)

(81) *B*([*Redbord*]*near a subway*, r_1=0.35)

In this example, the degree of *B*([α] *near a school*) refers to the probability of near a school after buying α. Taking into account our personal view, the agent's beliefs about buying houses near a school are as follows:

(82) *B*([*Grenpark*]*near a school*, r_2=0.85)

(83) *B*([*Blusky*]*near a school*, r_2=0.88)

(84) *B*([*Whitfish*]*near a school*, r_2=0.35)

(85) *B*([*Yellostar*]*near a school*, r_2=0.75)

(86) *B*([*Redbord*]*near a school*, r_2=0.46)

It is assumed that the desires are stochastically independent. Thus, we need to add the following inference rule for *BC*:

(87) if *B*([α] *near a school*, r_1) and *B*([α] *near a subway*, r_2), then *B*([α] *near a school* $\wedge$ *near a subway*, $r = r_1 \bullet r_2$)

Plan Contexts (*PC*): In this scenario, a series of atomic actions are the following:

(88) *action(Grenpark*, {dist-=3km}, {cost=7millions}, $Cost_\alpha$=0.54)
(89) *action(Blusky*, {dist-=5km}, {cost=5millions}, $Cost_\alpha$=0.32)
(90) *action(Whitfish*, {dist-=19km}, {cost=9millions}, $Cost_\alpha$=0.96)
(91) *action(Yellostar*, {dist-=16km}, {cost=8millions}, $Cost_\alpha$=0.89)
(92) *action(Redbord*, {dist-=6km}, {cost=6millions}, $Cost_\alpha$=0.48)

Now the agent can reason to determine which intention to adopt and which plan is associated with that intention. A brief schema of the steps in this process is the following:

Stage 1: The desires are passed from desire contexts to plan contexts.
Stage 2: Within plan contexts, plans for each desire are found.
The agent looks for a set of different houses plans in terms of the desires, and takes into consideration the beliefs about the possibilities of satisfying the goals near a school and a subway through the different actions. Using the restriction by the desire (75), that is, the distance no more than 10kms, the agent rejects plans (90) and (91), that is, the agent rejects to buy *Whitfish* and *Yellostar*. Therefore, plans are generated for each desire by (67). For example, for the most preferred desire, i.e. *near a school* ∧ *near a subway*, the generated plans are as follows:

(93) *plan(near a school∧near a subway*, *action(Grenpark*, {dist-=3km}, {cost=7million}, $Cost_\alpha$= 0.32), $r_1 \bullet r_2$ =0.85)
(94) *plan(near a school∧near a subway*, *action(Blusky*, {dist-=5km}, {cost=5million}, $Cost_\alpha$= 0.48), $r_1 \bullet r_2$ =0.396)
(95) *plan(near a school∧near a subway*, *action(Redbord*, {dist-=6km}, {cost=6million}, $Cost_\alpha$= 0.54), $r_1 \bullet r_2$ =0.161)

Stage 3: The plans determine the degree of intentions.
The function *f* is monotonically increasing with respect to *d* by (69). Therefore, it is enough to take into account the most preferred desire, i.e. *near a school* ∧ *near a subway*, which is preferred to a degree 0.96. In terms of (69), using $f(d, b, r) = r\,(b+(1-Cost_\alpha))/2 = r_1 \bullet r_2 \bullet (0.96+(1- Cost_\alpha))/2$, we successively have for $\alpha \in$ {*Grenpark*, *Blusky*, *Redbord*} as the following:

(96) *I(near a school∧near a subway*, $r_1 \bullet r_2 \bullet (0.96+(1- Cost_\alpha))/2 = 0.6035$)
(97) *I(near a school∧near a subway*, $r_1 \bullet r_2 \bullet (0.96+(1- Cost_\alpha))/2 = 0.32472$)
(98) *I(near a school∧near a subway*, $r_1 \bullet r_2 \bullet (0.96+(1- Cost_\alpha))/2 = 0.11914$)

The maximal degree of intention for *near a school∧near a subway* by the plan *Grenpark* is 0.6035.

Stage 4: A plan is adopted.
At last, the action α = buying *Grenpark* is selected and passed to the Communication context by (70).

10 Conclusions and Future Work

The graded BDI agent model in this paper explicitly represents the uncertainty of beliefs, desires and intensions using multi-context systems, and is general enough to specify different types of agents. We choose the infinite-valued Łukasiewicz Logic to formalize the degrees of each attitude: Belief, Desire and Intension. In order to represent the uncertainty behavior as probability, necessity and possibility, the corresponding axioms are added to the logic. The graded BDI agent's behavior is determined by the different measure of each context which is added by concrete conditions. This paper is to look for a possible axiomatic modeling of beliefs, desires and intensions, and to show how they influence the agent's behavior. This model can also easily be extended to include other mental attitudes.

As for future work, it would be interesting to add emotions, such as fear, anxiety and self-confidence, to the graded BDI agent in order to express the concepts of the Emotional Graded BDI model of agency, in the line of the work done by Pereira et al. in [16].

Acknowledgements. This research has been partially supported by the Postdoctoral Foundation of China (Grant No.2012M510101), and the Major Tender of National Social Science Foundation of China (Grant No.11&ZD088).

References

1. Wooldridge, M., Jennings, N.R.: Intelligent agents: theory and practice. The Knowledge Engineering Review 10(2), 115–152 (1995)
2. Rao, A., Georgeff, M.: Modeling rational agents within a BDI-architecture. In: Proceedings of the 2nd International Conference on Principles of Knowledge Representation and Reasoning (KR 1992), pp. 473–484 (1991)
3. Rao, A., Georgeff, M.: BDI agents: from theory to practice. In: Proceedings of the 1st International Conference on Multi-Agents Systems, pp. 312–319 (1995)
4. Bratman, M.E.: Intentions, Plans, and Practical Reason. Harvard University Press, Cambridge (1987)
5. Casali, A., Godo, L., Sierra, C.: Graded BDI Models for Agent Architectures. In: Leite, J., Torroni, P. (eds.) CLIMA 2004. LNCS (LNAI), vol. 3487, pp. 126–143. Springer, Heidelberg (2005)
6. Lang, J., van der Torre, L., Weydert, E.: Hidden uncertainty in the logical representation of desires. In: International Joint Conference on Artificial Intelligence, IJCAI 2003, Acapulco, Mexico, pp. 685–690 (2003)
7. Parsons, S., Sierra, C., Jennings, N.R.: Agents that reason and negotiate by arguing. Journal of Logic and Computation 8(3), 261–292 (1998)
8. Schut, M., Wooldridge, M., Parsons, S.: Reasoning about Intentions in Uncertain Domains. In: Benferhat, S., Besnard, P. (eds.) ECSQARU 2001. LNCS (LNAI), vol. 2143, p. 84–95. Springer, Heidelberg (2001)
9. Ghidini, C., Giunchiglia, F.: Local modal semantics, or contextual reasoning = locality + compatibility. Artificial Intelligence 127(2), 221–259 (2002)

10. Godo, L., Esteva, F., Hajek, P.: Reasoning about probabilities using fuzzy logic. Neural Network World 10, 811–824 (2000)
11. Hájek, P.: Mathematics of Fuzzy Logic. Kluwer (1998)
12. Harel, D., Kozen, D., Tiuryn, J.: Dynamic Logic. The MIT Press (2000)
13. Meyer, J.J.: Dynamic logic for reasoning about actions and agents. In: Workshop on Logic-Based Artificial Intelligence, Washington, D.C., June 14-16 (1999)
14. Goldblatt, R.: Logics of Time and Computation. CSLI Lecture, Notes 7 (1992)
15. Chagrov, A., Zakharyaschev, M.: Modal Logic. Clarendon Press (1997)
16. Pereira, D., Oliveira, E., Moreira, N.: Formal Modelling of Emotions in BDI Agents. In: Sadri, F., Satoh, K. (eds.) CLIMA VIII 2007. LNCS (LNAI), vol. 5056, pp. 62–81. Springer, Heidelberg (2008)

A Polymorphic Type System with Progress for Binary Sessions*

Zhenguo Yang, Farong Zhong**, and Jinfang Zhang

Department of Computer Science, Zhejiang Normal University,
Jinhua 321004, Zhejiang Province, P.R. China
{yangzhenguo1988,zjfcathy}@126.com, zfr@zjnu.cn,

Abstract. A static bounded polymorphic type system is presented in this paper, which ensures the progress property, i.e., the property that once a communication has been established, well-formed programs will never starve at communication points. The introduction of subtyping for session types and the relaxed duality relation increases the flexibility of the type system, and allows the participants in a conversation to follow different protocols that are nevertheless compatible in a sense defined by the subtyping relation. In addition, to keep progress in sessions, the type compliance is defined to associate with the relaxed duality relation, where the environment is balanced. Finally, the soundness and communication safety of the type system are proved, and some related work and possible future work in this area are discussed.

Keywords: Communication-Centered Programming, Session Types, Subtyping, Pi-calculus, Progress.

1 Introduction

New Internet technologies have boosted the amount of distributed applications and services, and communication-centered programming thus confronts us with challenges. The communication protocols between participants are often complicated, involving large numbers of states and state transitions caused by different types of messages. When implementing a process in an application, the sequence and structure of messages have to be confirmed correctly according to the protocol.

Session types [9-11] are one of the formalisms proposed to structure interactions and to explain communicating processes and their behaviors. They were first proposed in the context of a language based on Pi-calculus [22]. The flexibility and expressiveness of session types were enhanced by defining a notion of subtyping [5] for session types, whose main application is to make the participants following

---

* This work has been supported by the Natural Science Foundation of China under grant 60873234 and Top Key Discipline of Computer Software and Theory in Zhejiang Provincial Colleges at Zhejiang Normal University.

** Corresponding Author.

F.L. Wang et al. (Eds.): WISM 2012, LNCS 7529, pp. 451–461, 2012.

different protocols compatible in a dialogue. In addition, Girard [14] and Reynolds [13] presented the polymorphism in the polymorphic lambda calculus, and Turner [3] as well as Pierce and Sangiorgi [2] studied a polymorphism in Pi-calculus. Vasconcelos and Honda [23] and Gay [4] studied a weaker ML-style polymorphism in their language similar to Pi-calculus. Liu and Walker [16] presented a rather different style of polymorphism. Gay and Hole first integrated the bounded polymorphism in Pi-calculus into session types in [6-8], where the authors made the range of compatibility of subtyping clear. Yang et al. [24] extended the solution further by introducing a recursive session type to describe protocols with an indefinite number of repetitive behaviors and defined the relaxed duality relation.

However, a programmer expects that communicating programs should together realize a consistent conversation, but they are prone to failure to handle a specific incoming message or to send a message at correct timing, with no way to detect such errors before runtime. Dezani-Ciancaglini et al. proposed the language MOOSE [18], which integrated the object-oriented programming style and sessions, and established the progress property [19] that once a session has been initiated, well-typed processes will never starve at session channels.

In this work, we extend our previous type system in [24] with progress property by defining type compliance, which integrates the relaxed duality for both non-recursive and recursive session types. The definition of type compliance shows a good combination between the subtyping relation and progress property. In addition, the bilinearity condition [19] is introduced, ensuring that there are exactly two threads containing occurrences of the same live channel, which makes it easier to guarantee progress and prove Type Safety Theorem. We formalize the syntax, operational semantics and prove the soundness of the type system with progress for processes where the environment is balanced.

The remainder of the paper is organized as follows. In Section 2, session types, polymorphism and progress property are introduced through examples. In Section 3, the syntax and notation including the relaxed duality relation are defined. In Section 4, the type system with subtyping and relaxed duality relation is presented. In Section 5, the soundness of the type system and the progress property are proved. Finally, we conclude the paper.

2 Session Types and Polymorphism

2.1 Session Types, Subtyping and Bounded Polymorphism

A session, which is started after a connection has been established between two participants, is a sequence of messages exchanged through the connection following a fixed protocol. Proposed session types aim at characterizing such sessions, in terms of the types of messages received or sent by a participant. For example, the session type *begin*.![*int*].?[*bool*].*end* expresses that an integer will be sent, and then a boolean value will be expected to be received before the protocol is completed. Naturally, $Server_1 \mid Client_1$, can communicate well without an error, which reveals general client-server interactions. However, we have to face the case that the server has updated to the type $Server_2$ while the client stays the same.

$$Server_1 = \&\{go : ?[int].![int], quit : end\} \tag{1}$$

$$Client_1 = \oplus\{go : ![int].?[int], quit\} \tag{2}$$

$$Server_2 = \&\{go : ?[real].![nat], quit : end\} \tag{3}$$

Gay and Hole [5-8] addressed the problem by defining the subtyping for session type, in which the input and branch as well as continuation types are covariant, while the output and choice types are contravariant. The subtyping for session type could be a session type where receipt is replaced by sending with a smaller type, and vice versa. Such cases will be revealed when $Server_1$ upgrades to $Server_2$, and the communication $Server_2 \mid Client_1$ still proceeds well. Gay and Hole [6, 8] showed the bounded polymorphism to quantify the branch by a type variable with lower and upper bounds as the type $Server_3$,

$$Server_3 = \&\{go(?[nat] \le X \le ?[real]) : X.\overline{X}, quit : end\} \tag{4}$$

where the bounded polymorphism with a lower bound, in turn, offers an upper bound for a process using $\overline{X}$ to construct its typing derivation.

2.2 Progress Property

We have to distinguish *shared channels* and *live channels*, which are ignored in the above examples such that we can illustrate the session types and polymorphism more explicitly. Shared channels have not yet been connected; they are used to decide if two threads can communicate, in which case they are replaced by fresh live channels. After a connection has been created, the channel is live. And data may be transmitted through such active channels only. We present the progress by analyzing the following examples.

Example 1. (Bounded Shared Channels). A bound shared channel not having a dual to start a session can block the communication on live channels forever, as in

$$Server_4 = \nu a(a(x).?[real].end) \tag{5}$$

The problem does not arise if the shared channel a is free, since we can always compose with a dual process e.g. with

$$Client_2 = \overline{a}(x).![int].end \tag{6}$$

Note that the protocols followed by the two processes are different, yet they can be compatible in our type system by introducing subtyping for session types. This situation is solved well with definition of relaxed duality [24] for both non-recursive and recursive session types.

Example 2. (Type Compliance). Though we introduce the subtyping for session types, yet the protocols followed by the two processes can be incompatible.

$$Server_5 = a(x).b(y).(x![int]; y?[int].end) \quad (7)$$

$$Client_3 = \bar{a}(x).\bar{b}(y).(x?[nat]; y![real].end) \quad (8)$$

As $Server_5$ | $Client_3$ shows, such incompatibility prevents the composition from going. To cope with such kind of misbehaviour, Acciai et al. [15] incorporated *type compliance* in the theory of session types. Further, we integrate it with the relaxed duality for both non-recursive and recursive session types, which is defined in Definition 6. As a result, we consider $Server_5$ and $Client_4$ are in type compliance.

$$Client_4 = \bar{a}(x).\bar{b}(y).(x?[real]; y![nat].end) \quad (9)$$

Example 3. (Circularity of Channels). As we only focus on the synchronous communication, the order of session channels should be taken into account.

$$Server_6 = a(x).b(y).(x![int]; y?[real].end) \quad (10)$$

$$Client_5 = \bar{a}(x).\bar{b}(y).(y![real]; x?[int].end) \quad (11)$$

These processes use the channels bound by a and b in reverse order, hence $Server_6$ | $Client_5$ will result in a deadlock. This is prevented by type compliance in the type system, which allows instead to compose $Server_6$ e.g. with

$$Client_6 = \bar{a}(x).\bar{b}(y).(x?[real]; y![int].end) \quad (12)$$

Example 4. (Bilinearity). The $Server_7$ | $Client_7$ cannot complete the communication forever in synchronous communications, which is prevented by bilinearity condition, i.e. there are exactly two threads that contain occurrences of the same live channel.

$$Server_7 = a(x).b(y).(x![int]; y?[real].end) \quad (13)$$

$$Client_7 = \bar{a}(x).\bar{b}(y).(x?[real].end) \quad (14)$$

3 Syntax and Notation

3.1 Types

The language is built on a polyadic Pi-calculus with output prefixing [22], omitting delegation for simplicity. In the present work, $i \in \{1,\ldots, n\}$, $j \in \{1,\ldots, m\}$, $m \leq n$ and $i \leq j$, and recursive types are contractive, containing no subexpressions of the form $\mu X.\mu X_1\ldots\mu X_n.X$. Each session type S has a dual type $\bar{S}$, defined for non-recursive session types in Figure 2. We introduce the relaxed duality relation [24] as follows, where for any recursive type T, *unfold* (T) is the result of repeatedly unfolding the top level recursion until a non-recursive type constructor is reached.

Definition 5. *A relation $R_1 \subseteq Type \times Type$ is a relaxed duality relation if implies the following conditions:*
1, *If $unfold(T) = ?[T_1,\ldots,T_n].S_1$ and $unfold(U) = ![U_1,\ldots,U_n].S_2$, then $(S_1, S_2) \in R_1$ and $U_i \leq_c T_i$.*
2, *If $unfold(T) = ![T_1,\ldots,T_n].S_1$ and $unfold(U) = ?[U_1,\ldots,U_n].S_2$, then $(S_1, S_2) \in R_1$ and $T_i \leq_c U_i$.*
3, *If $unfold(T) = \&\{l_1{:}S_1,\ldots,l_n{:}S_n\}$, $unfold(U) = \oplus\{l_1{:}V_1,\ldots,l_m{:}V_m\}$, then $(S_i, V_i) \in R_1$.*
4, *If $unfold(T) = \oplus\{l_1{:}S_1,\ldots,l_m{:}S_m\}$, $unfold(U) = \&\{l_1{:}V_1,\ldots,l_n{:}V_n\}$, then $(S_i, V_i) \in R_1$.*
5, *If unfold $(T) = end$ then unfold $(U) = end$.*

Definition 6. *The coinductive relaxed duality relation $\perp_c$ is defined by $T \perp_c U$ if and only if there exists a type simulation R_1 such that $(T, U) \in R_1$*

Channels	c ::=	x, y, z	(variables)
	|	a, b	(shared channels)
	|	$k^{\bar{p}}, k^{p}$	(live channels)
Bounds	B ::=	$Bot \mid Top \mid M \mid S$	
	|	L	(lower bound)
	|	U	(upper bound)
	|	X	(type variable)
Message Types	M ::=	$bool \mid nat \mid int \mid real$	
Session Types	S ::=	$end \mid ?[\tilde{T}].S \mid ![\tilde{T}].S$	
	|	$\mu X.S$	(recursive session types)
	|	$\&\{l_i(L_i \leq X_i \leq U_i): S_i\}$	(branching)
	|	$\oplus\{l_i(L_i \leq X_i \leq U_i): S_i\}$	(selection)

Fig. 1. Types

$$\overline{X} = \overline{Y}\ (if\ Y \leq X) \quad \overline{end} = end \quad \overline{![\tilde{T}].S} = ?[\tilde{T}].\overline{S} \quad \overline{?[\tilde{T}].} = ![\tilde{T}].\overline{S}$$
$$\overline{Top} = Bot \quad \overline{\&\{l_i(L_i \leq X_i \leq U_i): S_i\}} = \oplus\{l_i(L_i \leq X_i \leq U_i): \overline{S_i}\}$$
$$\overline{Bot} = Top \quad \overline{\oplus\{l_i(L_i \leq X_i \leq U_i): S_i\}} = \&\{l_i(L_i \leq X_i \leq U_i): \overline{S_i}\}$$

Fig. 2. Relaxed duality for non-recursive session types

3.2 Processes

Process $x^p\,?[\tilde{y}:\tilde{T}].P$ inputs the names $\tilde{y}$ along the port x^p and then executes P. Process $x^p\,![\tilde{y}].P$ outputs the names $\tilde{y}$ along the port x^p and then executes P. Process $x^p \triangleright \{l_i(L_i \leq X_i \leq U_i): P_i\}$ offers a choice of the labels l_i and an accompanying type T ranging from the lower bound L_i to the upper bound U_i. Process

$x^p \triangleleft l(T).P$ selects and sends the label l and type T along port x^p, and then executes P. Lastly, the conditional expression where the boolean value b determines to select P or Q.

$$
\begin{array}{llll}
P ::= 0 & \text{(inaction)} & \mid P \mid Q & \text{(parallel)} \\
\mid a(x).P & \text{(accept)} & \mid \overline{a}(x).P & \text{(request)} \\
\mid x^p\,![\tilde{y}].P & \text{(data send)} & \mid x^p\,?[\tilde{y}:\tilde{T}].P & \text{(data receive)} \\
\mid (\nu x)P & \text{(hiding)} & \mid x^p \triangleright \{l_i(L_i \le X_i \le U_i): P_i\} & \text{(branch)} \\
\mid x^p \triangleleft l(T).\,P & \text{(selection)} & \mid \mathit{if}\ x\ \mathit{then}\ P\ \mathit{else}\ Q & \text{(condition)}
\end{array}
$$

Fig. 3. Processes

3.3 Operational Semantics

The operational semantics, denoted by $P \rightarrow Q$, is the smallest relation closed under the rules in Figure 4. R-COM is the standard rule for communication, where the channel on which communication takes place are polarized and names received are substituted in the continuation process P. R-SELECT is a rule for selection from labeled processes. R-TRUE and R-FALSE are standard, defining reduction in conditional expressions. R-PAR and R-CONG are also standard, defining reduction in parallel composition and structural congruence. Finally, R-NEW is standard, defining reduction for processes under ν bindings.

$$x^p\,?[\tilde{y}:\tilde{U}].P \mid x^{\overline{p}}\,![\tilde{z}].Q \rightarrow P\{\tilde{z}/\tilde{y}\} \mid Q \quad \text{R-COM}$$

$$x^p \triangleright \{l_i(L_i \le X_i \le U_i): S_i\} \mid x^p \triangleright l_i(T).Q \rightarrow P_i\{T/X_i\} \mid Q \quad \text{R-SELECT}$$

$$\mathit{if}\ \mathit{true}\ \mathit{then}\ P\ \mathit{else}\ Q \rightarrow P \quad \text{R-TRUE} \qquad \mathit{if}\ \mathit{false}\ \mathit{then}\ P\ \mathit{else}\ Q \rightarrow Q \quad \text{R-FALSE}$$

$$\frac{P \rightarrow P'}{P \mid Q \rightarrow P' \mid Q}\ \text{R-PAR} \qquad \frac{P \rightarrow P'}{(\nu x^{\pm}:S)P \rightarrow (\nu x^{\pm}:S)P'}\ \text{R-NEW}$$

$$\frac{P' \equiv P \quad P \rightarrow Q \quad Q \equiv Q'}{P' \rightarrow Q'}\ \text{R-CONG}$$

Fig. 4. Reduction rules

4 Type System

4.1 Subtyping

The subtyping rules for non-recursive types defined in Figure 5 are based on work [8]. The subtyping judgments are relative to an environment $\Delta = L_i \le X_i \le U_i, \ldots, L_n \le X_n \le U_n$, in which the order of bounded type variables is significant and any type variables in the environment are taken from $\{X_1, \ldots, X_n\}$.

What's more, to extend subtyping for recursive types, we introduce the coinductive subtyping relation [7] in a similar way to the coinductive duality relation.

Definition 7. *A relation $R_2 \subseteq Type \times Type$ is a type simulation if implies the following conditions*:
1, *If unfold(T)* = $?[T_1,...,T_n].S_1$ *and unfold(U)* = $?[U_1,...,U_n].S_2$, *then* $(S_1\ ,\ S_2) \in R_2$ *and* $(T_i, U_i) \in R_2$.
2, *If unfold(T)* = $![T_1,...,T_n].S_1$ *and unfold(U)* = $![U_1,...,U_n].S_2$, *then* $(S_1\ ,\ S_2) \in R_2$ *and* $(U_i, T_i) \in R_2$.
3, *If unfold(T)* = $\&\{l_1{:}S_1,...,l_m{:}S_m\}$, *unfold(U)* = $\&\{l_1{:}V_1,...,l_n{:}V_n\}$, *then* $(S_j\ ,\ V_j) \in R_2$.
4, *If unfold(T)* = $\oplus\{l_1{:}S_1,...,l_n{:}S_n\}$, *unfold(U)* = $\oplus\{l_1{:}V_1,...,l_n{:}V_n\}$, *then* $(S_j\ ,\ V_j) \in R_2$.
5, *If unfold (T) = end then unfold (U) = end.*

Definition 8. *The coinductive subtyping relation* $\le_c$ *is defined by* $T \le_c U$ *if and only if there exists a type simulation* R_2 *such that* $(T, U) \in R_2$.

$$\frac{}{\Delta \vdash end \le end}\text{S-END} \qquad \frac{\Delta \vdash V \le W \quad \Delta \vdash T_i \le U_i}{\Delta \vdash ?[\tilde{T}].V \le ?[\tilde{U}].W}\text{S-IN}$$

$$\frac{\Delta \vdash V \le W \quad (\Delta \vdash T_i \le U_i)}{\Delta \vdash ![\tilde{U}].V \le ![\tilde{T}].W}\text{S-OUT}$$

$$\frac{\Delta, L_j \le X_j \le U_j \vdash R_j \le S_i}{\Delta \vdash \&\{l_j(L_j \le X_j \le U_j) : R_j\} \le \&\{l_i(L_i \le X_i \le U_i) : S_i\}}\text{S-BRANCH}$$

$$\frac{\Delta, L_j \le X_j \le U_j \vdash R_j \le S_i}{\Delta \vdash \oplus\{l_i(L_i \le X_i \le U_i : R_i\} \le \oplus\{l_j(L_j \le X_j \le U_j) :: S_j\}}\text{S-CHOICE}$$

Fig. 5. Subtyping rules for non-recursive types

4.2 Typing Rules

Typing rules in Figure 6 are based on [7, 8], whose essence would be found in Piece and Sangiorgi's work [1]. T-NIL ensures that all interactions are finished when typing the nil process, **0**, by forcing the environment to be completed. T-PAR rule is for parallel composition. T-NEW forces two ports of a session channel to have dual types when ports are bound in order that sequences of communications on the channel will not get out of step. T-COND is standard. T-IN and T-OUT are on two ports of one session channel, showing the effect of lower and upper bounds. T-OFFER shows that each branch is typed with corresponding type for the channel, including appropriate upper and lower bounds for type variables. T-CHOOSE indicates a typing derivation for continuation process by appropriately-instantiated polymorphic channel type.

$$\frac{\Delta;\Gamma\ completed}{\Delta;\Gamma \vdash 0}\text{T-NIL} \quad \frac{\Delta;\Gamma_1 \vdash P \quad \Delta;\Gamma_2 \vdash Q}{\Delta;\Gamma_1,\Gamma_2 \vdash P \mid Q}\text{T-PAR}$$

$$\frac{\Delta;\Gamma,x^+:S,x^-:\overline{S} \vdash P}{\Delta;\Gamma \vdash (\nu x^{\pm}:S)P}\text{T-NEW} \quad \frac{\Delta;\Gamma \vdash P \quad \Delta;\Gamma \vdash Q \quad \Delta;\Gamma \vdash x \leq bool}{\Delta;\Gamma \vdash if\ x\ then\ P\ else\ Q}\text{T-COND}$$

$$\frac{\Delta;\Gamma_1,x^p:S_1,\tilde{y}:\tilde{U} \vdash P \quad \Delta \vdash T \leq ?[\tilde{U}].S_1}{\Delta;\Gamma_1,x^p:T \vdash (x^p?[\tilde{y}:\tilde{U}]).P}\text{T-IN} \quad \frac{\Delta;\Gamma_3,x^{\bar{p}}:S_2 \vdash Q \quad \Delta \vdash \overline{T} \leq ![\tilde{L}].S_2}{\Delta;\Gamma_2,x^{\bar{p}}:\overline{T},\tilde{z}:\tilde{L} \vdash x^{\bar{p}}![\tilde{z}].Q}\text{T-OUT}$$

$$\frac{\Delta, L_i \leq X_i \leq U_i;\Gamma,x^p:S_i \vdash P_i \quad \Delta \vdash T \leq \&\{l_i(L_i \leq X_i \leq U_i):S_i\}}{\Delta;\Gamma,x^p:T \vdash x^p \rhd \{l_i(L_i \leq X_i \leq U_i):P_i\}}\text{T-OFFER}$$

$$\frac{\Delta;\Gamma,x^p:S_i\{B/X_i\} \vdash P \quad \Delta \vdash T \leq \oplus\{l_i(L_i \leq X_i \leq U_i):S_i\} \quad \Delta \vdash L_i \leq B \leq U_i}{\Delta;\Gamma,x^p:T \vdash x^p \lhd l_i(B).P}\text{T-CHOOSE}$$

Fig. 6. Typing rules

5 Soundness of the Type System

5.1 Subject Reduction and Type Safety

Before we proceed for the soundness of the type system, we introduce two auxiliary results, Substitution Lemma and Subject Congruence from [21]. We prove the soundness for processes where the environment is balanced, showing that the result of a reduction from well-typed processes must be also well-typed, meanwhile, no errors will occur [21].

Lemma 9 (Substitution Lemma)
If $\Delta;\Gamma, x{:}U \vdash P$ *and* $\Delta \vdash T \leq_c U$ *and* $\Gamma, y{:}T$ *is defined, then* $\Delta;\Gamma + y{:}T \vdash P\{y/x\}$.

Lemma 10 (Subject Congruence)
If $\Gamma \vdash P$ *and* $P \equiv Q$, *then* $\Gamma \vdash Q$.

Theorem 11 (Subject Reduction)
If $\Gamma \vdash P$ *and* $P \rightarrow Q$ *and* Γ *is balanced, then* $\Gamma \vdash Q$.

Proof. The proof is the same as our previous work [24], and we omit the details.
Before we proceed, we need the following notion: an x-process is a prefixed process with subject x. Next, an x-redex is the parallel composition of a pair of dual x-processes. If the parallel composition of more than one x-processes does not result in an x-redex, then P is an error.

Theorem 12 (Type Safety)
A typable program never reduces to an error.

Proof. The proof is by reduction to absurdity, assuming error processes typable. Suppose that $\Delta;\Gamma \vdash P \mid Q$. There are two cases of an error in a typable program.

Case 1. $P=P_1|P_2$ is the parallel composition of two x-processes that do not form an x-redex. There are several subcases will occur, yet they are alike. Therefore, we only consider one subcases, for example, the pair output/label-select.

$P_1 = x^{\bar{p}}![\tilde{y}]$ while $P_2 = x^p \triangleright \{l_i(L_i \le X_i \le U_i): P_i\}$, from which we can conclude their types on the two ports of a channel are $x^{\bar{p}}: ![\alpha]$ and $x^p: \oplus\{l_i:\alpha_i\}$ respectively. Obviously, the type of the composition obtained by T-PAR is not defined, hence $P \mid Q$ is not typable.

Case 2. $P = (P_1|P_2)|P_3$ is the parallel composition of three or more x-processes, it will be against the bilinearity condition. If $P_1 \mid P_2$ is not an x-redex, it is just the Case 1. Otherwise, it must be the case that $x:\perp \in \Gamma_1$ and $x:T \in \Gamma_2$, thus the composition of the two environments is not defined, hence $P \mid Q$ is not typable.

In both cases, we conclude a contradiction.

5.2 Communication Safety

Definition 13. (Progress Property). *A process P has the progress property if* $P \to P'$ *implies that either* P' *does not contain live channels or* $P' \mid Q \to$ *for some* Q *such that* $P' \mid Q$ *is well-typed and* $Q \not\mapsto$*.*

A process P has the progress property if it is not blocked, and a process is blocked if it is some "bad" normal form. Even an irreducible process can have progress whenever it is able to interact in parallel with an irreducible process Q. In addition, The function *head*(P) is to get the head prefixed subprocess in P.

Definition 14. (Type Compliance). *Type compliance is the largest relation on types such that whenever S is compliant with T, written* $S \propto T$ *, it holds that either head(S) = ∅, or* $X \in head(S)$*,* $Y \in head(T)$*, and* $X \perp Y$ *(or* $X \perp_c Y$*) such that,* $S \xrightarrow{Y} S'$*,* $T \xrightarrow{X} T'$*, it holds that* $S' \propto T'$ *.*

The duality relation is relaxed, aiming to combine the subtyping for both recursive and non-recursive session types, which allows communications between participants following different protocols.

Lemma 15. *For communications occur in binary sessions, both non-recursive and recursive sessions must keep the following rules*:
1, *The shared channels are free, without using shared channel hiding.*
2, *The live channels on both participants satisfy the bilinearity condition.*
3, *The session types on the two ports satisfy the type compliance.*

Theorem 16. (Progress). *The parallel composition* $P \mid Q$ *keeps progress property if it follows the Lemma 15.*

Proof. Let $P_0 = P \mid Q$ be a process and $P_0 \to^* P_1$. There are three cases for process P_1:

Case 1. P_1 does not contain live channel or it $P_1 \to P'$. There is nothing to prove.

Case 2. P_1 is irreducible and its head prefixed subprocess is an accept/request on a free shared channel a. Naturally, we can build an irreducible Q_2 as a request process on a according to the relaxed duality for non-recursive or recursive session types. This is why we do not consider any irreducible process as blocked, rather we say that even a irreducible P_1 has the progress whenever it is able to interact in parallel with some Q_2 such that $P_1 \mid Q_2$ is well-typed.

Case 3. P_l does not contain an accept/request on a shared channel as head subprocess, but contains a live channel k^p. Since the live channels satisfy the bilinearity condition and type compliance, $k^{\overline{p}}$ must occur in P_l, which means it follows what P_l reduces.

6 Conclusions and Future Work

A general theme in this work has been the introduction of the progress property that once a session is started, the participants will be able to complete all the necessary communications without getting in a deadlock. The type compliance is defined to show a good combination between the subtyping relation and progress property. In addition, the introduction of bilinearity condition makes it easier to guarantee progress and prove Type Safety. However, the type system does not support session delegation [17-19], that means the transmissions of shared or live channels are forbidden. Clearly, there is scope for further work in this area.

On the other hand, the present work focuses on the case of two participants, yet there is nothing about the cases in multiparty sessions [12, 20], where session types of participants involved will not be dual any more. As a result, the whole process cannot be constructed from the two dual participants. We hope to touch on these aspects in our future work.

Acknowledgments. We would like to thank Simon Gay, Mario Coppo and Mariangiola Dezani-Ciancaglini for their valuable guidance on subtyping for session types and the progress property.

References

1. Pierce, B.C., Sangiorgi, D.: Typing and subtyping for mobile processes. Mathematical Structures in Computer Science 6(5), 409–453 (1996)
2. Pierce, B.C., Sangiorgi, D.: Behavioral equivalence in the polymorphic Pi-calculus. Journal of the ACM 47(3), 531–584 (2000)
3. Turner, D.N.: The polymorphic Pi-Calculus: theory and implementation. PhD thesis, University of Edinburgh (1996)
4. Gay, S.J.: A sort inference algorithm for the polyadic Pi-calculus. In: 20th ACM Symposium on Principles of Programming Languages, pp. 429–438. ACM Press (1993)
5. Gay, S.J., Hole, M.: Types and Subtypes for Client-Server Interactions. In: Swierstra, S.D. (ed.) ESOP 1999. LNCS, vol. 1576, pp. 74–90. Springer, Heidelberg (1999)
6. Gay, S.J., Hole, M.J.: Bounded polymorphism in session types. Technical Report, Department of Computing Science, University of Glasgow (2003)
7. Gay, S.J., Hole, M.J.: Subtyping for session types in the Pi-Calculus. Acta Informatica 42(2/3), 191–225 (2005)
8. Gay, S.J.: Bounded polymorphism in session types. Mathematical Structures in Computer Science 18, 895–930 (2007)
9. Honda, K.: Types for Dyadic Interaction. In: Best, E. (ed.) CONCUR 1993. LNCS, vol. 715, pp. 509–523. Springer, Heidelberg (1993)

10. Takeuchi, K., Honda, K., Kubo, M.: An Interaction-Based Language and Its Typing System. In: Halatsis, C., Philokyprou, G., Maritsas, D., Theodoridis, S. (eds.) PARLE 1994. LNCS, vol. 817, pp. 398–413. Springer, Heidelberg (1994)
11. Honda, K., Vasconcelos, V.T., Kubo, M.: Language Primitives and Type Discipline for Structured Communication-Based Programming. In: Hankin, C. (ed.) ESOP 1998. LNCS, vol. 1381, pp. 122–138. Springer, Heidelberg (1998)
12. Honda, K., Yoshida, N., Carbone, M.: Multiparty asynchronous session types. In: 32nd Symposium on Principles of Programming Languages, pp. 273–284. ACM Press, New York (2008)
13. Reynolds, J.C.: Towards a Theory of Type Structure. In: Robinet, B. (ed.) Programming Symposium. LNCS, vol. 19, pp. 408–425. Springer, Heidelberg (1974)
14. Girard, J.Y.: Interprétation fonctionnelle et élimination des coupures dans l'arithmétique d'ordre supétieur, PhD thesis, University of Paris VII (1972)
15. Acciai, L., Boreale, M.: A Type System for Client Progress in a Service-Oriented Calculus. In: Degano, P., De Nicola, R., Meseguer, J. (eds.) Concurrency, Graphs and Models. LNCS, vol. 5065, pp. 642–658. Springer, Heidelberg (2008)
16. Liu, X., Walker, D.: A Polymorphic Type System for the Polyadic Pi-Calculus. In: Lee, I., Smolka, S.A. (eds.) CONCUR 1995. LNCS, vol. 962, pp. 103–116. Springer, Heidelberg (1995)
17. Coppo, M., Dezani-Ciancaglini, M., Yoshida, N.: Asynchronous Session Types and Progress for Object Oriented Languages. In: Bonsangue, M.M., Johnsen, E.B. (eds.) FMOODS 2007. LNCS, vol. 4468, pp. 1–31. Springer, Heidelberg (2007)
18. Dezani-Ciancaglini, M., Mostrous, D., Yoshida, N., Drossopoulou, S.: Session Types for Object-Oriented Languages. In: Hu, Q. (ed.) ECOOP 2006. LNCS, vol. 4067, pp. 328–352. Springer, Heidelberg (2006)
19. Dezani-Ciancaglini, M., de'Liguoro, U., Yoshida, N.: On Progress for Structured Communications. In: Barthe, G., Fournet, C. (eds.) TGC 2007. LNCS, vol. 4912, pp. 257–275. Springer, Heidelberg (2008)
20. Castagna, G., Dezani-Ciancaglini, M., Padovani, L.: On global types and multi-party sessions. Logical Methods in Computer Science 8(1), 1–45 (2012)
21. Yoshida, N., Vasconcelos, V.: Language primitives and type disciplines for structured communication-based programming revisited. In: 1st International Workshop on Security and Rewriting Techniques. ENTCS, vol. 171, pp. 73–93. Elsevier (2007)
22. Milner, R., Parrow, J., Walker, D.: A calculus of mobile processes, I and II, Information and Computation, pp. 1–77 (1992)
23. Vasconcelos, V.T., Honda, K.: Principal Typing Schemes in a Polyadic Pi-Calculus. In: Best, E. (ed.) CONCUR 1993. LNCS, vol. 715, pp. 524–538. Springer, Heidelberg (1993)
24. Yang, Z., Zhong, F., Zhang, J.: A bounded polymorphic type system for client-server interactions. In: 4th International Conference on Networking and Digital Society; Journal of Digital Content Technology and its Application (to appear, 2012)

Energy Model of SARA and Its Performance Analysis

Bo Pang, Zhiyi Fang, Hongyu Sun, Ming Han, Xiaohui Nie

College of Computer Science and Technology of Jilin University, 130012
Changchun, Jilin, China
fangzy@jlu.edu.cn

Abstract. SARA algorithm is a self-adaptive wireless sensor network (WSN) routing algorithm. It can, depending on the parameter values, select a different network performance metrics for path selection. SARA algorithm takes three performance indicators into account: network life cycle, network delay, network reliability .By adjusting the proportion of each of the three performance indicators, SARA is able to fulfill the network requirements of a longer life cycle, shorter delay network and a higher reliability. In this paper, an experimental study of the SARA life is conducted and an energy model is formulated in order to prove the efficiency of the network performance. Furthermore, the established energy model proves the energy performance of SARA and its superiority in energy consumption by laboratory analysis, compared with LEACH algorithm.

Keywords: WSN, SARA, the energy model, life cycle.

1 Introduction

WSN (Wireless Sensor Network) is a novel sensor network. As one of the four pillar industries in the 21st century high-tech fields, it has been applied to fields such as military, agriculture, environmental monitoring, commercial, medical, marine exploration. Its application is of great significance to national security and socio-economic development.

There are four key technologies of WSN: routing, positioning[1], energy consumption and security issues. Among these four, routing is a core technology, as data is transmitted from the source node to destination node by the routing protocol. Meanwhile, routing algorithm can restrict the energy consumption and security problems. As a consequence, the research of infinite sensor network routing algorithm is of great value. Currently, there are four types of routing algorithms, namely, Energy Efficient Routing Protocol, query-based routing protocols, geographic routing protocols, QoS routing protocol [2], based on the different needs of the network. Among these, energy efficient routing protocol is a typical representative of the energy multi-path routing protocols; the main representative of the query routing protocol is directed diffusion routing protocol and rumor routing protocol; while location-based routing protocol is GEAR routing protocols. In recent years, QoS routing protocols are gaining an incremental attention and application in both business

F.L. Wang et al. (Eds.): WISM 2012, LNCS 7529, pp. 462–471, 2012.

and academic study, which, however, are usually at the expense of increased energy consumption.

Energy multi-path routing protocol [3] mainly concerns the life cycle of the network. Its goal is to extend the network lifetime to the greatest possible, while neglect other QoS parameters. However, it is not effective when there is long network delay. Directed diffusion routing protocol [4] mainly works through the release of interest in information related to the query data, rather than proactively sending data, so it has the disadvantages of poor real-time quality and weak ability to find new collection to real-time data and information. Rumor routing protocol [5] are likely to cause an empty network, thereby reducing the life cycle of the network. GEAR [6] focuses on the geographical coordination, with greater dependence on the top of the positioning algorithm, which makes it hard to transplant. In short, a common feature of the existing routing algorithms is that they are effective for a single destination routing; but not universally applicable.

This paper proposes an adaptive routing algorithm SARA Title (Self-the adaptive Routing Algorithm), as a response to the current routing algorithms. Depending on various the parameter settings, SARA can concentrate on different performance indicators and balance performance indicators. This modification increases router's adaption to various needs of the network, which expands the application scope and reduces the difficulty to transplant.

The remaining parts of the paper are organized as follows: first, SARA algorithms is discussed in Section 3, followed by energy modeling of SARA in Section 4; the performance of the simulation is proved in Section 5, after which there is a brief conclusion.

2 Sara Algorithm

SARA algorithms refer to the three performance indicators, namely, the life cycle of the network, the network transmission delay, and network reliability. According to the actual situation of the network, the life cycle of these three networks are modeled as follows:

2.1 The Life Cycle of the Network

The network life cycle refers to process from network deployment to the first node appears. The design of the algorithm aims to select a smaller energy consumption path routing in order to extend the life cycle. Formula 1 shows the energy consumption model:

$$C_i = \mathrm{C(s,l)} + \mathrm{Metric(l,j)} \quad (1)$$

C_i indicates the energy consumption of the its path; $\mathrm{C(s,l)}$ represents the consumption from node s to node 1; while $\mathrm{Metric(l,j)}$ is the path of energy from node 1 to node j. As shown in equation 3.1, the path of energy consumption takes the residual energy of nodes consumption and path energy consumption into account.

2.2 Network Transmission Delay

Transmission delay of the network path is represented by the transmission delay of the physical time, shown as in formula 2.

$$D_i = T_{V_n} - T_{V_s} \tag{2}$$

Among them, T_{V_s} is the time of packets sent from the node s, while T_{V_n} is the time destination node receiveing the packet. In order to obtain T_{V_s} and T_{V_n}, a stamp packets is designed, as shown in Figure 1.

The source node number	The destination node number
Packet type	
Optional fields	Forwarding the remaining energy of the packet node
Package creation time	Destination node received the packet time
Path information	

Fig. 1. Timestamp packet formats

2.3 The Reliability of the Network

This chapter defines the reliability of the proposed algorithm as the ratio of the distance of two nodes and communication radius, for instance, if the distance is within 2 times the communication radius, it is considered a reliable network. The network reliability modeling is shown in formula 3.

$$\mathrm{R}_k(\mathrm{i},\mathrm{j}) = \frac{\sqrt{(x_j-x_i)^2+(y_j-y_i)^2}}{2R} \tag{3}$$

k denotes the k-path reliability, as shown in Equation 3.3, the greater $\mathrm{R}_k(\mathrm{i},\mathrm{j})$, the higher the reliability.

2.4 Adaptive Parameter

SARA combines the three performance indicators above, and adjusts the relations of the constraints of the various performance indicators through parameters, as shown in formula 4.

$$\mathrm{Sa\_p}_i = \alpha \frac{1/C_i}{\sum_{k=1}^{m} 1/C_k} + \beta \frac{1/D_i}{\sum_{j=1}^{m} 1/D_j} + \gamma \frac{R_i}{\sum_{k=1}^{m} R_i} \tag{4}$$

In this formula, $\alpha+\beta+\gamma=1$, so it is clear that $\mathrm{sa\_p_i}$ and the path of energy consumption is inversely proportional, that is, a smaller energy consumption and less delay path routing is preferable. The reliability is proportional related, which means high credibility is ideal, and the higher path routing; the higher the probability the path is selected.

3 The Sara Energy Model

In order to analyze the energy consumption of the routing algorithm, energy consumption model is utilized, as is shown in the following figure:

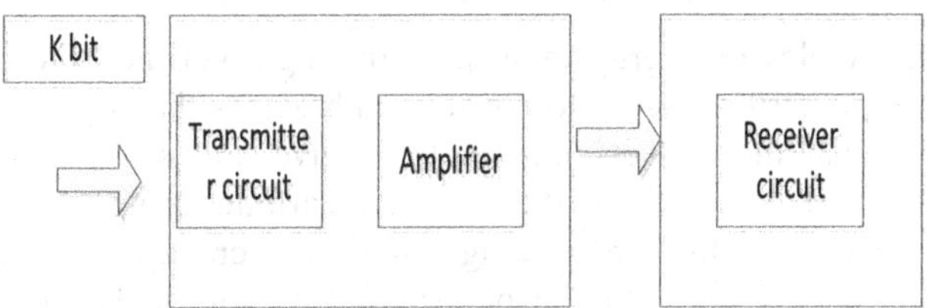

Fig. 2. The SARA energy model

As shown in Figure 2, the loss of the transmitter circuit energy consumption refers to the sending of a message. Assume that the length of the message sent is k bit, then energy consumption of sending the k-bit message is E_(Tx-elect) (k) . Power amplifier circuit loss is associated to transmission distance: if k is-bit message sent r m, denoted by the energy consumption of the amplifier circuit asE_(Tx-amp) (k,r), the energy consumption receiving the k-bit message asE_(Tx-elect) (k), and that the energy consumption of sending or receiving 1bit message asE_(Tx-elect) (k).
Assume that the radius of each node of the launch from the SARA algorithm to any one path is r, there is a total of n jump, so the average transmission distance is n × r; then the SARA algorithm energy consumption is:

3.1 Energy consumption of path is established.
3.2 Creating the path is based on the use of genetic algorithms, so there is no communication cost. Assume the implementation of the average energy consumed by an instruction node is E_cal, and that the implementation of genetic algorithm to generate a path average instruction book are m paths, the resulting computational overhead is as follows, shown in formula 5:
3.3 E_path=m*E_cal (5)
3.4 Updating the path of energy consumption.

After established, the path needs to update its value through energy delay request message. Assume that the updated message to k-bit, and then updated the loss E_(edit-path) is shown in formula 6:

$$E_{\text{edit-path}} = 2 * \begin{pmatrix} nE_{Tx\text{-elect}}(k) + (n-1)E_{Tx\text{-elect}}(k) + \\ nE_{Tx\text{-amp}}(\mathrm{k},\mathrm{r}) \end{pmatrix} \quad (6)$$

Transmitting k-bit data of energy consumption, such as shown in formula 7:

$$\mathrm{E}_{\text{edit-path}} = \mathrm{nE}_{\text{Tx-elect}}(\mathrm{k}) + (\mathrm{n}-1)\mathrm{E}_{\text{Tx-elect}}(\mathrm{k}) + \mathrm{nE}_{\text{Tx-amp}}(\mathrm{k},\mathrm{r}) \quad \ldots(7)$$

According to statistics, energy consumption is mainly concentrated in transmission. The 100m average energy consumption of 1-bit data transfer is the implementation of the 3000 instruction energy consumption, so the calculation of the energy consumption of an instruction with the message sending / receiving the energy consumption of the relationship is shown in Equation 8, as follows:

$$E_{cal} = \frac{E_{Tx-elect}(k)}{3000*rk} \ldots \tag{8}$$

Based on the principles of energy multi-path routing algorithm, the energy of the multi-path routing algorithm needs to create a path across the network broadcasting the path to establish the packets needed to receive and send large amounts of messages, rather than the path of the SARA algorithm using genetic algorithm. Assume that energy multi-path routing algorithm creates a path of energy consumption and energy consumption of the transmission of data is consistent (in fact, it is fairly different as it needs to establish paths through broadcasting), that is, the value of Equation 5.3. In order to facilitate the comparison, this article assumes that E_(Tx-elect) (k)≈E_(Tx-amp) (k,r)=E_elect, then the value of 5.3 is approximately equal to (3n-1)*E_elect. Empathy May Permit, E_cal≈E_elect/ (3000*rk), so the implementation of 3000 * (3n-1) r.

Instruction energy consumption is equivalent to the path of energy multi-path routing algorithm. Assume that in a small network, data sent each time is bits 4096bit, when r = 30m, the average number of hops in the 100 * 100m, the instructions executed by the hop count is 108, while the running of SARA requires 104, which is considerably different.

To sum it up, the overall energy consumption of SARA algorithm is lower than the energy of the multi-path routing algorithm.

4 Sara Performance Analysis

In order to verify the SARA algorithm' capacity to ensure the network life cycle basis, reduce network latency and increase network reliability, a simulation research is conducted towards SARA algorithm, MEAR, algorithm and LEACH algorithm under MATLAB2010 environment in this paper, and proves SARA's superiority from the following three aspects: life cycle, network the reliability and the network latency.

Simulation Environment

Under MATLAB environment, the simulation in this paper takes the matrix to represent. Assuming the random deployment of 100 nodes within the 100 * 100 m, and there are 100 nodes in laboratory in the region, the nodes of the initial situation are shown in Table 1.

Figure 3 and Figure 4 show the distribution of 100 nodes in 100 * 100 square meters, where the red stars nodes are the target node, and the blue circular represent the ordinary sensor nodes. Figure 1 indicates the network model without cluster network model, and Figure 2 shows the one with cluster network model. The blue stars represent the cluster head node. As shown in Figure 3 and Figure 4. The simulation in laboratory network is close to the realistic distribution situation which can represent reality to a large extent.

Table 1. The node in the initial situation

The deployment of regional	$100*100m^2$
Communication radius	R=20m
The number of nodes	N=100
The initial energy	E_{init}=0.5J
Energy consumption data is sent	E_{tx}=0.00000001J/bit
Data receiving energy consumption	E_{rx}=0.00000001J/bit
Data fusion energy consumption	E_{da}=0.0000000000001J/bit
Packet length	L=1024bit
Calculate the energy consumption	$E_{cal}=E_{tx}/30$
Leach algorithm cluster head probability	p=0.1

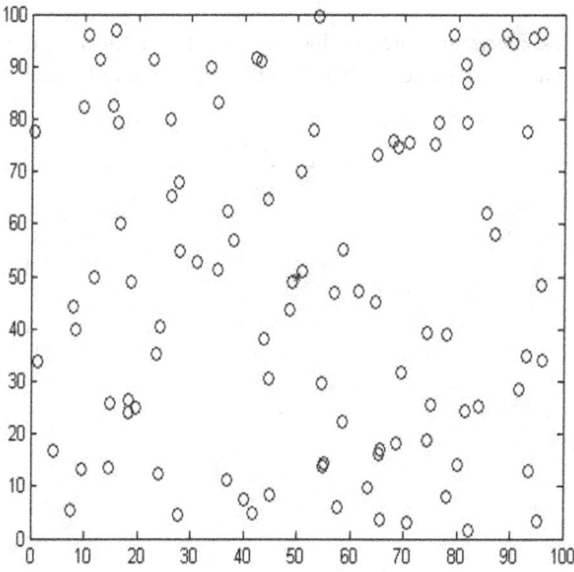

Fig. 3. Network distribution of plat network model

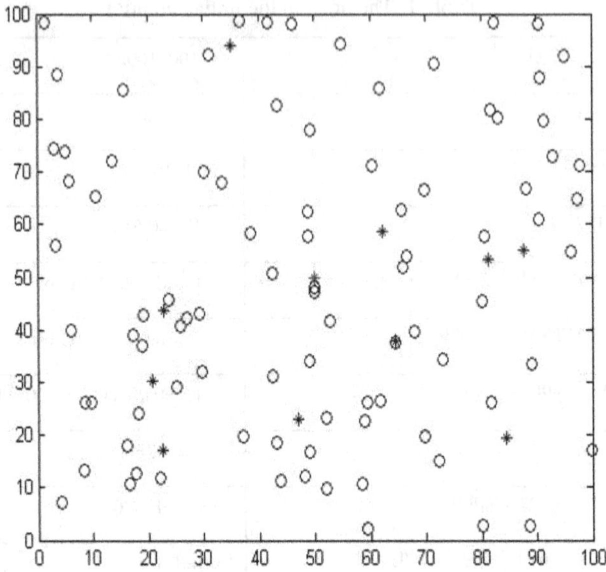

Fig. 4. Distribution of hierarchical network model

10 laboratory experiments are conducted under the models Figures 3 and 4 in the initial conditions shown in Table 1. The results are shown in Figure 5.

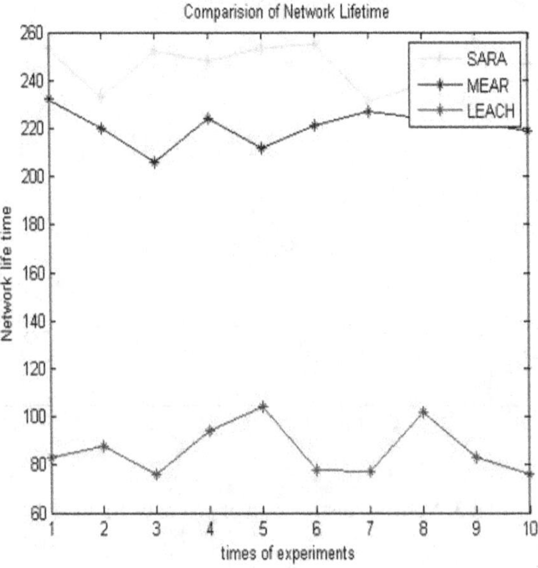

Fig. 5. Under the same conditions of the network lifecycle

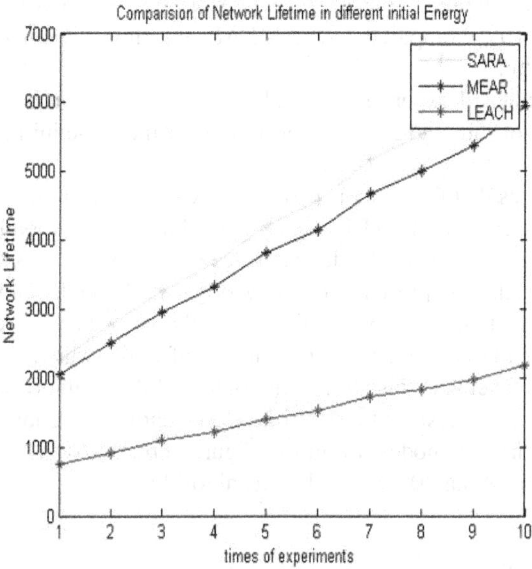

Fig. 6. Network life cycle comparison under different initial energy

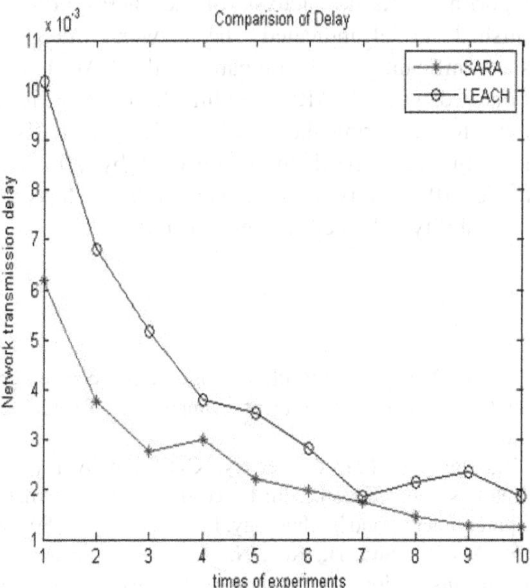

Fig. 7. Comparison of network delay

The network life cycle of the above three algorithms under different initial energy are shown in Figure 6.

As indicated in Figure 6, the average network life cycle of SARA algorithms delayed by about 10% compared to the MEAR algorithm, even in the case of the initial energy change.

This paper conducted experimental study on the comparison between the average delay of SARA algorithm and Leach algorithm under the same initial conditions, as is shown in Figure 7.

Figure 5 indicates that the delay is on a steady decrease with the increase of nodes, both in the SARA algorithm and LEACH algorithm. Consequently, the better the connectivity, the less the networks delay. However, the delay in SARA algorithm is significantly reduced, and presents a steady downward trend, compared to that of LEACH algorithm. This may be attributed to the fact that LEACH reduces the average latency of the network with the number of hops, which may vary with the dynamic changes of network instability. In contrast, SARA algorithm makes choices of the route with less physical time, so the delay shows a steady decline with the increase of the number of nodes. From the Figure above, SARA algorithm increases the average delay by about 20% over LEACH algorithm.

5 Conclusion

This paper makes a further analysis and established a network model of the previously proposed SARA algorithm. As the analysis of the energy efficiency of the SARA through the established model indicated, the network life cycle of the SARA algorithm presents a significant growth compared with LEACH algorithm, and shows a 10% increase compared with MEAR algorithm. In addition, it effectively reduces the network latency when ensuring the network life cycle basis at the same time. Through experiment, the network delay is lowered by 20% compared with the LEACH algorithm. Nevertheless, issues such as the relationship of the life cycles and the absence of accountability still need further research.

References

1. Guo, D., Jia, C.: A WSN routing algorithm using overpayment on complex network. In: 5th International Conference on Intelligent Computation Technology and Automation, ICICTA 2012 (2012)
2. Sun, H., Fang, Z., Wang, T., Ma, Y., Ren, N.: CDHL: A hybrid range-free localization algorithm in wireless sensor networks. In: Proceedings - 5th International Conference on Frontier of Computer Science and Technology, FCST 2010, pp. 180–183 (2010)
3. Fang, Z., Liu, Z., Ma, Y., Sun, H., Ren, N.: A new coordinate correction localization theory and its implementation mechanism in wireless sensor networks. Journal of Information and Computational Science 8(1), 23–28 (2011)
4. Hong, L., Yang, J.: An Energy-Balance Multipath Routing based on Rumor Routing for Wireless Sensor Networks. In: Fifth International Conference on Natural Computation, pp. 87–91 (2009)
5. Hun, X.-B., Leeson, M.S., Hines, E.L.: An effective genetic algorithm for network coding, pp. 952–963. Elsevier (2012)

6. Lei, Y., Shang, F., Long, Z., Ren, Y.: An energy efficient multiple-hop routing protocol for wireless sensor networks. In: Proceedings - The 1st International Conference on Intelligent Networks and Intelligent Systems, ICINIS 2008, pp. 147–150 (2008)
7. Yang, Y., Zhong, C., Sun, Y., Yang, J.: Network coding based reliable disjoint and braided multipath routing for sensor networks. Journal of Network and Computer Applications 33, 422–432 (2010)
8. Yin, L., Wang, C., Ien, G.E.: On the minimization of communication energy consumption of correlated sensor nodes. Wireless Personal Communications 50(1), 57–67 (2009)
9. Minhas, M.R., Gopalakrishnan, S., Leung, V.: An online multipath routing algorithm for maximizing lifespan in wireless sensor networks. In: Proceedings of the Sixth International Conference on Information Technology: New Generations, 2008, pp. 581–586 (2009)
10. Cheng, W., Xing, K., Cheng, X., Lu, X., Lu, Z., Su, J., et al.: Route recovery in vertex-disjoint multipath routing for many-to-one sensor networks. In: Proceedings of MobiHoc 2008, Hong Kong SAR, China, pp. 209–219 (May 2008)
11. Jung, S., Lee, J., Roh, B.: An optimized node-disjoint multi-path routing protocol for multimedia data transmission over wireless sensor networks. In: Proceedings of International Symposium on Parallel and Distributed Processing and Applications (ISPA), pp. 958–963 (December 2008)
12. Chao, H.L., Chang, C.L.: A fault-tolerant routing protocol in wireless sensor networks. International Journal of Sensor Networks 3(1), 66–73 (2008)
13. Wu, S., Candan, K.S.: Power-aware single- and multipath geopraphic routing in sensor networks. Ad Hoc Networks 5, 974–997 (2007)
14. Zhai, Y., Yang, O.W.W., Wang, W., Shu, Y.: Implementing multipath source routing in a wireless ad hoc network testbed. In: Proceedings of IEEE Pacific RIM Conference on Communications, Computers and Signal Processing (PACRIM), pp. 292–295 (2005)
15. Lou, W., Liu, W., Zhang, Y.: Performance optimization using multipath routing in mobile ad hoc and wireless sensor networks. Combinatorial Optimization in Communication Networks 18, 117–146 (2006)

Data Profiling for Semantic Web Data

Huiying Li

School of Computer Science and Engineering,
Southeast University, Nanjing 210096, P.R. China
huiyingli@seu.edu.cn

Abstract. Lots of RDF data have been published in the Semantic Web. For human users it is often rather difficult to get the big picture of a large RDF data exposed by Semantic Web applications. How to understand a large and unfamiliar RDF data becomes very important when the data schema is absent or different schemas are mixed. In this paper we describe a tool which can induce the actual schema, gather corresponding statistics, and present a UML-based visualization for the RDF data sources like SPARQL endpoints and RDF dumps. Experimental results, using six data sets from the Linked Data cloud, compare our approach and ExpLOD. The evaluations show that our approach is more efficient than ExpLOD.

Keywords: Semantic Web, RDF, Data profiling.

1 Introduction

As pointed out in [1], the Web is being extended with more and more RDF data sources and links between objects. The emergence of linked open data (LOD)[2] is being promoted by the Linking Open Data community project and is fostering the availability of many open RDF data sets, such as SwetoDBLP[3] and LinkedMDB[4]. These data sets are placed online by the community, fostering collaborative linkage of structured knowledge.

How to understand a large and unfamiliar RDF data becomes a key challenge when reusing it. For human users, it is often rather difficult to get the big picture of a large RDF data exposed by Semantic Web applications. An usual approach is to get familiar with the ontology which the RDF data conforms to. While it becomes complicated if the RDF data uses multiple ontologies. Even worse, the ontologies used by some RDF data set such as SPARQL endpoint can not be obtained. Inspired by early work on RDF usage presented in [5] enumerates RDF usages, we propose a SPARQL query based approach for RDF data profiling. For human users, data profiling is a cardinal activity when facing an unfamiliar data set. It can help to assess the importance of the data set as a whole, find out whether the data set or part of the data set can be easily reused, improve the user ability to query or search the data set, detect irregularities for improving data quality, etc.

In this paper, we propose an approach to obtain the actual schema of a SPARQL endpoint or RDF dump, gather corresponding statistics, present a visual, UML-based

F.L. Wang et al. (Eds.): WISM 2012, LNCS 7529, pp. 472–479, 2012.

notation for them. Such visualization can be considered as a customed schema for the RDF data behind SPARQL endpoint. We compare our approach to the similar approach ExpLOD. The evaluations show that our approach does not need to construct middle graph and is more efficient than ExpLOD.

The rest of the article is organized as follows. Section 2 introduces the related work. Section 3 introduces the SPARQL based approach to construct and visualize RDF data profiling. Section 4 details the experimental results of our approach. Section 5 concludes the study.

2 Related Work

Describing and understanding large RDF data can be enabled by statistics and summary. Recently, there exist some researches dealing with RDF data statistics and summary, such as Semantic sitemaps[6], RDFStats[7], SCOVO vocabulary[8], and ExpLOD[9]. Among the works above, ExpLOD is the most similar work to our approach. ExpLOD is a tool that supports constructing summaries of RDF usage based on bisimulation contraction mechanism. Before generating the summary graph, ExpLOD constructs a label graph by adding triples to describe the bisimulation label for each class, predicate, instance and literal. Then using a SPARQL-based approach, ExpLOD computes the instance blocks based on instances' predicate usages. Notice that the triple number of the middle label graph is much larger than the triple number of origin RDF graph, it will cost more storage space. Moreover, the number of instance blocks will be large when instances' predicate usages have different variations. Although ExpLOD enables coarse granularity summarization based on hierarchical bisimulation label. While the summarization grouped by high hierarchy label such as namespace is too general sometimes. Especially, ExpLOD does not care about the relation between instance blocks which is very important to help user understand the data structure.

3 Visualizing Semantic Web Data Profile

3.1 Semantic Web Data Profiling

We consider the RDF graph as a set of RDF triples. A triple is consisted of a subject, a predicate and an object[10]. Fig.1 shows the RDF graph of an sample snippet of RDF data. The snippet gives information about two music artists, their names, the records they have made and the record titles. In such a small snippet, there are four used ontologies, the FOAF ontology (with prefix label *foaf*), the Dublin Core ontology (with prefix label *dc*), the Music Ontology (with prefix label *mo*) and the RDF meta ontology (with prefix label *rdf*). When the ontologies are absent, it is rather difficult to get the big picture for human users. Our purpose is to obtain the actual schema of the SPARQL endpoint or RDF dump, gather corresponding statistics, and present a UML-based visualization.

```
eg:artist1 foaf:name "Cicada";
   rdf:type foaf:Person;
   rdf:type mo:MusicArtist;
   foaf:made eg:record1.
eg:artist2 foaf:name "Bobywan";
   rdf:type foaf:Person;
   rdf:type mo:MusicArtist;
   foaf:made eg:record2.
eg:record1 dc:title "Rebirth";
   rdf:type mo:Record.
eg:record2 dc:title "Live my dream";
   rdf:type mo:Record.
```

Fig. 1. RDF data snippet

Database profiling is the process of analyzing a data set to determine its structure and internal relationships. Common profiling methods and tools assume a starting point of relational data with a domain specific schema. This assumption does not hold for Semantic Web data published on the web, especially the SPARQL endpoint data. Such SPARQL endpoint only enables users to query a knowledge base via the SPARQL language. We propose a new SPARQL query based approach for Semantic Web data profiling. Six different kinds of descriptive statistics are used: (1) data structure, denotes the classes, their attributes (datatype property) and relations (object property) between classes; (2) class instantiation, the number of instances that are typed as a particular class; (3) property instantiation, the number of times a property is used to describe instances; (4) multiplicity of object, an inclusive interval of non-negative integers to specify the allowable number of objects of described property; (5) functional property, denotes whether a property have only one (unique) value for each instance or not; and (6) inverse-functional property, denotes whether a property is inverse-functional or not.

3.2 Visual Modeling for Semantic Web Data Profile

Since we construct the actual schema and gather the corresponding statistics, how to visualize them for human users becomes an important issue. We adopt the notations of properties and classes proposed by [11] and Ontology Definition Metamodel[12].

Classes. Usually, an OWL class can be described by six different ways. Among them, two ways are used in RDF data profiling. One is the simple named class, it is defined by a class identifier. The other is the property restriction class, it restricts the domain or range of a property for the context of the class. We use the *<<rdfsClass>>* stereotype to represent RDF/S and OWL class. Fig. 2 demonstrates the notations for named class and restriction class. The class name of Fig. 2(a) is *mo::MusicArtist* where *mo* is the prefix label of name space. Fig. 2(b) shows a restriction class which restricts the range of property *event::factor* for class *mo::Lyrics*.

Properties. The object properties and datatype properties are represented as UML associations and attributes respectively. Because properties can have multiple domains and ranges, we allow several associations with the same name. Fig. 3(a) demonstrates the notation for object properties, and Fig. 3(b) demonstrates the simplified notation for object properties and datatype properties.

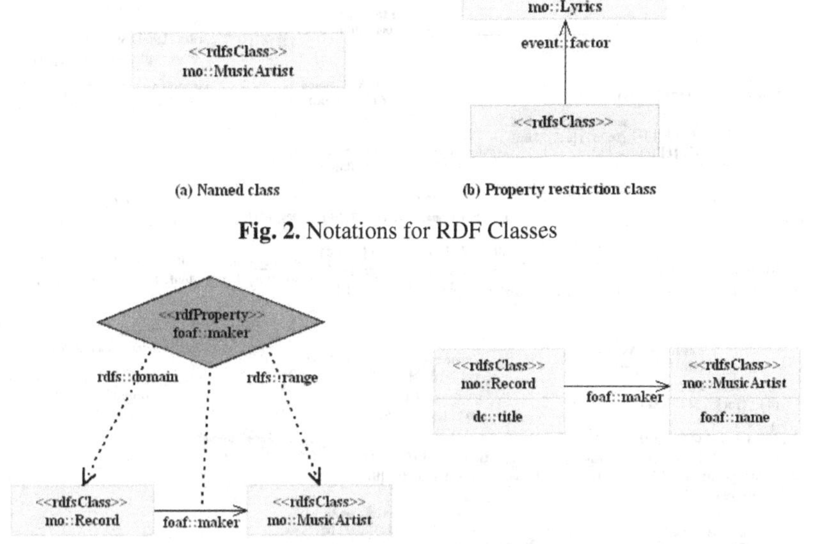

Fig. 2. Notations for RDF Classes

Fig. 3. Notations for RDF Properties

3.3 Implementation

This part describes the SPARQL-based approach to create the RDF data profiling. In order to compute the RDF data structure, we collect all classes used in the RDF data firstly. In RDF data, many classes are not declared to be an *owl:Class* or *rdfs:Class* explicitly. We use the following SPARQL query to statistic all the named classes: "*SELECT distinct ?c WHERE* {*?s rdf:type ?c.*}". Notice that the classes with *rdf*, *rdfs* and *owl* namespaces are considered as metaschema-level classes and are not collected (except for *rdfs:Seq* because many blank nodes are declared to be *rdfs:Seq*). With these named classes, the instances are belonged to different blocks according to their types.

After getting the named classes, we explore the property (including relation and attribute) usage of every class. For each class, we explore all its instances to collect their used properties. For example, the SPARQL query "*SELECT distinct ?prop WHERE*{*?s rdf:type mo:Record. ?s ?prop ?o.*}" queries the used properties of class *mo:Record*. Moreover, for the object property, we find out the range classes by the SPARQL query "*SELECT distinct ?rangeclass WHERE*{*?s rdf:type mo:Record. ?s mo:available_as ?o. OPTIONAL*{*?o rdf:type ?rangeclass.* }}".

There are still some instances without type declaration despite many typed instances. To solve this problem, the first solution is to collect all the non-type instances into one unnamed class. While, to provide more clear information for human user, the second solution is to divide these non-typed instances into different partitions according to their property restrictions. We collect the properties with non-type instance as subject or object in advance, then the SPARQL query is used to divide the non-type instances.

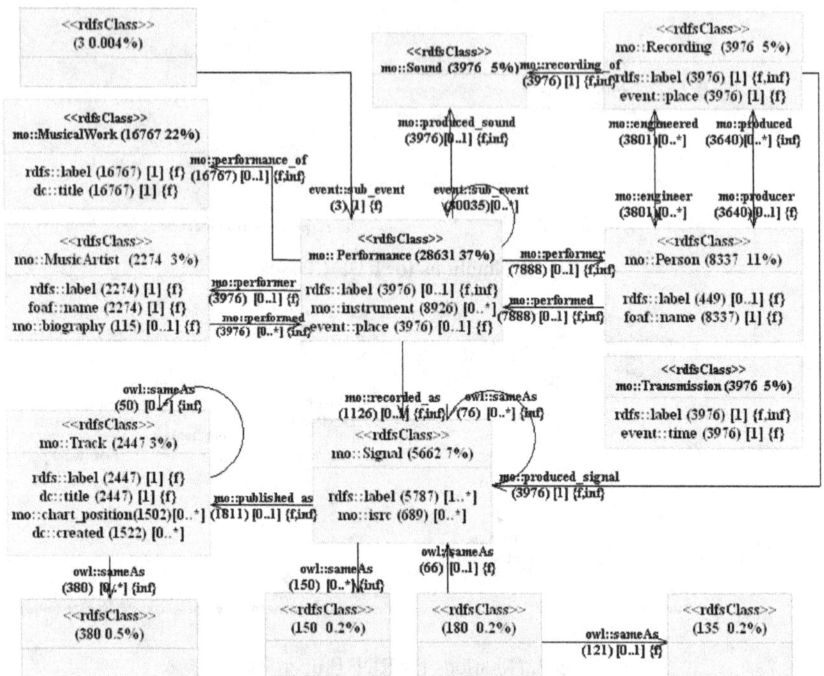

Fig. 4. Visualizing Peel data with UML-like Notation

Using the visual model introduced in Section 3.2, we visualize the data profiling of RDF data Peel, a data set from LOD cloud, which contains the information about music artists and their productions. Fig. 4 shows the structure and corresponding statistics about Peel. For each class, the number of instances belong to this class and the percent of all instances are provided. For every property, the instantiation and the multiplicity of object are provided, and whether the property is functional (denoted by *f*) or inverse-functional (denoted by *inf*) is also provided. Moreover, the unnamed block in Peel is divided into five restriction classes shown in Fig. 4. Such a graph provides the data structure and corresponding statistics for human users. With its help, users can understand the RDF data behind a SPARQL endpoint.

4 Experimental Study

In this section, we compare the performance of our approach with ExpLOD. ExpLOD is a tool that supports SPARQL-based summary creation of RDF usage, it constructs label graph by adding labels to RDF graph and computes the bisimulation contractions based on the label graph. We use Jena toolkit (jena.sourceforge.net) to manage RDF data for ExpLOD and our approach. All the experiments were developed within the Eclipse environment and on a PC with 4 GB of RAM (of which 1.5 GB was assigned to the JVM).

We chose six data sets which are shown in the Table 1. These data sets are considered from the LOD cloud because they vary in the amount and type of information they describe. Data Peel and Jamendo contain information about music artists and their productions. Data GeoSpecies contains information on biological orders, families, species as well as species occurrence records and related data. Data LinkedCT contains information about linked clinical trials. Data LinkMDB contains linked data about movies. Data SwetoDblp which focuses on bibliography data of publications from DBLP includes information about affiliations, universities, publications and publishers. Table 1 shows the following information about each data set: the name, the number of triples it contains, the number of instances, classes and properties.

Table 1. The statistics of RDF data sets

Data set	# of classes	# of properties	# of instances	# of triples
Peel	9	25	76,894	271,369
Jamendo	11	26	410,893	1,047,950
GeoSpecies	44	168	184,931	2,076,380
LinkedCT	13	90	1,169,905	9,804,652
LinkedMDB	53	222	1,326,001	6,147,995
SwetoDblp	10	145	2,394,479	13,041,580

To show the efficiency of our approach and ExpLOD, we conduct a performance evaluation. Performance is measured as the time taken by different methods. For ExpLOD, the premise of computing summary graph is to construct labeled graph. So, the time taken by ExpLOD is the sum of the labeled graph constructing time and the summary graph computing time. The running time of our approach is composed of three parts: the time to get all classes, the time to construct data structure and the time to compute corresponding statistics. We divide the six data sets into two groups, the Peel, Jamendo and GeoSpecies are in group one, these data sets are relative small and are fit into memory to deal with. The LinkedCT, LinkedMDB, SwetoDblp are in group two, these data sets are relative large and are put into database.

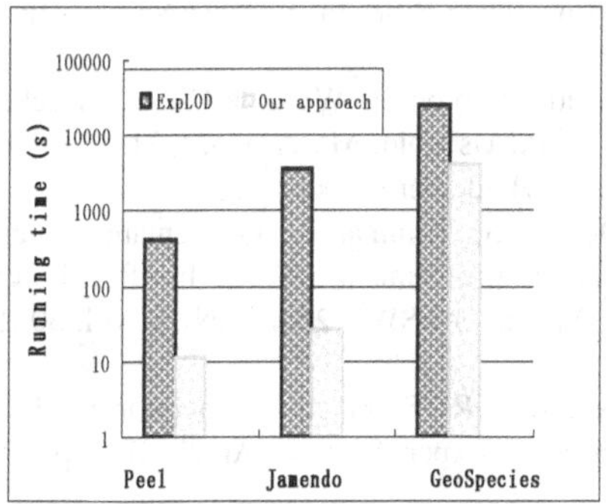

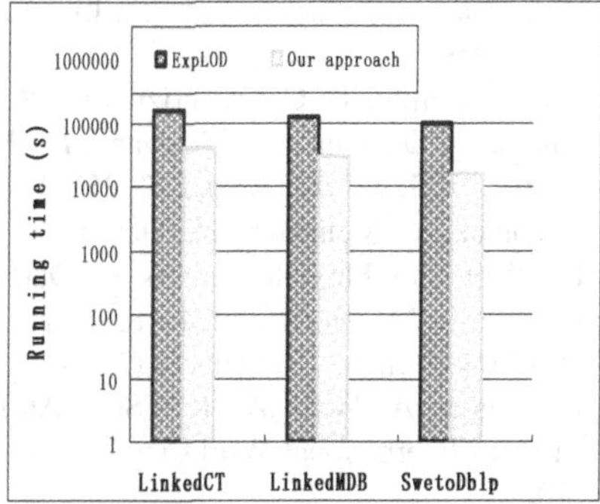

Fig. 5. a) Comparison for group one(in s) b) Comparison for group two(in s)

Fig. 5 shows the performance comparison. It can also be seen that the running time of ExpLOD is longer than our approach. For ExpLOD, before computing the summary graph, a bisimulation label must be assigned to each class, predicate, instance and literal. It will increase the triple number largely. Take data Jamendo for example, it has 1,047,950 triples. The triple number increase to 3,940,113 after adding bisimulation label, it is about four times more than the original triple number.

5 Conclusion

Lots of RDF data have been published in the Semantic Web. For human users it is often rather difficult to get the big picture of the information contained in a large RDF data, especially the data schema is absent or different schemas are mixed. We propose a SPARQL-based tool to construct the actual schema, gather corresponding statistics, and present a visual, UML-based notation for the RDF data sources like SPARQL endpoints and RDF dumps. It helps the human user to understand the data structure and construct suitable SPARQL query. Experimental results, using several collections from the Linked Open Data cloud, compare our approach and ExpLOD. The evaluations show that our approach does not need to construct middle graph and is more efficient than ExpLOD.

Acknowledgments. The work is supported by the National Natural Science Foundation of China under grant No. 60973024 and the National Natural Science Foundation of China under grant No. 61170165.

References

1. Hendler, J., Shadbolt, N., Hall, W., Berners-Lee, T., Weitzner, D.: Web Science: An Interdisciplinary Approach to Understanding the Web. Communications of the ACM 51(7), 60–69 (2008)
2. Bizer, C., Heath, T., Berners-Lee, T.: Linked Data - The Story So Far. IJSWIS 5(3), 1–22 (2009)
3. Aleman-Meza, B., Hakimpour, F., Arpinar, I.B., Sheth, A.P.: SwetoDblp Ontology of Computer Science Publications. Web Semantics: Science, Services and Agents on the World Wide Web 5(3), 151–155 (2007)
4. Hassanzadeh, O., Consens, M.P.: Linked Movie Data Base. In: I-SEMANTICS, pp. 194–196 (2008)
5. Ding, L., Finin, T.: Characterizing the Semantic Web on the Web. In: Cruz, I., Decker, S., Allemang, D., Preist, C., Schwabe, D., Mika, P., Uschold, M., Aroyo, L.M. (eds.) ISWC 2006. LNCS, vol. 4273, pp. 242–257. Springer, Heidelberg (2006)
6. Cyganiak, R., Stenzhorn, H., Delbru, R., Decker, S., Tummarello, G.: Semantic Sitemaps: Efficient and Flexible Access to Datasets on the Semantic Web. In: Bechhofer, S., Hauswirth, M., Hoffmann, J., Koubarakis, M. (eds.) ESWC 2008. LNCS, vol. 5021, pp. 690–704. Springer, Heidelberg (2008)
7. Langegger, A., Woß, W.: RDFStats-An Extensible RDF Statistics Generator and Library. In: 20th International Workshop on Database and Expert Systems Application, pp. 79–83 (2009)

8. Hausenblas, M., Halb, W., Raimond, Y., Feigenbaum, L., Ayers, D.: SCOVO: Using Statistics on the Web of Data. In: Aroyo, L., Traverso, P., Ciravegna, F., Cimiano, P., Heath, T., Hyvönen, E., Mizoguchi, R., Oren, E., Sabou, M., Simperl, E. (eds.) ESWC 2009. LNCS, vol. 5554, pp. 708–722. Springer, Heidelberg (2009)
9. Khatchadourian, S., Consens, M.P.: ExpLOD: Summary-Based Exploration of Interlinking and RDF Usage in the Linked Open Data Cloud. In: Aroyo, L., Antoniou, G., Hyvönen, E., ten Teije, A., Stuckenschmidt, H., Cabral, L., Tudorache, T. (eds.) ESWC 2010, Part II. LNCS, vol. 6089, pp. 272–287. Springer, Heidelberg (2010)
10. Klyne, G., Carroll, J. (eds.): Resource Description Framework (RDF): Concepts and Abstract Syntax, http://www.w3.org/TR/2004/REC-rdf-concepts-20040210/
11. Brockmans, S., Volz, R., Eberhart, A., Löffler, P.: Visual Modeling of OWL DL Ontologies Using UML. In: McIlraith, S.A., Plexousakis, D., van Harmelen, F. (eds.) ISWC 2004. LNCS, vol. 3298, pp. 198–213. Springer, Heidelberg (2004)
12. Documents Associated with Ontology Definition Metamodel (ODM) Version 1.0 (2009), http://www.omg.org/spec/ODM/1.0/

Checking and Handling Inconsistency of DBpedia

Zhaohua Sheng, Xin Wang*, Hong Shi, and Zhiyong Feng

School of Computer Science and Technology, Tianjin University, Tianjin, China
shengzhaoli24353@sina.com, {wangx, serena, zyfeng}@tju.edu.cn

Abstract. DBpedia is the hub of Linked Data, and there might be inconsistencies in it. Reasoning with inconsistent ontologies may lead to erroneous conclusions, so whether it is consistent is a critical issue. However, the current inference engine is only appropriate to reason lightweight ontologies, and the existing approaches to handle inconsistencies are unreasonable. In this paper, we check the inconsistency in DBpedia by rule-based distributed reasoning using MapReduce. The experimental results show that there are a number of inconsistencies in DBpedia. Furthermore, we should handle different types of inconsistencies respectively with different methods to improve data quality of DBpedia.

Keywords: inconsistency, large-scale ontology, DBpedia, MapReduce.

1 Introduction

The DBpedia project [1] is a community effort to publish data of Wikipedia infobox in the form of RDF. It has large volume of data and is relatively comprehensive. During the last few years, a great number of Linked Data datasets have been published in which entities are linked to their equivalent resource in DBpedia, making DBpedia a central interlinking hub for the Linked Data [2] datasets.

There might be inconsistencies in DBpedia. From data sources viewpoint, one's understanding of something may vary according to his own background. So, inconsistent data might exist in Wikipedia. From another viewpoint, when the data are extracted from Wikipedia, errors may occur during this process. For example, when we trace LibriVox to its Wikipedia page, we find that it is in fact an online digital library. However, in DBpedia it is asserted as a member of libraries in the real world. Reasoning with inconsistent ontologies may lead to erroneous conclusions. Hence, we need to check its consistency and handle inconsistencies with a reasonable method.

Since DBpedia is the hub of Linked Data and there are serious consequences if reasoning with inconsistent premises, we focus on checking and handling inconsistency of it in this paper. However, the TBox of the DBpedia ontology is relatively simple such that we cannot directly check its consistency. So, firstly we extend the DBpedia ontology with disjointness axioms. Ontologies, or generally speaking schemas, are an effective mean to improve the quality of Linked Data.

* Corresponding author.

F.L. Wang et al. (Eds.): WISM 2012, LNCS 7529, pp. 480–488, 2012.

In particular, the disjointness axioms are definitions of relation between two classes that share no instances, and can be used to check the consistency of ontologies. Thus, enriching ontologies with disjointness axioms is a hot topic in ontology engineering.

The remainder of this paper is organized as follows: after giving a brief overview of related work in Section 2, we present how to extend the DBpedia ontology using UMBEL in Section 3. Based on this, we show the consistency checking of DBpedia in detail in Section 4. In Section 5 we present our analysis and discussion of experimental results. Finally, there is our conclusion and future work in Section 6.

2 Related Work

The consistency is an important metric for quality of ontologies. Many researchers have been working on checking and handling inconsistencies of large-scale RDF data.

In [3] they define some types of inconsistency when measuring data quality of Linked Data. But they only give representations of inconsistencies in the natural language and there are no formal definitions. Pellet [4], one of the most popular OWL reasoners, could check the consistency of ontologies. However, Pellet is only appropriate to reason lightweight ontologies and there are memory overflow problems when reasoning large-scale ontologies. It is required to use a distributed method to solve this problem. WebPIE [5], proposed by J. Urbani, is a distributed inference engine based on MapReduce [6]. Its experimental results show that it has high efficiency and good scalability. But it only restricts to RDFS and OWL Horst rules.

When handling inconsistencies, the traditional approach is just removing or ignoring the corresponding axioms uniformly. In [7][8] they make efforts to obtain a collection of maximum consistent sub-ontologies. Both of the approaches have a shortcoming that they lack considering reasons causing inconsistencies.

The work in [9] is similar to ours. But it has two drawbacks. One is that they convert consistency checking to SPARQL query, so they do not tackle the essential large-scale ontology reasoning problem. The other is that they just simply list the number of inconsistencies in DBpedia, but they do not analyze the experimental results thoroughly and give no reasonable solutions for handling inconsistencies.

3 Ontology Extension of DBpedia

As shown in [1], the TBox of DBpedia is relatively simple such that we cannot directly check its consistencies. So, firstly we extend it with disjointness axioms using the relation between UMBEL and DBpedia.

UMBEL [10] (Upper Mapping and Binding Exchange Layer) is a lightweight ontology which has been created for interlinking Web content and data. It is a subset of OpenCyc[11]. OpenCyc is a common knowledge base and has a long development history. Now, it has been accepted around the world and there are many applications based on it. So, there is a strong persuasiveness to extend the DBpedia ontology using UMBEL. UMBEL has two valuable functions. First, it defines a base vocabulary which can be used for the construction of ontologies from other domains. Second, based on the vocabulary, it provides many reference concepts that can be used as standard links in external datasets, so different datasets could interoperate easily.

There are two reasons why we can use UMBEL to obtain disjointness axioms in the DBpedia ontology. First, the UMBEL ontology defines the disjointness relations between different classes. Second, the DBpedia ontology is linked to the UMBEL ontology by rdf:subClassOf. Thus, we could extend the DBpedia ontology using rule (1) and obtain 25534 disjointness axioms.

$$\frac{(RC1, \text{rdfs:subClassOf}, C1)(RC2, \text{rdfs:subClassOf}, C2)(C1, \text{owl:disjointWith}, C2)}{(RC1, \text{owl:disjointWith}, RC2)} \quad (1)$$

Note that RC_1 and RC_2 are classes from DBpedia and C_1 and C_2 are classes from UMBEL. All relationships are defined in files downloaded from the website[1].

4 Reasoning with MapReduce to Check Inconsistency

In this Section, we first give a brief introduction to MapReduce. Next, we summarize five types of inconsistencies in the form of inference rules in DBpedia based on [3]. Finally, we take a rule as an example to describe how we implement it using MapReduce.

4.1 The MapReduce Programming Model

The MapReduce programming model proposed by Google is a parallel processing model and suitable for handling large-scale data. A cluster of some commodity machines could achieve high efficiency and scalability using this model. In recent years, there is a lot of work using this model when consuming Linked Data.

Each job consists of two phases: a Map phrase and a Reduce phrase. When a job arrives, the input files are divided into many chunks. Then each chunk is processed respectively. The Map phase emits the input data in the form of a set of key/value pairs. All pairs with the same key will be collected to the same machine. Every machine processes intermediate values independently and outputs a new set of key/value pairs as output.

4.2 Definitions of Inconsistencies in DBpedia

INCONSISTENCY 1: Usage of undefined classes or properties

$$\frac{(s, \text{rdf:type}, X)}{(\text{inconsistency})} \ (X \text{ is not defined in TBox.}) \quad (2)$$

$$\frac{(s, p, o)}{(\text{inconsistency})} \ (p \text{ is not defined in TBox.}) \quad (3)$$

[1] http://code.google.com/p/umbel/source/browse/#svn%2Ftrunk%2Fv100

INCONSISTENCY 2: Usage of literals incompatible with ranges of data type properties

$$\frac{(p, \text{rdfs:range}, d_1)(s, p, o)(o, \text{rdf:type}, d_2)}{(\text{inconsistency})} \quad (d_1 \text{ is incompatible with } d_2.) \tag{4}$$

INCONSISTENCY 3: One class is subclass of and disjoint with another class

$$\frac{(X, \text{owl:disjointWith}, Y)(X, \text{rdfs:subClassOf}, Y)}{(\text{inconsistency})} \tag{5}$$

INCONSISTENCY 4: One class is subclass of two disjoint classes

$$\frac{(X, \text{rdfs:subClassOf}, Y)(X, \text{rdfs:subClassOf}, Z)(Y, \text{owl:disjointWith}, Z)}{(\text{inconsistency})} \tag{6}$$

INCONSISTENCY 5: Invalid entity definitions as members of disjoint classes

$$\frac{(s, \text{rdf:type}, X)(s, \text{rdf:type}, Y)(X, \text{owl:disjointWith}, Y)}{(\text{inconsistency})} \tag{7}$$

Note that *s* and *o* stand for instances; *p* stands for properties defined in the TBox; *X*, *Y* and *Z* stand for classes defined in TBox; d_1 and d_2 stand for data types from XML or DBpedia.

4.3 Example Rule Execution with MapReduce

In this Section, we take rule (7) as an example to illustrate how the inference rule is executed with MapReduce. The inputs of this job are documents of triples in the form (*s*, *p*, *o*). In the Map phrase, triples with rdf:type as predicates and classes as objects are emitted. The intermediate outputs are pairs of <*s*, *o*> as shown in Algorithm 1 and Fig. 1. After the Map phrase, values with the same key are collected to one machine. When all emitted triples are grouped for the reduce phase. The reduce function starts to work. In the Reduce phrase, we traverse values to check whether the two classes are disjoint. If they are, an inconsistency is output; if not, nothing is output.

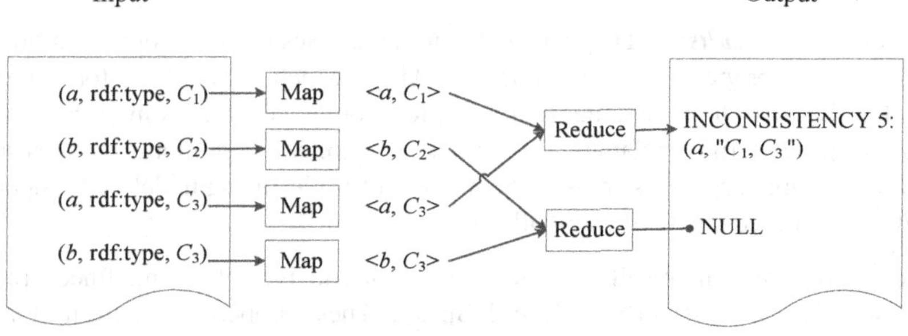

Fig. 1. Encoding rule (7) with MapReduce. C_1 and C_3 are disjoint classes; C_2 and C_3 are not

Algorithm 1. Encoding rule 7 with MapReduce

```
map (key, value)
        // key: line number (not relevent).
        // value:triples.
  if (value.p=rdf:type and value.o=owl:Class)
        EMIT (<value.s, value.o>)

reduce (key, iterator values)
        // key: resource.
        // values: classes which are types of the resource.
    for each v1 in values
      for each v2 in values
        if (v1 owl:disjointWith v2)
            new INCONSISTENCT 5 (key, "v1, v2")
            EMIT (<NullWritable, INCONCSISTENCY 5>)
```

5 Experiments

We conduct several experiments on the dataset provided by DBpedia[2]. So far, the latest version is 3.7. First we check consistencies in the original dataset, and then we experiment on the dataset enriched with triples obtained according to two RDFS reasoning rules (8, 9). Since RDF data have semantics and we often consume inferred data, it is necessary to check consistencies of it after reasoning.

In this section, we use the following prefixes to abbreviate resource URIs in the DBpedia dataset.

dbo: http://dbpedia.org/ontology.
dbr: http://dbpedia.org/resource.
dbd: http://dbpedia.org/datatype.

5.1 Data Inconsistency before Reasoning

(1) Usage of undefined classes or properties

***Experimental Results*:** In DBpedia, there are no classes used without definitions. But 15 undefined properties are used in the ABox which consists of foaf:based_near, foaf:familyName, foaf:surname, foaf:page, foaf:thumbnail, foaf:homepage, foaf:logo, foaf:name, foaf:givenName, foaf:nick, geo:lat, geo:long, skos:subject, purl:description, and georss:point. The number of triples using undefined properties is 3170567 and accounts for 18% of all.

***Discussion*:** From the result we conclude that although a lot of undefined properties are used, they are just not defined in DBpedia. These properties are from FOAF, the geospatial namespace, SKOS, Dublin Core, and GeoRSS.

---

[2] http://wiki.dbpedia.org/Downloads37

***Solution*:** We need to import ontologies which define these properties in the header of the DBpedia document.

(2) Usage of literals incompatible with ranges of data type properties

***Experimental results*:** There are 39632 triples whose literals are incompatible with ranges of data type properties. Table 1 enumerates the top five properties instantiated with incompatible values.

***Discussion*:** We find that ranges of these properties are all defined to xsd:double in the TBox. However, in the ABox, objects in triples with these properties are assigned to user-defined data types in DBpedia and they are not of the same type. So, there is no proper and unified data type for this type of properties.

***Solution*:** There are two solutions to solve this problem. First, we could make no definition for its range and the value of it can be of any data types. Second, we could define a new data type as superclass of these existing types as its range in the TBox. In the ABox, we could assert its value to the corresponding subclasses if necessary.

Table 1. Top five data type properties with literals incompatible with ranges

Data type properties	Number of occurrences
<http://dbpedia.org/ontology/budget>	8248
<http://dbpedia.org/ontology/revenue>	6960
<http://dbpedia.org/ontology/gross>	6447
<http://dbpedia.org/ontology/netIncome>	3502
<http://dbpedia.org/ontology/cost>	2871

(3) One class is subclass of two disjoint classes

***Experimental results*:** We find the following inconsistency of this type.
(dbo:Library , rdfs:subClassOf , dbo:Building)
(dbo:Library, rdfs:subClassOf, dbo:EducationalInstitution)
(dbo:Building , owl:disjointWith, dbo:EducationalInstitution)

***Discussion*:** When analyzed respectively they are all correct because of various characteristics of something. But there are real world libraries and digital ones. When it is the latter, the first axiom is wrong. Therefore, it is unreasonable to decide which of them should be removed only from the concept definitions.

***Solution*:** Classes defined in the DBpedia TBox are not comprehensive and the granularity of classification is relatively rough. We should improve quality of the DBpedia TBox. For example, we could define subclasses of the class dbo:Library.

(4) Invalid entity definitions as members of disjoint classes

***Experimental results*:** We find 960 inconsistencies of this type. All these inconsistencies are instances of the following two disjointness axioms.

(dbo:Building, owl:disjointWith, dbo:EducationalInstitution)
(dbo:Building, owl:disjointWith, dbo:Organisation)

***Discussion*:** As shown in Section 1, there are some reasons that cause inconsistencies. In order to handle inconsistencies reasonably we need to seek out what causes contradictions on earth. The first axiom is reasonable since we believe that buildings are disjoint with organizations. For example, we can assert dbr: Yale_University_Library to an instance of dbo:Building and dbo:Organization. Therefore, when handling this inconsistency we should ignore the disjointness axiom. But when we consider dbr:LibriVox the solution is not proper. It is in fact a digital library. There is an erroneous assertion when data are extracted from Wikipedia. So when handling this inconsistency we should remove the incorrect assertion.

***Solution*:** We divide this type of inconsistency into two categories based on reasons: (1) Inconsistencies that should be retained due to various characteristics of things. (2) Inconsistencies that should be removed. We should deal with each inconsistency respectively according to its category.

5.2 Data Inconsistency after Reasoning

As triples we obtain only relate to inconsistency 5, other experimental results are the same to results before reasoning. In this Section, we only show results of inconsistency 5.

***Experimental results*:** We obtain 2,483,391 triples with rules (8, 9) and find 55829 inconsistencies. Table 2 enumerates the top five disjoint classes.

***Discussion*:** The number and type of inconsistencies both increase after reasoning. As shown in Table 2, the top two pairs of disjoint classes are the same as results before reasoning. In addition, we also find apparent inconsistencies such as dbo:PopulatedPlace and dbo:Person. The results indicate that DBpedia cannot guarantee the correctness of reasoning. In other words, there are a lot of contradictions between the TBox and the ABox. The essential reason is that the domain or range of one property in the TBox is disjoint with types of an instance in the ABox.

***Solution*:** If we would like to develop applications based on the DBpedia dataset we should pre-process the data first of all. Fortunately, our results point out the direction in some extent. For example, we recommend to analyzing instances asserted to classes in Table 2.

$$\frac{(p, \text{rdfs:domain}, X)(s, p, o)}{(s, \text{rdf:type}, X)} \tag{8}$$

$$\frac{(p, \text{rdfs:range}, X)(s, p, o)}{(o, \text{rdf:type}, X)} \tag{9}$$

Table 2. Top five disjoint classes

Disjoint classes	Number of instances
Organisation & Building	10030
EducationalInstitution & Building	8559
School & Building	7910
MeanOfTransportation & Fungus	7692
PopulatedPlace & Person	1473

6 Conclusion and Future Work

In this paper, we focus on checking consistencies of DBpedia. We first extend it with disjointness axioms via UMBEL. Then we define five types of inconsistencies and present them in the form of six inference rules. Based on that, an algorithm using MapReduce is proposed to check consistencies for large-scale ontologies. After analyzing the experimental results we summarize three reasons that cause inconsistencies in DBpedia: (1) user-defined data types of DBpedia are inadequate; (2) there are a lot of incorrect assertions in the DBpedia ABox; and (3) the quality of the DBpedia TBox is low. For instance, the granularity of classification is relatively rough. Finally, we draw the conclusion that it is not wise just to remove or ignore corresponding axioms uniformly and we should classify inconsistencies according to the reasons and deal with them respectively.

In the future, we will give more rules of inconsistencies for more datasets such as GeoNames and Bio2RDF. Regards to handling inconsistencies, an automatic inconsistency classifier is needed.

Acknowledgements. This work was supported by the National Science Foundation of China (No. 61100049, 61070202) and the Seed Foundation of Tianjin University (No.60302022).

References

1. Bizer, C., Lehmann, J., Kobilarov, G., Auer, S., Becker, C., Cyganiak, R., Hellmann, S.: DBpedia - A Crystallization Point for the Web of Data. Web Semantics 7(3), 154–165 (2009)
2. Bizer, C., Heath, T., Berners-Lee, T.: Linked Data - The Story So Far. International Journal on Semantic Web and Information Systems 5(3), 1–22 (2009)
3. Hogan, A., Harth, A., Passant, A., Decker, S., Polleres, A.: Weaving the pedantic web. In: International Workshop on LDOW at WWW (2010)
4. Sirin, E., Parsia, B., Grau, B.C., Kalyanpur, A., Katz, Y.: Pellet: A practical owl-dl reasoner. Web Semantics 5, 51–53 (2007)
5. Urbani, J., Kotoulas, S., Maassen, J., Van Harmelen, F., Bal, H.: WebPIE: A Web-scale parallel inference engine using MapReduce. Web Semantics: Science, Services and Agents on the World Wide Web (2011)
6. Dean, J., Ghemawat, S.: MapReduce: Simplified Data Processing on Large Clusters. Communications of the ACM, 107–113 (2008)

7. Haase, P., Stojanovic, L.: Consistent evolution of OWL ontologies. The Semantic Web: Research and Applications, 91–133 (2005)
8. Meyer, T., Lee, K., Booth, R., Pan, J.Z.: Finding maximally satisfiable terminologies for the description logic ALC. In: Proceedings of the 21st National Conference on Artificial Intelligence, pp. 269–274. AAAI Press (2006)
9. Fleischhacker, D., Völker, J.: Inductive Learning of Disjointness Axioms. In: Meersman, R., Dillon, T., Herrero, P., Kumar, A., Reichert, M., Qing, L., Ooi, B.-C., Damiani, E., Schmidt, D.C., White, J., Hauswirth, M., Hitzler, P., Mohania, M. (eds.) OTM 2011, Part II. LNCS, vol. 7045, pp. 680–697. Springer, Heidelberg (2011)
10. UMBEL, `http://www.umbel.org/`
11. OpenCyc, `http://www.opencyc.org/`

Conceptual Representing of Documents and Query Expansion Based on Ontology

Haoming Wang, Ye Guo, Xibing Shi, and Fan Yang

School of Information, Xi'an University of Finance and Economics, Xi'an, Shaanxi 710100, P.R. China
hmwang@mail.xaufe.edu.cn, shiny_yang0000@sina.com, {guoyexinxi,xbshine}@126.com

Abstract. In vector space model, a document is represented by words. As the new words appear dramatically in the Internet era, this kind of method draws back the IR systems performance. This paper puts forward a new approach to present the concepts, query expressions, and documents based on the ontology. The approach has two levels, the Word-Concept level and the Concept-Document level. In the first level, the transition probability matrix is constructed by using the appearing times of word-word pairs in documents. The biggest eigenvector of matrix is computed, and it reflects the importance of words to the concept. In the second level, the distance matrix is constructed by using the distance between words in a given ontology, and the average variance value of elements is computed. It reflects the relevance of documents to concepts. In the last section, the query expansion is discussed by using the personal information profile of the user. It is proofed to be more effective than previous one.

Keywords: Ontology, Word-Concept level, Concept-Document level, Relevance Computing, Personal Information Profile.

1 Introduction

With the explosive growth of information in Internet, search engine (SE) is used in information retrieval (IR). Users of IR Systems expect to find the most relevant items to a certain query. Generally, the IR system does not feed back an ideal behavior. They feed back much more results to users, and users have to spend a considerable time to find these items which are really relevant to their initial queries.

One of the reasons is that the system does not know what the user wants to get actually. The search engine separates the terms in the query expression, and computing the page value by using the contents or the page-links or the both of the pages. The $top - n$ pages are feed back to the users. This kind of searching method neglects the relevant documents that do not contain the index terms which are specified in the users queries. In order to improve the effect of retrieval, the specific domain knowledge should be added to the queries.

F.L. Wang et al. (Eds.): WISM 2012, LNCS 7529, pp. 489–496, 2012.

Ontology is a conceptualization of domain knowledge. It is a concept set with the human understandable, machine readable format. It consists entities, attributes, relationship, and axioms. For an ontology based information retrieval system, when the user input the query expression, the system tries to insert the ontology knowledge to enhance to query expression in order to increase the probability of relevancy. In the concept level, documents having very different vocabularies could be similar in subject and, similarly, documents having similar vocabularies may be topically very different.

This paper is organized as follows: Section 2 introduces the concepts of Ontology and Query Expansion. Section 3 discusses new approach combining words, concepts and documents in two levels. Section 4 presents the methods of query expansion. Finally, a summary of this paper and directions for future work are discussed in Section 5.

2 Related Works

2.1 Ontology

In the traditional IR approaches, documents and query expression are represented as a vector of terms simply. One of the examples is VSM. The relevance of the document and the query is computed by the cosine distance between two vectors. This approach does not require any extraction or annotation phases. Therefore, it is easy to implement, however, the precision value is relatively low.

Ontology is explicit representations of a shared conceptualization, i.e., an abstract, simplified view of a shared domain of discourse. More formally, an ontology defines the vocabulary of a problem domain, and a set of constraints (axioms and rules) on how terms can be combined to model specific domains. An ontology is typically structured as a set of definitions of concepts and relations between these concepts. Ontology is machine-processable, and they also provide the semantic context by adding semantic information to models, thereby enabling natural language processing, reasoning capabilities, domain enrichment, domain validation, etc [1,2].

2.2 WordNet

WordNet, a manually constructed electronic lexical database for English, was conceived in 1986 at Princeton University, where it continues to be developed. WordNet is a large semantic network interlinking words and groups of words by means of lexical and conceptual relations represented by labeled arcs. WordNets building blocks are synonym sets (synsets), unordered sets of cognitively synonymous words and phrases. Each member of a given synset expresses the same concept, though not all synset members are interchangeable in all contexts. Each synset has an unique identifier (SynsetID). Synsets are interlinked by means of conceptual-semantic and lexical relations. The resulting network of meaningfully related words and concepts can be navigated with the browser [3].

In this paper, we select one ontology from the WordNet to discuss our approach.

2.3 Conceptual Representation

In VSM, the term is the word of the document normally. Due to the ambiguity and the limited expressiveness of single word, it is difficult to decide which word is more important for the document.

One way of improving the quality of similarity search is Term FrequencyInverse Document Frequency (TFIDF). The main idea of TFIDF is the more vocabulary entry in document set, the lower separate ability of document property, and then the weight value is small. On the other hand, the higher frequency for a certain vocabulary entry in a document, the higher separate ability, and then the weight value is big. The method is widely used in selecting text feature. But it has many disadvantages too. First, the method undervalues that this term can represent the characteristic of the documents of this class if it only frequently appears in the documents belongs to the same class while infrequently in the documents of the other class. Second TFIDF neglects the relations between the feature and the class [4].

The another way is Latent Semantic Indexing (LSI). The most improvement is mapping the document from the original set of words to a concept space. Unfortunately, LSI maps the data into a domain in which it is not possible to provide effective indexing techniques. Instead, conceptual indexing permits to describe documents by using concepts that are unique and abstract human understandable notions. After that, several approaches, based on different techniques, have been proposed for conceptual indexing.

One of the well-known mechanism for conceptual representation is conceptual graph (CG). In Ref. [5], two ontologies are implemented based on CGs: the Tendered Structure and the abstract domain ontology. And, the authors first survey the indexing and retrieving techniques in CG literatures by using these ontologies.

2.4 Query Expansion

In an IR system, the user inputs his query expression for his requirements. Normally, the query expression is not clear enough to let the IR system understands what the user wants. Query expansion is one of the ways to solve this problem [6].

Query expansion technology was brought forward in Ref. [7]. It expands a query expression with the addition of terms that are semantically correlated with the original terms of the query. Several works demonstrated the performance of IR system was improved by using it. As the terms, which are added to the query, play a decision rule in the query process, they should be selected carefully. Experimental results show that the incorrect choice of terms might harm the retrieval process by drifting it away from the optimal correct answer [8].

3 Document Representing

There are three tasks for the documents representing based on the concept:

(1) Labeling the terms in the document. The document consists of terms, and most of them should be belonged to one or more concepts. Some of the terms are not so closed to the concepts, and they are omitted in this step. The documents are represented by the remain terms.
(2) Computing the relevance. There are many terms in a concept. There is no experimental result shows that some terms are always more important than others. The importance of the term is different in different concept. For the document, we want to get a term list by the importance decrease order to the concepts.
(3) Deciding the attribution of the document to the concept. For a given document, it may have relation with two or more concepts, which concept the document belongs to finally? Sometime, we need to answer the question that if two documents had same terms but different term orders, do they have same importance for a query or in a concept?

In the following, we construct a new approach with two-level structures. The first level, called Word-Concept (W-C) level, reflects the relation between the words and the concepts. And the second level, called Concept-Document (C-D) level, reflects the relation between concepts and documents.

3.1 Labeling Document

In our discussion, the first task is labeling the terms in a document to the concept. The method of labeling is referencing the WordNet. For the ontology and the document, we can assume the facts:

(1) An ontology is a very large set of concepts, and there are several hundreds of concepts in it. Each concept is consisted of many terms and the relations between the terms, meanwhile each term may belong to more than one concept.
(2) For a document, it is impossible to include all of the terms in a special concept or a ontology. In other words, it is impossible that all of the terms in a document are belonged to one concept.
(3) Assuming the term d is one of the terms of a document $D(d \in D)$, d can be labeled to concept C_1 or C_2 according to the term-list of the concepts. Which concept should be selected for the term d finally?

The Word-Concept level of the new approach can be described as :

(1) Constructing the matrix UC for each concept C, it is:

$$\mathbf{UC} = \begin{pmatrix} uc_{11} & uc_{12} & \dots & uc_{1n} \\ uc_{21} & uc_{22} & \dots & uc_{2n} \\ \vdots & \vdots & \ddots & \vdots \\ uc_{n1} & uc_{n2} & \dots & uc_{nn} \end{pmatrix}$$

Where the element $uc_{ij}(i \neq j)$ is the times which word d_i and d_j appear synchronously in a paragraph, and uc_{ii} is the times which word d_i appears in a paragraph by itself. In the beginning, all of the elements are 0.

(2) Scanning the document D. We label the words to the different concepts. If a word was belong to two or more concepts, it was labeled to each of the concepts. After the scanning, we count the times, of which word d_i and d_j appear synchronously, and replace the uc_{ij} with the value.

(3) Dealing with the matrix UC. If the column i is all 0, it means the word d_i does not appear in document D. The column i and row i of this matrix should be deleted.

It is obviously that the UC is symmetric matrix. In order to decrease the amount of computation, we set a threshold for value of elements. Deleted the rows and columns synchronously, the matrix keeps the characters of symmetric.

The document D may have relevance with concepts $C_1, C_2, \ldots, C_k$. We denote the relevance by matrix $UC_p, p \in [1, k]$. In the following distribution , we indicate the matrix $UC_p, p \in [1, k]$ with Q for convenience.

3.2 Computing Relevance

The elements $q_{ij}(i, j \in [1, n])$ of matrix Q responds to the times of word pair d_i-d_j appeared in the same paragraph in a document D. Normalizing the matrix Q, we explain it as:

We have a word set $D = \{d_1, d_2, \cdots, d_n\}$, and we name each word in the set with the state. The process starts in one of these states and moves successively to another. Each moving is called a step. If the chain is currently in state d_i, then it moves to state d_j at the next step with a probability denoted by q_{ij} , and the probability does not depended upon which states the chain was in before . The word set $D = \{d_1, d_2, \cdots, d_n\}$ can be regarded as Markov Chain. The Q is row-stochastic matrix, and the elements q_{ij} is transition probabilities.

According to the Chapman-Kolmogorov equation [9] and characters of Markov chain [10], the $n - step$ transition matrix can be obtained by multiplying the matrix Q by itself n times.

In the Q , the elements are connected to others, and the matrix cannot be divided into two parts. So the Q is irreducible. Meanwhile the Q is aperiodic too. The Perron-Frobenius theorem guarantees the equation $x^{(k+1)} = Q^T x^{(k)}$ (for the eigensystem $Q^T x = x$)converges to the principal eigenvector with eigenvalue 1, and there is a real, positive, and the biggest eigenvector [11].

The biggest eigenvector means the importance of word d_i to the concept C.

3.3 Deciding Affiliation

In the above, we compute the importance of a word to a concept. In this section, we will discuss the relevance of the document to the concept. The relevance will be computed in Concept-Document(C-D) level of our new approach.

Assuming we have two documents D_1 and D_2 and a concept C, how can we consider the D_1 has more or less relevance than the D_2 to the C?

The graph G discussed previously is consisted by the nodes and the links. The node represents the word and the link represents the relation. According to the definition of ontology, there are four kinds of relation between the words, such as $part-of$, $kind-of$,$instancd-of$ and $attribute-of$. We define the relation between the words as,

Define 1. *Assuming w_i and w_j are the nodes of graph G. If w_i does not connect to w_j directly, there is a path from w_i to w_j. The distance between them is the minimum of the steps from w_i to w_j.*

$$distance(w_i, w_j) = Min(n|w_i \rightarrow w_1 \rightarrow w_2 \rightarrow \cdots \rightarrow w_n \rightarrow w_j)$$

Define 2. *If w_i connected to w_j directly, the distance between them is,*

$$distance(w_i, w_j) = \begin{cases} 1 \; Rela(w_i,w_j) \in \{part-of, attribute-of\} \\ 2 \; Rela(w_i,w_j) \in \{instance-of, kind-of\} \end{cases}$$

Here $Rele(w_i, w_j)$ is the one of the four relations between the words in a ontology.

According to the $Define1$ and $Define2$, we construct the distance matrix $Dis(C, D)$, which represents the distance of the document D to the concept C.

$$\mathbf{Dis(C, D)} = \begin{pmatrix} d_{11} & d_{12} & \ldots & d_{1n} \\ 0 & d_{22} & \ldots & d_{2n} \\ \vdots & \vdots & \ddots & \vdots \\ 0 & 0 & \ldots & d_{nn} \end{pmatrix}$$

In $Dis(C, D)$, column's order can be exchange in order to keep the column i is the hypernym of column j when $i < j$.

The sum of the row i, named it with $Dis(i) = \sum_j d_{ij}, j \in [1, n]$, means the ability of word w_i representing the concept. In general, the more the $Dis(i)$, the much irrelevance of the word w_i to the concept.

So, assuming the matrix $Dis(C, D_1)$ and $Dis(C, D_2)$ represent the relevance of document D_1 and D_2 to the concept C respectively, we compute the distance respectively just as fellows,

(1) Computing the Average Variance of each rows $AV_i, i \in [1, n]$;
(2) Sum the Average Variance value $V = \sum AV_i, i \in [1, n]$.

Hence, we get two Average Variance values V_1 and V_2 for the document D_1 and D_2 to the concept C respectively. We consider that the document with the less Average Variance of V_1 and V_2 has much relevance with the C.

4 Query Expansion

Search Engine (SE) plays the important role in finding information in Internet. The user inputs the query sentence to SE, and the most important thing for SE is to know what the user want to get exactly. In normal, the query sentence is not detail enough to be used to feed back the satisfactory results to the user. Query expansion is used to solve this problem.

There are many ways to expand the query sentence. But it is difficult to expand it without any other help, such as the domain information, surfing history or log records. In our approach, the user is requested to register for the personalized service. The personal information is used to construct the Personal Information Profile (PIP). After the IR system feed back the results to the user, he checks the results and estimates them. The IR system refine the PIP according to the estimation. In the constructing of PIP, the ontology play a key role. By using ontology, we can enrich the implication of query and to enhance the search capabilities of existing web searching systems. The method of expanding can be described as,

(1) Splitting the query to words and marking them in the ontology words pool. The weight of the word plus 1 for each time appeared in the query. It is obviously that the more times the word appear in the query, the more weight it is in the ontology words pool.
(2) Selecting the concept, which the query words are involved in, we order the words belong to the concept just as following steps,
   (a) Ordering two word-lists. The first one is that the words order by the relevance, which are computed in W-C level. We named it as,

$$M(w_{i1}, w_{i2}, ..., w_{im}).$$

   The second one is that the words order by the appearance times in user's query in a given period. We named it as,

$$N(w_{j1}, w_{j2}, ..., w_{jn}).$$

   (b) Setting the final word list as,

$$P = \alpha M(w_{i1}, w_{i2}, ..., w_{im}) + (1 - \alpha)N(w_{j1}, w_{j2}, ..., w_{jn}), \alpha \in (0, 1).$$

   (c) Setting the threshold, and selecting the $top-R$ words. The $top-R$ words have much relevance with the words appeared in the query sentence.

   In general, it can be imaged that the effect of this way is not ideal in the beginning as the limited of the words in query sentence. With the times of query input increased, the accuracy will be better.
(3) The R words selected in the last step will be submitted to the SE, and SE feed backs the results to user according to the these words. The user reviews the results, and he presents his owner opinion for the retrieval results. The opinion will be used to refine the parameter $\alpha \in (0, 1)$ in the formula.

5 Conclusion

The paper introduces the concepts of ontology, query expansion, and representing the document by using the ontology. We construct a new approach with two levels, the Word-Concept level and Concept-Document level, which reflects the relation among the words, concepts, documents and queries. By computing the biggest eigenvector of words matrix to determine the relevance of words appeared in document to concepts, and computing the average variance to determine the distance of document to the concept. By constructing the Personal Information Profile(PIP) of user to expand the query sentence. According to the forecast, the feedback results will be fine than before.

Acknowledgment. This work was supported by Scientific Research Program Funded by Shaanxi Provincial Education Department, P.R.China (Program No.09JK440), and Natural Science Foundation of Shaanxi Province of China (Program No.2012JM8034).

References

1. Kara, S., Alan, O., Sabuncu, O., Akpınar, S., Cicekli, N.K., Alpaslan, F.N.: An ontology-based retrieval system using semantic indexing. Information Systems 37(4), 294–305 (2012)
2. Kang, X., Li, D., Wang, S.: Research on domain ontology in different granulations based on concept lattice. Knowledge-Based Systems 27, 152–161 (2012)
3. Dragoni, M., da Costa Pereira, C., Tettamanzi, A.G.: A conceptual representation of documents and queries for information retrieval systems by using light ontologies. Expert Systems with Applications (2012) 10.1016/j.eswa.2012.01.188
4. Qu, S., Wang, S., Zou, Y.: Improvement of text feature selection method based on tfidf. In: International Seminar on Future Information Technology and Management Engineering, FITME 2008, pp. 79–81 (November 2008)
5. Kayed, A., Colomb, R.M.: Using ontologies to index conceptual structures for tendering automation. In: Proceedings of the 13th Australasian Database Conference, ADC 2002, vol. 5, pp. 95–101. Australian Computer Society, Inc., Darlinghurst (2002)
6. Kim, M.-C., Choi, K.-S.: A comparison of collocation-based similarity measures in query expansion. Information Processing and Management 35(1), 19–30 (1999)
7. Efthimiadis, E.N.: Query expansion. Annual Review of Information Science and Technology 31, 121–187 (1996)
8. Cronen-townsend, S., Zhou, Y., Croft, W.B.: A framework for selective query expansion. In: Proceedings of 13th International Conference on Information and Knowledge Management, CIKM 2004, pp. 236–237. ACM (2004)
9. Gardiner, C.: Stochastic Methods: A Handbook for the Natural and Social Sciences. Springer Series in Synergetics. Springer (2009)
10. Mian, R., Khan, S.: Markov Chain. VDM Verlag Dr. Muller (2010)
11. Serre, D.: Matrices: theory and applications. Graduate texts in mathematics. Springer (2010)

Robust Web Data Extraction: A Novel Approach Based on Minimum Cost Script Edit Model

Donglan Liu[1,2], Xinjun Wang[1,2,*], Zhongmin Yan[1,2], and Qiuyan Li[3]

[1] School of Computer Science and Technology, Shandong University, 1500 Shunhua Road, Jinan 250101, P.R. China

[2] Shandong Provincial Key Laboratory of Software Engineering, 1500 Shunhua Road, Jinan 250101, P.R. China

[3] Changchun Institute of Engineering Technology, Changchun, P.R. China

liudonglan@mail.sdu.edu.cn, {wxj,yzm}@sdu.edu.cn, lili000643@sohu.com

Abstract. Many documents share common HTML tree structure on script generated websites, allowing us to effectively extract interested information from deep webpage by wrappers. Since tree structure evolves over time, the wrappers break frequently and need to be re-learned. In this paper, we explore the problem of constructing robust wrappers for deep web information extraction. In order to keep web extraction robust when webpage changes, a minimum cost script edit model based on machine learning techniques is proposed. With the method, we consider three edit operations under structural changes, i.e., inserting nodes, deleting nodes and substituting nodes' labels. Firstly, we obtain the change frequencies of three edit operations for each HTML label according to the frequency of webpage change on real web data with machine learning method. Then, we compute the corresponding edit costs for three edit operations on the basis of change frequencies and minimum cost model. Finally, we choose the most proper data to extract the interested information by applying the minimum cost script. Experimental results show that the proposed approach can accomplish robust web extraction with high accuracy.

Keywords: Web Data Extraction, Wrapper, Robust, Minimum Cost Script, Machine Learning.

1 Introduction

Several websites use scripts to generate highly structured HTML from backend databases, such as: recruitment sites, shopping sites, academic sites and form-based websites. The structural similarity of script-generated webpages can help information extraction systems to extract information from the webpages using simple rules. These rules are called wrappers. Once a wrapper is learnt for a site, it will keep up-to-date information. Nowadays, wrappers are becoming a dominant strategy for extracting web information from script-generated pages.

[*] Corresponding author.

F.L. Wang et al. (Eds.): WISM 2012, LNCS 7529, pp. 497–509, 2012.

However, the extraction operation of wrappers is greatly depended on the structure of webpage. Since the information of webpage changes dynamically, even very slight change may lead to the breakdown of wrappers and require them to re-learn. This is the so-called Wrapper Breakage Problem. Therefore, it is very important for web data integration to effectively improve the adaptive capacity of web data extraction.

To illustrate, Fig. 1 displays an XML document tree of a script-generated job page from 51job.com. If we want to extract working place from this page, following XPath expression can be used:

$$W_1 \equiv \text{/html/body/div[2]/table/td[2]/text()}\ . \tag{1}$$

which is an instruction on how to traverse HTML DOM trees. However, there are several small changes which can break this wrapper. For instance, the order of "Working place" and "Position" is changed, the first div is deleted or merged with the second one, a new table or td is added under the second div, and so on.

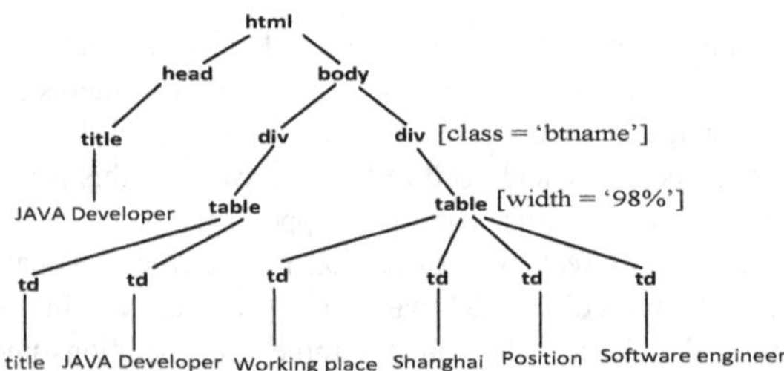

Fig. 1. An HTML webpage of 51job.com

In fact, the expression W_1 is one form of a simple wrapper, and the problem of robust web extraction has caught much attention and has been widely researched [1, 2, 3]. Jussi Myllymaki and Jared Jackson [1] proposed that certain wrappers are more robust than others, and the wrappers can have lower breakage. For instance, each of the following XPath expressions can be used as an alternative to W_1 to extract the working place.

$$W_2 \equiv \text{//div[@class = 'btname']/*/td[2]/text()}\ . \tag{2}$$

$$W_3 \equiv \text{//table[@width = '98\%']/td[2]/text()}\ . \tag{3}$$

Intuitively, these wrappers may be preferable to W_1 since they have a lesser dependence on the tree structure. For example, if the first div is deleted, W_1 will not work, while W_2 and W_3 might still work.

In this paper, we aim to design a wrapper that can achieve substantially higher robustness. Wrapper can complete and extract web data records from deep webpage accurately, which is denoted as "distinguished node" in this paper. Then, wrapper constructs a set of extraction rules according to website template, extracts information from webpage and translates into structured data automatically. In general, when webpage changes beyond the limitation of wrapper script, we can only re-locate the data by modifying wrapper scripts. Otherwise, the information extraction might be failed. When webpage evolves over time, we use the existing information of web data integrated system combining with other techniques to recognize and label the data elements and attribute tags. Then, we can generate an optimal training sample. Finally, we can rebuild a new wrapper by using the existing wrapper induction techniques, and make it possible for wrappers to cope with the changes in websites effectively. In order to keep web extraction robust when webpage changes, we propose a novel and highly efficient approach based on minimum cost script edit model with machine learning techniques for robust web extraction.

Our contributions involve four aspects in this work. Firstly, we propose a general framework for constructing robust wrapper to extract interested information from deep webpages. Secondly, we present a novel and efficient algorithm for enumerating all minimal candidate wrappers to speed up the robustness of the evaluation. Thirdly, we design a model that takes the archival data on real websites as input, and learns a model that best fits the data to extract the information from deep webpages. Finally, we perform an extensive set of experiments covering over multiple websites, and the experiments are able to achieve very high precision and recall. It demonstrates that our wrapper is highly effective in coping with changes in websites.

The rest of the paper is organized as follows. In Section 2, we introduce our robust web extraction framework, and generate minimal candidate wrappers with enumeration method in Section 3. The minimum cost script edit model for evaluating the robustness of wrappers is proposed in Section 4. Our experimental evaluation is presented in Section 5 and conclusion is provided in Section 6. Finally, acknowledgements and related references are given in the following sections.

2 Robust Web Extraction Framework

In this section, we give an overview of our robust web extraction framework on the basis of the framework which was recently proposed by Nilesh Dalvi et al. [2]. This framework is depicted in Fig. 2. Some related concepts about robust web extraction are from the reference [16]. We describe the main components as follows and the functions in italic indicate our new methods.

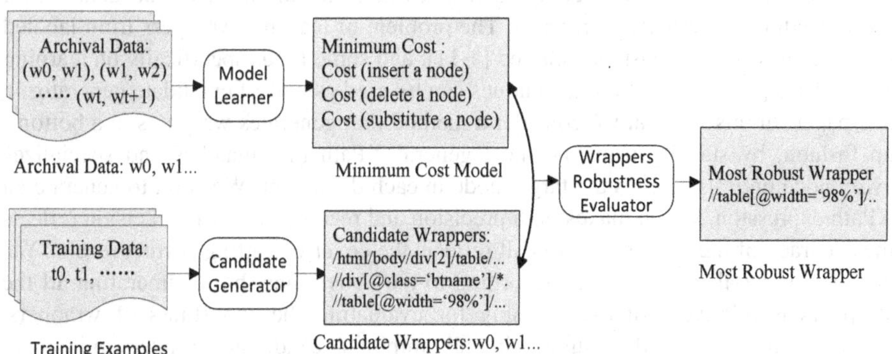

Fig. 2. Robust web extraction framework

In the framework, the *Archival Data* component contains various evolutions of webpage. Suppose a webpage w denotes the 51job page for the Java Developer, undergoing various changes across its time. Let w_0, w_1, … denote the various future versions of w. The archival data component is a collection of pairs (w_t, w_{t+1}) for the various future versions of w, i.e. it includes $\{(w_0, w_1), (w_1, w_2), \ldots, (w_t, w_{t+1})\}$. The archival data can be obtained by monitoring a set of webpages over time.

The *Model Learner* component is mainly responsible for minimum cost model. Model learner takes the archival data as input and learns a model that best fits the data. It learns the parameter values that minimize the cost of each edit operation, i.e.

$$\arg\min_{x} \prod_{(w_t, w_{t+1}) \in ArchivalDaa} \{ \cos t(insert: x), \cos t(delete: x), \cos t(substitute: x \to y) \} \,. \tag{4}$$

where x represents a node of webpage; *cost* (*insert:* x) represents the cost of inserting a node with the label x; *cost* (*delete:* x) represents the cost of deleting a node with the label x; *cost* (*substitute:* $x \to y$) represents the cost of substituting node x to node y.

The *Minimum Cost Model* component consists of costs of three edit operations: inserting nodes, deleting nodes and substituting labels of nodes. This is the salient component of our framework. The minimum cost model is specified by a set of parameters, computing the minimum cost of each edit operations by means of the parameter obtained by model learner. We intend to compute the change frequencies of three edit operations for each HTML label according to the frequency of webpage change on real web data with machine learning techniques. Then, we derive the cost of edit operations according to the frequency of webpage change combining with model learner. If the operations of webpages are frequent, then we need to make the corresponding cost as low as possible. Since the cost of each edit operation is minimal, the cost sum of a sequence of edit operations for webpage is minimum.

The *Training Examples* component for an extraction task consists of a small subset of the interested webpages that specify some fields. Such as: the working place, release date, education background and position for job information of the webpages from 51job and zhaopin sites.

The *Candidate Generator* component takes labeled training data and generates a set of candidate alternative wrappers. The problem of learning wrappers from labeled examples has been extensively studied [4-11], and some focus specifically on learning XPath rules [1, 4]. Any of the techniques can be used as part of candidate generator in this paper. In this section, we consider a method that generates wrappers in a bottom-up fashion, by starting from the most general XPath that matches and specializes every node until it matches the target node in each document. We want to generate an XPath expression, which makes both precision and recall equal to 1. Precision reflects the accuracy of the results, and Recall reflects the cover of getting correct results. We can enumerate all the wrappers according to the above idea, but enumerating all the wrappers is relatively time-consuming for evaluating the robustness of wrappers. Consequently, we consider enumerating all "minimal" candidate wrappers to improve our algorithm.

The *Wrappers Robustness Evaluator* component takes the set of candidate wrappers, evaluates the robustness of every one by the minimum cost model, and chooses the most robust wrapper. We define the robustness of wrappers as the minimum cost and it will continue to work on the future new versions of the webpage for extracting the distinguished node of interest when webpages evolve over time. The wrapper that has the most robustness is chosen among the set of candidate wrappers as the desired one.

3 Generating Minimal Candidate Wrappers

We intend to obtain a set of alternative wrappers for our extraction task to pick the most robust wrapper according to our model. The set of candidate wrappers should contain a variety of potentially robust ones. Previous work [11] on automatically learning XPath rules from labeled webpages works in a top-down fashion, namely, it starts from the specific paths in each webpage and generalizes them to a single XPath. But unfortunately, this results in the most specific XPath which contains all possible predicates across all webpage. For instance, in Fig 1, a full XPath expression W_1 is obtained by applying a top-down fashion when we extract the information of "Working place". For W_1, if the first div is deleted or merged with the second one, it will not work, namely the robustness of wrapper W_1 is extremely poor. Therefore, the resulting XPath is complex and not robust, thus, it is not a suitable candidate for a robust wrapper.

Nilesh Dalvi et al. [2] recently proposed an algorithm for enumerating wrappers; however, the algorithm is not complete. Besides, the problem of constructing the most robust wrapper is still unsolved. In this paper, we describe a more complete and effective method based on their algorithm. It generates wrappers in a bottom-up fashion, by starting from the most general XPath that matches and specializing every node until it matches the target node in each document.

Suppose D represents a set of labeled XML documents, i.e. it contains several distinguished nodes of interest in corresponding webpage. For each $d \in D$, we want to extract the target nodes coming from the set of labeled nodes of D ,which is written as $T(d)$. For an given XPath expression x, we intend to generate such an XPath expression x which satisfies the following rule: for each d, $x(d) = T(d)$, namely, the result of extraction is exactly equal to the target node. We define evaluation criteria according to the information retrieval as follows.

$$\text{Precision}(x) = \sum_{d} (x(d) \cup T(d)) / x(d) \quad (5)$$

$$\text{Recall}(x) = \sum_{d} (x(d) \cup T(d)) / T(d) \quad (6)$$

We intend to generate an XPath expression, which makes both precision and recall equal to 1. To illustrate, we use the following XPath expression:

$$x_0 \equiv \text{// html/body/div/ table/*/td/text()} \,. \quad (7)$$

We define a one-step specialization of x to be an XPath expression obtained by any of the following four operations on x:

1. Adding a predicate to some node in x. E.g.

$$x_1 \equiv \text{// html/body/div/ table[@}width = \text{'98\%']/*/td/text()} . \quad (8)$$

2. Adding child position information to some node in x. E.g.

$$x_2 \equiv \text{// html/body/div[2]/ table/*/td/text()} . \quad (9)$$

3. Adding a //* at the top of x. E.g.

$$x_3 \equiv \text{//*/ html/body/div/ table/*/td/text()} . \quad (10)$$

4. Converting a * to an HTML label name. E.g.

$$x_4 \equiv \text{// html/body/div/ table/tr/td/text()} . \quad (11)$$

We represent $x_0 \rightarrow x_1$ as a one-step specialization of x_0, and we identify $x_0 \xrightarrow{*} x_1$ as multi-step specializations, namely, x_1 can be obtained from x_0 using a sequence of specializations. The basic idea behind the proposed algorithm is maintaining a set P of partial wrappers. Each element of set P is an XPath expression which has a recall equal to 1 and a precision less than 1. Initially, set P contains the single XPath expression "//*" that matches every node in each document. The algorithm applies specialization steps to XPaths in P to obtain new ones repeatedly until the precision of XPaths reaches 1. Then, XPaths are removed from P and added to the set of output wrappers. By this method, we can enumerate all XPaths expressions. However, it is really time-consuming to evaluate the robustness of wrappers by enumerating all XPaths, besides, the efficiency is quite low. Thus, we consider enumerating all "minimal" candidate wrappers to improve our method.

For a set of given documents D and an XPath expression x, we say x is minimal if there is no other XPath x_0 fits: $x_0 \xrightarrow{*} x$, $Precision(x_0) = Precision(x)$ and $Recall(x_0) = Recall(x)$.

Suppose x is a wrapper expressed by XPath expression, namely its precision and recall equal to 1, but x does not meet the conditions of minimal wrapper. Then we need to discover a smaller XPath expression x_0 which is also a wrapper. Moreover, smaller XPaths expressions are less likely to break when extracting the information in the webpages.

Obviously, suppose X be any XPath expression and x a wrapper, such that $X \xrightarrow{*} x$. If x is minimal, X is also minimal. Thus, we can obtain all minimal wrappers by enumerating wrappers and discarding non-minimal XPath expressions in the set P after each specialization.

The algorithm includes three cases. Firstly, if x_0 in set P is not the minimal wrapper, we will remove it from P. Then, if the wrapper x_0 meets the conditions of minimal wrapper, the recall equals to 1 but the precision is less than 1, we will put x_0 in set P and continue looping until the precision reaches to 1. Finally, if the wrapper is the minimal one and the precision reaches to 1, then we remove the wrapper x_0 from P and add it to the set of resultset. In Table 1, the algorithm for enumerating minimal wrappers is described in detail. It can be seen that Algorithm 1 is very reasonable and complete, thus, it generates all minimal candidate wrappers and only minimal ones.

Table 1. Generating minimal candidate wrappers with enumeration method

Algorithm 1. Generating minimal candidate wrappers

Input: A set of labeled webpages, i.e. webpages include several distinguished nodes.
Output: *ResultSet*: a set of wrappers expressed by XPath expressions.

```
ResultSet = ∅      // Initially, ResultSet does not contain element.
P = {"//*"}              // P is the set of partial wrappers.
while P ≠ ∅ do
Suppose x be any XPath expression in P of partial wrappers, namely, each x∈P.
        P = P − x
for all x0 s.t. x → x0 do        // x to x0 is one-step specialization operation.
   if isMinimal(x0) = false then
        P = P − x0
   end if
   if isMinimal(x0) and Precision(x0) < 1 and Recall(x0) = 1 then
        P = P ∪ x0
   end if
   if isMinimal(x0) and Precision(x0) = Recall(x0) = 1 then
        ResultSet = ResultSet ∪ x0
   end if
  end for
 end while
 Return ResultSet
```

4 Minimum Cost Script Edit Model for Evaluating the Robustness of Wrappers

We have described edit costs for three edit operations (i.e. inserting nodes, deleting nodes and substituting labels of nodes). We now depict cost function for computing the cost corresponding to each operation in detail. Let L denote the set of all labels, l_i□L. We assume that there is a cost function *cost*(x) for computing the cost for each edit operation, such as *cost* ($\varnothing$, l_1) represents the cost of inserting a node with label l_1, *cost* (l_1, $\varnothing$) represents the cost of deleting a node with label l_1, and *cost* (l_1, l_2) represents the cost of substituting a node with label l_1 to another node with label l_2. Note that *cost* (l_1, l_1) = 0. In addition, we assume that cost functions are satisfied for the triangle inequality, i.e., $cost(l_1, l_2)+cost(l_2, l_3) \geqslant cost(l_1, l_3)$. S is an given edit script and the cost denoted as *cost*(S), which is simply the sum of costs of each of the edit operations in S as given by the cost function *cost*(x).

It is easy to compute the minimum cost scripts according to the cost model is trained in Section 2, since each cost of the operations is minimum which are trained in model.

Table 2. Minimum cost script edit algorithm for robust web extraction

Algorithm 2 Minimum Cost Script Edit Algorithm for Robust Web Extraction

Input: *W*: A webpage;
W': the future version of webpage *W*;
d(*W*): a distinguished node of interest.

Output: *New_d*(*W*)': new location of *d*(*W*) in *W'*;
Extr_Conf: confidence of extraction.

1. Compute the change frequencies of three edit operations for each HTML label according to the frequency of webpage change on real web data with machine learning techniques.
2. Compute the corresponding costs for three edit operations according to the change frequencies of edit operations and the minimum cost model.
3. cost(insert: x) := Cost(insert a node);
4. cost(delete: x) := Cost(delete a node);
5. cost(substitute: x→y) := Cost(substitute a node to another node);
6. Take webpage *W* to *W'* by a set of edit scripts, using the dynamic programming to compute the costs of all edit scripts and save the results with array.
7. Choose the minimum cost script S_1 such that *W'* are obtained by applying S_1 in *W*, namely $S_1(W)$ and *W'* are isomorphic, i.e.
   $S_1(W) \equiv W'$;
   $New\_d(W)' = S_1(d(W))$;
   $Extr\_C1 = \text{cost}(S_1)$;
8. Choose the minimum cost script S_2 such that *W'* are obtained by applying S_2 in *W*, namely $S_2(W)$ and *W'* are isomorphic but does not map *d*(*W*) to the node corresponding to $S_1(d(W))$, i.e.
   $S_2(W) \equiv W'$ and $New\_d(W)' \neq S_1(d(W))$;
   $Extr\_C2 = \text{cost}(S_2)$;
9. $Extr\_Conf = Extr\_C2 - Extr\_C1$;
10. **Return** *New_d*(*W*)', *Extr_Conf*

Confidence is a measure of how much we trust our extraction of the distinguished node for the given new version. Intuitively, if the page *w'* differs a lot from *w*, then our confidence in the extraction should be low. However, if all the changes in *w'* are in a distinct portion away from the distinguished node, then the confidence should be high despite those changes. Based on this intuition, we define the confidence of extraction on a given new version as follows [3].

Let S_1 be the smallest cost edit script that takes *w* to *w'*, namely $S_1(w)$ and *w'* are isomorphic trees, the node extracted by wrapper is denoted $S_1(d(w))$, and the corresponding cost is $cost(S_1)$. We also look at the smallest cost edit script S_2 that takes *w* to *w'* but does not map *d*(*w*) to the node corresponding to $S_1(d(w))$, and the corresponding cost is $cost(S_2)$. We define the confidence of extraction as $cost(S_2) - cost(S_1)$. Intuitively, if this difference between $cost(S_2)$ and $cost(S_1)$ is large, the extracted node is well separated from the rest of nodes, and the extraction is likely to be correct. If this difference is low, the extracted node is likely to be wrong.

Aditya Parameswaran et al. [3] proposed a method by enumeration method for computing the minimum cost scripts and extracting the distinguished node of interest. The method is simple but inefficient, for the efficiency of enumerative algorithm is very low. Next, we design a more efficient algorithm using dynamic programming to compute the costs of all edit scripts efficiently. The basic idea behind the more efficient algorithm is to pre-compute the costs of all edit scripts, and finally choose the most suited data to extract the interested information. In Table 2, the process is illustrated.

In Algorithm 2, output parameter *New_d(W)'* represents the new location of distinguished node of interest, expressed by XPath expression. Output parameter *Extr_Conf* represents the confidence of extraction. If *Extr_Conf* is large, the extraction is likely to be correct. The confidence of extraction can be used to decide whether to use the extracted results or not.

5 Experimental Evaluation

In our experiments, we evaluate the effect of our robust web extraction framework on a dataset of crawled pages on two real-world recruitment sites.

5.1 Data Sets

To test the robustness of our techniques, we use archival data from two sites: 51job and zhaopin sites. Each data set consists of a set of webpages from the above websites monitored over last 10 months respectively. We choose about 100 webpages acting as archival versions from each website, and we crawl every version once a week.

In each of our data sets, we manually select distinguished nodes which can be identified. We choose distinguished nodes including "Working place", "Position", "Release date" and "Education background" and so on.

5.2 Evaluation Criterion

In this paper, evaluation criteria from information retrieval [12] are adopted to evaluate the effect of this method.

We use the expressions (5) and (6) to evaluate the Precision and Recall, respectively. F1 measure is:

$$\mathrm{F1} = \frac{2 \times \mathrm{Precision} \times \mathrm{Recall}}{\mathrm{Precision} + \mathrm{Recall}} \tag{12}$$

Precision reflects the believe level of the results, and Recall reflects the cover of getting correct results, with F1 synthesizing precision and recall.

5.3 Experimental Results and Analysis

(1) The Change Frequencies and Costs of Edit Operations

In this section, firstly, we obtain the change frequencies of edit operations by using machine learning techniques, then edit costs are calculated.

Table 3 shows the top 10 change frequencies that we implement our approach on a dataset of crawled pages on two real-world recruitment sites. Intuitively, the edit cost for an operation should capture the difficulty for a website maintainer to make edit operation. For instance, inserting a new table should be more expensive than inserting a new row or column, so the frequency of inserting a new row is higher than inserting a new table. Therefore, if the operations of webpages are frequent, we need to make the corresponding cost as low as possible. We set the cost of edit operations as 1, same with the frequencies of edit operations. Table 4 shows the edit costs according to the frequency of webpage change combining with our minimum cost model.

Table 3. The change frequencies of edit operations

Insertion frequency		Deletion frequency		Substitution frequency	
Label name	Frequency	Label name	Frequency	Label name	Frequency
a	0.0286	a	0.1070	a	0.0518
span	0.0162	li	0.0779	td	0.0311
div	0.0154	td	0.0375	li	0.0199
li	0.0112	div	0.0269	div	0.0102
td	0.0081	img	0.0232	span	0.0077
b	0.0064	span	0.0143	img	0.0064
img	0.0051	tr	0.0129	b	0.0045
tr	0.0042	b	0.0125	tr	0.0040
br	0.0036	br	0.0110	br	0.0034
ul	0.0011	ul	0.0099	ul	0.0026

Table 4. The edit costs of edit operations

Edit operations	Edit costs
Insertion	0.5414
Deletion	0.1624
Substitution	0.2962

(2) Evaluating the Robustness of Wrappers

In our experiments, we implement two other wrappers for comparison. One uses the full XPath containing the complete sequence of nodes' labels from the root node to the distinguished node in the initial version of the webpage. This wrapper is often used in practice. The other one uses the probabilistic robust XPath wrappers from [2]. We call these wrappers FullXPath and ProRobustXPath respectively, and our wrapper MinCostScript.

The input of a wrapper consists of the old and new versions of a page as well as the location of the distinguished node in the page. As to each execution, we check whether the wrapper finds the distinguished node in the new version. We study how well our wrapper performs as a function of elapsed time between the old and new versions of the webpage. Also, we use *skip sizes* as a measure for the elapsed time. We say a pair of versions has a skip of *K* if the difference between the version numbers of the two versions is *K*. We evaluate the accuracy of wrappers by varying the skip size. We plot the results of three methods on 51job site in Fig. 3(a). As we can see, our wrapper performs much better. Meanwhile, our wrapper is also suitable for other websites, such as Zhaopin site in Fig. 3(b). For each skip size *K*, we run all the rappers and plot the results in Fig.4. We can find that the three schemes perform worse when skip size is increased, while our approach performs much better.

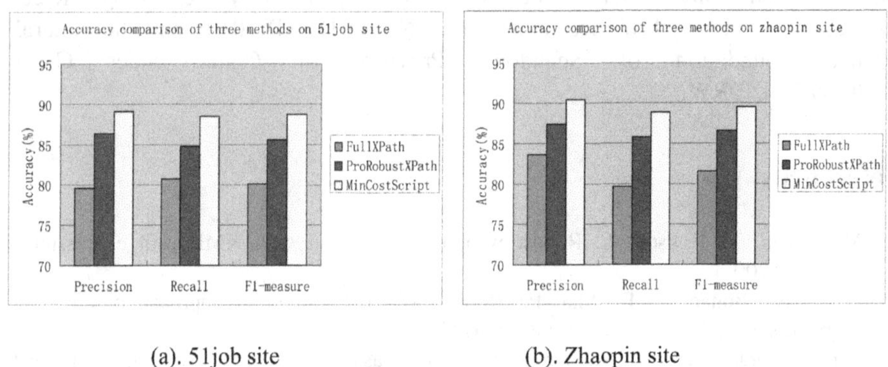

(a). 51job site (b). Zhaopin site

Fig. 3. Performance comparisons of three methods

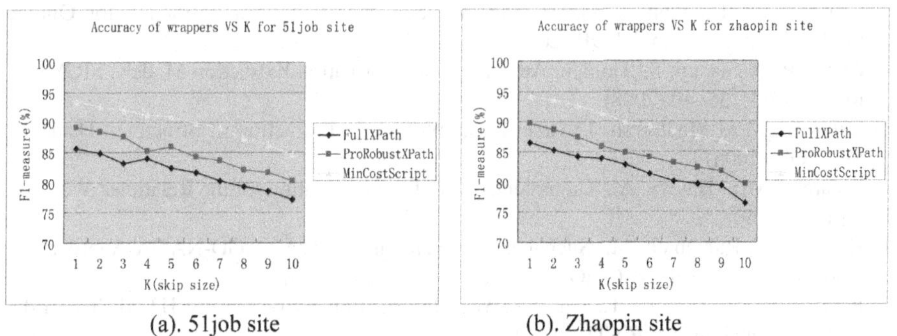

(a). 51job site (b). Zhaopin site

Fig. 4. Accuracy comparisons of three methods VS K(Skip Size)

6 Conclusion

In this paper, we propose a minimum cost script edit model based on machine learning techniques to construct robust wrapper for extracting interested information from deep webpages. The method considers three edit operations under HTML tree structural changes, namely, inserting nodes, deleting nodes and substituting labels of nodes. We obtain the change frequencies of three edit operations for each HTML label-name by applying our approach on a dataset of crawled pages on two real-world recruitment sites, i.e. 51job and zhaopin sites. The model takes archival data on the real-world recruitment sites as input and learns a model that best fits the data, such that the parameter values minimize the cost of each edit operation. Finally, the model chooses the most suited data to extract the interested information from deep webpages. By evaluating on real websites, it demonstrates that our wrapper is highly effective in coping with changes in websites.

Acknowledgments. This work is supported by Independent Innovation Foundation of Shandong University under Grant No.2010TS057, the Key Technology R&D Program of Shandong Province under Grant No. 2010GGX10108 and the Natural Science Foundation of Shandong Province of China under Grant No.2009ZRB019RW.

References

1. Myllymaki, J., Jackson, J.: Robust web data extraction with XML path expressions. CiteSeer (2002)
2. Dalvi, N., Bohannon, P., Sha, F.: Robust web extraction: an approach based on a probabilistic tree-edit model. In: SIGMOD (2009)
3. Parameswaran, A., Dalvi, N., Garcia-Molina, H., Rastogi, R.: Optimal Schemes for Robust Web Extraction. In: VLDB (2011)
4. Dalvi, N., Kumar, R., Soliman, M.: Automatic Wrappers for Large Scale Web Extraction. In: VLDB (2011)
5. Baumgartner, R., Gottlob, G., Herzog, M.: Scalable Web Data Extraction for Online Market Intelligence. In: VLDB (2009)
6. Gupta, R., Sarawagi, S.: Domain Adaptation of Information Extraction Models. SIGMOD Record 37(4), 35–40 (2008)
7. Cafarella, M.J., Madhavan, J., Halevy, A.: Web-Scale Extraction of Structured Data. In: SIGMOD (2008)
8. Cafarella, M.J., Halevy, A., Khoussainova, N.: Data Integration for the Relational Web. In: VLDB (2009)
9. Kasneci, G., Ramanath, M., Suchanek, F., Weikum, G.: The YAGO-NAGA Approach to Knowledge Discovery. SIGMOD Record 37(4), 41–47 (2008)
10. Kim, Y., Park, J., Kim, T., Choi, J.: Web Information Extraction by HTML Tree Edit Distance Matching. In: ICCIT (2007)
11. Anton, T.: Xpath-wrapper induction by generating tree traversal patterns. In: LWA, pp. 126–133 (2005)
12. van Rijsbergen, C.: Information Retrieval. Butterworths (1979)

13. Chidlovskii, B., Roustant, B., Brette, M.: Documentum ECI self-repairing wrappers: performance analysis. In: SIGMOD, pp. 708–717 (2006)
14. de Castro Reis, D., Golgher, P.B., da Silve, A.S.: Automatic web news extraction using tree edit distance. In: WWW, pp. 502–511 (2004)
15. Wang, W., Xiao, C., Lin, X., Zhang, C.: Efficient approximate entity extraction with edit distance constraints. In: SIGMOD, pp. 759–770 (2009)
16. Liu, D., Wang, X., Li, H., Yan, Z.: Robust Web Extraction Based on Minimum Cost Script Edit Model. Procedia Engineering 29, 1119–1125 (2012)
17. Hao, Q., Cai, R., Pang, Y., Zhang, L.: From one tree to a forest: a unified solution for structured Web data extraction. In: Proceeding of the 34th International ACM SIGIR Conference on Research and Development in Information Retrieval SIGIR, pp. 775–784 (2011)

Rule-Based Text Mining of Chinese Herbal Medicines with Patterns in Traditional Chinese Medicine for Chronic Obstructive Pulmonary Disease

Junping Zhan[1,2], Guang Zheng[2,3], Miao Jiang[3], Cheng Lu[3], Hongtao Guo[1,3], and Aiping Lu[1,⋆]

[1] Shanghai University of Traditional Chinese Medicine, Shanghai, 201203, China
[2] School of Information Science-Engineering, Lanzhou University, Lanzhou, 730000, China
[3] Institute of Basic Research in Clinical Medicine, China Academy of Chinese Medical Sciences, Dongzhimen, Beijing, 100700, China
lap64067611@126.com

Abstract. Through several thousands years of clinical research and theoretical thoughts, traditional Chinese medicine (TCM) has accumulated rich experiences on Chronic Obstructive Pulmonary Disease (COPD). However, the usage of Chinese herbal medicines in formulae are flexible in TCM clinical practice according to pattern differentiation, so, it is important to get the composition rules of Chinese herbal medicines through literatures. Based on the keyword list of Chinese herbal medicine, through the keyword filtering skill, we can get the lists of Chinese herbal medicines. However, for Chinese herbal medicines, they are not only mentioned in the plain format of herb names, but also densely described in the form of formulae (also called Chu-Fang, Yao-Fang or Tang-Yao in Chinese). As formulae are composed by Chinese herbal medicines according the theory of traditional Chinese medicine, so, it is necessary to filtering them out and de-compose them back into specified Chinese herbal medicines. In this study, take COPD for example, we explored the composition-rules of Chinese herbal medicines and the network of TCM pattern with them. By doing this, we might get a deeper understanding of composition-rules of Chinese herbal medicines in clinical practice which might be a better reference for both clinical practise and research.

Keywords: text mining, Chinese herbal medicine, composition rule, chronic obstructive pulmonary disease, traditional Chinese medicine.

1 Introduction

Through several thousand years of clinical practices and theoretical thoughts, traditional Chinese medicine has developed into several genres [1–3]. For even

⋆ Corresponding author.

F.L. Wang et al. (Eds.): WISM 2012, LNCS 7529, pp. 510–520, 2012.

the same disease, doctors of different genres might prescribed different formulae or different composition of herbal medicines, and what's more, patients with the same diseases can also be cured. This might be confusing [1], yet by text mining [4], we might get the objective composition rules of Chinese herbal medicines. In this study, focused on the disease of Chronic obstructive pulmonary disease (COPD), we got the rules of Chinese herbal medicines for treating this disease.

COPD, also known as chronic obstructive lung disease (COLD) [5–8], chronic obstructive airway disease (COAD), chronic airflow limitation (CAL) and chronic obstructive respiratory disease (CORD), is the co-occurrence of chronic bronchitis and emphysema, a pair of commonly co-existing diseases of the lungs in which the airways become narrowed (http://www.nhlbi.nih.gov/health/health-topics/topics/copd/). This leads to a limitation of the flow of air to and from the lungs, causing shortness of breath (dyspnea). In clinical practice, COPD is defined by its characteristically low airflow on lung function tests [9–14].

In traditional Chinese medicine, COPD belongs to asthma syndrome (Chuang-Zheng in Pinyin, or "喘证" in Chinese) [1]. In clinical, asthma syndrome can be demonstrated by symptoms of cough, sputum, chest stuffiness, palpitation, limb edema and so on. The therapeutic methods are facilitating the flow of blood ("活血" in Chinese)benefiting qi ("益气" in Chinese)reducing phlegm ("化痰" in Chinese)warming yang("温阳" in Chinese).

Based on TCM theory, on COPD, researchers have done some research work on it, i.e., syndrome [20], syndrome differentiation [21], and therapy effect[22, 23]. However, there is no work done on the analysis of Chinese herbal medicines used for COPD. In this study, based on the therapeutic methods, Chinese herbal medicines are classified into different functions according to their 'nature' characters [2]. What's more, in clinical practices, the usages of Chinese herbal medicines are flexible according to the healthy conditions of different patients [3][2]. So, it is important and necessary to mine out the composition-rules of Chinese herbal medicines for COPD.

By applying the algorithm set based on data slicing with the principle of co-occurrence, we mined the dataset of literature on cirrhosis. The text mining process is executed for Chinese herbal medicine, herb formula, and pattern in traditional Chinese medicine associated with them. Then, based on the rules of herb formulae, we decomposed formulae into Chinese herbal medicines and then built the networks of Chinese herbal medicines.

2 Material and Methods

2.1 Data Collection

The dataset were downloaded from SinoMed (http://sinomed.cintcm.ac.cn/index.jsp) with the query terms of "慢性阻塞性肺疾病" and "慢阻肺" (both are COPD in Chinese) on January 22, 2012. This dataset conatins 18,199 records of literatures on COPD, and in this dataset, each record/paper is tagged with an unique ID. These records contain the title, keywords, and abstract of published papers.

2.2 Data Filtering

Chinese Herbal Medicine in the Plain Text Format. Based on the keyword list of Chinese herbal medicines (both legal names and other popular names are included for calculation) [2], we filtered the Chinese herbal medicines in the plain text format, and then converted all popular names into legal names. All the Chinese herbal medicines were tagged with their unique paper ID. Based on the unique paper ID, we could construct the pairs of co-existed Chinese herbal medicines as they do co-existed in literature [4, 16–19]. For example, in one paper, Chinese herbal medicines of "Mahuang" ("麻黄" in Chinese), "Guizhi" ("桂枝" in Chinese), and "Xixin" ("细辛" in Chinese) are mentioned. Then, the pairs of co-existed Chinese herbal medicines of "Mahuang-Guizhi", "Mahuang-Xixin", and "Guizhi-Xixin" are constructed.

Chinese Herbal Medicine in Herb Formulae. As Chinese herbal medicines are densely described in herb formulae (also called FuFang or TangYao in Chinese) [1–3, 15], and they are composed in order of sovereign, minister, assistant and courier (Jun-Chen-Zuo-Shi in Pinyin, and "君臣佐使" in Chinese) [15] within a formula according to their functions which is based on the theory of traditional Chinese medicine [3]. So, in order to get a deeper understanding of the composition-rules of Chinese herbal medicines, it is necessary to decompose Chinese herbal medicine out from herb formulae. Based on this, we filtered the names of herb formulae. Then, according to their components of Chinese herbal medicines, we built another kinds of co-existed pairs of Chinese herbal medicines. That is to say, all the Chinese herbal medicines that existed within a formula are considered as co-existed. As for the calculation process, the formulae were also mined out with their unique paper ID.

Patterns and Symptom Sets of Traditional Chinese Medicine. Pattern differentiation is of key role in the clinical practise of traditional Chinese medicine [1, 3]. What's more, in traditional Chinese medicine, pattern can be identified by a set of symptoms. In clinical practices, patterns are usually as described with symptoms, herb formulae, Chinese herbal medicines, and other information associated with diseases. Based on the theory of traditional Chinese medicine, patterns should be associated with the usage of Chinese herbal medicines and the symptom sets of patients. Then, it is natural and intuitive to filter out the pattern, together with the association rules among Chinese herbal medicines, and symptom sets with them.

After calculation, the most outstanding TCM pattern are: *phlegm-damp obstructing lung* ("痰湿阻肺" in Chinese), and *qi-deficiency of lung* ("肺气虚"). For symptoms, the most frequently mentioned are: shortness of breath, cough, expectoration, chest stuffiness,lack of strength, and so on.

2.3 Calculating the Merged Order of Chinese Herbal Medicines

As we have the table of Chinese herbal medicines filtered from plain text format, together with those induced from herb formulae, we got the total order of them. This total results are listed in Table 1.

Table 1. Top 15 Chinese herbal medicines mentioned in plain text format

Pinyin Name	Chinese Name	Latin Name	Order of Merged	Order of Plain	Order of Formula
Renshen	人参	Radix Et Rhizoma Ginseng	1	1	3
Huangqi	黄芪	Radix Astragali	2	3	8
Gancao	甘草	Radix Et Rhizoma Glycyrrhizae	3	11	4
Banxia	半夏	Rhizoma Pinelliae	4	15	2
Maidong	麦冬	Radix Ophiopogonis	5	2	-
Baizhu	白术	Rhizoma Atractylodis Macrocephalae	6	-	1
Danshen	丹参	Radix Et Rhizoma Salviae Miltiorrhizae	7	4	-
Chenpi	陈皮	Pericarpium Citri Reticulatae	8	14	5
Fuling	茯苓	Poria cocos (Schw.) Wolf.	9	-	6
Guizhi	桂枝	Ramulus Cinnamomi	10	-	7
Mahuang	麻黄	Herba Ephedrae	11	7	-
Xixin	细辛	Radix Et Rhizoma Asari	12	8	-
Taoren	桃仁	Semen Persicae	13	-	14
Chuanxiong	川芎	Rhizoma Chuanxiong	14	5	-
Danggui	当归	Radix Angelicae Sinensis	15	13	-
Dongchongxiacao	冬虫夏草	Cordyceps	-	6	-
Xiebai	薤白	Bulbus Allii Macrostemonis	-	9	-
Shengdihuang	生地黄	Radix Rehmanniae	-	10	-
Gejie	蛤蚧	Gecko	-	12	-
Fuzi	附子	Radix Aconiti Lateralis Praeparata	-	-	9
Dongguazi	冬瓜子	Semen Benincasae	-	-	10
Dazao	大枣	Fructus Jujubae	-	-	11
Lugen	芦根	Rhizoma Phragmitis	-	-	12
Shengjiang	生姜	Rhizoma Zingiberis Recens	-	-	13
Wuweizi	五味子	Fructus Schisandrae Chinensis	-	-	15

From Table 1, we can see that the column of "Order of Merged", top 15 Chinese herbal medicines are selected from both "Order of Plain" and "Order of Formulae". Among the top 15 merged Chinese herbal medicines, 11 are from plain text format and 9 are from those induced from herb formulae. That is to say, there are 5 Chinese herbal medicines existing in both plain format and formulae.

By checking the nature of these Chinese herbal medicines according to their relationships with the major patterns of *phlegm-damp obstructing lung* and *qi-deficiency of lung*. They can be classified into three groups: group one is for pattern *phlegm-damp obstructing lung* which will be demonstrated in green, group two is for pattern *qi-deficiency of lung* which will be demonstrated in orange, and group three is belonging to other patterns which will be demonstrated in gray.

2.4 Network Construction

Rule Based Inducing Chinese Herbal Medicines Out from Herb Formulae. With respect to the theory of traditional Chinese medicine, each herb formula has its function structure for each consisting herbs. As the name of herb formulae can be filtered out from literatures tagged with paper IDs. Then, according to the unique paper ID, we built the co-existed pairs of Chinese herbal medicines. For example, herb formula of Xiao-Qing-Long decoction is one of the top two formulae which are associated with COPD. It is composed by seven Chinese herbal medicines, in PinYin, they are Mahuang ("麻黄" in Chinese), Shaoyao ("芍药" in Chinese), Xixin ("细辛" in Chinese), Ganjiang ("干姜" in Chinese), Zhigancao ("炙甘草" in Chinese), Guizhi ("桂枝" in Chinese), Wuweizi ("五味子" in Chinese), Banxia ("半夏" in Chinese). In this formula, co-existed pairs of Chinese herbal medicines can be constructed by rules which is demonstrated in Table 2 marked by checkmark.

From Table 2, we can see that only the different composition of Chinese herbal medicines within one herb formula (for example, Xiao-Qing-Long decoction) can be constructed into co-existed pairs. These co-existed pairs of Chinese herbal medicines are demonstrated by checkmark.

Chinese Herbal Medicine in the Plain Text Format. One natural and intuitive goal for text mining is to count the frequencies of Chinese herbal medicines existed in literatures. As Chinese herbal medicines were filtered out from literatures tagged with paper ID, and what's more, these herbal medicines

Table 2. Rules of co-existed pairs of Chinese herbal medicines constructed from herb formulae

Co-existed pairs	Mahuang (麻黄)	Shaoyao (芍药)	Xixin (细辛)	Ganjiang (干姜)	Zhigancao (炙甘草)	Guizhi (桂枝)	Wuweizi (五味子)	Banxia (半夏)
Mahuang	–	–	–	–	–	–	–	–
Shaoyao	√	–	–	–	–	–	–	–
Xixin	√	√	–	–	–	–	–	–
Ganjiang	√	√	√	–	–	–	–	–
Zhigancao	√	√	√	√	–	–	–	–
Guizhi	√	√	√	√	√	–	–	–
Wuweizi	√	√	√	√	√	√	–	–
Banxia	√	√	√	√	√	√	√	–

might be mentioned more than once in one particular paper. So, the filtering process which is based on the keyword filtering might get redundant Chinese herbal medicine within one particular paper. These redundant keywords tagged with paper ID are useless for further analysis. Based on this, the redundant list of Chinese herbal medicines must be cleared. After eliminating the redundant information, we got the clean list of Chinese herbal medicine with paper ID.

Based on the table of co-existed pairs of Chinese herbal medicines, we calculated the frequencies of herb pairs which existed in the whole dataset. Then, sorted them on descendent order of frequency, and this sorted list is the original data for network of Chinese herbal medicine filtered from plain text format.

Chinese Herbal Medicine Induced from Herb Formulae. As we have filtered out the formulae list from literature, together with the composition-rules of these formulae (demonstrated in Table 2), we can build the list of Chinese herbal medicines tagged with both formulae and paper ID. By then, as the table of Chinese herbal medicines induced from formulae was ready, we calculated the frequencies of Chinese herbal medicines.

Based on the table of co-existed pairs of Chinese herbal medicines which are induced from formulae, we calculated the frequencies of herb pairs which are existed and induced from formulae in traditional Chinese medicine. When that was done, these herb pairs were also sorted on descendent order of frequency. Again, these sorted list of co-existed pairs forms the original data of network of Chinese herbal medicine induced from herb formulae.

Chinese Herbal Medicine as a Whole. Now, we get the original data of networks of Chinese herbal medicines from two formats. One is filtered from plain text format, and the other is induced from herb formulae. Because both of them represent the usages of Chinese herbal medicines for COPD in TCM clinical practices yet from different formats, so, it is natural and reasonable to merge them into one.

The merging process for frequencies of Chinese herbal medicine is based on the formula of

$$Herb_{Merged} = Herb_{Plain} + Herb_{Formulae}$$

where the $Herb_{Merged}$ represents the frequencies of merged Chinese herbal medicines, $Herb_{Plain}$ represents the frequencies of Chinese herbal medicines filtered from plain text format, and $Herb_{Formulae}$ represents the frequencies of Chinese herbal medicines induced from herb formulae by rules.

3 Results

3.1 Network of Chinese Herbal Medicine

The results of Chinese herbal medicine networks are demonstrated in Figure 1. In this figure, nodes in orange area representing the Chinese herbal medicines for *tonifying qi* for pattern *qi-deficiency of lung*, nodes in green area representing

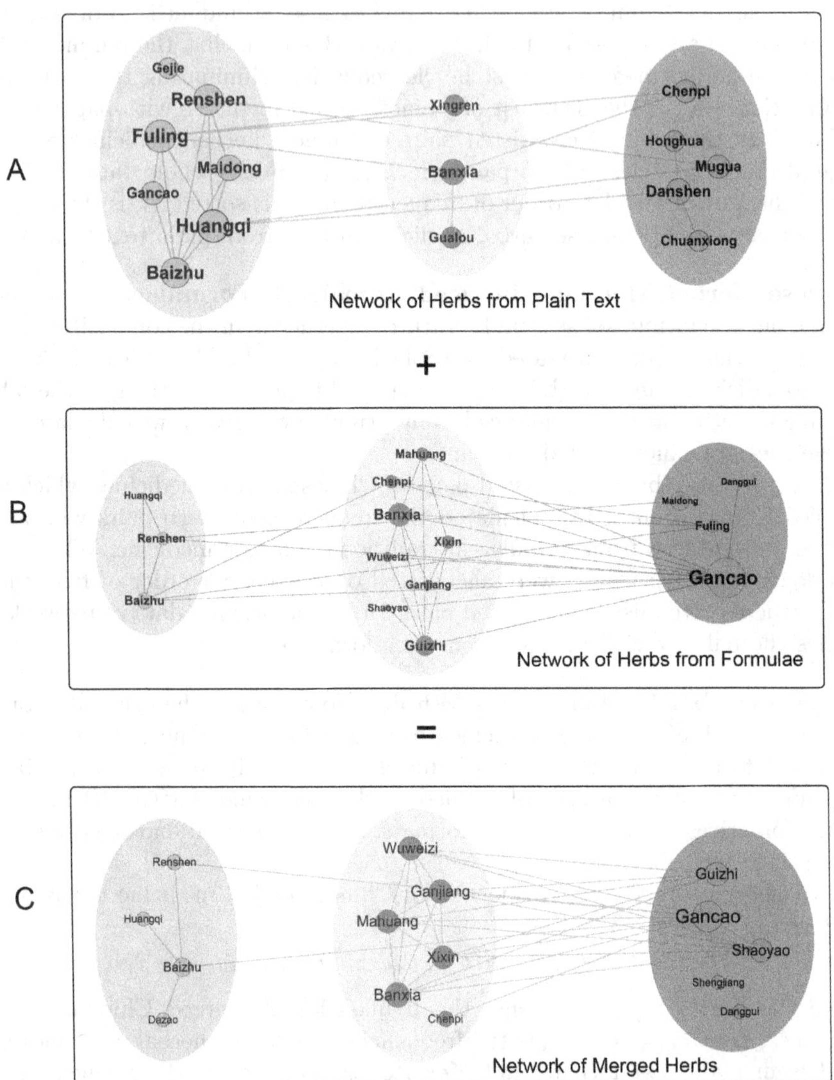

Fig. 1. Networks of Chinese herbal medicine on COPD from plain text, herb formulae, and the merged one. For part A, it demonstrates the network of Chinese herbal medicine mined from plain text format. For part B, it demonstrates the network of Chinese herbal medicine induced from herb formulae. For part C, it demonstrated the merged network of part A and B.

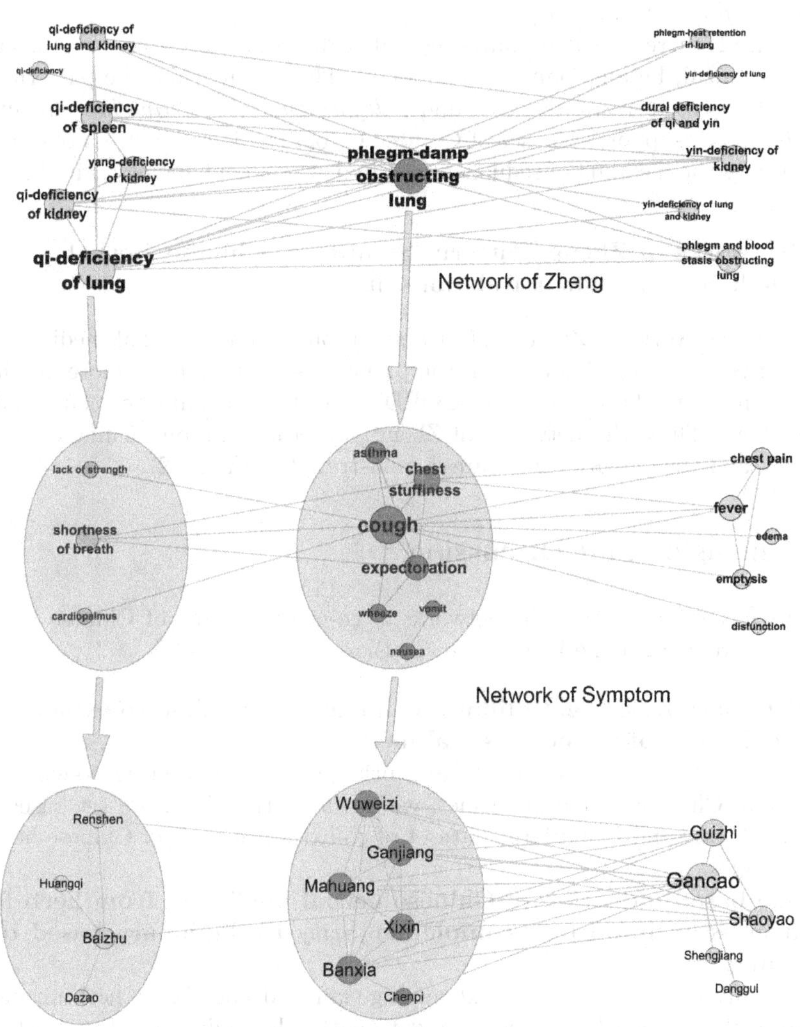

Fig. 2. Networks of Zheng, symptom and Chinese herbal medicine on COPD. The upper network is Zheng/syndrome of traditional Chinese medicine. The middle network is symptom which belongs to Zheng/syndrome in traditional Chinese medicine. The lower network is Chinese herbal medicine. The upper orange arrow points from *qi-deficiency of lung* to symptom set of it. The lower orange arrow points from the symptom set to the set of Chinese herbal medicines regulating them. As for the green arrows, the upper one points from *phlegm-damp obstructing lung* to the symptom set of it, and the lower one points from the symptom set to the set of Chinese herbal medicines regulating them.

the Chinese herbal medicines for *drying dampness to resolve phlegm* for pattern *phlegm-damp obstructing lung*.

In Figure 1, there are some Chinese herbal medicines in gray areas i.e., Gancao, Shaoyao, Guizhi, Fuling, Danggui, and so on. They do not aim the patterns of *drying dampness to resolve phlegm* and *phlegm-damp obstructing lung* directly. However, they "support" the set of Chinese herbal medicines in areas of orange and green by the meaning of reducing their toxicity and enhancing their efficacy.

3.2 Networks of Zheng/Pattern, Symptom, Chinese Herbal Medicine, and Their Relationships

We got the networks of Zheng/pattern, symptom, Chinese herbal medicine by executing the methods described in the previous section. What's more, as they are all came from the literature of COPD, then, there should be relationships among them. Thus, the networks of Zheng/pattern, symptom, Chinese herbal medicine, and their relationships are demonstrated in Figure 2.

4 Conclusion and Discussion

In this study, by analyzing the networks of co-existent pairs of Chinese herbal medicines, we can get the following conclusions:

1. **Dictionary based text mining is an effective method to explore the co-existence of Chinese herbal medicines**
   In this study, based on the mining principle of constructing co-existence pairs of Chinese herbal medicines, we executed the algorithm set based on data slicing, and found large dataset of co-existence pairs of Chinese herbal medicines.
2. **Rule based induction of Chinese herbal medicines from herb formulae is an important complementarity to dictionary based text mining**
   By applying the rule based text mining skill of decomposing herb formulae into Chinese herbal medicines, we got another large dataset of co-existence pairs of Chinese herbal medicines. These binary relationship of Chinese herbal medicines have been existing in the long history of clinical practise of traditional Chinese medicine. So, it is very importance to mine them out and demonstrated in networks.
3. **Merging results is a better option of finding co-existence of Chinese herbal medicines**
   Merging the results of dictionary based and rule based, we got more complete and deeper understanding of co-existence pairs of Chinese herbal medicines from literatures.
4. **"Supporting" set of Chinese herbal medicine for patterns in COPD are found**
   By checking the results of merged network of Chinese herbal medicines, we

not only found the set for patterns directly, but also found the "supporting" set of Chinese herbal medicines for those directly ones. According to the theory of Chinese herbal medicine, these "supporting" medicines have the functions of reducing their toxicity and enhancing their efficacy.

Acknowledgment. This work was partially supported by National Science Foundation of China (No. 30825047, 30902003, 30973975, 90709007, and 81072982). China Postdoctoral Science Foundation (No. 20110940553). CACMS Project (Z0134, Z0172).

References

1. Zhou, Z., et al.: Internal Medicine of Traditional Chinese Medicine. China Press of Traditional Chinese Medicine (2007) (in Chinese) ISBN 978-7-80156-313-2
2. Gao, X., et al.: Science of Chinese Pharmacology. China Press of Traditional Chinese Medicine (2008) (in Chinese) ISBN 978-7-80156-318-7
3. Deng, Z., et al.: Formulae of Chinese Medicine. China Press of Traditional Chinese Medicine (2008) (in Chinese) ISBN 7-80156-322-0
4. Zheng, G., Jiang, M., He, X., Zhao, J., Guo, H., Chen, G., Zha, Q., Lu, A.: Discrete derivative: a data slicing algorithm for exploration of sharing biological networks between rheumatoid arthritis and coronary heart disease. BioData Mining 4, 18 (2011), `http://www.biodatamining.org/content/4/1/18`
5. Blanc, P.D.: Occupation and COPD: a brief review. J. Asthma 49, 2–4 (2012)
6. Gaebel, K., et al.: Triple therapy for the management of COPD: a review. COPD 8, 206–243 (2011)
7. Grant, M., et al.: The impact of caring for those with chronic obstructive pulmonary disease (COPD) on carers' psychological well-being: A narrative review. Int. J. Nurs. Stud. (March 2, 2012)
8. Janaudis-Ferreira, T., et al.: How should we measure arm exercise capacity in patients with COPD? A systematic review. Chest 141, 111–120 (2012)
9. Lemmens, K.M., et al.: Chronic care management for patients with COPD: a critical review of available evidence. J. Eval. Clin. Pract. (December 2, 2011)
10. Lodewijckx, C., et al.: Impact of care pathways for in-hospital management of COPD exacerbation: a systematic review. Int. J. Nurs. Stud. 48, 1445–1456 (2011)
11. Reddel, H.K., et al.: Year-in-review 2009: Asthma, COPD and airway biology. Respirology 15, 365–376 (2010)
12. Rodrigo, G.J., Neffen, H.: Comparison of Indacaterol with Tiotropium or Twice-Daily Long-Acting Beta-Agonists for Stable COPD: A Systematic Review. Chest (March 1, 2012)
13. Rodrigo, G.J., et al.: Comparison of three combined pharmacological approaches with tiotropium monotherapy in stable moderate to severe COPD: a systematic review. Pulm. Pharmacol. Ther. 25, 40–47 (2012)
14. Steurer-Stey, C.: Management of COPD: a review. Praxis (Bern 1994) 93, 1673–1678 (2004)
15. WHO Health organization. WHO International Standard Terminologies on Traditional Medicine in the Western Pacific Region. WHO Library Cataloguing in Publication Data (2007) ISBN 978 92 9061 2484

16. Hotho, A., Nürnberger, A., Paaß, G.: A Brief Survey of Text Mining. LDV Forum - GLDV Journal for Computational Linguistics and Language Technology 20(1), 19–62 (2005)
17. Hand, D.J., Mannila, H., Smyth, P.: Principles of data mining. MIT Press (2001) ISBN 026208290X, 9780262082907
18. Tzanis, G., Berberidis, C., Vlahavas, I.P.: Biological Data Mining. Encyclopedia of Database Technologies and Applications, pp. 35–41 (2005)
19. Schmidt, S., Vuillermin, P., Jenner, B., Ren, Y., Li, G., Chen, Y.-P.P.: Mining Medical Data: Bridging the Knowledge Divide. In: Proceedings of eResearch Australasia (2008)
20. Yu, D.S., et al.: The revolving door syndrome: the Chinese COPD patients' perspectives. J. Clin. Nurs. 16, 1758–1760 (2007)
21. Zhang, W., et al.: Effects of TCM treatment according to syndrome differentiation on expressions of nuclear factor-kappaB and gamma-glutamylcysteine synthetase in rats with chronic obstructive pulmonary disease of various syndrome types. Zhongguo Zhong Xi Yi Jie He Za Zhi 27, 426–430 (2007)
22. Zheng, F., et al.: Effects of Chinese medicine internal-external combined therapy in treating chronic obstructive pulmonary disease in the stable period. Zhongguo Zhong Xi Yi Jie He Za Zhi 30, 369–372 (2010)
23. Zheng, J.P., et al.: The efficacy and safety of combination salmeterol (50 microg)/fluticasone propionate (500 microg) inhalation twice daily via accuhaler in Chinese patients with COPD. Chest 132, 1756–1763 (2007)

Fault Forecast of Electronic Equipment Based on ε −SVR

Lina Liu[1], Jihong Shen[1], and Hui Zhao[2]

[1] College of Automation, Harbin Engineering University, 150001, Harbin, Heilongjiang Province, China
[2] Department of Mathematics, Heilongjiang Institute of Technology, 150050, Harbin, Heilongjiang Province, China
{liulina,shenjihong,zhaohui}@hrbeu.edu.cn

Abstract. In order to ensure security and reliability of the equipment, so as to decrease the maintenance cost, combining with the characteristics of fault data, this paper adopts ε − support vector regression to establish a fault forecast model and evaluation system to prediction model effect which are proper to the electronic equipment. Selecting multi-electronic equipment and training on the ε − SVR with different kernel functions. It is demonstrated that the prediction effect is better and it is still of vital realistic significance for realizing condition-based maintenance of modern electronic equipment.

Keywords: Data Mining, ε -SVR, Kernel Function, Fault Forecast.

1 Introduction

Ship electronic equipment is applied to every field of naval industry. A lot of equipment needs repair, maintenance tasks is extremely difficult. Traditional maintenance mode can not keep with the requirements of development with high - speed.

So fault forecast should be done by data mining and other algorithms before the equipment out of work. Combining with available resource information, condition-based maintenance of modern electronic equipment can be realized, and also maintenance cost can be decreased.

In the related fields, for example, using adding-weight one-rank local-region method is applied in electric power system short-term load forecast[1]; using grey Markov model is applied in prediction of flight accident rate, annual precipitation and flood disasters[2-4]; using wavelet analysis and chaotic time series technology is applied in world oil price forecasting[5]. Chaos and data mining is applied in fault diagnosis for electronic system[6]. According to the characteristic of fault data and available maintenance information, this paper establishes a fault prediction model based on ε − SVR to forecast the fault of electronic equipment.

F.L. Wang et al. (Eds.): WISM 2012, LNCS 7529, pp. 521–527, 2012.

2 Support Vector Regression

Support vector machine (SVM) is proposed by Vapnik and some people, it is a new machine learning method based on the small sample statistics theory and has study ability. It solutes some problem, for example, dimension disaster, small sample, local minima and nonlinear problems, and so on. SVM is applied in the data classification and regression estimate at present[7].

The basic idea of support vector regression (SVR) is to map training data into a high dimensional feature space for the linear regression by a nonlinear transformation $\phi(x)$, it realize non-linear regression in the original space.

$\varepsilon-$SVR is an algorithm of SVR. The following is the algorithm:

(1) Let $T=\{(x_1,y_1),\cdots,(x_l,y_l)\}\in(X\times Y)^l$ be known training set, in which $x_i\in X=R^n$, $y_i\in Y=R$, $i=1,2,\cdots,l$;

(2) Select proper C and positive number ε and also proper kernel function $K(x,x')$;

(3) Structure and solve optimization problems

$$\min_{\alpha^{(*)}\in R^{2l}} \frac{1}{2}\sum_{i,j=1}^{l}(\alpha_i^*-\alpha_i)(\alpha_j^*-\alpha_j)K(x_i,x_j)+\varepsilon\sum_{i=1}^{l}(\alpha_i^*+\alpha_i)$$

$$-\sum_{i=1}^{l}y_i(\alpha_i^*-\alpha_i),$$

$$s.t \quad \sum_{i=1}^{l}(\alpha_i^*-\alpha_i)=0,$$

$$0\le\alpha_i,\alpha_i^*\le\frac{C}{l},i=1,2,\cdots,l$$

Optimal solution $\bar{\alpha}=(\bar{\alpha}_1,\bar{\alpha}_1^*,\cdots,\bar{\alpha}_l,\bar{\alpha}_l^*)^T$ can be obtained;

(4) Structure decision-making function

$$f(x)=\sum_{i=1}^{l}(\bar{\alpha}_i^*-\bar{\alpha}_i)K(x_i,x)+\bar{b}$$

In which $\bar{b}$ is calculated by the way as follows: select $\bar{\alpha}_j$ or $\bar{\alpha}_k^*$ which are in the open interval $\left(0,\frac{C}{l}\right)$. If $\bar{\alpha}_j$ is selected,

$$\bar{b} = y_j - \sum_{i=1}^{l}\left(\bar{\alpha}_i^* - \bar{\alpha}_i\right)\left(x_i \cdot x_j\right) + \varepsilon ;$$

If $\bar{\alpha}_k^*$ is selected,

$$\bar{b} = y_k - \sum_{i=1}^{l}\left(\bar{\alpha}_i^* - \bar{\alpha}_i\right)\left(x_i \cdot x_k\right) + \varepsilon .$$

The following are kernel functions:

(1) Linear kernel function

$$K(x,x') = (x \cdot x')$$

(2) Polynomial kernel function

$$K(x,x') = ((x \cdot x') + c)^m$$

in which, $c \geq 0$, m is arbitrary positive integer

(3) RBF kernel function

$$K(x,x') = \exp\left(-\|x - x'\|^2 / \sigma^2\right)$$

(4) Sigmoid kernel function

$$K\left(x,x'\right) = \tanh\left(\rho(x \cdot x') + b\right)$$

3 Empirical Analysis

This paper selects fault data from February 2009 to December 2010 of electronic equipment for the fault diagnosis and detection. The following is prediction model:

$$y_t = f\left(y_{t-1}, y_{t-2}, \cdots, y_{t-n}\right) + b$$

In which, n is the numbers of continuous month before month t, and y_t is fault conditions of electronic equipment in the month t.

$$y_t = \begin{cases} 0, & non-failure \\ 1, & failure \end{cases}$$

This paper adopts $\varepsilon -$ SVR with Linear kernel function, Polynomial kernel function, RBF kernel function and Sigmoid kernel function, compared with BP neural network.

RMS error is used as the standard of prediction model in which $n=3$. Select anterior 14 group data from 19 group data samples as training samples and the rest 5 group data as testing samples. Figure.1 is simulation diagram of $\varepsilon-$ SVR and Figure.2 is simulation diagram of BP neural network.

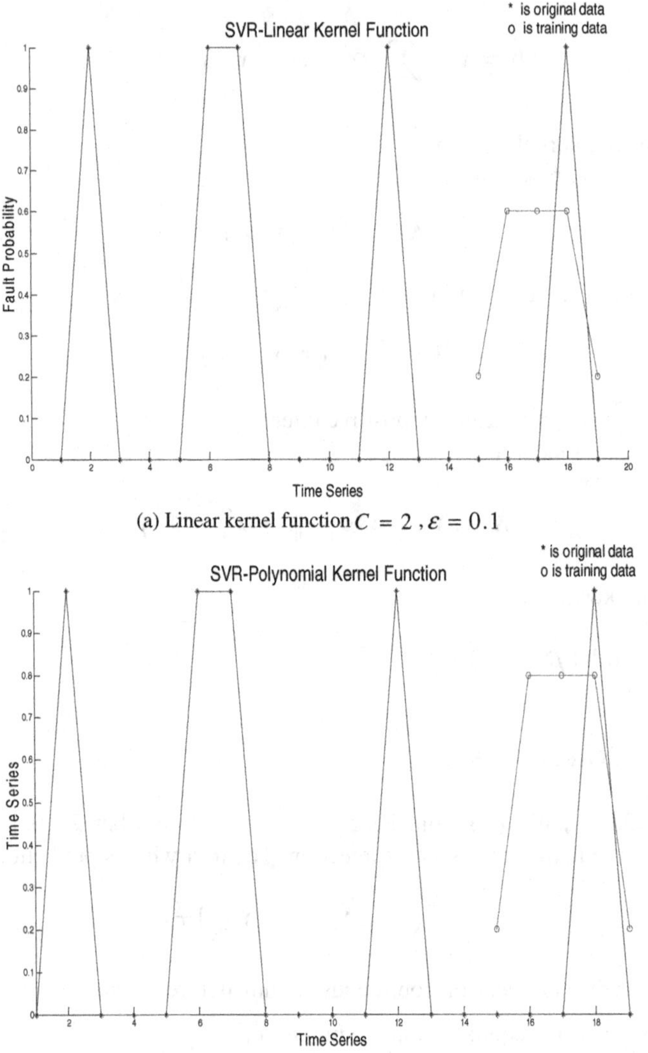

(a) Linear kernel function $C = 2$, $\varepsilon = 0.1$

(b) Polynomial kernel function $C = 2$, $\varepsilon = 0.2$, $m = 3$

Fig. 1.

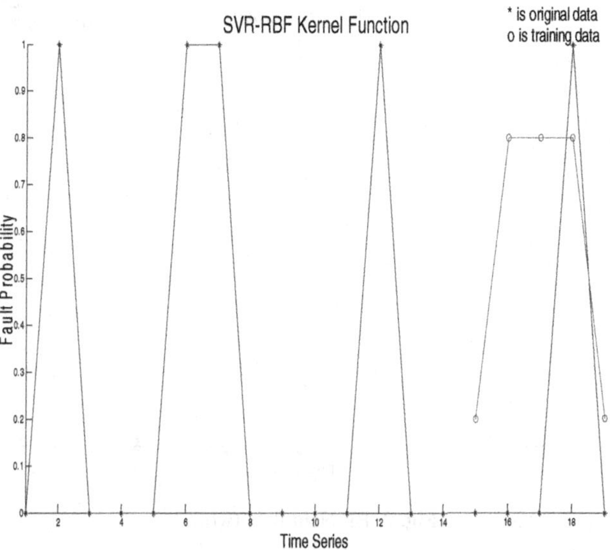

(c) RBF kernel function $\sigma = 2$, $C = 2$, $\varepsilon = 0.2$

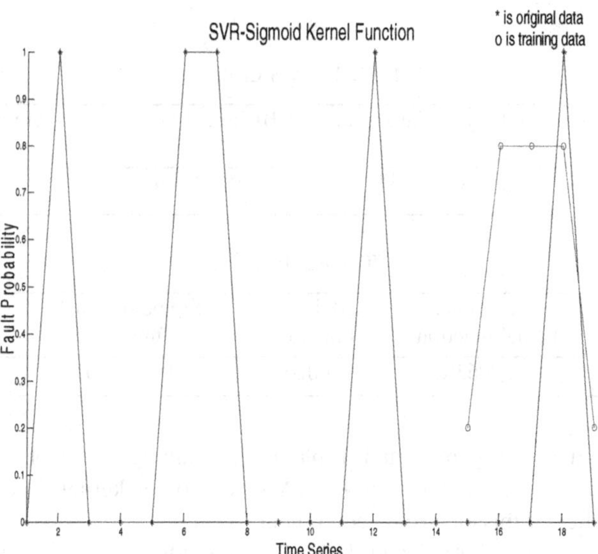

(d) Sigmoid kernel function $\sigma = 20$, $C = 2$, $\varepsilon = 0.2$, $b = 0$

Fig. 1. *(Continued)*

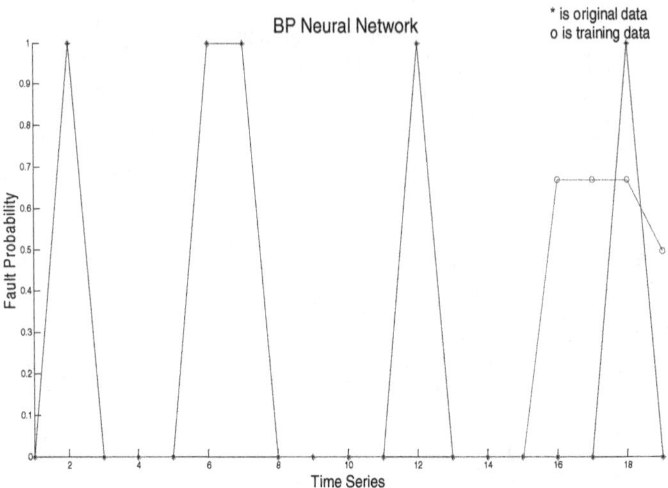

Fig. 2. BP Neural Network

Training effects of $\varepsilon-$ SVR are shown as Table 1, and prediction results are shown as Table 2.

Table 1. RMS Error

Linear kernel function	Polynomial kernel function	RBF kernel function	Sigmoid kernel function
0.166772	0.239734	0.239909	0.239999

Table 2. Error

Linear kernel function	Polynomial kernel function	RBF kernel function	Sigmoid kernel function	BP neural network
0.360410	0.200630	0.200200	0.200000	0.331880

From the Figure 1, Figure 2 and Table 1, the training effect of $\varepsilon-$ SVR with Linear kernel function is better than $\varepsilon-$ SVR with other kernel functions for data with characteristics of this electronic equipment.

From the figure 1, Figure 2 and Table 2, the prediction results of $\varepsilon-$ SVR is better than BP neural network. The comparison study of the prediction results of $\varepsilon-$ SVR with four kernel functions, $\varepsilon-$ SVR model with Sigmoid kernel function is the best.

4 Conclusion and Prospect

This article does explorative research on the fault data in time series based on $\varepsilon-$ SVR with different kernel functions. The experiment results of $\varepsilon-$ SVR for fault prediction demonstrates effectiveness, and the prediction effect is better than BP neural network. It is still of vital realistic significance for realizing condition-based maintenance of modern electronic equipment. For the future study, this approach is prospectively suggested in other kinds of fault prediction.

References

1. Lv, J.-H., Zhang, S.-C.: Application of Adding-weight One-rank Local—region Method in Electric Power System Short-term Load Forecast. Control Theory and Applications 19(5), 767–770 (2002)
2. Li, D.-W., Xu, H.-J., Liu, D.-L., Xue, Y.: Improved Grey Markov Model and Its Application in Prediction of Flight Accident Rate. China Safety Science Journal 19(9), 53–57 (2009)
3. Wei, D.-J., Xie, M.-Y.: Application of Grey Markov Model to Forecast Annual Precipitation. Journal of Huazhong Normal University (Nat. Sci.) 41(1), 23–26 (2009)
4. Li, Y., Wang, X.-Y., Li, Y.-L., Zhang, G.-S.: Forecasting Flood Disasters in the Chaohu Lake Basin Based on Grey-Markov Theory. Journal of China Hydrology 26(4), 43–46 (2006)
5. Ge, G., Wang, H.-L., Xu, J.: World Oil Price Forecasting Based on Waevelet Analyze and Chaotic Time Series Technology. Systems Engineering-Theory & Practice 29(7), 64–68 (2009)
6. Ma, H.-G., Han, C.-Z., Wang, G.-H., Xu, J.-F., Zhu, X.-F.: Chaos and Data Mining Based Fault Diagnosis for Electronic System. Journal of Data Acquisition & Processing 19(3), 273–277 (2004)
7. Deng, N.-Y., Tian, Y.-J.: Support Vector Machine—a New Method in Data Mining. Science Press, Beijing (2004)

Analysis and Design of Internet Monitoring System on Public Opinion Based on Cloud Computing and NLP

Hui Wen, Peiguang Lin, and Yaobin Hu

School of Computer & Information Engineering, Shandong University of Finance, Jinan 250014
wenhuisdufe@126.com

Abstract. This paper analyzes and designs a monitoring system on public topic based on cloud computing and NLP technology. The system solves the internet massive data processing and computational complexity based on Hadoop platform; it realizes the analysis on the web page, extraction of public opinion and tracking technology based on NLP techniques and machine learning technology; it also can analyze the feelings on the users' comments and further determine the trend of public topic based on emotional thesaurus; finally, it provides a visual interface and the retrieval interface for users to use this system. Implementation of the system will improve the efficiency and quality of public topic.

Keywords: Cloud computing, Hadoop, NLP, Public topic.

1 Introduction

At the present time, network media has become the main carrier of public topic. In recent years, at home and abroad there will cause a strong reaction and a high degree of public attention after the occurrence of major events on the network media immediately, this phenomenon has brought tremendous impact to the different social groups and social structures. Because of development of information technology, especially the emergence of computer networks and social complexity, rapid increase in levels of interaction, the contemporary social sciences, management science and information science has been unable to deal with the modeling, analysis, management and control challenges brought by complex social dynamics network. While the network is providing a convenient way of information exchange, it also constitutes a threat that cannot be ignored on political security and cultural security.

Facing on the massive network information, it is very important to collect and process them only rely on traditional manual method. Therefore, we need the more scientific and effective information analysis technology to analyze the public topic quickly, thereby enhancing the regulatory capacity and the ability to handle emergencies. So it is an important task currently to use the scientific method to research public topic and master the development and the law of the social impact.

F.L. Wang et al. (Eds.): WISM 2012, LNCS 7529, pp. 528–535, 2012.

In order to efficiently process vast amounts of information on the network, this paper proposes cloud computing platform and stores large amounts of web information based on Hadoop and HBase. To improve the computational efficiency, we can realize distribute computing based on Map/Reduce framework. In addition, in order to effectively identify and describe network events, it uses NPL to analyze and process the text information, and classify and track network events using the related clustering method, finally it will realize the timely detection, tracking and monitoring on public topic.

2 Related Works

The research on public topic on abroad started earlier and the development was more rapid. Some universities, research establishments and commercial companies have close cooperation, the former have a professional research team, the latter can quickly research the formation of products, the practical application can return to the promotion of research and create a good recycling industry chain. The professional public topic network of European Commission, the world's public topic network are the professional bodies to research the network public topic.

Domestic research on network public topic started late, because of the difference between Chinese and Western culture, we cannot draw their advanced technologies such as text mining technology. At the same time, as the application institutions and research institutions are out of line, our public topic products are more expensive and the development is relatively slow. Throughout the areas of domestic public topics, public topic monitoring room of People's Daily Network Center is the earliest one of the professional bodies, which began as early as 2006. Founder Technology, Beijing Institute of Technology Networks and Distributed Computing Laboratory, respectively, also developed a proprietary network monitoring of public topic. January 10, 2009 China's first network security research organization of public topic--Beijing Jiaotong University, Center for Security Studies Network was formally established.

At present, we have formed many networks monitoring system products, the majority of them are based on keyword frequency or co-occurrence frequency. In this paper, we can fully extract the overall event through the monitoring system based on NLP, it has good usability.

3 The Analysis and Design of Network Monitoring System on Public Topic Based on Cloud Computing and NLP

3.1 System Analysis

In order to monitor public topic on the network, we must first collect the related web page information , and filter, classify them; the main task of filtering is to delete the non-essential information such as navigation and advertisements; the main task of classification is the original classification on web pages, such as news, forums, reviews, blog, etc;

Then, based on NLP we extract the related information from the web pages we collect, including a time, place, characters and their relevant information;

Third, we should cluster all the events pages, and find the theme of network public topic. In order to achieve tracking and analysis to the theme of public topic, the system needs to continue to download and analyze the relevant pages, and get the new theme or update existing information on the subject.

Finally, based on the above results, we can provide the relevant visual presentation and retrieval services.

Figure 1 shows the basic flow of the system and the main task of each step.

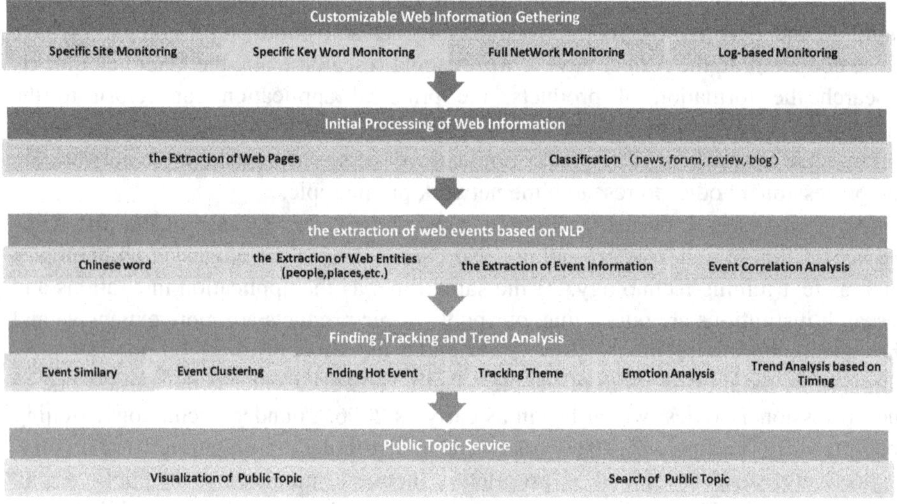

Fig. 1. The Flow of Public Topic

Taking into account the complexity of the implementation in dealing with the massive information, this system will use Hadoop to realize distributed storage and distributed processing on related technology based on MapReduce. Hadoop is a similar technology platform to the Google "cloud computing", which is a distributed system focusing on the mass data storage, processing. And it provides the Map/Reduce framework based on Java for distributed applications can be deployed to a large low-cost cluster. Map/Reduce is a parallel computing model that can efficiently handle task scheduling and load balance. Based on Hadoop Amazon company launched Amazon S3(Amazon Simple Storage Service), to provide reliable, fast network storage services, while IBM's cloud computing project - "Blue Cloud Scheme", Hadoop is also an important basis for software. At the same time, Hadoop has been applied in the library, for example, Nutch's distributed search, index, and distributed Lucene.

Map/Reduce is a parallel computing model that can efficiently handle task scheduling and load balance. Programming model is one of the core technology of cloud computing, it extract the simple business logic from the complex implementation details, provides a series of simple and powerful interfaces. It enables developers not to need a large number of parallel or distributed developments experience to use efficiently distributed resources.

In summary, the overall structure of the system is as shown in Figure 2.

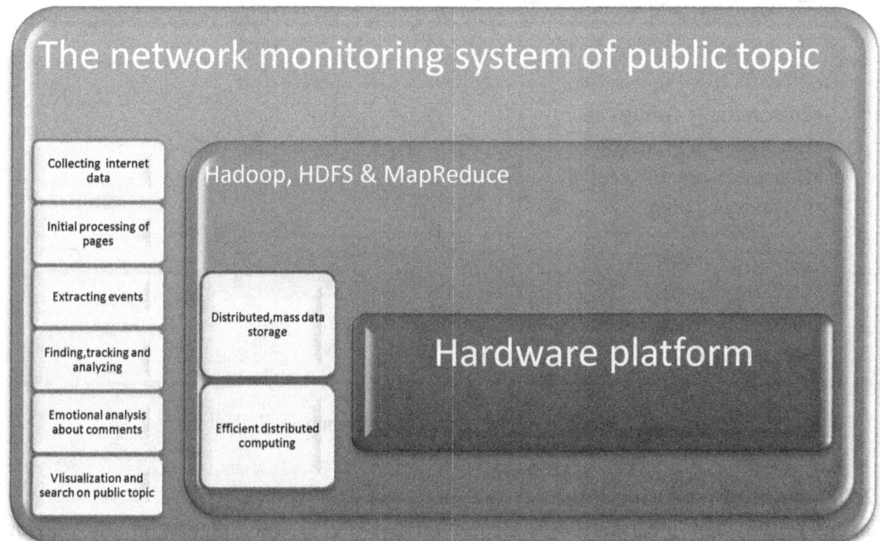

Fig. 2. The Overall Structure of the System

3.2 Detailed Design

Based on these preliminary findings, this section describes the system for further details

3.2.1 The Collection of Internet Information Can Be Customized

The module mainly achieves the automatic collection of Internet information. In order to enhance the adaptability of the system, the user can customize the collection of targeting data that the user can make specify system to search the specified site, also can use search engine by specific keyword to get search engine results, or search the entire Internet and information in Gateway-based record log.

In order to achieve the collection of Web information, Nutch provides a good solution. Nutch is an open source search engine framework, based on a modular architecture design ideas, working to build a query and index based on Lucene open source search engine achieved. It uses the Java language to develop, with the advantages of cross-platform applications. Therefore, the system uses Nutch to achieve collection of Internet information.

Finally, taking into account the mass of Internet information and the actual needs of the system, the module will be distributed based on Hadoop technology. The specific structure of the module is as shown in Figure 1Figure 3.

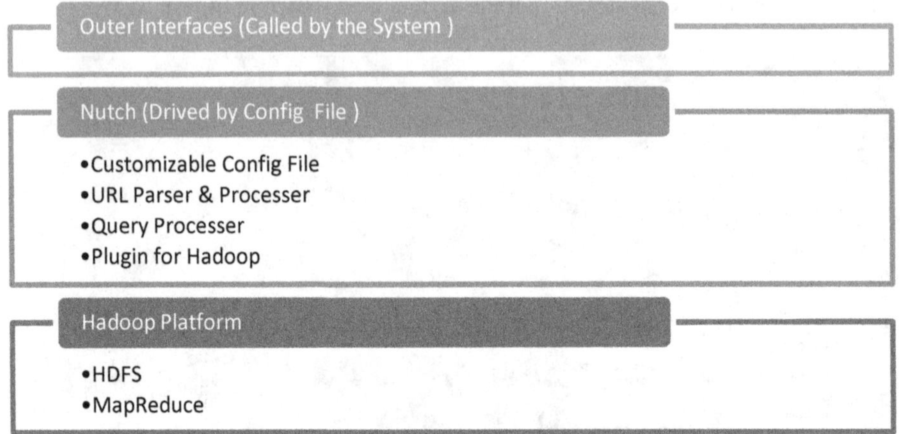

Fig. 3. the Module of Nutch-based Information Collection

3.2.2 Initial Processing Page

Currently, the web information is described by HTML language, and the current HTML specification cannot distinguish between the content of Web pages and other navigation information, advertising, irrelevant information. In order to achieve the extraction of the events described on the page, we need to filter out irrelevant information on web page.

As Web browser admit that HTML Web pages have some faults, most HTML pages do not comply with W3C standards, such as the lack of related marks, nesting and unreasonable. To properly analyze the structure and features of pages ,the structure of pages must comply with HTML standards of pages. To parse all the pages according to HTML standards, you must scan the Web page and restore defective pages. The system uses JTidy repair defects on the web. JTidy is a powerful tool for open-source can be used to correct common errors in the HTML document and generate good formatting documents .Web pages will take pretreatment by JTidy, and finally an equivalent HTML document will convert into XHTML document.

After the standardization of web pages, it is necessary to preliminary classify web pages that to determine pages belong to which? Press reviews, forum or blog and so on? The specific classification is based on the combination of rules and machine learning. Rules include the rules of the URL and sensitive words on web pages; machine learning methods achieve the classification of web pages by clustering or classification.

Through observation, the body of a page usually appears in the middle of the page, the top or left of a page generally is related navigation information, the bottom of a page is advertising or other links. Based on this assumption, we use Microsoft

Research Asia's VIPS (Vision-based Page Segmentation) algorithm to solve problem of web content extraction. At the same time, access to standard information of web page, we use the method based on DOM tree to extract the comment.

Figure 4shows the main tasks and process in the initial processing of web pages.

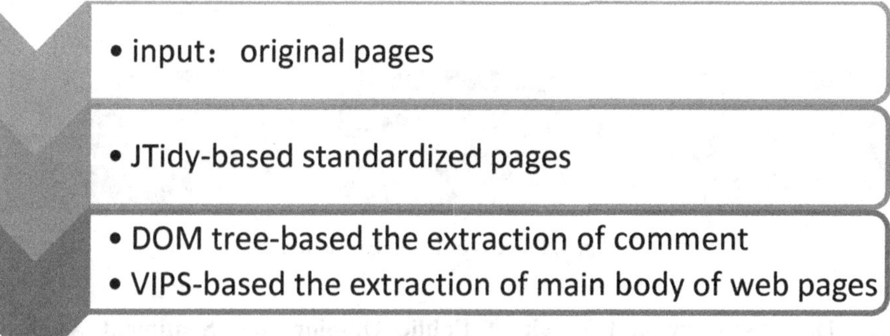

Fig. 4. the Main Tasks and Process in the Initial Processing of Web Pages

3.2.3 NLP-Based Event Extraction

Information extraction is the rise of the past 20 years in the field of text processing is an important direction. Despite no or less information extraction system using natural language technology (such as through meta-data, HTML tags or according to key words).But the more in-depth linguistic analysis, the more stronger will be when information extraction system for the extraction of targets in different areas of adaptability and portability.

A complete information extraction system need to go through the text block, preprocessing, filtering out irrelevant statements, analysis, fragment combination, semantic interpretation, lexical disambiguation, template generation. To improve the efficiency of the system, the system accept 8 event categories and 33 seeds types as event template defined by ACE2005 and use HMM-based machine learning methods to achieve the extraction of relevant events.

Hidden Markov Model is a double Markov random process, including a state transition probability of Markov chain and the output observations of the random process, its status is showed only by observing the sequence of random process Hidden Markov model provides a probability based on training data automatic identification system .This system has shown good momentum of development in the field of information extraction.

On the basis of pretreatment of web pages, we first create a hidden Markov model, and then train Hidden Markov Model with the manual tagging data sources. When the training reaches a certain accuracy, new web pages could be extracted. The system uses open source HMM framework jahmm as a model.

The processing module is shown in Figure 5.

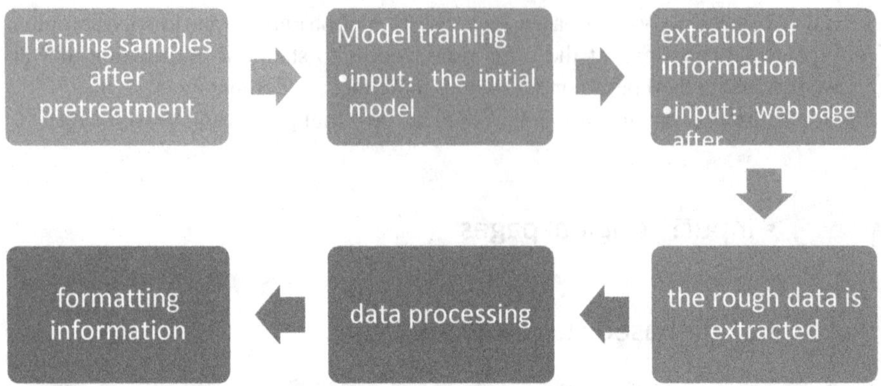

Fig. 5. Process of Event Extraction

3.2.4 The Discovery and Track of Public Opinion and Sentiment Analysis Comments

Access to the events described in each page, we cluster all the events, and statistic the number of samples in each category, the size of the sample is most that is taken as the current hot topic. At the same time, we also achieve the track of public opinion by this method, the event tracking. We achieve Sentiment analysis of the comments information with thesaurus of emotional word.

3.2.5. Public Opinion Service

Based on the above analysis, the system provides event timing change of plans, information retrieval interfaces to help users view the results of public opinion analysis.

4 Summary

This article analysis and design a system of public opinion monitoring which is based NLP and NLP and cloud computing. The core of each module technology and solutions is given in this article. The implementation of the system will effectively improve the time efficiency of the Internet public opinion monitoring, and provide efficient public opinion monitoring services for government departments and others related departments. Meanwhile, after the system is transformed, it could provide products of automatically track, intelligence collection and analysis services for related companies.

Acknowledgement. This work was supported by MOE Project of Humanities and Social Sciences (Project No.10YJC880076) and Natural Science Foundation of Shandong Province (Project No. ZR2010FL008).

References

1. Apache Hadoop (EB/OL) (April 15, 2011), `http://hadoop.apache.org/`
2. Allan, J., Carbonell, J.: Topic Detection and Tracking Pilot Study: Final Report. In: Proceedings of the DARPA Broadcast News Transcription and Understanding Workshop, pp. 194–218. Morgan Kaufmann Publishers, Inc., San Francisco (1998)
3. Brin, S., Page, L.: The Anatomy of a Large-scale Hypertextual Web Search Engine. Computer Networks and ISDN Systems, 107–117 (2009)
4. Qu, W.-G., Chen, X.-H., Dong, Y.: Chinese WSD based on context calculation model. Journal of Guangxi Normal University 24, 179–182 (2006)
5. Fisher, D.H.: Knowledge Acquisition via Incremental Conceptual Clustering. Machine Learning 1, 139–172 (2007)
6. Hulten, G., Domingos, P.: A General Framework for Mining Massive Data Streams. Journal of Computational and Graphical Statistics 12 (2003)
7. Charikar, M., Chen, K., Farach-Colton, M.: Finding Frequent Items in Data Streams. In: Widmayer, P., Triguero, F., Morales, R., Hennessy, M., Eidenbenz, S., Conejo, R. (eds.) ICALP 2002. LNCS, vol. 2380, pp. 693–703. Springer, Heidelberg (2002)
8. Kossinets, G., Watts, D.J.: Empirical analysis of an evolving social network. Science 311(5757), 88–90 (2006)
9. Yu, M., Luo, W., Xu, H., Bai, S.: Research on hierarchical topic detection in topic detection and tracking. Journal of Computer Research and Development 43(3), 489–495 (2006)
10. Tang, Y., Zheng, J.: Linguistic Modeling Based on Semantic Similar relation among linguistic labels. Fuzzy Sets and Systems, 1662–1673 (2006)
11. Zaharia, M., Konwinski, A., Joseph, A.D., Katz, R., Stoica, I.: Improving MapReduce Performance in Heterogeneous Environments. In: Proceedings of OSDI (2008)
12. Song, C., Luo, Q., Shi, F.: A Bayesian Dynamic Forecast Model Based On Neural Network. In: Proceedings of the International Symposium on Intelligent Information Technology Application Workshops, pp. 130–132 (2008)

Using Similes to Extract Basic Sentiments across Languages

Bin Li[1,2], Haibo Kuang[2], Yingjie Zhang[1], Jiajun Chen[1], and Xuri Tang[3]

[1] State Key Lab for Novel Software Technology, Nanjing University, 210046 Nanjing, P.R. China
{lib,zhangyj,chenjj}@nlp.nju.edu.cn

[2] Research Center of Language and Informatics, Nanjing Normal University, 210007 Nanjing, P.R. China
mchypocn@hotmail.com

[3] Foreign Language School, Huazhong University of Science and Technology, 430074 Wuhan, P.R. China
xrtang@126.com

Abstract. People often use similes of pattern "as adjective as noun" to express their feelings on web medias. The adjective in the pattern is generally the salient property and strong impression of the noun entity in the speaker's mind. By querying the simile templates from search engines, we construct a large database of "noun-adjective" items in English and Chinese, which highlight the same and different basic sentiments on the same entity in the two languages. The approach is a fast and efficient way to extract people's basic sentiments and feelings across languages.

Keywords: sentiment extraction, simile, natural language processing.

1 Introduction

Sentiment extraction is a heated topic in natural language processing. People's attitudes towards a person, product, organization and other objects are potentially of great value to business intelligence and decision making systems[1]. Most of the previous researches focused on the supervised or unsupervised machine learning algorithms for the constructions of monolingual or multilingual lexical resources, corpus and systems[2]. However, there is one important and simple phenomenon that is not fully explored but able to provide new access to the knowledge. People would use the salient sentiments of an object in simile expressions, which could be used to detect people's sentiments. As shown in example(1,2,3), the three real sentences express strong attitudes on *BlackBerry apps*, *Mike* and *he*, which are the key topics to extract in sentiment research. But we want to emphasize that, the *iPhone*, *Kobe* and *pig* are more basic and important topics with their salient properties *polished*, *selfish* and *lazy*. These properties are the speaker's sentiments on the nouns.

(1) BlackBerry apps tend not to look as ***polished*** as ***iPhone***.
(2) Mike was just as ***selfish*** as ***Kobe***.
(3) He is as ***lazy*** as a ***pig***.

F.L. Wang et al. (Eds.): WISM 2012, LNCS 7529, pp. 536–542, 2012.

It is obvious that, the adjective used in a simile template "X be as ADJECTIVE as VEHICLE" precisely describes the property of the vehicle. In linguistics, the template can serve as comparison (example 1, 2) or simile(example 3) or irony(as fast as a snail), but the vehicle has the property expressed by the adjective in both the first two cases. So the simile templates are very useful in acquiring the vehicle nouns' properties. If tens of thousands of simile expressions are collected, a great number of vehicle-adjective items will be gathered. Then it is easy to obtain the different sentiments and properties of the same vehicle like *iPhone*-polished, coveted, fancy, etc.

The similes do not often occur in news and technique texts but in web blogs, micro blogs and forums. Thus, using search engine is a good way to get the similes fast and efficient. For the purpose of extracting basic attitudes cross languages, we employ simile sentences to collect "vehicle-adjective" items in English and Chinese. In this way, we construct a large database which reveals the shared and dependent attitudes on entities in the two languages.

2 Related Work

Cross language sentiment extraction has been experimented on parallel texts[3], translated texts[4] and comparable corpus[5]. All these work employed machine learning methods to classify or extract the sentiments in the corpus of comments, reviews and blogs.

Turney put forward a fast method to obtain the polarity of a word by querying the word from search engine with good and bad words[6]. Point-wise mutual information(PMI) is calculated as the polarity score. But it cannot supply the detailed attitudes of people on the entity word.

To collect similes from search engines is the interest of the metaphor researchers. Veale collects English "noun-adjective" pairs by querying the simile templates "as adjective as *" and "as * as noun" from Google(google.com) with WordNet to construct the English lexical metaphor knowledgebase "sardonicus", which contains about 10,000 items of "noun-adjective" pairs[7]. Similarly, Jia collects 20,000 items of Chinese "noun-adjective" pairs with similes from Baidu(Baidu.com)[8]. Thus, querying search engine is an efficient way to collect "noun-adjective" pairs. However, they concentrate on the adjectives of the common nouns in WordNet-like dictionaries, but not the entity words like "iPhone" and "Kobe". Second, they separately do the collecting work in one language. Third, they don't pay enough attention the frequencies of the items. Therefore, we want to extend the research to multi-languages and further the research with stress on the frequencies to show the sentiments distributions.

3 Simile Extraction

As the methods applied in [7] and [8], we use the specific simile templates to collect English and Chinese "noun-adjective" pairs by querying the search engines. The words in WordNet[9] and HowNet[10] are used for querying the search engines.

3.1 English Simile Extraction

We use the 21,479 adjectives in WordNet to fill in the simile template "as ADJ as". Google advanced search is queried with 3 limitations to refine the search results: exact search, English language and up to 100 results for each query. Then, 585,300 "as…as…" items(1,054,982 tokens) are obtained, many of which are nonsense, noisy and error items. We simply use the adjectives in HowNet as the filter to trim these items. Functional words, pronouns and the vehicles longer than 2 words are trimmed off. Only 98,057 types(178,622 tokens) of "noun-adjective" pairs are left, covering 12,468 adjectives in WordNet and 68,752 vehicles. Table 1 gives the top 10 most frequent pairs with their frequencies.

Table 1. Top 10 most frequent vehicle-adjective pairs in English

Id	Vehicle	Adjective	Freq
1	blood	red	628
2	twilight	gay	466
3	grass	perennial	413
4	ice	cold	392
5	mustard	keen	385
6	**Barack Obama**	**Irish**	358
7	snow	white	340
8	sea	boundless	314
9	feather	light	289
10	night	black	280

The "blood-red" is the most frequent item in English. And "Barack Obama-Irish" gets a high rank in the results. The frequency tells the salience of people's feelings of the vehicles. "Barack Obama" also have ten other adjectives like inexperienced(4 times), socialistic(once), etc. The short name "Obama" gets many more results like liberal(26 times), funny(14 times), and black(6 times), etc. These items are very useful as they take the basic and strong impressions in people's minds.

The frequencies here are not the exact data on the web. They are only the statistical situation in the collected items. And the frequency of the item can be over 100, because the item also occurs in the querying results of other words.

3.2 Chinese Simile Extraction

For Chinese, 3 simile templates "像(as)+NOUN+一样(same)", "像(as)+VERB+一样(same)", "像(as)+一样(same)+ADJ" are filled with the 51,020 nouns, 27,901 verbs and 12,252 adjectives in HowNet to query the Chinese search engine Baidu(baidu.com). Verbs are considered as vehicles, because Chinese words do not have inflections and some of verbs may serve as a noun in some context. We submit 91,173 queries to Baidu advanced search, setting that up to 100 results returned for each query. As a result, 1,258,430 types(5,637,500 tokens) of "vehicle-adjective" pairs are gathered. Then, adjectives in HowNet are used to filter these items, leaving only 75,336 items. The database of the Chinese filtered items is already available for search at http://nlp.nju.edu.cn/lib/cog/ccb_nju.php. Table 2 shows the top 10 most frequent items with their frequencies.

Table 2. Top 10 most frequent vehicle-adjective pairs in Chinese.

Id	Vehicle	Adjective	Freq
1	苹果apple	时尚fashionable	1445
2	宝钗lady name	懂事reasonable	998
3	可卿lady name	漂亮pretty	943
4	美玉fine jade	美丽beautiful	840
5	呼吸breath	自然natural	758
6	晨曦sun rise	朝气蓬勃spirited	750
7	纸paper	薄thin	660
8	雨点rain drop	密集dense	557
9	自由freedom	美丽beautiful	543
10	雪snow	白white	521

It is surprising to see that the products of "apple (Inc.)" have taken the first place in Chinese eyes on the web media. And the two ladies named "宝钗" and "可卿" in the famous Chinese novel "A Dream of Red Mansions" get the second and third place. The rest of the vehicles in top10 items are common nouns. The rest of the adjectives of "apple" are 红(red,68 times) and 可爱(lovely, 25times), etc. Most of them do not refer to the company but the fruit.

In next section, we will compare the basic sentiments based on the collected data from Google and Baidu.

4 Bilingual Comparison of Basic Sentiments

In the last 2 sections, we've got the basic sentiments in English and Chinese. Now, it is natural to see if the sentiments are the same between them. However, we still lack a large dictionary or ontology of entity words and the sense tagging tool for disambiguation the senses of the vehicles. So what we can supply is to search these named entity words in the database bilingually with Google (translate.google.cn) or Bing (dict.bing.com.cn) translation service.

We randomly select 13 words of famous persons, products and companies to see their sentiments in people's eyes bilingually. The nouns in table 3 are common topics in English, including 3 famous persons, 3 products of apple and 7 enterprises. Most of the properties are correct, with few polysemous words like "Kobe-marbled"[1] and "apple-round". The bold adjectives are the same property shared by English and Chinese speakers.

[1] "Kobe" is also the name of a city in Japan.

Table 3. Bilingual comparison of sentiments on 13 entities

Noun_# of adjs	Top8 adjs with frequencies and Chinese translations
Obama_35	liberal_26,funny_14,magnanimous_6,black_6,ineligible_4,incompetent_4,phony_4, fake_4
奥巴马_5	横 peremptory_15,单纯 pure minded_5,健美 good look_3,**好 good_2**,优秀 excellent_1
Messi_6	talented_8,acceptable_4,midget_2,reliant_5,skilful_2
梅西_7	伟大 great_16,灵活 nimble_4,过人 outstanding_3,致命 deadly _2,低调 low-key _2,好 good_1,灵便 nimble_1,高 high level_1
Kobe_7	cocky_6,[marbled_6],arrogant_2,**selfish_2**,disliked_1,emphatic_1,streaked_1
科比_8	强 strong_11, 坚强 strong heart_5,成功 successful_3, 准 accurate_2,出色 outstanding_1,孤独 lonely_1,勇猛 brave_1,自私 **selfish_1**
iPod_9	portable_4,**common_2**,swarthy_1,preferred_1,noteworthy_1,intuitive_1,important_1,dinky_1
iPod_8	**流行 popular_10**,多彩 colorful_7,容易划伤 delicate _4,简单 simple_2,豪华 luxury_2,完胜 success_1,容易使用 easy to use_1,火 heated_1
iPad_24	lambent_8,diverse_5,slippery_5,intuitive_4,less_4,fast_3,nascent_2,telepathic_2
iPad_4	流行 popular_5,持久 long haul_4,精致 exquisite_2,轻松 easy_1
iPhone_80	useless_18,polished_6,mature_5,responsive_5,coveted_4,modal_4,popular_4,insecure_4,smooth_4,discreet_3,
iPhone_8	方便 convenient_2,成功 successful_2,供不应求 short supply_1,薄 thin_1,坚挺 strong_1,帅 cute_1,拉风 cool_1, 炫 dazzle_1
apple_62	**round_49**,modern_12,cool_6,light_6,**loved_6**,**pretty_6**,closed_5,**crisp_4**
苹果_30	**时尚 fashionable_1445**,红 red_68,**可爱 lovely_25**,甜 sweet_20,圆 round_17,坚实 firm_12,**鲜嫩 fresh_8**,鲜红 red_7
Microsoft_22	**sneaky_16**,commercialized_4,rugged_4,arrogant_2,inadequate_2,evil_2,ephemeral _2, monopolistic_2
微软_4	过时 outdated_3,成功 successful_2,**强大 powerful_1**,**无耻 sneaky_1**
Google_28	mighty_16,permanent_4,acquisitive_3,omniscient_2,relevant_2,international_2,intelligent_2,helpful_2
谷歌_6	简洁 simple_19,**家喻户晓 well known_3**,年轻有为 bright young_1,强大 powerful_1,好 good_1,财大气粗 rich_1
Facebook_56	social_40,bitchy_6,boring_6,creepy_6,indispensable_6,malign_6,erudite_4, extraordinary_4
Facebook_2	成功 successful_2,好 good_1
KFC_4	sly_4,fried_2,fortified_1,synergistic_1,
肯德基_2	好吃 delicious_2,恶心 sick_1
McDonalds16	fattening_10,inhumane_6,mammoth_2,patronized_2, prevalent_2,caloric_2,corporate_2,dying_2
麦当劳_5	方便 convenient_8,强大 powerful_8,正常 nomal_6,火 heated_3,好吃 delicious_1
Walmart_7	classy_4,scummy_2,sanctified_2,predatory_1,exploitive_1,communist_1,avaricious _1
沃尔玛_3	大 large_5,多 many_3,实惠 boon_1

It is obvious that, most of the properties are different in the two languages. They are language or culture dependent, and the number of adjectives in English is almost larger than in Chinese. For the famous persons, people have positive and negative sentiments on them. *Obama* is somewhat welcomed in China, *Messi* is perfect without negative comments. *Kobe* is a great player, but his selfish is criticized cross languages. For apple's products, people love them very much. To have an apple product is a fashion in China, while the west users are not very satisfied with it. The three great companies *Apple*, *Google* and *Facebook* are enjoyed by the 2 language users, while *Microsoft* is getting worse. The two fast food companies and supermarket company *Walmalt* are still welcomed in China, but get more criticisms in the west.

The evaluation of the sentiment data is not easy to conduct. First, it is almost impossible to score the recall rate of the "vehicle-adjective" items, because we cannot know exactly how many adjectives people will use to describe the objects. Second, due to the large scale of the data, we can only manually check some samples to see the accuracy of the items. And the accuracy is high on the randomly selected 13 nouns. In the future, we need to design better methods for the evaluation.

5 Conclusion and Future Work

Extracting multi-lingual sentiments from web is a useful but hard task in natural language processing, because it has to face the efficiency and accuracy in lexical analysis, parsing, word sense disambiguation and machine translation. To avoid the complexity and low efficiency of the traditional methods, we put forward an easy and fast way to extract basic sentiments from search engines. Simile templates are accurate to obtain the adjectives expressing people's feelings and search engines are easy to query. Using Google and Baidu, we collected a database of tens of thousands "vehicle-adjective" pairs and then conducted filtering by phrase length and adjectives. The database constructed in the process supply the bilingual basic sentiments of a given topic.

In the future, we will collect bilingual entity dictionaries and ontologies to make full analysis of the basic sentiment database and supply online search service. We also want to expand our method to collect basic sentiments in more languages. At that time, it will become easy to browse the basic attitudes of things all over the world.

Acknowledgments. We are grateful for the comments of the anonymous reviewers. This work was supported in part by National Social Science Fund of China under contract 10CYY021, 11CYY030, China PostDoc Fund under contract 2012M510178, State Key Lab. for Novel Software Technology under contract KFKT2011B03, Jiangsu PostDoc Fund under contract 1101065C.

References

1. Pang, B., Lee, L.: Opinion Mining and Sentiment Analysis. Foundations and Trends in Information Retrieval 2(1-2), 1–135 (2008)

2. Liu, B.: Sentiment Analysis and Subjectivity. In: Handbook of Natural Language Processing, 2nd edn., Chapman and Hall/CRC (2010)
3. Ni, X., Sun, J.-T., Hu, J., Chen, Z.: Mining Multilingual Topics from Wikipedia. In: Proceedings of the 18th International Conference on World Wide Web, New York, USA (2009)
4. Wan, X.: Co-training for Cross-lingual Sentiment Classification. In: Proceedings of the Joint Conference of the 47th Annual Meeting of the ACL and the 4th International Joint Conference on Natural Language Processing of the AFNLP, Stroudsburg, PA, USA, vol. 1 (2009)
5. Boyd-Graber, J., Resnik, P.: Holistic Sentiment Analysis across Languages: Multilingual Supervised Latent Dirichlet Allocation. In: Proceedings of the 2010 Conference on Empirical Methods in Natural Language Processing, pp. 45–55 (2010)
6. Turney, P.D.: Thumbs Up or Thumbs Down? Semantic Orientation Applied to Unsupervised Classification of Reviews. In: Proceedings of the 40th Annual Meeting of the ACL, pp. 417–424 (2002)
7. Veale, T., Hao, Y.F.: Learning to Understand Figurative Language: From Similes to Metaphors to Irony. In: Proceedings of CogSci 2007, Nashville, USA (2007)
8. Jia, Y.X., Yu, S.W.: Instance-based Metaphor Comprehension and Generation. Computer Science 36(3), 138–141 (2009)
9. Miller, G.A., Beckwith, R., Fellbaum, C.D., Gross, D., Miller, K.: WordNet: An Online Lexical Database. Int. J. Lexicography 3(4), 235–244 (1990)
10. Dong, Z.D., Dong, Q.: HowNet and the Computation of Meaning. World Scientific Press, Singapore (2006)

Automatic Summarization for Chinese Text Using Affinity Propagation Clustering and Latent Semantic Analysis

Rui Yang, Zhan Bu, and Zhengyou Xia

College of Computer Science and Technology, Nanjing University of Aeronautics and Astronautics, China
zhengyou_xia@nuaa.edu.cn

Abstract. As the rapid development of the internet, we can collect more and more information. it also means we need the abitily to search the information which really useful to us from the amount of information quickly. Automatic summarization is useful to us for handling the huge amount of text information in the Web. This paper proposes a Chinese summarization method based on Affinity Propagation(AP)clustering and latent semantic analysis(LSA). AP is a new clustering algorithm raised by B. J. Frey on science in 2007 that takes as input measures of similarity between pairs of data points and simultaneously considers all data points as potential exemplars. LSA is a technique in natural language processing, in particular in vectorial semantics, of analyzing relationships between a set of sentences. Experiment results show that our method could get more comprehensive and high-quality summarization.

Keywords: Summarization, AP, LSA, clustering.

1 Introduction

With the rapid development of Internet, it has become increasingly important to find ways for getting useful information. The automatic summarization [1] can generate a short summary expressing the main meaning of a text, which helps us to find the information we really need quickly.

Summarization can be divided into different categories along several different dimensions. Based on whether or not there is an input query, the generated summary can be query-oriented or generic; based on the number of input documents, summarization can use a single document or multiple documents; in terms of how sentences in the summary are formed, summarization can be conducted using either extraction or abstraction — the former only selects sentences from the original documents, whereas the latter involves natural language generation. Overall, automatic summarization systems aim to generate a good summary, which is expected to be concise, informative, and relevant to the original input.

In this paper, we consider extractive summarization. We propose a new clustering algorithm called Affinity Propagation [2,3,4]. In the sentence extraction strategy,

F.L. Wang et al. (Eds.): WISM 2012, LNCS 7529, pp. 543–550, 2012.

clustering is frequently used to eliminate the redundant information resulted from the multiplicity. Clustering for text summarization mainly take the sentences into some clusters according to the distance or similary between each two senteces, and select one or more sentences from each cluster until fitting the length of the summary.

The popular k-centers clustering technique [5] begins with an initial set of randomly selected exemplars and iteratively refines this set so as to decrease the sum of squared errors. k-centers clustering is quite sensitive to the initial selection of exemplars, so it is usually rerun many times with different initializations in an attempt to find a good solution. However, this works well only when the number of clustersis small and chances are good that at leastone random initialization is close to a good solution. AP simultaneously considers all data points as potential exemplars. By viewing each data point as a node in a network, AP transmits real-valued messages along edges of the network until a good set of emplars and corresponding clusters emerges. As described later, messages are updated on the basis of simple formulas that search for minima of an appropriately chosen energy function. At any point in time, the magnitude of each message reflects the current affinity that one data point has for choosing another data point as its exemplar, so we call this method "affinity propagation".

Affinity propagation takes as input a collection of real-valued similarities between data points, where the similarity between two sentences calculated by Latent semantic analysis (LSA) here. LSA is a technique in natural language processing, in particular in vectorial semantics, of analyzing relationships between a set of sentences. LSA assumes that words that are close in meaning will occur close together in text. This method could mitigate the problem of identifying synonymy and the problem with polysemy [6].

This paper is organized as follows: We first give a brief survey on previous work in Section 2. Then, we detail our proposed automatic summarization method in Section 3. Following that, we present the experiments in Section 4. Finally, conclusions are given in Section 5.

2 Related Work

Automatic summarization has received a lot of attentions in recent years, and various approaches exist for extractive summarization, including the use of word frequency [7], cue words or phrases [8], machine learning [9], lexical chains [10] and sentence compression through syntactical or statistical restrictions [11].

Wang et al. [12] proposed a Chinese automatic summarization method based on thematic sentence discovery. They utilized terminology rather than traditional word as the minimal semantic unit, computed terminology weight with its length and frequency to extract keywords, and discovered thematic sentences using an improved k-means clustering method.

Rada Mihalcea and Paul Tarau [13] introduced TextRank – a graph-based ranking model for text processing. Graph-based ranking algorithms are essentially a way of deciding the importance of a vertex within a graph, based on global information recursively drawn from the entire graph. The basic idea implemented by a graph-based ranking model is that of "voting" or "recommendation". When one vertex links to another one, it is basically casting a vote for that other vertex. The higher the number of votes that are cast for a vertex, the higher the importance of the vertex. Moreover, the importance of the vertex casting the vote determines how important the vote itself is, and this information is also taken into account by the ranking model. Hence, the score associated with a vertex is determined based on the votes that are cast for it, and the score of the vertices casting these votes.

Jen-Yuan Yeh, Hao-Ren Ke [14] proposed a approach to address text summarization: LSA based and T.R.M.approach(LSA+T.R.M.).). They used latent semantic analysis(LSA)to derive the semantic matrix of a document or a corpus and uses semantic sentence representation to construct a semantic text relationship map.

Shasha Xie [15] investigated several approaches to extractive summarization on meeting corpus. Their thesis first proposed two unsupervised frameworks: Maximum Marginal Relevance (MMR) and the concept-based global optimization approach. Second, They treat extractive summarization as a binary classification problem, and adopt supervised learning methods and propose using various sampling techniques and a regression model for the extractive summarization task. Third, They focus on speech specific information for improving the meeting summarization performance.

Lucas Antiqueira, Osvaldo N. Oliveira Jr. [16] employed concepts and metrics of complex networks to select sentences for an extractive summary. The graph or network representing one piece of text consists of nodes corresponding to sentences, while edges connect sentences that share common meaningful nouns. This method uses a simple network of sentences that requires only surface text pre-processing, thus allowing us to assess extracts obtained with no sophisticated linguistic knowledge.

3 Process of LSA+AP

The process of LSA+AP consists of five phases: (1) preprocessing, (2) calculating similarities between sentences by LSA, (3) clutering the senteces in the text by AP algorithm, (4) sentences selection for summary.

3.1 Preprocessing

Preprocessing delimit each sentence by punctuation. Furthemore, it segments each sentence into words based on dictionary. In addition, we remove the meaningless words like "de", "a", etc.

3.2 Calculating Similarities between Sentences by LSA

LSA was patented in 1988 (US Patent 4,839,853) by Scott Deerwester, Susan Dumais, George Furnas, Richard Harshman, Thomas Landauer, Karen Lochbaum and Lynn Streeter [17]. In the context of its application to information retrieval, it is sometimes called Latent Semantic Indexing (LSI).

We select the LSA to get the similarities between sentences because that LSA could analyzing relationships between a set of sentences by producting a set of concepts related to the sentences. LSA assumes that words that are close in meaning will occur close together in text. So it can mitigate the problem of identifying synonymy and the problem with polysemy.

The specific steps of LSA to calculate similarities between sentences are as follows:

Occurrence Matrix. The following elucidates how to construct the word-by-sentence matrix for the single-document.

Let D be a document, W(|W| = M)be the set of keywords in D, and S(|S| = N)be the set of sentences in D. A is the word-by-sentence matrix whose every row Si indicates a sentence and every column Wi indicates a keyword.

In A, $a_{i,j}$ is defined as Eq.(1), where L_{ij} is the local weight of W_i in S_j, and G_i is the global weight of W_i in D.L_{ij} is defined as $L_{ij} = \log(1 + \frac{aij}{nj})$, and G_i is defined as $G_i = 1 - E_i$, where c_{ij} is the frequency of W_i occurring in S_j, n_j is the number of words in S_j, and E_i is the normalized entropy of W_i , which is defined as (Bellegarda, Butzberger, Chow, Coccaro, & Naik, 1996).

$$aij = Gi \times Lij \tag{1}$$

Rank Lowering. We then perform singular value decomposition (SVD) to A. The SVD of A is defined as AUZV ^{T}, where U is an $M \times N$ matrix of left singular vectors, Z is an $N \times N$ diagonal matrix of singular values, and V is an $N \times N$ matrix of right singular vectors.

Finally, the process of dimension reduction is applied to Z by deleting a few entries in it, and the result of dimension reduction is a matrix Z'. A new matrix, A', is reconstructed by multiplying three component matrixes. A' is defined as Eq. (2), where Z' is the semantic space that derives latent semantic structures from A, U' is a matrix of left singular vectors whose ith vector u_i' represents W_i in Z', and V' is a matrix of right singular vectors whose jth vector v_j' represents Sj in Z'.

$$A' = U'Z'V'T \approx A \tag{2}$$

Each column of A' denotes the semantic sentence representation, and each row denotes the semantic word representation.

Calculating Similarities between Sentences. In our method, a sentence S_i is represented by the corresponding sementic sentence representation, instead of the original keyword-based frequency vector. The similarity between a pair of sentences S_i and S_j is evaluated to determine if they are semantic ally related. The similarity is defined as Eq.(3).

$$\text{sim(Si,Sj)} = \frac{\vec{s_1} \cdot \vec{s_2}}{|\vec{s_1}| \cdot |\vec{s_2}|} \tag{3}$$

3.3 Clutering the Senteces in the Text by AP

Up to now, we have got the similarity between each sentence, which is the input of the AP clustering algorithm.

Rather than requiring that the number of clusters be prespecified, affinity propagation takes as input a real number s(k,k) for each sentence k so that the sentences with larger values of s(k,k) are more likely to be chosen as exemplars. These values are referred to as "preferences". The number of identified exemplars(number of clusters) is influenced by the values of the input preferences, but also emerges from the message-passing procedure. The preferences should be set to a common value— this value can be varied to produce different numbers of clusters. The shared value could be the median of the input similarities (resulting in a moderate number of clusters) or their minimum (resulting in a small number of clusters).

Because of that we need to generate a summary, so we want the number of clusters should be small, we select the minimum of the input similarities as the " preferences".

In addition, we knows that the first sentence in a paragragh always be important than other sentences in a chinese text. So we add a weight to a sentence if it is the first sentence of a paragragh as Eq.(4).

$$\text{P(i)} = \begin{cases} Pvalue + \gamma \times Pvalue, if \text{ Si is the first sentence} \\ Pvalue, if \text{ not} \end{cases} \tag{4}$$

Where the Pvalue is the minimum of the input similarities, and the γ is 0.9.

There are two kinds of message exchanged between sentences, and each takes into account a different kind of competition. Messages can be combined at any stage to decide which sentence are exemplars and, for every other sentence, which exemplar it belongs to. The "responsibility" r(i,k), sent from sentence i to candidate exemplar sentence k, reflects the accumulated evidence for how well-suited sentence k is to serve as the exemplar for sentence i, taking into account other potential exemplars for sentence i. The "availability" a(i,k), sent from candidate exemplar sentence k to sentence i, reflects the accumulated evidence for how appropriate it would be for sentence i to choose sentence k as its exemplar, taking into account the support from other points that sentence k should be an exemplar. r(i,k) and a(i,k) can be viewed as log-probability ratios. To begin with, the availabilities are initialized to zero: a(i,k) = 0. Then, the responsibilities are computed using the rule:

$$\text{r(i,k)} = \text{s(i,k)} - \max_{k' s.t. k' \neq k} \{a(i,k') + s(i,k')\} \tag{5}$$

Whereas the above responsibility update lets all candidate exemplars compete for ownership of a sentence, the following availability update gathers evidence from sentences as to whether each candidate exemplar would make a good exemplar:

$$
a(i,k) = \begin{cases} \min\{0, r(k,k) + \sum_{i' s.t. i' \neq i, i' \neq k} \max\{0.r(i',k)\}\}, if\ i \neq k \\ \sum_{i' s.t. i' \neq k} \max\{0, r(i',k)\}, if\ i = k \end{cases} \tag{6}
$$

For sentence i, the value of k that maximizes a(i,k) + r(i,k) either identifies sentence i as an exemplar if k = i, or identifies the sentence that is the exemplar for sentence i. The message-passing procedure may be terminated after a fixed number of iterations, after changes in the messages fall below a threshold, or after the local decisions stay constant for some number of iterations. We select in our process that if the exemplar have not changed for 20 times that the procedure could be terminated.

When updating the messages, it is important that they be damped to avoid numerical oscillations that arise in some circumstances. Each message is set to λ times its value from the previous iteration plus 1–λ times its prescribed updated value, where the damping factor λ is between 0 and 1. In all of our experiments, we used a default damping factor of λ = 0.5, and each iteration of affinity propagation consisted of (i) updating all responsibilities given the availabilities, (ii) updating all availabilities given the responsibilities, and (iii) combining availabilities and responsibilities to monitor the exemplar decisions and terminate the algorithm when these decisions did not change for 20 iterations.

Finally, we put the each sentence i which is not the exemplar into the corresponding exemplar which max the value of a(i,k) + r(i,k).

3.4 Sentences Selection for Summary

After clutering, we could select sentences from each cluster orderly until the length of summary is suitable. Morris A.H etal [18] tested many short text from GMAT reading comprehension and determined that the best length of the summary of the reported text was the 20% - 30% of the original length of text, and found that at the length the information of the summary is similar to the original text.

4 Experiments and Results

In this section, we report our experimental results.

4.1 Data Corpus

We select 100 news and articles from the web as our test text for generate the corresponding summary.

4.2 Evaluation Methods

There are two sorts of methods to evaluate the performance of text summarization: extrinsic evaluation and intrinsic evaluation (Mani & Bloedorn, 1999; Mani & Maybury, 1999). Extrinsic evaluation judges the quality of a summary based on how it affects other task(s), and intrinsic evaluation judges the quality of a summary based on the coverage between it and the manual summary. We chose intrinsic evaluation and used recall (R), precision (P) and F -measure (F) to judge the coverage between the manual and the machine-generated summaries. Assume that T is the manual summary and S is the machine-generated summary, the measurements are defined as Eq. (7) (Baeza-Yates & Ribeiro-Neto, 1999).

$$\mathrm{P}=\frac{|S\cap T|}{|S|},\quad \mathrm{R}=\frac{|S\cap T|}{|T|},\quad \mathrm{F}=\frac{2PR}{R+P} \tag{7}$$

4.3 Results

We select the TextRank and k-means clustering method for comparing to our method. Table 1 shows the average performance of each method.

Table 1. The average performance of each method

	R	P	F_measure
TextRank	0.377	0.392	0.376
k-means clustering	0.330	0.326	0.330
AP+LSA	0.665	0.489	0.548

From the table 1, we can see that our method could get better accuracy than the other two method.

5 Conclusions

We used affinity propagation to cluster chinese sentences for summary,and from the results we can see that this method could get a summary which has higher accuracy. LSA not only calculate the similary between two sentences effective and mitigates the problem of identifying synonymy and the problem with polysemy. Affinity propagation cluster need not define the number of the cluters in advance and it only need simple update rules and could get high precision.

There are also any problems in our proposed method, such as the definition of the preferences and the the damping factor λ .We will continue to research how to define these two parameters more effectively.

References

1. Sicui, W., Weijiang, L., Feng, W., Hui, D.: A Survey on Automatic Summarization. In: International Forum on Information Technology and Applications, IFITA (2010)
2. Frey, B.J., Dueck, D.: Clustering by passing messages between data points. Science 315, 972 (2007)
3. Mézard, M.: Where Are the Exemplars? Science 315, 972 (2007)

4. Eiler, J.M.: On the Origins of Granites. Science 315, 972 (2007)
5. Kummamuru, K., Lotlikar, R.: A hierarchical monothetic document clustering algorithm for summarization and browsing search results. In: Proceedings of the 13th International Conference on World Wide Web. ACM, New York (2004)
6. Ai, D., Yuchao, Z., Dezheng, Z.: Automatic text summarization based on latent semantic indexing. In: Artificial Life and Robotics (2010)
7. Edmundson, H.P.: New methods in automatic abstracting. Journal of the Association for Computing Machinery 16(2), 264–285 (1969)
8. Paice, C.D.: The automatic generation of literature abstracts: an approach based on the identification of self-indicating phrases. In: Proceedings of the Third Annual International ACM SIGIR Conference on Research and Development in Information Retrieval, pp. 172–191 (1981)
9. Kupiec, J., Pedersen, J., Chen, F.: A trainable document summarizer. In: Proceedings of the 18th Annual International ACM SIGIR Conference on Research and Development in Information Retrieval, pp. 68–73 (1995)
10. Barzilay, R., Elhadad, M.: Using lexical chains for text summarization. In: Advances in Automatic Text Summarization, pp. 111–121. MIT Press (1999)
11. Zajic, D., Dorr, B.J., Lin, J., Schwartz, R.: Multi-candidate reduction: sentence compression as a tool for document summarization tasks. Information Processing & Management 43(6), 1549–1570 (2007)
12. Meng, W., Chun-gui, L., Pei-he, T., Xiao-rong, W.: Chinese Automatic Summarization Based on Thematic Sentence Discovery. Computer Engineering 33(8), 180–181 (2007)
13. Mihalcea, R., Tarau, P.: TextRank: Bringing Order into Texts. In: Proceedings of the Conference on Empirical Methods in Natural Language Processing (EMNLP), Barcelona, Spain (2004)
14. Yeh, J.-Y., Ke, H.-R.: Text summarization using a trainable summarizer and latent semantic analysis. Information Processing & Management (2005)
15. Xie, S., Liu, Y., Hansen, J.H.L., Harabagiu, S.: Automatic Extractive Summarization on Meeting Corpus (2010)
16. Antiqueira, L., Oliveira Jr., O.N., da Fontoura Costa, L.: A complex network approach to text summarization. Information Sciences (2009)
17. Landauer, T.K., Foltz, P.W., Laham, D.: An introduction to latent semantic analysis. Discourse Processes (1998)
18. Morris, A.H., et al.: The effects and limitations of automated text condensing on reading comprehension performance. Information Systems Research (1992)

A Webpage Deletion Algorithm Based on Hierarchical Filtering

Xunxun Chen[1,*], Wei Wang[2], Dapeng Man[2], Sichang Xuan[2],

[1] National Computer Network Emergency Response Technical Team Coordination Center
Beijing China
13911226679@139.com
[2] School of Computer Science and Technology
Harbin Engineering University
Harbin China
{w_wei,mandapeng,xuanshichang}@hrbeu.edu.cn

Abstract. Duplicate webpages can affect the user experience of search engine. This paper proposed webpage deletion algorithm based on hierarchical filtering according to the features of duplicate webpage. The webpage feature extraction is divided into three layers, which are paragraphs, sentences and words. The webpage features are formed by layer filtering redundant information. In the sentence layer paragraph sentences are extracted according to the sentence semantics, while in the word layer the sentences are denoised filtering based on statistics of the part of speech in them. This algorithm improves the noise immunity and the original coverage of the feature extraction. The experiments show that the proposed method can accurately filter out duplicate webpage.

Keywords: webpage deletion, hierarchical filter, paragraph sentences, denoise.

1 Introduction

The internet developed rapidly for its openness and interoperability. There are a large number of duplicate webpages among some websites, especially in news and blog webpage. This case is formed by reproducing or copying the same content. These duplicate webpages will be indexed repeatedly by search engines, such as Google or Baidu, which waste the storage and lower the retrieve efficiency. Users may pay more time to read the same content, which is a bad practice. Therefore, how to get rid of these duplicated webpages is an urgent problem to be solved. Webpage deletion is an effective way to eliminate the defects.

Webpage deletion algorithm can be classified based on URL and text content and the latter is concerned in this paper. Generally, the algorithm based on text contents includes two steps, including feature extraction and similarity computation. Lots of related works are concerned[1][2][3]. There are two categories of methods to extract feature, they are rule based and weight based method. The former study on the position of feature extraction [4] or the part of speech[5], which rely on the stop

---

* This work is supported by National Natural Science Foundation of China(61170242)), Fundamental Research Funds for the Central Universities (HEUCF100601).

F.L. Wang et al. (Eds.): WISM 2012, LNCS 7529, pp. 551–557, 2012.

words selection. For the semantic of original text content is ignored completely, the feature matching process will be interfered by noise. The latter based on term weight calculated by some ways and terms with heavy weight are selected as keywords. TF-IDF[6] is an popular method to calculate term weight. Meng W. C.[7] and Matsuo Y.[8] propose keyword weight calculation method respectively, the former is concerned with term similarity and the latter consider the word co-occurrence probability. Other algorithms proposed by Jie Luo[9] and Jun Fang[10][11] introduce many text factors into weight calculation, but the features retrieved can't cover the original content for ignoring the text structure and word appearance order. The similarity calculation commonly utilize Jaccard Coefficient, angle cosine of feature vector or Longest Common Substring(LCS). The first two methods will deal with data with high dimension, and LCS limit the repeat form of feature string, which can't handle the partly duplicate contents.

As the analysis above, some issues of the current webpage deletion algorithm as follows: feature extraction granularity is too simple; noise immunity and adaptability in feature extraction are poor; similarity calculation can't consider text structure and semantic at same time with low accuracy rate; weight calculation can't retrieve keywords from partially duplicate content. A webpage deletion algorithm based on hierarchical filtering is proposed in this paper by combining rule extraction with weight extraction, which can extract feature richly and cover the original content uniformly.

2 Hierarchical Filtering Algorithm

The main idea of Hierarchical Filtering Algorithm(HFA) is to considers feature extraction as the process of the filtering information in different layers. Initial input of HFA is some webpage texts with some noisy information. During the feature exaction, the output of one layer is treated as the input of next layer, and features to eliminate duplicate content are retrieved finally. Three layers are defined in HFA, they are paragraphs, sentences and terms respectively. Paragraphs are retrieved by screening in webpage text content. For each paragraph, sentences are extracted by sentence screening strategy. For sentence sets, feature terms are generated term layer filtering. The input of each layer above is independent, which can't interfere with each other to avoid noise. The overall process of HFA is shown as Fig. 1.

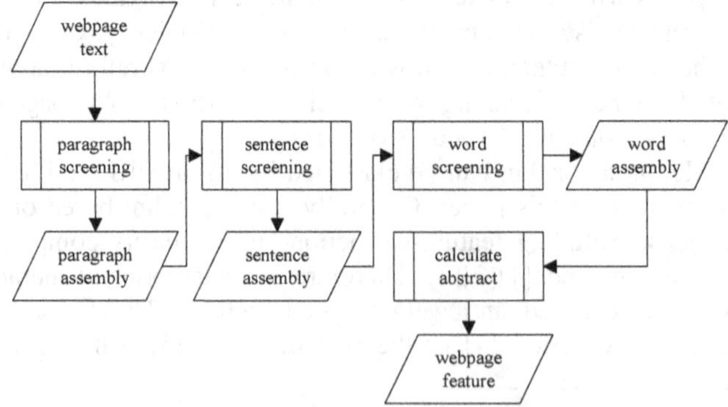

Fig. 1. The overall process of HFA

2.1 Paragraph Filtering

According to analysis of large number of webpage text content related to news and blog, two main interferences in paragraph level filtering are found, short paragraphs and news explanation. Short paragraphs include webpage explanation or copyright information. News explanation is generally located in the start or end of one paragraph, three forms as follows:

1 The paragraph is around by " ( )", such as "(Sina sports in Guangzhou Asian Games combined reported group of reporter Liang Xiaochen)".

2 The font is different from others in the same paragraph, such as bold, italic or different colors.

3 The paragraph is separated with others by spaces.

Paragraph filtering steps are as follows:

1. According to the webpage text content A, calculate the number of paragraph n and length of each paragraph l_i.

2. Calculate the average paragraph length l in text A with $l = \sum_{i=1}^{n} l_i / l$.

3. For each paragraph, calculation its weight $w_i = l_i / l$.

4. If w_i is greater than the threshold, the corresponding paragraph is reserved, go to step 5; otherwise, the paragraph is abandoned.

5. Remove the description information in the paragraph reserved, go to step 3.

2.2 Sentence Filtering

The input of the sentence filtering is the paragraph collections reserved. Sentence filtering process includes two steps, key words extraction and sentence extraction

1. Keywords extraction

Chinese terms are segmented and the average frequency of each terms, *freq*, are calculated in each paragraph. Set $p = freq \times 0.7$, $q = freq \times 1.3$ and reserve those terms as candidate keywords whose term frequency is between p and q. Select verbs and adverbs from the candidate keywords and calculate their weight as formula (1).

$$w_i = a \times w_{i1} + (1-a) \times w_{i2} \tag{1}$$

Formula (1) is composed of two parts: word frequency w_{i1} and semantic w_{i2}. w_{i1} is derived by the traditional TF-IDF, but the strategy is improved, that is when the both terms and their speech are same, they can be considered as same. w_{i2} is defined as whether the term appear in title, w_{i2} is determined as (2)

$$w_{i2} = \begin{cases} 1 & appear\,in\,title \\ 0 & never\,appear\,in\,title \end{cases} \tag{2}$$

The n terms with largest weight are extracted as the paragraph key words. Semantic weight is defined as whether or not is in the title.

In formula (1), the value of a is between 0 and 1, Here its value is 0.8.

2. Sentence Extraction
Sentence extraction process refers to long sentence extraction[12]which is improved in this paper. If the number of sentences in a paragraph is too small, no sentence will be extracted. In order to avoid this case, different extraction threshold is used for different number of sentences in different paragraph. Other feature weights of sentence are also referred. Sentence extraction process as follow:

(1) Calculate the number of sentence in a paragraph. if the number is greater than two, then go to step 2, otherwise the sentences are extracted in their original order in the paragraph, and go to step 5.

(2) For those sentences whose length is greater than threshold, calculate their weight according to their own features.

(3) For a paragraph, sort the sentences according to their weights.

(4) Select 1/3 of the total number of sentences which has the larger weight and output them in their original order, go to step (5).

(5) The end.

In step 2, sentence weight is calculated as formula (3)[13]

$$W_i = \sum_{j=1}^{n} a_j * w_{ij} \tag{3}$$

In formula (3), n is the number of the weight factor, here n is four, including sentence position in paragraph, sentence classification, whether the sentence contains contain keywords and sentence length; $0<a_j<1$; w_{ij} is the weight of the factors of sentence i and $\sum_{j=1}^{n} a_j = 1$. The weight values of the factors of sentence are show in Table 1.

Table 1. The weight values of the factors of sentence

Sentence attribute	Attribute value	weight
Position	The begin or the end of a segment	0.5
	The begin of the paper	1
	others	0
Sentence class	The statement / imperative sentences	1
	Interrogative sentence	0
	Exclamatory sentence	0.5
If it contains the key words	Yes	W(sum of the keywords weight contained in the sentence)
	No	0
Sentence length	m	m/max(length of the sentence)

2.3 Term Filtering

The input of term filtering is the complete sentences with term segmentation, and the output is the keywords screened. Term filtering can eliminate the noise in sentence and extract the core terms. In order to reduce the computation overhead, some candidate terms are selected from sentences based on part of speech, further filtering will be done for the candidate words

Candidate term set definition:

$$set_{can}(w)=\{w \mid w_{pos} \in S\} \text{ , } S=\{a \text{ , } an \text{ , } d \text{ , } n \text{ , } nr \text{ , } ng \text{ , } ns \text{ , } v \text{ , } vg \text{ , } vn \text{ , } t\}$$

Candidate term filtering rules:

1. adverb/ adjective+(verb/noun): these adverb/adjective are eliminated.
2. in addition to adverb+verb and adverb+Auxiliary word, other adverb are eliminated.
3. verb and gerund appear at the end of sentence:gerund are eliminated.
4. verb+"front/middle/behind":verb are eliminated.
5. preposition+noun: noun are eliminated.

3 Experimental Results and Analysis

3.1 Experiment Data

A web crawler called Larbin is used to download lots of webpages, and a number of duplicate webpages are saved by utilize Google and Baidu. These webpages acquired above constitute the basic data set.

3.2 Experiment and Result Analysis

The following experiments verify that HFA is better than Long Sentence based Algorithm(LSA)[12]. The basic data set is divided into group3 according to the number of terms in webpages: G_1, smaller than 4000 bytes; G_2, between 2000 and 4000 bytes; G_3, smaller than 2000 bytes. The proposed algorithm and LSA are executed with G_1, G_2 and G_3 to obtain their precision, recall and storage overhead comparison. The value of paragraph ratio is set 0.55. The comparison results are shown in Fig.2, Fig.3, and Fig.4.

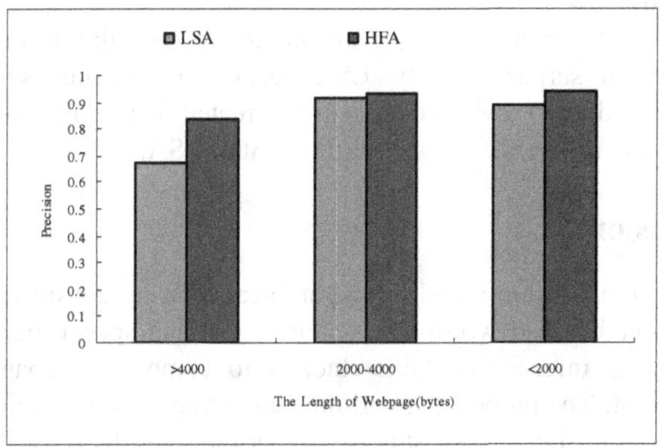

Fig. 2. Precision comparison

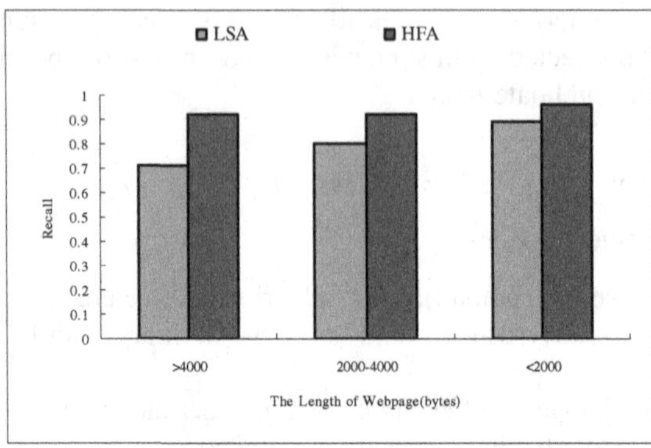

Fig. 3. Recall comparison

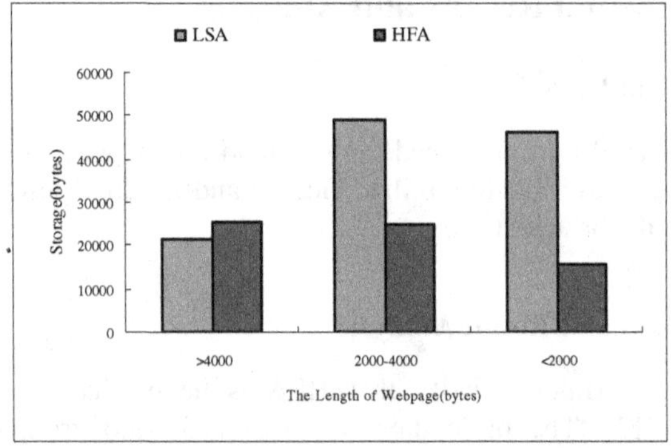

Fig. 4. Storage overhead comparison

From Fig. 2 and Fig. 3 we can see that the precision and recall of HFA are better than that of LSA. Especially for G_1, the precision of LSA is 0.675 and 0.885 for HFA. We also can see that for G_1, the precision of both of algorithms are not good for fewer sentences are extracted.

Fig.4 shows that for the same precision and recall, the storage overhead of proposed algorithm surpass that of LSA because no surplus sentences will be extracted, which induct to fewer nodes will be created. Especially for G_3, the storage overhead of our algorithm is only one third of that of LSA.

4 Conclusion

Webpage deletion algorithm based on hierarchical filtering considers the association of paragraph, sentence and word. The feature word and speech pairs are obtain by screening webpage information hierarchically to combine webpage structure and information weight. The proposed algorithm can extract features which can cover the whole webpage uniformly and resist noise. Experiment results show that the proposed algorithm can delete duplicate webpage with better performance.

References

1. Ukkonen, E.: Constructing Suffix Trees On-Line in Linear Time. In: Proceedings of the IFIP 12th World Computer Congress on Algorithms, Software, Architecture, pp. 484–492 (1992)
2. Broder, A.: Identifying and Filtering Near-Duplicate Documents. In: Giancarlo, R., Sankoff, D. (eds.) CPM 2000. LNCS, vol. 1848, pp. 1–10. Springer, Heidelberg (2000)
3. Zhang, G., Liu, T., Zheng, S.F.: Fast Deletion Algorithm for Large Scale Duplicated Web Pages. The First Student Computational Linguistics Seminar 2002 (2002)
4. Elhadi, M., Al-Tobi, A.: Webpage Duplicate Detection Using Combined POS and Sequence Alignment Algorithm. In: 2009 World Congress on Computer Science and Information Engineering, pp. 630–634 (2009)
5. Theobald, M., Siddharth, J., Paecke, A.: SpotSigs: Robust and Efficient Near Duplicate Detection in Large Web Collections. In: Proceeding of the 31st Annual International ACM SIGIR Conference on Resesrch and Development in Information Retrieval, pp. 563–570 (2008)
6. Chowdhury, A., Frieder, O., Grossman, D., et al.: Collection Statistics for Fast Duplicate Document Detection. ACM Transactions on Information System 20(2), 171–191 (2002)
7. Meng, W.C., Liu, L.C., Dai, T.: A Modified Approach to Keyword Extraction Based on Word-similarity. In: Proceedings of the Sixteenth International Florida Artificial Intelligence Research Society Conference, pp. 392–396 (2003)
8. Matsuo, Y., Ishizuka, M.: Keyword Extraction From a Single Document Using Word Co-occurrence Statistical Information. International Journal on Artificial Intelligence Tools 13(1), 157–169 (2004)
9. Luo, J., Chen, L.: Research on Fast Text Classifier Based on New Keywords Extraction Method. Application Research of Computers 4, 32–34 (2006)
10. Cheng, L.L., He, P.L., Sun, Y.H.: Study on Chinese Keyword Extraction Algorithm Based on Naive Bayes Model. Journal of Computer Applications 25(12), 2780–2782 (2005)
11. Fang, J., Guo, L., Wang, X.D.: Semantically Improved Automatic Keyphrase Extraction. Computer Science 35(6), 148–150 (2008)
12. Liu, S.W., Zhang, Y., Xia, Y.M.: Webpage Deletion Algorithm Based on HTML Mark and Long Sentence Retrieve. Microcomputer Applications 25(8), 30–32 (2009)
13. Fan, Y., Zheng, J.H.: Research on Elimination of Similar Web Pages Computer. Engineering and Application 45(12), 141–143 (2009)

Research in Keyword Extraction

Bing Wu and Pingping Chen

School of Economics and Management, Tongji University
200092 Shanghai, China

Abstract. This study is a productivity review on the literature gleaned from science citation index expanded (SCI-EXPANED) database on web of science, concerning knowledge transfer research with network. The result indicates that the number of literature productions on this topic mainly distributes in recent 10 years, reaching climax of 4 in 2010 and then followed by 2008 of 3. The main research development country is USA, accounting for 40% meanwhile China is on the second place. And from the analysis of institutions, UNIV TORONOTO and UNIV WASHINGTON are in parallel top one for almost one fourth of the total. As for the analyze results of source title, it has close relation with computer science and management. As a result, expert system, management and methods become the three main points of research focus.

Keywords: knowledge transfer, keyword, computer science, management, expert system.

1 Introduction

Keywords can be considered as brief summaries of a text. Although a summary of a text is capable of providing more information about the text than keywords of the text, the summary may not be suitable for some applications due to the complex structure of sentences. Keywords are not replacements for summarization but alternative summary representations that could be consumed by other applications more easily. Since they are concise representations of the underlying text, it is possible to use them in different applications such as indexing in search engines or text categorization.

With the currently growing interest in the Semantic Web, keywords/metadata extraction is coming to play an increasingly important role. Keywords extraction from documents is a complex task in natural languages processing. Ideally this task concerns sophisticated semantic analysis. However, the complexity of the problem makes current semantic analysis techniques insufficient. Automatic keyword extraction has attracted much attention due to the many applications in information retrieval, text mining, information processing, etc. To extract desired keywords from documents automatically, other than lexical relations among words or sentences, the content-based measures are critical at the same time.

The main objective of this paper is to analyze keyword extraction research on Science Citation Index Expanded (SCI-E) database from web of science. As a result the related work in this area can be thoroughly explored. The rest of this article is

F.L. Wang et al. (Eds.): WISM 2012, LNCS 7529, pp. 558–563, 2012.

organized as follows: Section 2 surveys the keyword extraction research. Section 3 briefly introduces research focuses of them. Last, we conclude our article in Section 4.

2 Analyze Result of Keyword Extraction Research

Keywords are a fundamental part of information retrieval (IR) and as such they have been studied extensively. They are used for everything from searching to describing a document. A Keyword extraction algorithm can be defined as a combination of a keyword representation and a selection/weighting scheme. The most common selection/weighting schemes are based on collection statistics or using supervised machine learning algorithms. In these cases, keywords can, typically, only be extracted from documents that belong to a collection or using a large amount of annotated training data. Unfortunately, a great portion of existing documents available today does not have keywords available for them. Keyword extraction is highly related to automated text summarization. In text summarization, most indicative sentences are extracted to represent the text. In keyword extraction, most indicative keywords are extracted to represent the text.

According to Science Citation Index Expanded and Social Sciences Citation Index Database in web of science, only 29 records were found in related discipline when "keyword extraction" as search title.

2.1 Country/Territory Analyze

According to country/territory, analyze results can be shown as table 1. People R China accounts for 34.483%, ranking in the top one, following that is Japan accounting for 17.241%.

Table 1. Analyze Results of Country/Territory

Country/Territory	Record Count	% of 29
People R China	10	34.483%
Japan	5	17.241%
Canada	3	10.345%
South Korea	2	6.897%

2.2 Publication Year

Table 2. Analyze Results of Publication Year

Publication Year	Record Count	% of 29
2006	6	20.690%
2007	4	13.793%
2003	3	10.345%
2011	3	10.345%
2001	2	6.897%

Almost all the papers were published mainly in last ten years. As shown in table 2, there are 6 papers in publication year 2006, twice the number of that in year 2003 and 2011. Publication year 2007 is in the second place with 4 records, twice that of the rest publication year. And accordingly citations per year are presented in Fig 1, citations in 2009 reach climax.

2.3 Institutions

As far as institutions are concerned, as shown in table 3, TSING HUA UNIV ranks top one, following by BEIJING UNIV POSTS TELECOMMUN, UNIV TEXAS DASSLAS AND UNIV TOKUSHIMA, accounting for 6.897% respectively.

Table 3. Analyze Results of Institutions

Institutions	Record Count	% of 29
TSING HUA UNIV	3	10.345%
BEIJING UNIV POSTS TELECOMMUN	2	6.897%

2.4 Source Title

There are four main sources, lecture notes in computer sciences is in the priority place, accounting for 27.586%. Following that are lecture notes in artificial intelligence. The four sources have the half number of all papers, as shown in Table 4.

Table 4. Analyze Results of Source title

Source Titles	Record Count	% of 29
LECTURE NOTES IN COMPUTER SCIENCEs	8	27.586%
LECTURE NOTES IN ARTIFICIAL INTELLIGENCE	3	10.345%
COMPUTATIONAL INTELLIGENCE	2	6.897%
INFORMATION PROCESSING MANAGEMENT	2	6.897%

2.5 Subject Areas

As shown in Table 5, subject areas are mainly in computer science, accounting for 79.310%.

Table 5. Analyze Results of Subject Areas

Subject Areas	Record Count	% of 15
Computer Science	23	79.310%
Engineering	3	10.345%
Mathematical Computational Biology	3	10.345%
Biochemistry Molecular Biology	2	6.897%
Biotechnology Applied Microbiology	2	6.897%

3 Focuses of Keywords Extraction Research

3.1 Applications of Keyword Extraction

Interest Topic. Messages posted by micro blogging users, are usually noisy and full of new words, which is a challenge for keyword extraction. The mining of the interests of micro blogging users via keyword extraction have been proposed by Liu Zhiyuan, et al (2012) [1]. There are still several open problems that can be further investigated, including other sophisticated word alignment models can be explored for keyword extraction, to build a more practical system and to build recommender systems.

Contextual Advertising. Keyword extraction is a key step in this kind of advertising, through which appropriate advertising keywords are extracted from Web pages so that corresponding ads can be triggered. Liu Jianyi et al (2010) describes a system that learns how to extract keywords from web pages for advertisement targeting [6].

Meeting Transcript. As meeting transcripts is significantly different from written text or other speech domains, a supervised framework for extracting keywords from the meeting transcripts is investigated by Liu Fei, et al (2011) [5]. A variety of features are proposed beyond the traditional frequency and position-based features, including term specificity features, decision-making sentences-related features, prosodic prominence score, and a set of features extracted from the system generated summaries. Different reasons for errors in the current keyword extraction systems point out interesting directions for future studies.

Press Release. Samejima Masaki, et al.(2010) propose a method of automatically extracting numbers for particular objects related to product trends and the corresponding time stamps from press releases on a company's web pages, and graphing the extracted time series numerical data [7]. Focusing on the fact that the same words are used for the same objects in the press releases from one company, words included in the topics for extraction can be regarded as keywords.

Scientific Text. Domain keywords of scientific text play a primary role in scientific text classifying, clustering and personalized services. Luo Xiangfeng et al (2008) propose a new method called TDDF to extract domain keywords from multi-texts [8]. The quantity and the quality of domain keywords can be flexibly controlled by the rate of word common possession. TDDF is only limited to the extraction of domain keywords from multi-texts that belong to one domain, which is not suitable for the extraction of textual keywords from single text.

Internet News Articles. The importance of extracting keywords without a document collection has been gradually increasing due to the Internet. A keyword extraction algorithm designed with news in mind that requires neither a document collection nor training data is presented by Bracewell David B., et al (2008) [9]. It uses noun phrases as its keyword representation and takes in document statistics to derive its weighting scheme.

3.2 Keyword Extraction Methods

Supervised Classification. High Relevance Keyword Extraction (HRKE) facility is introduced to Bayesian text classification to perform feature/keyword extraction during the classifying stage, without needing extensive pre-classification processes. In order to perform the task of keyword extraction, HRKE facility uses the posterior probability value of keywords within a specific category associated with text document [2]. The experimental results show that HRKE facility is able to ensure promising classification performance for Bayesian classifier while dealing with different text classification domains of varying characteristics. This method guarantees an effective and efficient Bayesian text classifier which is able to handle different domains of varying characteristics, with high accuracy while maintaining the simplicity and low cost processes of the conventional Bayesian classification approach.

Ranking. A method for ranking the words in texts by means of nonextensive statistical mechanics is put forth by Mehri Ali and Darooneh Amir H (2011) [3]. The nonextensivity measure is a convenient tool used to classify the correlation range between word-type occurrences in a text. Therefore, a q measure is applied to order the words of a given text. It is clear that q values for most relevant words are greater than their values for irrelevant words. Consequently, this method can be applied systematically to keyword extraction and summarization processes.

Supervised Learning. From the perspective of information retrieval, multiple content-based measures for the weights, contributions to document classifications and coverage of words are proposed by Yue Kun et al. (2011) to facilitate effective and intuitively-interpretable methods for automatic keyword extraction [4]. The measure based on the Laplace's law to determine the weight of a word in documents and the measure based on the concept of average mutual information to determine the influence of a word on document classifications were given. Furthermore, the concept of semantic coacervation degree as the measure for the coverage of words on documents was proposed, and a branch and bound algorithm to find the minimal set of keywords that maximize the coverage was given.

The problem of automatic extraction of keywords from documents can be treated as a supervised learning task. A lexical chain holds a set of semantically related words of a text and it can be said that a lexical chain represents the semantic content of a portion of the text. Although lexical chains have been extensively used in text summarization, their usage for keyword extraction problem has not been fully investigated. A keyword extraction technique that uses lexical chains is described by Ercan Gonenc and Cicekli Yas (2007) [10], and encouraging results are obtained to investigate the benefits of using lexical chain features in keyword extraction.

Machine learning methods can support the initial phases of keywords extraction and can thus improve the input to further semantic analysis phases. A machine learning-based keywords extraction for given documents domain is proposed [11], namely scientific literature. More specifically, the least square support vector machine is used as a machine learning method. The proposed method takes the advantages of machine learning techniques and moves the complexity of the task to the process of learning from appropriate samples obtained within a domain. Preliminary experiments show that the proposed method is capable to extract keywords from the domain of scientific literature with promising results.

4 Conclusions

Keywords can be considered as condensed versions of documents and short forms of their summaries. Keywords enable readers to decide whether a document is relevant for them or not. They can also be used as low cost measures of similarity between documents. The written word is one of the important manifestations of natural language.

Keywords are a fundamental part of information retrieval (IR) and as such they have been studied extensively. Various universal features characterize text as a complex system. The existence of a long-range correlation in the spatial distribution of terms in a given text is another universal feature. The methods borrowed from statistical mechanics are the best candidates for finding universal regularities in complex systems.

Acknowledgement. This work is supported by the National Nature Science Foundation of China (No. 71071117).

References

1. Liu, Z., Chen, X., Sun, M.: Mining the interests of Chinese microbloggers via keyword extraction. Frontiers of Computer Science in China 6(s1), 76–87 (2012)
2. Hong, L.L., et al.: High Relevance Keyword Extraction facility for Bayesian text classification on different domains of varying characteristic. Expert Systems with Applications 39(1), 1147–1155 (2012)
3. Ali, M., Darooneh Amir, H.: Keyword extraction by nonextensivity measure. Physical Review E 83(5) (2011)
4. Yue, K., Liu, W.-Y., Zhou, L.-P.: Automatic keyword extraction from documents based on multiple content-based measures. Computer Systems Science and Engineering 26(2), 133–145 (2011)
5. Liu, F., Liu, F., Liu, Y.: A Supervised Framework for Keyword Extraction From Meeting Transcripts. IEEE Transactions on Audio Speech and Language Processing 19(3), 538–548 (2011)
6. Liu, J., Wang, C., Yao, W.: Keyword Extraction for Contextual Advertising. China Communications 7(4), 51–57 (2010)
7. Masaki, S., et al.: An Extraction Method of Time-Series Numerical Data from Press Releases by Using Co-Occurrence Conditions of Numbers and Time Stamps Related to Target Business Keywords. Electronics and Communications in Japan 93(3), 61–69 (2010)
8. Luo, X., et al.: Experimental study on the extraction and distribution of textual domain keywords. Concurrency and Computation-Practice & Experience 20(16), 1917–1932 (2008)
9. Bracewell David, B., et al.: Single document keyword extraction for Internet news articles. International Journal of Innovative Computing Information and Control 4(4), 905–913 (2008)
10. Gonenc, E., Yas, C.: Using lexical chains for keyword extraction. Information Processing & Management 43(6), 1705–1714 (2007)
11. Wu, C., et al.: Machine learning-based keywords extraction for scientific literature. Journal of Universal Computer Science 13(10), 1471–1483 (2007)

Tuple Refinement Method Based on Relationship Keyword Extension

Xiaoling Yang, Jing Yang[2,*], and Chao Chen[3]

Department of Computer Science and Technology, East China Normal University
200241 Shanghai, China
{51101201023,51101201029}@ecnu.cn
jyang@cs.ecnu.edu.cn,

Abstract. Entity relation extraction is mainly focused on researching extraction approaches and improving precision of the extraction results. Although many efforts have been made on this field, there still exist some problems. In order to improve the performance of extracting entity relation, we propose a tuple refinement method based on relationship keyword extension. Firstly, we utilize the diversity of relationships to extend relationship keywords, and then, use the redundancy of network information to extract the second entity based on the principle of proximity and the predefined entity type. Under open web environment, we take four relationships in the experiments and adopt bootstrapping algorithm to acquire the initial tuple set. Three tuple refinement methods are compared: refinement method with threshold set, refinement method with relation extension and refinement method without relation extension. The average F-scores of the experimental results show the proposed method can effectively improve the performance of entity relation extraction.

Keywords: Relation Extraction, Tuple Refinement, Relationship Keyword Extension, Proximity Principle.

1 Introduction

With the rapid development of the Internet, extracting structured data from the vast amounts unstructured data called as information extraction (IE) has been the present research hotspot. The major research directions of IE [10, 11] are named entity recognition, anaphora resolution and relation extraction (RE). Since RE can be used in many applications such as semantic web, automatic answering and social relationship network, it is very important in the IE.

Generally speaking, RE has two methods based on knowledge engineering and machine learning. Knowledge engineering methods require a lot of professional knowledge, which waste a lot of time and manpower. Meanwhile they lack the area portability. Machine learning methods need to apply a lot of deep natural language processing (NLP) technologies which inevitably produce a lot of noise. SVM methods

* Corresponding author.

F.L. Wang et al. (Eds.): WISM 2012, LNCS 7529, pp. 564–571, 2012.

use syntactic analysis and semantic analysis [5, 6, 7] to construct feature vectors, kernel approaches [4, 12] introduce dependency trees to build the shortest tree kernels, while weakly supervise bootstrapping methods require the entity recognition technology. Machine learning methods are more extensible and applicable than knowledge engineering methods in general.

With the continuous improvement of extraction approaches, RE needs more and more support of language technology platform. However deep NLP technologies face a lot of problems and bootstrapping methods have the cyclic dependence problem. Both of them lower precision of RE.

As the complexity of natural languages, a relationship has different expressions. It makes RE become a big challenge while it makes refining extracted entity tuples become possible. In this paper, we present a tuple refinement method according to the diversity relationships and the redundancy of web information to further refine the extracted tuples, thereby improving the precision of entity relation extraction.

The remainder of this paper is organized as follows. Section 2 introduces the related work of tuple refinement methods and section 3 introduces the tuple refinement method based on relation keyword extension. The experimental procedure and analysis of the experimental results are described in section 4. A discussion of the proposed method in the paper and future work are given in section 5.

2 Related Work

Recently, researchers make so many efforts on the study of searching extract methods, but ignoring tuple refinement makes the precision of RE low in general. So finding effective and useful refinement methods is very meaningful.

Now, tuple refinement methods are mainly concentrated on the credibility evaluation. DIPRE [2] uses bootstrapping methods to generate new tuples. Since it lacks a tuple refinement method, the precision is very low. Baseline tuple refinement method [3] simply uses co-occurrence statistics, which also have low precision. Snowball [1] designs a tuple credibility evaluation method (see Formulas 1 and 2) without human intervention. It only preserves the most credible patterns and tuples whose credibility value exceeds the threshold T into the next iteration. Since the credibility evaluation depends only on one key attribute of the relationship, so it has small scalability. HIT [8] designs an automatic credibility evaluation method (see Formula 3 and 4). Firstly it uses the initial seeds to extract multiple iteration patterns; secondly, it uses the initial seeds to instance each pattern; thirdly, it puts each instance pattern into search engine for extracting the matched estimated number, then ranks the tuples according to the estimated number of the search engine returned results; finally, it takes the first N patterns as the evaluate pattern set called as EvalPSet. HIT [8] just uses the seed tuples to evaluate, and it doesn't take into account the diversity expressions of tuples and patterns, so the precision is not high.

$$\mathrm{Conf(P)} = \mathrm{P}_{\cdot\mathrm{Positive}} / (\mathrm{P}_{\cdot\mathrm{Positve}} + \mathrm{P}_{\cdot\mathrm{negative}}) \ . \tag{1}$$

$$\mathrm{Conf(T)} = 1 - \prod_{i=0}^{|P|} (1 - (\mathrm{Conf}(P_i) \cdot \mathrm{Match}(C_i, P_i))) \ . \tag{2}$$

$$C(T) = \log|\Sigma_{i=1}^{n}(T + P_i) + \lambda| / \log|\Sigma_{i=1}^{n}\text{Hits}(T) + \gamma| \times \Sigma_{i=1}^{n}\delta_{P_i} \, . \quad (3)$$

While

$$\delta_{p_i} = \begin{cases} 1, & \text{if Hits}(T + p_i) > 0 \\ 0, & \text{if Hits}(T + p_i) = 0 \end{cases} . \quad (4)$$

In order to improve the performance of RE, we propose a new tuple refinement method based on extension of relationship keyword. Since an entity relation has different expressions, so we can use a different keyword to express a same relation. For example, the words of 夫妇 and 夫妻 can be used to express the relation of couple. In order to use this feature, we construct an extended relational keyword list according to the relation diversity. And then, we use the entity and keyword co-occurrence features to extract the second entity with the predefined type.

The relation keyword extended method has two advantages. First it is independent of extraction methods. It just has been given initial evaluated tuples, while the HIT [8] relies on initial seed tuples and seed patterns, and snowball [1] must use the last patterns and tuples in each iteration. Second it can be used after any per iteration of RE, at the same time it doesn't use deep NLP technologies to avoid noises.

In experiments of this paper, the initial tuples are acquired by HIT [8] method without refinement. HIT [8] proposed a bootstrapping method of automatically obtaining entity relation tuples based on web mining. According to cutting rules, it uses seed tuples and relationship keywords to obtain multiple context patterns, and puts the patterns into web to get more tuples in each iteration.

3 Proposed Tuple Refinement Method

In this paper, a novel tuple refinement method based on relationship keyword extension [9] is proposed. Firstly it extends the relationship by an extended relationship keyword list, followed by using the entity and keyword co-occurrence features to extract the second entity. Finally, it uses the new extracted tuples to evaluate the candidate evaluated tuple for improving the relation extraction results.

3.1 Problem Definition

Evaluated Tuple: ET(*e1*, *e2*, *k*, *r_type*), where *e1* represents first entity, *e2* represents second entity, *k* represents relationship keyword, and *r_type* represents entity relationship type.

Initial Tuple: IT(*e1*, *k*, *r_type*), where *e1* represents first entity, *k* represents relation keyword, *and r_type* represents entity relationship type.

Extended Relation List: KeywordsList(*k*, *ki*), where *k* represents relationship keyword, and *ki* represents the extended relationship keyword.

Co-occurrence Sentence Set: CSS(*e1*, *ki*), where *e1* represents first entity, and *ki* represents the extended relationship keyword.

Extended Tuple: ET(*e1*, *ki*, *r_type*), where *e1* represents first entity, *ki* represents the extended relationship keyword, and *r_type* represents entity relationship type.

As a relationship has various expressions, we build a relational keyword extended list to extend relation, for example, the relationship of couple could be extended to (夫妻、夫妇、伉俪、伴侣、两口子).

ET gets from the HIT [8] without tuple refinement. IT comes from ET. KeywordsList is firstly acquired by the existing knowledge base of TongYiCi CiLin and then artificial extended. For example, the relationship of couple is firstly extended to（夫妻、夫妇、伉俪、终身伴侣、小两口、老两口、两口子、家室）by TongYiCi CiLin, then we remove manually some of the uncommonly used words and contradictory words, finally the relationship of couple can be extended to (夫妻、夫妇、终身伴侣、两口子、伉俪).

3.2 Tuple Refinement Method Based on Extended Relationship Process

Refinement method detailed process:

1. Extend IT into a set of ET by KeywrodsList, for instance, (姚明，夫妻，人物与人物) is extended to {(姚明，夫妻，人物与人物), (姚明，夫妻，人物与人物), …}.

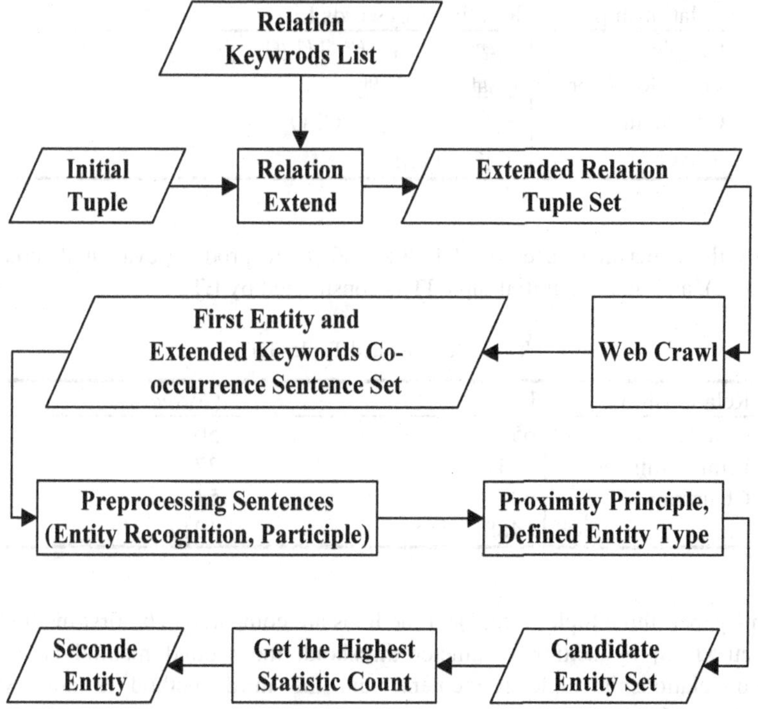

Fig. 1. The process of tuple refinement method based on relation keyword extension.

2. Crawl the co-occurrence sentences CSS of *e1* and *ki*.
3. Take word segmentation and POS tagging of CSS.
4. Extract candidate words from CSS according to the predefined relation type *r_type* and the principle of proximity
5. Take the largest statistical value of the candidate words as *e2*.
6. Evaluate ET by the new acquired tuple (*e1*, *e2*): remove those different and unreturned tuples and then same tuples are reserved.

Our method makes use of the relationship's diversity and the redundancy of the web information, according to extended relationship grabs a lot of web information, and then uses proximity principle to extract second entity after simple pretreatment.

4 Experiment and Analysis

4.1 Experiment Set

Four common relationships that are couple (夫妻), chairman of the company (董事长), mayor (市长) and prime minister (总理) are selected. The relationship keywords list is shown in table 1.

Table 1. Relation Keywords List

Relationship	Relation Keywords List
Couple	夫妻、夫妇、终身伴侣、两口子、伉俪
Prime Minister	总理、 代总理
Chairman	董事长、老总、CEO
Mayor	市长、代市长

Firstly the experiment uses the HIT method [8] to produce evaluated tuples ETs (see table 2), and then the initial tuple IT is constructed by ET.

Table 2. Evaluated Tuples ETs

Relationship	ET	Correct ET
Couple	65	50
Prime Minister	114	27
Chairman	63	51
Mayor	1488	1214

In this paper, three tuple refinement methods are compared. The first method is the HIT method [8] by setting parameter threshold, the second method is our non-extended relationship tuple refinement, and the third method is our extended relationship tuple refinement.

4.2 Assessment Methods

As different relationship extract different number of tuples, we evaluate all of the tuples less than 100 and randomly select 100 tuples from more than 100 tuples.

In order to assess the experiment result, we use three common evaluation indexes: precision P, recall R and F-score. Here, CRT represents the correct refined tuples, ART represents the all refined tuples, and CTD represents the correct evaluated tuples (ETs).

$$P = CRT/ART . \quad (5)$$

$$R = CRT/CTR . \quad (6)$$

$$F = 2*P*R/(P+R) . \quad (7)$$

4.3 Experimental Results and Analysis

A method: HIT method [8] of setting parameter threshold, B method: tuple refinement method without relational keyword extending, C method: tuple refinement with relational keyword extending. The experiment results are shown in table 3.

Table 3. Experiment Results

Relationship	Method	P	R	F
Couple	A	79.63%	86.00%	82.69%
	B	82.00%	82.00%	82.00%
	C	87.23%	82.00%	84.53%
Prime Minister	A	26.74%	85.18%	40.70%
	B	42.85%	77.77%	55.26%
	C	45.60%	77.70%	57.50%
Chairman	A	85.10%	90.10%	87.60%
	B	89.70%	86.20%	88.00%
	C	90.00%	88.20%	89.10%
Mayor	A	80.52%	83.44%	81.19%
	B	93.49%	70.00%	80.05%
	C	93.82%	75.04%	83.38%

To fully evaluate the experimental results, we compare the average F-score, the average F-score of A method is 73.04%, the average F-score of B method is 76.32%, and the average F-score of C method is 78.62%. According to the comparison of A and B, we can find our non-extended method is valid, meanwhile we can find our extended method is more effective by comparing B and C methods.

5 Conclusion

In this paper, we use the redundancy of web information and the relationship's diversity to design a tuple refinement method. We firstly construct a keyword list based on the diversity of relation to extend the entity relation keyword, crawl the co-occurrence of entity and relation keywords by the redundant web information, then use the principle of proximity and the predefined entity type to extract entities as the candidate entities, and finally save the highest statistical value of candidate entities as the second entity. The proposed method only uses shallow language processing technologies to avoid noise. The experimental results show our method is relatively effective.

However, in our proposed method, relationship extension still needs some manual participation, so our future work will focus on extending relationship automatically.

Acknowledgments. This material is based upon work supported by the Opening Project of Shanghai Key Laboratory of Integrate Administration Technologies for Information Security and the Shanghai Science and Technology commission Foundation (No. 10dz1500103, No. 11530700300 and No. 11511504000). We also thank ICA members for much discussion.

References

1. Eugene, A., Luis, C.: Snowball: Extracting relations from large plain-text collections. In: 5th ACM International Conference on Digital Libraries, New York, pp. 85–94 (2000)
2. Brin, S.: Extracting Patterns and Relations from the World Wide Web. In: Atzeni, P., Mendelzon, A.O., Mecca, G. (eds.) WebDB 1998. LNCS, vol. 1590, pp. 172–183. Springer, Heidelberg (1999)
3. Eugene, A.: Confidence Estimation Methods for Partially Supervised Relation Extraction. In: Proc. of SIAM Intl. Conf. on Data Mining, SDM 2006 (2006)
4. Frank, R., Hannes, K., Gerhard, P.: Semantic Relation Extraction With Kernels Over Typed Dependency Trees. In: The 16th ACM SIGKDD International Conference on Knowledge Discovery and Data Mining, New York, pp. 773–782 (2010)
5. Yuan, F.K., Chen-Chuan, C.: Searching Patterns for Relation Extraction over the Web: Rediscovering the Pattern-Relation Duality. In: The 4th ACM International Conference on Web Search and Data Mining, New York, pp. 825–834 (2011)
6. Jeoghee, Y., Neel, S.: Mining the web for acronyms using the duality of patterns and relations. In: The 2nd International Workshop on Web Information and Data Management, New York, pp. 48–52 (1999)
7. Danushka, B., Yutaka, M., Mitsuru, L.: Relational Duality: Unsupervised Extraction of Semantic Relations between Entities on the Web. In: The 2nd International Conference on World Wide Web, New York, pp. 151–160 (2010)
8. Li, W.G., Liu, T., Li, S.: Automated Entity Relation Tuple Extraction Using Web Mining. Acta Electronica Sinica (11), 2111–2116 (2007)
9. Yao, C.L., Di, N.: A relation extraction method based on large-scale characters web social. Pattern Recognition and Artificial Intelligence (6), 740–744 (2007)

10. Grishman, R.: Information Extraction: Techniques and Challenges. In: Pazienza, M.T. (ed.) SCIE 1997. LNCS, vol. 1299, pp. 10–27. Springer, Heidelberg (1997)
11. Aron, C., Andrew, M.: Confidence estimation for information extraction. In: 2004 HLT-NAACL, pp. 109–112 (2004)
12. Shubin, Z., Ralph, G.: Extracting relations with integrated information using kernel methods. In: 43rd Annual Meeting on ACL, pp. 419–426 (2005)

An Agent Based Intelligent Meta Search Engine

Qingshan Li and Yingcheng Sun

Software Engineering Institute, Xidian University, 710071 Xi'an, China
qshli@mail.xidian.edu.cn

Abstract. Addressing the problems that available search engines seldom consider the personalized needs of users with low precision rate and the discrete retrieval results, an Agent-based intelligent meta-search engine model is proposed. Agent technology is used, which makes the system more intelligent. In order to achieve personalized retrieval analysis, the model uses a four-tuple user interest model and improved text classification model. A retrieval result synthesis strategy is proposed based on the factors of initial positions, related degree of retrieval queries and abstracts, and weight of individual search engines. And result consistence sorting is also realized. The experimental results show that the proposed model has a preferably performance.

Keywords: Agent, Meta search engine, Personalized retrieval.

1 Introduction

Search engine is an important tool for people to find information online. But traditional search engines have exposed various problems since network information resources are expanding greatly and needs of users are increasing quickly. It is very difficult to locating users' specific information needs accurately; different retrieval resources of various search engines lead to diverse and distributed retrieval results, so users have to seek different search engines to get a more comprehensive result. Besides, most available search engines work in an identical mode and return the same results for different users with the traditional retrieval model without considering their personalized needs.

Addressing the problems above, a new search model needs to be developed, which should not only retain the advantages of traditional search engines such as with simple operations and quick responses, but also solve the problems above to improve users' searching experiences and fit their needs. The meta-search model using Agent technology is an effective and prospect method. Agent is an autonomous, initiative and intelligent entity, which can perceive environment information autonomously, carry on intelligent learning and then make reactions instantly. With meta-search model, meta-search engines convert user's search queries to a unified format which most individual search engines can recognize, and search the individual search engines at the same time, and then compare and analyze all the retrieval results. After that the final results are integrated with a unified format and presented to users. Combination of Agent technology and meat-search engine technology can solve the

F.L. Wang et al. (Eds.): WISM 2012, LNCS 7529, pp. 572–579, 2012.

problems above to a certain extent. Based on the above research and analysis, an Agent-based intelligent meta-search engine model is proposed in this paper.

2 An Agent-Based Intelligent Meta-search Engine Model

As an instinct feature, meta-search engines don't need to consider the low level retrieval processes such as crawling web pages and analyzing them and so on for all of these processes are done by individual search engines. So more efforts can be made to meet users' high-level needs and improve the intellectualization level. An Agent-based meta-search engine model proposed by Amir [1] has been proved effective in the Webfusion meta-search engine. But the model is a simplified one, in which intellectualization has not been much considered. This paper makes a further analysis based on it and proposes a new meta-search engine model. It adds several intelligent modules like user personalization module, retrieval results feedback module and so on, which make the procedure of the model more specific and efficient.

The entire model procedure can be divided into three phases: The first phase is user query personalization analysis phase. Personalization Agents group intelligently and analyze user queries based on users' personal information in the personalization information database and then form new retrieval queries which reflect users' personal needs. The second phase is to schedule and retrieve member search engines. Scheduling Agent will choose several most appropriate search engines in the individual search engine list and then request to the corresponding retrieval Agents. The third phase is to synthesize retrieval results. Synthesizing result Agent deals with various retrieval results according to weight of each member search engines, and sorts and presents them to user in a unified format. In addition, after the entire retrieval process it tracks users' options and clicks. Feedback Agent will return information about users' final options and update the personalization information Agent and individual search engines knowledge Agent. A concrete structure is presented in Fig.1.

Functions of each part of the model are illustrated as follows.

1) Personalization Agent Groups

Personalization Agent Groups contain a group of related Agent personalizing for individual users, including recording and analyzing Agent which mines and analyzes users' browsing histories, segmentation and classification Agent which initializes user interest model, interest renewal Agent which renews users' interests; query and analysis Agent which personalizes users when they are searching.

2) Personalized Information Database

Personalized Information Database is used to store users' personal information, including their different kinds of interests and corresponding weights according to the user interest model. In addition, it includes classification reference lexicon and words with their occurrence frequencies by removing low frequent words.

3) Individual Search Engine Knowledge Database

Individual Search Engine Knowledge Database stores different capabilities of different topics and languages (Chinese/English) of different individual search engines. Information is stored in form of a weight matrix. The database also includes different search formats of different individual search engines to ensure it can retrieve as requested.

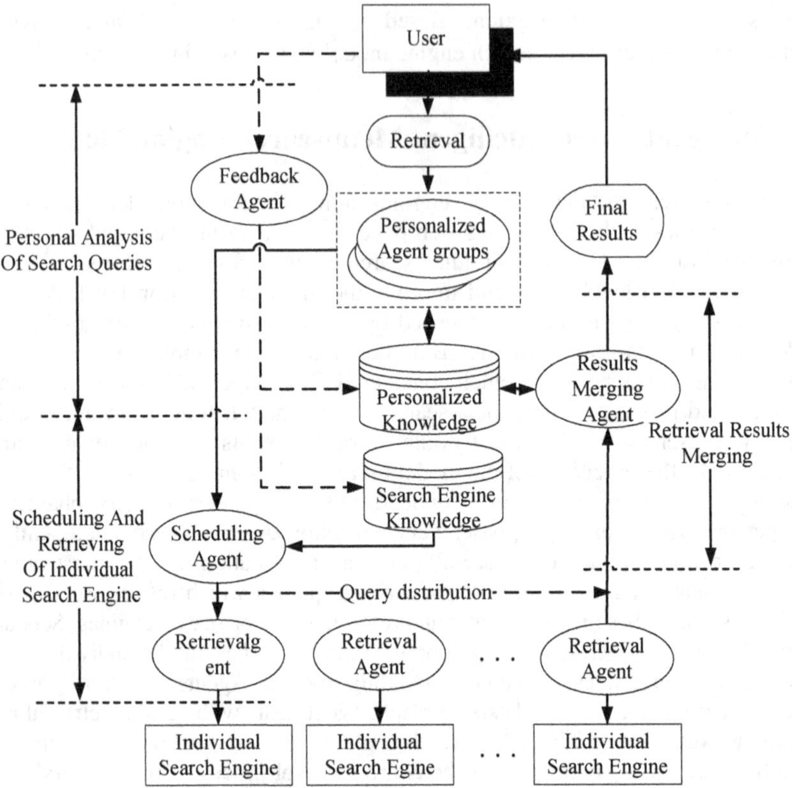

Fig. 1. The structure of the Agent-based intelligent meta-search engine model

4) Scheduling Agent

Scheduling Agent's main function is to choose appropriate individual search engines based on the search queries. After analyzing queries' topic classes and using languages, it chooses several individual search engines and informs corresponding retrieval Agent by synthetic consideration of searching information capabilities of all individual search engines in individual search engine knowledge database.

5) Retrieval Agent

Retrieval Agent's main responsibility is to receive messages from scheduling Agent and individual search engines with retrieval queries provided by Scheduling Agent and take corresponding coordinative actions. Different individual search engines are accessible through their own corresponding interface by which retrieval Agent submits queries for retrieval.

6) Synthesizing Results Agent

Synthesizing Results Agent gets all retrieval results and calculates each retrieval result's final weight on factors of initial results, related degree of retrieval queries and abstracts, and weight of individual search engines, and sorts the results. After reduplication removal, the results are presented to users.

7) Feedback Agent
Feedback Agent tracks users' options and clicking behaviors and returns information according to the options' topic and information of the individual search engine which provides it. Then feedbacks are converted to personalization information Agent and individual search engine knowledge database to update information stored in them.

3 Personalization Analysis of Retrieval Queries

Personalization analysis of queries is the first phase of the model. And there are two main problems in the phase: One is how to construct the user interest model, and the other is how to do text classification, judging which class it belongs to. The most frequently used method of constructing users' interest model is the spatial vector modeling which is improved in this paper based on the research of Park YW [2] and in this paper a four-tuple spatial vector model is proposed. This paper also makes an improvement on the Naive-Bayesian model and proposes a refined word-frequency-based Naive-Bayesian model about the classification.

3.1 A User Interest Model

For the user interest model, users' potential behavior features and interests are found out through analyzing users' browsing histories firstly to construct the fundamental interest model. Then users' browsing behaviors when using search engines and are analyzed to update the user interest model. The user interest model consists of a series of interest feature vectors which can be denoted as a four-tuple spatial vector as follows.

$$\{k_i, c_i, d_i, w_i\} \tag{1}$$

In the formula above, k_i denotes a vocabulary entry referring to users' interest. c_i denotes classification referring to users' classified category. d_i denotes the updated date and time meaning the latest updated date and time. If users haven't searched for a certain topic for a long time, its weight should be decreased. w_i denotes weight, meaning the vocabulary entry k_i's weight in its displayed updated date and time. The TF algorithm is used to calculate the weights with the following formula.

$$w_i = TF\ (k_i,\ d) = \sum_{d=1}^{n} (f_{id} / \sum_{v=1}^{m} f_v) \tag{2}$$

In the formula above, d denotes the document name where the keyword k_i appears in, where 1<=p<n fid denotes the number of the keyword w_i occurring in a document. m denotes the vocabulary entry number in the document d, where 1<=w<m, meaning the v_{th} word occurrence times in the document d. The algorithm first calculates the number of a certain keyword occurrence in a document, and let it divided by the sum of all vocabulary entries occurrence to get the word frequency in this document. Then the keyword occurrence frequencies in other documents of the category which the

keyword is classified to are calculated and the frequencies are summed up. The sum is used as the key word's weight.

3.2 A Text Classification Model

Text classification model uses the improved word-frequency-based Naive-Bayesian model. In this paper, the Text Classification Corpus provided by the Sogou Lab is preprocessed to be used as a training library. There are nine categories in the Text Classification Corpus: IT, Finance, Health, Travel, Education, Recruitment, Culture and Military. There are 1990 documents in each category. The preprocesses includes classifying all documents, ruling out useless words by word characteristic and length, using the remaining words as characteristic results of their own category and saving them into text files. The model uses the following Naïve-Bayesian formula.

$$C_{NB} = \text{argMax}\ (P\ (c_j)*\prod_1^c P\ (x_i|c_j)) \tag{3}$$

CNB denotes text classification. P (c_j) is the prior probability of the category j. P $(x_i|c_j)$ is the class condition probability that the feature vector xi belongs to the category c_j. Take the maximum as the text category after calculating its probabilities of all sorts of categories.

$$P\ (c_j) = \sum_{k=1}^{v} TF\ (X{=}x_i,C{=}c_j) / \sum_{m=1}^{w}\sum_{k=1}^{v} TF\ (X{=}x_k,C{=}c_m) \tag{4}$$

Formula (4) is a prior probability calculation formula based on the word-frequency-based Naïve-Bayesian formula. And v denotes feature vocabulary, meaning the number of words/attributes of c_j category in the treated training library. TF($X=x_k$, $C=c_m$) denotes attribute x_i's sum of occurrence numbers in the category c_j. w denotes the sum of the categories. The formula calculates the sum of all feature words occurrence in a certain category divided by the sum of the numbers of all feature words occurrence in all categories as the certain category's prior probability.

$$P\ (x_i|c_j) = (TF(X{=}x_i,C{=}c_j)+1)/(v+\sum_{k=1}^{v} TF\ (X{=}x_i,C{=}c_j)) \tag{5}$$

Formula (5) is a class condition probability calculation formula based on the word-frequency-based Naïve-Bayesian formula. In the formula, the increase is to avoid the class condition probability is 0. In this paper, we improve the class condition probability with considering the situation where a feature word occurs in a to-be-classified document many times. The boosted-formula is as follows.

$$P\ (x_i|c_j) = TF\ (k_i,d)*(TF(X{=}x_i,C{=}c_j)+1)/(v+\sum_{k=1}^{v} TF\ (X{=}x_i,C{=}c_j)) \tag{6}$$

In the formula TF (k_i, d) is the occurrence number of the keyword k_i in a to-be-classified document d. Take P (c_j) and boosted P $(x_i|c_j)$ as input parameters of the (3-3) formula and we can get the text's classification.

4 Retrieval Result Synthesis Strategy

This paper designs a retrieval result synthesis strategy consists of three factors of the initial positions, the related degree of retrieval queries and abstracts, and the weight of individual search engine, referring to the view that a synthesis algorithm based on the retrieval results and abstracts is more efficient than other algorithms addressed by Lu Y[3]. Concrete related functions are completed by synthesizing result Agent.

The feedback position relations of retrieval results and individual search engine should be considered first. Different individual search engines return different positions with different results, which are calculated using the following formula.

$$T_i(\text{result})=1-r_i/N \tag{7}$$

In the formula, result denotes a certain retrieval result. r_i denotes the result's position in feedback result list of the individual search engine i. N is the number of all results retrieved by individual search engine i. If a certain result hasn't been appeared in the results retrieved by i, T_i(result)=0. Those reduplicate retrieval results' should be deleted after calculation. The related degree of retrieval results position in different individual search engine list can be calculated with the following formula.

$$Rel_{rank}\ (\text{result}) = \sum_{i=1}^{n} T_i\ (\text{result}) \tag{8}$$

Then consider the related degree of the query and result abstract and the body topic. The query sentence is segmented to words $Q_{div}= \{q_1,\ldots, q_m\}$. The segmented query sentence Q_{div} and abstract can be calculated with the following formula.

$$Rel\ (Q_{div},\text{abstract})=\sum_{j=1}^{m} \text{Rank}(q_j,\text{abstract}) \tag{9}$$

$$\text{Rank}(q_j,\text{abstract})=\text{occ}(q_j,\text{abstract})\sum_{i=1} \ln(\text{len}(\text{abstract})/\text{loc}(q_j,i,\text{abstract})) \tag{10}$$

In the formula above, occ (q_i, abstract) denotes the keyword q_j occurrence times in the abstract. If q_j dosen't occur, Rank (q_j, abstract) =0. len (abstract) denotes the length of the abstract. loc (q_j,i,abstract) denotes the i_{th} occurrence position in the abstract. The body topic is always short, and related degree of retrieval queries position in the body topic has a small effect. The important consideration is whether the word has appeared in the body topic. If it has appeared, the topic has a close tie with the retrieval result. The related degree of the retrieval result and the body topic can be calculated with the following formula.

$$Rel\ (Q_{div},\text{topic})=(n_{q1}+n_{q2}+\ldots+n_{qm})/\text{len (topic)} \tag{11}$$

In the formula above, the nq1 denotes the keyword q_j occurrence number in the body topic ($1 \le j \le m$). len (top) denotes the length of the body topic. m denotes the keyword number in the retrieval queries. Considering the two factors above, the related degree of the retrieval queries and the body topic and abstracts can be calculated with the following formula. In the formula, α, β denotes the weight coefficients of body topic and abstracts, and α+β=1.

$$Rel\ (Q,result)=\alpha Rel\ (Q_{div},\ topic)+\beta Rel\ (Q_{div},abstract) \quad (12)$$

Finally, the effect of the weight of individual search engines on results synthesis is considered. A higher weight an individual search engine has, a relatively higher quality its retrieval results will have. So the w_{final} calculated by scheduling Agent should be a factor taken into account of sorting retrieval results. The final weights of individual search engines can be calculated based on the three factors above.

$$w\ (result,\ q)=\ \alpha Rel_{rank}\ (result)\ +\beta Rel(Q,\ result)\ +\gamma w_{final} \quad (13)$$

In the formula above, α, β, γ are weight factors and α+ β+ γ=1. Retrieval results are sorted by w (result, q) in descending order and the top N are be used to as the final retrieval results.

5 Experimental Results

In order to verify the meta-search engine model proposed in this paper, an Agent-based meta-search engine model prototype system ASE is designed. We choose the popular search engines Baidu, Google, youdao, soso, meta-search engine Webcrawler, and metacrawler as comparative search engines. Then it is practically tested in web environment. Because the analysis on the searching habits shows that most users only browse the top 10~30 retrieval results [4], the top 30 results are collected, and according to the related degree of each retrieval result documents, the nine categories degrees are averaged. Then the results are evaluated using evaluating methods proposed by literature [5].

$$R=\ (2r+u)/2t \quad (14)$$

In the formula, r, u and t denote related documents, uncertain documents and retrieval documents. The experiments test the average correlation between queries and results provided by prototype of meta-search engine ASE and compare them with different individual search engines. The related degree distribution of different Chinese search engines can be found in Fig.2.

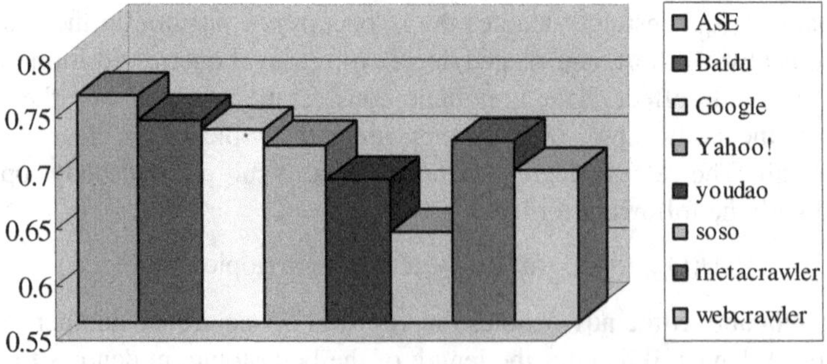

Fig. 2. The distribution of related degree of Chinese search engines

From Fig.2 we can find that the average related degree of ASE is better than the others'. Compared with the domestic search engines, ASE shows some other certain advantages as well.

6 Conclusion

This paper provides an Agent-based meta-search engine model, which implements the strategies and algorithms of the three phases of model procedure, reflecting intellectualization. The personalization Agent groups implement users' personal retrieval. And in retrieval result synthesis the initial positions, related degree of retrieval queries and abstracts, and weight of individual search engines are considered to reach a satisfying synthetic effect. Through experiments based on the prototype of ASE system it shows that the model has a higher personalization level than those individual search engines.

Acknowledgments. This work is supported by the National Natural Science Foundation of China (61173026), the National High Technology Research and Development 863 Program of China (2012AA02A603), the Fundamental Research Funds for the Central Universities of China, and the Defense Pre-Research Project of the 'Twelfth Five-Year-Plan' of China (513***301).

References

1. Keyhanipour, A.H., Moshiri, B., Kazemian, M., Piroozmand, M., Lucas, C.: Aggregation of web search engines based on users' preferences in WebFusion. Knowledge-Based Systems 20, 321–328 (2007)
2. Park, Y.W., Lee, E.S.: A New Generation Method of a User Profile for Information Filtering on the Internet. In: Proceedings of the 13th International Conference on Information Networking, pp. 261–264. IEEE Computer Society, Washington, DC (1998)
3. Lu, Y., Meng, W., Shu, L., Yu, C., Liu, K.-L.: Evaluation of Result Merging Strategies for Metasearch Engines. In: Ngu, A.H.H., Kitsuregawa, M., Neuhold, E.J., Chung, J.-Y., Sheng, Q.Z. (eds.) WISE 2005. LNCS, vol. 3806, pp. 53–66. Springer, Heidelberg (2005)
4. Spink, A., Jansen, B.J., et al.: Real Life Information Retrieval: A study of results overlap and uniqueness among major web search engines. Information Processing and Management, 1379–1391 (2006)
5. Keyhanipour, A.H., Moshiri, B., Piroozmand, M., et al.: WebFusion: Fundamentals and Principals of a Novel Meta Search Engine. In: Proceedings of the 2006 International Joint Conference on Neural Networks, pp. 4126–4131. IEEE Press, Vancouver (2006)

A Novel Image Annotation Feedback Model Based on Internet-Search

JianSong Yu[1,2], DongLin Cao[1,2,*], ShaoZi Li[1,2], and DaZhen Lin[1,2]

[1] Cognitive Science Department, Xiamen University, Xiamen, China, 361005
[2] Fujian Key Laboratory of the Brain-like Intelligent Systems, Xiamen, China, 361005
yujswawj@hotmail.com,
{another,szlig,dzlin}@ xmu.edu.cn

Abstract. We propose an Internet-search-based automatic image annotation feedback model, combining content-based and web-based image annotation, to solve the relevance assumption between the image and text and the limited volume of the database. In this model, we extract candidate labels from search results using web-based texts associated with the image, and then verify the final results by using Internet search results of candidate labels with content-based features. Experimental results show that this method can annotate the large-scale database with high accuracy, and achieve a 5.2% improvement on the basis of web-based automatic image annotation.

Keywords. automatic image annotation, Internet search, the verification of feedback.

1 Introduction

Automatic image annotation technology has great value and significance on image retrieval. The technology can be widely used in image search engine and image data management. Common Web-based automatic image annotation approach research presumes certain relevance between the text and the corresponding image in the page, and is highly subject to the scale of the image dataset.

As a training set, the size, scope and accuracy of the database are the key to the evaluation of the automatic image annotation system. For example, Corel5K [1], LabelMe[2], PASCAL[3] have been widely used in the field of computer vision. However, due to its limited number of pictures, these databases are lack of diversity and practical application. The appearance of Flicker makes up these shortcomings, and covers most of the content that related. Since Flicker can be added by users, so the quality of the data is unevenness.

It's effective to use the image-to-word relation (IWR), image-to-image relation (IIR), or word-to-word relations (WWR) to annotate images. Using a training set of annotated images, we learn the joint distribution of blobs and words which we call a cross-media relevance model (CMRM [4]) for images. CRM [5] and MBRM [6] make use of feature values directly, together with kernel based techniques as a distance measure. In contrast, as to classification-based methods, each semantic keyword or

F.L. Wang et al. (Eds.): WISM 2012, LNCS 7529, pp. 580–588, 2012.

concept is treated as an independent class and corresponds to one classifier. Taking into account the semantic relations between words, the WWR will often be added to the annotation process, such as the coherent language model [7] based on WordNet[8]. Then Liu etc.[9] augment the classical model with generic knowledge-based WordNet. It employs image-based image learning to achieve basic image annotation to fix annotation results combined with the text-based image learning. The text-based annotation method relies on the specific context of the images. Using the text related to the web images to reveal its semantic content is an important method of automatic Web image annotation.

Most of the content-based automatic image annotation methods are based on the training set. Since most of the training sets are limited and difficult to update, and every day there will be a lot of image resources and new keywords appear, the majority of automatic image annotation methods only achieves good results in a particular small database in the laboratory. While facing the large quantity images with updating, it is hard to achieve the desired results. Web related text-based method is often confined to a separate page where the picture in and also based on the assumption of the relevance between the images and text [10]. Therefore, the use of the text is very limited, and vulnerable to the current text.

We propose an Internet - search based automatic image annotation feedback model combining the content based and web-based automatic image annotation. First, we use the search engine and the nature language process to extract keywords. The point of this process is using the results that from several different pages by the search with image, reducing the impact of a single text associated with the image in one page. Then we use the image content features of the image, combined with powerful database of search engine, for analysis the initial keywords and revise, resulting in the final keywords. Taking advantage of the huge Internet data, without using the training set, the novel model can annotate the large-scale image set, and then uses the Internet data again to the analysis and re-rank.

2 Image Annotation Feedback Model

2.1 The Main Idea of the Image Annotation Feedback Model

In the paper, we obtain the relevant text and images using the image-to-image search supported by the search engine online. Comprehensive utilization of the text and image's visual features, corresponding keywords will be extracted as the initial result. Then image content features will be used to revise the keywords. The basic algorithm is as follows:

Input: image g

Output: keyword set $\{fw_1, fw_2, \ldots, fw_n\}$

1. Images similar with image g obtaining through search engine serves as the image set. $G=\{(g_1,h_1,r_1),(g_2,h_2,r_2),\ldots,(g_m,h_m,r_m)\}$, where g_i is the related image, h_i is the text associated with image g_i, and r_i is the position of output sequence by the search engine.

2. Count the weight of all words in the associated text $\{h_1,h_2,\ldots,h_m\}$,and take the first n words as the candidate word, $\{w_1,w_2,\ldots,w_n\}$.
3. Put the candidate annotation word set into the search engine to obtain the image set $\{G_{w_1},G_{w_2},\ldots,G_{w_n}\}$, where $G_{w_i}=\{g_{i_1},g_{i_2},\ldots,g_{i_s}\}$ is the image set related to the word w_i.
4. Calculate $Sim(g,G_{w_i})$ to achieve the similarity value of the HSV color, texture and SIFT features between the original image g and the image in G_{w_i}. Finally, reorder candidate label word $\{w_1,w_2,\ldots,w_n\}$, and get the corresponding keyword set $\{fw_1,fw_2,\ldots,fw_n\}$.

2.2 The Selection of Candidate Words Based on Multi-page

We use the image search engine, comprehensive use of multiple texts where the images exist, to get the candidate label words. It reduces the impact of a single text, being more robust compared to the traditional single text-based annotation.

Input the original image g, and upload it to search using the image-to-image search module of search engine. Get the websites that the similar images in and open them for the text information. Keep the Chinese and then we will get the textual information $T=\{h_1,h_2,\ldots,h_n\}$, and $h_i\,(i=1,2,\ldots n)$ corresponds to each text message of one similar image. Each piece of text information contains the title and content, $h_i=\{t,c\}$, where t is the text title and c is the content. We then do word segmentation, tagging, retain the term and filter the abstract nouns to get a noun set, $NW=\{nw_1,nw_2,\ldots,nw_n\}$.

Set individual weight on each paragraph to measure that the front of the text in the search results is more accurate than the latter:

$$t\_weight_i = Title\_Weight - \frac{Para\_Num_i \times m}{10} \tag{1}$$

$$c\_weight_i = Text\_Weight - \frac{Para\_Num_i \times n}{10} \tag{2}$$

where t_weight_i, c_weight_i are weight of the title and the body. $Title\_Weight$, $Text\_Weight$ are the initial weight set by the man, $Para\_Num_i$ is the position in the search result sequence, m and n are the related parameters.

Because of the entity object the image content reflects, the label word must be a noun. For each noun nw_i, there is N_{nw_i} weight value while the word appears N_{nw_i} times in the text. Each weight could be denoted by $W_{nw_i,k}\,(k=1,2,\ldots,N_{nw_i})$. The problem that whether the word nw_i exists in the title or the text, $X_{nw_i,k}$, which appears at the k-th time, can be estimated as follows:

$$X_{NWi,k} = \begin{cases} 1 & \text{when } X_{nw_i,k} \text{ is in the title} \\ 0 & \text{when } X_{nw_i,k} \text{ is in the text} \end{cases} \tag{3}$$

While one word appears at somewhere, the weight value is set as follows:

$$W_{NWi,k} = t\_weight_i \times X_{nw_i,k} + c\_weight_i \times (1 - X_{nw_i,k}) \tag{4}$$

We integrate the weight of each word $nw_i (i = 1,2,\ldots,n)$ in NW to calculate W_{nw_i}, the formula is as follows:

$$W_{nw_i} = \sum_{k=1}^{N_{nw_i}} W_{nw_i,k} \tag{5}$$

Calculate the weight of all words in the NW, and normalize the weight as follows:

$$WI_{nw_i} = \frac{W_{nw_i} - MIN}{MAX - MIN} \tag{6}$$

where MAX and MIN is the biggest and smallest weight of the words in the set NW.

We set the weight t and c in the h_i, organize the text, and do word segmentation, tagging, retain the term and filter the abstract nouns, then we get the final sort of the text through the sequencing as follows:

$$F = [(w_1, WT_1), (w_2, WT_2), \ldots (w_n, WT_n)] \tag{7}$$

where w_i is the keyword, WT_i is the final value of the keyword calculated by the text T. Order F by descending order of WT_i value.

2.3 Feedback Based on Image Content

The candidate label word is extracted according to the text associated with the image. The final result is based on the keywords sequence sorted by WT_i, and the source is the relevant text without considering the actual content of the image. In this paper, we search the image using the candidate label word, and then revise the initial result with the content based Image annotation method.

For the initial label sequence of words $w_1, w_2, \ldots, w_n$ (n is the number of selected keywords), upload to the word-to-image search module and get the result search of the word w_i, $G_{w_i} = g_{i_1}, g_{i_2}, \ldots, g_{i_s}$, then the final image set is:

$$G_w = \{ g_{1_1}, g_{1_2}, \ldots, g_{1_s}, g_{2_1}, g_{2_2}, \ldots, g_{2_s}, \ldots, g_{n_1}, g_{n_2}, \ldots, g_{n_s} \} \tag{8}$$

We select the HSV color, texture and SIFT [11] feature as the lower visual features to match the distance between the original image and the images in G_w.

We extract the HSV feature of the image, then rank the Bhattacharyya distance between the three channel histogram of the images in ascending order, and we will obtain the sequence $a_1, a_2, a_3, \ldots, a_n$. Extract the Energy, Contrast, Entropy and Correlation as eigenvalues of the texture feature with gray level co-occurrence matrix[12], and describe the parameters as the vector $\text{Tex} = (En, Ct, Er, Cr)$. We make the Euclidean distance of the texture feature in ascending order to get $b_1, b_2, b_3, \ldots, b_n$. Score each image as follows:

$$GD_{hsv\_i} = a_1 / a_i \quad \text{When } a_1 \neq 0 \tag{9}$$

$$GD_{hsv\_i} = \begin{cases} 10 \times \dfrac{a_2}{a_{i+1}} & i = 1 \\ \dfrac{a_2}{a_i} & i \neq 1 \end{cases} \quad \text{When } a_1 = 0 \tag{10}$$

$$GD_{gray\_i} = b_1 / b_i \quad \text{When } b_1 \neq 0 \tag{11}$$

$$GD_{gray\_i} = \begin{cases} 10 \times \dfrac{b_1}{b_{i-1}} & i = 1 \\ \dfrac{b_2}{b_i} & i \neq 1 \end{cases} \quad \text{When } b_1 = 0 \tag{12}$$

where GD_{hsv_i} and GD_{gray_i} is the HSV feature and texture feature match score of the i-th image.

We calculate the Euclidean distance between the 128-dimensional feature vectors of different points. When it is less than a certain threshold, these two points will be marked as a pair of match point. The number of the paired points of the original image and the match image set should be sorted in ascending order, $c_1, c_2, c_3, \ldots, c_n$, and the score the SIFT feature respectively:

$$GD_{sift\_i} = \begin{cases} 0 & c_i = 0 \\ \dfrac{c_i}{c_n} & c_i \neq 0 \end{cases} \tag{13}$$

where GD_{sift_i} is the image's sift feature match score.

Based on the three results above, the final score can be calculated as follows:

$$GD_i = GD_{hsv\_i} \times W_{hsv} + GD_{gray\_i} \times W_{gray} + GD_{sift\_i} \times W_{gray} \tag{14}$$

where W_{hsv}, W_{gray}, W_{sift} are the weights of HSV feature, the texture feature and the sift feature respectively. Then normalized the total score to get the score MP_{Gik}, which means the normalized results of the match score between the k-th image and the original image g associated with w_i. By the end, we sort $\{w_1, w_2, \ldots, w_n\}$ by MP_{Gik} to obtain the final annotation results $\{fw_1, fw_2, \ldots, fw_n\}$.

3 Experimental Results and Analysis

3.1 Experiments Setting

The image-to-image search module and word-to-image search module used in the experiments is from Baidu. We take a random way to select the images as the image data set online (use the JPG as the image format). In the experimental process, we get 1000 images after the image-to-image search, whose search result is not null. As the example shows in Table (1), the four images are the original images, and there are some outputs after uploaded to the image-to-image search. After statistics, the original images are classified as "Animal", "Landscape", "Transport", "Characters", "Groceries", "Food", "Plants", and "The Other" eight categories, and the number of images of each type is shown in Figure (1). Also it can be shown indirectly that, the larger the number of some category is, the higher the probability of receiving the return value through the search engine, while the smaller means that the probability of the images in this category is relatively lower or even hardly to have a return value. From the Figure (1), we can see that the "Characters" and "Landscape" categories are the top 2 biggest.

Table 1. Examples of the image data set

the images				
Human annotated words	plane	Mona Lisa	watermelon	lily

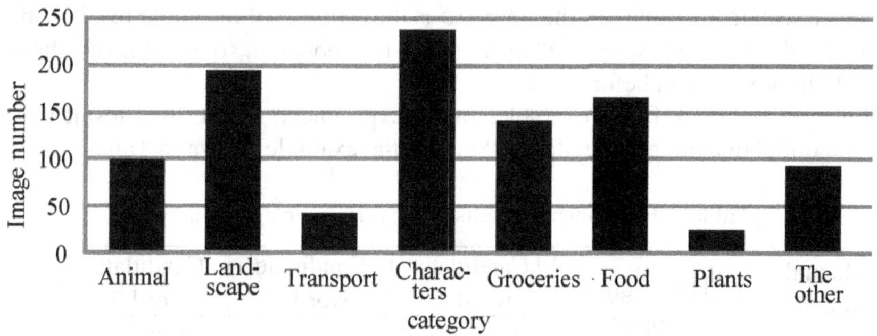

Fig. 1. Experimental image classification result

In this experiment, the Average Precision value, AP, is used to evaluate a single result. The AP is the average of the precision rate of the recall at different point, and in this paper it is defined as:

$$AP = \frac{1}{r}\sum_{i=1}^{r}\frac{i}{Po_i} \tag{15}$$

where r is the number of the right label word, Po_i is the position of the i-th right label word. When there is no right label word, $AP = 0$.

At the same time, we use the MAP (MeanAP) to estimate the whole experiment which integrate all single experimental results and calculate the macro average. It's defined in the text as:

$$MAP = \frac{\sum_{i=1}^{n} AP_i}{n} \tag{16}$$

where AP_i is the evaluation results of a single experiment, n is the number of the single experiments.

For each image, we choose the first visual label as the artificial label word result. Considering the influence resulting from image size while extracting SIFT feature, the images will be scaled to 150 pixels or less both in width or height before extracting SIFT. But the image will be kept the original size as extracting the texture feature and the HSV histogram feature, to alleviate the impact of image scaling.

In the experiment, the $Title\_Weight$, $Text\_Weight$ in formula (1) (2) are set to 0.6 and 0.4 empirically. m, n are set to 0.15 and 0.1 and $W_{hsv}, W_{gray}, W_{sift}$ are set to 1 in formula (14). We choose the first 3 or at most 3 and the first 2 or at most 2 words in the sequence F as the candidate keywords respectively to do the experiment. For each candidate label word in the initial result, we download three images to build a comparison image set. The experiment uses 1000 original images to analyze which has a return value after search, so the n in formula (16) is 1000.

3.2 Experiment Result Analysis

As can be seen from Table (2), the $AP = 0.5$ is the estimate of the initial result, and after the feedback, the word "straw hat" move to the first position, so the estimate change to $AP = 1$, improved than before.

However, there are failed examples in the experiment, whose final result is worse than the initial one after the feedback. Such as the example shown in Table (3).

Table 2. Experimental results of the original image "straw hat"

Manual annotation results:	The original image	Candidate word 1	Candidate word 2	Candidate word 3	AP
		Material	Straw hat	Still life	0.5
Straw hat		word after feedback 1	word after feedback 2	word after feedback 3	AP
		Straw hat	Still life	Material	1

The final result of the initial label in comparison to the results after feedback is shown in Figure (2). While there are 3 keywords at most, there are 90 images whose AP is 0.33, the 0.5 ones are 155, and the 1 ones are 487. After the final feedback, the 0.33 images are 55, the 0.5 ones are 140, and the 1 ones are 537. After the experiment,

Web-based image annotation at the initial result has $MAP = 0.5942$, and the MAP reaches 0.62515 after the feedback using the image content, increasing by 5.2%.

Table 3. The results of the original image "Grape"

Manual annotation results:	The original image	Candidate word 1	Candidate word 2	Candidate word 3	AP
		Grape	Flower shop	Tonic	1
Grape		word after feedback 1	word after feedback 2	word after feedback 3	AP
		Tonic	Grape	Flower shop	0.5

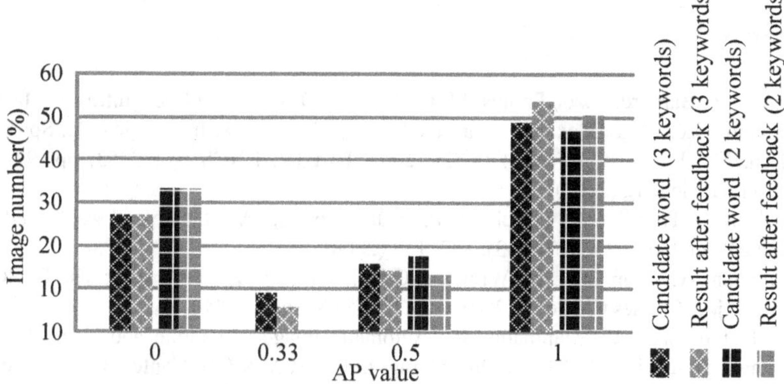

Fig. 2. The comparison of the results of the candidate label words and the ones after feedback with at most 3 keywords and 2 keywords

When there are 2 keywords at most, there are only three different estimate values. At the point 0.5 the number of images reduced to 136 from 179, and it rises to 523 from 480 at point 1. The MAP of the final result with at most 2 keywords has a 3.8% increased from $MAP = 0.5695$ to $MAP = 0.591$.

The Figure (2) show that where there is 0, there is no change, because the final label is revised based on the initial results. The final annotation results at point of 0.33 and 0.5 are slightly less than the initial ones, and at the point of 1 there is a significantly raised. The experiment proves that the feedback on the initial annotation has a good effect.

4 Conclusions

In this paper, we develop an internet-search-based automatic annotation framework which is able to achieve annotating a large-scale image set automatically. We employ the powerful database from Internet search to process text where the similar images are, and then get the result of initial label words as the candidate words. After putting the initial result in the search engine again, we obtain a series of images related to the word and then revise the initial result with the visual features as the final result. Analysis and correction to the image content based on the Internet-search ensures the accuracy of the image content annotation. Access to image information based on the Internet, this method can work without a certain training set, and thus overcomes the limitations of the traditional training set. A series of experiments with the image data set built by ourselves verify the validity of this automatic image annotation method.

Acknowledgements. This work was supported by National Nature Science Foundation of China(No.60873179), the Nature Science Foundation of Fujian Province(No.2011J01367)Doctoral Program Foundation of Institutions of Higher Education of China(No.20090121110032), and Shenzhen Science and Technology Research Foundation(No.JC200903180630A).

References

1. Duygulu, P., Barnard, K., de Freitas, J.F.G., Forsyth, D.A.: Object Recognition as Machine Translation: Learning a Lexicon for a Fixed Image Vocabulary. In: Heyden, A., Sparr, G., Nielsen, M., Johansen, P. (eds.) ECCV 2002, Part IV. LNCS, vol. 2353, pp. 97–112. Springer, Heidelberg (2002)
2. Russell, B.C., Torralba, A., Murphy, K.P., et al.: Labelme: A database and web-based tool for image annotation. IJCV 77(1-3), 157–173 (2008)
3. Everingham, M., Van Gool, L., Williams, C.K.I., Winn, J., Zisserman, A.: The PASCAL Visual Object Classes Challenge 2008 (VOC 2008) Results (2008)
4. Jeon, J., Lavrenko, V., Manmatha, R.: Automatic image annotation and retrieval using cross-media relevance models. In: Proceedings of the ACM SIGIR Conference on Research and Development in Information Retrieval, Toronto, Canada, pp. 119–126 (2003)
5. Lavrenko, V., Manmatha, R., Jeon, J.: A model for learning the semantics of pictures. In: NIPS (2004)
6. Feng, S., Manmatha, R., Lavrenko, V.: Multiple bernoulli relevance models for image and video annotation. In: CVPR, pp. 1002–1009 (2004)
7. Jin, R., Chai, J.Y., Si, L.: Effective automatic image annotation via a coherent language model and active learning. In: ACM SIGMM, pp. 892–899 (2004)
8. Tufiş, D., ŞtefĂnescu, D.: Experiments with a differential semantics annotation for WordNet 3.0. In: Proceedings of the 2nd Workshop on Computational Approaches to Subjectivity and Sentiment Analysis, ACL-HLT 2011, Portland, Oregon, USA, pp. 19–27 (2011)
9. Liu, J., Li, M., Liu, Q., Lu, H., Ma, S.: Image annotation via graph learning. PR 42(2), 218–228 (2009)
10. Tseng, V.S., Su, J.H., Wang, B.W., Lin, Y.M.: WEB Image Annotation by Fusing Visual Features and Textual Information. In: Proeeedings of the 2007 ACM Symposium on Applied Computing, pp. 1056–1060. ACM Press, New York (2007)
11. Lowe, D.G.: Distinctive Image Features from Scale-Invariant Key points. International Journal of Computer Vision 60(2), 91–110 (2004)
12. Gadelmawla, E.S.: A vision system for surface roughness characterization using the gray level co-occurrence matrix. NDT & E International 37(7), 577–588 (2004)

The Design and Application of an Ancient Porcelain Online Identification Analysis System*

Wei Li[1], Xiaochuan Wang[1], Rongwu Li[2], Guoxia Li[1,**], Weijuan Zhao[1], Wenjun Zhao[3], and Peiyu Guo[3]

[1] School of Physical Engineering, Zhengzhou University, Zhengzhou 450052, China
liwei658@163.com, liguoxia@zzu.edu.cn
[2] Department of Physics, Beijing Normal University, Beijing 100875, China
[3] Institutes of Cultural Relic and Archaeology of Henan Province, Zhengzhou 450000, China

Abstract. This paper puts forward an ancient porcelain system that can be identified and analyzed online. It can put discriminant analysis of ancient porcelain technology identification use to network platform. In addition, the system can provide efficient ancient porcelain identification service and collect more ancient porcelain samples data. The system exercises JSP, JavaBean and SQL Server 2000 technologies to establish an ancient porcelain database management system, and to achieve data exchange between JSP pages and database, and uses Java program to invoke Matlab program to achieve the identification analysis function. Test analysis results show that the system can be used to identify Liu Jiamen celadon porcelain, Ruguan porcelain and Zhang Gongxiang celadon porcelain.

Keywords: porcelain, identification analysis, JSP.

1 Introduction

China has a long history of porcelain production and the craftsmanship is exquisite [1]. People have always been strongly interested in the research of ancient porcelain production technology and culture. Ancient porcelain identification has a special meaning for the research. Traditional identification methods are facing challenges in highly copy porcelain, while technology workers make certain progress by combining nuclear analysis technology with multivariate statistical analysis [2, 3].

Taking the problems of ancient porcelain identification into account, several science research institutions have tested ancient porcelain samples, and have set up database management systems in China [4, 5]. But most of these systems are standalone versions, which are not beneficial to the communication and cannot meet the needs of normal users. Using multivariate statistical analysis method to analyze the ancient

* Project Funds: National Natural Science Foundation(No. 51172212, No. 50772101), The Chinese academy of sciences nuclear analysis technology key laboratory fund(No. B0901), Henan province foundation and frontier technology research plan(No. 102300410168)

** Corresponding author.

F.L. Wang et al. (Eds.): WISM 2012, LNCS 7529, pp. 589–594, 2012.

porcelain nuclear analysis data, it is expected to find some scientific criterions to distinguish ancient porcelain [6]. Now, there are a variety of tools to help people analyze the data, such as Matlab, SPSS and so on, but it is difficult to grasp these tools for non-professional peoples on data processing and analyzing.

The purpose of this paper is to design an ancient porcelain online analysis system. We establish the ancient porcelain database, and add analysis methods into the system. The task of the system is to provide identification analysis service or research for a wider range of people. We use JSP and SQL Server 2000 technologies constructing network platform, and make some attempts to implement the analysis functions by using mixed programming of Java and Matlab.

2 System Analysis and Function Design

2.1 The System Function Modules Analysis

According to the requirement analysis, the system uses the B/S structure. In order to ensure the safety of a large number of ancient porcelain nuclear analytical data, the system should design a user management subsystem to manage user permissions. To access the ancient porcelain data, users need to register and login the system. After successfully applying for certain permission, users can use the identification analysis function. According to the quantity of samples, the users can choose single-sample identification by manually entering a group data or multi-sample identification by submitting a excel file. If the data were analyzed, the samples data will be temporarily stored in a table of system database. With approved by administrators, the data can be shifted to the ancient porcelain standard sample data table and enlarge the amount of the standard samples information in the database. By means of the above analysis, we develop three subsystems: users management system, ancient porcelain data management system and identification analysis system. Fig.1 lists the subsystems and main function.

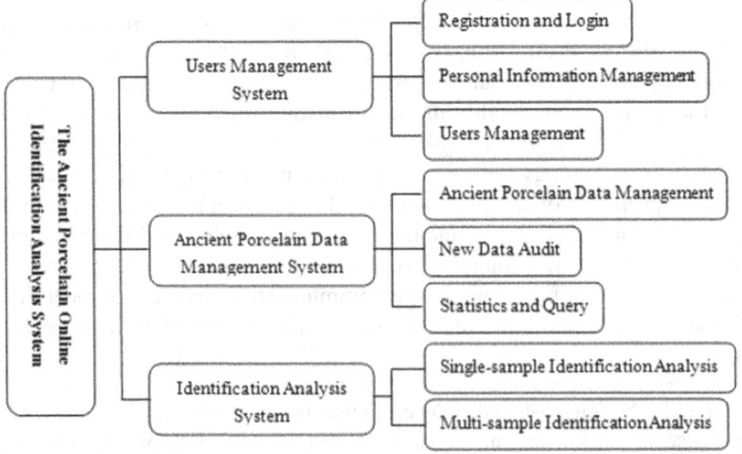

Fig. 1. Ancient Porcelain Identification Analysis System Function Modules

2.2 Database Design

Taking into account the amount of samples information and data security, we select SQL Server 2000 as a backend database.

The database consists of the following tables:

(1)The user table (tb_usertable): It mainly includes user number, username, password, permission and E-mail.

(2)The ancient porcelain categories table (tb_category): The category number, name and remarks are the main components.

(3)The chemical composition table (tb_PIXE): It is the standard samples library to identify porcelain samples, and mainly covers sample number, sample name, category, properties, remarks, the content of Al_2O_3, SiO_2, K_2O, CaO, TiO_2, MnO, Fe_2O_3 and so on.

(4)The temp table(tb_temp): Its content is alike tb_PIXE. The table is mainly used to temporarily store the porcelain samples data submitted when users use the identification function.

3 The Implementation of the System

3.1 Development Technology Sketch

We use JSP and JavaBean technologies to develop dynamic pages. JSP technology has some advantages, such as writing once and running everywhere, supporting multi-platform. JavaBean is a reusable cross-platform component written in Java. In fact, it is a reusable public class packed by Java. JavaBean can achieve database connection and data operation, while JSP mainly focuses on generating dynamic pages. Thus, it can not only make full use of the reusable JavaBean components, but make the code easy to maintain and improve development efficiency. In addition, we use Matlab compiler that can pack Matlab code as JavaBean components to achieve the data analysis function.

3.2 Using Database Connection Pool

Because distribution, management and release of database connections straightly influence the performance of web applications, we use database connection pool (DBCP) to manage the database connection. DBCP is a Java connection pool project and a connection pool component what tomcat use. When tomcat starts, DBCP can produce a sufficient number of database connections for the program, improve the speed of the program and save the memory. Program can get one connection from the memory when it needs to establish a database connection, and return to the memory after using [7]. These connections can make it efficient and safe to reuse, and avoid the overhead of database connections that frequently are established and closed [8].

3.3 Implementation of the Analysis Model

3.3.1 To Compile Discriminant Analysis Function File

The identification analysis function of the system is achieved by Fisher discriminant analysis. Matlab has powerful data processing capability, and Fisher discriminant analysis function is in Matlab's toolbox, so we can use Matlab Builder for Java to compile Matlab program to .jar file which can call by Java codes [9].

With deploytool command in Matlab, We can open deployment tool, and create a project, then add and compile a .m file, while we can get a .jar file and a .ctf file from fiher.m file. In order to import the package to MyEclipse web project, we can use the component in the web project. And then we can use Beans to take porcelain samples data from database. Finally, it can invoke the component to identify samples.

3.3.2 The Single Sample Identification

According to the number of samples, the identification analysis module is divided into the single-sample identification and the multi-sample identification. If the porcelain sample number is small, users can enter the data of one sample's chemical composition of Al_2O_3, SiO_2, K_2O, CaO, TiO_2, MnO, Fe_2O_3 and so on, input introduction of its origin, and select its properties, then the system will identify it and display the analysis result.

If the information of the sample is entered, the result may be affected. So the system should check it when we click on the "Identify" button. What's more, it will alert the user if the information is not complete. The function is achieved by JavaScript.

The system will add the data submitted successfully to tb_temp, and then invoke the identification analysis module to identify. The single-sample identification analysis call classify()—a member function of fisherclassify class. The function can read the

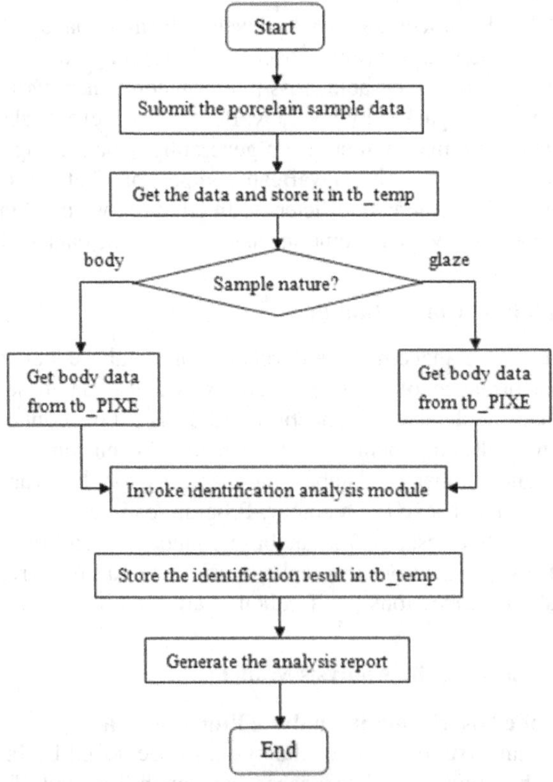

Fig. 2. Flow chart of single-sample identification analysis

data form tb_temp table, get its property, read the data with the property from tb_PIXE table to identify and store analysis result. The program flow chart of the identification analysis is shown in Fig. 2.

3.3.3 The Multi-sample Identification

If several or more samples are to be tested, it is inconvenient to select the single-sample identification analysis, while users can select multi-sample identification analysis via submitting a specific format Excel file. The system will add the samples data in the file to tb_temp table, and then invoke the module to analyze and store their analysis results. The module is more complex than the single-sample module, but the process is somewhat similar, so there is no longer described.

3.4 To Increase the Standard Samples in the Database

The more standard samples are in the database, the more reliable identification results are. So it is an important work to increase the number of standard samples in the database by rationally using users' samples data. However, this data is not necessarily reliable. Therefore, we cannot add them to standard samples table, but store in temporary table—tb_temp. If this data is reliable, the administrator can shift them to the standard samples table.

4 Running Instances

According to certain criteria, to develop the discriminant equation, discriminant analysis can test the unknown sample's category based on the known types of samples data. As a result, we cannot correctly distinguish the samples unless there are standard samples which have the same category with the samples in the database. To test identification performance of the system, we can make use of the single-sample and the multi-sample identification analysis to test the samples of known category. Now we use the former to test a Liu Jiamen celadon porcelain sample, and its identification result is shown in Fig. 3.

Sample Identification Analysis Report	
The main information of the sample you entered is as follows:(%)	
Al2O3	9.52
SiO2	73
K2O	3.9
CaO	9.5
TiO2	0.237
MnO	0.035
Fe2O3	2.37
Remarks	Liu JM
Result	Liu Jiamen celadon porcelain

Fig. 3. The report of the single-simple identification analysis

Next, to submit 4 samples data through an Excel file, the identification report shows that their test results are all correct. It is shown in Fig. 4.

Samples Identification Analysis Report									
After the test, the Excel file have 4 effective samples, and the identification results are as follow:									
No.	Sample ID	Al2O3	SiO2	K2O	CaO	TiO2	MnO	Fe2O3	Results
1	R600G	13.33	71.67	3.67	9.13	0.15	0.05	1.50	Zhang Gongxiang celadon porcelain
2	R611G	13.90	68.15	4.22	10.18	0.20	0.07	1.79	Zhang Gongxiang celadon porcelain
3	R334g	16.72	63.78	4.79	10.97	0.17	0.14	1.93	Ruguan porcelain
4	R335g	15.56	64.03	4.12	12.63	0.19	0.12	1.86	Ruguan porcelain

Fig. 4. The report of the multi-simple identification analysis

5 Summarization

Through the design and application of the ancient porcelain online identification analysis system, we have the following harvest: Firstly, to build the system platform by using JSP and SQL Server 2000 database, we can effectively manage ancient porcelain data, and provide users with more convenient services through the Internet. Secondly, the combination of the JSP and JavaBean enables separation of the HTML and Java program. What's more, it is easy to program development and code maintenance. Finally, the identification analysis function is applied to network by mixed programming of Java and Matlab. The running instances show that the system can be used effectively identify Liu Jiamen celadon porcelain, Ruguan porcelain and Zhang Gongxiang celadon porcelain, and satisfy users' need of identifying porcelain samples from these categories. However the system still need to add more variety of types of ancient porcelain samples and other nuclear analytical data and to use other methods to scientifically analyze the data to validate each other, and then the system will be more practical. Furthermore, this is the next work we still need to do.

References

1. Zhu, Z.F.: The Present Situation and Development of China Ceramics Industry. China Ceramic Branch to 2003 Silicate Conference Proceedings (2003)
2. Xiao, P.X., Zhao, H.M., Li, R.W., Zhao, W.J., Li, G.X., Zhao, W.J.: Multivariate Statistical Analysis of Ruguan Porcelain, Junguan Porcelain and Liu jiamen Kiln Celadon Porcelain. Bulletin of the Chinese Ceramic Society 2, 312–315 (2011)
3. Wang, S., Li, G.X., Zhao, W.J., Sun, H.W., Li, R.W., Xie, J.Z.: Multivariate statistical analysis of the major chemical compositions of Ru official ware and Jun official ware. Sciences of Conservation and Archaeology 3, 1–5 (2007)
4. Yang, Y., Guo, L.W.: The Establishment of Ancient Porcelain Structure Database. Journal of Cdramics 1, 53–56 (2005)
5. Liu, Q., Wang, X.C., Yue, C.Y., Tian, H.Y., Wang, Y.F., Li, G.X.: Visual Design for Ancient Porcelain Database Management System. Computer Knowledge and Technology 28, 7887–7888 (2010)
6. Liu, J.L., Wang, X.C., Li, G.X., Zhao, W.J., Li, R.W., Zhao, Q.Y.: Differentiate Analysis of Ru Guan Porcelain, Jun Guan Porcelain and Folk Jun Porcelain Made By LiuJiamen Kiln. China Porcelain 10, 75–77 (2010)
7. Zhang, Y.F., Li, S.Y., Xu, L.: The Application of JSP Database Accessing with Connection Pool. Microprocessors 1, 29–32 (2005)
8. Liang, W.: Research on Database Connection Technology of Web Application Based on JDBC. Journal of Hefei University (Natural Sciences) 4, 29–32 (2010)
9. Fan, Z., Gao, Z.Y., Huang, Z.H., Wang, L.: Research and Realization of Web-Based Face Recognition System. JiSuanji Yu XiaDaihuan 3, 96–99 (2010)

A Self-adaptive Strategy for Web Crawler in In-Site Search

Rui Sun, Peng Jin, and Wei Xiang

Laboratory of Intelligent Information Processing and Application,
Leshan Normal University, Leshan, China
{dram_218,lstcxw}@163.com, jandp@pku.edu.cn

Abstract. This paper analyzes some characteristics of in-site search and proposes a self-adaptive strategy for web crawler. This strategy is polite and the number of concurrent threads is automatically adjusted according to the analyses of pages' average download time in different time units. Some factors such as web server load and network bandwidth are synthetically considered. The experimental results show that our strategy can achieve higher performance than some other strategies. It objectively reflects the practical crawling course of web crawler and fully exploit local and network resources.

Keywords: In-site search, Web Crawler, Multithreading, Self-adaptive.

1 Introduction

Web crawler provides data source for search engine as an essential component. The performance of crawler directly affects the richness and timeliness of source data. Since the first crawler was implemented by Matthew Gray Wanderer in the spring of 1993 [1], the study of web crawling has never been interrupted in both academia and industry.

No matter what kind the crawler is, its efficiency is inevitably influenced by some factors such as the quality of network environment and the operational efficiency of local environment. Many approaches and strategies have been used to sufficiently utilize network bandwidth and increase the crawling speed. Meng et al. (2008) proposed a bandwidth control strategy to predict download speed and maximum request frequency for crawling different sites. Yin et al. (2008) present efficiency bottleneck analysis and built an improved schema for web crawler.

In order to improve operational efficiency of local environment, concurrent technique is widely applied in crawlers. Most research on web crawler concentrate on internet search where distributed and cloud computing are required because of vast amounts of information. Those studies rarely involved in-site search.

This paper analyzes various characteristics of web crawler in in-site search, and presents a self-adaptive strategy for downloading web pages from specific site. This strategy is to consider the average download time of web page, the number of concurrent threads, network bandwidth and web server load, etc. The number of

F.L. Wang et al. (Eds.): WISM 2012, LNCS 7529, pp. 595–602, 2012.

threads can be adjusted to adapt to the actual operating environments. We carry out several experiments to compare different thread settings of crawler.

The remainder of the paper is structured as follows. We first describe our related works in Section 2. The next Section is concerned with the self-adaptive strategy in detail. Section 4 presents our experiments and analyses the experimental results and performances. At last we conclude our present achievements and possible future works.

2 Related Works

2.1 Operational Strategy of Web Crawler

The essential algorithms, such as breadth first search (BFS) and depth first search (DFS) [4, 5], are adopted to visit each node in network topology diagram.

Normal crawlers only focus on efficiently downloading useful web pages and ignore their topics. Focused crawlers needs to consider the relevance of web pages and the specific topic. Valuable resources should be given priority, and low-value resources should be put off even ignored [4]. Incremental crawlers must consider the frequent updated pages and regularly maintain the freshness of local copies [6].

2.2 Characteristics of In-Site Search

In-site search engine searches a specific website for content that matches the visitors' query. It is similar to internet search engine on basic search concepts, but there are some characteristics which need to be concerned in it. In this paper, we are mainly concerned about their influences in crawling system.

Simple Network Topology. In a crawling system, we often only care about the hyperlinks in pages. Many irrelevant links should be thrown off by URL parsing. In in-site search, the network topology may be approximately viewed as a tree structure, as shown in figure 1, and this tree can be traversed in level-order.

Polite Crawling Strategy. Similarly, concurrent technique ought to be used in in-site search. Thread synchronization implies that the thread waiting time will be increased. If the shared data is frequently accessed, a lot of overhead will be spent to achieve the management of critical section. So, the multithreading may result in an efficiency bottleneck of crawling system.

Inevitably, the crawler will access web server for a long time. The connection established to server is bound to occupy some server resources. In some violence crawling strategy, because thread number is too large, the internet user's normal access to the website is seriously impacted. In extreme cases, web server even may collapse because of too much burden. So, crawling in-site should adopt courteous crawling strategy. This polite way does not affect the normal operation of the site [1, 2, and 5].

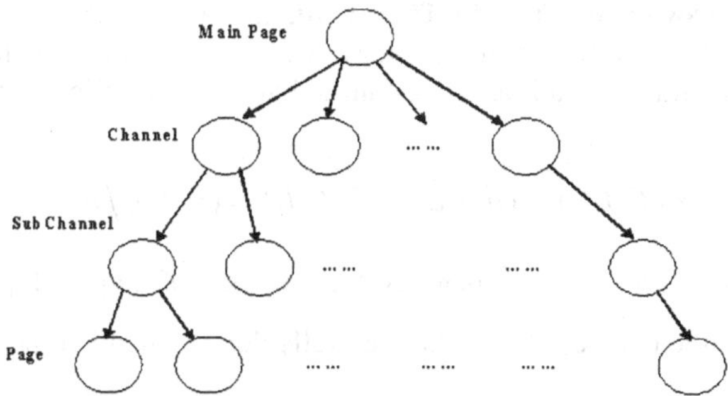

Fig. 1. Network Topology of Website

Crawling Strategy for Sub-periods. Generally, the frequency of accessing website in different time periods is different. The number of connections that web server allows is limited. The performance of crawler in the different time periods is also different. Therefore, the number of threads should be increased to take advantage of network resources in server idle time, and be reduced in server busy periods.

We propose a self-adaptive algorithm in next section. These factors above-mentioned are concerned in our crawling system.

3 Self-adaptive Strategy

3.1 Basic Concept

Pages' Average Download Time. The main factors which affect the page's download speed are its size and its download time. There is certain randomness in page size on the same web server. Considering that paging and other techniques are used in most of web sites, the page size may be approximately looked as the average size in this paper. We uses average model to calculate pages' download time.

In the single-threaded mode, web pages' average download time is the mean of time which is used to download all pages, as shown in (1).

$$\hat{t} = \frac{1}{n}\sum_{i=1}^{n} t_i \tag{1}$$

Where $\hat{t}$ denotes average download time, t_i denotes the actual web pages' download time, n denotes the number of pages downloaded.

Pages' Download Precision

Some web pages may be lost during the downloading. If the pages' download precision is lower, the local resources are wasted; otherwise the performance of crawler is better. The pages' download precision is defined by

$$p = td/d \tag{2}$$

where d denotes the number of pages which will be downloaded, td denotes the number of pages which are successfully downloaded.

Pages' Average Download Time in Time Unit. Considering that a lot of crawling tasks are executed in multithreading strategy, we simply think that these tasks are started in the same time and all tasks are finished at end time. Web pages' download time in unit time is defined by

$$t_i = (1/p_i)\times(T/td_i) = (d_i/td_i)\times(T/td_i) = (d_i \times T/td_i^2) \tag{3}$$

where T denotes a time unit, d_i denotes the number of required pages in time unit i, td_i denotes the number of pages successfully downloaded in time unit i.

Pages' Average Download Time in Window. The number of threads is dynamically adjusted according to the tendency of the changing of average download time. So, a window-like prediction mechanism is used to predict this tendency. Using n to denote the window size, we can calculate the average download time $\overline{t}$ in a window. It is defined by

$$\overline{t} = \sum_{k=i}^{i-n+1}(p_k \times t_k) = \sum_{k=i}^{i-n+1}((thrd_k / \sum_{l=i}^{i-n+1} thrd_l)\times t_k) \tag{4}$$

where t_k denotes average download time in time unit k, p_k is defined as the weight of time unit k and calculated according to the difference of thread number in windows, $thrd_k$ denotes thread number which is set in the time unit k.

Comparing $\overline{t}$ to the average download time in prior time unit t_{k-1}, we can predict the number of threads should be set in time unit k. In formula (5), x denotes the number of thread which should be added or reduced in next time unit, namely step size.

$$thrd_k = \begin{cases} thrd_{k-1} + x & \overline{t} - t_{k-1} > \alpha \\ thrd_{k-1} & |t_{k-1} - \overline{t}| \leq \alpha \\ thrd_{k-1} - x & t_{k-1} - \overline{t} > \alpha \end{cases} \quad (k \geq 3) \tag{5}$$

3.2 Algorithm Description

1) Set the initial number of threads to 1, and calculate average download time t_1;

2) Set the number of threads to $x+1$, calculate t_2;

3) According to the formula 4, calculate average download time in window $\overline{t}$;

4) If $\overline{t} - t_{k-1} > \alpha$, we predict that web server load or network bandwidth is limited. The number of threads is too high and should be decreased by step size;

5) If $t_{k-1} - \overline{t} > \alpha$, the download speed of pages is climbing, the number of threads should be increased by step size;

6) If $\left|t_{k-1} - \overline{t}\right| \leq \alpha$, the current download speed of pages is stable, and the number of threads should not be changed;

7) In the next new time unit, go back to step 3, and loop for the course above mentioned. The algorithm is over until all pages are downloaded or t_k is equal to 0.

4 The Experiments and Discussion

4.1 Experimental Settings

A large portal website, York BBS (www.yorkbbs.ca), which is one of biggest Chinese website in Canada, is the crawl target in our experimental system. Considering the efficiency of a web crawler is tightly related to its operating environment, all experiments are carried out on same computer system. The network bandwidth is limited to 10Mbps.

4.2 Experiments

We design several experiments to compare various implementations on different strategies respectively, such as different number of threads, self-adaptive and fixed number of threads, polite crawling and violent crawling.

The frequency of accessing website in different time periods is also different and the number of allowed connections to the website is limited, so the performances of web crawler at different time periods are uneven. So our experiments are carried out at the same time everyday (e.g. midday).

In each experiment, the time unit is set to 5 minutes, and the total crawling time is 2 hours, and the step size is 5. We respectively record the number of the requested pages, the number of the successfully downloaded pages, and calculate the pages' average download time.

4.3 Experimental Results and Discussion

Different Number of Concurrent Threads. The fixed number of threads in experiments is set to 1, 10, 20, 30, 40, 50, 60, 80, 100, and 120 respectively. We compare with the pages' download precision and the pages' average download time in different strategies. As shown in figure 2, the pages' download precision gradually declines when the number of threads is increased to 50. There is small change before 50. Therefore, it is not true that the bigger the number of threads, the better the performance of crawler is. Similarly, the pages' average download time doesn't always decline due to the increasing number of threads.

So, we think that the performance of crawler is divided into three stages, as shown in figure 3.

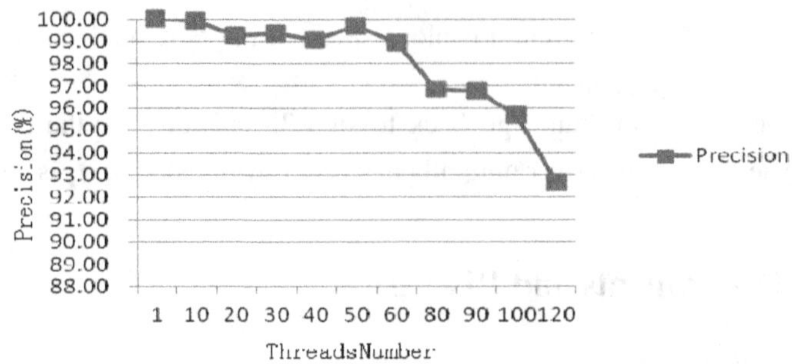

Fig. 2. Change of precision values in different thread number

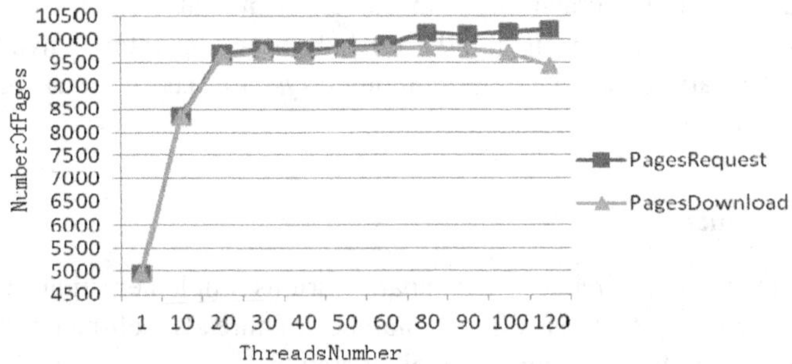

Fig. 3. Comparison of the numbers of downloaded pages in different thread number

The first is a rising stage. With the increase of the number of threads, the pages' average download time is declining. The advantage of multithreading has been shown undoubtedly. At this time, the performance of crawler is quickly improved.

When the number of threads is increased to a specific range, the precision and average download time is stabilized because of the influence of the local operational environment and the network environment. This is the stable stage.

In the downward stage, the pages' download precision starts to decline. Though the number of requested pages is increasing, the number of successfully downloaded pages decline sharply. At this time, the increase of the number of threads will not bring any efficiency gains. On the contrary, it will expend more system overhead, and even cause network congestion. So, the violence crawling strategy will bring bad or even fatal effects on the web server.

Obviously, the number of threads should be added in the rising stage and vice versa in the downward stage. In our experiments, when the number is approximately set in the range between 30 and 50, the performance of crawler can remain relatively high.

Self-adaptive Strategy. Use self-adaptive crawling algorithm to record the average download time of each time unit, and compare to other strategies where the different number of threads is fixed. Figure 4 illustrate the comparison of the pages' average download time of different strategies in a certain sequence of time units.

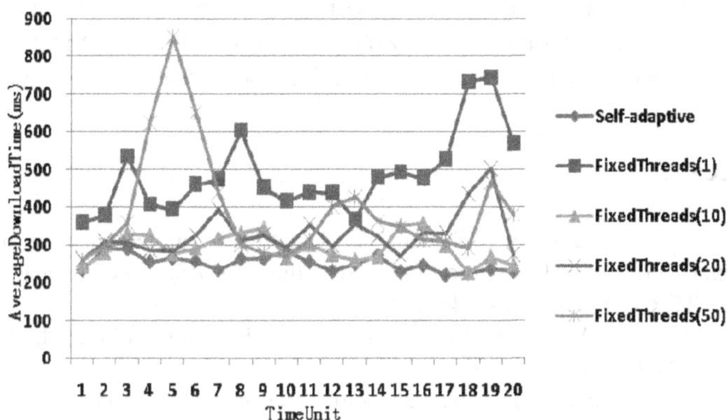

Fig. 4. Comparison of the pages' average download time of different strategies

Some significant phenomena can be found from this above figure.

The pages' average download time is relatively stable in self-adaptive strategy. On the contrary, there are some fluctuations the other crawling strategies. Our strategy is little influenced by the actual operational environmental.

The average download time is also relatively small in self-adaptive strategy where the number of threads is neither fixed nor limited by specific site. So it can effectively reduce the influence of local operational efficiency and network environment.

Parameters' Settings of Self-adaptive Strategy. We carry out several experiments where the windows size is set to 2,3,5,8 individually. Through comparing the precision and the average download time, we find that the performance is better when the size is set to 3. What's more, the value of step above-mentioned is set to 5 presently.

5 Conclusions and Future Works

This paper presents a self-adaptive crawling strategy based on polite crawling policy. According to the statistics of pages' average download time and pages' download precision in different periods of crawling process, the number of concurrent threads is dynamically adjusted in crawling system.

Experiment results show that our strategy is better than other fixed number strategies. It is simple but effective, and can be applied in some fields, such as general crawling, deep web crawling, focused crawling, etc. It should be noted that the rationality of some parameters like the window size still needs to be testified in the algorithm.

In future work, we will deeply study our algorithm and improve the performance of web crawler. At the same time, we will combine with distributed and cloud computing in our system.

Acknowledgments. This work is supported by Research Fund of Sichuan Provincial Education Department (Grant No. 11ZZ020) and Youth Fund of Sichuan Provincial Education Department (Grant No. 11ZB134). Peng Jin is the corresponding author.

References

1. Bouras, C., Poulopoulos, V., Thanou, A.: Creating a polite, adaptive and selective incremental crawler. In: Procedings of the IADIS International Conference (WWW/INTERNET), pp. 307–314 (2005)
2. Meng, X.-Q., Ye, Y.-M.: A Fast Iris Localization and Recognition Method. Microcomputer Information 24(33), 76–77 (2008)
3. Yin, J., Yin, Z.-B., Huang, H.: Efficiency bottlenecks analysis and solution of Web crawler. Computer Applications 28(5), 1114–1116 (2008)
4. Zhou, D.-M., Li, Z.-J.: Survey of High performance Web Crawler. Computer Science 36(8), 26–29 (2009)
5. Zhang, X., Zhang, Z.-J., Liu, Y.: Research on crawler module of BBS Public Opinion System. Railway Computer Application 19(12), 18–20 (2010)
6. Wolf, J.L., Squillante, M.S., Yu, P.S., Sethuraman, J., Ozsen, L.: Optimal crawling strategyes for Web search engines. In: Proceedings of the International Conference on World Wide Web (WWW), pp. 136–147. ACM, New York (2002)

A Novel Shark-Search Algorithm for Theme Crawler*

Lei Luo, Rong-bo Wang, Xiao-xi Huang, and Zhi-qun Chen

Institute of Computer Application Technology,
Hangzhou Dianzi University, Hangzhou, 310018, China
royluo7@gmail.com, wangrongbo@hdu.edu.cn,
itshere@zju.edu.cn, chenzq@hdu.edu.cn

Abstract. The shark-search algorithm is a classical content-based theme crawling algorithm. However, it has some disadvantages on crawling scope, including the *viscousness* phenomenon. To avoid this shortcoming of the original shark-search algorithm, an improved shark-search algorithm combining URL-analysis algorithm and host-control strategy is proposed in this paper. The accessed frequency of a host is considered in this new algorithm. The experimental results show that the proposed algorithm can overcome shortages of the original shark-search algorithm and improve the efficiency of a theme crawler.

Keywords: theme crawler, shark-search algorithm, crawling strategy, vertical search.

1 Introduction

With the developments of Internet industry in recent years, the amount of web information increases greatly. So it is more difficult to gather web information for traditional search-engines. To avoid this problem, researchers proposed the vertical search-engine which becomes a new development direction of search-engine technology in this information explosion era [1].

One vertical search engine focuses on a specific domain which can generate more accurate and professional search results than a traditional search engine. For a vertical search-engine, the most important module is a well-designed theme crawler [2]. Some algorithms are integrated into a theme crawler to filter the information unrelated to a theme. So a theme crawler can reduce greatly I/O cost and the number of indexed documents compared to a traditional search engine which makes a vertical search-engine more efficient. The core part of a theme crawler is the crawling strategies. The efficient crawling strategies can improve the performance of a theme crawler.

The Shark-Search algorithm [3] is a classical content-based algorithm of a theme crawler. However, in current Internet situations, the Shark-Search algorithm works only in one host or a small number of hosts because the breadth of Internet nodes

* This research is supported in parts by Youth Fund Project of Humanities and Social Sciences Research from the Chinese Ministry of Education(No.12YJCZH201) and National Natural Science Fund (No.61103101).

F.L. Wang et al. (Eds.): WISM 2012, LNCS 7529, pp. 603–609, 2012.

structure is neglected in the algorithm which leads to the obtained data concentrative excessively.

The main contributions of this paper can be described as follows. (1) A schema is proposed on how to evaluate the *viscousness* phenomenon of a theme crawler algorithm (The *viscousness* phenomenon is defined in section 2). (2) Two strategies are introduced for reducing the *viscousness* of a crawler. (3) A new algorithm that combined these two strategies is proposed. (4) Experiments were performed to confirm whether the new algorithm can lessen the *viscousness* phenomenon effectively.

2 The Existing Theme Crawling Strategies

There are two kinds of common used crawling strategies: the heuristic strategies based on the evaluation of text content and the strategies based on the evaluation of web links [4].

The early heuristic strategies based on the evaluation of text content include Fish-Search [5] and Shark-Search [3]. These algorithms evaluate a URL link by calculating the similarity between the contents of the web page it locates and the themes, and the similarity between its anchor text and the themes. The main idea of Shark-Search algorithm is as follows.

The Shark-Search algorithm adopts a *similarity engine* to compute the score of a web page.

$$\begin{aligned} potential\_score(child) &= \gamma * inherited(child) + \\ &(1-\gamma) * neighborhood(child) \quad (0 \le \gamma \le 1) \end{aligned} \tag{1}$$

where *inherited()* is used to compute the *inherited score*, which is defined as equation (2):

$$inherited\ (child) = \begin{cases} \sigma * sim(q, parent) & (sim(q, parent) \ge \phi) \\ \sigma * inheritd\ (parent) & (sim(q, parent) < \phi) \end{cases} \tag{2}$$

The decay factor δ is a threshold value in a range of 0 and 1. The *neighborhood score* is computed by function *neighborhood()* with the similarity between the anchor text of a hyperlink and the theme.

$$neighborhood(child) = \beta * sim(q, anchor) + (1-\beta) * sim(q, anchor\_text) \tag{3}$$

Based on above equations, the vector space model (VSM) is used to compute the similarity.

$$sim\ (d_1, d_2) = \frac{\sum_{k=1}^{n} (w_{ik} * W_{jk})}{\sqrt{(\sum_{k=1}^{n} w_{ik}^2)(\sum_{k=1}^{n} W_{jk}^2)}} \tag{4}$$

Note that n means the dimension of the vector. $(w_{i1}, w_{i2}, \ldots, w_{ik})$ is feature vector of the context text and $(w_{j1}, w_{j2}, \ldots, w_{jk})$ is feature vector of the theme keywords.

The main advantages of the strategy based on text content evaluation include the well-formed theoretical basis, low computing complexity and high accuracy. However, it cannot obtain the web link structure effectively because it neglects the link information. When these algorithms were proposed, most of web pages in a website belong to static (inactive) web pages. Moreover, the total number of web pages at that time is very small compared to the current number. Therefore, these algorithms presented a good performance at that time. However, there is currently a huge number of websites and web pages which are updated frequently. In current Internet environment, these original algorithms can make one crawler focus on some websites containing much information related to one theme or several themes, which is called as *viscousness* phenomenon. For example, from the investigation in WOW-bar of Baidu Tieba (Baidu Tieba is a famous website of theme-BBS groups in China), we can know that about 200 posts are replied in one minute at rush hour. New information of one theme is generated at a high speed, which does beyond the limitation of a simple theme crawler. A traditional theme crawler may meet *viscousness* phenomenon that keeps itself always access this website. The *Viscousness* phenomenon is harmful to a vertical search engine, because it limits the source websites in a finite range. Obviously, a vertical search-engine with a little number of indexed websites is not a satisfied search engine.

The representatives of strategies based on web-link evaluation are PageRank[6] and Hits[7]. In these algorithms, the importance of web pages is evaluated through the URL links among them, and the important web pages are downloaded preferentially. One of the important advantages of these algorithms is that the structure of a web page can be recognized effectively, so that the information in a web page can be obtained pointedly. However, there are some obvious disadvantages in these algorithms, including high computing complexity and topic excursion, which are caused by the lack of content evaluation.

The combination of the two kinds of strategies becomes a hot research topic in consideration of their properties. References [8-11] presented the investigation results of all crawling strategies. A link structure analysis method combined with Shark-Search based algorithm is proposed in literature [8], in which the web-link evaluation is computed in a low complexity. Unfortunately, it still cannot avoid the *viscousness* phenomenon. In literature [9], a method similar to PageRank's link analysis was presented to calculate the importance of a web page. Reference [10] introduced a genetic algorithm (GA) and the mutation of URLs to alleviate the *viscousness*. LIU Shu-mei etc. proposed a theme transfer method [11] to improve the original Shark-Search algorithm. All of these methods managed to combine the two kinds of strategies and to reduce the *viscousness* phenomenon. But their emphasis was to enhance recall and precision rate.

In reference [8], Liu proposed a new method to control a theme crawler, which is inspired by the idea of access speed control strategies for crawlers and can avoid *viscousness* phenomenon effectively. In this method, the number of accessed hosts and the concentration degree are considered as the measurement parameters. Experimental results show that this method can reduce *viscousness* phenomenon efficiently.

3 The Improved Shark-Search

3.1 The Core Idea of Proposed Algorithm

The proposed algorithm is based on Shark-Search algorithm, combining link weights analysis with web-link structure analysis and hosts access control method with Fish-Search algorithm. In the proposed algorithm, every access of a crawler to each host is recorded, and the weight of a host is decreased if this host is accessed frequently. Therefore, the crawler can avoid *viscousness* phenomenon effectively.

3.2 Shark-Search Algorithm Combined with Web-Link Evaluation

In order to add the web-link evaluation to the traditional Shark-Search algorithm, we define the link score as follows:

$$link\_score(url) = 1/(L-2) \tag{5}$$

where L is the number of symbol "/" appearing in a URL, the more "/" in a URL, the harder this web page can be found by a crawler, and the less important the web page is. Because a HTTP URL always starts with string "http://", L minus 2 operation and its reciprocal transformation should be performed. *link_score* is a value between 0 and 1.

The new value score is defined as follows:

$$\begin{aligned} now\_score(child) &= \rho * potential\_score(child) \\ &\quad + (1-\rho) * link\_score(child) \quad (0 \le \rho \le 1) \end{aligned} \tag{6}$$

Note that *potential_score(child)* represents the score computed by traditional Shark-Search algorithm, ρ is the weight of a Shark-Search score.

3.3 Adding Host Control Mechanism to the Algorithm

In Fish-Search algorithm, there is a response time evaluation in order to reduce the time cost. We apply it to the Shark-Search algorithm.

$$\begin{aligned} final\_score(child) &= \theta * now\_score(child) \\ &\quad + (1-\theta) * time\_score(child.host) \quad (0 \le \theta \le 1) \end{aligned} \tag{7}$$

where the *time_score* is defined as follows:

$$time\_score = \begin{cases} 1 & if\,(Average\_access\_time(child.host) \le 1) \\ 1/Average\_access\_time(child.host) & if\,(Average\_access\_time(child.host) > 1) \end{cases} \tag{8}$$

In equation (8), *Average_access_time(child_host)* is the average access time of the host containing current URL.

For *viscousness* problem of the traditional Shark-Search algorithm, a host control strategy is introduced as follows:

(1) Set three threshold values n, m and t, and set a counter c for every host for counting a crawler's visit.

(2) For every visiting page, perform following operations according the value of c:

a) If c meets the condition $c \leq m$, then compute the *time_score* of a web page with formula *Average_access_time(child.host)* = *Average_access_time(child.host)* + $t * c$, and do c++.

b) If c satisfies $m<c<n$, then compute the *time_score* of a web page with formula *Average_access_time(child.host)* = *Average_access_time(child.host)* + $t * m$, and do c++.

c) If c satisfies $c=n$, then compute the *time_score* of a web page with formula *Average_access_time(child.host)* = *Average_access_time(child.host)* + $t * c$, and reset c to 0.

If a host is frequently visited, the counter value c will rise and *Average_access_time (child.host)* will increase too. According to equation (7), the weights of web pages in this host will reduce, and then the frequency of this host to be accessed will be limited. When the counter value c reaches the threshold value m, a host is in a maximum inhibition; the negative influences caused by the host control method meet the maximum value and the host will keep stable. When c rises and reaches the threshold value n, the host has been inhibited for a long time and finishes a cycle period. We reset c to 0 and prepare for a next cycle.

4 The Experimental Results and Analysis

In section 3, three different kinds of theme crawlers are designed. In the first crawler, only the traditional Shark-Search algorithm is used. The Shark-Search algorithm with web-link evaluation is used in the second one. The last one is our proposed algorithm in which a host control mechanism is added compared to the second crawler.

In order to investigate the ability to reduce *viscousness* phenomenon of a crawler, we record the number of accessed hosts and concentration degrees. The concentration degree is defined as the proportion of the number of indexed pages from three most frequent hosts to the total number of indexed pages. When the number of obtained web pages is given, the more hosts a crawler accesses, and the lower concentration degree it has, the better performance for the crawler to reduce *viscousness*.

The threshold values are set as follows: $\gamma = 0.2$, $\varphi = 0.5$, $\beta = 0.7$, the maximum crawling depth is 2, $\theta = 0.7$, $m = 100$, $n = 1000$, $t = 0.1$. The number of web pages to be crawled is set as 5000, which means a crawler could gather 5000 web pages in each experiment. The keywords used in the experiments are as follows, *HangZhou, ZhouJieLun, BianXingJinGang, BaiXueZheng*. The entrance of a crawler is a list of URLs obtained from the previous data by a search engine. We consider these keywords as an input in Baidu.com search engine and collect the returned URLs. The researching on News is included in nearly all main search engines, but the results of News-searching are influenced by time strongly. In order to guarantee the stability of the experiments, URLs from News-search engine are removed from the URLs list. The experimental results are showed in Table 1.

Table 1. The results of three theme crawlers

Keywords	Theme related pages	Not theme related pages	Visit error pages	Precision ratio	Obtained hosts number	Pages from frequent hosts	Aggregation rate
HangZhou [1]	4301	642	57	86.02%	276	2809	65.31%
HangZhou [2]	4217	633	150	84.34%	346	2961	70.22%
HangZhou [3]	3888	956	156	77.76%	1782	1653	42.52%
ZhouJieLun [1]	4412	576	12	88.24%	12	3808	86.31%
ZhouJieLun [2]	4567	415	18	91.34%	13	3936	86.18%
ZhouJieLun [3]	4346	577	77	86.92%	149	3083	70.94%
BianXingJinGang [1]	2126	2720	154	42.52%	48	1906	89.65%
BianXingJinGang [2]	1644	3233	123	32.88%	60	1456	88.56%
BianXingJinGang [3]	1313	3470	217	26.26%	1435	706	53.77%
BaiXueZheng [1]	1306	3599	95	26.12%	6	1293	99.00%
BaiXueZheng [2]	1315	3602	83	26.30%	8	1305	99.24%
BaiXueZheng [3]	1089	3766	145	21.78%	372	852	78.24%

[1]: The crawler with algorithm 1.
[2]: The crawler with algorithm 2.
[3]: The crawler with algorithm 3.

From Table 1, we can know that more hosts can be accessed by algorithm 2 than those by algorithm 1, but the concentration degree has not reduced obviously. Generally, both algorithm 1 and algorithm 2 have not shown a good performance in the number of accessed hosts and the concentration degree. Especially, the concentration degree can reach 99% for uncommon keywords, such as *BaiXueZheng*, which means 99% of web pages indexed are from one identical host. This is a typical example case of *viscousness*. If algorithm 1 and algorithm 2 are applied in a vertical search engine, then the information sources obtained may be duplicated.

The number of obtained hosts increased obviously by using algorithm 3, together with the reduced concentration degree. It shows that the new algorithm can obtain more hosts and gather information from more hosts. At the same time the *viscousness* reduced, which makes the data sources of a vertical search engine diversified. However, the precision of our algorithm is lower than those of the other two algorithms duo to the work of avoiding *viscousness*.

5 Conclusion

In this paper, in order to solve the *viscousness* problem of the original Shark-Search algorithm, a new improved algorithm is proposed by integrating web-link evaluation and host control mechanism to the original one. Experimental results show that the new algorithm is efficient and feasible. From the results, we can know that the new algorithm reduces the precision ratio of a crawler in a certain extent. In future work, we will manage to overcome this shortage.

References

1. Panidis, A., Poulos, G.K.C., Pitas, I.: Combining Text and Link Analysis for Focused Crawling-an Application for Vertical Search Engines. Information System 32(6), 886–908 (2007)
2. Menczer, F., Pant, G., Srinivasan, P.: Topical web crawlers: evaluating adaptive algorithms. ACM Transactions on Internet Technology 4(4), 378–419 (2004)
3. Herseovici, M., Jacov, M., Maarek, Y.S.: The Shark-Search Algorithm-An Application: Tailored Web Site Mapping. Computer Networks and ISDN Systems 30, 317–326 (1998)
4. Ouyang, L.-B., Li, X.-Y., Li, G.-H., et al.: A survey of web spiders searching strategies of topic-specific search engine. Computer Engineering 30(13), 32–46 (2004)
5. Bra, D.P., Post, R.: Searching for arbitrary information in the WWW: the fish-search for mosaic. In: Second WWW Conference, pp. 45–51. ACM Press, Chicago (1994)
6. Page, L., Brin, S., Motwani, R.: The PageRank Citation Ranking: Bring Order to the Web. Stanford University (1998)
7. Kleinberg, J.: Authoritative Sources in A Hyperlinked Environment. Journal of the ACM 46(5), 604–632 (1999)
8. Liu, Y.-F.: Focus crawler researching in search engine. SUN Yat-Sen University, Guangzhou (2005)
9. Liu, P., Lin, H., Gao, D.-W.: Research on crawling strategy of subject searching spider by content-based and hyperlink-based analysis. Computer & Digital Engineering, 22–24 (January 2009)
10. Chen, Y.-F., Zhao, H.-K., Yu, X.-Q., Wan, W.-G.: Improvement of focused crawling strategy based on genetic algorithm. Computer Simulation 27(17), 87–90 (2010)
11. Liu, S.-M., Xia, L., Xu, N.-S.: Search strategy and achieve of the topic search engine crawler. Computer System & Applications 19(3), 49–52 (2010)

A Framework of Online Proxy-Based Web Prefetching*

Zhijie Ban and Sansan Wang

College of Computer Science, Inner Mongolia University, China
banzhijie@imu.edu.cn, sswang@ustc.edu

Abstract. Web prefetching is emerging as a promising technology and attracts many attentions from industry and research domains. Researches have intensively explored many web prefetching related problems such as prediction model, prefetching control and performance indexes. However, existing approaches emphasize on how to make use of users' browsing information to predict the next request, the technologies on the framework of online web prefetching are not intensively explored. It's obvious that these fundamental technologies are also very important for the successful application of web prefetching. To address this problem, this paper presents a framework for developing the online proxy-based web prefetching, analyzes the technologies included in this framework, and presents the design decisions and solutions. Finally, a prototype system based on the framework is implemented and proves the framework is effective.

Keywords: web prefetching, web caching, prediction model.

1 Introduction

When users browse web pages, it generally exists idle time between two requests because users often read some parts of one document before jumping the next one. Web prefetching makes use of the idle time to pre-fetch some web pages which will be requested and get them into the cache in the background before an explicit request is made for them. The main advantages of employing prefetching are that it prevents bandwidth underutilization and hides part of the latency. At present, the research works of web prefetching mostly focus on prediction models, prefetching control, prefetching performance measurement and so on.

Various prediction models have been proposed to represent users' behavior patterns and predict the user's next request page. These models can be divided into five types. The first type is URL-based related graph where an arc from node A to B means that B is likely to be accessed within a short interval after an access to A[1]. Each arc was labeled the conditional probability. The second type is Markov model, e.g., traditional Markov model[2], multi Markov chains-based model[3], Hybrid-Order Tree-like Markov model[4], All-k^{th}-Order-Markov Model[5], Selective Markov Model[6],

* Supported by higher education institution's science research project of Inner Mongolia Autonomous Region(NJZZ12005).

F.L. Wang et al. (Eds.): WISM 2012, LNCS 7529, pp. 610–620, 2012.

standard PPM (Prediction by Partial Matching), and LRS (Longest Repeat String) PPM[7]. Zipf-like law shows that the probability of one object accessed again is the larger if its access number is the larger. According to this law, some popularity-based algorithms are proposed such as Top-10[8], Top-n[9], Good-Fetch[10], APL[11] and Objective-Greedy[12]. Data mining-based approaches try to find user browsing patterns and the motivation behind behaviors by analyzing server logs. Young[13] et al. provided comparative study on different kinds of sequential association rules for web document prediction which showed the antecedents of rules and criterion for selecting prediction rules. Nanopoulos[14] et al. identified the factors which affected the performance of Web prefetching algorithms and proposed one algorithm called WMo, which was proven to be a generalization of existing ones. It is observed that client surfing is often guided by some keywords in the anchor text of Web objects. Thus, Xu[15] et al. presented a keyword-based semantic prefetching approach to predict future requests based on semantic preferences of past retrieved Web documents.

In order to reduce web prefetching adverse network effect, Wang[16] et al. presented one measurement-based web prefetching control. Based on analyzing the www's traffic characteristics,a new method to control prefetching is proposed[17].

To identify which indexes are the meaning ones, Jose Domenech[18] et al. proposed a taxonomy based on three categories and studied the relation among them. Shi[19] et al presented one stochastic Petri nets based on integrated web prefetching and web caching model, and discussed the performance metrics, access latency, throughput, hit rate.

From the above-mentioned, we see that it is easy to ignore others if only one side's problem is considered for web prefetching, which influences its total application. Thus, based on the foundation of an important amount works in the open literature, we believe it is time to emphasize the total framework of web prefetching and take into account its related problems. Based on the thinking, this paper presents a framework for developing the online proxy-based web prefetching. The framework includes three research contents such as web prediction model, cache management and prefetching control engine.

The rest of the paper is organized as follows. Section 2 presents the related background. Section 3 describes the framework for online web prefetching. Section 4 gives the details of information storage. Section 5 describes prototype system and section 6 elaborates the experiments. Section 7 is the summary and conclusions.

2 Related Works

There is an important set of research works concentrating on prefetching techniques to reduce the user perceived latency. Various prediction models have been proposed to model and predict a user's browsing behavior on the web. Markatos and Chronaki [8] proposed a Top-10 approach which combined the popular documents of the servers with client access characteristics. Web servers regularly pushed the most popular documents to web proxies, and then proxies pushed those documents to the active clients. But the approach only made use of page access frequency. In order to solve the problem, the study in [1] presented a prefetching algorithm based on prefetching in the context of file systems. But the model was not very accurate in predicting the user browsing behavior because it only considered first-Order dependency [14] and didn't look far into the past to correctly discriminate the different observed patterns. Thus, the studies in [20] described the use of a kth-Order Markov model for user request represented the sequence of k previous requests, and had conditional probabilities to each of the next possible states. However, it is likely that there will be instances in which the current context is not found in a kth-Order Markov model if the context is shorter than the order of the model. Therefore, the PPM (Prediction by Partial Matching) model [5][7]which originates in the data compression community, overcomes the problem. It trained varying order Markov models and used all of them during the prediction phase. Bouras[21] et al. studied prefetching's potential in the Wide Area by employing two prediction models. Fan[22] et al. studied how user access latency could be reduced for low-bandwidth users by using compression and PPM prediction model between clients and proxies. The LRS (Longest Repeated Subsequences) PPM[7] was used to store long branches with frequently accessed pages. It made use of the longest repeated sequence and conditional probability threshold to predict next request. These PPM models don't implement the online update and timely reflect the changing user request patterns. An online PPM with dynamic updating is presented[23]. Some literatures studied prefetching performance and prefetching control.

These above-mentioned researches provided a foundation for studying the framework of web prefetching. Based on our previous researches[23][24][25], We study the framework of the online proxy-based web prefetching system.

3 Framework of Online Proxy-Based Web Prefetching

We design and implement one proxy-based online web prefetching prototype system, called POLP. Based on our previous researches[23][24][25], POLP makes use of PPM model to mine user browsing patterns, incremental inserts new user requests and deletes outdated access information using non-suffix tree to reflect users' current access interests[23]. According to the model, it uses stochastic gradient descent rule to predict the next user request and updates prediction functions' weights by feedback information[24]. In order to decide what user requests to be deleted from the model, we make use of two-window approach to control samples[25]. By the techniques of the online updating prediction model, comprehensive considering access characteristics and the changing of prediction precision, POLP improves web prefetching performance and reduce its adverse effects. On the other hand, POLP integrates web prefetching and web caching which initiative prefetches the possible requested pages and caching those accesses a moment ago.

Figure 1 gives the structure of the online web prefetching framework about the proxy side. POLP includes three parts which is respectively prediction model, prefetching control and cache management. It is composed of the following modules: OT (offline training), GUIP(Gathering user information and preprocess), OP(online prediction), OU(online update), PCE(prefetching control engine), CC(cache control), RA(replacement algorithm), CM(construction model), LPP(log preprocess), APD(access pattern database), DG(data gather), DPP(data preprocess), DM(data management), PM(precision monitor), PDA(prediction decision algorithm), STP(selection training period), swc(sliding window control), II(incremental insertion) and ID(incremental deletion). In the following sections, we will specify mainly modules' function.

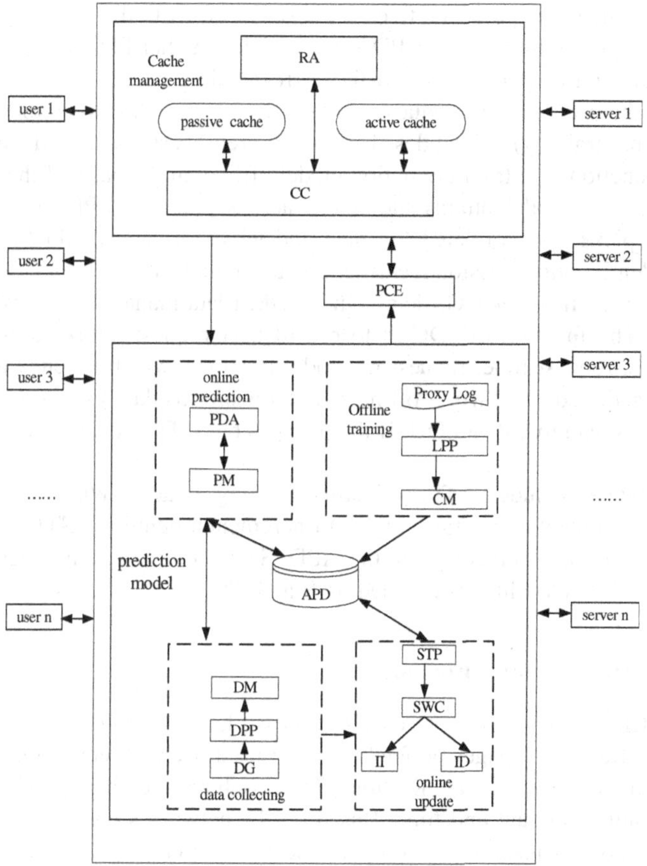

Fig. 1. System structure of POLP

3.1 Cache Management

Active cache stores web pages prefetched and passive cahce stores user access requests a moment ago. CC chooses appropriate replacement algorithm to weed out some web pages when cache is full. For active and passive cache, CC may use different replacement policy because web caching makes use of the temporal locality and web prefetching adopts spatial locality. At the same time, CC comprehensively considers active and passive cache to prevent repeated storage.

3.2 Prediction Model

Prediction model includes online prediction, offline training, data collecting and online update.

The online prediction contains PM (precision monitor) and PDA (prediction decision algorithm). The function of PM is to monitor the prediction precision. In order to avoid the lower precision to cause negative effect to the network system, PM reflects user access information to PDA and decide whether PDA to predict the next page. The function of PDA is to find the corresponding access patterns in the APD according to the user's current request and uses decision method to predict.

OT (offline training) includes LPP (log preprocess) and CM (construction model). Its function is to train prediction model offline on the basis of the proxy's log so that prediction model attains the initial access patterns. LPP preprocesses the proxy's log, picks up the related log data and wipes off noise, which includes log clear, user identification, session identification and data format.

Data collection includes DG (data gather), DM (data management) and DPP (data preprocess). The function of DG is to record user request information such as IP address, URL, request time, request method and so on, which it sends to DPP. For information gathered, DPP wipes off noise and transforms data format. If the proxy is idle, DM sends data to OP and OU. If the proxy is busy, DM temporary stores data or puts away.

Online update includes STP (selecting training data), SWC (sliding window control), II (incremental insertion) and ID (incremental deletion). STP is responsible for selecting data blocks included in the APD. SWC controls the number of deletion and insertion requests, which is finished by II and ID.

3.3 Prefetching Control Engine

Web prefetching may reduce user's access latency and improves network performance. But if web pages prefetched aren't used, web prefetching has an adverse influence on other network's applications. Thus, web prefetching system must control web prefetching's amount and time. Based on the network's condition and system's resource, the cost of resource is computed so that prefetching threshold is attained. When the network is congestion or the server's load is larger and the cost of web prefetching is higher, web prefetching system reduces the prefetching's amount. By doing this, it can prevent system resource's massive consumption and reduce the adverse effect on the network.

The function of PCE (prefetching control engine) is to receive web page candidates from OP module, compute the cost of system and resource, decide the prefetching's opportunity and quantity, then prefetche web pages and send them to CM.

4 Information Storage

The data quantity of user access information is larger. Suppose the throughput of one proxy is 10 requests per second, the accumulated access number of the proxy is 36000 per second. Thus one important problem of information storage is to better organize and store access characteristics so that web prefetching system can rapidly access related user request patterns.

User access interests are changing as time goes on, so storage's information changes along with user interests. At the same time, we must find the similar access patterns to prefetch some web pages before they are really requested. Thus, its operations to user access information mainly include insertion, deletion and locating. In order to improve access speed, we choose non-compact suffix tree to store the user access patterns because we can make use of the structure's suffix pointer to incremental insert and incremental delete. In order to insert the newest user request, we partly use the result of prediction's location so that it saves a large amount of time. Meanwhile, we do index from every user request to non-compact suffix tree so that the time of deleting one user request tends to zero.

5 Prototype System Function Test

We almost realize all modules in the figure 1. We only introduce prediction model partition because of space limit.

In order to test the function of POLP, we do experiments using three PC machines whose configuration is AMD Athlon X2 Dual 2GHZ. One PC, called PC_1, is one proxy and is also one client side. Other machines are client side, respectively called P_2 and P_3. The machines are connecting with local area network and P_1 is connecting to the internet by campus network. We orderly browse http://www.sina.com.cn, http://edu.sina.com.cn, http://edu.sina.com.cn/foreign and http://edu.sina.com.cn/toefl/index.shtml in the P_1. We orderly browse http://www.sina.com.cn, http://edu.sina.com.cn, and http://edu.sina.com.cn/foreign in the P_2. We orderly browse http://edu.sina.com.cn and http://edu.sina.com.cn/foreign in the P_3. The table 1 gives the relation between web pages and machines.

In order to test POLP's functions such as incremental insertion, incremental deletion, prediction, sliding window control and prefetching, we don't train the model. With the arrival of user requests, we construct prediction model and predict the user's next request. Suppose the sliding window's length is 5 and ignore web caching. According to the orderly user requests, POLP makes use of incremental insertion, incremental deletion and sliding window control to realize dynamic changing of prediction tree. Table 2 shows the prediction tree after the fifth request and table 3 gives the result after the eighth request. POLP inserts the new user request and deletes the requests beyond the sliding window.

Table 1. Successively relation of user requests in the proxy

Num	PC	URL
1	P_1	http://www.sina.com.cn/
2	P_1	http://edu.sina.com.cn/
3	P_2	http://www.sina.com.cn/
4	P_2	http://edu.sina.com.cn/
5	P_3	http://edu.sina.com.cn/
6	P_3	http://edu.sina.com.cn/foreign/
7	P_1	http://edu.sina.com.cn/foreign/
8	P_1	http://edu.sina.com.cn/toefl/index.shtml
9	P_2	http://edu.sina.com.cn/foreign/

Table 2. Prediction tree after the fifth user request

Node	Son Node	Num
Root	www.sina.com.cn	2
	edu.sina.com.cn	3
edu.sina.com.cn	edu.sina.com.cn	2

Table 3. Prediction tree after the eighth user request

Node	Son Node	Num
root	edu.sina.com.cn	2
	edu.sina.com.cn/foreign	2
	edu.sina.com.cn/toefl/index.shtml	1
edu.sina.com.cn	edu.sina.com.cn/foreign	1
edu.sina.com.cn/foreign	edu.sina.com.cn/toefl/index.shtml	1

When the the third user request arrives, POLP inserts it into the prediction and predicts the next request. The prediction result of P_1 is http://edu.sina.com.cn/ because has requested http://www.sina.com.cn/. POLP analyses http://www.sina.com.cn/ and prefetches embedded objects.

6 Experiments

In order to test the performance of POLP, we make use of log-derived simulation because of few machines. In the experiments, each log is partitioned into two parts. One part is training data and other is test data. We use two PC machines. One is client side which reads test log and sends user requests to the proxy according to the real user's requests. Other is the proxy. Firstly, it does offline training using training log and stores the training result. Then it is waiting for accepting user's requests. Once it receives the user's request, the proxy do corresponding operations. The replacement method of active and passive web caching is LRU.

6.1 Log

In the experiments, we use the following logs. 1) NLANR logs: National Lab of Applied Network Research provides pasted seven days web log and users may download them by authorized users' name and password. We collect proxy logs such as BO2, PA, SV and NY, respectively called NLANR_BO2, NLANR_PA, NLANR_SV and NLANR_NY. 2)College log: We collect one proxy log using proxy software from some medium-size education institution, called CU. The access record varies from Jan. 1, 2005 to 2 Feb., 2005. Table 4 shows some information about logs.

We remove all dynamically generated files. These files can be in types of ".asp", ".php", ".cgi" and so on. We also filter out embedded image files such as ".gif" and ".jpg" because we believe the image file is an embedded file in the HTML file. Access request sequence of each log file is partitioned into user sessions. One user session is one orderly access sequence from the same user. If a user has been idle for more than two hours, we assume that the next request from the same user starts a new user session. We recognize that the time interval of partitioning sessions may introduce some inaccuracy in the simulator, but it will not affect the evaluation of different models.

Table 4. Some information about logs

Log file	Days	Type	Time
NLANR_BO2	24	Proxy	2007/06/03-2007/06/26
NLANR_PA	23天	Proxy	2007/06/04-2007/06/26
NLANR_SV	20	Proxy	2007/06/05-2007/06/24
NLANR_NY	23	Proxy	2007/06/03-2007/06/28
CU	43	Proxy	2005/01/01-2005/02/12

6.2 Predictin and Updating Time

The aim of this experiment is to test the time of update and prediction about POLP, which is called prediction time as a matter of convenience. We respectively choose CU log from Jan. 1, 2005 to Jan. 10, 2005 and NLANR_BO2 log from JUN. 3,2007 to JUN. 14, 2007 to train prediction model. Then we make use of follow-up logs to update prediction model and predict the next user's requests. In the experiment, probability's threshold is 0, the predictin number every time is 3 and the model's order is 8. Figure 2 shows the total time of prediction, where abscissa is the number of times about prediction and vertical coordinate represents prediction time.

From the figure 2, we can see that the total time about 3000 predictions is 3 seconds and the average prediction time is 1 millisecond for CU log. The total time about 3000 predictions is 2 seconds and the average prediction time is 0.67 millisecond for NLANR_BO2 log.

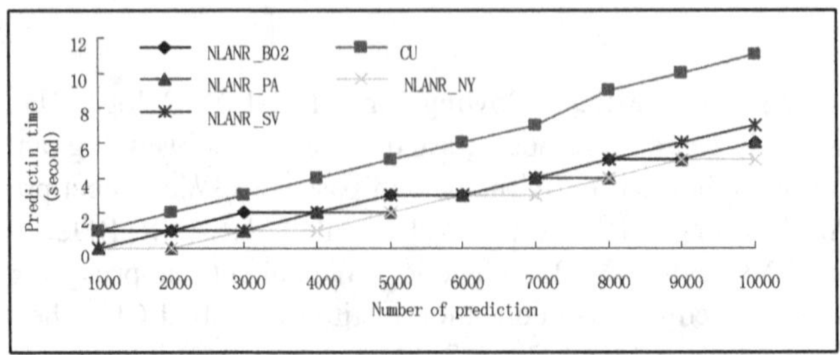

Fig. 2. Relation between prediction time and number

6.3 Prediction Number for Each User Request

In the experiment, we mainly analyze the web page's number of the proxy's prediction for each user request how to affect prefetching performance which is defined in [25]. Suppose that all predicted web pages are prefetched, put them in the active web caching and the new web page replaces the old web page if the web caching is full. We choose CU log from Jan. 1, 2005 to Jan. 10, 2005 for training prediction model and take it from

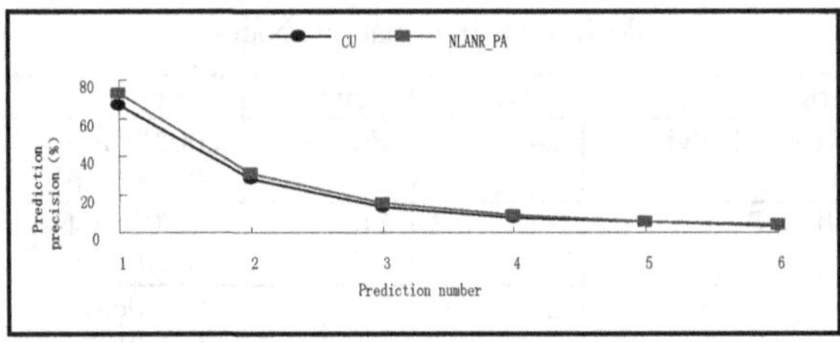

(a) prediction precision and prediction number

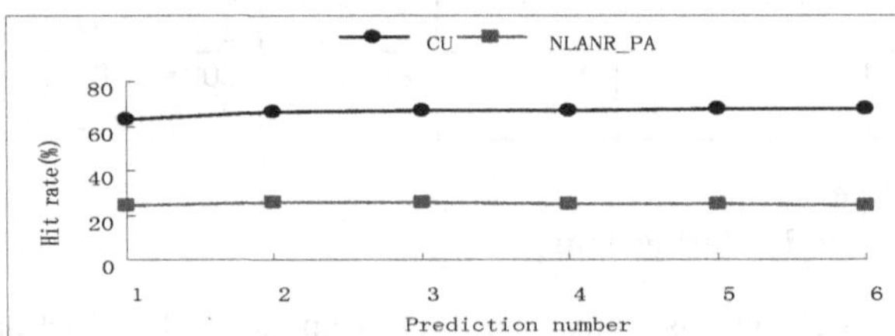

(b) hit rate and prediction number

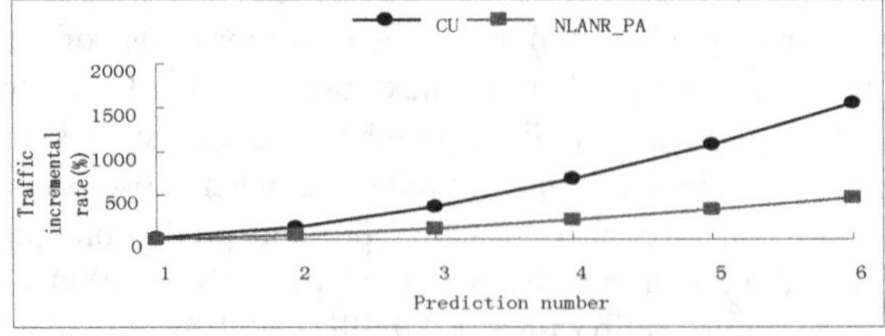

(c) traffic incremental rate and prediction number

Fig. 3. Prefetching performance vs prediction number

Jan 11, 2005 to 20 as test data. We choose NLANR_PA log from Jun. 4, 2007 to Jun. 14, 2007 for training prediction model and take it from Jun. 15, 2007 to 26 as test data. For each log, the model's order is 3 and the probability threshold is 0. The size of web cache is one percent of each log's requests about one day. In another words, the active and passive web cache is 0.5% of one day's requests. The figure 3 from (a) to (c) respectively gives POLP's different performance index at the condition of different prediction number for two logs. In the figure, abscissa is the prediction number for user's current request, which means the system provides the large prediction number according to the prediction model, and the vertical coordinate respectively represents prediction precision, hit rate and traffic incremental rate.

We can see that curves' variation tendency of two logs is almost no difference. With the prediction's increase, POLP's prediction precision is gradually decreasing. For instance, the prediction precision of NLANR_PA drops 69.16% when the prediction number varies from 1 to 6. The reason is that the total prediction number is increasing with the increasing of the prediction number while at most one prediction is correct every time. At the same time, the prefetching quantity increases with the increasing of prediction volume so that the traffic from prefetching is increasing like the figure 3(c) showing. In addition, we can deduce that hit rate slightly rises from the figure 3(b) and then it keep steady. Because the active cache accommodates the more new prefetching pages and the hit's probability will increases in the prefetching cache with the increasing of prediction quantity. However, with its continuous increasing, the size of active web cache is constant and new web pages replace the old ones so that it is possible that the web pages which may be visited are replaced.

7 Conclusion

This paper designs and realizes one online web prefetching system based on the proxy server. It specifies its architecture and each part's role. We test the POLP's some functions and prove the framework is effective by series of experiments about prefetching performance using log-driven simulations.

References

1. Padmanabhan, V.N., Mogul, J.C.: Using Predictive Prefetching to Improve World Wide Web Latency. Computer Communication Review 26(3), 22–36 (1996)
2. Nicholson, A.E., Zukerman, I., Albrecht, D.W.: A Decision-Theoretic Approach for Pre-sending Information on the WWW. In: Lee, H.-Y., Motoda, H. (eds.) PRICAI 1998. LNCS, vol. 1531, pp. 575–586. Springer, Heidelberg (1998)
3. Xing, Y., Ma, S.: Modeling User Navigation Sequences based on Multi-Markov Chains. Chinese Journal of Computer 26(3), 1510–1517 (2003)
4. Xin, D., Shen, J.: A New Markov Model for Web Access Prediction. Computing in Science & Engineering 4(6), 34–39 (2002)
5. Palpanas, T., Mendelzon, A.: Web Prefetching Using Partial Match Prediction. In: 4th Web Caching Workshop, San Diego, California (1999)
6. Deshpande, M., Karypis, G.: Selective Markov Models for Predicting Web Page Accesses. ACM Transactions on Internet Technology 4(2), 163–184 (2004)

7. Pitkow, J., Pirolli, P.: Mining Longest Repeating Subsequences to Predict World Wide Web Surfing. In: USENIX Annual Technical Conference, Monterey, USA, pp. 139–150 (1999)
8. Markatos, E.P., Chronaki, C.E.: A Top-10 Approach to Prefetching the Web. In: 8th Annual Conference of the Internet Society, Geneva, Switzerland (1998)
9. Shi, L., Gu, Z., Wei, L., Shi, Y.: Quantitative Analysis of Zipf's Law on Web Cache. In: Pan, Y., Chen, D.-X., Guo, M., Cao, J., Dongarra, J. (eds.) ISPA 2005. LNCS, vol. 3758, pp. 845–852. Springer, Heidelberg (2005)
10. Venkataramani, A., Yalagandula, P., Kokku, R., et al.: The Potential Costs and Benefits of Long-Term Prefetching for Content Distribution. Computer Communications 25(4), 367–375 (2002)
11. Jiang, Y., Wu, M., Shu, W.: Web Prefetching: Costs, Benefits and Performance. In: 11th World Wide Web Conference, Hawaii, USA (2002)
12. Wu, B., Kshemkalyani, A.D.: Objective-Greedy Algorithms for Long-Term Web Prefetching. In: 3rd IEEE International Symposium on the Network Computing and Applications, Cambridge, pp. 61–68 (2004)
13. Yang, Q., Li, T.Y., Wang, K.: Building Association-Rule Based Sequential Classifiers for Web-Document Prediction. Data Mining and Knowledge Discovery 8(3), 253–273 (2004)
14. Nanopoulos, A., Katsaros, D., Manolopoulos, Y.: A Data Mining Algorithm for Generalized Web Prefetching. IEEE Transactions on Knowledge and Data Engineering 5(5), 1155–1169 (2003)
15. Xu, C., Ibrahim, T.I.: A Keyword-Based Semantic Prefetching Approach in Internet News Services. IEEE Transactions on Knowledge and Data Engineering 16(5), 601–611 (2004)
16. Wang, L., Zhang, L., Shu, Y., et al.: Study of Measurement-Based Web Prefetch Control. In: 2000 Canadian Conference on Electrical and Computer Engineering, Nova Scotia, Canada, vol. 1, pp. 204–208 (2000)
17. Zhu, P., Lu, X.: Traffic Smoothing of WWW Presending. Chinese Journal of Computers 22(6), 668–671 (1999)
18. Domènech, J., Gil, J.A., Sahuquillo, J., et al.: Web Prefetching Performance Metrics: A Survey. Performance Evaluation 63(9), 988–1004 (2006)
19. Shi, L., Han, Y., Ding, X., et al.: A SPN-based Integrated Model for Web Prefetching and Caching. Journal of Computer Science and Technology 21(4), 482–489 (2006)
20. Su, Z., Yang, Q., Lu, Y., et al.: WhatNext: A Prediction System for Web Requests Using N-Gram Sequence Models. In: 1st International Conference on Web Information Systems Engineering, pp. 200–207 (2000)
21. Bouras, C., Konidaris, A., Kostoulas, D.: Predictive Prefetching on the Web and its Potential Impact in the Wide Area. World Wide Web: Internet and Web Information Systems 7(2), 143–179 (2004)
22. Fan, L., Cao, P., Jacobson, Q.: Web Prefetching Between Low-Bandwidth Clients and Proxies: Potential and performance. In: ACM SIGMETRICS 1999, Atlanta, Georgia (1999)
23. Ban, Z., Gu, Z., Jin, Y.: An Online PPM Prediction Model for Web prefetching. In: 9th ACM International Workshop on Web Information and Data Management, Lisboa, Portugal (2007)
24. Ban, Z., Gu, Z., Jin, Y.: A PPM Prediction Model Based on Stochastic Gradient Descent for Web Prefetching. In: IEEE 22nd International Conference on Advanced Information Networking and Applications
25. Ban, Z., Gu, Z., Jin, Y.: Selection of Training Period Based on Two-Window. In: 10th International Conference on Advanced Communication Technology, pp. 2043–2047. IEEE Computer Society Press, Gangwon-Do (2008)

Mapping the Intellectual Structure by Co-word: A Case of International Management Science

Hongjiang Yue[1,2]

[1] School of Management,Nanjing University,
Nanjing,China, 210093
[2] Institute of Public Administration and Performance Evaluation,
Nanjing Audit University, Nanjing, China, 210029
yuehj@nau.edu.cn

Abstract. The aim of this study is to mapping the intellectual structure of the field of international management science during the period of 1997-2006. Text mining and co-word analysis was employed to reveal patterns and trends in the international management science field by measuring the association strengths of terms representative of relevant publications. Data were collected from selected 52 kinds of international journal papers collected from Social Science Citation Index (SSCI). In addition to the keywords added by SSCI databases. Observe the co-word try to distinguish its dynamic development and evolution, and with the help of discipline knowledge to explain these information, mine the knowledge mode of its research field. The results show that the international management science field has some established research themes and it also changes rapidly to embrace new themes.

Keywords: intellectual structure, co-word analysis, international management science.

1 Introduction

Along with the development of computer, co-word analysis becomes one of the effective methods for mining lexical semantics and text content architecture, also becomes a useful method for extracting text knowledge. Co-word analysis mainly counts the co-occurrence frequency of the technical terms (words or phrases), with this as basis to take datamation processing to these technical terms, accordingly proclaim the similarity or dissimilarity of these technical terms, and then analysis the structure alteration of discipline and subject presented by these technical terms.

From the words in the literatures, and take the high co-occurance frequency keywords as research object, the understanding to the relationship of the conceptions reflected by different words can be obtained, and a "more meaningful ubstance" can be formed. Take these "meaningful ubstance" as subject, sub-subject or specific tools, accordingly build up a network desciption to knowledge structure[1], [2]. The words in literatures can generally describes text knowledge unit relationship and text research topic relationship. Accordingly forms description to the discipline structure

F.L. Wang et al. (Eds.): WISM 2012, LNCS 7529, pp. 621–628, 2012.

and its relationship with relative fields, furtherly analysis the discipline and subject structure presented by these words, by comparing the discipline structure in different periods and space, to obtain acknowledgement to the discipline's develop, intersect, seep and vicissitude tendency[3].

2 Research Method and Data

The general procedure for co-word analysis research can be summarized as: confirm research object, collect and extract relative literature set, mine text property set from literature, build up formalized matrix and take statistical analysis, mapped visualization chart display and the result analysis and explaination.

Word is an important index for scientometrics and bibliometrics research. There are two essential methods can be used to extract words and phrases from technical texts, one is the keywords, subjects and summaries cited by academic journal papers, conference papers or research reports; the other one is the words or phrases extracted from complete technical texts[4], [5].

Keyword has already becomes an essential constituent for the papers published by academic journals, is a necessary supplement for dissertation topic, and is the substantive terms or phrases which can furthest explain the subject and reflect centric content. Keyword is the natural language word obtained by direct random selection from textual topic, summary or full-text, owns material meaning, and processed by normative or non-normative approach. As the common mark for databank recruits and retrieves literature, the effect of keywords is to make readers to be clear with the works' main content at a glance.

Intercept high frequency keywords according to the variation of high frequency keywords' accumulative frequency. Take the front 150 keywords whose accumulative frequency reached up to 62.5% of the total keyword frequency as research object; because these keywords own high occurrence frequency in the management science research, they present the hotspot of current management science research. In order to furtherly reflect the relationship of these words, build up a co-word matrix for these high frequency keywords: doubly count their occurrence frequency in the same dissertation, and name it as co-occurrence frequency, accordingly forms a 150×150 co-word matrix. This matrix is named as symmetric matrix, in off-diagonal, the unit value is the co-occurrence frequency of the high frequency keywords, the data of leading diagonal is defined as missing values.

Law and Whittaker thought, firstly, the keywords cited by papers can reflect the current situation of science research without any doubt, secondly, the viewpoints accepted by the other scientists can effect the science papers which published with similar keywords[6], [7]. Co-word analysis method is based on such hypothesises. If these presuppositions are tenable, it is probable to reflect the text conception with the co-occurrence frequency of the word pairs in the papers.

With co-word matrix as base to analysis the connect strength of these words, firstly utilize the Pearson's correlation coefficient to transfer the initial matrix into correlation matrix, then takes hierarchical clustering. The adoptive method for the visualization of co-word structure is multidimensional scaling and social network map, through the visual map, the structure and variation of discipline and subject of management science can be directly seen.

In the social network map of each word group, the nodal point link presents the co-occurrence strength of the keywords, the co-occurrence strength takes use of Salton index[8],

$$\text{Salton} = C_{ij} / \sqrt{N_i N_j} \tag{1}$$

presents the co-occurrence frequency of keyword i and j, presents the occurrence frequency of keyword i, presents the occurrence frequency of keyword j. Salton value is between 0 and 1, the bigger is the value, the higher is the co-occurrence frequency strength of the keywords, this index can excellently proclaim the co-occurrence strength of these keywords. In this section, the nodal point link strength of the social network map is defined as Salton≥0.2.

In the visual social network map, the nodal points present the co-corrence keywords, the link of the nodal points presents the two nodal points exist strong co-occurrence relation. It can be thought that, the nodal points lie in the network center are the center conceptions of word groups, the other keywords associated with the center conceptions reflect the research content of the word groups. The research content with keywords as presentation get detailed and full-scale display in the two-dimensional space.

In order to understand the inner connection and reciprocity of different word groups, can measure the centrality and density of word groups. The centrality is used to measure the interrelation between one and the other word groups, the bigger is the interrelation strength between one and the other word groups, the nearer to the centre stage is the word group in the whole discipline. Density is used to measure the interrelation strength of the keywords in the same word group [9].

In order to fully understand the developed discipline structure of international management science in 21 century, this dissertation selected 52 kinds of international journal papers published in ten years from 1997 to 2006 as data source. With co-word analysis method to proclaim the subject and knowledge structure of the international management science research, with visualization method to directly understand the co-relation between content architecture and thematic structure of the discipline knowledge. Observe the co-word try to distinguish its dynamic development and evolution, and with the help of discipline knowledge to explain these information, mine the knowledge mode of its research field.

3 Result Analysis

3.1 High Frequency Keywords

Through statistics and analysis to the dissertation keyword frequency, can quantize the attended degree of some research object or the utilized degree of some research method, accordingly analysis the hotspot of some discipline or some research field. Take statistics to the occurrence frequency of all keywords in the selected journal papers; find 1001 different keywords in the utilized databank.

On the whole, in the past 10 years, the hottest keyword for the international management science is performance. From 1997-2001 to 2002-2006, this word lied in the first position. After the performance conception was brought up, researchers continue to open the intension and extension of this word, performance management is becoming a human resource management procedure with extensive approval. Then is innovation. Many scholars continue to take research to the innovation issue; it is still a hot topic in management science field before and after the 21st century. If add up the two keywords model and models, its occurrence frequency· is only next to performance. If translate model(s) into model, which is also one of the high frequency words used to describe the research method. Strategy is treated as an important research field is started at the 1960th. The conventional research issues of management science including knowledge, organizations, firm, industry, behavior, which are still attractive research objects.

3.2 Discipline Structure Based on Co-word

Utilizing Hierarchical clustering method to cluster the 150 high frequency keywords in 1997-2006, according the cluster result to divide the high frequency keywords of this period into four word groups. In order to display the positional concepts of each word group, select 5 high frequency keywords from the MDS visualization histogram, respectively present each word group; the word groups' different positions display the relations of keywords in each word group.

Figure 1 displays the MDS histogram of the four word groups in 1997-2006; respectively take the front 5 high frequency keywords of each word group as its representation and display.

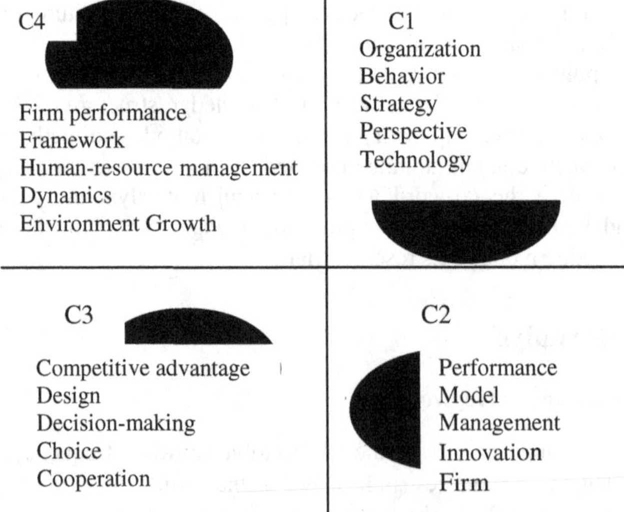

Fig. 1. MDS histogram of the four word groups

Word group C1 contains high frequency keywords including organization, behavior, strategy, perspective and technology, etc; word group C2 contains high frequency keywords including performance, model, management, innovation and firm, etc; word group C3 contains high frequency keywords including competitive advantage, design, decision-making, choice and cooperation, etc; word group C4 contains high frequency keywords including firm performance, framework, human-resource management, dynamics and environment growth , etc.

In order to display the detailed network structure of the four word groups, figure 2-5 respectively display the inner network connect visualization chart of each word group. If you have more than one surname, please make sure that the Volume Editor knows how you are to be listed in the author index.

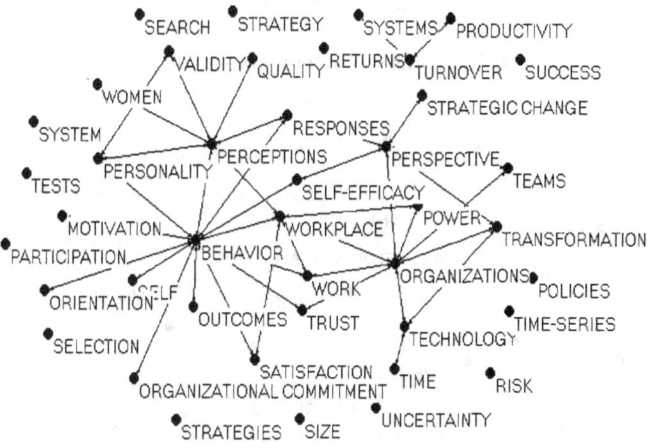

Fig. 2. The visualization for the word group (C1)

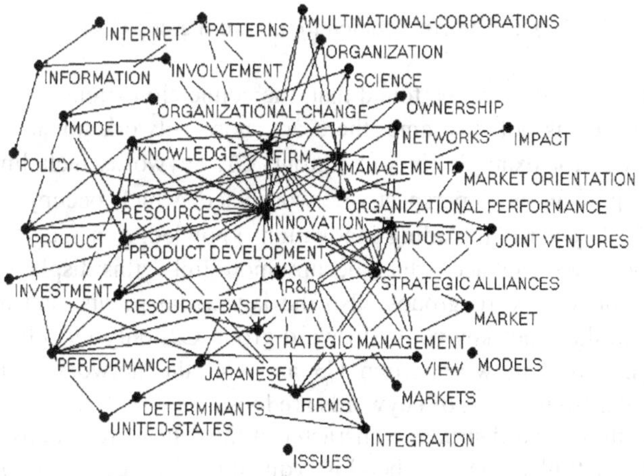

Fig. 3. The visualization for the word group (C2)

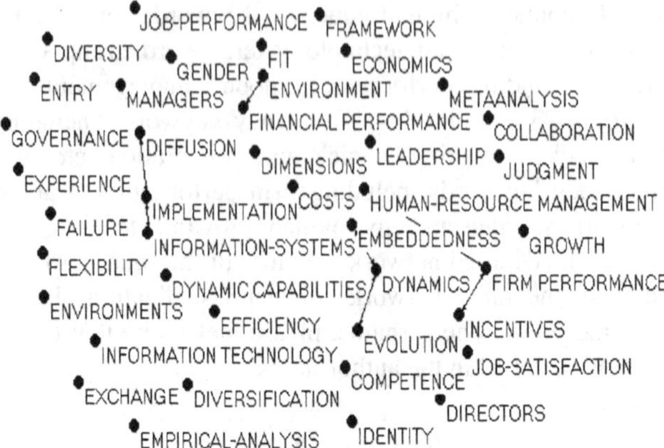

Fig. 4. The visualization for the word group (C3)

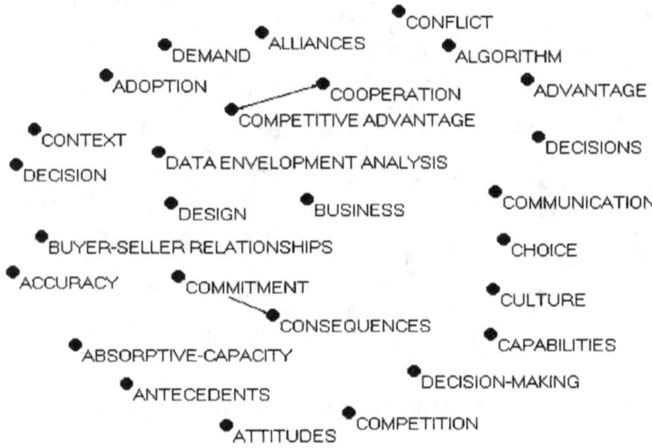

Fig. 5. The visualization for the word group (C4)

From the above chart it can be found that, the connection strength of the keywords in each word group is not the same, there are more network nodal points in word group C1 and C2, for example, in word group C1, keywords including behavior、organizations、perceptions、perspective、workplace, which occupy more links; in word group C2, keywords including innovation, knowledge, management, industry, resources, firm(s), performance, which occupy more links, this displays that, they are the active keywords in word groups, also the key nodal points, and own extensive connections with the other keywords. In word group C2, only models and issues are isolated points. While in word group C3 and C4, the network nodal points are relatively less, the connections of keywords are loose.

Table 1 lists the co-word structure character of the four word groups in 1997-2006. In this table, outlinks means the link quantity between one keyword and the other keywords in a word group. Intraconnection means the link quantity of the keywords in a word group. Intraconnection percentage means the ratio between the intraconnection and the intraconnection and outlink sum, and outlink percentage means the ratio between the outlink and intraconnection and outlink sum. There are many different calculation methods for centrality and density, in this dissertation, the centrality takes the ratio between outlink quantity and keyword quantity; density takes the ratio between intraconnection and keyword group.

Table 1. Co-word structure of the four word groups

	C1	C2	C3	C4	Average
Keyworde	42	40	41	27	37.5
Inlinks	158	162	57	61	109.5
Outlinks	80	216	14	4	78.5
Outlinks (%)	66.4	42.9	80.3	93.8	58.2
Inlinks (%)	33.6	57.1	19.7	6.2	41.8
Centrality	1.05	1.08	0.38	0.41	-
Density	1.91	5.40	0.34	0.15	-

From table 1, it can be found that, the intraconnection strength of word group C1 and C2 is bigger than their outlink strength, while word group C3 and C4 are on the contrary.

Word group C2 owns the highest centrality and density in the whole research network, its density is much higher than centrality. This word group not only owns intense inner connection, also owns extensive connection with the other word groups. This explains that, this word group lies in the core of all research subjects, and the research subjects composed by this word group are tending to maturate. From figure 4, it also can be found that, the main nodal points for this word group are the keywords including innovation、knowledge、management、industry、resources、firm(s)、performance, accordingly it is not difficult to understand the positions occupied by the word groups composed by these high frequency words in the international management science research.

In the whole research network, word group C1 occupies higher density but lower centrality, the inner connection of this word group is intense, this explains that, the research of these fields have already been cared, and have got good researchment. This word group is a network mainly composed by the high frequency keywords including behavior、organizations、perceptions、perspective、workplace.

Centrality and density of the word group C3 and C4 are low, their research subjects lie in the edge of the whole research field, and the research is immature. The two word groups almost don't have important network nodal points.

4 Conclusion and Discussion

This study demonstrates the feasibility of co-word analysis as a viable approach for extracting patterns from, and identifying trends, in large corpora where the texts collected are from the same domain or sub-domain and are divided into roughly equivalent quantities. Hence, this examination points the way for further research into international management science field and well-restricted sub-domains. On the other hand, co-word analysis successfully visualizes the inter-relations of the keywords and sub-fields of international management science, while the importance of visualizing methods in the convincing presentation of results has not been sufficiently understood in the past. Co-word analysis opens a new opportunity for cartography of science and information visualization. The co-word results have produced a great deal more than statistical artifact. This study aimed to exploit the visualization effect of the co-word maps to the aid of searchers in the international management science domain. Overall, this study has led us to an increased confidence in the co-word analysis.

Acknowledgments. This work was supported by supported by the Humanities and Social Sciences Planning Fund of Chinese Ministry of Education under grant No. 11YJC630272, supported by China Postdoctoral Science Foundation, under grant No. 20110491375, supported by the Social Science Foundation of Jiangsu Province, China, under grant No. 10ZZC005. supported by A Project Funded by the Priority Academic Program Development of Jiangsu Higher Education (Auditing Science and Technology), supported by Jiangsu Province College of Humanities and Social Science Fund, under grant No. 2011SJB630037.

References

1. Rapp, R.: The computation of word associations: comparing syntagmatic and paradigmatic approaches. In: Proceedings of the 19th International Conference on Computational Linguistics, vol. 1, pp. 1–7. Association for Computational Linguistics, Morristown (2002)
2. Mittendorf, E., Mateev, B., Schauble, P.: Using the Co-Occurrence of Words Forretrieval Weighting. Information Retrieval 3, 243–251 (2000)
3. He, Q.: Konwledge Discovery Through Co-Word Analysis. Library Trends 48, 133–159 (1999)
4. Braam, R.R., Moed, H.F., van Raan, A.F.J.: Mapping of Science by Combined Co-Citation and Word Analysis. I. Structural aspects. Journal of the American Society for Information Science 42, 233–251 (1991)
5. Braam, R.R., Moed, H.F., van Raan, A.F.J.: Mapping of Science by Combined Co-Citation and Word Analysis. II: Dynamical aspects. Journal of the American Society for Information Science 42, 252–266 (1991)
6. Börner, Chen, C., Boyack, K.: Visualizing Knowledge Domains. In: Cronin, B. (ed.) Annual Review of Information Science & Technology, vol. 37, ch. 5, pp. 179–255. Information Today, Inc./American Society for Information Science and Technology, Medford, NJ (2003)
7. Law, J., Whittaker, J.: Mapping Acidification Research: a Test of the Co-Word Method. Scientometrics 23, 417–461 (1992)
8. Zhang, J., Rasmussenb, M.: Developing a New Similarity Measure from Two Different Perspectives. Information Processing & Management 37, 279–294 (2001)
9. Callon, M., Courtial, J.P., Laville, F.: Co-Word Analysis as a Tool for Describing the Network of Interactions between Basic and Technological Research: The Case of Polymer Chemistry. Scientometrics 22, 155–205 (1991)

Study on Multi-sensor Information Fusion Technology in the Dynamic Monitoring of Coal Mine Roof*

Yong Zhang[1], Zhao-Jun Liu[2], and Yun-Fu Cheng[1,**]

[1] Dept of radioactivity, Taishan medical college, China
[2] Department of Information Science, Taishan medical college, China
gczkyanchen@126.com

Abstract. China is the world's largest coal producer and consumer countries. With the continuous progress of the modernization drive, the demand for coal is more and more, but coal mine safety situation is still grim. All of various types of coal mine accidents in production, the roof accident continued to hold the forefront. As we known, coal production is a dynamic process, a variety of parameters and variables are subject to change, yet all variables are not isolated, they are interconnected. These factors are from a different side to reflect a change or movement of the roof, so they are multi-information fusion integration. In this study, we take Xing-Cun village coal mine, which locate on Shandong province, for an example. In E3207 working face, underground 1000 meters, five kinds of sensors are installed. They are micro seismic sensors, fully mechanized working face powered support sensor, roof-to-floor convergence, two sides convergence of roadway, and borehole stress sensor. The ground control central can collected above five kinds of data in real-time. Using data fusion method, we setup up the model of fuzzy logic. Through data analysis, the five membership function are given, finally we get the fusion curve. By the curve we can identify the roof movement consistent with the law production practices, analysis the overlying rock strata, with good guidance to production. Facts have proved that the information can be fused to fully reflect the comprehensive and complementary data, improve the accuracy of the warning of roof movement, has positive significance to ensure safety in production.

Keywords: information fusion, mine roof, multi sensor, safety warning, fuzzy logic.

1 Introduction

With the development of China, the demand for coal mine is more and more. And coal-related security incidents frequently occur, and the security situation remains grim. All of various types of coal mine accidents in production, the roof accident continued to hold the forefront. As we known, Coal production is a dynamic process, a variety of parameters and variables are subject to change, but all these variables are

---

* This work is supported by the Natural Science Foundation of Shandong Province, China(Number: ZR2011EL019).

** Corresponding author.

F.L. Wang et al. (Eds.): WISM 2012, LNCS 7529, pp. 629–636, 2012.

not isolated, they are interconnected. These factors are from a different side to reflect a change in the roof, is a question of multi-information fusion integration. In recent years, scholars have made a lot of research work of information fusion. Some applications had also been in the coal field.

The multi-sensor information fusion technology is a new discipline leads by modern warfare as early as the 1970s. Multi-sensor information fusion refers to data from multiple sensors and multi-level, multi-faceted, multi-level processing, resulting in new and meaningful information, and this new information can not be obtained in any single sensor [1]. [2]Set up roof integrated dynamic monitoring system, but it only applied statistical pattern recognition to deal with the real-time information of roof. Model for safety of coal mine roof based on evidence fusion theory discussed the method, applying combination rule of Dempster-Shafer (D-S) evidence theory, showed the mine roof safety status [3]. [4] Using information fusion theory predicts the mine floor water irruption.

From above we can see those method does not using real-time data extraction and analysis; Failed to identify the nature of links between the various variables. Data fusion method could be improved. In this paper, the author proposes a multi-sensor information fusion method based on fuzzy logic, the method using for dynamic monitoring and control of the roof. It includes five sections.

2 Individual Systems and Monitoring Principles

In this study, we take Xing-Cun village coal mine for an example, which locate on Shandong province. In E3207 working face, five kinds of sensors are installed. They are micro seismic sensors, fully mechanized working face powered support sensor, roof-to-floor convergence, two sides convergence of roadway, and borehole stress sensor. The control central can collect real-time mining parameters and analysis and processing. Next we will introduce the working principle of five independent sensors.

2.1 Micro Seismic System

The new generation SOS micro seismic monitoring instrument purchased from Portland. It was design and manufacture by Polish Mining Research Institute of Mining Institute of Seismology. The main purpose of micro seismic is predicting rock burst.

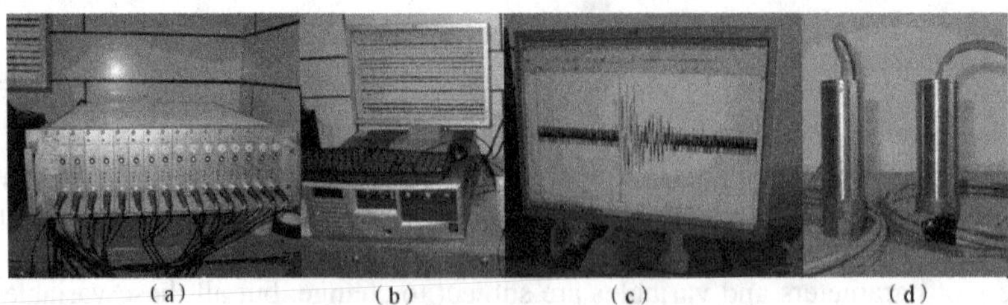

Fig. 1. It is the working principle of micro seismic monitoring system. It includes four sub graphs, (a) is 16-channel DLM-SO signal acquisition station; (b) is recorder system; (c) is Analyzer and (d) is detector measuring probe.

The SOS micro seismic monitoring system can be achieved, including the rock burst on mine earthquake signals over long distances no more than 10km in real time, dynamically and automatically monitoring, getting full vibration waveform of the rock burst and mine earthquake signal. Software can accurately calculate the energy greater than 10^2J of the coal-rock shock occurred at a time, include energy and three-dimensional space coordinates, to determine each mine earthquake shock type, to determine the vibration of the power source rock mine pressure level of risk assessment and forecast. And through the application of the SOS micro seismic monitoring system, engineer can analyze the mine overburden fracture, describe the migration of the space rock structure motion and stress field evolution for coal mine safety production. Fig. 1 is the working principle.

2.2 Fully Mechanized Working Face Resistance of Support

The resistance is connected with shield powered support. Using strain gauge pressure sensors or vibrating wire pressure sensor. The emulsion of the pillars of the internal pressure is delivered to the sensor, the sensor output of the analog signal converted by the circuit after the computer acquisition. The total collection real-time pressure information includes three parts: the anterior column pressure, posterior column pressure and the probe beam pressure.

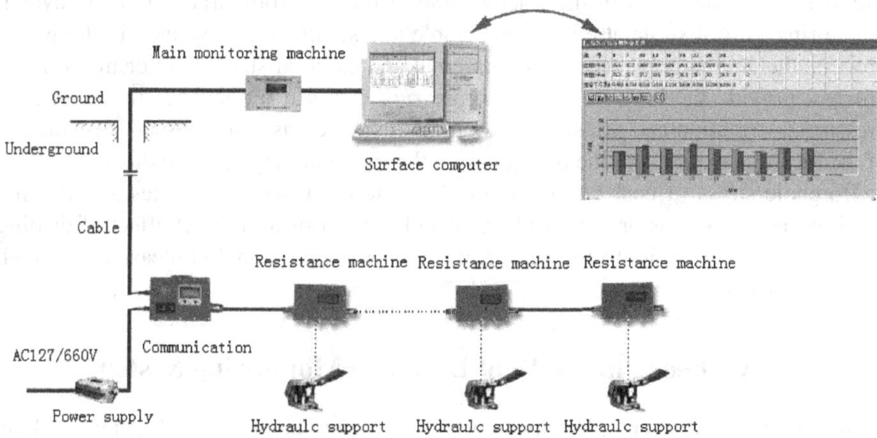

Fig. 2. is the structure of mining shield working resistance system. The sensors are installed in the stope. With the promoting of the working surface, the working resistances of supports are observation, The size of the total pressure provide the basis for research and rational bracket-wall rock relations, roof control and management.

Step and strength of the working face support resistance monitoring role mainly in: master the law of cycle pressure and strata behavior; analyze and verify the adaptability of the supports on the roof conditions; adjust supports reasonable control top state; roof accident prediction that may occur; bracket failure rate of the hydraulic system monitoring, and easy to correctly guide the production.

2.3 Roof-to-Floor Convergence

The sensors are used to detect roof-to-floor convergence, usually the convergence ratio are recorded everyday. Detection sensor includes: inductive displacement sensor, a capacitive displacement sensor, magneto resistive displacement sensor and so on .General instrument perpendicular to the roof and floor installation, wood braze or drill rod set the reference point can be used in the installation of a relatively soft floor. It can also be tilted installation. The resolution of the measurement is of 0.1-0.01mm. From the moving speed, the staff can master the law of strata movement, it can also inferring the stability of surrounding rock and strata behaviors.

2.4 Two Sides of Roadway Convergence

Tunnel convergence generally refers to the relative displacement between the two opposite sides, it common first set of basis points, then measure the deformation by displacement method. The displacement is different from geological condition. Similar to the roof-to-floor convergence, the convergence ratio are also recorded everyday.

2.5 Borehole Stress

The system is located in the face ahead of the roadway, the sensor is located in the stress field on both sides of the fracture line of the face roof strata end and beyond, monitoring of roof strata at a relatively stable and significant movement in the whole process, the abutment pressure in the internal and external stress field changes in the process, provide the basis for the rock pressure is based on changes in the bearing pressure distribution and transfer decision-making, such as peak pressure distribution of the location, extent, and the roof pressure the mid-and long-term prediction.

Borehole stress of coal or rock is used to internal vertical stress testing. Playing level in the coal seams or rock drilling, install the sensor to a direct effect of drilling deep, coal or rock stress to the sensor, the sensor output signal is measured through the secondary instrument.

3 Fuzzy Theory in the Roof Dynamic Monitoring System

Coal mining is a very complex, dynamic process. The factors that affect the working face stability are much, such as geological conditions, mining conditions, equipment conditions, management level and so on. These conditions have fuzzy characteristics. Different mines have different feature. It is not possible to quantitatively describe the precise variable or algorithm. Therefore, in this paper, the author proposes the application of fuzzy mathematics approach to create the roof dynamic response model.

3.1 The Set Up of Information Fusion Model Base on Fuzzy Logic

The model shows in Fig.3, with detailed descriptions of the modeling methods and ideas. From Fig.3 we can see the model has five inputs, micro seismic data, fully mechanized working face powered support, roof-to-floor convergence, two sides convergence of roadway, and borehole stress. The value comes from the same roadway; weighted dimensionless value of between 0 and 1, and I called it the eigenvalue of the tunnel roof movement. The size of the eigenvalue reflects the state of the roof movement. The model has an output, to reflect the state of motion of roof characteristic; its value can change with time.

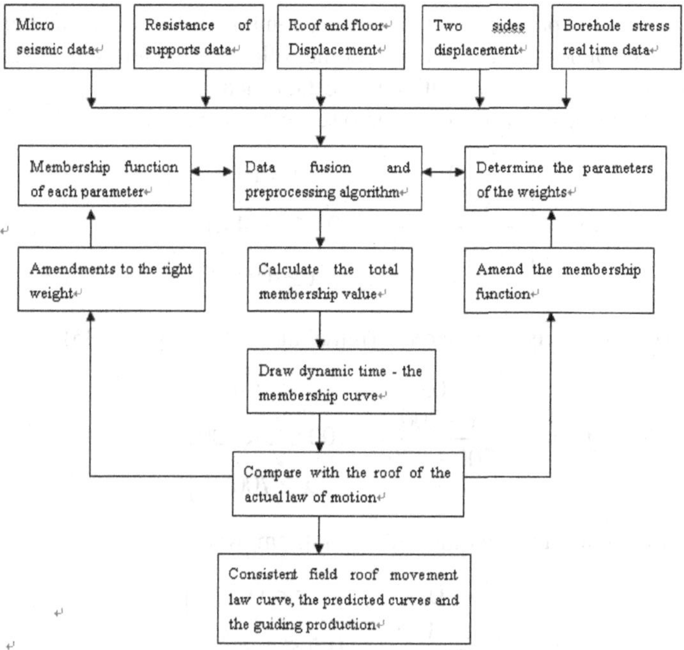

Fig. 3. Instruct the model of information fusion of monitoring system base on fuzzy logic

3.2 The Membership Function to Determine the Method of Classification

Membership function of fuzzy control applications based on one of the keys to the membership function of a properly constructed is the ability to make good use of fuzzy control. The membership function to determine the process, in essence, should be objective and understand their differences, but everyone with a vague understanding of the concept, therefore, the membership function to determine and subjective. Fuzzy set membership function curve by fuzzy statistics based on indicators of the score distribution, according to the shape of the curve to select the appropriate function expression, we can get the corresponding membership function. Commonly it used membership functions of fuzzy statistics, example of law, and expertise in law, and dual contrast sort.

4 Experiments

Let us take Xing-Cun village Coal Mine, the E3207 working face, for an example. The face is located underground 1000 meters. The experiment time was from

3/13/2011 to 3/28/2011, the data from underground 1000m, five kinds of sensors parameters are collected. According to data fusion method, we get the fusion curve; they are shown on Fig.4 and Fig.5.

4.1 Membership Function

According to working principle of the sensors, combination of expert and engineering and technical personnel, field staffs, we get the following membership function.Semi-trapezoidal and trapezoidal distribution was established as follows.

(1) The membership function of the micro seismic system (initial unit J, weights 0.35)

$$f_1(x)=\begin{cases} 0, & x<10^3 \\ \dfrac{x-10^3}{4\times10^4-10^3}, & 10^3\le x<4\times10^4 \\ 1, & x\ge 4\times10^4 \end{cases} \tag{1}$$

(2) Resistance of supports data function (initial unit KN, weights 0.35)

$$f_2(x)=\begin{cases} 0, & x<200 \\ \dfrac{x-200}{700-200}, & 200\le x<700 \\ 1, & x\ge 700 \end{cases} \tag{2}$$

(3) Roof-to-floor convergence ratio (initial unit mm, weights 0.1)

$$f_3(x)=\begin{cases} 0, & x<0.5 \\ \dfrac{x-0.5}{1.5-0.5}, & 0.5\le x<1.5 \\ 1, & x\ge 1.5 \end{cases} \tag{3}$$

(4) The two sides convergence of roadway membership function (the initial units of mm, weights 0.1)

$$f_4(x)=\begin{cases} 0, & x<0.5 \\ \dfrac{x-0.5}{1.5-0.5}, & 0.5\le x<1.5 \\ 1, & x\ge 1.5 \end{cases} \tag{4}$$

(5) Borehole stress function (initial units of Mpa, weights 0.1)

$$f_5(x)=\begin{cases} 0, & x<4 \\ \dfrac{x-4}{7-4}, & 4\le x<7 \\ 1, & x\ge 7 \end{cases} \tag{5}$$

4.2 Data Fusion Method

According to the above membership functions formula to calculate the membership value of each parameter:

$$U = \{U_1, U_2, U_3, U_4, U_5\} \tag{6}$$

1) According to the five parameters on the roof dynamic monitoring results, selected the values of various parameters of the right, on this study, the weights are 0.35, 0.35, 0.1, 0.1 and 0.1.

$$W = \{W_1, W_2, W_3, W_4, W_5\} \tag{7}$$

2) Calculation of the total membership model. The result of A is between 0 and 1. Weight can be modified according to the specific circumstances and working field data.

$$A = \{U_1 \times W_1 + U_2 \times W_2 + U_3 \times W_3 + U_4 \times W_4 + U_5 \times W_5\} \tag{8}$$

3) Generate time-varying data fusion curve and optimization model

4.3 Experimental Results

The results are shown on Fig.4 and Fig.5. Fig.5 highlights the fusion curve. On the graph, A is micro seismic data; B is resistance of fully mechanized powered support; C is roof-to-floor convergence; D is two sides convergence of roadway; E is borehole stress and F is the result of fusion curve. All data have been converted into the value between 0 and 1.

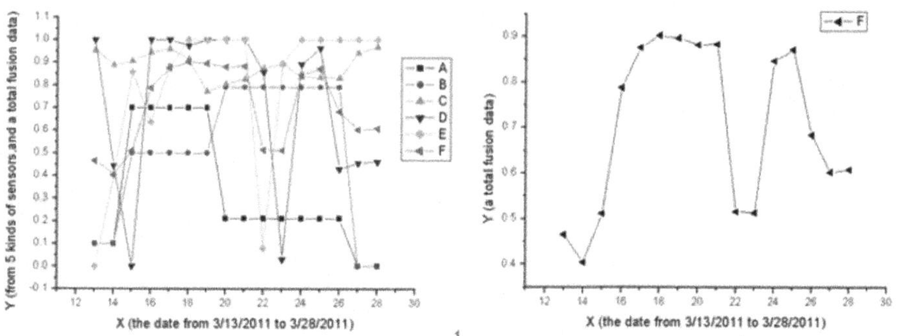

Fig. 4. The data form five sensors and fusion curve

Fig. 5. The fusion curve

5 Conclusion

From above graph we can see the roof experienced a cycle pressure process. During the period from 3/17/2011 to 3/21/2011, the direct roof pressure occurred, strata behavior markedly enhanced than usual, and micro seismic events and vibration energy significantly enhanced. On the day 3/21, direct roof has been breaking down. But these are normal monitoring situation, without beyond the pre-set warning value. Through Fig.5 we can got the movement law of roof.

In fact, in the early stages of model building, there are no ready-made curves and the actual mining conditions consistent and, we had to find the membership function, the weights based on site conditions and experience. Finally we must determine whether the fusion curve we obtained line with actual production. If not, the weights of each parameter, must be recalculated, the model must be modify and optimize, and thus infinite approximation of the actual exploitation of the dynamic model.

In conclusion, in this paper, information fusion method was used to deal with fives different sensors data, finally finding a curve matching of the dynamic variation of the roof. From this we can draw a conclusion: information fusion model can reflect the roof moving behavior, has positive significance to guide mine coal safety production.

References

1. Llinas, J., Waltz, E.: Multi sensor Data Fusion. Artech House, Morwood (1990)
2. Zhang, Y.: Study on integrated technology based on roof dynamic monitoring. Shan Dong University of Science and Technology (2009)
3. Jia, R.-S., Sun, H.-M., Yan, X.-H.: Model for safety evaluation of coal mine roof based on evidence fusion theory. Journal of China Coal Society, 1496–1500 (2010)
4. Li, L., Cheng, J.-L.: Floor water irruption prediction based on information fusion. Journal of China Coal Society, 623–626 (2006)
5. Zhou, H., Li, S.-H.: Combination of support vector machine and evidence theory in information fusion. Chinese Journal of Sensors and Actuators, 1566–1570 (2008)

Detecting Hot Topics in Chinese Microblog Streams Based on Frequent Patterns Mining*

Weili Xu[1], Shi Feng[1,2], Lin Wang[1], Daling Wang[1,2], and Ge Yu[1,2]

[1] School of Information Science and Engineering, Northeastern University
[2] Key Laboratory of Medical Image Computing (Northeastern University), Ministry of Education, Shenyang 110819, P.R.China
xuweili_18@163.com, wanglin0204@qq.com, {fengshi,wangdaling,yuge}@ise.neu.edu.cn

Abstract. Microblog plays a more and more important role on the emerging and propagation of the public opinion on the Web. Although topic detection has long been a hot research topic, the characteristics of microblog make it a non-trivial task. In this paper, we propose a novel hot topic detection approach based on keyword extraction and frequent patterns mining. We analyze the characteristics of hot topic microblogs and the topical keywords are extracted according to the increasing rate and frequency in Chinese microblog streams. Different from traditional clustering based detection methods, in this paper we treat the short texts of microblogs as transaction items, and apply Apriori algorithm to generate the hot topics. The experiments in the real dataset verify the efficiency and effectiveness of our proposed methods.

Keywords: Microblogs, topical keyword extraction, hot topic detection.

1 Introduction

As an emerging Web 2.0 application, the microblog has become a new platform for information sharing and dissemination. Microblog users can publish short texts, pictures through the Web and mobile clients, which ensures that the users can receive the shared information immediately. Each microblog is mainly restricted to 140 characters, which makes the published microblogs become fast streaming short texts. Therefore, hot topics can spread fast. However, huge numbers of users publish their microblogs with little hot topics information. They mainly talk about personal things.

The major challenges about detecting hot topics in microblog streams include:

Mixed Topics. Since the microblogs containing hot topics are mixed up with the ones expressing personal feelings and opinions, it is necessary to design efficient algorithms to filter out irrelevant microblogs for further processing steps.

* Project supported by the State Key Development Program for Basic Research of China (Grant No. 2011CB302200-G), State Key Program of National Natural Science of China (Grant No. 61033007), National Natural Science Foundation of China (Grant No. 61100026, 60973019,), and the Fundamental Research Funds for the Central Universities (N100704001).

F.L. Wang et al. (Eds.): WISM 2012, LNCS 7529, pp. 637–644, 2012.

Limited Length. Different from traditional long news articles and blogs, the short texts in Chinese microblogs bring new obstacles to the hot topic detection problems.

To tackle the challenges, we analyze the characteristics of microblogs that contain hot topic and design algorithms to filter out the microblogs without hot topics. Some words without topical information are employed by users frequently, which interfere with algorithm. We filter out these words, and extract potential hot topics keywords.

Based on the burst and wide-spread of hot topics in microblog streams, a method to extract hot topic keywords is proposed, which considers both frequency and growth.

Hot topics can be inferred by several keywords, which appear in the similar context. We apply an improved Apriori algorithm to compute the frequency of every word (as itemset) in the high score words, and the hot topics with the maximal frequency itemset are inferred from microblog streams.

The framework of the proposed approach is shown in Figure 1. Firstly, we eliminate the most irrelevant topic microblogs. Secondly, we preprocess the topic keywords. Then we compute the keyword score and extract the high ranked keywords. Finally, we discover maximal frequent patterns from the extracted topic keywords to generate the hot topics.

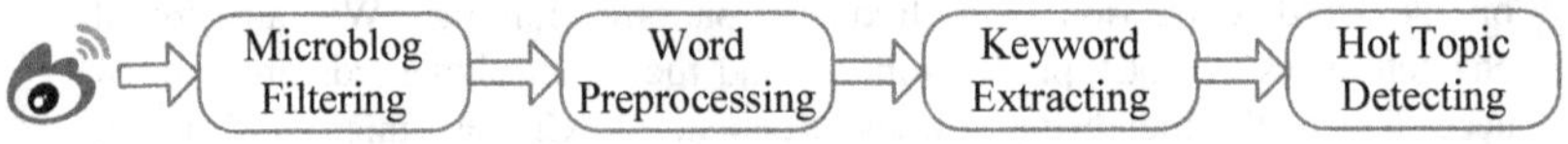

Fig. 1. The framework of the proposed approach

2 Related Work

Recently, the microblog has attracted more and more attentions in the research communities. Wayne Zhao compared Twitter [4] and traditional media using topic models, and developed a new Twitter-LDA model for short tweets [5]. Pal et al. summarized a set of features for characterizing social media authors [6]. He showed how probabilistic clustering over the new feature space, followed by a within-cluster ranking procedure, could yield a final list of top authors for a given topic. Welch et al described a detailed model of Twitter as a graph, and demonstrated important distinctions between edge types in the graph [7].

Our work focuses on the hot topics keywords extraction and hot topics detection in Chinese microblogs. Lu et al choose a two-level *K*-means-hieratical hybrid clustering method to cluster the selected tweets into news [8]. Zheng Fei-ran combined the word frequency and the increasing rate to measure the possibility of a keyword, and they constructed the topics by the conditional probability [9].

3 Microblog Preprocessing

There are large numbers of useless texts in the microblog data streams, which bring obstacles to our research. To tackle these challenges, we propose a preprocessing method based on the characteristics of the Chinese microblogs.

3.1 Microblog Filtering

We analyze the differences between hot topic microblogs and the common ones. Based on the differences, we eliminate some microblogs considering 5 factors:

(1) **Number of words.** The distributions of the number of words in common (30,745 items) and hot topic microblogs (2,000 items) are shown in Figure 2. The number of microblogs with less than 60 words in common microblogs is much larger than in topic ones. We filter out microblogs with less than 60 words.

(2) **Emoticons and smileys.** Microblogs containing emoticons and smileys usually talk about personal feelings. So we delete microblogs with emoticons and smileys.

(3) **#topic name#.** There are large numbers of microblogs containing hashtags. These kinds of microblogs are usually initiated by the microblog platform, and contain a lot of human factors. Therefore, we also filter out these microblogs.

(4) **@username.** We ignore the microblogs including "@username", because these microblogs mostly appear in the conversational contexts. They have directionality. However, the hot topic microblogs have no directionality, and have no possibility of including the "@username". We ignore these microblogs containing @username.

(5) **URL.** Many companies publish the advertisement microblogs with the form of URL. Hot topic microblogs rarely contain URL. So, we filter out these microblogs.

The examples of the last four kinds of microblogs are shown in Table 1.

Table 1. The examples of the filtered microblogs

Characteristic	Example
Emoticons, smileys	今天好开心啊！ (What a happy day!)
#topic name#	#男女距离#，50厘米(#distance between boys and girls#, 50 cm)
@username	@Xuweili，好想你啊!(@Xuweili, I miss you so much!)
URL	工具不错：http://t.cn/zOKyq9C (good tool : http://t.cn/zOKyq9C)

3.2 Hot Topic Keywords Preprocessing

We make the word segmentation on the remaining microblogs. Each microblog is represented as a word vector, and each word is labeled with part-of-speech tags. Since the nouns and verbs can express hot topic information, we select the nouns and verbs.

The most frequent words in microblogs always have nothing to do with the hot topics. The statistical distribution of word-frequency in one day is showed in Figure 3. We manually tagged the top ranked words without topic information, and finally 297 words are selected. These words spend much virtual memory and much time when computing their weights. So we delete these common words as well as the stop words.

4 Hot Topic Detection

Topic detection has long been a hot research area, and many research works have been done. The characteristics of microblogs make it difficult in microblog text streams. In this section we propose a new keyword extraction approach.

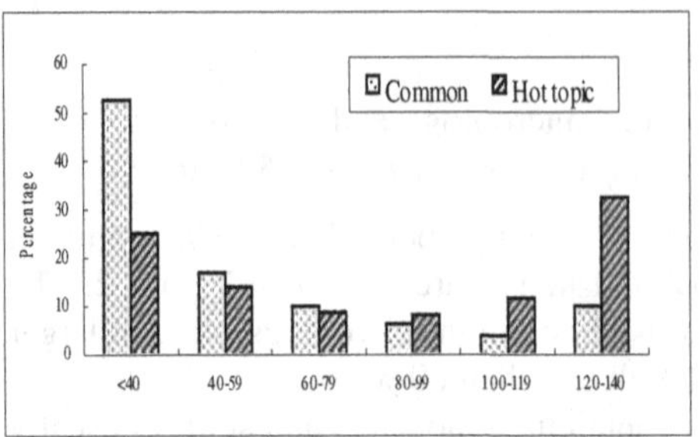

Fig. 2. The distributions of the number of words

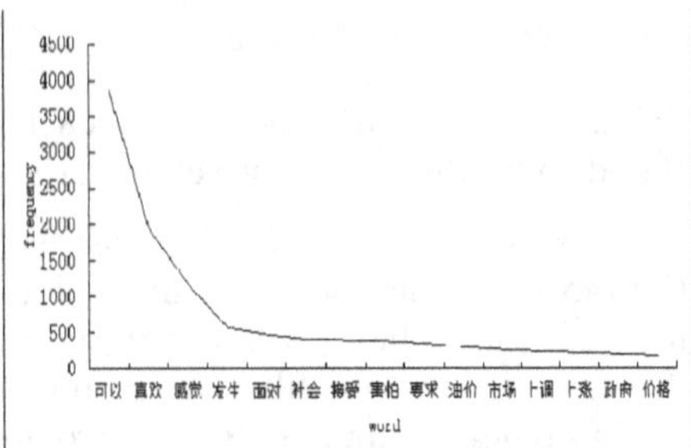

Fig. 3. The distribution of word-frequency

4.1 Hot Topic Detection in Microblogs and Other Datasets

There are many previous researches focusing on hot topic detection. The topic can be defined as an event or live activity and their related events or activities [1]. James Allan et al. divide the topic detection into two branches [2]: the first one is the retrospective topic detection; the other branch is the new online event topic detection. The characteristics of microblogs make them as the fast streaming texts. We can also divide the microblogs into streaming documents by the time interval.

The traditional topic detection methods focus on finding new clustering algorithm or improving the existing algorithms. Based on the characteristics of the microblogs, we employ a frequent pattern mining based algorithm to generate hot topics.

Generally, the hot topics microblogs have two characteristics. The first one is that the contents are always changing. There are few previous microblogs contents are similar to current hot topic microblogs. They will appear in a short period of time. Another one is that once the hot topics happened, they will acquire much discussion.

4.2 Hot Topic Keywords Extraction

According to the first characteristic mentioned above, we compute the increasing rate of each word (*Growth*) as follow:

$$Growth(w_i) = \log \frac{F(w_{ci}) - F(w_{pi})}{1 + F(w_{pi})} \tag{1}$$

where $F(w_{ci})$ presents the number of times word w_i used in the current time period, $F(w_{pi})$ presents the number of times word w_i appear in prior time period.

According to the second characteristic, we get another ranking metric, i.e. the traditional TF-IDF method. The formula is defined as follow:

$$TD(w_i) = \frac{F(w_{ci})}{F_{max}} * \log \frac{1 + D}{D(w_i)} \tag{2}$$

where F_{max} is the maximum number of unique words in documents, $D(w_i)$ is the number of documents which contain the word w_i. D is the number of the documents.

We now propose the measurement of weighting each word and define it as follow:

$$Score(w_i) = \alpha * TD(w_i) + (1-\alpha) * Growth(w_i) \tag{3}$$

where α is a parameter between 0 and 1 to control the probability of the growth.

According to Formula (1)~(3), we give the algorithm extracting hot topic keywords as Algorithm 1.

Algorithm 1: Extracting hot topic keywords

Input: *MD* //microblog set of all time period including current and before current

α //parameter for controlling the probability of the growth; *k* //a parameter

Output: *HW* //high score words set (words with top-*k* score)

Method:

1. for all $md \in MD$ preprocess *md* with the method in Section 3;
2. for every word $w \in MD$
3. {compute *Growth*(*w*) with Formula (1);
4. compute *TD*(*w*) with Formula (2);
5. compute *Score*(*w*) with Formula (3);}
6. for all $w \in MD$ Rank *w* by *Score*(*w*);
7. *HW*= word set with top-*k* score;

The purpose of line 1~5 is to compute the *Score* in Formula (3) for every words, and line 6~7 is to rank the scores for discovering top-*k* words with high score.

4.3 Hot Topic Detection

The previous algorithms of hot topic detection are based on the words' semantics. However, these methods are not appropriate for hot topic keywords generation. According to the characteristics of microblogs, based on the score in Formula (3), we detect hot topics by applying Apriori algorithm. Algorithm 2 describes the process.

Algorithm 2: Detect hot topics

Input: *MD*, *HW* // see Algorithm 1; *min_sup* // minimum support threshold;

Output: *HT* //hot topic set detected;

Method:

1. L_1=*HW*; //set the high score words as the frequent_1-itemsets
2. Apply Apriori algorithm for mining all frequent itemsets set *L*;
   //Apriori algorithm description see [3]
3. *HT*=maximal(*L*) //$\forall ht \in HT$, $\neg\exists ht' \in HT \wedge ht' \supset ht$

In the algorithm, we apply classical Apriori algorithm. Differing from the process of generating frequent 1-itemsets in Apriori algorithm, we assign *HW*, i.e. hot topic keyword set obtained in Algorithm 1 to frequent 1-itemsets (line 1), and then continue to call Apriori algorithm to get other frequent itemsets with various length (line 2). Finally, we discover the maximal frequent itemsets which mean that $ht \in HT$, and there exists no super-itemset ht' such that $ht' \supset ht$ and $ht' \in HT$, as the hot topic (line 3).

An example for above algorithm is shown in Table 2. There are eight microblogs (transactions) in the microblog set as shown in Table 2.

Table 2. The Transaction List

No.	Text
1	房价上涨是为了让我们好好工作，油价上涨是为了让我们好好节约
2	上调汽柴油600元，汽油0.44元，柴油0.51元。油价上涨以后，价格突破8元大关。
3	房价上涨是为了让我们好好工作
4	由于油价上涨，车流虽有所下降，但人流大幅上升并出现交通堵塞的情况
5	汽油上涨，车流量下降，但行人流量增加，交通压力增大
6	单眼皮单日出行，双眼皮双日出行，一单一双夜间出行，戴墨镜出行者按故意遮挡号牌
7	今天好开心，好开心呀，好开心
8	因93号汽油上涨至8元大关，车流量下降，行人流量陡增，交通压力增大。

We set the minimum support as 0.25. After preprocessing and computing, three words ("油价" (oil price), "汽油" (oil), "上涨" (rise)) are extracted from a document. The maximal frequent itemset {"上涨油价" (oil price rises), "上涨汽油" (oil rises)} is acquired. We can infer that the hot topic is 'the price of oil rises'.

5 Experiment

For the purpose of this study, the microblog data was crawled from Weibo.com from April 9th 2012 to April 15th 2012. There are 3,634,150 microblogs in total. After filtering, there remains 159,942 items. We manually tag each day's hot topics.

5.1 Hot Topic Keywords Extraction in the Dataset

The preprocessing can improve both the time and the space efficiency. After filtering microblogs, the dataset reduce to about 10%. The time of processing words from 1 minute 50 seconds reduces to 11 seconds. For the two days' dataset, it took 30 minutes to weight the words before preprocessing, and after that it took 3 minutes.

In order to find the most efficient function, we conduct the experiment by varying the parameter α. For example, the hottest topic on April 11th is 'earthquake occurs at the 8.6 level in Indonesia'. Topic keywords are '海啸' (tsunami), '震感' (earthquake feelings), '里氏' (Richter), '海域' (waters). These words are denoted as set S_h.

Assume l is a ranked word list computed by the Formula (3) with different α, w_i is a word in S_h. Let $l(w_i)$ be the rank position of w_i in l (a higher ranked word corresponds to a lower value of $l(w_i)$). The efficiency of the function is measured by the metric $Q(\alpha)=\sum_{w_i \in S_h} l(w_i)$. The lower value $Q(\alpha)$ is, the higher efficiency of the corresponding function is. Table 3 lists the $Q(\alpha)$ for different value of parameter α.

The function achieves the optimal performance when α=0.7 or α=0.8 from the results in Table 3. However, the other keywords of the topics rank higher when α=0.8. So we set α=0.8 for further steps. Table 4 shows part of the ranked words when α=0.8.

Table 3. The $Q(\alpha)$ values with different α

α	0	0.4	0.5	0.6	0.7	0.8	0.9	1
海啸(tsunami)	1	1	1	1	1	1	2	16
里氏(Richter)	10	4	10	2	2	2	3	13
海域(waters)	5	5	5	5	4	11	19	429
震感(earthquake feelings)	105	58	47	21	20	13	34	132
$Q(\alpha)$	121	68	62	29	**27**	**27**	58	590

Table 4. Part of the top ranked words and their corresponding scores

word	score	word	score	word	score	word	score
海啸(tsunami)	0.6022	火线(fire wire)	0.4386	书本(book)	0.4204	特产(specialty)	0.4003
里氏(Richter)	0.5443	放弃(give up)	0.4314	海域(waters)	0.4168	学生(student)	0.3997
姑息(tolerate)	0.4400	碟子(dish)	0.4204	变味(sour)	0.4008	国旗(flag)	0.3890
条例(rules)	0.4393	拥护(support)	0.4204	震感(quake feel)	0.4008	辛苦(hard)	0.3880

5.2 Hot Topics Detection in the Dataset

We use *F-score* to measure the performance as defined in formula (4). *S* presents the number of hot topics in keyword set extracted. *C* is the number of the elements in frequent itemset. We denote the number of the manual tagged hot topics as *R*.

$$Precision = \frac{S}{C} \qquad Recall = \frac{S}{R} \qquad F-score = \frac{2 \times Precision \times Recall}{Precision + Recall} \tag{4}$$

5.2.1 The Influence of the Number of Extracted Words

The experiments results on different number of the extracted words are shown in Figure 4. *F-score* reaches the highest value at the top 500 ranked words. Here Maximal F-set represents the maximal frequent itemset. The result shows that, the maximal frequent itemset can express the hot topics the most effectively.

5.2.2 The Influence of the Minimum Support

The experiment results with different minimum supports are shown in Figure 5. *F-score* achieves the highest value at 0.0001.

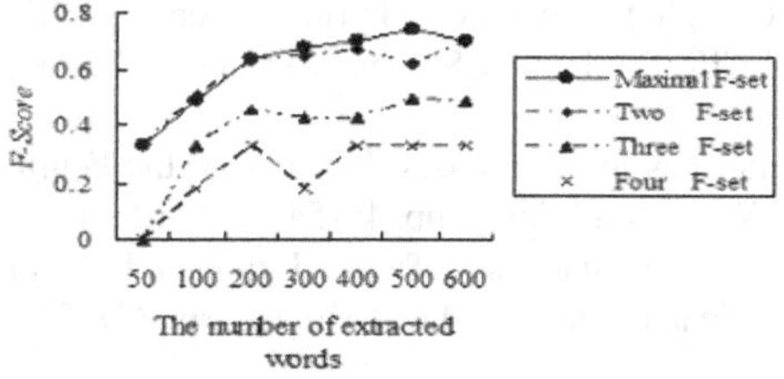

Fig. 4. Influence of the number of keywords

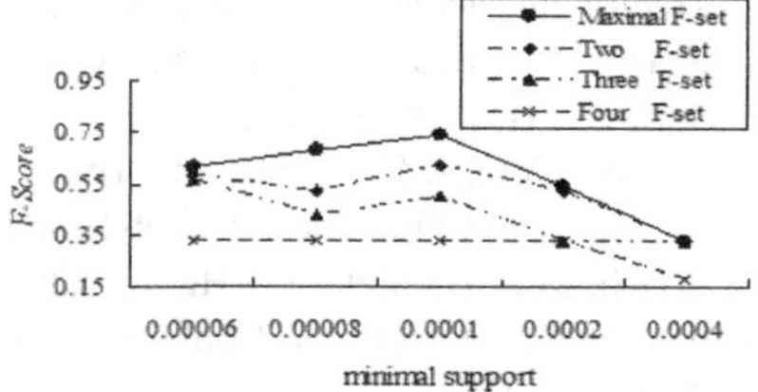

Fig. 5. Influence of minimum support

5.2.3 Case Study

Table 5 lists part of the detected hot topics.

Table 5. The experiments results of hot topic detection

Date	Hot Topic keywords	Corresponding microblog with hot topic
2012/ 04/11	海啸|海域|里氏|震感 (tsunami| waters| Richter| earthquake feel)	印尼亚齐省附近海域11日发生8.6级地震，当地已发出海啸预警。据美国地质勘探局网站消息，北京时间4月11日16时38分，北苏门海岸发生里氏8.6级地震。震源深度33公里。
	军舰|对峙|海域|渔民 (warship| confrontation| waters| fishermen)	据法新社4月11日报道，菲律宾政府宣布，菲律宾的一艘军舰11日“在该国海域试图抓捕非法捕捞的中国渔民”时，与数艘中国海监船发生对峙。
2012/ 04/14	单身|贵族|过节 (single| nobility| holiday)	4月14日是“黑色情人节（Black Day）”，是属于单身贵族们的情人节。打算过节的人享受黑咖啡的苦涩原味。

6 Conclusion

In this paper, we propose a hot topic detection approach for Chinese microblog based on keyword extraction and frequent pattern mining. Firstly we analyze the hot topic microblogs' characteristics, based on which to eliminate most topic irrelevant microblogs. Then we extract the hot topic keywords considering both the increasing rate and the frequency of the words. Finally, we introduce a frequent pattern mining algorithm to form hot topics in the extracted words. The experimental results show that the proposed methods can effectively find hot topics in Chinese microblogs.

References

1. Fiscus, J., Doddington, G.: Topic Detection and Tracking Evaluation Overview. The Information Retrieval Series, vol. 12, pp. 17–31. Kluwer, Norwell (2002)
2. Allan, J., Carbonell, J., Doddington, G., Yamron, J., Yang, Y.: Topic Detection and Tracking Pilot Study: Final Report. In: Proc. of the DARPA Broadcast News Transcription and Understanding Workshop, pp. 194–218. Morgan Kaufmann, San Francisco (1998)
3. Agrawal, R., Srikant, R.: Fast algorithms for mining association rules. In: Proceedings of International Conference on VLDB, pp. 487–499 (1994)
4. Twitter, `http://twitter.com`
5. Zhao, W.X., Jiang, J., Weng, J., He, J., Lim, E.-P., Yan, H., Li, X.: Comparing Twitter and Traditional Media Using Topic Models. In: Clough, P., Foley, C., Gurrin, C., Jones, G.J.F., Kraaij, W., Lee, H., Mudoch, V. (eds.) ECIR 2011. LNCS, vol. 6611, pp. 338–349. Springer, Heidelberg (2011)
6. Pal, A., Counts, S.: Identifying Topical Authorities in Microblogs. In: Proc. of the Fourth International Conference on Web Search and Web Data Mining, pp. 45–54. ACM (2011)
7. Welch, M., Schonfeld, U., He, D., Cho, J.: Topical Semantics of Twitter Links. In: Proc. of the Fourth International Conference on Web Search and Web Data Mining, pp. 327–336. ACM (2011)
8. Lu, R., Xiang, L., Liu, M., Yang, Q.: Extracting News Topic from Microblogs based on Hidden Topics Analysis and Text Clustering. In: Proc. of the CCIR 2010 (2010)
9. Zheng, F., Miao, D., Zhang, Z., Gao, C.: News Topic Detection Approach on Chinese Microblog. Computer Science 39(1), 138–141 (2012)

KACTL: Knowware Based Automated Construction of a Treelike Library from Web Documents*

Ruqian Lu[1,2], Yu Huang[1], Kai Sun[2], Zhongxiang Chen[1], Yiwen Chen[3], and Songmao Zhang[1,2]

[1] CAS Key Lab of IIP, Institute of Computing Technology
[2] CAS Key Lab of MADIS, Academy of Mathematics and Systems Science
[3] Tianjin University
{rqlu,smzhang}@math.ac.cn, sunkai@amss.ac.cn,
{hy.hyperborean,czx139,jacky19900606}@gmail.com

Abstract. This paper proposed a knowware based supervised machine learning technique for domain specific regression and classification of Web documents. It is simple because it is only based on word counting techniques without natural language understanding and complicated statistic techniques. Starting from constructing a domain sub-division tree and assigning a training set of documents to its nodes, the algorithm produces a labeled classification tree with a characteristic vector for each node. This tree is used to classify any number of documents in that particular domain. A tool for developing Web portal is also provided to build a Web station for displaying the final treelike library of documents.

Keywords: Web portal, document classification, supervised learning, knowware.

1 On Knowledge Acquisition from World Wide Web

The World Wide Web has become a major source of knowledge acquisition for experts and also for the general public. However, we still suffer from the insufficient capability of currently available browsers in acquiring Web knowledge. Generally, a browser provides only a quite huge set of Web pages upon the key words inputted by the user. They (the browsers) are incapable of extracting the needed knowledge from this huge set of Web pages and let the users take the burden of reading, searching and extracting knowledge from each Web page. What even worse is, the Web pages pushed forwards to the users by the browsers are usually not sorted carefully to easy the users during knowledge searching. The Web pages are often listed in a messy order with lots of repeats and junk documents, together with the massive sets of advertisements. Our experience of using different Web browsers has shown that the claim made by them like 'we have found 1234567890 Web pages for your search' is often meaningless.

The motivation of the research presented in this paper is to create a knowware based DIY technique, which, when inputted an appropriate search request, can develop a

* Partially supported by 973 Program 2009CB320701,NSFC 61073023 and Tsinghua-Tengxun Lab Project.

F.L. Wang et al. (Eds.): WISM 2012, LNCS 7529, pp. 645–656, 2012.

library of documents acquired from the search results of any browser. These documents will be classified and presented in tree form to easy users' work of finding the documents containing the information or knowledge they need. Knowware is a technique proposed by the first author[15-16], which aims at separating domain knowledge from application software to make it an independent product to be used interchangeably with different products of software. The use of knowware in this technique is essential because only knowledge based approach will make the search and classification successful.

Roughly speaking, the KACTL approach consists of the following parts:

Step 1: Prepare the initial domain knowledge in form of knowware (a classification tree of domain/sub domain organization);

Step 2: Using a set of documents to train the classification tree to get a labeled classification tree;

Step 3: Using the labeled classification tree to classify a large set of documents to get a treelike library of documents;

Step 4: Establish a Web portal of this library for public use.

Step 1 shows that the KACTL approach is knowware based. Step 2 shows that it is also based on supervised learning. Step 3 shows that the learning result is effective. Step 4 shows that our approach can help people develop domain specific Web portals in a basically automated and even DIY way.

2 The Main Idea of KACTL Approach

More precisely, the (unlabeled) classification tree is a repeated division of user interested points with respect to some well defined domain. Fig. 1 shows such a tiny classification tree regarding 'computing technology'.

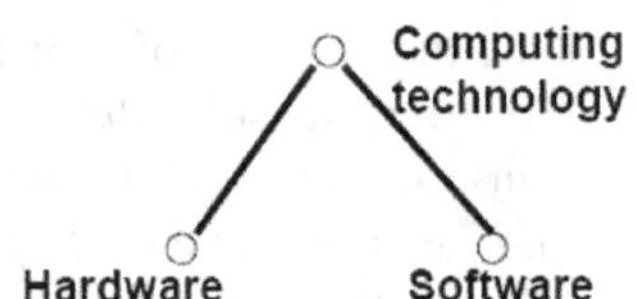

Fig. 1. Unlabeled Classification Tree for the domain IT

Once the classification tree is built, we have to train it. In this context training means telling the computer which documents should belong to which node(*s*) of the classification tree. Given a small set of training documents, the knowledge engineer is required to assign each of them to some nodes of the tree, meaning that this document belongs to sub-domains represented by these nodes. Note that it is possible that a document belongs to several sub-domains. Just considering the case of a survey paper is enough for convincing oneself. However, each document belongs to at most one node along each path from root to leaf. Now assume we have 4 documents {P1, P2, P3,

P4} as the trainings set, where P2 contains information about both hardware and software, P1 contains neither hardware information nor software information. They are attached by the knowledge engineer to the tree as shown in Fig. 2.

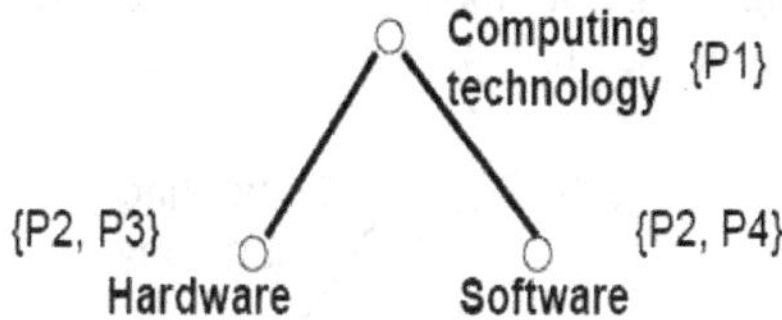

Fig. 2. Classification tree with Training Documents

The second part of training is generalization. An algorithm is provided to extract common characteristics of documents assigned to the same node. We decided to take frequency vectors (consisting of 'most frequent words' contained in documents) and strength vectors (strength of a word equals to frequency of the word divided by length of the document containing it) as common characteristics of documents belonging to the same node. A classification tree with its nodes equipped with strength vectors is called a labeled classification tree and is ready for use in document classification. For simplicity we assume the frequency vectors and strength vectors are all of length 2. Assume the training algorithm calculates the following results:

Frequent words: P1: {IT, market}, P2:{chip, compiler}, P3:{chip, CPU}, P4:{ service, compiler};
Frequency vectors: P1: (9, 4), P2: (7, 5), P3: (8, 4), P4: (6, 3);
Strength vectors: P1: (9/10, 2/5), P2: (7/8, 5/8), P3: (8/11, 4/11), P4: (3/4, 3/8);

A short note to explain the above data: the term 'IT' appears 6 times in document P1. The length of P1 is 10.

The algorithm then calculates the strength vectors for the nodes. They are composed of 'most strong' terms of each node as shown in Fig. 3:

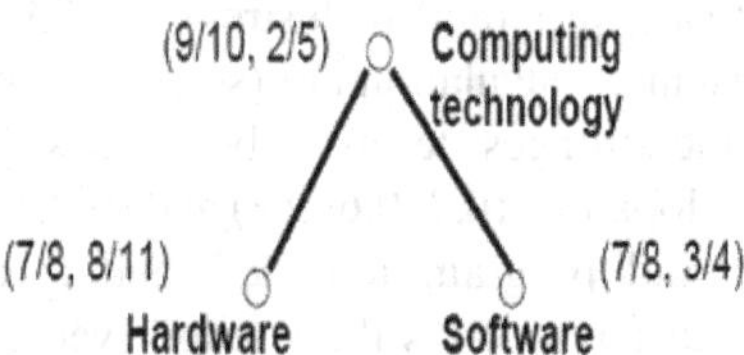

Fig. 3. Classification Tree with Basic Strength Vectors

Computing technology: (9/10, 2/5), hardware: (7/8, 8/11), software: (7/8, 3/4).

When calculating these strength vectors only the documents in the relevant node are taken into account (for example P2, P3 and P4 are not taken into account when calculating the strength vector of 'Computing technology'). They are therefore called basic strength vectors. However, the son nodes of each node are supposed to belong to the sub-domains of the father node. They should be also considered when classifying new documents. The algorithm therefore calculates the 'enhanced' strength vectors by taking into account also the son nodes. Thus, the enhanced strength vectors of 'hardware' and 'software' nodes are the same as their basic strength nodes, while that of the 'computing technology' gets the value (9/10, 7/8).

By the above reasoning we get a labeled classification tree with enhanced strength vectors attached to each node, see Fig. 4.

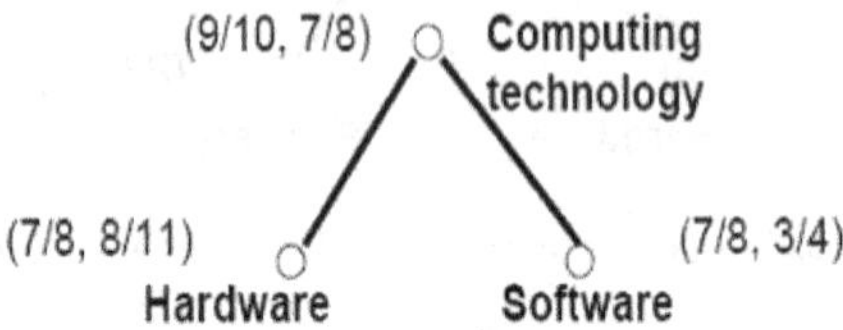

Fig. 4. Labeled Classification tree with Enhanced Strength Vectors

During document classification, the enhanced strength vectors will be used to calculate the distance between a new document and a node of the labeled classification tree. We have tried two different ways to calculate this distance. The first one is the usual tf-distance of the vector space model. The second choice is the extended tf-distance, which is a generalization of the conventional tf-distance of the vector space model. We call it the extended tf-distance. For the conventional tf-distance model the query term vector decides the terms to be queried, while the document term vector checks the existence of the queried terms in the document. But for the extended tf-distance model proposed by this paper, the query term vector and the document term vector decide the terms in a cooperative way by merging the terms of both side's strength vectors. For example, if the terms (computer, program) are to be queried, the classical query term vector may have the value (0.2, 0.5), where the numbers are weights. Searching the (same) terms (computer, program) in the document provides the document term vector (0.3, 0.8). The tf-distance between (0.2, 0.5) and (0.3, 0.8) will be calculated. But in our definition, if the strongest words of node N are (computer, program) with enhanced strengths vector (0.6, 0.4), while the strongest words of the new document P are (software, program) with strengths (0.9, 0.5), then the terms to be checked in extended tf-distance calculation are (software, computer, program), which is generated by merging the strongest terms of both sides. The node term vector and document term vector may look like (0.2, 0.6, 0.4) and (0.9, 0.3, 0.5), where 0.2 and 0.3 are strengths of 'software' and 'program' in node N and document P, respectively.

Note that according to different ways the strength vectors are produced, we have designed and tested four variants of the learning algorithm. The variance 1 algorithm produces strength vectors of fixed length 10 (the 10 'most strong' words). The variance 2 algorithm produces strength vectors with a maximal length of 10 $\times$ number of documents in that node. The variance 3 algorithm produces strength vectors of a length no more than the number of documents in that node. Experiments have shown that the three variants of the algorithm provided different accuracies in classification.

Note also that the classifying of new documents is not based on mutual exclusion of document sets on different nodes. The same new document may belong to different nodes of the tree, provided that no document belongs to two different nodes on the same path of the tree. When classifying a new document, the procedure starts from the root node in a top down way. The principle is to attach new documents to nodes as close to the leaves as possible. The intuitive meaning is to classify new document as 'concrete' as possible. Whenever a new document *P* 'arrives' at some node *N* of the tree its tf-distances (or extended tf-distances) with *N* and with all son nodes of *N* will be calculated. *P* will be moved to (some) son nodes if the calculation shows that *P* is closer to these son nodes than to *N*. Otherwise *P* will remain in *N*. We call such documents as rear documents of *N* to mean that *N* is one of the 'best fit' nodes for *P*.

Now we can explain the problem on where and how the knowware concept is applied in the KACTL methodology. We provide a three level approach of knowware application to support the learning and classification task. The most basic level of knowware used in it is the classification tree, where the hierarchical sub-division of a domain has already been made. The second level of knowware application is the labeled classification tree with enhanced strength vectors or characteristic vectors. The third level of knowware application is the treelike library of classified documents. One can start from any level of knowware application to construct the library, as shown by Table 1.

Table 1. Application of Knowware in Web Portal Establishment

Embedded Knowware	System Modules remain to be produced
nothing	Classification tree + labeled classification tree + Treelike library of documents + Web portal
Classification tree	labeled classification tree + Treelike library of documents +Web portal
labeled classification tree	Treelike library +Web portal
Treelike library	Web portal

3 Procedure and Algorithms of KACTL

This section transforms the main ideas described in last section into concrete procedure and algorithms. In this section we differentiate between procedure and algorithm. An algorithm is such a list of instructions, which can be programmed in computer languages and performed by a computer automatically. On the other hand, a procedure can not be performed in a totally automated way. A major part of it must be performed by a human being, either a knowledge engineer or a programmer.

In the following we provide the tree construction procedure as the first module of our KACTL approach.

Procedure 1: Constructing the Classification Tree

1.Given a (unlabeled) classification tree *T*(*D*), where *D* is a specific domain;
2.Given a training set *W* of documents supposed to belong to the domain *D*;
3.Mark all nodes of *T*(*D*) as 'inactive' but the root node, which is marked as 'active';
4.Attach all documents of W to the root node of *T*(*D*)
5.**While** there is at least one 'active' node in *T*(*D*)

Do

1) Take any 'active' node N, remark it as 'inactive';
2) **For** all documents P attached to N **do**
3) **Case of**

Case1: If P is supposed to belong to some sub-domains represented by some son nodes $\{N_1, N_2, \ldots, N_k\}$ of N, then make k copies of P and attach them to $\{N1, N2, \ldots, N_k\}$ respectively. Mark $\{N_1, N_2, \ldots, N_k\}$ as active and remove P from N;

Case 2: If P is supposed to belong to some sub-domain D' of D, which is not represented by any son node of N, then generate a new son node N' of N, mark N' as 'active' and remove P from N;

Case3: If neither case 1 nor case 2 is true, then mark P as a rear document of N

End of Case;

End of Do

Now comes the central part of the approach—the learning algorithm.

Algorithm 1: Learning the Labeled Classification Tree

1. Given the (unlabeled) classification tree built in **Procedure 1** with the training documents attached to its nodes;
2. Calculate the frequency vector $V(P)$ of each document P. It consists of 10 most frequent words appearing in P together with their frequency, where the length of each frequent word should be larger than 1;
3. Mark all nodes of $T(D)$ as 'unprocessed';
4. **While** there is at least one 'unprocessed' node of $T(D)$

Do // *calculate the (basic and enhanced) strength vector of each node in a bottom up way)*

1) Take any 'unprocessed' node N, which has no 'unprocessed' son nodes, remark it as 'processed';
2) Calculate the basic strength vector $BV(N)$ as follows: let the set of rear documents in node N be S. Using the formula:

strength of w in P = frequency of w in P / length of P

to calculate the strength of each word w in each rear document P in S. The ten words with highest strength with respect to all documents in S are selected to form the basic strength vector $BV(N)$;
3) Calculate the enhanced strength vectors $EV(N)$ in a bottom up way as follows: If N has no son nodes, then $EV(N) = BV(N)$, otherwise let the set of enhanced strength vectors of all son nodes of N be S'. The ten words with highest strength are selected from $S' \cup \{BV(N)\}$ to form the enhanced strength vector $EV(N)$;

End of Do;

5. Call the classification tree with enhanced strength vectors at all its nodes as labeled classification tree.

Finally, we present the algorithm for applying the above result (to classify new documents):

Algorithm 2: Build the Treelike Document Library

---

1. Given a labeled classification tree T and a set U of documents to be classified;
2. Mark all documents in U as 'unclassified';
3. Set up a set S of (*document*, *node*) pairs. Its initial content is empty;
4. **While** there is at least one 'unclassified' document in S

   Do

   1) Build a copy W of T;
   2) Mark all nodes of W as 'unprocessed';
   3) Take any 'unclassified' document P from U and remark it as 'classified';
   4) Calculate the strength vector $V(P)$;
   5) Assign P to the root node R of W;
   6) Calculate the extended tf-distance between the strength vector $V(P)$ and the enhanced strength vector $EV(R)$ as follows: Combine $V(P)$ with $EV(R)$ to form a new vector $NV(P, R)$, which consists of (and only consists of) elements of $V(P) \cup EV(R)$. Using $NV(P, R)$ to calculate the distance between P and R based on the vector space model.
   7) Remark R as 'undeveloped';
   8) **While** there is at least one 'undeveloped' node of W, where the set of its son nodes is not empty and all its (direct) son nodes are 'unprocessed'

      Do

      1)) Take any 'undeveloped' node N from W and remark it as 'developed';
      2)) Calculate the extended tf-distance between P and Z for all son nodes Z of N in the same way as the extended tf-distance between P and R is calculated in step 6) above;
      3)) **For** all such nodes Z **do**

      Begin

      If the extended tf-distance between P and Z is less or equal than that between P and N

      Then

      Begin

      1))) Remark Z as 'undeveloped' and put the pair (P, Z) in M;
      2))) Attach the document P to Z;

      End

      Else

      remark Z as 'processed';

      End;

      4)) **If** P is attached to at least one of the son nodes Z of N

      Then

      remove P from N;

      End of Do

   End of Do

---

4 Experiments and Results

We take the domain 'cloud computing' as the target of experiment. In the first step, we construct a classification tree for this domain according to our personal feeling about how to subdividing this now very hot domain. The resulted tree is shown in Table 1 of appendix. In the second step we use the Chinese browser Baidu to search Web pages with the keywords 'cloud computing'. The browser reported in total about 100,000,000 Web pages. Based on the assumption that all these Web pages are 'cloud computing' related, we selected the first 730 Web pages and used a document extractor to extract all documents from these Web pages as our training documents. The number of documents we got after removing repetitions was 103. In the third step we attach each of the 103 documents to some node(s) of the tree. Note that according to our approach a document can be attached to several nodes of the tree. The number of documents attached to each node is shown in Table 2. In the fourth step we apply the learning algorithm to the tree with attached documents to calculate the enhanced strength vectors. We take a sub-tree and show the strength vectors of all its nodes in Table 3. After all these four steps are done, the labeled classification tree is built and the learning process is completed. Then we test the labeled classification tree by using the training set as test set, in order to see how well the obtained classification will meet the original classification made manually. We used three measures for the comparison: the Jaccard Coefficient, recall and precision, all based on the vector space distances tf and extended tf.

For the Jaccard Coefficient, we just calculate the value of

$$|HD(N) \wedge CD(N)| \,/\, |HD(N) \vee CD(N)| \tag{1}$$

for each node, where *HD*(*N*) and *CD*(*N*) are the sets of documents assigned to node N by human assignment and by computer classification respectively. The average values for the three measures are shown in Table 2 for the learning algorithm and both models of tf-distance.

Table 2. Statistical comparison of human and computer classification

Comparison	Jaccard Coefficient	Recall	Precision
tf-distance	0.2211	0.4782	0.2947
Extended tf-distance	0.1432	0.7050	0.1492

The values reported in Table 2 are not very high because a general statistical comparison does not reflect the real quality of the classification precisely. One of the reasons is that this test does not take account of the data structure of the classification tree. In fact, whether to assign a document to some node N or to its father node is not a strict contradiction, but a problem of inheritance. Usually, in this case neither assignment should be considered as a mistake. On the other hand, human classification usually suffers from casualness, randomness and incompleteness. Similar situation appears in particular during assigning a document among several brother nodes of a tree. For example, a document reporting some conference on cloud computing may be assigned by the programmer to the node "conference". But the computer would assign the same document to the node 'technique' as well because many technical problems are discussed on this conference. In this case, the computer's behavior can be considered as a complement to the programmer's work.

Table 3. The Classification Tree

Labeled Classification Tree	Tree-like Document Library
Cloud Computing (0)	云计算(2)
CC Concept (0)	云计算概念(10)
CC Basic Principles (8)	云计算基本原理(0)
CC Definition (22)	云计算定义(0)
CC Discrimination Criteria (8)	云计算判断准则(1)
CC Publication (0)	云计算出版物(4)
CC Journals (1)	云计算刊物(0)
CC Web sites (0)	云计算网站(0)
CC Books (3)	云计算专著(5)
CC Activity (0)	云计算活动(0)
CC Conference (0)	云计算大会(0)
CC Domestic Conference (6)	云计算国内大会(0)
CC International Conference	云计算国际大会(0)
	世界云计算大会(0)
	美国云计算大会(0)
	欧洲云计算大会(0)
	亚太云计算大会(0)
CC Training Course (0)	云计算培训班(0)
CC Workshop (0)	云计算研讨会(0)
CC Technique (7)	云计算技术(12)
CC Basic Technique (0)	云计算基本技术(11)
CC and Map Reduce Software (1)	Map Reduce技术(0)
CC and Big Table Technique (1)	Big Table技术(0)
CC and Distributed File System (1)	分布式文件系统(0)
CC New Technique (0)	云计算新技术(0)
CC Classification (10)	云计算分类(2)
Public Cloud (2)	公有云(2)
Private Cloud (5)	私有云(9)
Mixed Cloud (2)	混合云(2)
CC Application (14)	云计算应用(32)
CC and Browser (6)	云计算与浏览器(1)
CC and e-Commerce (5)	云计算与电子商务(35)
CC and Mobile Communication (6)	云计算与移动计算(53)
CC Product (1)	云计算产品(3)
CC Hardware (1)	云计算硬件(2)
CC Software (4)	云计算软件(11)
CC Platform (5)	云计算平台(0)
Hadoop (2)	Hadoop平台(0)
GAE (0)	GAE平台(0)
EC2 (0)	EC2平台(0)
Blue Cloud (4)	蓝云(1)
CC Develop Tendency (14)	云计算发展趋势(39)
CC Market (22)	云计算市场(14)
CC Device Provider (9)	云计算设备供应商(0)
CC Service Provide (12)	云计算服务供应商(4)
CC Consultation Provider (2)	云计算咨询公司(4)
CC Stock Market (7)	云计算股市(10)
CC Concept Stock (3)	云计算概念股(19)
CC Policies (3)	云计算政策(11)
CC Security (12)	云计算与安全(59)
CC History (5)	云计算历史(21)
CC Current Issues (10)	云计算现状(23)
CC Problems (11)	云计算的问题(16)
CC Characteristics (8)	云计算的特点(11)
CC Practice (1)	云计算实践(52)
CC Standards (2)	云计算标准(34)
CC Comments (4)	云计算的相关评价(60)

Our technique is already put in use in practice. The reader can visit our experimental Web portal with the address http://159.226.47.103/sct/demo.jsp, wheresome 3500 documents on 'cloud computing' are organized as a treelike library by algorithm 2. To visit the Web portal, one needs browsers such as IE 9, Firefox or other browsers that comply with the W3C standard. Two pieces of images cut down from this Web portal, together with an English translation of the classification tree, are shown in Table 2. They are the original classification tree with training data (103 documents) and the same classification tree with re-classified documents obtained after using the learning algorithm. The numbers attached to each node in these trees are the numbers of documents belonging to this node.

5 Related Work

Categorizing Web pages automatically is an active research area in information retrieval, usually aiming at maximizing the intra-category similarity and minimizing inter-category similarity. Most previous work focused on grouping the results returned by search engine into clusters flatly or hierarchically and assigning a label to every cluster.

The first commercial clustering search engine named Northern Light, based on predefined set of categories, was released in the late 1990s. Vivisimo, founded in 2000 by a trio of computer scientists at Carnegie Mellon University, started to generate clusters and cluster labels dynamically and won "The Best Meta-search Engine Award" assigned by SearchEngine.com from 2001 to 2003 and "InfoWorld Technology of the Year" from 2006 to 2008 by Best Enterprise Search.

In addition, a large number of published papers also discussed specific issues and systems in this research area. Suffix Tree Clustering [1-2], proposed by Zamir et al, was a linear time clustering algorithm and used in the demonstrated system called Grouper. Zheng et al [3] reformulated the clustering problem as a salient phrases ranking problem. In [4-5], Osiński et al developed a novel algorithm called Lingo, which took advantage of algebraic transformations of the term-document matrix and frequent phrase extraction using suffix arrays to cluster the search result and accomplish the cluster label induction. This algorithm was implemented as the fundamental component in the open source project Carrot2[6-7], another state-of-art clustering search engine besides Vivisimo. Moreover, one recent and more detailed survey in this area can be found in [8].

In recent years, more advanced clustering algorithms have been developed and applied in several practical areas including clustering the documents and Web pages. One approach was called Spectral clustering and was already successfully used in image segmentation [9] and search result clustering [10]. Roughly speaking, it was a technique of partitioning the rows of a matrix according to their components in the top few singular vectors of the matrix [11]. In [12], Frey et al devised a method called "affinity propagation," which took as input measures of similarity between pairs of data points and then exchanged real value messages between data points until a high-quality set of exemplars and corresponding clusters gradually emerged. Hierarchical affinity propagation was also proposed in [13]. Another different modern approach, based on

Bayesian nonparametric model and Chinese restaurant random process, was proposed in [14]. Given a collection of documents, a hierarchy of topics and words allocation to these topics could be generated by this algorithm. However, it may not be directly used in Web page clustering in an online system because of its efficiency problem and practical issues.

In this paper, as mentioned above, we used a supervised learning algorithm based on classification tree and strength vector instead of traditional unsupervised algorithms for clustering. As a result, it will bring us significant improvement in efficiency and enable us to evaluate the results more accurately. Moreover, our algorithm can provide more delicate taxonomies, which means our algorithm can classify the Web pages or documents into categories more accurately than existing algorithms and systems. More important than all advantages mentioned above, the most salient feature of our work is to provide a DIY technique which is missing in any other related work.

6 Conclusion and Future Work

In this paper we proposed a new approach of classifying Web documents in form of a treelike library. This approach has the following characteristics. First, it provides a tool for setting up Web portals of documents by non-experts in a DIY way. Everybody can set up a Web library in tree form by himself, just by using the KACTL software with a minimum of knowledge (in form of knowware). No extra programmer or knowledge engineer is required. Second, the KACTL approach is knowware based. Since the first author proposed the knowware concept and its methodology seven years ago[15-16], this is the first time to have the knowware technique implemented in a software/knowware complex in practical use. The knowware used in this approach is exchangeable to meet new user's need. Third, the KACTL approach is domain independent. Though the example given in this paper is cloud computing, actually any domain is possible. This is a consequence of using knowware technique in the approach. Fourth, the result of Web document classification is developer dependent. Even with the same title such as cloud computing, different developers will provide different classification trees and produce different libraries of Web documents. Therefore, developing document libraries on the same domain but of different flavors are always possible. Fifth, the knowware applied in this approach is scalable. We provide a three level approach of knowware application in the KACTL methodology.

Our future work will be improving the document classification accuracy by making a deep study on the non-deterministic and noisy nature of Web page content. Many problems are difficult to solve by currently existing techniques. It is our goal to look for techniques or to create new techniques appropriate for use in such kind of tasks.

References

1. Zamir, O., Etzioni, O.: Grouper: A Dynamic Clustering Interface to Web Search Results. In: Proceedings of the Eighth International World Wide Web Conference (WWW8), Toronto, Canada (May 1999)

2. Zamir, O., Etzioni, O.: Web Document Clustering: A Feasibility Demonstration. In: Proceedings of the 19th International ACM SIGIR Conference on Research and Development of Information Retrieval (SIGIR 1998), pp. 46–54 (1998)
3. Zeng, H.-J., He, Q.-C., Chen, Z., Ma, W.-Y., Ma, J.: Learning to cluster web search results. In: Proceedings of the 27th Annual International ACM SIGIR Conference on Research and Development in Information Retrieval (SIGIR 2004), pp. 210–217. ACM, New York (2004)
4. Osiński, S., Stefanowski, J., Weiss, D.: Lingo: Search Results Clustering Algorithm Based on Singular Value Decomposition. In: Advances in Soft Computing, Intelligent Information Processing and Web Mining, Proceedings of the International IIS: IIPWM 2004 Conference, Zakopane, Poland, pp. 359–368 (2004)
5. Osiński, S., Weiss, D.: Conceptual Clustering Using Lingo Algorithm: Evaluation on Open Directory Project Data. In: Advances in Soft Computing, Intelligent Information Processing and Web Mining, Proceedings of the International IIS: IIPWM 2004 Conference, Zakopane, Poland, pp. 369–378 (2004)
6. Stefanowski, J., Weiss, D.: Carrot^2 and Language Properties in Web Search Results Clustering. In: Menasalvas, E., Segovia, J., Szczepaniak, P.S. (eds.) AWIC 2003. LNCS (LNAI), vol. 2663, pp. 240–249. Springer, Heidelberg (2003)
7. Osiński, S., Weiss, D.: Carrot^2: Design of a Flexible and Efficient Web Information Retrieval Framework. In: Szczepaniak, P.S., Kacprzyk, J., Niewiadomski, A. (eds.) AWIC 2005. LNCS (LNAI), vol. 3528, pp. 439–444. Springer, Heidelberg (2005)
8. Carpineto, C., Osiński, S., Romano, G., Weiss, D.: A survey of Web clustering engines. ACM Computing Surveys (CSUR) 41(3), Article No. 17 (July 2009) ISSN:0360-0300
9. Shi, J., Malik, J.: Normalized cuts and image segmentation. IEEE, Trans. on Pattern Analysis and Machine Intelligence 22, 888–905 (2000)
10. Cheng, D., Kannan, R., Wang, G.: A Divide-and-Merge Methodology for Clustering. In: Proc. of the ACM Symposium on Principles of Database Systems (2005)
11. Kannan, R., Vetta, A.: On clusterings: good, bad and spectral. Journal of the ACM (JACM) 51(3), 497–515 (2004)
12. Frey, B.J., Dueck, D.: University of Toronto. Science 315, 972–976 (2007)
13. Givoni, I.E., Chung, C., Frey, B.J.: Uncertainty in Artificial Intelligence (UAI) (July 2011)
14. Ghosh, S., Ungureunu, A., Sudderth, E., Blei, D.: Spatial distance dependent Chinese restaurant processes for image segmentation. In: Neural Information Processing Systems (2011)
15. Lu, R.: From hardware to software to knowware: IT's third liberation? IEEE Intelligent Systems, 82–85 (March/April 2005)
16. Lu, R.: Knowware, the Third Star after Hardware and Software. Polimetrica Publishing Co., Italy (2007)

Associating Labels and Elements of Deep Web Query Interface Based on DOM

Baohua Qiang[1,2], Long Shi[1], Chunming Wu[2], Qian He[1], and Chao Shen[1]

[1] School of Computer Science and Engineering, Guilin University of Electronic Technology, Guilin, 541004, P.R. China
[2] College of Computer and Information Science, Southwest University, Chongqing, 400715, P.R. China
{Qiangbh,Shilong,Heqian,Shenc}@guet.edu.cn,
Chunmingnone@163.com

Abstract. Query interface schema extraction is an important issue for Deep Web data acquisition and integration. In order to obtain the query interface schema, it is firstly required to associate elements and labels of Deep Web query interface correctly. Due to the fact that query interface on HTML page can be parsed as well structured DOM, we proposed an effective algorithm for associating elements and labels of Deep Web query interface based on hierarchical DOM. Our algorithm mainly adopted the nearest-neighbor-distance and other two useful heuristic rules to associate the most related label of a given control element. The experimental results on real query interfaces show that our proposed algorithm is highly effective.

Keywords: Deep web, element and label association, DOM, query interface.

1 Introduction

As the largest web data resources on the Internet, Deep Web, which can be accessed through form-based query interfaces, has become the main channel for us to obtain the valuable information [1-3]. Logically, elements and their associated labels of Deep Web query interface together form different attributes of the underlying database. Using such a schema-based interface, users can specify complex and precise queries to Web databases [4]. So the accurate extraction of query interface schema is essential and it is firstly required to associate labels and elements of the Deep Web query interface correctly.

Generally, two approaches can be used to associate labels and elements of the Deep Web interface schema: 1) Label-centered association, i.e., to scan the context control elements based on the labels and associate semantically related labels and elements to form logical attributes. 2) Control element-centered association, i.e., to scan the context labels based on the control elements and associate semantically related elements and labels to form logical attributes. ATTACH [5] is a control element-centered algorithm in which the query interface is divided as some blocks and the labels are associated with given elements by considering the vision-distance in each block. This algorithm has a feasible theoretical basis as well as high precision.

F.L. Wang et al. (Eds.): WISM 2012, LNCS 7529, pp. 657–663, 2012.

But the shortcoming of the algorithm is the low performance due to highly computational complexity. Literature [6] presented a label-centered approach to associate the elements with given labels. This algorithm can solve some limitations in literature [5], but the computational complexity still exists and the results are unstable. Literature [7] proposed an improved N-Gram approach to associating the elements and labels by measuring the similarity of string. But this algorithm only applies to English characters and the Chinese characters of Deep Web query interface should be translated into English characters firstly. So whether the Chinese and English can be mapped correctly determine the results. More researches about this topic can be found in the literatures [8-10].

The present approaches excessively rely on theoretical computation to associate the elements and labels and neglect the characteristics of the HTML code of query interface. According to our observations and analysis, the Web browser parses the page to generate DOM according to the sequence from top to bottom and from left to right. We can obtain the desired results by considering the characteristics of Web DOM nodes. In this paper we proposed an effective algorithm for associating labels and elements of the Deep Web query interface based on DOM. The experimental results showed that our proposed approach not only has higher precision, but also can maintain higher recall.

The main body of this paper is organized as follows. We firstly describe the DOM of Deep Web query interface in Section 2. Then the algorithm for associating labels and elements of Deep Web query interface based on DOM is presented in Section 3. In Section 4 the experiments of our proposed approaches are given. Section 5 concludes the paper.

2 DOM for Representing Query Interfaces

Deep Web query interface is an HTML form with some control elements and labels. The Deep Web control element refers to those form elements, such as textbox, radio button, checkbox and selection list, for receiving users' parameters. The Deep Web label refers to those text elements semantically associated with the control elements. For example, *E1* and *E2* in Fig.1 are two control elements and *From* and *To* are their corresponding labels in the domain of airline, respectively. The correct combination of each control and its corresponding label can reflect the semantic relationship of Deep Web query interface, which will help to improve the accuracy of subsequent schema matching and query processing. So the effective association between the control elements and their corresponding label texts is critical.

According to the principle of web browser, web browser adopts DOM to parse and represent the Web page. DOM is a platform and language-neutral interface that will allow programs and scripts to dynamically access and update the content, structure and style of documents. DOM defines the logical structure of documents and the way a document is accessed and manipulated. So the web page can be expressed as a set of elements such as TagNode representing control elements, TextNode representing label texts and CommentNode representing comments. Due to the control elements and label texts are our concerned objects, we will focus on the TagNode denoted as INPUT, SELECT and RADIO, and TextNode denoted as #TEXT. DOM has good hierarchical structure as Fig.2. By adopting DOM, the irregular HTML codes of web pages can be represented as well structured interface as Fig.1 visually. So the DOM can be employed to associate labels and elements of the Deep Web query interface effectively.

Fig. 1. A flight query interface example

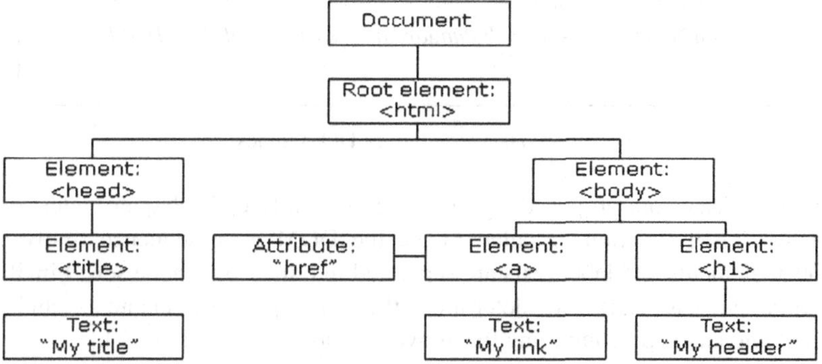

Fig. 2. The hierarchy of DOM

3 Algorithm for Associating Elements and Labels of Deep Web Query Interface

Given a control element, its corresponding label usually locates within a certain range before or after this control element. We denote the DOM sequence of a Form as DomNodes:

$$DomNodes = \{Tag_1, Txt_2, Tag_3, Txt_4, \ldots, Tag_n\}$$

Tag_i denotes the control element node, and Txt_j denotes the text element node. The probability of Node tag_3 associated with text node Txt_2 or Txt_4 is far greater than with other text nodes by our large number of experiments, which enhances our inference. So one important heuristic rule, the most related label of a given control element is the minimum distance label text node, will be adopted to improve the association accuracy. This heuristic rule is called as the nearest-neighbor-distance rule. Due to the diversity of HTML coding style, some other useful heuristic rules should be used to filer unrelated information before associating the control element and its label.

Other two useful heuristic rules are: (1) Generally, the control element should be associated with the label text in front of it; (2) The corresponding labels of elements such as radio button, checkbox are always located behind them in the sequence of DOM nodes. So the control elements such as radio button, checkbox should be associated with the label text behind it.

In order to capture the semantic information of DOM, we define the structure of DomNodes as follows:

```
DOMNodes
{
   int  ParserSequence;     //Parsing order
   String  NodeName;     //Name of node such as INPUT
   String  LocalName;     //Attribute Name of NodeName
   String  TextContent;  // Label text
   String  Type;      //Node type
   w3c.dom.Node W3CNode;  //Standard data structure of W3C DOM
}
```

Fig. 3. Data structure of DOMNodes

The basic idea of our proposed algorithm is: given a Deep Web query interface, the web browser API function is called to get the DOM node sequence firstly. Then the control elements and label texts are separated into their own node sets. Finally, the association process starts to determine the corresponding elements and their corresponding labels by comparing the above two node sets (See Fig.4.).

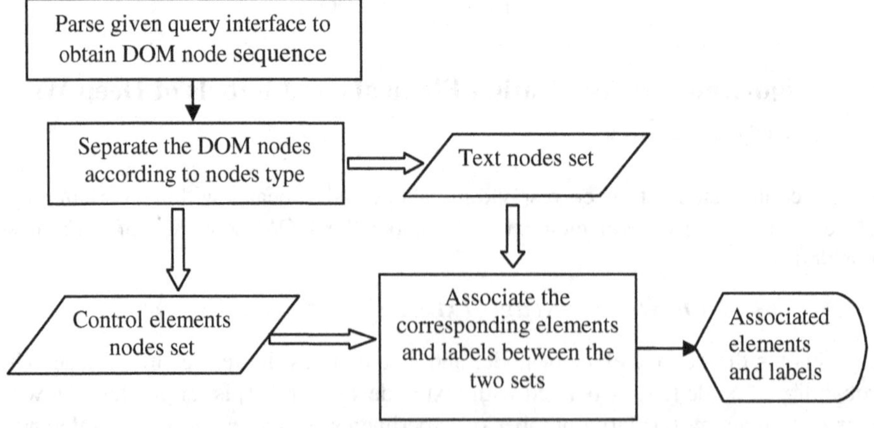

Fig. 4. Flow chart for associating controls and labels

According to the idea of our approach, the detailed algorithm description for Associating labels and elements of Deep Web query interface is presented below.

$RELATER(Taglist,Txtlist) \rightarrow List < RelatePair >$

Input: List<DOMNode> Taglist *//To be associated control element node list*
List<DOMNode> Txtlist *//To be associated text element node list*
Output: List<RelatePair> Outputlist *//Associated control and text element node list*

```
Algorithm Begins:
  List< RelatePair> Outputlist=null;
  While(Taglist.hasNext)      // Traverse the control node list
  Do
     Tag=taglist.next();
     tagSequence=Tag. ParserSequence;
    Txt_Before= Txtlist.getBefore(tagSequence);  //Obtain the nearest text node
                                                   before tagSequence
    Txt_After= Txtlist.getAfter(tagSequence);    //Obtain the nearest text node after
                                                   tagSequence
    Txt_ToRelate=NULL;     // Text node pointer associated with Tag
    If(Tag.Type=={"radio" ,"checkbox"})
       Txt_ToRelate= Txt_After;
    Else
       SeqDistance_bef=Absolute(tagSequence - Txt_Before.ParserSequence);
              //Compute the distance between the tag and the nearest text node before it
       SeqDistance_aft=Absolute(tagSequence - Txt_After.ParserSequence);
              //Compute the distance between the tag and the nearest text node after it
       SeqDistance_bef= SeqDistance_bef /2;
       If(SeqDistance_bef >SeqDistance_aft)
          Txt_ToRelate=Txt_After;
       Else
           Txt_ToRelate= Txt_Before;
       End if
    End if
    Relatepair =new RelatePair (Tag, Txt_ToRelate); //Associate the Tag and
                                                      Txt_ToRelate
    Outputlist.add(Relatepair);
    Txtlist.remove(Txt_ToRelate);
  End Do
Reurn Outputlist.
```

4 Experiments

In order to evaluate the algorithm proposed in this paper, we randomly submitted queries on Google and Baidu to obtain different numbers of search interfaces from five application domains: flights (15), book (19), hotel (33), job (24), and music (9). The total number of search interfaces used is 100 from 100 different sites. The detailed characteristics of data set and experimental results are listed in Table 1.

Table 1. Results for associating elements and labels of Deep Web query interface

Domain	Flight	Book	Hotel	Job	Music
Nubmer of query interfaces	15	19	33	24	9
Nubmer of control elements	169	199	231	275	88
Number of correctly associated elements	154	162	206	247	83
Accuracy	92.0%	82.1%	88.5%	89.3%	93.9%

From Table.1, we note that our algorithm can obtain high accuracy and the average overall accuracy of five domains is over 89.2%. Specially, the accuracy in the music domain can reach 94%. By analyzing the reason of relatively lower accuracy in domain Book, we observe that some of the attributes of Book domain, such as the date control elements which is composed of year element, month element and day element are not regularly arranged, and this results in the wrong association between control elements and labels. Now we are trying to tackle this problem by referring to ontology.

5 Conclusions

In this paper, we proposed an effective algorithm for associating labels and elements of the Deep Web query interface, which adopted DOM to parse Form query interface and used two useful heuristic rules and the nearest neighbor distance between control elements and labels to obtain the associated elements and labels. The experimental results show that our proposed algorithm is highly effective. In the future work, we will make further research to improve the accuracy by combining our proposed approach with semantic information of query interface.

Acknowledgments. This work is supported by the National Natural Science Foundation of China (61163057), Guangxi Nature Science Foundation (2012jjAAG0063), Open Fund of Guangxi Key Laboratory of Trusted Software (KX201117), Chongqing Nature Science Foundation (CSTC2009BB2126). The author would like to thank Long Shi, Qian He from Guilin University of Electronic Technology, and Chunming Wu from Southwest University for their suggestions and system evaluation.

References

1. Liu, W., Meng, X., Meng, W.: A Survey of Deep Web Data Integration. Chinese Journal of Computers 30(9), 1475–1489 (2007)
2. Chang, K.C., He, B., Li, C., Patel, M., Zhang, Z.: Structured database on the Web: Observations and Implications. SIGMOD Record, 61–70 (2004)
3. Jayant, M., Jeffery, S.R., Cohen, S., et al.: Webscale Data Integration: You Call Only Afford to Pay as You Go. In: Proceedings of the 3rd Biennial Conference on Innovative Data Systems Research, Asilomar, pp. 342–350 (2007)
4. He, H., Meng, W., Yu, C., Wu, Z.: Constructing Interface Schemas for Search Interfaces of Web Databases. In: Ngu, A.H.H., Kitsuregawa, M., Neuhold, E.J., Chung, J.-Y., Sheng, Q.Z. (eds.) WISE 2005. LNCS, vol. 3806, pp. 29–42. Springer, Heidelberg (2005)
5. Wu, W.: Integrating Deep Web data sources. University of Illinois at Urbana-Champaign (2006)
6. Liang, H., Zuo, W., Ren, F.: Attribute extraction of Deep web query interface based on heuristic rule. Computer Research and Development (46), 48–54 (2009)
7. Wang, H., Yu, J.: Attribute extraction of Deep web interface based on N-Gram. Computer and Modernization 12, 135–138 (2010)
8. He, H., Meng, W., Yu, C.T., Wu, Z.: WISE—integrator: An automatic integrator of Web search interfaces for e-commerce. In: Proceedings of the 29th International Conference on Very Large Data Bases, Berlin, pp. 357–368 (2003)
9. Wang, Y., Peng, T., Zuo, W., Zhu, H.: Schema Extraction of Deep Web Query Interface. In: International Conference on Web Information Systems and Mining, Shanghai, pp. 391–395 (2009)
10. Wu, W., Doan, A., Yu, C.: WebIQ: Learning from the Web to match Deep-Web query interfaces. In: Proceedings of the 22nd IEEE International Conference on Data Engineering, Atlanta, pp. 44–53 (2006)

Design and Implementation of the Online Shopping System

Guoyong Zhao and Zhiyu Zhou

College of Information and Electronics, Zhejiang Sci-Tech University, Hangzhou, China
550118999@qq.com, zhouzhiyu1993@163.com

Abstract. In today's society, online shopping has adapted to the fast-paced lifestyle, making ccustomers enjoy the convenience of choosing and buying their favorite products at home. This system is based on MVC architecture and adopts ASP.NET, Dreamweaver, SQL Server 2005, ADO.NET Entity Framework and other related technologies. The foreground system achieves some functions including the user registration and login, checking and buying commodities, the shopping cart, the personal order management, the customer complaint and personal information management, etc. And the background system achieves functions including the administrator login, the commodities category management, the commodities management, the order management, the news and information management, and so on. When released, this system will be dynamic and interactive, and become an online shopping system which is operated easily and has many functions.

Keywords: management information systems, online shopping, ASP.NET.

1 Introduction

In recent years, with the improvement of people's living standard and the popularization of personal computers, online shopping [1-3] has become an indispensable part of people's life. Therefore, more and more stores online are opened to expand the business scale and market influence, effectively reducing the operating costs for enterprises and improving the work efficiency. Compared to the traditional shopping style with features of high-cost, low efficiency and extensiveness and a variety of waste and corruption happening in the intermediate links, online shopping, with features of the "directness" and "transparency" in the business activities, can effectively reduce the economic costs and set up a good economic order. Online shopping system has a powerful interactive function which makes businessmen and users transfer information easily.

In literature [4], on the basis of the analysis of system requirements, the overall programs of the system and the business process and database of the system, are designed and Visual Studio. NET 2005 and SQL Server 2005 are adopted to implement the online shopping system on the basis of the analysis of system requirements. In literature [5], the online shopping system model, based on the colored Petri net-based method, is proposed, and the layered characteristics of CPN tools are used to overcome the shortcomings of the bigness and the complexity in the

F.L. Wang et al. (Eds.): WISM 2012, LNCS 7529, pp. 664–670, 2012.

established CPN model. In literature [6], using the technology of the ASP.NET is to implement online shopping system. In this paper, the ASP.NET MVC2 architecture design, the database connection technology and the ADO.NET Entity Framework 4.1 are adopted to implement the online shopping system.

2 Key Technologies

This system, based on B / S mode, uses the ASP.NET MVC2 in the architecture design, the Dreamweaver as a front-end development tool, the SQL Server 2005 in the management of the database, the ADO.NET the Entity Framework 4.1 technology as the database-driven, the Linq as the data query technology. Meanwhile, the front desk of this system adopts div + css, jQuery, Ajax and other related developmental technologies; the integrated development environment of this system is the Visual Studio 2010 and the operating platform is Window 7.

ASP.NET is a kind of developmental technology of dynamic websites and a part of microsoft.COM framework. ASP.NET can be written by Visual Basic.NET、C#、J# and any other languages that are compatible to .NET framework. This system is written by C#. ASP.NET is the compiled common language runtime code running on the server, and the ASP.NET Framework complements the designer integrated development environment in the Visual Studio.NET. ASP.NET, using the text of the layered configuration system, simplifies the server environment and the application of the web program. At the same time, ASP.NET considers the scalability, and increases the function of improvement used in the gathered environment and the multi - processor environment.

The MVC architecture is a widely used mature procedural framework, which divides the application architecture as a model (Model), views (View), controller three major units [7-8]. The model layer is responsible for the definition of the data structure, communicating with databases, data processing and processing of all data-related tasks. In the view layer, the ASP.NET technology is used. With ASP.NET, ASP.NET can use some of JavaScript. NET Framework, custom tags and other technologies.The control layer is mainly used to forward data and requests coming from the view to the model layer.

3 System Design

The main function modules of the system are divided into the front desk function module and the backstage function module: the front part is presented to the users including tourists and login users with different permissions and the back part is operated by the system administrator.

3.1 The Function Module Design of the Foreground Users

Through a browser on the network, the foreground tourists can receive the user registration, login, view product information, browse the site news, etc. with the user registration module, the user login module, the news-browsing module and the product-browsing module. The login users can manage personal information, purchase, view and modify individual orders, review products, initiate complaints, etc. with the personal information management module, the purchasing module, the personal order management module, the commodity comments module, and the customer complaint module. As shown in figure 1

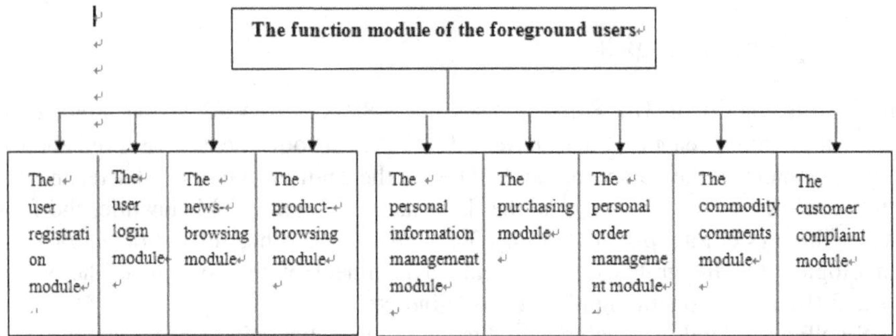

Fig. 1. The picture of the function module of foreground users

3.2 The Function Module Design of the Background Administrator

The administrator manages the website in the function module of the background administrator which consists of the administrator login module, the personal information management module, the products category management module, the product information management module, the customer order management module, the customer complaint management module, the news and information management module. As shown in figure 2

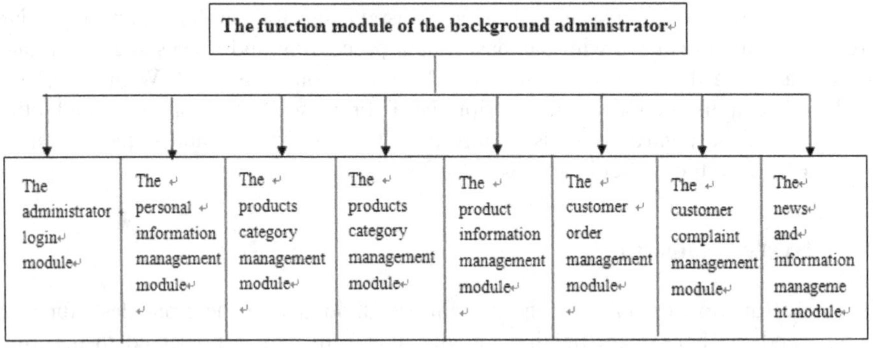

Fig. 2. The picture of the function module of the background administrator

4 System Implementation

4.1 Front System Implementation

A. User registration

The user registration entrance is located in the upper right corner of the site, the user should click on the "Register" link to enter the member registration page where the user need to fill out information about himself or herself. The page has been done with the input validation. As what is submitted does not meet the requirements of the input, the submission will fail. And the page will prompt the related error about the user's filling.

The user registration interface can save the member registration information to the Member of the database table, so that the user can manage his own data, which is convenient for the site administrator's management. When the user submits the right registration information, the Register action in the Account Controller with Http Post attribute will accept the treatment.

B. User login

The user login entrance is also located in the upper right corner of the site, the user should click on the "Login" link to enter the member login page. When login, the user needs to fill in the correct user name, password and verification code and then submit the login. If the login is successful, the page will turn to the home page, if not, the page will prompt the error information.

The user using the verification code to login effectively prevent the possibility of the user's password being cracked with the brute force. At the same time, using the Linq technology to query the membership information in database greatly improves the security of user information. After the user submits the right login information, the Login action in the Account Controller with Http Post attribute will accept the treatment. First it will verify whether the user exists, if it exists, then will judge whether the current user is an administrator, if the user is the administrator, the page will jump to the main interface of the management of background, otherwise go to the home page.

C. News and information browsing

The user will open web news list by clicking the "News bulletin" at the top of the website which is linked. Through browsing, the user will know the latest news on the website, as well as other things.

D. Products browsing

The home page provides the list of product information browsing, click on a single commodity, the user can browse the detailed information of the products.

E. Personal information management

If the user is registered as a member, he or she can login to the page of user's information to modify the related data, in order that the user's information can get updated.

After modifying the information, the user should click "Edit information" to submit the modification. Then the modification should be verified whether it meets the requirement, if meets, the information will be updated and saved to the database, and the user's information gets updated. If the user wants to change his or her password, he or she needs to fill in the current password, the new password and repeat the new password to change.

F. Commodity purchasing

The commodity will be added to the shopping cart after the user clicks on the "Buy". In the interface of the shopping cart, the user also can change the quantity of commodities and delete the commodities he or she doesn't want to buy. Then, the user is going to choose the method of payment and fill in the receipt information to place an order, and then the user finishes the shopping. The default receipt information is the personal information which the user fills in when registration. After placing an order successfully, the Order database will increase a piece of order information and the commodities in the shopping cart will be added to the Order Goods table.

G. Personal order management

The user can check all his or her orders in "My orders" which is presented in a list form. Meanwhile, the user can check the detailed information of orders. In the personal order management module, the user can cancel the unfilled orders, confirm the receipt. The operation of changing the order made by the user should come before the commodities are not shipped. Once shipped, the user can do nothing to the order other than confirming the receipt.

H. Commodity comments

In the interface of the detailed order information, the user will find the entrance of the commodity comments module in the "deal". Just click it, the user will go to the commodity comments interface. Comments on the commodity will be displayed directly under the detailed information of the commodity. And the comments will be added to the "Comment" table in the database after the user submits the comments. Comments on the commodity are shown as the Figure 3.

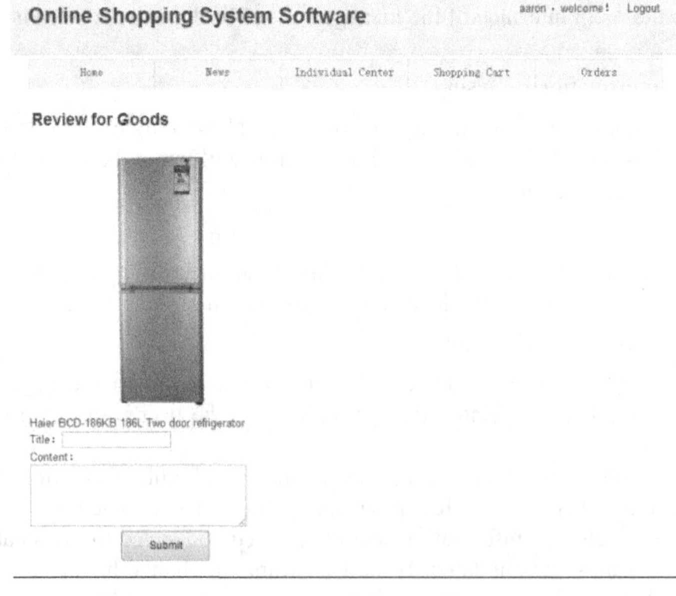

Fig. 3. The interface of Commodity comments

I. Customer complaint

At the bottom of the detailed order information interface, the user can find the entrance of the customer complaint module. Click it, the user will turn to the complaint-suggestion interface, by which the user can communicate with the website administrator directly. And the complaint from the user will be saved in "Complaint" table of the database.

4.2 Background System Implementation

The background system is mainly about the administrator's management on the site in which the administrator's login and the personal information management share the function of front users. When eloigning, the administrator will directly enter the background management system through the value judgment of the field is admin in Member table.

A. Products category management
The entrance of the products category management is located on the left side of the background system interface, where the administrator can add new categories, delete and change the existing categories. Commodities could be categorized in the products category management module, which speeds up the user's looking for commodities and presents the user commodities he or she is caring about, improving the user's shopping experience.

B. Product information management
In the product information management module, commodities on line can be added, modified and put off the shelf, so that the information of commodities can get increased and updated in time. All the products on the website are presented by the form of list in the interface of the product information management. Adding and modifying commodities, made in the same interface, are two different operations which are distinguished by different IDs. If there is the ID of the commodity in the database, commodities are modified; if not, new commodities are added. The management of the product information is one of core background system managements of online shopping and the convenience of updating commodities information can increase the volume.

C. Customer order management
The customer order management module aims at managing the orders from users. In this module, the detailed information of orders can be checked and the status of orders can be set. Entering the interface of the customer order management, you will find the customer orders appear in lists. Clicking on the "check" of the specified order, the administrator can enter the interface of the detailed information of orders and set the orders status. The management of orders is mainly about setting orders status, cancelling Invalid orders, make shipment according to effective orders, and so on.

D. Customer complaint management
The customer complaints management is the communicative medium between users and administrators. In this module, complaints from users about orders are displayed, and the administrator can check and respond to the users complaints.

E. News and information management
The news and information management is the primary means of releasing the latest news and information to users. By doing this, the administrator can make users know

the latest news and information on the website. The management of the news and information includes adding, modifying and viewing news.

5 Conclusion

The system of online shopping is a kind of commercial information system with the interactive function. It can make the commodities information gets updated in time and users have a better understanding of the trade. This system is developed as an e-commerce platform based on the MVC pattern, adopting the ASP.NET technology, SQL Serer 2005 as the DBMS and IIS7 as the application server. The functions of the front system include the user registration, checking commodities, buying commodities, the shopping cart, the personal order management, the customer complaint and personal information management, etc. The functions of the background system include the administrator login, the commodities category management, the commodities management, the order management, the news and information management, and so on. At the same time, it is using the jQuery, Ajax, Repository encapsulation technology that improves the performance and maintainability of the system. When users are shopping on line, the application of Ajax-no-refurbish modification technology greatly speeds up the purchase and the user's shopping experience.

References

1. Hsu, C.H., Yang, C.M., Chen, T.C., et al.: Applying AHP method select online shopping platform. In: 2010 7th International Conference on Service Systems and Service Management, Proceedings of ICSSSM 2010, pp. 912–916 (2010)
2. Wang, Y.L.: China online shopping markets and innovation of marketing. In: 2011 International Conference on E-Business and E-Government, pp. 8691–8694 (2011)
3. Chen, H.: The impact of comments and recommendation system on online shopper buying behaviour. Journal of Networks 7, 345–350 (2012)
4. Zhang, Z.W., Xu, H.L., J, E., et al.: Online shopping simulation experiment system design and development. China Modern Educational Equipment 1, 22–23, 29 (2012)
5. Peng, J., Li, S.Z., Yang, S.X.: Online Shopping System Modeling and its Soundness Analysis Based on CPN Tools. Journal of Jiangxi University of Science and Technology 32, 49–52 (2011)
6. Peng, L., Liang, J.I., Zhang, L.Y.: The Implement of On-Line Shopping System Based on .net. Journal of Wuhan Engineering Institute 23, 29–33 (2011)
7. Huang, B.X., Chen, X.J., Li, Y.: ASP.NET MVC2 Development practices. Publishing House of Electronics Industry (2011)
8. Li, J.J., Fu, H., Zhang, L., et al.: Key Technology Research of Web Framework Based on MVC Pattern. Computer Knowledge and Technology 7, 2308–2309, 2332 (2011) (received date: May 24, 2012)

System Development of Residence Property Management Based on WEB

Hanbin Cui and Zhiyu Zhou

College of Information and Electronics, Zhejiang Sci-Tech University, Hangzhou, China
554796870@qq.com, zhouzhiyu1993@163.com

Abstract. In order to making owner' life convenience and property management standardized, the B/S structure MVC model and SSI frame etc. are employed. System of residence property management based on WEB is empoldered combining with Java, JavaScrip, HTML and SQL language. The prospect of this system achieves register module, log in module, complaint service module, repair service module, parking space module, building information module and PIM module. The background of this system receives log in module, system user module, resident management module, complaint management module, repair management module and payment management module. The software application of residence property management based on WEB is vastly convenient the proprietors' lives and the proprietors' demand service from the property management company. The property management standard is developed.

Keywords: Residence Property Management, WEB, System Development.

1 Introduction

With the development of network techniques, the residence property management [1-4] has already grown to network. Compared with the traditional property management and property companies, the residence property management system software can administrate several residential quarters based on Wed. It needs no work place in each residential quarter. After the servers are set up, the information of proprietors in the residential quarters can be checked and administrated by the property companies, and the work efficiency is improved greatly.

Literature [5] gives research on MapX technology to develop the system with the actual examples. It gives out the system's main functions and the implementation techniques of some graphic function, which achieves the seamless connect of MIS and GIS. Literature [6] designs and realizes the residence property management using the Delphi7.0 and MSSQL2000.This system implements the functional module of house property set, customer data set, charge management, administrative personnel, report inquiry, community management and system setup. This paper employs Java, JavaScript, HTML, SQL language to programme. The residential property management platform is user-friendly control and rich function.

F.L. Wang et al. (Eds.): WISM 2012, LNCS 7529, pp. 671–677, 2012.

2 Key Technology

B/S is the application model of Browser/Server which often has the aid of IE browser to work. MySQL is SQL date base server with multi-user and multi-threading. It is composed with a server demons mysql and many different customer program libraries. MySQl allows you to save and record files or images flexibly and it can dispose the large-scale very fast. SSI technology means the Strtus, Spring, iBatis technology.The aim to use the Struts is to decrease the time to exploit the Web applied through the MVC design module. Spring frame is in layered architecture constituting with 7 well defined modules. The core point of Spring frame supports binding to reusable business and data access object of specific J2EE server. MVC module enforces function division of Web system. Model storey realizes system business logic. View storey achieves the user interaction using JSP and HTML.

3 System Model Design

A. Foreground proprietor module design

The foreground browser system is an operating system that permits proprietors to register, log in, modify and browse. It consists of user register and log in module, housing information module, repair service module, complaint service module, payment inquiry module and carport information module. The foreground user example is as Fig.1.

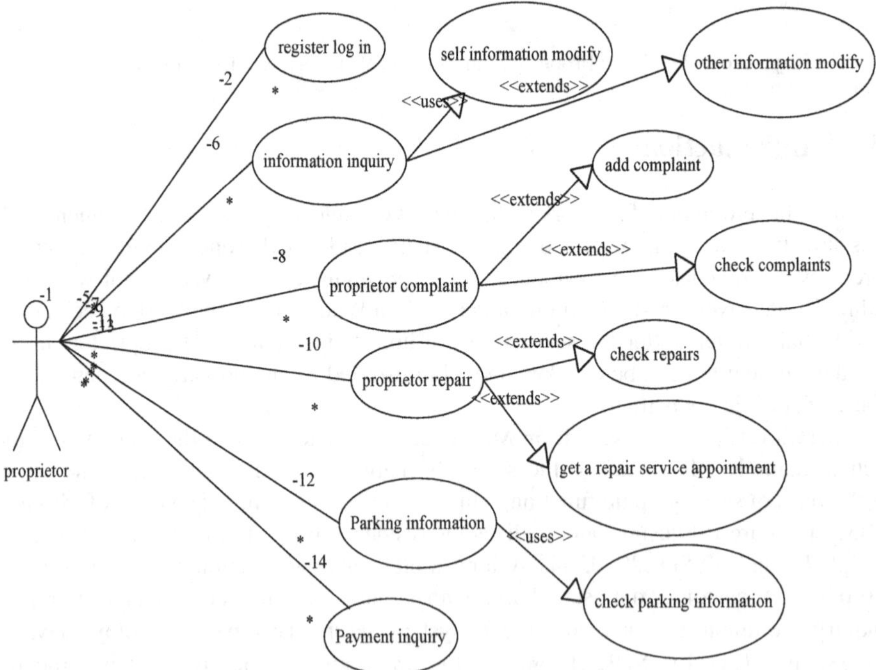

Fig. 1. Foreground user test cases

User register log in: first, the proprietors should register as the website's user, which can only promise them to log in the website. It can protect user's privacy and also to the website information.

Complaint service module the user can add complaint and track the progress of existing complaint. For instance, the audit is pass or not and who processes it.

Repair service module: the user can get a repair service appointment and track the progress that the audit is pass or not or who deal with it.

Payment inquiry module: the user can check the payment and the payment.

Parking information module: Parking information is checked.

Building information module: the user log in the website and check the information of his residential quarter.

B. Background administrator module design

The background website administrate system is a system that allows the administrator to manage the web. It consists of building information module, user management module, complaint management module, repair service module, payment management module and carport management module. The test cases are as Fig.2.

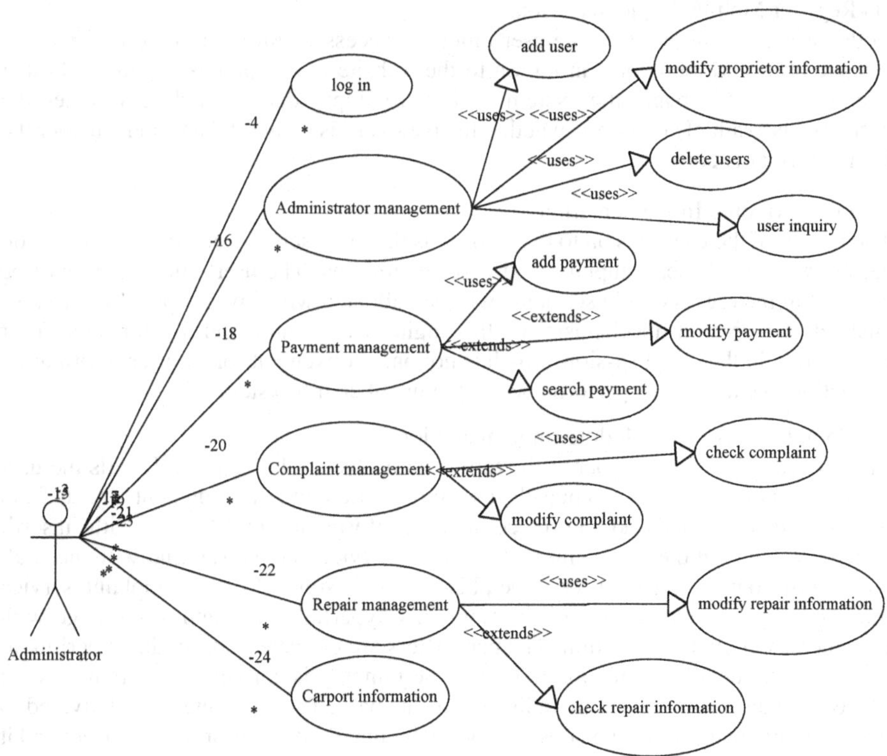

Fig. 2. Webmaster module test cases

Administrator management module: The administrator can do the operation to delete the user's account number and give the administrator privileges to some users.

Complaint management module: the administrator can check, search complaint, delete or modify complaint information, and guarantee the complaint information's update timely.

Repair management module: The manager can browser, search, and delete or modify the repair information to ensure the information to update timely.

Payment management module: according to the user payment state, the manager can browser, search, and delete or modify payment.

Carport information module: The manager can browse, search the carport, delete or modify the carport information, and ensure the carport timely update.

Building information module: the building information is promised to add and modify the building information.

4 System Module Implementation

4.1 System Foreground Implementation

(1) Register Module Implementation

In the user registration page, the user input the necessary information to register to a website user, and then you can log on to the website to complaints, repair, and other operations. In the page, the system will verify input data, including whether the username is available or empty, whether the password is greater than 6, and whether the E-mail format is correct.

(2) Login Module Implementation

The barrier of the user logs on to the website is the login screen, only by logging in you can browse, purchase, complaint and other operations. The user's input data will be checked in the registry and user name will be validated with JavaScript. </script> also enabled the session mechanism in the login background and it stores all user information in the client session, which is not only convenient for the user information extraction, but also helps performance of the membership system.

(3) Complaints Service Module Implementation

There is complaint service bar at the top of the website's home page. It calls the user information in the session and puts the user information into the complaint objects from the website page finally adding the complaints information to data base with 'insert'. When you click on the complaint records in the service, the background will check all the records depending on users' member Id in session. On the complaint service interface, the user can click on the complaints hyperlink and jump to the page with complaints detailed information. The parameters are delivered during the hyperlink to the complaint detail information to ensure the jumping is correct. The parameters are the record number of the complain list in the data base. The parameters are delivered to the complaint page, which ensures the complaint detail information show correct and it makes the user have a further learning about complaint.

(3) Repair Service Module Implementation

The complaint information is checked according the complaint names. The implementation technology of the module is similar to complaints module. The repair module is added to the residential user rights. When users add repair appointment, the background will take out a house Id from user objects from session to store it to repair objects. And finally the object stores repair appointment by inserted into background database. Repairing and adding interface are shown in Figure 3.

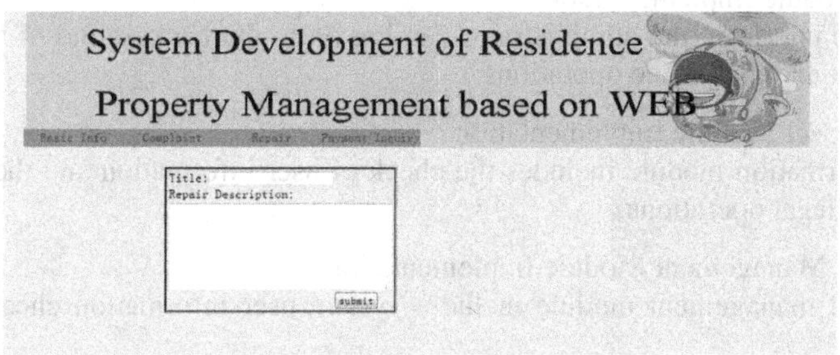

Fig. 3. Repairing and adding interface

The repair list can help users to find his record quickly and check the status in time. When clicking on repair record in record service, the background will check all records based on the user house Id in session which are only users have.

In the repair record interface, the user can click on the repair hyperlink and jump to the repair detailed information page, in order to ensure the jumping is correct. The parameters are the record number of repairs list. The parameters are transmitted to the repair page, which makes the repair information correct show and let user have a further learning about repair.

(5) Payment Query Module Implementation
According to owner number, the payment information is checked. The module technology is similar to the repair module. The payment list is helpful for users to find their payment records to check the status quickly. When the button is clicked, the background will find all records based on user's house Id in session. It's the rights only the users have.
In the payment record interface, the user can click on the payment hyperlink and jump to the payment detailed information page. To ensure the jumping is correct and let user have a further learning about the payment, when jumped to the page, the parameters are also passed to the detail page.

(6) Packing Module Implementation
Depending on the corresponding parking information, all user numbers are checked .The module implement technology is similar to payment module.

(7) Building Information Module Implementation
Depending on the corresponding building information, all user numbers are checked .The module implement technology is similar to the parking module.

(8) Personal Information Management Module Implementation
Basic information module includes basic information modify, E-mail changing and password changing. The input data test is the same as that of the foreground system,

including whether the user name is empty, whether the length of password is longer than 6, and whether the E-mail format is correct.

4.2 System Background Implementation

(1) Login Module Implementation
Through the interface, the administrator can log on to the background of the website management and give some operations.

(2) System User Module Implementation
System information module includes the check of user information and the removing users with illegal operations.

(3) Resident Management Module Implementation
The resident management module includes system user information check and user adding.

(4) Complaint Management Module Implementation
The complaint module is allowed to check, modify and delete operation, which is convenience for manager to update the complaint information in time. The complaint management is divided into verify waiting, verify losing record and verify passing record. The complaint detail information is checked and status change by click the check button to mean the property management company starts to process the information. The manger also can select and delete the records in the listing.

(5) Repair Management Module Implementation
The repair management module allows managers to search, modify and delete, which is same as complaint management module.

(6) Payment Management Module Implementation
The manager can modify and add payment. In this module, the payment status is most important. Managers change the status according the users' payment. By clicking the detail, manage can log in the payment detail interface to operate the payment status. The manager can add a proprietor's payment bill. The realization of the payment bill is as Fig.4.

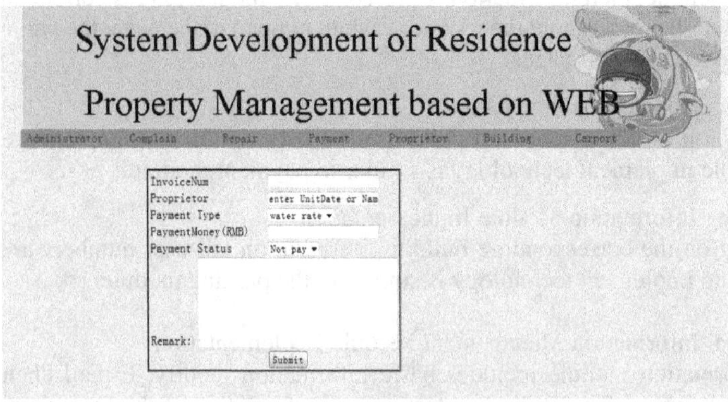

Fig. 4. Payment add interface

(7) Other Modules Implementation

Other modules management is constituted of carport management and building information. The building information is the management of the building. The manager should add and modify the information in time. The carport management is the parking space management. When the owner bought the house, manages should add connect of the parking space and the proprietor' information.

5 Conclusion

Combining the actual demand of current property management, on the basis of residence property management design, this paper develop the residence property management system based on WEB and finish the basic function for the users and administrators. The software application of residence property management based on WEB is vastly convenient the proprietors' lives and the proprietors' demand service from the property management company. It is convenience for the property management company to keep and check the user's information. It moves forward a single step to regulate the property management and the standard is developed.

References

1. Sun, Y.X., Liang, Y.: Design and implementation of a residential community management system based on GIS. In: International Conference on Information Engineering and Computer Science, ICIECS 2009 (2009)
2. Zheng, H.F., Yang, Z.X.: Urban community property management research. In: 2nd International Conference on Information Science and Engineering, ICISE 2010, pp. 3345–3347 (2010)
3. Tian, J.L.: Application design of community property management system based on Delphi. In: 2010 International Conference on Computer and Communication Technologies in Agriculture Engineering, vol. 1, pp. 85–88 (2010)
4. Li, Y.X., Cao, L.J., Qian, Y., et al.: A property management system using WebGIS. In: Proceedings of the 2010 International Conference on Computer Engineering and Technology, vol. 2, pp. 2683–2685 (2010)
5. Li, H.Q., Ren, Y.K.: Study of Property Management Information System Based on MapX. Journal of Henan Institute of Engineering (Natural Science Edition) 22(3), 38–41 (2010)
6. Chen, K.H., Pan, G.N., Ling, Y.X.: The Design and Implement on Residence Property Management System in Yulin. Journal of Yulin Normal University (Nature Science) 30(3), 130–134 (2009)
7. Han, X.C., Li, J., Du, X.F.: The Design of New Property Management System. Journal of Zhongyuan University of Technology 20(3), 66–68, 75 (2009) (received date: May 24, 2012)

An Intelligent Metadata Extraction Approach Based on Programming by Demonstration

Binge Cui[1,2] and Jie Zhang[1]

[1] The First Institute of Oceanography. SOA. 266061 Qingdao, Shandong, P.R. China
[2] College of Information Science and Engineering, Shandong University of Science and Technology, 266510 Qingdao, Shandong, P.R. China
cuibinge@yahoo.com.cn

Abstract. Metadata extraction is an important prerequisite for remote sensing images management and sharing. Business users urgently need to extract metadata automatically and quickly. However, existing metadata extraction applications extract only one single type of image metadata. In this paper, we proposed a generic and extensible metadata extraction approach based on Programming by Demonstration (PbD). Data owners can specify the metadata items to be extracted and the extraction methods in an interactive and visual user interface. Such knowledge will be stored in the rule base in the form of rules. The advantage of this approach is that users do not need any programming knowledge, but they can handle new types of images themselves.

Keywords: Metadata Extraction, Demonstration by Programming, Rule-based System.

1 Introduction

With the rapid development of remote sensing technology, more and more remote sensing images can be acquired from satellite data receiving stations. To facilitate the management of remote sensing images, data providers usually generate an ASCII image metadata in a separate file or in the image file header. However, after investigation we found that almost every kind of image metadata formats are not the same. Some metadata is a text file, and some metadata is an XML files. Even if in the same file type, metadata description and organization are also different. As the number of metadata items is very large (from dozens to hundreds) and the extraction means is very complex, the programming effort is very large.

The traditional method of extracting the metadata can be described as follows: Business users tell programmers what metadata items need to be extracted; Business users tell programmers the extraction approach of each metadata item; Programmers write code to achieve the extraction of metadata. Can this process be done in one step? Business users tell the computer what to extract and how to extract, and programmers do not need to participate in the process. This is the basic idea of programming by demonstration (PbD). PbD defines a way to define a sequence of

F.L. Wang et al. (Eds.): WISM 2012, LNCS 7529, pp. 678–685, 2012.

operations without having to learn a programming language. The PbD paradigm is first attractive to the robotics industry due to the costs involved in the development and maintenance of robot programs. In this field, the operator often has implicit knowledge on the task to achieve (he/she knows how to do it), but does not have usually the programming skills (or the time) required to reconfigure the robot. Demonstrating how to achieve the task through examples thus allows the robots to learn the skill without explicitly programming each detail.

In this paper, we want to apply PbD to remote sensing image metadata extraction. To achieve this goal, we must address some key technical issues. Such issues include: How users interact with computers? How users pass their knowledge to the computer? How computer organize and store this knowledge? How computer use the knowledge during metadata extraction process? To solve these issues, we propose to use XML as the intermediate file format. Each original header file should be converted to an XML file before further processing. Through the introduction of file tree and operation wizard, business users can tell the computer how to extract and process each metadata item. These steps will be stored in the form of rules in the rule base. At last, computer can complete the metadata extraction work based on these rules.

2 Related Works

In metadata extraction, there are two main technical routes: statistical model-based route and rule-based route.

2.1 Statistical Model-Based Metadata Extraction

The basic idea of this route is as follows: find an appropriate statistical model, change the parameters and training sample set, apply the trained model for metadata extraction. This route has greater learning ability and adaptability compared with rule-based route. HMM (Hidden Markov Model) and SVM (Support Vector Machine) are the typical statistical model-based approaches.

HMM-based metadata extraction regards the document as a phrase sequence generated by some hidden states. The state sequence that can output the document with maximum probability will be selected. The model parameters can learn from the samples. Seymore et al used HMM to extract the paper header metadata, and the total accuracy has reached 90.1% [1]. McCallum et al proposed MEMM (Maximum Entropy Markov Model) by adding the text context into HMM [2], which achieved good result in the FAQ metadata extraction.

2.2 Rule-Based Metadata Extraction

Rules can be obtained in two ways: manual define and automated generation. Manual define is a very heavy physical labor, and it requires a high level of knowledge engineers to complete. Automated generation does not need the involvement of knowledge engineers, thus it is applicable to the diversity of information sources format. The recognition accuracy of the generated rules is generally between 80% and 90%. Some systems have been developed to automatically generate extraction rules in different domains, such as Web site information [3], citation metadata [4], journal metadata [5], and biomedical document images [6], digital documents [7], etc. As the accuracy of the automatically generated rules is not high, and it is difficult to define a number of rules manually, people want to use a new programming technology for the definition of rules.

PbD is based on the idea: the user operation sequence on one instance can be applied to other instances. IBM Almaden Research Center developed a Firefox-based browser extension: Koala/ Coscripter [8], which support users to record operational processes into the script. R. Tuchinda et al developed a Mashup build tool "Karma" based on PbD [9], which proposed one unified interactive framework to support users accomplish the Mashup build process through demonstration. Wang et al designed a table-based programming environment: Mashroom [10], which has implemented the data operations as visual actions on nested tables and expressed business logic by performing actions on current data. Use cases and analysis show that compare with Yahoo! Pipes, Mashroom can reduce time cost (by 90%) and learning cost of developing Situational Applications.

PbD can convert the user's operation into business rule. However, Existing work is mainly aimed at Web data manipulation. As we know, HTML file can be easily converted to XML file to locate the path for each metadata item. Remote sensing image header file may be a plain text file. Therefore, it is difficult to locate metadata items. Moreover, the extracted metadata items need to be further processed to be able to meet the needs of users. Thus, the rule description and generate methods should also be re-enacted.

3 Convert Original Data into XML Data

As remote sensing satellites were launched by different countries or organizations in different historical periods, metadata storage format is diverse and heterogeneous. Before the introduction of our work, let us look at several typical header files, which are shown in Figure 1.

From Figure 1 we can see, although these header files are not standard XML format, they still have some structural features. However, it is very difficult to extract the value of a metadata item easily. For example, we want to extract the upper left corner latitude of the remote sensing images: 38.379949. How can we locate the value uniquely? If string "GRINGPOINTLATITUDE" above it is unique in the header file, we can search that string first, and then read 3 lines backward. If not, we can only find the value by line number. Moreover, To isolate that value, we need to further break down the string by equal sign, parentheses, commas, etc. Thus, To extract one metadata item from the header file, we should write a separate program for it. This not only makes the program complex and unstructured, but the program is difficult to extend and maintain. Is there an easy way to extract the metadata?

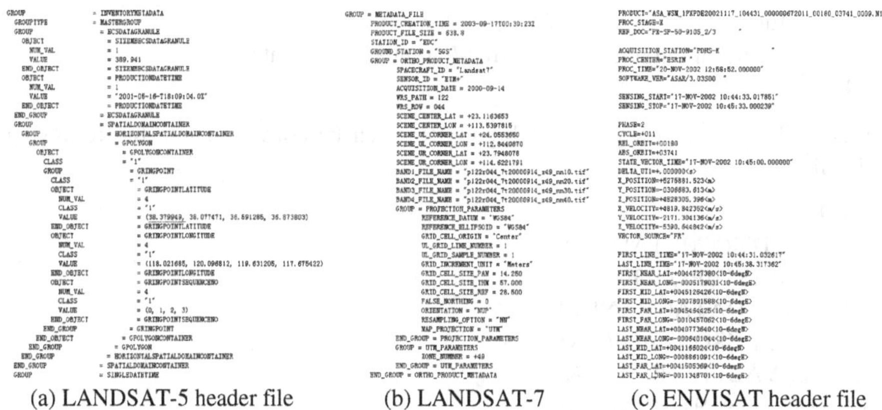

(a) LANDSAT-5 header file (b) LANDSAT-7 (c) ENVISAT header file

Fig. 1. Three Typical Remote Sensing Images Header File Fragment

To improve the readability of header file, metadata are generally organized through newlines and indentations, which allows them to have the characteristics of tree. Thus, using such layering visual features, we can convert the header file into an XML file. The conversion principle can be described as follows:

1) If next line indentation is greater than current line indentation, then next line is the child element of current line.

2) If next line indentation is equal to current line indentation, then next line is the sibling element of current line.

3) If next line indentation is less than current line indentation, then next line is the closing tag of current line's parent element.

4) For leaf element, the left side of equal sign is the element name, and the right side of equal sign is the element value. For non-leaf element, the left side of equal sign can be omitted, and the right side of equal sign is the element name.

5) For non-leaf element, the element name of the opening tag should be equal to that of the closing tag.

Assume that the indentation of the ith line is denoted as Indentation(i); the total number of rows equal to the Num. The conversion algorithm is described below.

Algorithm 1. Converting original data format into XML format

```
1.  Create an XML document object, set the CurrentParentNode as the root
    node;
2.  Read each line of the header file, and store them into one array lines;
3.  CurrentNode = lines[0];
4.  CurrentParentNode. AddChildNode(CurrentNode);
5.  For (int i = 0; i < Num; i++) {
6.    If (Indentation(i+1) > Indentation(i))        // Child Element
7.      CurrentParentNode = CurrentNode; CurrentNode = lines[i+1];
8.      CurrentParentNode. AddChildNode(CurrentNode);
9.    Else If (Indentation(i+1) == Indentation(i))        //Sibling Element
10.     CurrentNode = lines[i+1];
11.     CurrentParentNode. AddChildNode(CurrentNode);
```

```
    Else //Closing Tag of Parent Element
        CurrentParentNode = CurrentParentNode.ParentNode
}
```

For example, metadata segment in Figure 1(a) can be converted into the following XML file.

```
<ROOT>
  <INVENTORYMETADATA>
      <MASTERGROUP>
      <ECSDATAGRANULE>
  <SIZEMBECSDATAGRANULE>
  <NUM_VAL>1</NUM_VAL>
  <VALUE>389.941</VALUE>
  </ SIZEMBECSDATAGRANULE>
  ......
  </ECSDATAGRANULE>
  <SPATIALDOMAINCONTAINER>
  ......
      < GRINGPOINTLATITUDE>
          < NUM_VAL>1</ NUM_VAL>
          <CLASS> "1"</CLASS>
          <VALUE> (38.379949, 38.077471, 36.591285, 36.873803) </VALUE>
      </GRINGPOINTLATITUDE>
  ......
  </SPATIALDOMAINCONTAINER>
  </INVENTORYMETADATA>
</ROOT>
```

Now, the path for the element that contains value "38.379949" can be described using XPath:

ROOT/INVENTORYMETADATA/SPATIALDOMAINCONTAINER/HORIZONTALSPATIALDOMAINCONTAINER/GPOLYGON/GPOLYGONCONTAINER/GRINGPOINT/GRINGPOINTLATITUDE/VALUE, which is unique and easily obtained. Thus, the element positioning problem has been solved.

4 Extract the Metadata from the Raw Data Based on PbD

Due to historical reason, the value of metadata item is mixed and non-standard. For example, the four latitude values are organized into a string (38.379949, 38.077471, 36.591285, 36.873803) in Figure 1(a). We must break down the string to get each latitude value. The value for production data time is "2001-05-16-T18:09:04.0Z", which must be converted to standard time format before storing it into the database. The numeric data presentation is also diverse. For example, the first near latitude is equal to +0044727380<10-6degN> in Figure 3, which means that total is 10 digits, fractional part is 6 digits. So the real value is 44.727380. Some values are expressed in scientific notation, e.g., the line time interval is equal to +1.13293054E-02<s>. In summary, non-standard metadata exist everywhere, which are too numerous to mention.

Generally, to convert such non-standard data into standard data, we should write a separate program for each of them. This not only makes coding workload very large, but the code is difficult to reuse and maintain. Moreover, programmers do not quite understand the implications and handling approach of these metadata, which should be introduced by business users firstly. To improve the programming model, we proposed a group of data cleaning and processing approach. There are six commonly used means which are defined as follows:

1) Redundant characters removing: remove redundant character, such as quotes, left and right parentheses, etc.

2) Long string splitting: Common delimiters include comma, semicolon, etc. After the split, each separate metadata can be mapped to one standard metadata item.

3) Substring: By taking the substring to remove irrelevant information, such as <10-6degN>, <s>, etc. The start position and end position should be specified.

4) Date format conversion: Specify the date format by business users. For example, only business users know whether 08-09-2011 means August 9, 2011 or September 8, 2011. Recognize the date and time according to special characters. For example, "-" is usually used to separate year, month and day; ":" is used to separate hour, minute and second.

5) Scientific notation: Specify whether scientific notation is used in numerical representation. If true, the real number in front of E(e) is the coefficient, and the integer behind e is the exponent.

6) Decimal representation: Specify the location of the decimal point. It is no doubt that the front part is the integer, and the back part is the decimal.

For example, To obtain the upper left corner latitude 38.379949, we need to perform the following steps: 1) remove the redundant character "(" and ")"; 2) split the string into an array, which contain four decimals; 3) pick out the first decimal from the array. To get the first near latitude 44.72738 from string +0044727380<10-6degN>, we need to perform the following steps: 1) take the substring from the 1st character to the 11th character; 2) set the decimal point in front of the last 6th characters; 3) parse the front part into an integer, parse the back part into a decimal, and add them together to get the final value 44.72738. To get line time interval 0.0113293054 from string +1.13293054E-02<s>, we need to perform the following steps: 1) take the substring from the 1st character to the 15th character; 2) convert the scientific notation into decimal format.

From the above description we can see through a combination of some of the basic operation, we can implement the cleaning and conversation of the metadata value. The operations collection and operations order can be specified by business users. Thus, coding was replaced by an operation combination. Each operation is assigned zero to multiple operands. This combination approach is easy to understand for business users, which allows them to define or expand metadata extraction items by themselves.

On the basis of the above work, we proposed the novel metadata extraction approach based on PbD. The demonstration process is omitted due to page limit. A typical configuration file is as follows (Some non-essential information is deleted):

```
<Satellite satName="LANDSAT-5" sensorName="TM">
        <Parameter>
                <FieldName>PRODUCTIONDATETIME</FieldName>
                <DataType>datetime</DataType>
                <ExtractRule
        path="ROOT/INVENTORYMETADATA/ECSDATAGRANULE/PRODU
        CTION DATETIME/ VALUE" operations=" redundantChar =‘"’;
        dateFormat=‘Y-M-D’; " />
        </Parameter>
        <Parameter >
                <FieldName> UpperLeftLatitude </FieldName>
                <DataType>float</DataType>
                <Length>8</Length>
                <DecimalPlaces>6</DecimalPlaces>
                <ExtractRule
        path="ROOT/INVENTORYMETADATA/SPATIALDOMAINCONTAINE
        R/HORIZONTALSPATIALDOMAINCONTAINER/GPOLYGON/GPOLY
        GONCONTAINER/GRINGPOINT/GRINGPOINTLATITUDE/VALUE"
        operations="redundantChar=‘()’; separator=‘,’; itemIndex=‘0’" />
        </Parameter>
</Satellite>
```

From the above example we can see, the extraction rule is described using two attributes. One attribute is the "path", which is the exact and unique path for us to get the raw metadata item. Another attribute is the "operations", which contains zero to multiple operations separated using ";". Each operation are in the form "name = value". Detailed description of the usage of these operations is as follows:

1) redundantChar: its value includes ‘"’, ‘(’, ‘)’, etc.
2) separator: its value includes ‘,’, ‘;’, etc.
3) itemIndex: its value is the index of the metadata sub item.
4) dateType : its value includes ‘Y-M-D’, ‘M-D-Y’, ‘D-M-Y’, etc.
5) scientificNotation: its value is true or false.
6) subString: its value includes two integer separated by ‘,’.
7) decimalPoint: its value is the decimal point position.

Through these rules, the image metadata can be extracted automatically. Firstly, the header file will be converted into an XML file using algorithm 1. Secondly, the raw metadata item will be extracted using the "path" attribute in the extract rule. Thirdly, the cleaned target metadata value will be extracted based on the "operations" attribute in the extract rule. Due to space limitations, we won't discuss the implementation in detail.

5 Conclusion

Metadata extraction is always the core issues of metadata sharing and integration. Due to the diversity and complexity of metadata, existing extraction method is difficult to expand into new types of images. In this paper, we proposed the PbD based metadata extraction approach. Before the demonstration, the original header file should be converted into XML file using our algorithm described in section 3. Business users can demonstrate how to extract metadata through graphical interface. The demonstration result is stored in the configuration file, which will be used in future metadata extraction. Our approach builds a bridge between business users and the computer. Thus, it greatly reduces the workload of the programmer. Our approach is very simple and intuitive to use for business, so it will have a good prospect in the remote sensing domain.

Acknowledgments. This research is supported by foundation for outstanding young scientist in Shandong province (No. BS2009DX011) and Special Funds for Basic Scientific Research Project of the First Institute of Oceanography, S.O.A (No. 2008T29).

References

1. Seymore, K., McCallum, A., Rosenreid, R.: Learning hidden Markov model structure for information extraction. In: Califf, M.E., Freitag, D., Kushmerick, N., Muslea, I. (eds.) Proc. of the AAAI 1999 Workshop on Machine Learning for Information Extraction, pp. 37–42. MIT Press, Cambridge (1999)
2. McCallum, A., Freitag, D., Pereira, F.: Maximum entropy Markov models for information extraction and segmentation. In: Langley, P. (ed.) Proc. of the Int'l Conf. on Machine Learning (ICML 2000), pp. 591–598. Morgan Kaufmann Publishers, San Francisco (2000)
3. Chidlovskii, B.: Wrapping Web Information Providers by Transducer Induction. In: De Raedt, L., Flach, P.A. (eds.) ECML 2001. LNCS (LNAI), vol. 2167, pp. 61–72. Springer, Heidelberg (2001)
4. Hitchcock, S., Carr, L., et al.: Developing services for open eprint archives: Globalisation, integration and the impact of links. In: Proc. of the 5th ACM Conf. on Digital Libraries (ACMDL 2000), pp. 143–151. ACM Press, New York (2000)
5. Klink, S., Dengel, A., Kieninger, T.: Rule-Based document structure understanding with a fuzzy combination of layout and textual features. Int'l Journal on Document Analysis and Recognition 4(1), 18–26 (2001)
6. Kim, J., Le, D.X., Thoma, G.R.: Automated labeling algorithms for biomedical document images. In: Proc. of the 7th World Multiconference on Systemics, Cybernetics and Informatics, pp. 352–357. IIIS, Orlando (2003)
7. Zhao, P.X., Zhang, M., et al.: Automatic extraction of metadata from digital documents. Computer Science 30(10), 217–204 (2003) (in Chinese with English abstract)
8. Leshed, G., Haber, E.M., et al.: CoScripter: Automating & Sharing How-To Knowledge in the Enterprise. In: Proc. of the CHI 2008 Proceedings · End-Users Sharing and Tailoring Software, pp. 1719–1728. ACM Press, New York (2008)
9. Tuchinda, R., Szekely, P., Knoblock, C.A.: Building Mashups by Example. In: Proc. of the 13th International Conference on Intelligent User Interfaces (IUI 2008), Gran Canaria, Spain, pp. 139–148 (2008)
10. Wang, G.L., Yang, S.H., Han, Y.B.: Mashroom: End-User Mashup Programming Using Nested Tables. In: Proc. of the 18th International World Wide Web Conference (WWW), Madrid, Spain, pp. 861–870 (2009)

OF-NEDL: An OpenFlow Networking Experiment Description Language Based on XML*

Junxue Liang[1], Zhaowen Lin[1], and Yan Ma[1,2]

[1] Research Institute of Network Technology,
[2] Beijing Key Laboratory of Intelligent Telecommunications Software and Multimedia
Beijing University of Posts and Telecommunications
100876, Beijing, China
{liangjx,linzw,buptnic}@buptnet.edu.cn, mayan@bupt.edu.cn

Abstract. OpenFlow is a promising future Internet enabling technology and has been widely used in the network research community to evaluate new network protocols, applications, and architectures. However, most of these research activities or experimentations are lack of a uniformed description so can be repeated by other researchers. In this paper, we investigate the general model of an OpenFlow networking experiment and propose a language, OF-NEDL, which aspires to bridge this gap by providing a simple, comprehensive and extensible language for describing OpenFlow networking experiment. OF-NEDL allows the researcher to write a script to control every aspect of an OpenFlow networking experiment, including the hierarchical network topology description, the OpenFlow network devices configuration, the experiment software deployment, the experiment process control, monitoring and output collection. Our preliminary usage scenario shows that it has the ability to describe simple but extensible networking experiment, and we expect to refine considerably its design to make it more practical in the future work.

Keywords: Future Internet, OpenFlow, Network Experiment, XML, OF-NEDL.

1 Introduction

The Internet has grown so successfully that it plays a crucial role in today's society and business. It interconnects over a billion people, running a wide range of applications and is regarded as a great success by almost every measure. However, the Internet comes with important shortcomings because it originally not been designed for this wide range of usage scenarios. It can't handle new requisites such as end-to-end quality of service, security, and mobility. The network research community is generally more interested in fundamental questions about the structure of the Internet, including the design of new networking algorithms, protocols, and services, even construct the Internet completely new from scratch [1]. Of course, any such ideas must be rigorously tested and evaluated before being deployed in the actual Internet.

* Foundation Items: The National High Technology Research and Development Program of China (863 Program) (2011AA010704)

F.L. Wang et al. (Eds.): WISM 2012, LNCS 7529, pp. 686–697, 2012.

A network testbed is simply a collection of networking and systems resources that share a common infrastructure, and provides environment which allows researchers to develop, test, and evaluate new ideas. During recently years, many kinds of network testbed have been developed and some of them were widely used in network research community, such as PlanetLab [2] and Onelab [3]. However, most of these testbeds have only PC nodes as user configurable resources, and provide only application level networking experiment facility for the researchers. Hence, it is almost no practical way to experiment with new network protocols (e.g., new routing protocols, or alternatives to IP) in sufficiently realistic settings. To address the problem, Software-Defined Network (SDN) [4] is proposed and getting a lot of attention. SDN restructures network and exposes network APIs so that any software can configure the network as they want. Recently, OpenFlow [5] has been developing and is considered as a promising candidate technology to realize SDN. In brief, OpenFlow is based on a model that provides a physical separation of the network forwarding function and the network control function. The forwarding elements are dummy elements that have a generic shared forwarding table, which is accessed and configured by the independent control element called controller. In this situation, it is much easier to change or add new features to an OpenFlow enabled network testbed.

To start experiment on an OpenFlow network testbed, the first problem is how to describe the experiment, including the resources will be consumed, the control of experiment processes, the result or output of the experiment and so on. In a sense, the testbed is similar to the network simulation tools (NS2 etc.) and needs an experiment description language to tell the testbed system how to execute the experiment. For example, PlanetLab has implemented its experiment description language based on XML. However, these languages are lack of support to describe the OpenFlow network devices and the characteristics of OpenFlow enabled experiments. So we need to explore the principles and design a uniform and systematic framework to describe an OpenFlow networking experiment.

In this paper, we present the preliminary prototype of the OpenFlow Networking experiment Description Language (OF-NEDL), which intends to make it possible to write a script to control every aspect of an OpenFlow networking experiment, including the hierarchical network topology description, the OpenFlow network devices configuration, the experiment software deployment, the experiment process control, monitoring and output collection. This script file may be used by other researchers to repeat the experiment. The open question we are trying to answer is how to define a uniform abstract framework that can be used to describe arbitrary OpenFlow-based networking experiment, while keep it easy for the new user and extensible to design more complex experiments.

The remainder of this paper is organized as follows. Section 2 reviews some of the related works. Section 3 outlines the model of OpenFlow experiment and the Web-based experiment infrastructure. Section 4 presents the design principle and general architecture of OF-NEDL, and then demonstrates a simple usage scenario to evaluate it in section 5. Finally, section 6 concludes this paper and presents some future work.

2 Related Work

NS2[6] is a widely used network simulation tool in network research community. It is a discrete event simulator that has native support for common networking protocols and components. The simulation description language for NS2 is an extension of the Tcl (Tool Command Language). Using the various available NS2 commands, it can define network topology, configure traffic sources and sinks, program arbitrary actions into the configuration, collect statistics and invoke simulation.

NS3[7] is another popular network simulator with more advanced feature than NS2, but not an extension of NS2. Its simulation script language is C++ or optional Python, not the Tcl. Recently, the NS3 community are developing an XML-based language called NEDL (Ns3 Experiment Description Language) and NSTL (Ns3 Script Templating Language) [8], which provide a foundation for the construction of better interfaces between the user and the NS3 simulator.

Emulab [9] is a large network emulator, based on a set of computers which can be configured into various topologies through emulated network links. The Emulab control framework supports three different experimental environments: simulated, emulated, and wide area networks. It unifies all three environments under a common user interface, definitely, it uses an extend ns script written in Tcl to configure an experiment. For the researchers familiar with ns, it provides a graceful transition from simulation to emulation. The cost of this feature, however, is that the specification language and the relationship between virtualized resources is complex which makes it difficult to parse at the syntactic level.

The Open Network Laboratory (ONL) [10] is another emulation testbed that is similar in some respects to Emulab. ONL provides researcher with an user interface named Remote Laboratory Interface (RLI). The primary goal of the RLI is to make user interaction with testbed resources as easy as possible. In RLI, the most important abstraction is the resource type abstraction, which is represented in an XML file and consists of two types: the base description and the specialization description. The base description represents the physical device, while the specialization description represents one set of functionality supported by a particular type. It allows the testbed to include highly configurable and programmable resources that can be utilized in a variety of ways by the users.

Plush (PlanetLab user shell) [11] is a set of tools for automatic deployment and monitoring of applications on large-scale testbeds such as PlanetLab. Using Plush, the researcher can specify the experiment through an XML-RPC or a web interface and then remotely execute and control the experiment. In Plush, the experiment description or application specification consists of five different abstractions, that is, the process, barrier, workflow, component, and application blocks. The combinations of these blocks provide abstractions for managing resource discovery and acquisition, software distribution, and process execution in a variety of distributed environments. However, Plush focus solely on the application-level experiment in an overlay network, and is hard to extend to OpenFlow network.

Gush (GENI user shell) [12] is an experiment control system that aims to support software configuration and execution on many different types of resources. It leverages prior work with Plush, and is developed as part of the GENI (Global Environment for Network Innovations) project [13]. Gush accomplishes experiment control through an easily-adapted application specification language and resource.

Similar to Plush, the main goal of Gush project is used to simplify the tasks associated with application deployment and evaluation in distributed environments, but is more adaptable with other GENI control frameworks such as ORCA (Open Resource Control Architecture) [14] and ProtoGENI [15].

The cOntrol and Management Framework (OMF) [16] provide tools and services to support and optimize the usage of testbeds on a global scale, through systematic experiment description and execution, resource virtualization, and testbed federation. OMF is heavily focused on supporting repeatable experiments through OEDL [17], a domain-specific language based on Ruby fully capturing the experiment setup, its orchestration, and all relevant contexts.

Others, such as NEPI (Networking experiment Programming Interface) [18] and Expedient [19], proposed alternative frameworks to OMF to describe, execute and instrument networking experiments. NEPI proposes a framework based on a unified object model to describe a networking experiment which could subsequently be executed on different environments (e.g. simulations, emulations, and testbeds). Expedient is an ongoing effort to provide a testbed control framework that aggregates heterogeneous APIs to the underlying resources such as PlanetLab and Emulab, allowing for a user-friendly Web-based configuration and control of the experiment slices.

3 The OF-NEDL Design

3.1 Principles

In the previous section we investigated the model and the process of an OpenFlow experiment executing on the Web-based experiment infrastructure. Based on the discussion in the previous section, we now extract some general principles for an OpenFlow network experiment description framework.

Simple. It should be simple and easy to use, so that the user can concentrate his attention on the experiment software design itself but not the experiment description. The XML tag names and element structures should be easy enough to write and edit by hand as well as automatically generated by some other module (i.e., a front-end GUI). Keep it simple also means that it is easy to parse or handle by the Backend, so that the development complexity is to be reduced.

Comprehensive. It should be comprehensive and capture the abstract nature of the experiment, identify all aspects of the execution and environment needed to successfully deploy, control, and monitor the experiment. Basically, an OpenFlow network experiment description should include following components:

a description of the resources needed and network topology;
a description of the OpenFlow experiment configuration and deployment;
a description of the experiment control flow;
a description of the expected output.

Extensible. It should be extensible and provide explicitly defined procedures and system interfaces, making it easy to incorporate additional resources and technologies, even including those that do not exist today. In particular, it should be designed to minimize the effort of adding new node and link types, and ideally it should be handled without the need to modify any of the experiment infrastructure software.

The main challenge OF-NEDL attempts to overcome is the trade-off between simplicity and extensibility, which means that it should be easy to understand and write, while providing flexible constructs that allow for more complex experiments design.

3.2 The Architecture

According to the principles mentioned above, we have defined an OpenFlow Networking Experiment Description Language (OF-NEDL) to describe every task of a typical OpenFlow networking experiment workflow. As depicted in Figure 1, OF-NEDL is hierarchical. Experiment is defined at the highest level, and it contains information, topology, deployment, control, output components.

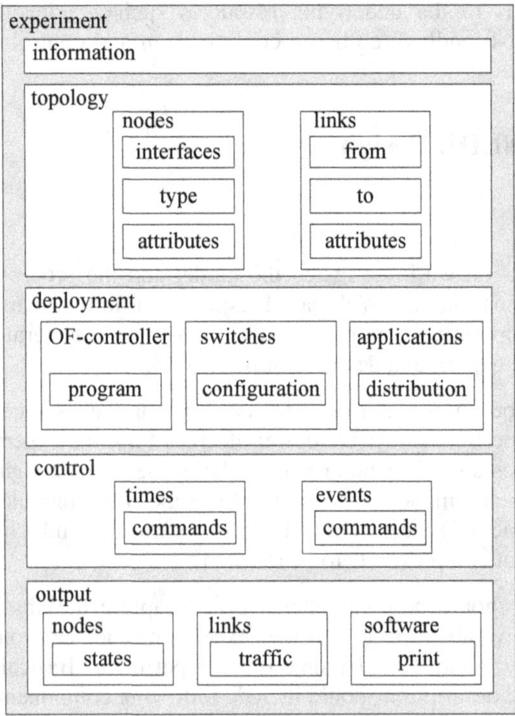

Fig. 1. General Structure of OF-NEDL

Information Component. Experiment information may include information about the user, group, experiment and global parameters. The most important is the setup of global parameters. These basic parameters, such as the start, duration, end times of experiment, the policy of error handle, will tell the Backend how to "treat" this experiment.

Topology Component. It is static representation of a group of nodes connected by links. Hence, it consists of two types of component, node and link. In communication networks, a node is either a redistribution equipment such as router, switch, hub, or a communication endpoint like a PC or sensor device. Each node has one or more interfaces to communicate with each other, and exhibit some attributes to materialize itself. To better illustrate the logical structure of the node component, we make use of UML class diagram. As show in Figure 2, a node has some general entries and attributes, which is similar to the Base Class in OOP (Object-Oriented Programming). Other special entries and attributes are stored in a Type structure whose type is one of the special devices (Router, Switch, KVM or openVZ virtual node, etc.) Another component, link, is the communication channel that connects two node through interfaces. To provide more configurable characteristics, the link is designed to have "from" and "to" endpoints to define a single direction communication. The attributes of link, such as delay, bandwidth, and loss-ratio are common for all kinds of physical channel, so there is no need to define "type" like node component.

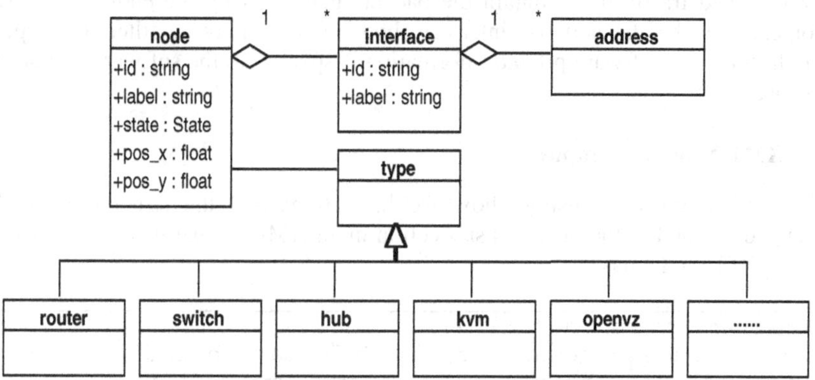

Fig. 2. UML Class Diagram for the Node Component

Deployment Component. The process of experiment deployment involves preparing the physical resources with the correct configuration and software. This typically includes copying and installing the OF program to OpenFlow controller, configuring the OF switches, copying and installing the software to all of the endhosts. According to the analysis of OpenFlow experiment in the subsection 3.2, the OF program and OF switch configuration are dedicated to themselves. Otherwise, the endhosts and software may have different combinations, so there may be two different ways to deploy the endhosts software, one is

```
SW_A: (EH_A, EH_B,... EH_X)
```

and another is

```
EH_A: (SW_A, SW_B,... SW_X)
```

where SW and EH represent the software and endhost. The main consideration is efficiency and compactness, especially when the number of endhosts and software increase to high, so we will provide the options to the user to decide which way to use.

Control Component. Experiment control includes start/stop the OF program and endhosts software, change the OF switches configuration, and issue commands on endhosts. All of these control actions may depend on times and events as following example:

```
AllReady: (ALL: CMD_A, CMD_B,... CMD_X)
   Wait_30S: (ALL_EXCEPT_EH_A•CMD_A)
```

The means of these expressions are literal, so it is easy for a novice user to get started. One problem is the control granularity provided. Fine granularity may provide user more control flexibility, but involve more complexity in the experiment infrastructure. We will refine the design details in the future work.

Output Component. The expected outputs in an experiment may vary widely due to the different goals and design. In principle, we categorize these outputs as node-related states, link-related traffic and software printed information. In general, the node-related states include the physical resources' states such as CPU, memory usage; the link-related traffic may contain the packets generated by the endhosts software and/or captured by the network interface, both of them may be filtered by special packet header; the software printed information is specific to the software and defined by the users.

3.3 XML Schema Samples

In this subsection we illustrate how we have translated the structure and class diagrams described in the previous subsection in an XML format that we describe by means of XML Schemas.

```
<xs:schema version="0.1">
  <xs:include schemaLocation="Information.xsd"/>
  <xs:include schemaLocation="Topology.xsd"/>
  <xs:include schemaLocation="Deployment.xsd"/>
  <xs:include schemaLocation="Control.xsd"/>
  <xs:include schemaLocation="Output.xsd"/>
  <xs:element name="experiment">
    <xs:complexType>
    <xs:sequence>
     <xs:element name="infomation" type="Information"
/>
     <xs:element name="topology" type="Topology"/>
     <xs:element name="deployment" type="Deployment"/>
     <xs:element minOccurs="0" name="control"
type="Control"/>
  <xs:element minOccurs="0" name="output"
type="Output"/>
  </xs:sequence>
  </xs:complexType>
  </xs:element>
</xs:schema>
```

Fig. 3. Experiment XML Schema

Figure 3 shows the format that we have defined for an XML document describing an OpenFlow networking experiment. Such a Schema is stored in as XSD file, which refers to five external XSD files. These files contain the components of an experiment that we have described above.

```
<xs:schema version="0.1">
<xs:include schemaLocation="Node.xsd"/>
<xs:include schemaLocation="Link.xsd"/>
 <xs:complexType name="Topology">
   <xs:sequence>
      <xs:element maxOccurs="unbounded" minOccurs="0"
name="node" type="Node"/>
      <xs:element maxOccurs="unbounded" minOccurs="0"
name="link" type="Link"/>
   </xs:sequence>
 </xs:complexType>
</xs:schema>
```

Fig. 4. Topology XML Schema

```
<xs:include schemaLocation="Interface.xsd"/>
<xs:include schemaLocation="Type.xsd"/>
 <xs:complexType name="Node">
    <xs:element  maxOccurs="unbounded"  minOccurs="0"
  name="interface" type="Interface"/>
    <xs:element name="type" type="Type" use="required"/>
    <xs:attribute       name="id"       use="required"
  type="xs:nonNegativeInteger" >
   <xs:attribute name="label"
      <xs:simpleTyep>
         <xs:restriction base="xs:string">
         <xs:pattern   value="   ([a-zA-Z](_)?)*   "/>
      </xs:restriction>
      </xs:simpleType> use="optional"/>
  <xs:attribute name="pos_x" type="xs:float" use="optional" />
  <xs:attribute name="pos_y" type="xs:float" use="optional" />
  <xs:attribute name="state" type="State" use="required" />
   <xs:simpleType name="State">
      <xs:restriction base="xs:string">
         <xs:enumeration value="CREATED"/>
         <xs:enumeration value="AVAILABLE"/>
         <xs:enumeration value="UNAVAILABLE"/>
         </xs:restriction>
      </xs:simpleType>
 </xs:complexType>
</xs:schema>
```

Fig. 5. Node XML Schema

Figure 4 shows the structure of the Topology part of a experiment description. This document also refers to external files (e.g. Node.xsd). For the sake of brevity, we just present the Node structure in Figure 5. These XML Schema samples imply that the design of OF-NEDL if a from-top-to-bottom process. The detail of each component can be redefined without affecting other component. This property provides the capacity to extend the language with minimal effort in the future.

4 Usage Scenario

To demonstrate the potential use of the OF-NEDL framework, we consider a simple but realistic networking experiment, the ping application. A minimal setup is shown in Figure 6. It consists of one switch (s1) and two endhosts (h1 and h2), each of which is configured with an IP address and connected by physical link. The logical model and XML description of the topology are shown in Figure 7. Each physical link is represented by two logical links with specified attributes. To be brief, we only show the description of s1 and ulink1.

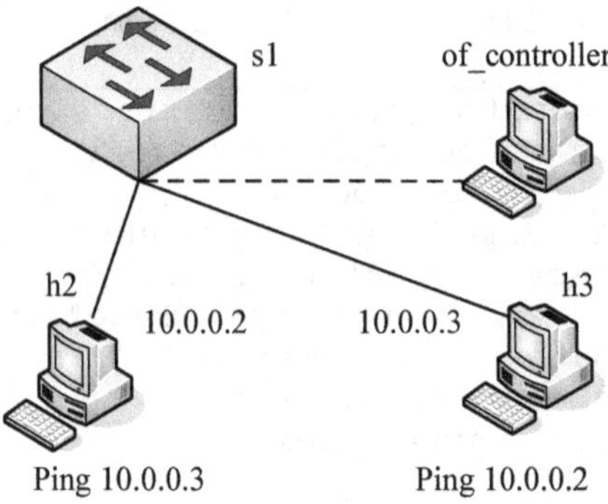

Fig. 6. The Ping Experiment Setup

The main goal of the ping application is to determine the connectivity of the network. In the experiment, the user need to implement an OF program with some basic function (ARP, STP etc.), a ping software (of course, user can use the existing software), and configure the OF switch properly (e.g. send the whole packet other than the first 128 bytes to the controller at the beginning). The deployment of the ping experiment is shown in Fig. 8. The two “config” components will add flow tables on the OF switch, so it can forward the packets correctly. By the way, the actions of manipulating the flow tables are all through the OF controller, so this configuration can be moved to the OF program.

The control of the ping experiment is simple (Fig. 9). When all the nodes are ready to run, endhosts h1 and h2 execute the commands to ping each other, and after 10 seconds, all the processes of the experiment will stop. The last part is the output component and we can imagine that the user want to show the information printed by the ping application, so the user only need to specify the “software print” as “ping” software in the output component.

```
<topology>
 <node id="s_0001" label="switch1"
   pox_x="400" pos_y="200" state="AVALIABLE">
    <interface id="s_0001_001" label="s1-eth1">
    <macAddr>fa:7a•c3:ea:c6:d1</macAddr></interface>
    <interface id="s_0001_002" label="s1-eth2">......
    </interface>
    <switch> ...... </switch>
 </node>......
   <link  id="ul_00001"  label="ulink1"    delay="0.0"
lossratio="0.0" bandwidth="100000.0">
    <from>h1-eth0</from>
    <to>s1-eth1</to>
 </link>......
</topology>
```

Fig. 7. Topology of the Ping Experiment

```
<deployment>
 <ofp>
    <name>example</name>
    <location>./example.py</location>
 </ofp>
 <ofs>
    <name>s1</name>
     <config> in_port=1,actions=output:2</config>
     <config> in_port=2,actions=output:1</config>
 </ofs>
 <software>
    <name>ping</name>
    ......
 </software>
 <ehs>
    <sw>ping</sw>
    <endhost>h2</endhost>
    <endhost>h3</endhost>
 </ehs>
</deployment>
```

Fig. 8. Deployment of the Ping Experiment

```
<control>
 <event name="AllReady">
    <command>ofnedl h2 ping c3 h3
             ofnedl h3 ping c3 h2</command></event>
    <time wait="10s"><command> ofnedl killall</command>
 </time>
</control>
```

Fig. 9. Control of the Ping Experiment

5 Conclusions and Future Work

In this paper we have investigated the general model of OpenFlow networking experiment, presented the design principle and architecture of an XML description language OF-NEDL, which aims to describe every aspects of an OpenFlow networking experiment. Since OF-NEDL is still in early design and prototyping stages, it is lack of full evaluations in the real experiment infrastructure. We will refine the design of OF-NEDL through ample evaluation in practice so it can be more practical for the researchers in the future.

References

1. Standford University: Clean Slate for the Internet, `http://cleanslate.stanford.edu/`
2. PlanetLab: An open platform for developing, deploying, and accessing planetary-scale services, `http://www.planet-lab.org/`
3. Onelab website, `http://www.onelab.eu/`
4. Yap, K.-K., Huang, T.-Y., Dodson, B., Lam, M., Mckeown, N.: Towards software-friendly networks. In: Proc. the First ACM Asia-Pacific Workshop on Systems, APSys 2010 (August 2010)
5. McKeown, N., Anderson, T., Balakrishnan, H., Parulkar, G., Peterson, L., Rexford, J., Shenker, S., Turner, J.: Openflow: Enabling innovation in campus networks. SIGCOMM Computer Communication Review 38(2), 69–74 (2008)
6. NS2 network simulator, `http://www.isi.edu/nsnam/ns/`
7. NS3 network simulator, `http://www.nsnam.org/`
8. Hallagan, A.W.: The Design of XML-based Model and Experiment Description Languages for Network Simulation. Bachelor Thesis, Bucknell University (June 2011)
9. Emulab website, `http://www.emulab.net/`
10. Open Network Laboratory, `https://onl.wustl.edu/`
11. Albrecht, J.R., Braud, R., Snoeren, A.C., Vahdat, A.: Application Management and Visualization with Plush. In: Proceedings of Peer-to-Peer Computing, pp. 89–90 (2009)
12. Albrecht, J., Huang, D.Y.: Managing Distributed Applications Using Gush. In: Magedanz, T., Gavras, A., Thanh, N.H., Chase, J.S. (eds.) TridentCom 2010. LNICST, vol. 46, pp. 401–411. Springer, Heidelberg (2011)
13. GENI: Global Environment for Network Innovations, `http://www.geni.net`
14. ORCA website, `http://www.nicl.cs.duke.edu/orca/`
15. ProtoGENI website, `http://www.protogeni.net`
16. Rakotoarivelo, T., Ott, M., Seskar, I., Jourjon, G.: OMF: a control and management framework for networking testbeds. In: SOSP Workshop on Real Overlays and Distributed Systems (ROADS 2009), Big Sky, USA, p. 6 (October 2009)
17. The OMF Testbed Control, Measurement and Management Framework, `http://www.omf.mytestbed.net`
18. Lacage, M., Ferrari, M., Hansen, M., Turletti, T.: NEPI: Using Independent Simulators, Emulators, and Testbeds for Easy Experimentation. In: Workshop on Real Overlays and Distributed Systems, ROADS (2009)
19. Expedient: A Pluggable Centralized GENI Control Framework, `http://www.yuba.stanford.edu/~jnaous/expedient`

20. OpenFlow in Europe: Linking Infrastructure and Applications, `http://www.fp7-OFELIA.eu/`
21. CHANGE project, `http://www.change-project.eu`
22. Kanaumi, Y., Saito, S., Kawai, E.: Deployment of a Programmable Network for a Nation wide R&D Network. In: IEEE Network Operations and Management Symposium Workshops, pp. 233–238 (2010)
23. Park, M.K., Lee, J.Y., Kim, B.C., Kim, D.Y.: Implementation of a Future Internet Testbed on KOREN based on NetFPGA/OpenFlow Switches. In: NetFPGA Developers Workshop, Stanford, CA, August 13-14 (2009)
24. Gude, N., Koponen, T., Pettit, J., Pfaff, B., Casado, M., McKeown, N., Shenker, S.: NOX: Towards an Operating System for Networks. SIGCOMM Comput. Commun. Rev. 38, 105–110 (2008)
25. Beacon, `http://www.openflowhub.org/display/Beacon/`
26. SNAC OpenFlow Controller, `http://snacsource.org/`

Structural Similarity Evaluation of XML Documents Based on Basic Statistics

Chen-Ying Wang, Xiao-Jun Wu*, Jia Li, and Yi Ge

Teaching and Research Division of Health Service,
Logistics Academy of the Chinese PLA, Beijing 100858, China
wangchenying2010@gmail.com

Abstract. The similarity evaluation between XML documents is the basis of XML structural mining, and it is a crucial factor of the mining result. After introducing the popular XML tree edit method and frequent pattern method for XML data mining, in this paper, we use 10 basic statistics to describe the structural information of the XML documents, and then using the improved Euclidean distance to evaluate the similarity of XML documents. Moreover, in order to verify the performance of the proposed evaluation method, it is applied to XML documents clustering. The experimental results show that our method is superior to the methods based on edit tree or frequent pattern.

Keywords: XML, basic statistics, similarity evaluation, cluster.

1 Introduction

With the growing popularity of XML for data representation and exchange, large numbers of XML documents have emerged. There is a pressing need for discovering valuable information from the massive documents. XML data mining is an important application of knowledge discovery technology, and similarity evaluation plays a crucial role in XML data mining.

XML data mining consists of contextual mining and structural mining. XML documents are semi-structural data, and the structure is an important characteristic of XML documents. Structural mining could excavate the knowledge among the structures of XML documents, and facilitate XML data extraction, integration and other applications. Classification and clustering are commonly used methods in XML structural mining. No matter classification or clustering, it is the first problem to be solved to evaluate the distance (or the similarity) between XML documents.

In the XML arena, similarity evaluation between documents has been receiving a lot of attention and many approaches have been developed. At present, there are two main categories of approaches, namely approaches based on XML tree edit and approaches based on frequent pattern. Modeling XML documents as trees is an ordinary representation. Such as DOM [1], which is a W3C proposed standard, presents XML documents as a hierarchy tree of node objects. A common way of

---

* Corresponding author.

F.L. Wang et al. (Eds.): WISM 2012, LNCS 7529, pp. 698–705, 2012.

evaluating the similarity of two trees is by measuring their tree edit distance [2], and there are three classical algorithms of tree edit distance: Selkow [3], Chawathe [4] and Dalamagas [5]. The tree edit methods contain all information of the XML documents and the tree edit operations are time-consuming operations, thus the efficiency of the evaluation algorithms based on tree edit is considered as a challenge work. Frequent patterns [6] are another category of important models transformed from tree models, such as frequent paths model [7], frequent sub tree model [8] and element sequence pattern [9]. The evaluation algorithms based on frequent patterns can evaluate the similarity between XML documents efficiently, but it is difficult to find out all the frequent patterns, and the accuracy of the evaluated similarity is not satisfied because they miss lots of structural information.

In this paper, we use 10 basic statistics to describe the structural information of the XML documents, and then using the improved Euclidean distance to evaluate the distance (or the similarity) of XML documents. Moreover, in order to verify the performance of the proposed evaluation method, it is applied to XML documents clustering. The experimental results show that our method is superior to the methods based on tree edit or frequent pattern considering either the processing cost or the clustering accuracy.

2 Basic Statistics of XML Documents

In an XML document, elements, attributes and text nodes are the three most common types of nodes, and the elements can be regarded as structural nodes and the others as contextual nodes. In this paper, we only focus on the structural similarity, and content similarity can be done a good job by the traditional information retrieval technology, so we use 10 basic statistics to describe XML documents' structural information. The notations used in this paper are shown in Table 1. If node v is the root node, then $h(v) = 1$; otherwise, $h(v) = h(\varphi(v))+1$. Using $|S|$ to denote the count of elements in a set S, then the fan out $f(e)=|\pi(e)|+|\rho(e)|$.

Table 1. The notations used in this paper

Notation	Description
D	An XML document or an XML document tree
v, e, a	A node, an element node or an attribute node of D
V, E, A	Nodes set, elements set or Attributes set of D
Σ	Different tokens set used as names by E and A of D
$\varphi(v)$	The parent node of the node v
$h(v)$	The height of the node v
$\pi(e)$	The child nodes set of the element e
$\rho(e)$	The attribute nodes set of the element e
$f(e)$	The fan out of the node e

Definition 1 (Statistics of the XML document) *Given an XML document D, then the structural information of document D can be presented as statistics X_1~X_{10}, where X_1 is the ratio of the number of contextual nodes and the number of structural nodes which is defined as:*

$$X_1 = \frac{|V| - |E|}{|E|}. \qquad (1)$$

X_2 is the number of different tokens used by the names of elements and attributes in D which is defined as:

$$X_2 = |\Sigma|. \qquad (2)$$

X_3, X_4, X_5 and X_6 are the mean, the standard deviation, the standard error and coefficient of variation of the height of the nodes in V, which are defined as follows:

$$X_3 = \frac{1}{|V|}\sum_{i=1}^{|V|} h(v_i),\ X_4 = \sqrt{\frac{1}{|V|-1}\sum_{i=1}^{|V|}(\mathrm{h(vi)} - \mathrm{X3})\ 2}\ ,\ X_5 = \frac{X_4}{\sqrt{|V|}}\ ,\ X_6 = \frac{X_4}{X_3}. \qquad (3)$$

X_7, X_8, X_9 and X_{10} are the mean, the standard deviation, the standard error and coefficient of variation of the fan out of the elements in E, which are defined as follows:

$$X_7 = \frac{1}{|E|}\sum_{i=1}^{|E|} f(e_i),\ X_8 = \sqrt{\frac{1}{|E|-1}\sum_{i=1}^{|E|}(f(e_i) - X_3)^2}\ ,\ X_9 = \frac{X_8}{\sqrt{|E|}}\ ,\ X_{10} = \frac{X_8}{X_7}. \qquad (4)$$

According to the definition 1, given an XML document *D*, the structural information of *D* is presented as a vector (x_1, x_2, ⋯, x_{10}), where x_i ($1 \le i \le 10$) is the value of the statistic X_i ($1 \le i \le 10$) of *D*. Now, we can use the basic statistics to evaluate the similarity between XML documents.

3 XML Documents Clustering

The first step of XML document clustering is to define the similarity or to calculate the distance between documents. Euclidean distance is the most widely used method in data mining, but the influence of variable's dimension and different variance among variables are not considered by the Euclidean distance. Apparently, different variance of variables will produce different effects to the Euclidean distance. Moreover, the variable with bigger variance will produce greater contribution to the distance. Therefore, given *n* XML documents, the weighted Euclidean distance is introduced to evaluate the similarity among the *n* XML documents.

Definition 2 (Weighted Euclidean Distance of XML Documents) *Given two XML documents D_i and D_j, ($x_1(D_i)$, $x_2(D_i)$, ⋯, $x_{10}(D_i)$) and ($x_1(D_j)$, $x_2(D_j)$, ⋯, $x_{10}(D_j)$) are the vectors describing the structural information of D_i and D_j, and $\bar{X}_k$ denotes the mean of X_k, where $1 \le i \ne j \le n$ and $1 \le k \le 10$. Then the structural similarity between document D_i and D_j is defined as d_{ij}:*

$$d_{ij}=\sqrt{\sum_{k=1}^{10}\left(\frac{x_k(D_i)-x_k(D_j)}{s_k}\right)^2}, \quad where \ s_k=\sqrt{\frac{1}{n-1}\sum_{t=1}^{n}(x_k(D_t)-\overline{X_k})^2}. \tag{5}$$

The second step is to choose the clustering algorithm, and the hierarchical clustering algorithm is used in this paper. Firstly, the n XML documents are considered as n cluster; secondly, the two clusters with the minimum distance are combined into a cluster, and then we get n-1 clusters; finally, the process of combining clusters with the minimum distance is repeated until all the XML documents are in a cluster. Therefore, it is the key problem to evaluate the distance between two clusters in the hierarchical clustering method. Three classic methods to evaluate distance between clusters are applied to verify our idea.

Let d_{ij} denote the distance between D_i and D_j, where $1\leq i\neq j \leq n$, C_1, C_2, ... denote the clusters, and S_{pq} denote the distance between cluster C_p and C_q. When cluster C_p and C_q are combined into cluster C_r, S_{rk} denotes the distance between the new cluster C_r and other clusters C_k. Then the minimum distance method defines:

$$S_{pq}=min\{d_{ij} \mid D_i \in C_p \wedge D_j \in C_q\},\ S_{rk}=min\{\ S_{pk}, S_{qk}\}. \tag{6}$$

The maximum distance method defines:

$$S_{pq}=max\{d_{ij} \mid D_i \in C_p \wedge D_j \in C_q\},\ S_{rk}=max\{\ S_{pk}, S_{qk}\}. \tag{7}$$

The third method is the centroid method, and the distance between clusters is defined as the distance between the centroids. Given C_p and C_q contain n_p and n_q XML documents, and cluster C_p and C_q are combined into cluster C_r with $n_r=n_p+n_q$ XML documents. Then S_{rk} denoting the distance between the new cluster C_r and other clusters C_k is defined as:

$$S_{rk}^2 = \frac{np}{nr}\cdot S_{pk}^2 + \frac{nq}{nr}\cdot S_{qk}^2 - \frac{np}{nr}\cdot\frac{nq}{nr}\cdot S_{pq}^2 \ . \tag{8}$$

4 Experiments

4.1 Experiment Setup

In order to verify our idea, we have implemented the hierarchical clustering algorithm based on three classic methods for distance evaluation. All the algorithms were implemented in C# and all the experiments are carried out on a 2.8 GHz Pentium processor with 2 GB RAM running Windows XP.

In order to apply the algorithms to XML documents clustering for performance evaluation, two kinds of datasets were used, namely ACM Sigmod Record [10] and INEX 2007 datasets [11]. To test the sensitivity of the algorithms on datasets with different sizes, we prepared 4 sub datasets shown in Table 2 for our experiments. After preparing the datasets, all the 48 XML documents no matter from ACM Sigmod or Shakespeare are encoded from D_{01} to D_{48}.

Table 2. Datasets used in our experiments

ID	Datasets	Sources	Documents(*.xml)
D_{01}~D_{12}	*Proceedings Page*	Sigmod-2002	152,162,172,182,192,202, 212,222,232,242,252,262.
D_{13}~D_{24}	*Ordinary Issue Pages*	Sigmod-1999	211,213,214,221,223,224, 231,233,234,241,243,244.
D_{25}~D_{36}	*Ordinary Issue Pages*	Sigmod-2002	251,253,254,261,263,264, 271,273,274,281,283,284.
D_{37}~D_{48}	140358	INEX 2007	3810,4036,4530,4649, 4904,5206,6104,6548, 6777,6945,3802,3439.

4.2 Results of Preprocessing Stage

In order to cluster XML documents, the algorithm are divided into two stages. One is the preprocessing stage including reading the documents and turning their structural information into the vector of 10 basic statistics. The other is the clustering stage including the similarity evaluation of any pair of documents and clustering them by the hierarchical clustering algorithm.

Table 3. The results of the 48 XML documents after preprocessing stage

ID	X_1	X_2	X_3	X_4	X_5	X_6	X_7	X_8	X_9	X_{10}
D_{01}	1.179	24	4.665	0.028	0.845	0.182	2.177	0.108	2.199	1.010
D_{02}	1.191	24	4.685	0.025	0.839	1.179	2.190	0.103	2.334	1.066
D_{03}	1.186	24	4.688	0.024	0.835	0.178	2.185	0.102	2.410	1.103
D_{04}	1.184	24	4.680	0.026	0.843	0.180	2.182	0.104	2.274	1.042
…	…	…	…	…	…	…	…	…	…	…
D_{45}	2.036	27	5.672	0.044	1.335	0.235	3.033	0.196	3.412	1.125
D_{46}	3.056	20	4.041	0.064	0.778	0.193	4.028	0.547	3.282	0.815
D_{47}	2.388	20	5.767	0.121	1.818	0.315	3.373	0.321	2.628	0.779
D_{48}	3.038	14	4.038	0.096	0.980	0.245	4.000	0.590	3.007	0.752

Table 3 shows the results of the 48 XML documents after preprocessing stage. It can be seen that the observations of 10 basic statistics from the same dataset have small differences, and the observations of 10 basic statistics from the different datasets have obvious differences.

4.3 Clustering Accuracy

Among the different quality measures for clustering accuracy, we used the precision and recall rates as clustering accuracy measure. The measure assumes that a cluster is the result of a query and a category is the desired set of the documents for the query.

Let C_i and R_i denote the cluster set and category set respectively, where $1 \leqslant i \leqslant 4$ in our experiments, $R_1=\{D_{01}\sim D_{12}\}$, $R_2=\{D_{13}\sim D_{24}\}$, $R_3=\{D_{25}\sim D_{36}\}$ and $R_4=\{D_{37}\sim D_{48}\}$. Then the precision and recall rates of that cluster-category pair can be computed as follows:

$$Precision = \frac{|C_i \cap R_i|}{|C_i|}, Recall = \frac{|C_i \cap R_i|}{|R_i|}. \tag{9}$$

where $|C_i \cap R_i|$ is the number of the documents of category R_i falling into the cluster C_i, $|C_i|$ is the number of documents in the cluster C_i, and $|R_i|$ is the number of documents in the category R_i.

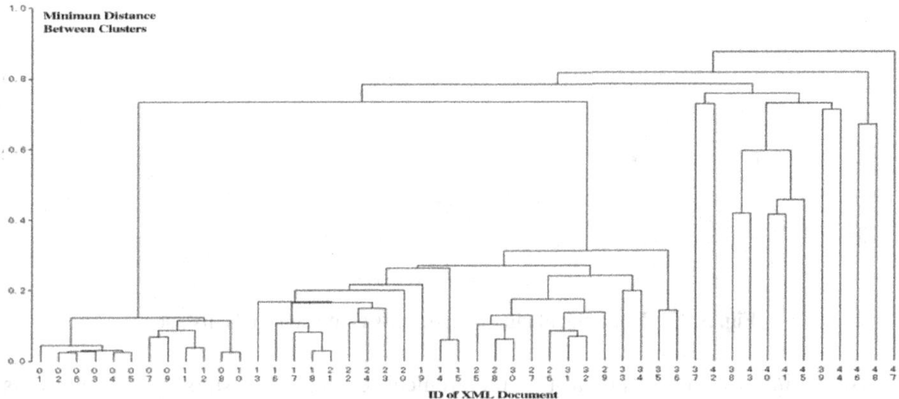

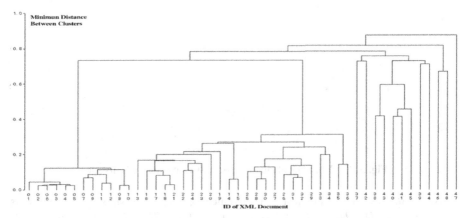

Fig. 1. The hierarchical clustering results using minimum distance

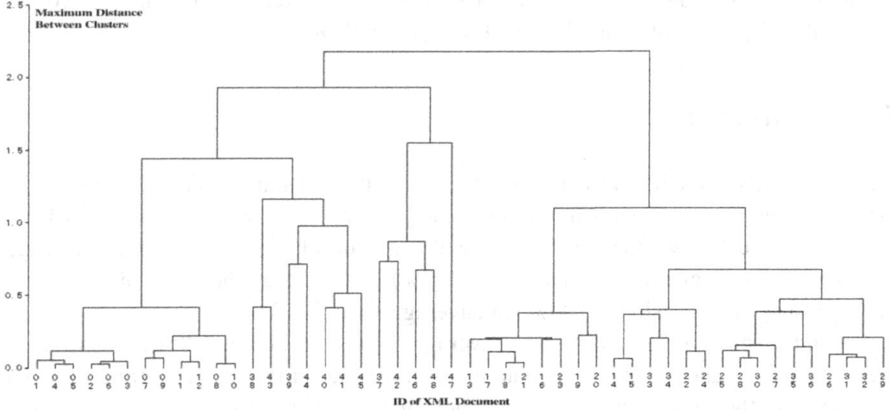

Fig. 2. The hierarchical clustering results using maximum distance

According to Fig.1 shown the hierarchical clustering results using minimum distance, $C_1=\{D_{01}\sim D_{12}\}$, $C_2=\{D_{13}\sim D_{24}\}$, $C_3=\{D_{25}\sim D_{34}\}$, $C_4=\{D_{37}\sim D_{45}\}$, $C_5=\{D_{35}, D_{36}\}$, $C_6=\{D_{46}, D_{48}\}$ and D_{47} is an isolated document. Therefore, the precision of the 4 clusters is 100%; the recall of C_1 and C_2 is 100%, the recall of C_3 is 83.33%, the recall of C_4 is 75%.

According to Fig.2 shown the hierarchical clustering results using maximum distance, C_1={D_{01}~D_{12}}, C_2={D_{13}, D_{16}~D_{21}, D_{23}}, C_3={D_{25}~D_{32}, D_{35}, D_{36}}, C_4={D_{38}~D_{41}, D_{43}~D_{45}}, C_5={D_{37}, D_{42}, D_{46}~D_{48}}, and C_6={D_{14}, D_{15}, D_{22}, D_{24}, D_{33}, D_{34}}. Therefore, the precision of the 4 clusters is 100%; the recall of C_1 is 100%, the recall of C_2 is 66.67%, the recall of C_3 is 83.33%, and the recall of C_4 is 58.33%.

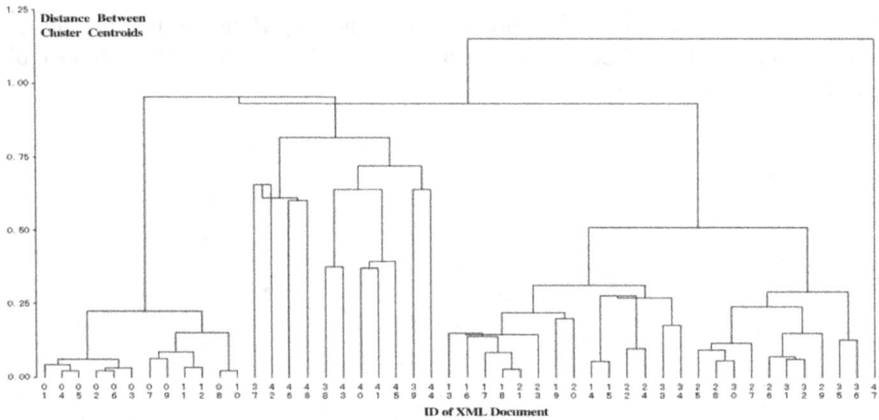

Fig. 3. The hierarchical clustering results using centroids distance

According to Fig.3 shown the hierarchical clustering results using centroids distance, C_1={D_{01}~D_{12}}, C_2={D_{13}~D_{24}, D_{33}, D_{34}}, C_3={D_{25}~D_{32}, D_{35}, D_{36}}, C_4={D_{37}~D_{46}, D_{48}}, and D_{47} is an isolated document. Therefore, the precision of C_1, C_3 and C_4 is 100%, and the precision of C_2 is 85.71%; the recall of C_1 and C_2 is 100%, the recall of C_3 is 83.33%, and the recall of C_4 is 91.67%.

5 Conclusions

The experimental results show that: Firstly, in the three distance calculation methods, the clustering accuracy result using centroids distance (average precision is 96.43%, and average recall is 93.75%) is the best, and the result of minimum distance (average precision is 100%, and average recall is 89.58%) is better than the result of maximum distance (average precision is 100%, and average recall is 77.08%).

Secondly, on synthetic data, the average precision and recall of Chawathe's method are 71% and 90%; on real data the average precision and recall are 100% and 95% [5]. By the report in [7], the accuracy of the PBClustering based on frequent pattern is affected by the different minimum support. When the minimum support is from 70% to 100%, the accuracy of the PBClustering algorithm is around 40%, the accuracy becomes over 50% when the minimum support is from 35% to 70%, and the accuracy becomes around 80% when the minimum support is from 2% to 35%. Therefore, our method using centroids distance is superior to the methods based on tree edit or frequent pattern.

Thirdly, in order to clearly draw the clustering hierarchy tree, the sample size of XML documents is 48; with the sample size increasing, the system error will be reduced and the clustering accuracy rate will be better.

Future work will focus on: using the increment of the cluster's variance and the variance between two clusters to evaluate the distance between the clusters, and designing basic statistics to describe the contextual information of XML documents to further improve the clustering accuracy.

References

1. W3C Recommendation, Document Object Model (DOM) Level 3 Core Specification (2004), http://www.w3.org/TR/DOM-Level-3-Core/
2. Bille, P.: A survey on tree edit distance and related problem. Theoretical Computer Science 337, 217–239 (2005)
3. Selkow, S.M.: The tree-to-tree edit problem. Information Processing Letter 6, 184–186 (1997)
4. Chawathe, S.S.: Comparing Hierarchical Data in External Memory. In: Proceedings of the 25th VLDB, pp. 90–101 (1999)
5. Dalamagas, T., Cheng, T., Winkel, K., Sellis, T.K.: A Methodology for Clustering XML Documents by Structure. Information Systems 31(3), 187–228 (2006)
6. Nayak, R., Iryadi, W.: XML schema clustering with semantic and hierarchical similarity measures. Knowledge-Based Systems Archive 20(4), 336–349 (2007)
7. Leung, H.-P., Chung, F.-L., et al.: XML Document clustering using Common XPath. In: Proc. of the Internation Workshop on Challenges in Web Information Retrieval and Integration, pp. 91–96 (2005)
8. Hwang, J.H., Gu, M.S.: Clustering XML Documents Based on the Weight of Frequent Structures. In: Proc. of the 2007 International Conference on Convergence Information Technology, pp. 845–849 (2007)
9. Zhang, H., Yuan, X., Yang, N., Liu, Z.: Similarity Computation for XML Documents by XML Element Sequence Patterns. In: Zhang, Y., Yu, G., Bertino, E., Xu, G. (eds.) APWeb 2008. LNCS, vol. 4976, pp. 227–232. Springer, Heidelberg (2008)
10. SIGMOD Record Datasets (2007), http://www.sigmod.org/record/xml/
11. INEX. INitiative for the Evaluation of XML Retrieval (2007), http://inex.is.informatik.uni-duisburg.de/

An XML Data Query Method Based on Structure-Encoded

Zhaohui Xu[1,2], Jie Qin[2], and Fuliang Yan[2]

[1] School of Computer, Wuhan University of Technology, Wuhan, China
[2] Grain Information Processing and Control Key Laboratory of Ministry of Education, Henan University of Technology, Zhengzhou Henan, 450001, China
{morsun,qinjie}@haut.edu.cn

Abstract. Aiming the special tree-structure of XML data, a triples-encoded method for searching XML documents is proposed. By representing both XML documents and XML queries in triples-encoded sequences, querying XML data is equivalent to finding subsequence matches. Triples-encoded method uses tree structures as the basic unit of query to avoid expensive join operations. The proposed method also provides a unified index on both content and structure of the XML documents; hence it has a performance advantage over methods indexing either just content or structure. Experiments show that the proposed method is more effective, scalable, and efficient in supporting structural queries than traditional XML data query ways which based on Join operations.

Keywords: XML, structure-encoded, query, index, ternary-code.

1 Introduction

XML (eXtensible Markup Language) has emerged as the standard format for web data expression and exchange, which leads to requirements involving XML data management. Data described by XML have semi-structure characteristics and make challenges to the traditional data management[1],[2],[3].

In order to improve the query efficiency of the XML data, researchers made a lot of different index technologies [1],[2],[3],[4]. These ways need JOIN operation to merge the middle results, and affect the query efficiency. Moreover, existing ways are inefficient when they deal with XML data with "*" or "//"[2].In the existing methods, typical methods are XISS[2] and Index Fabric[1].

Index Fabric takes path as the basic query units, and establishes indexes from the root node to all the other nodes to different paths to accelerate XML data query speed. This method suits single paths. If the query statement structure with branches, it will be divided into some single paths and separately execute its queries. Then get the final results through JOIN operation of every query result of single paths.

In fact, XML-based semi-structured data query is a matching procedure of the tree (the query) and the graph (the data object) [2], XML data and XML queries can be transformed to specific character strings. And a query based on XML data becomes to search the continuous substring (the discontinuous substring) to match in a set of strings. Based on such consideration. This paper presents a new method to query XML data based on the structural features: each of the basic unit of XML data use

F.L. Wang et al. (Eds.): WISM 2012, LNCS 7529, pp. 706–713, 2012.

Triples-encoding to express, (Name, Path, Branch) ,and in this encoding method, XML data query is transformed into a string matching problem. The content as following: The second part introduces the relevant knowledge. The third part shows the algorithm for the XML data query based on structural features encoding. The fourth part is experimental results and the last is conclusions.

2 Preliminary Knowledge

XML data query format has 4 kinds: the signal path query (the logical structure of the query without branches), the signal path query with wildcard character asterisks (*), the signal path query with character "//" and the query with a branch. As a simple example: Fig. 1 is a product record of an expression tree that contains buyer and purchases.

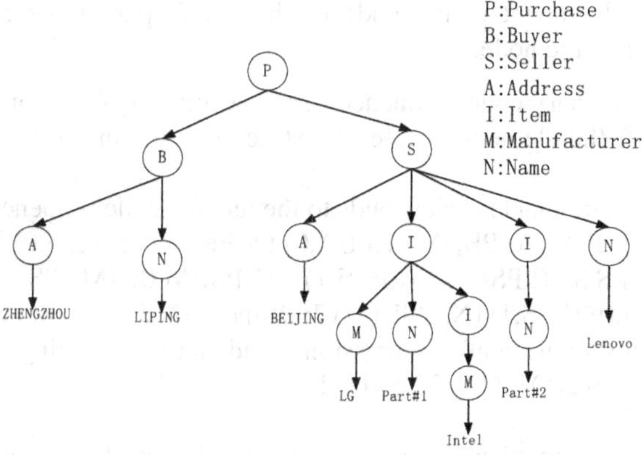

Fig. 1. A tree expression of a product purchased records

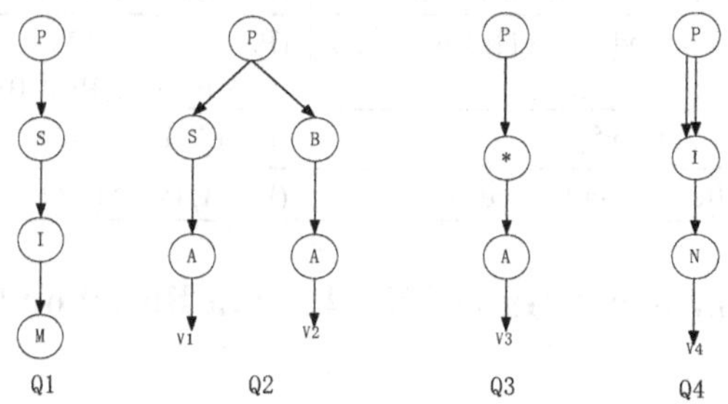

Fig. 2. A tree expression of a product purchased records

The tree representation of the several fundamental queries:

Q1: Find all manufacturers;

Q2: Find all the orders which sellers' addresses are in v1 and buyers' addresses are in v2;

Q3: Find all the orders contain sellers in v1 and buyers in v2;
Q4: Search for all the orders contains product v4;

Fig. 2. shows the query tree expressions corresponding to them, which are the single-path query (Q1), with a branch of the query (Q2), with the wildcard (*) single-path query (Q2) and with the "/ / "symbol of the single-path query (Q4).

The XML documents with tree structure can be expressed by preorder traversal, for the records in Fig. 1., corresponds to the preorder traversal sequence is:

$$\text{PBAv1Nv2SAv3}\underline{\text{IIMv6Mv4Nv5I}}\text{Nv7Nv8} \tag{1}$$

In which, V1 V2 …V8 represent ZHENGZHOU、LIPIN、…Lenovo.

Definition 1: Ternary-code is a triple (Name, Path, Branch), where Name denotes a node of the XML document tree, it can be the name of the element or attribute and can also be the specific values of the elements or attributes; Path shows a path from the root of the document to the node of the node's parent; Branch indicate the children number of the node.

Definition 2: Ternary-code sequence is a string sequence of triples which is composed by all the elements of the tree structure adopting triples encoding as pre-order traversal.

For example, formula(1) corresponds to the ternary-code sequences as follows:

D= (P,ε,2) (B, P,2) (A, PB,1) (v1,PBA,0) (N,PB,1) (v2, PBN,0) (S, P,4) (A, PS,1) (v3, PSA,0) (I,PS,3) (I,PSI,1) (M,PSII,1) (v6,PSIIM,0) (M, PSI,1)(v4,PSIM,0) (N, PSI,1) (v5,PSIN,0) (I,PS,1) (N, PSI,1) (v7; PSIN,0) (N,PS,1) (v8; PSN,0) (2)

Table 1 shows the path expressions and ternary coding sequences which correspond to queries Q1, Q2, Q3 and Q4.

Table 1. The path expressions of queries Q1, Q2, Q3, Q4 and the ternary coding sequences

	Path expressions	Corresponding ternary coding sequences
Q1	/Purchase/Seller/Item/Manufacturer	(P,ε)(S,P)(I,PS)(M, PSI)
Q2	/Purchase[Seller[add = v1]]/Buyer[add = v2]	(P,ε,2)(S,P)(A,PS)(v1, PSA,0) (B, P)(A,PB)(v2, PBA,0)
Q3	/Purchase/*[add = v5]	(P,ε)(A, P*)(v5, P*A,0)
Q4	/Purchase//Item[Manufacturer = v3]	(P,ε)(I, P//)(M, P//I)(v3,P//IM,0)

3 The Query Algorithm of XML Data Based on Structural Features

3.1 The String Matching Algorithm

According to the previously described the way of ternary encoding, the XML data is converted to ternary coding sequences. Now the query is equivalent to searching a specific sub-string (the XML query) in a long string (the XML data). Therefore the traditional string matching algorithms can complete the task, such as KMP algorithm, BM algorithm [7] etc. Of course, the key issue of XML data query need to address is how to process the queries with branches. Here focus on the queries with branches. Fig. 3. shows the MatchSearch algorithm of the XML data queries based on the string matching principles.

```
Input: Q =q1….qn,
S= S1…. Sm,
Output: the results of all matching Q in S.
MatchSearch(S•root, 1);
Function MatchSearch(m, i)
  If i ≤ n
  then
    for each node j is a descendent of node m do
      if (Namei = Namej)and (Pathi ≤ Pathj) and
      (Branchi≤Branchj) //sj matching qi
      then MatchSearch(j, i + 1)
      else
        if i > n
        then
          Output matched results;
      i=1 ;
    end
  end
```

Fig. 3. The MatchSearch algorithm of the XML data

Where: Q is a query sequence expressed by ternary-code q_i = ($Name_i$, $Path_i$, $Branch_i$); S is a data sequence expressed by ternary-code S_j = ($Name_j$, $Path_j$, $Branch_j$)

The algorithm's time complexity is O(m + n). Where m is the length of the data string and n represent the length of the corresponding query.

3.2 The indexed Matching Algorithm

After adoption above ternary encoding the data alignments in the string has two relations: the logical ancestor – successor(A-D) relation (such as the elements (B,P,2) and (v1,PBA,0) in the sequence (3)), the priority order (B-A) (such an the elements (B,P,2) and (S,P,4) in the (3)). Clearly the elements have the priority orders don't mean they have logical ancestor - successor relations. On the contrary, it is the same. In the query process, after it had found the element X matching with q_{i-1}, when it was needed to find the successor element of X matching with q_i, sometimes X and Y having the relationship A-D.

For any two data elements in the sequence, by examining the Path of each other, it can determine whether there is a logical relationship A-D between them. However, to determine whether they have the order relationship B-A, more information is needed.

Here we add a binary identifier <n_X, $size_X$> for the each element X of the data sequence, where n_X is the sequence number of the ternary coding sequence of X, $size_X$

is the number of subsequent elements of the ternary coding sequence of X. The binary identifier can be obtained by depth-first traversal of data elements. For example, the binary sequence identifier corresponding to the sequence (2) is as follows:

(P,ε,2) <1,21>(B, P,2)<2,20> (A, PB,)<3,19> (v1,PBA,0)<4,18> (N,PB)<5,17> (v2, PBN,0)<6,16> (S, P,4) <7,15>(A, PS,)<8,14>(v3, PSA,0) <9,13> (I,PS,3) <10,12> (I,PSI)<11,11> (M,PSII)<12,10> (v6,PSIIM,0) <13,9>(M, PSI) <14,8> (v4,PSIM,0)<15,7> (N, PSI)<16,6> (v5,PSIN,0)<17,5>(I,PS)<18,4> (N, PSI)<19,3> (v7; PSIN,0)<20,2> (N,PS)<21,1> (v8; PSN,0)<22,0>

By introducing the binary identifier, it is easily to determine the B-A relation between any two elements in the indexes. If the binary identifier of X is $<n_X, \ size_X>$ and the binary identifier of Y is $< n_Y, \ size_Y >$, then:

Theorem 1:X and Y satisfy the relationship B-A if nY∈ (nX, nX + sizeX) .

The correctness of the theorem is clearly and the proof is omitted.

Here we introduce the ways to build up indexes.

First, we can build up indexes in (Name, Path) for the keywords to the data objects by the use of B + tree. Through the indexes we can directly find the set {Y} of elements, which matching to the X and meeting to A-D with X, in the data objects. Of course, between X and some (or all) part of {Y} may not the B-A relationship of the actual sequence, the elements of this part shall be excluded. Therefore, for the same (Name, Path) elements, B+ trees are built up by the use of their own n_X. So the B-A relationship between the elements can be determined.

This set includes two-level B+ trees: it is used to determine the A-D relationship between the elements of B+ trees by the keyword of (Name, Path) or n_X.

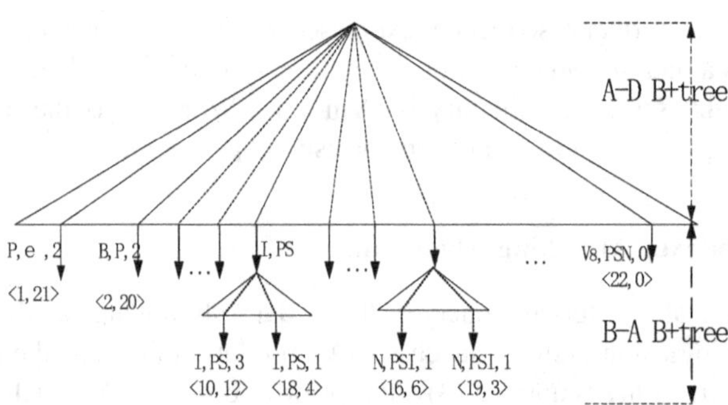

Fig. 4. The index structure corresponding to the ternary-code sequences of D

Here B+ trees built by the (Name, Path) keyword are A-D trees, and by the n_X are B-A trees. Figure 4 is the index structure corresponding to the Figure 2. Through the establishment of two-level B+ trees, it realized the indexing to the structure and content of XML data, and avoided to build up the indexes to the structure and contents of the semi-structured data, making the index structure is relatively simple.

After building up the index structure, query can be achieved as follows: Suppose that node X has been matched to the query sequence $q_1 \ldots q_{i-1}$, in order to find the next element q_i matching to the element Y, first, to use (Name, Path) of q_i as the key words to search in the A-D tree, if the elements satisfy the condition, it will obtain a pointer to point a B-A tree as ($Name_y$, $Path_y$) for its root. The B-A tree covers the set {Y} of the elements which meet the A-D relationship to the matching to the X and possibly to the Y. As between X and some elements of {Y} might have no the B-A relationship, it is needed to remove such elements from {Y}. In the B-A tree queries are executed according to the range of $n_x < n_Y \leq n_x + size_x$ so as to find the real subsequent {Y} of the X. Also , when looking for the qualified Y, if the Branch of q_i is more than 1, the Y satisfying $Branch_i \leq Branch_y$ is the real matching elements of q_i . So it is needed to search for Y satisfying $Branch_i \leq Branch_y$ in the {Y} of the real subsequent of X. Similarly you can continue to look for the elements matching to q_{i+1}.

Figure 5 shows the matching algorithm based on query string in this index structure.

```
Input: Q =q1....qn,
  Data S is the B+ tree index whose keywords is (Name,
  Path) // the A-D tree
  B+ tree index whose keywords is nx ; // B-A tree
Output: the results of all matching to  Q in  S;
BMS (S→root, 1) ;
Function  BMS (m, i)
   If i ≤ n
      then
        if qi ='//'  then i=i+1 ;
        Search (Namei, Pathi) in the A-D tree, and assign
        the results to T; // T is a pointer which point
        to the root of B-A.
        while T ≠nil do
          Looking for the elements within (nm, nm +size)
          in the B-A tree pointed by T, and assign the
          results to N;
        for each node j ∈ N do
          if Branchi ≤ Branchj
             then BMS(m+1, i+1)
          end;
          T=T.NEXT;
        end
     else
        if i > n
          then
            Output matched results;
            i=1 ;
          end
     end
```

Fig. 5. BMS query algorithm

Assuming the length of XML data is M and the data length of the query is N. The core factor of the algorithm is how to find the next matching element. The algorithm accomplishes the task by twice indexes. First, through the statement 4 it searches indexes to the B+ tree satisfying the A-D relationship, $O(\log 2^{M})$ of time is required and then it does the search based on the scope to the B-A tree to find the match meeting the conditions, it takes $O(\log 2^{M})$. So the time to find a matching element is $O(\log 2^{M})$. Since the data length of query is N, it takes $O(\log 2^{M})$ to find a query result satisfying the conditions. Known by the KMP algorithm, $O(M+ N\log 2^{M})$ is required to find all matches in the XML data, of which the length is M.

As the indexes space of A-D tree and B-A tree is O(M), the indexes of the algorithm is O(M).

4 Performance Test

In order to verify the actual performance of the algorithm, we achieved the BMS algorithm, and compared to the XISS algorithm, the Index Fabric algorithm and the RISI algorithm. Experimental environment: Linux Operating System and through the API of B+ tree provided by the Berkeley DB library [8] build up the indexes. Two test data sets, DBLP [9], XMARK [10] are chosen to compare.

The queries shown in Table 2. Selected the DBLP size is 30MB, contains 289,627 records, 2,934,188 elements and 364,348 properties. Size of the XMARK data set is 108MB.

Table 2. Queries and the corresponding data sets

	Path Expressions	Data sets
Q1	/in proceedings/title	DBLP
Q2	/book/author[text=`David']	DBLP
Q3	//author[text=`David']	DBLP
Q4	/book[key=`books/bc/MaierW88']/author	DBLP
Q5	/site//item[location=`US']/mail/date[text=`02 /18/2000']	XMARK
Q6	/site//person/*/city[text=`Pocatello']	XMARK

As shown in Table 3, the query based on string matching method is better than the traditional Join-based query method when executing the complex query with braches.

Table 3. The comparison of performances on BMS, RIST and XISS

	BMS	RIST	Index Fabric	XISS
Q1	1.8	1.2	0.8	9.7
Q2	1.7	2.4	4.6	53.2
Q3	1.7	1.9	22.1	29.7
Q4	1.7	2.8	7.5	18.1
Q5	2.3	3.7	17.2	22.1
Q6	2.4	2.5	36.3	26.9

The experiments show that the BMS algorithm has good scalability, and better query performance than RIST which also belongs to string matching method.

5 Conclusions

An advantage of using structural coding to realize XML data query is the query as a whole processing, and not need to decompose a complex query into some units to execute the query, and then obtain the final result by JOIN operation. It promotes the query efficiency. This paper adopts a triples-encoded method to describe the XML data and it clearly shows the structural features of the data elements. Using similarly string matching to query XML data. It can make full use of the existing information retrieval technology. It is more efficient when executing the query with branches.

Acknowledgment. This research was supported by the Scientific and technological project of Henan Province(0424220184) and Henan Province, Department of Finance food special(ZX2011-25).

References

1. Brian, F.C., Neal, S., Michael, F., Moshe, S.: A fast index for semistructured data. In: VLDB, pp. 341–350 (2001)
2. Wang, H.X., Park, S., Fan, W., Yu, P.S.: ViST: A Dynamic Index Method for Querying XML Data by Tree Structures. In: ACM SIGMOD (2003)
3. Jie, Q., Shu, M.Z.: Querying XML Documents on Multi-processor. In: Sixth International Conference on Fuzzy Systems and Knowledge Discovery, pp. 327–331 (2009)
4. Bruno, N., Koudas, N., Srivastava, D.: Holistic Twig Joins: Optimal XML Pattern Matching. In: ACM SIGMOD (2002)
5. McCreight, E.M.: A space-economical suffix tree construction algorithm. Journal of the ACM 23(2), 262–272 (1976)
6. Shasha, D., Wang, J.T.L., Rosalba, G.: Algorithmics and Applications of Tree and Graph Searching. In: ACM Symposium on Principles of Database Systems (PODS), pp. 39–52 (2002)
7. Leeuwen, J.V.: Algorithms for finding patterns in strings. In: Handbook of Theoretical Computer Science. Algorithms and Complexity, vol. A, pp. 255–300. Elsevier (1990)
8. Sleepycat Software, The Berkeley Database (Berkeley DB), `http://www.sleepycat.com`
9. The DBLP Computer Science Bibliography (2009), `http://www.informatik.uni-trier.de/~ley/db`
10. XMark-An XML Benchmark Project (2009), `http://www.xml-benchmark.org/`

Author Index